Current Trends and Open Problems in Computational Mechanics

Prof. Dr.-Ing. habil. Dr. h.c. mult. Dr.-Ing. E. h Peter Wriggers

Fadi Aldakheel · Blaž Hudobivnik ·
Meisam Soleimani · Henning Wessels ·
Christian Weißenfels · Michele Marino
Editors

Current Trends and Open Problems in Computational Mechanics

Springer

Editors
Fadi Aldakheel
Institute of Continuum Mechanics
Leibniz Universität Hannover
Garbsen, Germany

Meisam Soleimani
Institute of Continuum Mechanics
Leibniz Universität Hannover
Garbsen, Germany

Christian Weißenfels
Institute of Materials Resource Management
University of Augsburg
Augsburg, Germany

Blaž Hudobivnik
Institute of Continuum Mechanics
Leibniz Universität Hannover
Garbsen, Germany

Henning Wessels
Institute of Computational Modeling in Civil Engineering
Technical University of Braunschweig
Braunschweig, Germany

Michele Marino
Department of Civil Engineering and Computer Science
University of Rome Tor Vergata
Rome, Italy

ISBN 978-3-030-87311-0   ISBN 978-3-030-87312-7 (eBook)
https://doi.org/10.1007/978-3-030-87312-7

This Springer imprint is published by the registered company Springer Nature Switzerland AG
The registered company address is: Gewerbestrasse 11, 6330 Cham, Switzerland

Preface

This Festschrift is dedicated to Prof. Dr.-Ing. habil. Dr. h.c. mult. Dr.-Ing. E.h. Peter Wriggers on the occasion of his 70th birthday. Peter Wriggers achieved the degree of Dipl.-Ing. and obtained his Ph.D. and Habilitation at the Universität Hannover. In 1990, he was appointed Full Professor of Mechanics at the Technische Universität Darmstadt. Afterward, he moved back to Hannover, being in charge of the Institute of Mechanics and Computational Mechanics (IBNM) for almost ten years and then being appointed as Head of the Institute of Continuum Mechanics (IKM) in 2008. Moreover, Peter Wriggers has been an important collaborator of a number of internationally recognized institutes in Computational Mechanics, and among them he has been Visiting Professor at the University of California at Berkeley in the USA and the University of Newcastle in Australia. In his career, Peter Wriggers has received several Prizes, among which we mention the IACM Computational Mechanics Award and the ECCOMAS Euler Medal, as well as three Honorary Doctorates at TU Darmstadt, ENS Cachan and the University of Technology Poznan. Peter Wriggers has played a leading role in the academic community, actively participating in important scientific and administration boards.

As a scientist, Peter Wriggers has achieved outstanding results and an extraordinary international reputation. He has contributions in a wide range of topics, such as contact mechanics, meshless methods, novel finite element technologies, micromechanical approaches, damage mechanics, and many others. This wide range of topics is also covered in this book, where friends, collaborators, former and current Ph.D. students have enthusiastically contributed. Almost every continent and every scientific area in Mechanics is represented, highlighting the relevance Peter Wriggers has played in the field of Computational Mechanics. To many in our scientific community, Peter Wriggers has been a genuine supporter or, in other words, a friend. For most of the contributors, and surely for the Editors of this Festschrift, the encounter with Peter Wriggers has represented a professional game changer. The observational skills, creativity, as well as the innovativeness and professional attitude of Peter Wriggers have been, are and will be a source of inspiration.

We are glad to be in the position to edit this Festschrift and would like to thank Springer Verlag for the collaboration regarding this project. In the name of all authors, we congratulate Peter Wriggers on his 70th birthday. We wish him happiness, health and success. We are looking forward to his next successes and to learn more about Computational Mechanics in his forthcoming inspiring works.

Garbsen, Germany — Fadi Aldakheel
Garbsen, Germany — Blaž Hudobivnik
Garbsen, Germany — Meisam Soleimani
Braunschweig, Germany — Henning Wessels
Augsburg, Germany — Christian Weißenfels
Rome, Italy — Michele Marino
May 2021

Contents

Multiphysics Computation of Thermomechanical Fatigue in Electronics Under Electrical Loading

Bilen Emek Abali, Fadi Aldakheel, and Tarek I. Zohdi

Dedicated to Peter Wriggers, a real gentleman and scholar, who is a role model to us all (B. A. Abali, F. Aldakheel and T. I. Zohdi).

Abstract Miniaturization increases more complexity in Integrated Circuits (IC) as well as circuit boards. Several electronic components are assembled on a Printed Circuit Board (PCB). The board is a composite material of a fiber reinforced thermosetting polymer based matrix. Within and on the board, conducting traces and vertical interconnect access (via) carry out electrical signals. So-called JOULE's heat causes a cyclic occurrence of thermal stresses on vias and traces leading to plastic deformation and fatigue related damage. Multiphysics simulations make a study of this coupled and nonlinear system possible. We demonstrate an attempt to compute an electro-thermo-mechanical system and a damage problem by using finite element method in space and finite difference method in time.

Keywords Multiphysics · Damage · Electro-thermo-mechanics · FEM

1 Introduction

Metals fail under monotonous loading after reaching the ultimate tensile strength (UTS) by showing a ductile behavior. Even if the UTS is not reached, under a cyclic

---

B. E. Abali (✉)
Department of Materials Science and Engineering, Uppsala University, Uppsala, Sweden
e-mail: bilenemek@abali.org

F. Aldakheel
Institute of Continuum Mechanics, Leibniz University Hannover, Garbsen, Germany
e-mail: aldakheel@ikm.uni-hannover.de

T. I. Zohdi
Department of Mechanical Engineering, University of California, Berkeley, CA, USA
e-mail: zohdi@berkeley.edu

F. Aldakheel et al. (eds.), *Current Trends and Open Problems in Computational Mechanics*, https://doi.org/10.1007/978-3-030-87312-7_1

deformation, a failure is also possible known as fatigue. In electronic devices, as a natural consequence of switching on and off, where signals are sent in bits, the fatigue failure is an indispensable phenomenon characterizing the lifetime of an electronic system. There are different methods to estimate the lifetime by using empirical relations [1–3] as well as computations [4–7]. One possible experimental setting, called an active test, is as follows. A so-called bare circuit board—a board without components on it—is put under a cyclic electric current. The current produces JOULE's heat that alters the temperature [8–10]. The composite PCB has adhering copper and fiber reinforced plastics, both of different coefficients of thermal expansion. Hence, the thermal strain mismatch between the adhering materials creates thermal stresses. These stresses are high enough to deform the traces and vias irreversibly and the plastic deformation is accumulated in every cycle of the electric current loading.

In this work we present the method of solving a simple system in thermo-electro-damage mechanics by using a numerical algorithm by means of the open-source packages developed under the FEniCS project, see [11–13]. Coupled and nonlinear field equations are computed by using the finite element method in space and finite difference method in time. For preprocessing we use Salome CAD and Netgen meshing algorithm, for visualization we use ParaView. An accelerated reliability test is performed by applying the electric loading on a single thru-hole via on PCB.

2 Governing Equations

The system is composed of an anisotropic, linear elastic composite material and ductile, conductive copper. On both ends of the copper trace, as in a real experiment, we set the electric potential, ϕ in V(olt), as a harmonic function. Effected by the electric potential difference, an electric field, $\boldsymbol{E}$ in V/m(eter), leads to an electric current, $\boldsymbol{\mathcal{J}}$ in A(mpere)/m^2, measured on the material frame. This current is the effective motion of charges with respect to the continuum body and is modeled by OHM's law:

$$\mathcal{J}_i = \varsigma E_i \;, \tag{1}$$

where the electrical conductivity, ς, is a material dependent parameter. We ignore magnetization related effects in this study, thus, the electric field is given by the ansatz function,

$$E_i = -\phi_{,i} \;. \tag{2}$$

We use a comma notation $(-)_{,i}$ for partial derivative in space. All tensors are expressed in Cartesian coordinates. The continuum body deforms with respect to the laboratory frame. The electric current in the laboratory frame is given by

$$J_i = \mathcal{J}_i + v_i \rho z \;, \tag{3}$$

where ρ denotes the mass density in k(ilo)g(ram)/m^3, z the specific charge in C(oulomb)/kg, and $\boldsymbol{v}$ the velocity of the continuum body, which is the rate of displacement, $\boldsymbol{v} = \boldsymbol{u}^{\bullet}$. By assuming small deformations we refrain from distinguishing between reference and current placement, for an elaborate discussion we refer to [14]. Hence, the time rate $(\cdot)^{\bullet}$ is simply the partial time derivative. The starting point is the balance of electric charge:

$$\frac{\partial(\rho\, z)}{\partial t} + J_{i,i} = 0 \,, \tag{4}$$

where and henceforth we understand the EINSTEIN summation convention over repeated indices. By using MAXWELL's equation:

$$\rho\, z = D_{i,i} \,, \tag{5}$$

with the charge potential (electric displacement) $\boldsymbol{D}$ in C/m^2, we acquire

$$\frac{\partial D_{i,i}}{\partial t} + J_{i,i} = 0 \,. \tag{6}$$

This governing equation will be used to compute the electric potential, ϕ. Copper is a conductor, so its electric polarization is not significant. Also for the composite board, the polarization is neglected as well. Hence, we assume that both materials for the printed circuit board are unpolarized. For the charge potential we use the MAXWELL–LORENTZ aether relation:

$$D_i = \varepsilon_0 E_i \,, \tag{7}$$

with the universal constant $\varepsilon_0 = 8.85 \cdot 10^{-12}$ C/(V m). The electric current produces so-called JOULE's heat in addition to plastic deformation. These effects alter temperature T, see [15, 16]. For computing the temperature distribution, we use the balance of entropy:

$$\rho\eta^{\bullet} + \Phi_{i,i} - \rho\frac{r}{T} = \Sigma \,, \tag{8}$$

where the specific (per mass) entropy, η, its flux term, $\boldsymbol{\Phi}$, and its production term, Σ, needs to be defined. The entropy supply (radiant heat) r vanishes in this application. By referring to the thermodynamical study for unpolarized materials in [13, Sect. 3.3], we apply the following entropy flux and entropy production:

$$\Phi_i = \frac{q_i}{T} \,, \quad \Sigma = -\frac{q_i}{T^2}T_{,i} + \frac{1}{T}\mathscr{J}_i E_i + \frac{1}{T}\sigma_{ij}\, {}^{\mathrm{p}}\varepsilon^{\bullet}_{ij} \,. \tag{9}$$

Second term of the entropy production is related to the JOULE's heat and the third term is the internal friction caused production in plasticity, by the mechanical stress, $\boldsymbol{\sigma}$, and plastic strain, ${}^{\mathrm{p}}\boldsymbol{\varepsilon}$. For the heat flux we use FOURIER's law:

$$q_i = -\kappa T_{,i} \,, \tag{10}$$

with the material parameter κ called the thermal conductivity. For the entropy, we use a simple material with the specific heat capacity, c, and coefficients of thermal expansion, $\boldsymbol{\alpha}$, as follows:

$$\eta = c \ln\left(\frac{T}{T_{\text{ref.}}}\right) + \frac{1}{\rho}\alpha_{ij}\sigma_{ij} \,. \tag{11}$$

Mechanical stress, $\boldsymbol{\sigma}$, is modeled by HOOKE's law with DUHAMEL–NEUMANN extension,

$$\sigma_{ij} = C_{ijkl}(\varepsilon_{kl} - {}^{\text{p}}\varepsilon_{kl} - {}^{\text{th}}\varepsilon_{kl}) \,, \tag{12}$$

where we use the initial temperature as the reference (no strains condition) value, $T_{\text{ref.}} = 300\,\text{K}$ in the linear thermal strain:

$${}^{\text{th}}\varepsilon_{ij} = \alpha_{ij}(T - T_{\text{ref.}}) \,. \tag{13}$$

Plastic strain, ${}^{\text{p}}\boldsymbol{\varepsilon}$, is modeled by the PRANDTL–REUSS approach. Linear strain measure,

$$\varepsilon_{ij} = \frac{1}{2}(u_{i,j} + u_{j,i}) \,, \tag{14}$$

is the kinematic relation to displacements, $\boldsymbol{u}$. For computing the displacement field, we use the balance of linear momentum with the electromagnetic interaction governed by the production term, $\mathscr{F}$, on the right-hand side,

$$\rho u_i^{\bullet\bullet} - \sigma_{ji,j} - \rho f_i = \mathscr{F}_i \,, \tag{15}$$

where the specific body force, $\boldsymbol{f}$, is simply the gravitational acceleration, which is negligible regarding the thermal stress caused deformation. For unpolarized systems, the production term is the LORENTZ force density:

$$\mathscr{F}_i = \rho\, z\, E_i + (\boldsymbol{J} \times \boldsymbol{B})_i = D_{j,j}\, E_i \,, \tag{16}$$

since we have assumed that magnetic flux, $\boldsymbol{B}$, vanishes. The displacement has to fulfill Eq. (15) such that we obtain

$$\rho \frac{\partial^2 u_i}{\partial t^2} - \sigma_{ji,j} - D_{j,j}\, E_i = 0 \,. \tag{17}$$

Additive decomposition of strain, $\boldsymbol{\varepsilon} = {}^{\text{r}}\boldsymbol{\varepsilon} + {}^{\text{p}}\boldsymbol{\varepsilon}$, requires the aforementioned small displacements. Reversible part of strains, ${}^{\text{r}}\boldsymbol{\varepsilon}$, contains the elastic as well as thermal strains.

In time, we use a finite difference scheme for the discretization. So-called time step, $\mathrm{d}t$, is kept constant. In order to calculate current (unknown) plastic strain, ${}^{\text{p}}\boldsymbol{\varepsilon}$, we utilize the (known) plastic strain from the last time step, ${}^{\text{p}}\boldsymbol{\varepsilon}^0$, incrementally,

$${}^{\text{p}}\varepsilon_{ij} = {}^{\text{p}}\varepsilon_{ij}^0 + \mathrm{d}t\, {}^{\text{p}}\varepsilon_{ij}^{\bullet} \,. \tag{18}$$

By exploiting the PRANDTL–REUSS approach with kinematic hardening, by so-called back stress, $\boldsymbol{\beta}$, we obtain

$$ {}^{\mathrm{p}}\varepsilon^{\bullet}_{mn} = \langle\gamma\rangle \frac{(\sigma^0_{|ij|} - \beta^0_{ij})C_{ijkl}(\varepsilon^{\bullet}_{kl} - {}^{\mathrm{th}}\varepsilon^{\bullet}_{kl})}{\frac{4}{9}h\sigma_{\mathrm{Y}}^2 + (\sigma^0_{|ij|} - \beta^0_{ij})C_{ijkl}(\sigma^0_{|kl|} - \beta^0_{kl})}(\sigma^0_{|mn|} - \beta^0_{mn}) \,, \tag{19} $$

where the material parameters h and σ_{Y} are determined from a uniaxial tensile testing. The slope of stress versus plastic strain is given by h. The yield stress σ_{Y} represents the threshold for plastic deformation. MACAULAY brackets in $\langle\gamma\rangle$ defines a conditional parameter as being 1 or 0 depending on the VON MISES equivalent stress, σ_{eq}, defined by the deviatoric stress, $\sigma_{|ij|}$, as follows:

$$ \sigma_{\mathrm{eq}} = \sqrt{\frac{2}{3}\sigma_{|ij|}\sigma_{|ij|}} \,, \quad \sigma_{|ij|} = \sigma_{ij} - \frac{1}{3}\sigma_{kk}\delta_{ij} \,, \tag{20} $$

such that it becomes

$$ \langle\gamma\rangle = \begin{cases} 1 & \text{if } \sigma_{\mathrm{eq}} \geq \sigma_{\mathrm{Y}} \\ 0 & \text{otherwise} \end{cases} \,. \tag{21} $$

For the back stress, $\boldsymbol{\beta}$, we use a linear approach for its evolution,

$$ \beta_{ij} = \beta^0_{ij} + \mathrm{d}t\beta^{\bullet}_{ij} \,, \quad \beta^{\bullet}_{ij} = \bar{c}\;{}^{\mathrm{p}}\varepsilon^{\bullet}_{ij} \,, \tag{22} $$

where we choose $\bar{c} = 2h/3$ in the simulations. Under the usual assumption of 1-D power is equivalent to 3-D power, we obtain the equivalent strain rate:

$$ {}^{\mathrm{p}}\varepsilon^{\bullet}_{\mathrm{eq}} = \sqrt{\frac{3}{2}\,{}^{\mathrm{p}}\varepsilon^{\bullet}_{ij}\;{}^{\mathrm{p}}\varepsilon^{\bullet}_{ij}} \,. \tag{23} $$

The plastic strain accumulates in each cycle (of loading) with the latter rate of equivalent strain, ${}^{\mathrm{p}}\varepsilon_{\mathrm{acc}} = \int_{\mathrm{cycle}} {}^{\mathrm{p}}\varepsilon^{\bullet}_{\mathrm{eq}}\mathrm{d}t$, therefore, we use $e_{\mathrm{acc}} = \sigma_{\mathrm{Y}}\,{}^{\mathrm{p}}\varepsilon_{\mathrm{acc}}$ as an accumulated (irreversible) energy density (per volume). Indeed, the reversible energy density is composed of the elastic part

$$ e_{\mathrm{rev}} = \frac{1}{2}C_{ijkl}(\varepsilon_{ij} - {}^{\mathrm{th}}\varepsilon_{ij} - {}^{\mathrm{p}}\varepsilon_{ij})(\varepsilon_{kl} - {}^{\mathrm{th}}\varepsilon_{kl} - {}^{\mathrm{p}}\varepsilon_{kl}) \,. \tag{24} $$

For computing a damage variable, we use the standard phase-field approach [17–25] with an order parameter, ω, which is 0 for "virgin" material and 1 for fracture. A deformation energy, ψ, which degrades by the degradation function $g(\omega)$, and a dissipated energy because of forming the crack is used, For the deformation energy, we use a history variable, $\psi_{\max} := \max(\psi, \psi_{\max})$, where $:=$ means an algorithmic update (not a mathematical equality). In this way, $\psi_{\max}$ is monotonously growing and enables a non-healing fracture propagation by

$$\psi = \begin{cases} e_{\text{acc}} & \text{if } e_{\text{acc}} \geq G_{\text{c}} \text{ and } e_{\text{rev}} + e_{\text{acc}} \geq \psi_{\max} \\ \psi_{\max} & \text{otherwise} \end{cases} . \tag{25}$$

The fracture toughness, G_{c}, is determined by experiments, for copper we use $K_{Ic} = 15$ MPa $\sqrt{\text{m}}$ and $C_{1111} = 169.1$ GPa in order to obtain $G_{\text{c}} = K_{Ic}^2/C_{1111}$ in the simulation. The choice of the degradation function, $g(\omega)$, is of importance, we used the quadratic form as $g(\omega) = (1-\omega)^2$. Starting with the GRIFFITH's theory and approximating the discrete surface by a so-called crack density function, $\psi_{\text{c}}/G_{\text{c}}\big(2\omega + \ell^2\omega_{,i}\omega_{,i}\big)$ with $\psi_{\text{c}} = 3G_{\text{c}}/(8\sqrt{2}\ell)$. It is governed by the regularized length scale parameter ℓ which accounts for gradient-damage mechanics.

3 Model Problem

We model a circuit board shown in Fig. 1. The used electroplated copper has a grain size of 0.5 µm according to experiments in [26]. We use a realistic wall thickness of 35 µm and propose to model the copper as a cubic material. For this length-scale, a polycristalline structure may be assumed, but we observe that the miniaturization makes such an assumption invalid in the near future. For the composite board, we use an orthotropic material, namely (glass) fiber-reinforced epoxy known as FR4 in electronics industry. The stiffness tensor in VOIGT's matrix notation

$$C_{IJ} = \begin{pmatrix} C_{1111} & C_{1122} & C_{1133} & C_{1123} & C_{1113} & C_{1112} \\ C_{2211} & C_{2222} & C_{2233} & C_{2223} & C_{2213} & C_{2212} \\ C_{3311} & C_{3322} & C_{3333} & C_{3323} & C_{3313} & C_{3312} \\ C_{2311} & C_{2322} & C_{2333} & C_{2323} & C_{2313} & C_{2312} \\ C_{1311} & C_{1322} & C_{1333} & C_{1323} & C_{1313} & C_{1312} \\ C_{1211} & C_{1222} & C_{1233} & C_{1223} & C_{1213} & C_{1212} \end{pmatrix} , \tag{26}$$

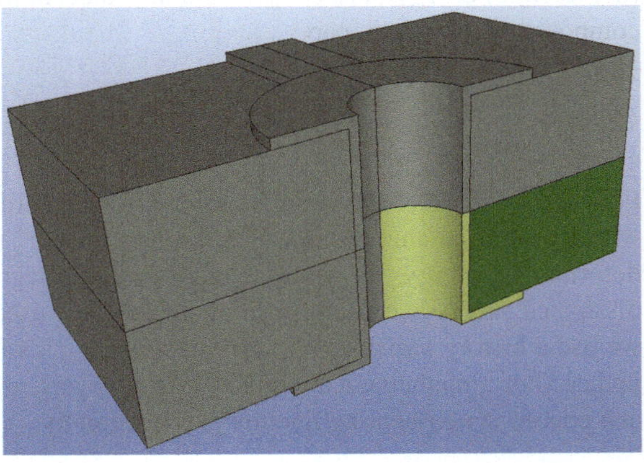

Fig. 1 Geometry used to model a copper via (yellow) embedded in a composite board (green), the gray parts are for the sake of visualization, how the model corresponds to a realistic case, copper wall thickness is 35 µm and the board height is 0.8 mm

or its inverse, the compliance matrix,

$$C_{IJ} = (S_{JI})^{-1}\ ,\quad S_{IJ} = \begin{pmatrix} \frac{1}{E_x} & -\frac{\nu_{xy}}{E_y} & -\frac{\nu_{xz}}{E_z} & 0 & 0 & 0 \\ & \frac{1}{E_y} & -\frac{\nu_{yz}}{E_z} & 0 & 0 & 0 \\ & & \frac{1}{E_z} & 0 & 0 & 0 \\ & & & \frac{1}{G_{yz}} & 0 & 0 \\ & \text{sym.} & & & \frac{1}{G_{zx}} & 0 \\ & & & & & \frac{1}{G_{xy}} \end{pmatrix}\ , \tag{27}$$

is used for implementing engineering constants (moduli). Analogously for the coefficients of thermal expansion, we use

$$\alpha_{ij} = \begin{pmatrix} \alpha_x & 0 & 0 \\ & \alpha_y & 0 \\ \text{sym.} & & \alpha_z \end{pmatrix}\ . \tag{28}$$

All parameters are taken from [27, Table 10] to [28, 29]. For copper we use

$$\begin{aligned} C_{IJ} &= \begin{pmatrix} 169.1 & 122.2 & 122.2 & 0 & 0 & 0 \\ & 169.1 & 122.2 & 0 & 0 & 0 \\ & & 169.1 & 0 & 0 & 0 \\ & & & 75.42 & 0 & 0 \\ & \text{sym.} & & & 75.42 & 0 \\ & & & & & 75.42 \end{pmatrix} \text{GPa}\ , \\ \alpha_{ij} &= \begin{pmatrix} 17 & 0 & 0 \\ 0 & 17 & 0 \\ 0 & 0 & 17 \end{pmatrix} \cdot 10^{-6}\,\text{K}^{-1}\ , \\ \sigma_\text{Y} &= 100\,\text{MPa}\ ,\quad h = 615\,\text{MPa}\ ,\quad \rho = 8\,940\,\text{kg/m}^3\ , \\ c &= 390\,\text{J/(kg K)}\ ,\quad \kappa = 385\,\text{W/(K m)}\ ,\quad \varsigma = 5.8 \cdot 10^7\,\text{S/m}\ . \end{aligned} \tag{29}$$

In the simulation, we reduce the resistivity by one hundred for modeling an electronic circuit with components on it. Of course, the chosen reduction is not enough, on purpose, we obtain high electric current such that the damage occurs in only a few cycles in this so-called "accelerated" test. In reality, these configurations hold up to 10k cycles or more. The lifetime estimation from such accelerated tests is often handled by empirical fit functions with a relatively low accuracy in general.

For the FR4 board, we apply the classical laminate theory and obtain the effective parameters, we refer to [30], and use in the simulation for the board of mass density, $\rho = 2500\,\text{kg/m}^3$, with its thickness along z-axis, for mechanical response,

$$\begin{aligned} E_x = E_y &= 34\,\text{GPa}\ ,\quad E_z = 6\,\text{GPa}\ , \\ \nu_{xy} &= 0.05\ ,\quad \nu_{xz} = \nu_{yz} = 0.08\ , \\ G_{yz} = G_{zx} &= 1.6\,\text{GPa}\ ,\quad G_{xy} = 1.5\,\text{GPa}\ , \end{aligned} \tag{30}$$

and for thermal response,

$$\alpha_x = \alpha_y = 13\,\mu\text{m/(m K)}\ ,\quad \alpha_z = 28.2\,\mu\text{m/(m K)}\ ,\\ c = 800\,\text{J/(kg K)}\ ,\quad \kappa = 1.3\,\text{W/(m K)}\ . \tag{31}$$

Indeed, the board is an insulator, $\varsigma = 0$.

4 Simulation and Results

The primitive variables, ϕ, $\boldsymbol{u}$, T, ω are continuous functions in space and time. We use a staggered scheme such that damage variable ω is solved separately and then ϕ, $\boldsymbol{u}$, T are solved at once. This approach relies on the fact that the damage variable is solved by using a solver with constraints where $\omega = [0;\,1]$ has an upper bound of 1 and a lower bound of 0 as value. For other variables, no constraints are implemented. We approximate space by means of the finite element method (FEM) and time by using finite difference method (FDM). As aforementioned, we discretize in time by a constant time step, $\mathrm{d}t$, and utilize EULER backward method with superscript zeros, $(\cdot)^0$ and $(\cdot)^{00}$, indicating already computed values from the last and second last time steps, respectively. In order to approximate the functions in a discretized space, we multiply the governing equations by appropriate test functions, integrate by parts, and obtain a variational form for each primitive variable,

$$\mathrm{F}_\phi = -\int_{\Omega^e} \big(D_i - D_i^0 + \mathrm{d}t\,J_i\big)\delta\phi_{,i}\mathrm{d}V + \int_{\partial\Omega^e} \big(D_i - D_i^0 + \mathrm{d}t\,J_i\big)\delta\phi N_i \mathrm{d}A\ , \tag{32}$$

$$\mathrm{F}_{\boldsymbol{u}} = \int_{\Omega^e} \Big(\rho\frac{u_i - 2u_i^0 + u_i^{00}}{\mathrm{d}t\mathrm{d}t}\delta u_i + \omega\sigma_{ji}\delta u_{i,j} - D_{j,j}E_i\delta u_i\Big)\mathrm{d}V - \int_{\partial\Omega^e} \sigma_{ji}\delta u_i N_j \mathrm{d}A\ , \tag{33}$$

$$\mathrm{F}_T = \int_{\Omega^e} \big(\rho(\eta - \eta^0)\delta T - \mathrm{d}t\,\Phi_i\delta T_{,i} - \mathrm{d}t\,\Sigma\delta T\big)\mathrm{d}V + \int_{\partial\Omega^e} \mathrm{d}t\,\Phi_i\delta T N_i \mathrm{d}A\ , \tag{34}$$

with N_i being the plane normal pointing outward from Ω^e that is one finite element. The whole computational domain, Ω, consists of two different materials, each material is divided by finite elements satisfying $\mathrm{F}_1 = 0$ with

$$\mathrm{F}_1 = \mathrm{F}_\phi + \mathrm{F}_{\boldsymbol{u}} + \mathrm{F}_T\ . \tag{35}$$

We can assemble by summing over all elements. Adjacent elements have opposed N_i, so continuous primitive variables leads to vanishing boundary terms within one material. Over the interface, $\partial\Omega^I$, between different materials (herein copper and reinforced epoxy), there may occur jumps since the material parameters have different values. First weak form reads

$$\mathrm{F}_\phi = -\int_\Omega (D_i - D_i^0 + \mathrm{d}t\, J_i)\delta\phi_{,i}\mathrm{d}V + \int_{\partial\Omega^I} [\![D_i - D_i^0 + \mathrm{d}t\, J_i]\!] \delta\phi N_i \mathrm{d}A + \\ + \int_{\partial\Omega} (D_i - D_i^0 + \mathrm{d}t\, J_i)\delta\phi N_i \mathrm{d}A\,, \tag{36}$$

where, $N_i [\![D_i]\!] = 0$, since ϕ is continuous. Moreover, no electric current along the surface normal occurs, $N_i [\![J_i]\!] = 0$, as epoxy is an insulator and also on the boundaries we regulate the voltage by DIRICHLET boundary conditions. Second weak form becomes

$$\mathrm{F}_{\boldsymbol{u}} = \int_\Omega \Big(\rho \frac{u_i - 2u_i^0 + u_i^{00}}{\mathrm{d}t\mathrm{d}t} \delta u_i + \omega\sigma_{ji}\delta u_{i,j} - D_{j,j} E_i \delta u_i \Big) \mathrm{d}V - \\ - \int_{\partial\Omega^I} [\![\sigma_{ji}]\!] \delta u_i N_j \mathrm{d}A - \int_{\partial\Omega} \sigma_{ji}\delta u_i N_j \mathrm{d}A\,, \tag{37}$$

where the traction vectors, $t_i = N_j\sigma_{ji}$, are equal in magnitude between neighboring elements—action is equal reaction. In this example, we have free surfaces, i.e. no mechanical loading is applied on boundaries. Third weak form reads

$$\mathrm{F}_T = \int_\Omega \big(\rho(\eta - \eta^0)\delta T - \mathrm{d}t\,\Phi_i \delta T_{,i} - \mathrm{d}t\,\Sigma\delta T \big) \mathrm{d}V + \\ + \int_{\partial\Omega^I} \mathrm{d}t [\![\Phi_i]\!] \delta T N_i \mathrm{d}A + \int_{\partial\Omega} \mathrm{d}t\,\Phi_i \delta T N_i \mathrm{d}A\,. \tag{38}$$

Temperature at the boundary is modeled by using a ROBIN boundary condition such that a deviation from the reference temperature causes a heat flux, $q_i N_i = \bar{h}(T - T_{\text{ref.}})$, depending on the convective heat transfer coefficient $\bar{h}$ in J/(s m^2 K). Finally, we acquire the weak form to be implemented

$$\mathrm{F}_1 = \int_\Omega \Big(-(D_i - D_i^0)\delta\phi_{,i} - \mathrm{d}t\,J_i\delta\phi_{,i} + \rho\frac{u_i - 2u_i^0 + u_i^{00}}{\mathrm{d}t\mathrm{d}t}\delta u_i + \omega\sigma_{ji}\delta u_{i,j} - \\ - D_{j,j}E_i\delta u_i + \rho(\eta - \eta^0)\delta T - \mathrm{d}t\,\Phi_i\delta T_{,i} - \mathrm{d}t\,\Sigma\delta T \Big) \mathrm{d}V + \\ + \int_{\partial\Omega^I} \mathrm{d}t [\![\Phi_i]\!] \delta T N_i \mathrm{d}A + \int_{\partial\Omega} \mathrm{d}t\,\bar{h}(T - T_{\text{ref.}}) \frac{\delta T}{T} \mathrm{d}A\,. \tag{39}$$

For the damage variable, we use the weak form

$$\mathrm{F}_2 = \int_\Omega \Big(\frac{\partial g(\omega)}{\partial\omega} \psi\delta\omega + 2\psi_{\mathrm{c}}\big(\delta\omega + \ell^2 \omega_{,i}\delta\omega_{,i}\big) \Big) \mathrm{d}V\,, \tag{40}$$

where the chosen degradation function delivers, $\frac{\partial g(\omega)}{\partial\omega} = 2(\omega - 1)$.

For all primitive variables, we choose standard, continuous elements with linear shape functions, from the same HILBERTian SOBOLEV space, $\mathcal{H}^1$, we refer to [31] for details of the finite element method. In the case of F_1 in Eq. (39), the polynomial finite elements construct a vector space

$$\mathcal{V} = \left\{ \{\phi, u_i, T\} \in [\mathcal{H}^1(\Omega)]^5 : \{\phi, u_i, T\}\Big|_{\partial\mathcal{B}_D} = \text{given} \right\}, \tag{41}$$

for the computational domain, Ω and its closure, $\partial\Omega$. Values are given on DIRICHLET boundaries, $\partial\Omega_D$. Specifically, the electric potential is given on both ends, by a sinusoidal function of period 10 s. Displacements on the symmetry surfaces are prevented along the surface normals. As known as the GALERKIN procedure, test functions stem from the same type of space

$$\tilde{\mathcal{V}} = \left\{ \{\delta\phi, \delta u_i, \delta T\} \in [\mathcal{H}^1(\mathcal{B})]^5 : \{\delta\phi, \delta u_i, \delta T\}\Big|_{\partial\mathcal{B}_D} = 0 \right\}. \tag{42}$$

The weak form F_1 in Eq. (39) is nonlinear and coupled, therefore, this type of monolithic approach is useful, although computationally costly. We use a triangulation by Netgen algorithms and solve F_1 with around 100k degrees of freedom (DOFs) transiently on a mesh with around 20k nodes. In each time step, a linearization is performed by a standard NEWTON–RAPHSON solver, where the derivative for the JACOBIan is computed by a symbolic computation.

In the case of F_2 in Eq. (40), for the damage variable, ω, we use the same type of linear space with purely zero NEUMANN condition. Physically, this choice facilitates that the damage is not "migrating out" across boundaries. Indeed, for the irreversible crack formation, this choice is adequate. Computationally, a NEWTON–RAPHSON solver may cause instabilities for no DIRICHLET boundaries proble, so we use a line search method instead. Specifically, we use a component from Scalable Nonlinear Equations Solvers (SNES) as part of the PETSc library, which is a semi-smooth solver for variational inequalities (vinewtonssls) based on NEWTON's method [32]. With this method, we solve the minimization problem with additional constraints that the damage parameter is restricted, $\omega \in [0; 1]$. We use a finer mesh of 100k nodes for the damage problem and project the solutions between different meshes by using the generalized minimal residual method (gmres) iterative solver as part of the PETSc library as well.

Choosing a small resistivity on the trace, the electric current under 0.2 V electric potential difference is so high that temperature increase after a quarter of the first cycle is around 350 °C as seen in Fig. 2. The temperature increase is indeed lower in reality, mainly because of shorter signals, higher resistance due to electronic components [33], as well as partly because of using an active ventilation and thus an increased flux across the boundaries, given by the heat transfer coefficient, $\bar{h}$. In an accelerated test, this higher increase is acquired such that the damage starts earlier. Of course, correlation is challenging to determine between small number of cycles in an accelerated test to a realistic lifetime for guaranteed durability of many years. As we start the simulation with $T_{\text{ref.}} = 300\,\text{K}$, which is also the reference value for van-

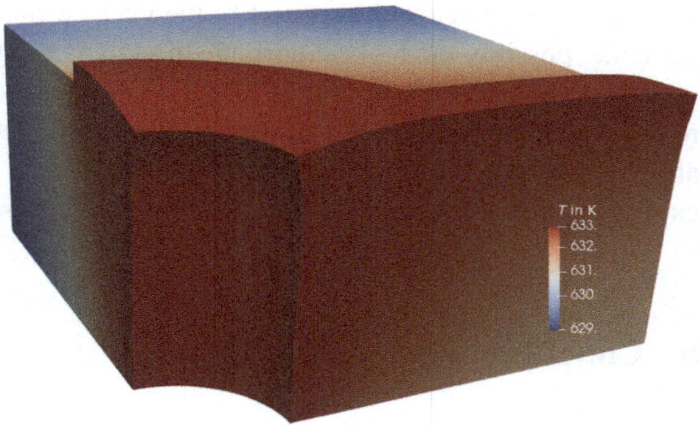

Fig. 2 Temperature distribution on the board, via, and trace at 2.5 s, the quarter of one period of sinusoidal electric loading, visualized on a deformed configuration (deformation is scaled by 50)

ishing thermal strains, temperature increase cause thermal strains. This deformation is different in copper and FR4 board effected by differences of thermal expansion coefficients. The adherence between copper and epoxy is modeled ideally—we are unaware of experiments showing a fracture along the interface. This condition causes extreme thermal strains and a plastic deformation with a released energy higher than the fracture toughness. The damage is steered by this accumulated dissipative energy as well as the reversible deformation energy such that the crack formation is different than the accumulated plastic strains. Already in the second cycle in Fig. 3, we observe

Fig. 3 Damage initiation (top) and propagation (bottom) within the second cycle, visualized in the reference configuration

a crack initiation below the corners of the via as well as roughly in the middle of the trace, i.e. possibly the highest normal stress in analogy with a bending beam.

We emphasize that the model is not axisymmetric because of orthogonal fiber reinforced epoxy board as well as the trace in the geometry. Thus, the crack initiation and propagation happens in a 3-D framework, very challenging to detect in experiments and recently possible by using multiphysics computations.

5 Conclusion

Especially in electronics industry, electro-thermo-mechanical systems pose multiphysics problems to be computed [34, 35]. We have developed a robust numerical scheme, based on [36] verified in [37], for solving electric potential, temperature, and displacement monolithically in transient systems. All electronic devices use a motherboard holding the electric circuitry. The connection of electronic components by copper traces and vias gets damaged by subsequent use. This fatigue related fracture is of interest to estimate the reliability of the circuitry. Since the multiphysics problem is difficult to analyze in full detail, semi-empirical approaches are established, where their prediction capabilities are in question. We have demonstrated a possible, comprehensive method for solving the damage problem by the phase-field approach in order to increase the understanding of this coupled phenomenon. Even a phase-field model incorporating fatigue [38] may be utilized for this problem. In order to allow a transparent and efficient exchange, the developed code under the GNU Public license [39] is made available in [40].

References

1. Solomon, H. D. (1991). Predicting thermal and mechanical fatigue lives from isothermal low cycle data. In *Solder Joint Reliability* (pp. 406–454).
2. Ridout, S., & Bailey, C. (2007). Review of methods to predict solder joint reliability under thermo-mechanical cycling. *Fatigue & Fracture of Engineering Materials & Structures, 30*(5), 400–412.
3. Roellig, M., Dudek, R., Wiese, S., Boehme, B., Wunderle, B., Wolter, K.-J., & Michel, B. (2007). Fatigue analysis of miniaturized lead-free solder contacts based on a novel test concept. *Microelectronics Reliability, 47*(2), 187–195.
4. Atli-Veltin, B., Ling, H., Zhao, S., Noijen, S., Caers, J., Weifeng, L., Feng, G., & Yuming, Y. Thermo-mechanical investigation of the reliability of embedded components in pcbs during processing and under bending loading. In *13th International Conference on Thermal, Mechanical and Multi-Physics Simulation and Experiments in Microelectronics and Microsystems (EuroSimE), 2012* (pp. 1–4). IEEE.
5. Emek Abali, B., Lofink, P., & Müller, W. H. (2014). Variation of plastic materials data of copper and its impact on the durability of Cu-via interconnects. In R. Aschenbrenner & M. Schneider-Ramelow (Eds.), *Microelectronic packaging in the 21st century* (Chapter 7.2, pp. 305–308). Fraunhofer Verlag.

6. Kpobie, W., Martiny, M., Mercier, S., Lechleiter, F., Bodin, L., des Etangs-Levallois, A. L., & Brizoux, M. (2016). Thermo-mechanical simulation of PCB with embedded components. *Microelectronics Reliability, 65*, 108–130.
7. Abali, B. E., Müller, W. H., & Walter, H. (2016). Gekoppelte Berechnung der Durchkontaktierungen in Leiterplatten unter elektrischer Belastung. Volume GMM-Fachbericht 84 *Elektronische Baugruppen und Leiterplatten - EBL 2016* (pp. 206–210). Berlin: VDE Verlag.
8. Zohdi, T. I. (2011). Joule-heating field phase-amplification in particulate-doped dielectrics. *International Journal of Engineering Science, 49*(1), 30–40.
9. Zohdi, T. I. (2012). Estimation of electrical heating load-shares for sintering of powder mixtures. *Proceedings of the Royal Society A: Mathematical, Physical and Engineering Sciences, 468*(2144), 2174–2190.
10. Zohdi, T. I. (2014). On cross-correlation between thermal gradients and electric fields. *International Journal of Engineering Science, 74,* 143–150.
11. FEniCS project. Development of tools for automated scientific computing, 2001–2016. http://fenicsproject.org.
12. Alnaes, M. S., & Mardal, K.-A. (2012). *Automated solution of differential equations by the finite element method, the FEniCS book*, chapter 15 Syfi and SFC: symbolic finite elements and form compilation. Springer.
13. Abali, B. E. (2017). *Computational Reality Advanced Structured Materials* (Vol. 55). Singapore: Springer Nature.
14. Abali, B. E., & Queiruga, A. F. (2019). Theory and computation of electromagnetic fields and thermomechanical structure interaction for systems undergoing large deformations. *Journal of Computational Physics, 394*, 200–231.
15. Zohdi, T. I. (2014). Modeling electrical power absorption and thermally-induced biological tissue damage. *Biomechanics and Modeling in Mechanobiology, 13*(1), 115–121.
16. Ince, C.-V., Chugreeva, A., Böhm, C., Aldakheel, F., Uhe, J., Wriggers, P., et al. (2021). A design concept of active cooling for tailored forming workpieces during induction heating. *Production Engineering, 15,* 177–186.
17. Aldakheel, F. (2016). *Mechanics of nonlocal dissipative solids: Gradient plasticity and phase field modeling of ductile fracture*. Stuttgart: Institut für Mechanik (Bauwesen), Lehrstuhl I, Universität Stuttgart.
18. Dittmann, M., Aldakheel, F., Schulte, J., Schmidt, F., Krüger, M., Wriggers, P., & Hesch, C. (2020). Phase-field modeling of porous-ductile fracture in non-linear thermo-elasto-plastic solids. *Computer Methods in Applied Mechanics and Engineering, 361*, 112730.
19. Aldakheel, F. (2020). A microscale model for concrete failure in poro-elasto-plastic media. *Theoretical and Applied Fracture Mechanics, 107*, 102517.
20. Krüger, M., Dittmann, M., Aldakheel, F., Härtel, A., Wriggers, P., & Hesch, C. (2019). Porous-ductile fracture in thermo-elasto-plastic solids with contact applications. *Computational Mechanics*, 1–26.
21. Aldakheel, F., Noii, N., Wick, T., & Wriggers, P. (2020). A global-local approach for hydraulic phase-field fracture in poroelastic media. *Computers & Mathematics with Applications*. https://doi.org/10.1016/j.camwa.2020.07.013.
22. Yang, S., Aldakheel, F., Caggiano, A., Wriggers, P., & Koenders, E. (2020). A review on cementitious self-healing and the potential of phase-field methods for modeling crack-closing and fracture recovery. *Materials, 13*(22), 5265.
23. Wriggers, P., Aldakheel, F., Lohaus, L., & Heist, M. (2020). Water-induced damage mechanisms of cyclically loaded high-performance concretes. *Bauingenieur, 95*(4), 126–132.
24. Kienle, D., Aldakheel, F., & Keip, M.-A. (2019). A finite-strain phase-field approach to ductile failure of frictional materials. *International Journal of Solids and Structures, 172,* 147–162.
25. Wriggers, P., Aldakheel, F., & Hudobivnik, B. (2019). Application of the virtual element method in mechanics. *GAMM-Rundbriefe, 1*(2019), 4–10.
26. Song, J.-M., Wang, D.-S., Yeh, C.-H., Wei-Chih, L., Tsou, Y.-S., & Lin, S.-C. (2013). Texture and temperature dependence on the mechanical characteristics of copper electrodeposits. *Materials Science and Engineering: A, 559,* 655–664.

27. Ledbetter, H. M., & Naimon, E. R. (1974). Elastic properties of metals and alloys. II. Copper. *Journal of Physical and Chemical Reference Data,* **3**(4), 897–935.
28. Kupferinstitut, D. (2014). Kupfer in der Elektrotechnik—Kabel und Leitungen. Retrieved from www.kupferinstitut.de.
29. Srikanth, N., Premkumar, J., Sivakumar, M., Wong, Y. M., & Vath, C. J. (2007). Effect of wire purity on copper wire bonding. In *2007 9th Electronics Packaging Technology Conference. EPTC 2007* (pp. 755–759). IEEE.
30. Abali, B. E. (2017). Computational study for reliability improvement of a circuit board. *Mechanics of Advanced Materials and Modern Processes, 3*(1), 1–11.
31. Zohdi, T. I. (2018). *Finite element primer for beginners*. Springer.
32. Benson, S.J., & Munson, T. S. (2006). Flexible complementarity solvers for large-scale applications. *Optimization Methods and Software, 21*(1), 155–168.
33. Abali, B. E., & Zohdi, T. I. (2018). On the accuracy of reduced-order integrated circuit simulators for computing the heat production on electronic components. *Journal of Computational Electronics, 17*(2), 625–636.
34. Zohdi, T. I. (2010). Simulation of coupled microscale multiphysical-fields in particulate-doped dielectrics with staggered adaptive FDTD. *Computer Methods in Applied Mechanics and Engineering, 199*(49–52), 3250–3269.
35. Abali, B. E., & Zohdi, T. I. (2020). Multiphysics computation of thermal tissue damage as a consequence of electric power absorption. *Computational Mechanics, 65*(1), 149–158.
36. Abali, B. E., & Reich, F. A. (2017). Thermodynamically consistent derivation and computation of electro-thermo-mechanical systems for solid bodies. *Computer Methods in Applied Mechanics and Engineering, 319,* 567–595.
37. Abali, B. E., & Reich, F. A. (2020). Verification of deforming polarized structure computation by using a closed-form solution. *Continuum Mechanics and Thermodynamics, 32*(3), 693–708.
38. Schreiber, C., Kuhn, C., Müller, R., & Zohdi, T. (2020). A phase field modeling approach of cyclic fatigue crack growth. *International Journal of Fracture, 225*(1), 89–100.
39. GNU Public. (2007). Gnu general public license. Retrieved June, 2007, from http://www.gnu.org/copyleft/gpl.html.
40. Abali, B. E. (2020). Supply code for computations. Retrieved 2020, from http://bilenemek.abali.org/.

Phase-Field Modeling of Fatigue Crack Propagation in Brittle Materials

Fadi Aldakheel, Chistoph Schreiber, Ralf Müller, and Peter Wriggers

This paper is dedicated to my Habilitation advisor Professor Peter Wriggers in acknowledgement of his pioneering contributions to computational mechanics and modern element technologies. I had the great fortune of being a post-doctoral student, group leader and currently chief engineer at his institute. Thank you Peter very much for your confidence in my work and the continuous support in the scientific community. Congratulations and many more successful, healthy and happy years to come! (F. Aldakheel).

Abstract In this contribution, the phase-field PF approach to brittle fracture is extended to model fatigue failure in the high cyclic regime. Fatigue is the primary failure mode for more than 90% of mechanical failures. It occurs when a structure is subjected to repeated loading at stress levels that are below the yield stress of the material. On the modeling side, a local energy accumulation variable which takes the loading history of a structure into account is introduced within the PF formulation. This is inserted into a fatigue degradation function which degrades the fracture material properties. To this end, only *one* additional parameter is proposed, that enables the reproduction of main material fatigue features. The model performance is demonstrated by two representative numerical examples.

F. Aldakheel (✉) · P. Wriggers
Institute of Continuum Mechanics, Leibniz University Hannover, Garbsen, Germany
e-mail: aldakheel@ikm.uni-hannover.de

P. Wriggers
e-mail: wriggers@ikm.uni-hannover.de

C. Schreiber · R. Müller
Institute of Applied Mechanics, Technische Universität Kaiserslautern, Kaiserslautern, Germany
e-mail: schreibc@rhrk.uni-kl.de

R. Müller
e-mail: ram@rhrk.uni-kl.de

F. Aldakheel et al. (eds.), *Current Trends and Open Problems in Computational Mechanics*, https://doi.org/10.1007/978-3-030-87312-7_2

1 Introduction

Engineering materials which are subjected to time-varying loads can fail at a stress level that is significantly lower than yielding or ultimate strength. This phenomenon is known as *fatigue failure*. It accounts for the vast majority of mechanical engineering failures worldwide. This happens without any warning compared with static loading; i.e. it does not form a neck; thus its failure is unpredictable. The bolts in an office chair, the crank arm on the bicycle, the pressurized oil pipelines are just a few examples of components that are subjected to time-varying loads and may be at risk of fatigue failure. Three stages process can describe material fatigue: (i) Crack formation which usually occurs at free surfaces and stress concentrations; (ii) the crack grows in size in a stable and continuous manner; and in stage (iii) after the crack has grown to a critical size, abrupt and rapid fracture occurs. It is common to differentiate between high cycle (HCF) and low cycle (LCF) fatigue. This work focus on the high cyclic fatigue that occurs when the applied cyclical stresses are low and failure occurs after a large number of cycles, typically more than $10,000$ cycles. Because of the low stresses, only elastic deformation will be considered before failure.

A great number of macro-meso-micro-nano-mechanically motivated approaches exist in the literature to model fatigue failure behavior, see for example [1] and the citations therein. Specifically, the continuum phase-field model for high cyclic fatigue life prediction is employed in this work. Due to its simplicity, this methodology has gained wide interest and started to be used in the engineering community since 2008. From there on many scientists have worked in this field and developed phase-field approaches for finite elements, isogeometric analysis, and lately also for the virtual element technology. The main driving force for these developments is the possibility to handle complex fracture phenomena within numerical methods in two and three dimensions. In recent years, several brittle [2–4] and ductile [5–7] phase-field fracture formulations have been proposed in the literature.

The phase-field fracture framework has been very recently extended to the fatigue crack propagation problems in [8–11]. In this work, a phenomenological fatigue fracture model able to reproduce the main features of high-cycle fatigue is presented. Hereby, a new energy accumulation variable is introduced to account for the cyclic loading history. The presented PF extension contains only one additional material parameter.

2 Phase-Field Modeling of Brittle Fracture

In this section, we outline a variational approach to brittle fracture in elastic solids at small strains.

2.1 Basic Kinematics at Small Strains

Let $\Omega \in \mathcal{R}^\delta$ with $\delta = 2, 3$ be a solid domain. The response of fracturing solid at material points $\boldsymbol{x} \in \Omega$ and time t is described by the displacement field $\boldsymbol{u}(\boldsymbol{x}, t)$ and the crack phase-field $d(\boldsymbol{x}, t)$ with $\dot{d} \geq 0$. Here, $d(\boldsymbol{x}, t) = 0$ and $d(\boldsymbol{x}, t) = 1$ represent the unbroken and fully broken state of the material, respectively. The gradient of the displacement field defines the symmetric strain tensor of the geometrically linear theory as

$$\boldsymbol{\varepsilon} = \nabla_s \boldsymbol{u} := \frac{1}{2}\left(\nabla \boldsymbol{u} + \nabla \boldsymbol{u}^T\right). \tag{1}$$

For the phase-field problem, a sharp-crack surface topology $\Gamma \rightarrow \Gamma_l$ is regularized by the crack surface functional

$$\Gamma_l(d) = \int_\Omega \gamma_l(d, \nabla d)\, dV \quad \text{with} \quad \gamma_l(d, \nabla d) = \frac{1}{2l} d^2 + \frac{l}{2} |\nabla d|^2 \tag{2}$$

based on the crack surface density function γ_l per unit volume of the solid and the fracture length scale parameter l that governs the regularization. Hence, the combination of elasticity with the first-order gradient damage modeling focuses on the set

$$\text{Constitutive State Variables: } \boldsymbol{C} := \{\boldsymbol{\varepsilon}, d, \nabla d\}. \tag{3}$$

2.2 Constitutive Work Density Function

The constitutive work density function W is assumed to depend on the constitutive state variables $\boldsymbol{C}$ introduced in Eq. (3). It consists of the sum

$$W(\boldsymbol{C}) = W_{bulk}(\boldsymbol{\varepsilon}, d) + W_{frac}(d, \nabla d) \tag{4}$$

of a degrading elastic bulk energy W_{bulk} and a contribution due to fracture W_{frac}, which contains the accumulated dissipative energy. The elastic bulk contribution in Eq. (4) is assumed to be a quadratic function as

$$W_{bulk}(\boldsymbol{\varepsilon}, d) = g(d)\ \psi(\boldsymbol{\varepsilon}) \quad \text{with} \quad \psi(\boldsymbol{\varepsilon}) = \frac{\lambda}{2}(\text{tr}\boldsymbol{\varepsilon})^2 + \mu\ \text{tr}(\boldsymbol{\varepsilon}^2) \tag{5}$$

where, $\lambda > 0$ and $\mu > 0$ are the elastic Lame constants. The function $g(d) = (1 - d)^2$ models the degradation of the stored elastic energy of the solid due to fracture. The fracture contribution in Eq. (4) is defined as

$$W_{frac}(d, \nabla d) = [1 - g(d)]\ \psi_c + 2\frac{\psi_c}{\zeta}\ l\ \gamma_l(d, \nabla d) \tag{6}$$

where, $\psi_c > 0$ is a critical fracture energy and ζ controls the post-critical range after crack initialization.

2.3 Governing Equations

2.3.1 Balance of Linear Momentum

The first equation is the stress equilibrium or the quasi-static form of the balance of linear momentum defined as

$$\text{Div}\,[\boldsymbol{\sigma}] = \mathbf{0}\,, \tag{7}$$

by neglecting volume forces. The stress tensor is obtained from W in (4) by

$$\boldsymbol{\sigma} := \partial_{\boldsymbol{\varepsilon}} W = (1-d)^2 \widetilde{\boldsymbol{\sigma}} \quad \text{with} \quad \widetilde{\boldsymbol{\sigma}} = \lambda \text{tr}[\boldsymbol{\varepsilon}]\mathbf{1} + 2\mu\boldsymbol{\varepsilon} \tag{8}$$

with the effective stress tensor $\widetilde{\boldsymbol{\sigma}}$.

2.3.2 The Fracture Phase-Field Equation

Evolution of the regularized crack surface functional (2) can be driven by the constitutive functions as outlined in [4, 6], postulating a global evolution equation of regularized crack surface as

$$\frac{d}{dt}\Gamma_l(d) =:= \frac{1}{l}\int_{\Omega} [\,(1-d)\mathcal{H} - \eta\dot{d}\,]\,\dot{d}\,\text{d}v \geq 0\,, \tag{9}$$

where $\eta \geq 0$ is a material parameter that characterizes the artificial/numerical viscosity of the crack propagation. The crack driving force

$$\mathcal{H} = \max_{s\in[0,t]} D(\boldsymbol{x}, s) \geq 0 \quad \text{with} \quad D{:=}\zeta\left\langle \frac{\psi^+(\boldsymbol{\varepsilon}^+)}{\psi_c} - 1 \right\rangle_+ \tag{10}$$

is introduced as a local history variable that accounts for the irreversibility of the phase-field evolution by filtering out a maximum value of what is known as the crack driving state function D. This is achieved by introducing the Macaulay bracket $\langle x \rangle_+ := (x + |x|)/2$. Note that only the tensile/positive part of the *elastic* energy in $(5)_2$ is considered for computing the crack driving force. It is defined in terms of the positive strain tensor $\boldsymbol{\varepsilon}^+ {:=} \sum_{a=1}^{\delta} \langle \varepsilon_a \rangle_+ \boldsymbol{N}_a \otimes \boldsymbol{N}_a$. Here, $\{\varepsilon_a\}_{a=1..\delta}$ are the principal elastic strains and $\{\boldsymbol{N}_a\}_{a=1..\delta}$ are the principal strain directions. Then the evolution statement (9) provides the local equation for the evolution of the crack phase-field in the domain Ω along with its homogeneous Neumann boundary condition as

$$[d - l^2 \Delta d] + \eta \dot{d} + (d-1)\mathcal{H} = 0 \tag{11}$$

with $\nabla d \cdot \boldsymbol{n} = 0$ on $\partial\Omega$. Here, $\boldsymbol{n}$ represents the outward normal on $\partial\Omega$.

3 Phase-Field Model for Fatigue Crack Growth

The above introduced phase-field model for brittle fracture is now extended to account for fatigue phenomena. This extension contains only one additional material parameter, namely the critical fatigue energy ψ_f. This is achieved by modifying the fracture energy degradation due to the repeated externally applied loads. To this end, the constitutive work density function (4) is modified as follows

$$W(\boldsymbol{C}) = W_{bulk}(\boldsymbol{\varepsilon}, d) + f(\bar{\psi}) W_{frac}(d, \nabla d) \tag{12}$$

with fatigue degradation function that reduce fracture material properties as

$$f(\bar{\psi}) := \begin{cases} 1 & \text{if} \quad \bar{\psi} < \psi_f \\ \left(\frac{2\,\psi_f}{\bar{\psi}+\psi_f}\right)^2 & \text{if} \quad \bar{\psi} \geq \psi_f \end{cases} \tag{13}$$

defined in terms of the accumulated energy $\bar{\psi}$. This is accumulating (active) only during the loading and for the unloading kept constant. The local energy accumulation function $\bar{\psi}$ is chosen such that it depends only on the positive part of the bulk elastic energy $(5)_2$. The incremental evolution of $\bar{\psi}$ is given through the following choice:

$$\bar{\psi}_{n+1} = \bar{\psi}_n + \Delta\bar{\psi} \quad \text{with} \quad \Delta\bar{\psi} = \left|\psi^+_{n+1} - \psi^+_n\right| \cdot H\left(\frac{\psi^+_{n+1} - \psi^+_n}{\Delta t}\right) \quad \text{and} \quad \Delta t = t_{n+1} - t_n \tag{14}$$

where $H(*)$ is the Heaviside function and Δt is the time increment. The subscribe $(\cdot)_n$ denotes quantities evaluates at time t_n. Following same derivation procedure for the governing equations as outlined in Sect. 2.3, it is obvious that such a fatigue extension only changes the crack driving state function D in $(10)_2$ as

$$D := \zeta \left\langle \frac{\psi^+}{f(\bar{\psi}) \cdot \psi_c} - 1 \right\rangle_+ \tag{15}$$

The above modified equations along with Sect. 2.3 for the phase-field formulation of fatigue failure are then converted into a numerical model using the finite element method (FEM).

4 Numerical Examples

We now demonstrate the performance of the proposed fatigue phase-field model by means of representative boundary value problems. On the computational side, a robust and efficient *monolithic scheme* is employed using the software tool ACEFEM in the numerical implementation to compute the unknowns (displacement $\boldsymbol{u}$ and crack phase-field d). Within each load step a quadratic convergence is achieved, due to the fact that all formulations are linearized in a consistent manner using ACEGEN.

4.1 Single-Edge Notched Test

The first benchmark test considers a square plate ($L = 1$ mm) with a horizontal notch placed at the middle height. The prescribed cyclic loading is plotted in Fig. 1, representing HCF with 10^5 cycles. The elastic parameters are chosen as $E = 21$ kN/mm^2 and $\nu = 0.3$, the viscosity of the crack propagation $\eta = 10^{-6}$ kNs/mm^2, the critical fracture energy $\psi_c = 1.35 \times 10^{-3}$ kN/mm^2, the critical fatigue energy $\psi_f = 40$ kN/mm^2 and the fracture length scale $l = 0.004$ mm. The specimen is discretized using FEM with linear triangles, as depicted in Fig. 2 (left). A mesh refinement in the expected fracture zone is applied. Furthermore, Fig. 2 shows the contour-plot of d for different cyclic loading states up to final rupture. Herein, the crack phase-field initiates at the notch-tip for 20000 cycles and propagates horizontally till separation reaching 10^5 cycles. Figure 3 shows the force associated with the displacement (left) and the length of a developing crack (right) as a function of the cycles number. As expected, the force decreases during cyclic loading. In turn, the crack length grows with increasing the number of cycles until the complete failure.

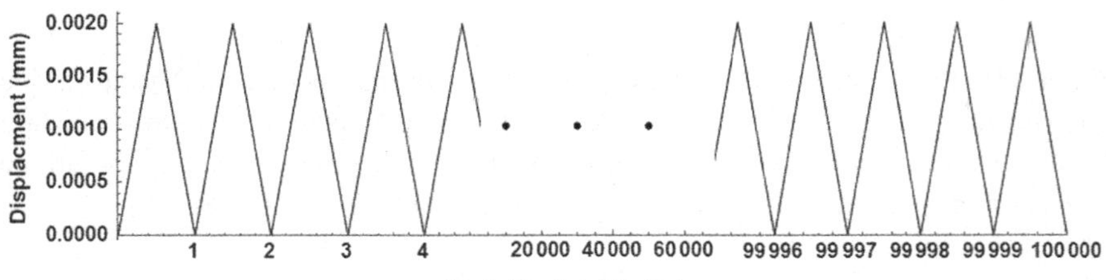

Fig. 1 Single-edge notched test. Applied loading for high cycle fatigue (HCF)

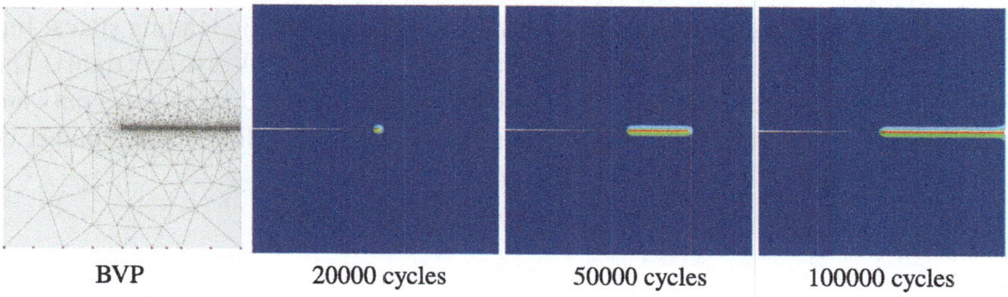

Fig. 2 Single-edge notched test. Contour plots of the fracture phase-field d for different cyclic loading states up to final failure

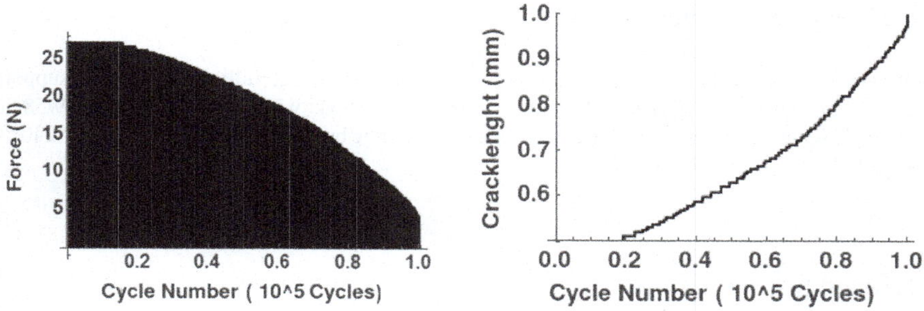

Fig. 3 Single-edge notched test. Degradation of force (left) and increasing crack length (right) depending on the number of cycles for HCF (10^5 cycles)

4.2 *Block with Multiple Holes Under Cyclic Loading*

The second model problem is concerned with analyzing brittle failure of a block with multiple holes subjected to cyclic tension/compression loading. As a geometrical setup, a square plate with length $L = 1$ mm is considered. To test the formulations performance, arbitrarily distributed circular holes are inserted inside the block, ranging between $0.01 - 0.1$ mm. We fix the bottom edge of the block and apply a cyclic tension/compression loading at the top edge until the plate is fully broken. Figure 4

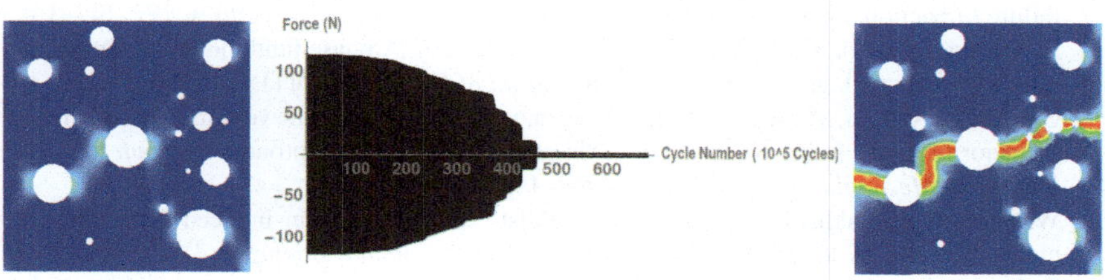

Fig. 4 Block with multiple holes. Contour plots of the phase-field d along with the force degradation for different cyclic loading states up to final failure

demonstrates the phase-field d evolution for two different cyclic loading states: Initiation at around 10^7 cycles (left) and final failure at 4.5×10^7 cycles (right). The crack starts to initiates around the bigger hole in the center and propagates outward until rupture. Furthermore, the degradation of the force depending on the number of cycles (HCF) is also illustrated in Fig. 4 (middle).

5 Conclusion

In this contribution a numerical model was developed, which allows simulations of fatigue crack propagation in brittle materials.

Acknowledgements The *corresponding author* Fadi Aldakheel gratefully acknowledges support for this research by the "German Research Foundation" (**DFG**) within **SPP** 2020-WR 19/58-2. Christoph Schreiber and Ralf Müller acknowledge the funding by **DFG** within **IRTG** 2057-2524083 and **SPP** 1748-255846293.

References

1. Radaj, D., & Vormwald, M. (2013). *Advanced methods of fatigue assessment*. Berlin: Springer. https://doi.org/10.1007/978-3-642-30740-9
2. Kuhn, C., & Mueller, R. (2010). A continuum phase field model for fracture. *Engineering Fracture Mechanics, 77*(18), 3625–3634.
3. Aldakheel, F., Noii, N., Wick, T., & Wriggers, P. (2020). A global-local approach for hydraulic phase-field fracture in poroelastic media. *Computers & Mathematics with Applications*. https://doi.org/10.1016/j.camwa.2020.07.013
4. Noii, N., Aldakheel, F., Wick, T., & Wriggers, P. (2020). An adaptive global-local approach for phase-field modeling of anisotropic brittle fracture. *Computer Methods in Applied Mechanics and Engineering, 361*, 112744.
5. Aldakheel, F., Hudobivnik, B., & Wriggers, P. (2019). Virtual element formulation for phase-field modeling of ductile fracture. *International Journal for Multiscale Computational Engineering, 17*(2), 181–200.
6. Aldakheel, F. (2020). A microscale model for concrete failure in poro-elasto-plastic media. *Theoretical and Applied Fracture Mechanics, 107*, 102517.
7. Kienle, D., Aldakheel, F., & Keip, M. A. (2019). A finite-strain phase-field approach to ductile failure of frictional materials. *International Journal of Solids and Structures, 172*, 147–162.
8. Schreiber, C., Kuhn, C., Mueller, R., & Zohdi, T. (2020). A phase field modeling approach of cyclic fatigue crack growth. *International Journal of Fracture, 225*(1), 89–100.
9. Carrara, P., Ambati, M., Alessi, R., & De Lorenzis, R. (2020). A framework to model the fatigue behavior of brittle materials based on a variational phase-field approach. *Computer Methods in Applied Mechanics and Engineering, 361*, 112731.
10. Wriggers, P., Aldakheel, F., Lohaus, L., & Heist, M. (2020). Water-induced damage mechanisms of cyclically loaded High-performance concretes. *Bauingenieur, 95*(4), 126–132.
11. Seiler, M., Linse, L., Hantschke, P., & Kaestner, M. (2020). An efficient phase-field model for fatigue fracture in ductile materials. *Engineering Fracture Mechanics, 224*, 106807.

A Non-intrusive Global/Local Cycle-Jumping Techniques: Application to Visco-Plastic Structures

Olivier Allix, Maxime Blanchard, and Pierre Gosselet

I've had the privilege to interact closely with Peter Wriggers for fifteen years or so, as for the IRTG 1627 [11]. I now have the great pleasure and honor to work with him on exciting projects, for example on the Virtual Element Method. I am every year more impressed by the depth and breadth of his research, but also by his avant-gardiste winning choices as investing in tools like AceGen. Peter is a gentleman and a very close friend. I hope to have the chance to continue working with him, it is such a pleasure!—O. Allix.

Abstract This paper is a first attempt to make possible the computation of the cycling response of a complex elastoviscoplastic structure up to its possible limit cycle. For this, we try to couple the non-invasive global/local iterative coupling technique with the cycle-jumping method. The different issues regarding the method are discussed on the basis of a 2D examples.

1 Introduction

In order to correct the drawbacks of the submodeling technique while keeping its simplicity and flexibility, a non-invasive method was proposed which allows exact *local/global* analysis while making use of the same basic tools inside an iterative

---

O. Allix (✉)
LMT, École normale Paris-Saclay/CNRS, Gif-sur-Yvette, France
e-mail: olivier.allix@ens-paris-saclay.fr

M. Blanchard
Safran Aircraft Engines, Courcouronnes, France
e-mail: maxime.blanchard@safrangroup.com

P. Gosselet
LaMcube, Université de Lille/CNRS/Centrale Lille, Lille, France
e-mail: pierre.gosselet@univ-lille.fr

F. Aldakheel et al. (eds.), *Current Trends and Open Problems in Computational Mechanics*, https://doi.org/10.1007/978-3-030-87312-7_3

procedure [8]. It was extended in several directions, notably in [7, 9, 10, 14, 15]. Safran Aircraft Engines was interested in exploring its potential for the detailed analysis of complex structures undergoing viscoplastic strains that can spread over the whole structure.

The quality of the integration of viscoplastic models is very sensitive to the time step. The global and local models do not need the same time steps to achieve the same precision. To handle this issue, we proposed in [2] a space/time global/local non-invasive coupling strategy that we try to extend to cycle-jumping.

For the class of material that we study, one may observe locally accommodation (elastic response around the hardened position), adaptation (stable viscoplastic cycle), or even ratcheting (the viscoplastic cycle evolves continuously). In general, the structure experiences large variations during the firsts cycles, with potentially strong stress redistribution. After that, the evolution is much smoother. The principle to reduce the computational cost is, after the first cycles, to extrapolate the response of the structure over a certain number of cycles [6, 16]. As such, the technique has the potential for high efficiency but depends strongly on the choice of the criteria to decide the number of cycles which can be skipped. In [1, 4] such drawback is alleviated by controlling the time interpolation using the Latin Method but the latter is quite intrusive.

The paper is organized as follows. In Sect. 2, the constitutive law, the structural example and the reference model are presented. The main lines of the proposed cycle-jumping method are given in Sect. 3. The principle of the coupling with the two-scale in time and space global/local approach is discussed and illustrated in Sect. 4. Conclusion and perspectives are provided in Sect. 5 on the basis of the cycle-jumping procedure applied to a 3D example.

2 Reference Problem and Solution

The material model used is the one proposed in [13], adapted from the Marquis-Chaboche's behavior [3]. The elasticity itself is linear and isotropic. The nonlinear part of the model is ruled by the yield function based on von Mises criterion. The plastic strain tensor ϵ^p is split into a fast part (f) and a slow part (s) with the associated values of the cumulated plasticity p as follows:

$$\dot{\epsilon}^p = \dot{\epsilon}^p_f + \dot{\epsilon}^p_s \quad and \quad p_i = \int_0^t \sqrt{\frac{3}{2}\dot{\epsilon}^p_i : \dot{\epsilon}^p_i}\, d\tau \quad \text{with for } i \in \{f, s\} \tag{1}$$

The kinematic hardening $X = X_f$ is only related to the fast cumulated plasticity and follows an Armstrong-Frederick's formulation. The fast plasticity dominates for strain rates in the range $\left[10^{-5},\ 10^{-2}\right] s^{-1}$ whereas the slow one dominates in the range $\left[10^{-9},\ 10^{-5}\right] s^{-1}$. The material of interest being confidential, we make use of the parameters given in [12] for a nickel based superalloy IN100 evolving at 800 °C which is the mean temperature value during a flight.

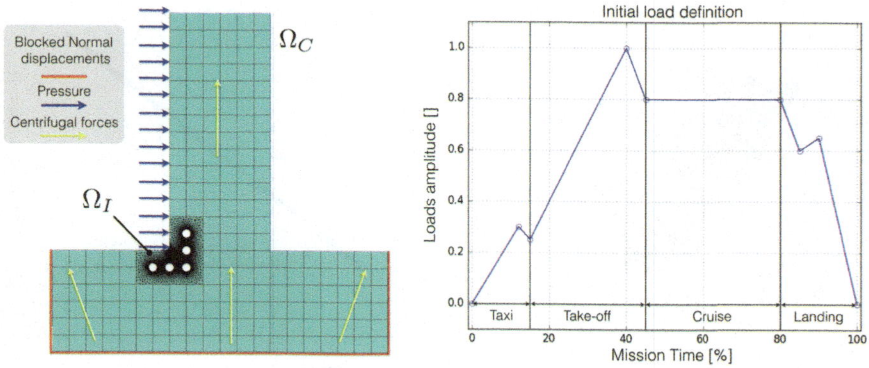

Fig. 1 Monolithic approach: reference model and definition of one cycle

In order to precisely assess the lifespan of the structure, a damage criterion should be added as in [17]. The 2D example of Fig. 1 is used in the paper.

The problem outlines a turbine blade. The external loads (pressure, temperature, centrifugal force) evolve in time according to the curve of Fig. 1 which mimics the main phases experienced by an engine during a flight. Asymptotic elastic behaviors like perfect elasticity or accommodation lead to high-cycle fatigue whereas plastic shakedown (adaptation) or ratcheting conduct to low-cycle fatigue.

The example of this paper corresponds to fatigue with adaptation because the viscous effects due to large plasticity dominate the response at the considered high temperature regime. As a first approximation, we consider that the structure is stabilized when the maxima of both total plastic strain and displacement increments are respectively below 10^{-6} and 10^{-5}, which occurs at about 150 cycles in this example.

3 Summary of the Chosen Cycle-Jumping Technique

In what follows, $Y_i(c)$ denotes the value of the internal variable Y_i at the end of cycle number c, $Y_i'(c) = Y_i(c) - Y_i(c-1)$ approximates its "derivative" and $Y_i''(c) = Y_i'(c) - Y_i'(c-1)$ its "second derivative".

In this section, we try to summarize the chosen cycle-jumping technique which is adapted from [5]. Note that the straight procedure as described in that paper fails in our examples, which stresses the difficulty of setting a proper cycle-jumping procedure:

1. The first aspect is the choice of a pertinent internal variable, or set of internal variables, to estimate the number of cycles Δc which can be skipped. In our case, it appears that the best choice is the fast cumulated plasticity p_f computed at the end of cycles (the evolution of this quantity constitutes the shakedown curve as shown in Fig. 2).
2. At any stage of the process, the computation of the possible jump is made after the full computation of three cycles. To compute the jump Δc, one first

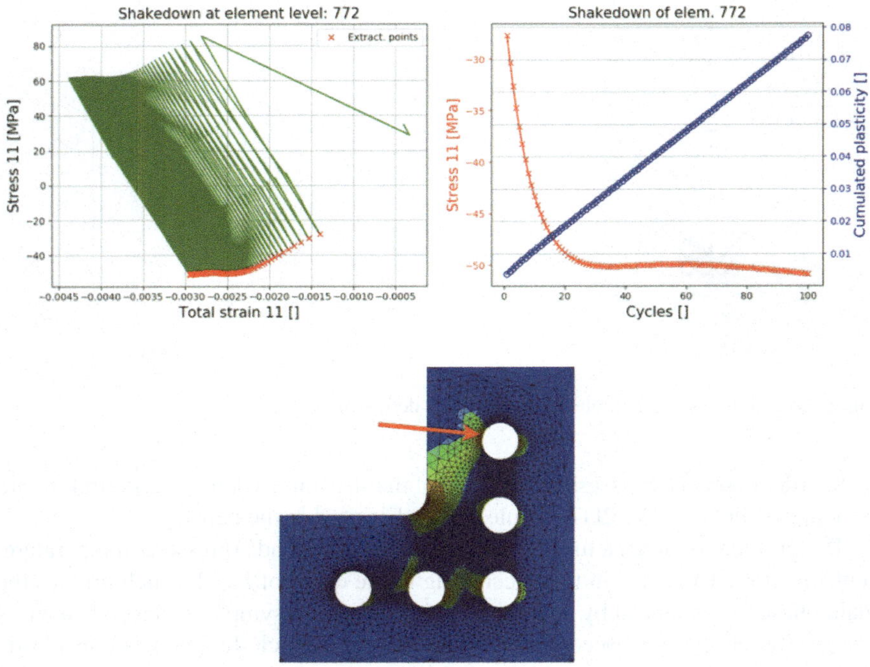

Fig. 2 The values at the end of cycles (top left, red dots), constitute the shakedown curve (top right), for the selected Gauss point (bottom)

eliminates the Gauss points that are stabilized, that is for which $p''_f < 10^{-12}$. The remaining set of Gauss points is denoted by GP. The chosen value is then $\Delta c = q_{Extr} Mean_{GP} \left\lfloor \frac{p'_f[c]}{p''_f[c]} \right\rfloor$.

3. The quality factor q_{Extr} is computed once for all from the previous formulae by forcing a first jump of fixed length $\Delta c = 2$ after the first three cycles.
4. To extrapolate the whole set of internal variables, we consider a first approximation, denoted $\tilde{Y}'_i$, of the slope after the jump of any internal variable $\tilde{Y}'_i \cong Y'_i(c) + Y''_i(c)\Delta c$. The slope used to extrapolate Y_i is defined as $0.7Y'_i + 0.3\tilde{Y}'_i$ that is $Y_i(c + \Delta c) \equiv Y_i(c) + (0.7Y'_i + 0.3\tilde{Y}'_i)\Delta c$.
5. After the jump, one computation is needed to rebalance the structure starting from the extrapolated values of the set of internal variables, before computing the next three cycles.

Figure 2 illustrates the first three aspects of the process.

4 Coupling with the Global/Local Method

The nonlinear global/local non-invasive coupling proposed in [2] is used in order to insert structural details given by local models into the global coarse representation of the whole structure, see Fig. 1. Thanks to an iterative coupling the solution obtained by the separate models converges toward the reference. In order to improve the performance, as explained in the introduction, the solutions are weakly coupled in time: the synchronization only occurs at the time steps necessary for the global model, see [2] for more details.

The quality of the integration of viscoplastic models is very sensitive to the size of the time steps. The global and local models do not need the same time step to achieve the same precision regarding the integration. Moreover, if one does not achieve a compatible precision between the two models, the local-global procedure may not converge to the reference. To handle this problem, we proposed and compared in [2] various space/time global/local non-invasive coupling strategies. The most efficient one, called weak time-coupling, which allowed us to achieve the desired precision, is schematised in Fig. 3.

After the computation of three consecutive cycles by the previous coupling algorithm, the question is which model to choose for the estimation of the jump length. The critical areas being located in the local model, a common skipped length, computed on the local model, is used in the global and local models. Then the extrapolation strategy is applied to the global and local models and the weak coupling is applied for the next 3 cycles. Figure 4 summarises the cycle-jumping procedure with about 40 cycles computed over the 150 cycles simulated.

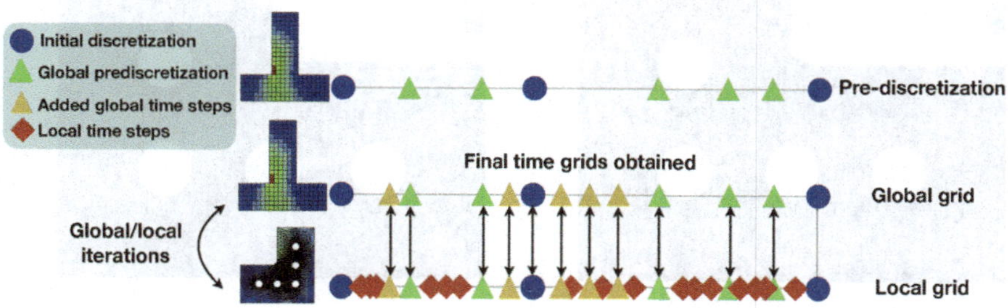

Fig. 3 Schematisation of the weak-time coupling strategy

Fig. 4 Overall view of the computed and skipped cycles

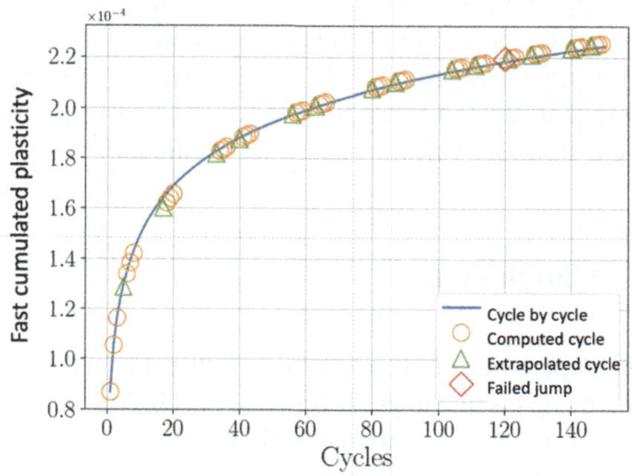

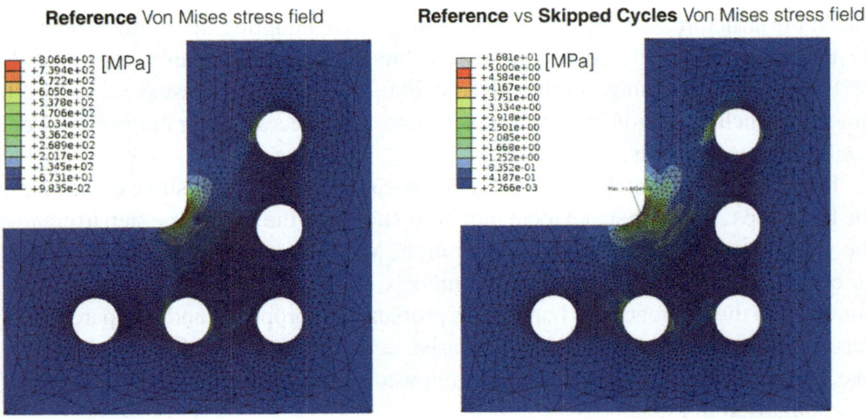

Fig. 5 Accuracy obtained with the cycle-jumping method: von Mises stress

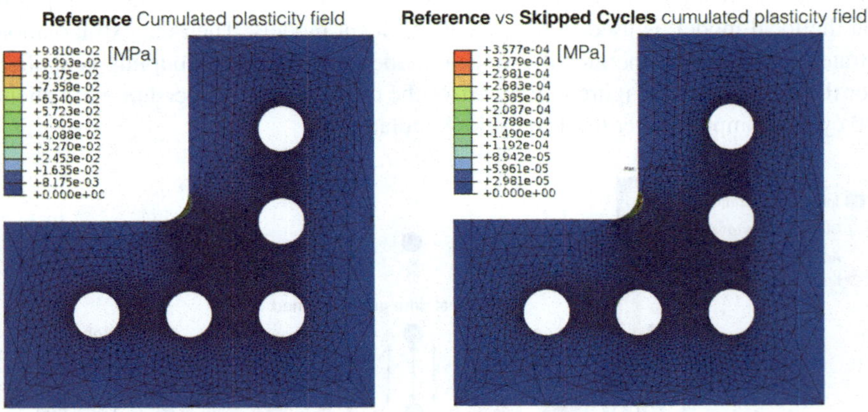

Fig. 6 Accuracy obtained with the cycle-jumping method: total cumulated plasticity

Figures 5 and 6 compare the von Mises stress and the cumulated plasticity at the last cycle for the reference and cycle-jumping procedures. The maximum relative errors are respectively about 2 and 0.4% compared to a fine monolithic computation. Note that the computation is never stopped just after a jump, but after few normal cycle computations in order to let plastic redistribution smooth the solution.

5 Conclusion

Our experience of cycle-jumping is that it is still an art to define a proper procedure. Maybe this is due to the severity of the chosen example which reaches nearly perfect viscoplasticity on some critical areas of the structure. For too large jumps or not precise enough extrapolations, rebalance may not even be possible. Our proposed

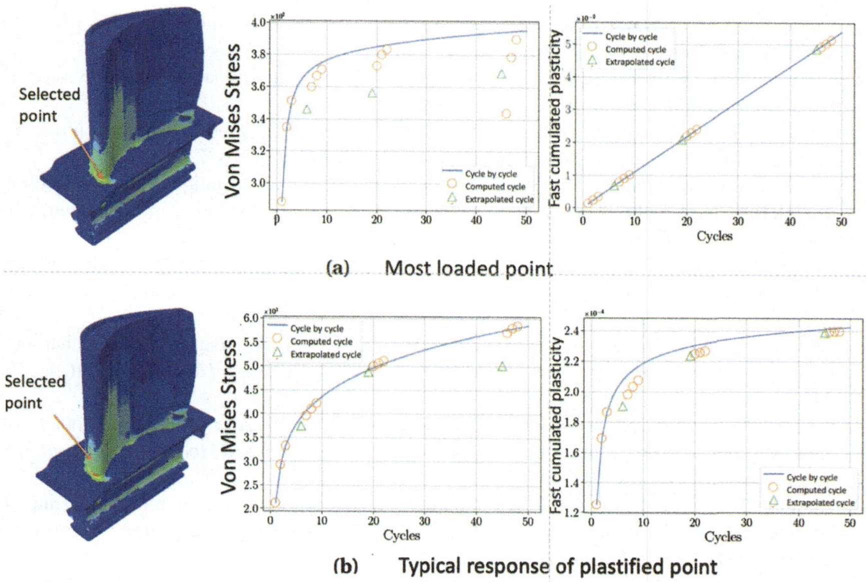

Fig. 7 Application of the proposed cycle-jumping procedure to a 3D global model

procedure leads to satisfactory results in terms of precision even if the factor of 4 for the gain that we obtained is not as high as what can be observed on examples from the literature about plasticity.

Let us note that it seems that 2D cases are more severe than 3D cases. This may be due to the fact that in 3D examples, a whole section never fully evolves in a plastic manner. This is at least what appears from our first experiments concerning the application of the proposed cycle-jumping procedure on a global 3D example of 1.510^6 degrees of freedom over 50 cycles (see Fig. 7). In this example, the exact same procedure as the one defined in Sect. 3 was applied.

What is still lacking, and which would probably imply a more intrusive procedure, is to include an evaluation of the quality of the solution within the process.

Acknowledgements This work was supported by Safran Aircraft Engines as part of the MAIA-MM1 research program.

References

1. Bhattacharyya, M., Fau, A., Desmorat, R., Alameddin, S., Neron, D., Ladevèze, P., & Nackenhorst, U. (2019). A kinetic two-scale damage model for high-cycle fatigue simulation using multi-temporal Latin framework. *European Journal of Mechanics A-Solids, 77,* 1880–1892.
2. Blanchard, M., Allix, O., Gosselet, P., & Desmeure, G. (2019). Space/time global/local noninvasive coupling strategy: Application to viscoplastic structures. *Finite Elements in Analysis and Design, 156,* 1–12.

3. Chaboche, J.-L. (1989). Constitutive equations for cyclic plasticity and cyclic viscoplasticity. *International Journal of Plasticity, 5*(3), 247–302.
4. Cognard, J.-Y., & Ladevèze, P. (1993). A large time increment approach for cyclic viscoplasticity. *International Journal of Plasticity, 2*(9), 141–157.
5. Cojocaru, D., & Karlsson, A. M. (2006). A simple numerical method of cycle jumps for cyclically loaded structures. *International Journal of Fatigue, 28,* 1677–1689.
6. Dunne, F. P. E., & Hayhurst, D. R. (1994). Efficient cycle jumping techniques for the modelling of materials and structures under cyclic mechanical and thermal loading. *European Journal of Mechanics A-Solids, 5*(13), 639–660.
7. Duval, M., Passieux, J.-C., Salaün, M., & Guinard, S. (2016). Non-intrusive coupling: Recent advances and scalable nonlinear domain decomposition. *Archives of Computational Methods in Engineering, 23*(1), 17–38.
8. Gendre, L., Allix, O., Gosselet, P., & Comte, F. (2009). Non-intrusive and exact global/local techniques for structural problems with local plasticity. *Computational Mechanics, 44*(2), 233–245.
9. Gerasimov, T., Noii, N., Allix, O., & de Lorenzis, L. (2018). A non-intrusive global/local approach applied to phase-field modeling of brittle fracture. *Advanced Modeling and Simulation in Engineering Sciences, 5,* 1.
10. Guguin, G., Allix, O., Gosselet, P., & Guinard, S. (2014). Nonintrusive coupling of 3D and 2D laminated composite models based on finite element 3D recovery. *International Journal for Numerical Methods in Engineering, 98*(5), 324–343.
11. Leibniz Universität Hannover and Ècole Normale Supèrieure de Cachan. (1627). Internation Research and Training Group (IRTG 1627). https://www.irtg1627.uni-hannover.de/.
12. Lemaitre, J., Chaboche, J. L., Benallal, A., & Desmorat, R. (2009). *Mécanique des matériaux solides 3 °édition*. Dunod.
13. Longuet, A., Burteau, A., Comte, A., & Crouchez-Pilot, A. (2013). Incremental lifing method applied to high temperature aeronautical component. In *Actes du 11eme colloque national en calcul des structures* (pp. 703–708).
14. Noii, N., Aldakheel, F., Wick, T., & Wriggers, P. (2020). An adaptive global-local approach for phase-field modeling of anisotropic brittle fracture. *Computer Methods in Applied Mechanics and Engineering, 361*, 766–777.
15. Passieux, J.-C., Réthoré, J., Gravouil, A., & Baietto, M.-C. (2013). Local/global non-intrusive crack propagation simulation using a multigrid x-fem solver. *Computational Mechanics, 52*(6), 1381–1393.
16. Saï, K. (1993). *Modèles à grand nombre de variables internes et méthodes numériques associées*. Ph.D. thesis, École Nationale Supérieure des Mines de Paris.
17. Yue, P., Ma, J., Zhou, C., Jiang, H., & Wriggers, P. (2020). A fatigue damage accumulation model for reliability analysis of engine components under combined cycle loadings. *Fatigue and Fracture of Engineering Materials and Structures, 43*, 1880–1892.

VEM Approach for Homogenization of Fibre-Reinforced Composites with Curvilinear Inclusions

Edoardo Artioli

I first met Prof. Wriggers through his books, when I was a Ph.D. student. Subsequently, I had multiple opportunities to attend his courses and seminars at prestigious international research institutions such as University of California at Berkeley where we first met personally in 2010. I was more recently involved in collaborative research with Prof. Wriggers and his co-workers during a research stay in Hannover at Leibniz University for the development of VEM methods in 2D contact mechanics. I personally find Peter a brilliant scientist and a great computational mechanicist, and a very charming and friendly person to go along with. My sincere wishes on his 70th birthday.

Abstract We propose a curvilinear virtual element method (VEM) for the asymptotic homogenization of fibre-reinforced composites with straight long fibres having general curvilinear cross sections. This technique is able to exactly represent the microstructural curvilinear geometry still granting all the standard features of VEM methods for elliptical boundary value problems. The method is here applied to doubly periodic fibre arrangements. Accuracy and computational efficiency of the proposed homogenization procedure is confirmed by numerical examples by comparison with semi-analytical solutions.

1 Introduction

Composite materials are extensively used materials in many engineering applications due to their interesting properties, as, for instance, high strength-to-weight ratio and tunable features of the constituents.

The present communication focuses on *fibre reinforced* composite materials analysed via *asymptotic homogenization method*. In particular, the analysis is here devel-

E. Artioli (✉)
Department of Civil Engineering and Computer Science, University of Rome - Tor Vergata, Rome, Italy
e-mail: artioli@ing.uniroma2.it

F. Aldakheel et al. (eds.), *Current Trends and Open Problems in Computational Mechanics*, https://doi.org/10.1007/978-3-030-87312-7_4

oped for composites with long fibre-like inclusions having random size and shape of the cross section, and doubly periodic space distribution within the hosting medium. In this latter case, the computation of homogenized quantities will require solving a boundary value problem at the microscale on the unit cell domain [1–3].

In this framework, a major issue of micro scale computational modeling is represented by meshing curved fibre/matrix subdomains and relevant interfaces thus requiring efficient discretization for any realization and any given loading condition for a composite.

Recently, the Virtual Element Method (VEM) has been introduced and proved an efficient alternative to standard finite element method [4, 5]. It represents a generalization of the FE method with the capability of dealing with very general polygonal/polyhedral meshes. The VEM has already been successfully adopted to solve linear elasticity problems [6–8], as well as with complex material nonlinearity such as plasticity, viscoelasticity, damage and shape memory problems, see, e.g. [9–13] for a short representative list of related works. In the framework of computational homogenization, VEM based procedures with straight edges have been proposed in [14, 15], for evaluating homogenized material moduli of a doubly periodic composite material reinforced by cylindrical circular inclusions, either with linear elastic or inelastic material behavior, while the same problem with random inclusion has been tackled with a VEM procedure in [16].

In this communication we present a curvilinear VEM method (i.e. with the possibility of using curvilinear polygonal elements [17–20]) for the antiplane shear homogenization problem of doubly periodic composites with fibres having general cross section. In particular, VEM elements characterized by linear and higher order polynomial approximation are proposed. Homogeneous and functionally graded constitutive laws are considered for the fibre constituents of the composite. Numerical applications are developed to assess the effectiveness of the proposed VEM elements by comparisons with more established techniques showing efficiency of the proposed methodology.

2 Asymptotic Homogenization of Doubly Periodic Fibre Reinforced Composite Materials

We here consider a composite material with two material components, a surrounding matrix with long cylindrical fibre-like inclusions, embedded into it according to a doubly periodic grid characterized by an angle ϕ, as can be seen in Fig. 1a. The bimaterial microstructure in the plane orthogonal to the ffibres consists of a two dimensional array of unit cells, developing periodically along the x_1 and ϕ directions, see Fig. 1b. The cell sides measure L_1 and L_2 respectively, being ϕ the cell angle.

In order to compute the effective material shear moduli of the composite via asymptotic homogenization a family of problems is introduced, indexed by a param-

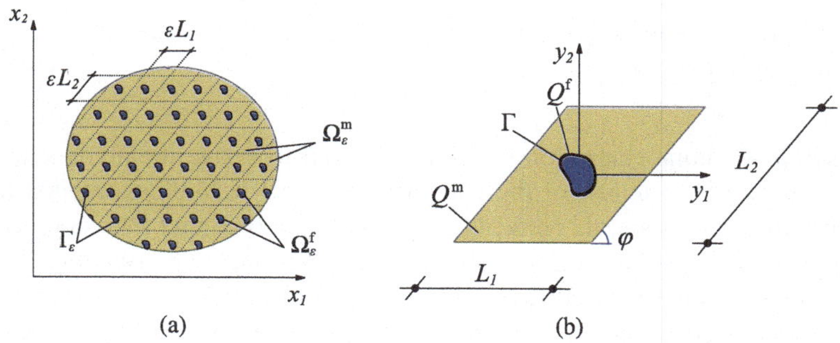

Fig. 1 Composite material with long cylindrical inclusions. **a** Microstructure lattice with cell doubly periodic arrangement. **b** Unit cell fibre/matrix geometric features

eter $\boldsymbol{\varepsilon}$: the ratio of the microstructure size to the total size of the analysis region (Fig. 1a). The homogenization limit is obtained by letting $\boldsymbol{\varepsilon}$ go to zero.

In the framework of antiplane shear deformation, the problem of determining the longitudinal displacement field w_{ε} in the composite domain is stated as follows:

$$\operatorname{div}(\boldsymbol{G}\nabla w_{\varepsilon}) = 0\,, \qquad \text{in } \Omega_\varepsilon^{\mathrm{f}} \cup \Omega_\varepsilon^{\mathrm{m}}\,; \tag{1}$$

$$[\![\boldsymbol{G}\nabla w_{\varepsilon} \cdot \boldsymbol{\nu}]\!] = 0\,, \qquad \text{on } \Gamma_{\varepsilon}\,; \tag{2}$$

$$\boldsymbol{G}\nabla w_{\varepsilon} \cdot \boldsymbol{\nu} = \frac{1}{\boldsymbol{\varepsilon}} D[\![w_{\varepsilon}]\!]\,, \qquad \text{on } \Gamma_{\varepsilon}\,. \tag{3}$$

Here $\Omega_\varepsilon^{\mathrm{f}}$ and $\Omega_\varepsilon^{\mathrm{m}}$ denote fibre and matrix domains respectively, Γ_{ε} is the union of fibre/matrix interfaces, $\boldsymbol{\nu}$ is the normal unit vector to Γ_{ε} pointing into $\Omega_\varepsilon^{\mathrm{m}}$, and square brackets $[\![\,\cdot\,]\!]$ denote the jump of the enclosed quantity across the interface, defined as extra-fibre value minus intra-fibre value.

Equation (1) is the field equilibrium equation; Eq. (2) represents the continuity of the normal-to-interface component of the shear stress hence equilibrium at fibre/matrix interface; (3) describes the interface constitutive law, being D a material parameter characterizing fibre/matrix strength. These equations must be complemented by suitable boundary conditions on the boundary of the domain $\Omega = \Omega_\varepsilon^{\mathrm{f}} \cup \Gamma_{\varepsilon} \cup \Omega_\varepsilon^{\mathrm{m}}$.

Fibres and matrix are assumed to be linear elastic, and their shear moduli are collected in the constitutive tensor $\boldsymbol{G}$, which specializes in

$$\boldsymbol{G} = \boldsymbol{G}^{\mathrm{f}} \;\text{ in } \Omega_\varepsilon^{\mathrm{f}}\,, \qquad \boldsymbol{G} = \boldsymbol{G}^{\mathrm{m}} \;\text{ in } \Omega_\varepsilon^{\mathrm{m}}\,. \tag{4}$$

Fibre/matrix interfaces are assumed to have zero-thickness and can encompass a spring-layer model, with linear relation for the displacement discontinuity $[\![w_{\varepsilon}]\!]$ and interface traction $\boldsymbol{G}\nabla w_{\varepsilon} \cdot \boldsymbol{\nu}$, with D a given spring constant parameter [21–23]. According to this model, interfaces have a physical thickness t which, though much smaller than the microstructural length scales L_1 and L_2, rescales as the latter ones in the homogenization process.

2.1 Homogenized Equilibrium Equation and Effective Material Moduli

The asymptotic homogenization method employed to derive the homogenized or effective constitutive tensor of the composite material is briefly recapped in this section. More details and theoretical background may be found for example in [1, 3] and in [24, 25] for the specific problem of antiplane shear deformation.

As shown in Fig. 1a, two different length scales characterize the problem under consideration. Hence, two different space variables are introduced: the macroscopic one, x, and the microscopic one, $y = x/\varepsilon$, $y \in Q$, being Q the unit cell (see Fig. 1b), whose intra-fibre space, extra-fibre space and fibre-matrix interface are denoted by Q^{f}, Q^{m} and Γ, respectively. Accordingly, the divergence and gradient operators are given by the following relations:

$$\mathrm{div} = \mathrm{div}_x + \frac{1}{\varepsilon}\,\mathrm{div}_y\,, \qquad \nabla = \nabla_x + \frac{1}{\varepsilon}\nabla_y\,. \tag{5}$$

An asymptotic expansion of the unknown displacement field is considered in the form:

$$w_\varepsilon(x, y) = w_0(x, y) + \varepsilon w_1(x, y) + \varepsilon^2 w_2(x, y) + \cdots\,, \tag{6}$$

where w_0 is the macroscopic or average value of the field variable, w_1, w_2 are Q-periodic functions in y representing perturbations in the field variable due to the microstructure, with zero integral average over Q.

Introducing the cell function $\boldsymbol{\chi}(y)$, the function w_1 is represented in the following form [1, 3]:

$$w_1(x, y) = -\boldsymbol{\chi}(y) \cdot \nabla_x w_0(x)\,, \tag{7}$$

where the components χ_h, $h = 1, 2$, are the unique, null average, Q-periodic solutions of the ensuing cell problem [24, 25].

The problem for w_2 hence results:

$$\mathrm{div}_y[\boldsymbol{G}(\nabla_y w_2 + \nabla_x w_1)] = -\mathrm{div}_x[\boldsymbol{G}(\nabla_y w_1 + \nabla_x w_0)]\,, \qquad \text{in } \mathrm{d}^{\mathrm{f}} \cup \mathrm{d}^{\mathrm{m}}\,; \tag{8}$$

$$[\![\boldsymbol{G}(\nabla_y w_2 + \nabla_x w_1) \cdot \boldsymbol{\nu}]\!] = 0\,, \qquad \text{on } \Gamma; \tag{9}$$

$$\boldsymbol{G}(\nabla_y w_2 + \nabla_x w_1) \cdot \boldsymbol{\nu} = D[\![w_2]\!]\,, \qquad \text{on } \Gamma\,. \tag{10}$$

Integrating (8) both in Q^{f} and in Q^{m}, using the Gauss-Green Lemma, adding the two contributions and exploiting (9), the following equation is obtained:

$$\frac{1}{|Q|}\int_{\mathrm{d}^{\mathrm{f}} \cup \mathrm{d}^{\mathrm{m}}} \mathrm{div}_x[\boldsymbol{G}(\nabla_y w_1 + \nabla_x w_0)]\mathrm{d}a = 0\,, \tag{11}$$

where $\mathrm{d}a$ is the area element of $\mathrm{d}^{\mathrm{f}} \cup \mathrm{d}^{\mathrm{m}}$ and $|\cdot|$ is the Lebesgue measure. Substituting (7) into (11), the homogenized equation for the macroscopic displacement w_0 is finally derived:

$$\mathrm{div}_x(\boldsymbol{G}^{\#}\nabla_x w_0) = 0\,. \tag{12}$$

Here $\nabla_x w_0$ is the macroscopic shear strain, and

$$\boldsymbol{G}^{\#} = \frac{1}{|Q|}\int_{\mathrm{d}^{\mathrm{f}}\cup\mathrm{d}^{\mathrm{m}}} \boldsymbol{G}(\mathbf{I} - \nabla_y^{\mathsf{T}}\boldsymbol{\chi})\,\mathrm{d}a \tag{13}$$

are the effective shear moduli, where the superscript ^{T} denotes the transpose.

Equation (13) yields the *effective shear moduli* of the composite material in terms of the cell function $\boldsymbol{\chi}$, solution of the cell problem. In the following section, a curvilinear virtual element methodology to solve the above problem for various is presented.

3 C^0 Curved Virtual Element Method

A weak formulation for the cell problem is provided by the virtual work principle [15, 16]. In this regard, the space of the admissible auxiliary cell functions $\tilde{\boldsymbol{\chi}}$ which are shift d-periodic is introduced, i.e. for $s \in \{1, 2\}$:

$$\tilde{\mathbf{V}} = \Big\{\tilde{\chi} \in L^2(\mathrm{d}) \text{ such that } \tilde{\chi}\,|_{\mathrm{d}^{\mathrm{f}}} \in H^1(\mathrm{d}^{\mathrm{f}}) \text{ for },$$
$$\tilde{\chi}\,|_{\mathrm{d}^{\mathrm{m}}} \in H^1(\mathrm{d}^{\mathrm{m}}),\ \tilde{\chi}(y_1, y_2) + y_s \text{ is d-periodic}\Big\}.$$

We denote by $\mathbf{V}$ the space of the admissible d-periodic variations of $\tilde{\mathbf{V}}$. The bilinear form characterizing the variational formulation is:

$$a(\tilde{\chi}_s, \delta\chi_s) = -\int_{\mathrm{d}} \mathrm{div}_y[\boldsymbol{G}(\nabla_y\tilde{\chi}_s)]\,\delta\chi_s\,\mathrm{d}\mathbf{x} \tag{14}$$

which, applying Gauss-Green lemma, considering the constitutive equation and that unit normal vectors to $\partial\mathrm{d}^{\mathrm{m}}$ on opposite sides of the unit cell are opposite, becomes:

$$a(\tilde{\chi}_s, \delta\chi_s) = \int_{\mathrm{d}} \nabla_y\delta\chi_s \cdot \boldsymbol{G}(\nabla_y\tilde{\chi}_s)\mathrm{d}\mathbf{x} + \sum_{j=1}^{F}\int_{\Gamma} [\![\delta\chi_s]\!]\, D\, [\![\tilde{\chi}_s]\!]\,\mathrm{d}. \tag{15}$$

The form $a(\cdot, \cdot)$ is symmetric, continuous and coercive on $\tilde{\mathbf{V}}$, hence the variational problem is well posed.

3.1 The Virtual Element Space

In order to devise a discretization of the boundary value problem under consideration adopting virtual elements with curved edges, we exploit the construction outlined in [14, 16, 17]. Let $\mathcal{T}_h$ be a *simple polygonal mesh* on d, i.e. any decomposition of d in a finite set of simple polygons e, without holes and with boundary given by a finite number of edges. Whenever an element has an edge lying on an interface Γ, such edge is then allowed to be curved in order to describe exactly the geometry of the problem. We assume that each interface Γ is parametrized by an invertible C^1 mapping γ from an interval in the real line into Γ. It is not restrictive to assume that each curved edge is a subset of only one Γ and therefore regular. In order to simplify the notation in the following we sometimes drop the index j, simply use Γ and

$$\boldsymbol{\gamma} \,:\, [0, L] \longrightarrow \Gamma$$

to indicate a generic curved part of the fibre/matrix interface and its associated parametrization.

The virtual element space is built elementwise. Indicating with $E \in \mathcal{T}_h$ a generic polygonal element of. Note that E may have some curved edge, laying on some curved interface Γ ($j \in \{1, 2, .., F\}$). For any of such curved edges e, let $\boldsymbol{\gamma}_e : [a, b] \rightarrow e$ denote the restriction of the parametrization describing Γ to the edge e. Then we indicate the space of mapped polynomials (living on e) as

$$\widetilde{\mathcal{P}}_k(e) = \left\{ p \circ \boldsymbol{\gamma}_e^{-1} \,:\, p \in \mathcal{P}_k[a, b] \right\}.$$

The local virtual element space on E is then defined as

$$\begin{aligned} \mathbf{V}_h(E) = \Big\{ & v \in H^1(E) \cap C^0(E) \,:\, v|_e \in \mathcal{P}_k(e) \text{ if } e \text{ is straight}, \\ & v|_e \in \widetilde{\mathcal{P}}_k(e) \text{ if } e \text{ is curved}, -\Delta v \in \mathcal{P}_{k-2}(E) \Big\}. \end{aligned} \tag{16}$$

The associated degrees of freedom are (see [17] for the simple proof)

- pointwise evaluation at every vertex of polygon E;
- pointwise evaluation at $k - 1$ distinct points lying on every edge of E;
- area-averaged moments $\int_E v p_{k-2}$ for all $p_{k-2} \in \mathcal{P}_{k-2}(E)$.

The global space is obtained by a standard procedure preserving interelement C^0—-continuity:

$$\tilde{\mathbf{V}}_h = \left\{ v \in \tilde{\mathbf{V}} \,:\, v|_E \in \mathbf{V}_h(E) \; \forall E \in \mathcal{T}_h \right\},$$

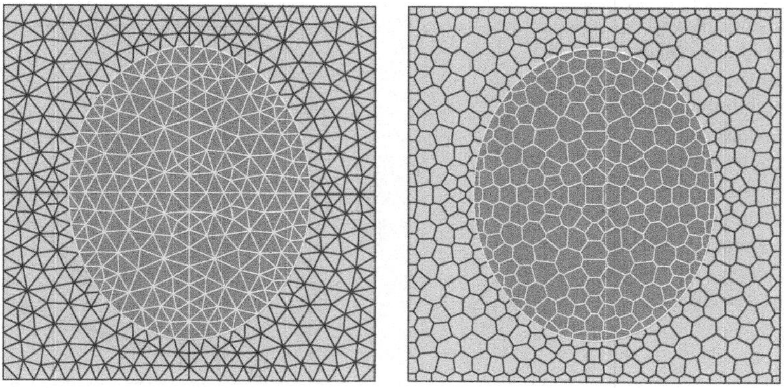

(a) (b)

Fig. 2 Square unit cell with elliptical inclusion. Curvilinear meshes. **a** Triangles. **b** Voronoi-like polygons

Global degrees of freeedom are the obvious extension of the local ones. The discretization of the problem is a combination of the scheme proposed in [14] for the case with standard straight edges and the curved-edge technology introduced in [17] for a model linear diffusion problem. Implementation details can thus be found in the aforementioned references.

3.2 Numerical Test

A composite arrangement with elliptical inclusions in square matrices is considered, cf. Fig. 2 with maximum/minimum axis ratio of 2 [26]. Fibre/matrix shear stiffness contrast factor is here $G^f/G^m = 18$, with perfect interfaces. The solution for the shear moduli are computed for the Tri-mesh and Poly-mesh discretizations, as can be appreciated from Fig. 2, and compared to a reference solutions obtained with quadratic triangular displacement-based finite elements on a very fine mesh. The homogenized principal shear moduli are reported in Fig. 3 confirming the accuracy of the method even in case of complex curvilinear fibre cross section geometry. The accuary of the present computation opens the door for the proposed methodology to even more involved geometries of the composite constituents, i.e. when cross fibre sections may present sharply curved edges which may be selected as to taylor material specific features [26].

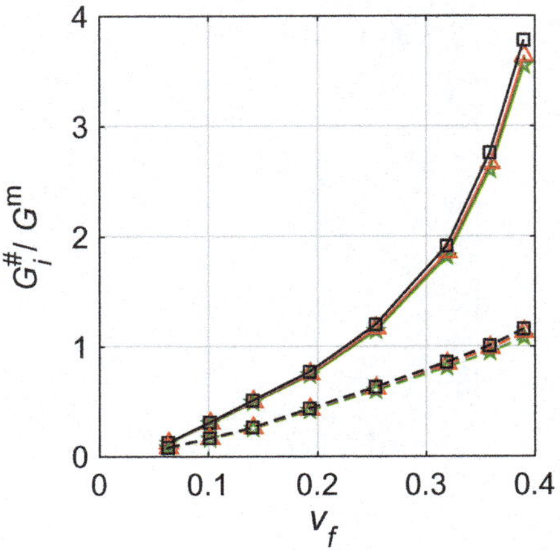

Fig. 3 Unit cell with elliptical fibre inclusion. Homogenized principal shear moduli for isotropic homogeneous constituents. Red triangle: Tri-mesh; green-pentagon: Poly-mesh; black squares: Quad-mesh for $Q4$ reference solution. Aspect ration $\kappa = 2$, shear constrast $\xi = 20$

4 Conclusion

In this contribution we have presented a curvilinear VEM for homogenization of unidirectional fiber-reinforced composite materials with inclusion curvilinear cross section. The procedure proves efficient and accurate as confirmed by several numerical results.

Acknowledgements The author gratefully acknowledges the partial financial support of PRIN 2017 project "3D PRINTING: A BRIDGE TO THE FUTURE (3DP_Future). Computational methods, innovative applications, experimental validations of new materials and technologies.", grant 2017L7X3CS_004, and of project "Innovative Numerical Methods for Advanced Materials and Technologies", University of Rome Tor Vergata—Beyond Borders call, (CUP): E89C20000610005.

References

1. Bensoussan, A., Lions, J. L., & Papanicolau, G. (1978). *Asymptotic Analysis for Periodic Structures*. Amsterdam: North-Holland.
2. Lions, J. L. (1980). Asymptotic expansions in perforated media with a periodic structure. *Rocky Mountain Journal of Mathematics, 10,* 125–140.
3. Sanchez-Palencia, E. (1980). *Non-Homogeneous media and vibration theory*. Lecture notes in physics. Berlin: Springer.
4. Beirão da Veiga, L., Brezzi, F., Marini, L. D., & Russo, A. (2014). The hitchhiker's guide to the virtual element method. *Mathematical Models and Methods in Applied Sciences, 24*(08), 1541–1573.
5. Beirão da Veiga, L., Brezzi, F., Cangiani, A., Manzini, G., Marini, L. D., & Russo, A. (2013). Basic principles of virtual element methods. *Mathematical Models and Methods in Applied Sciences, 23*(1), 199–214.
6. Beiro da Veiga, L., Brezzi, F., & Marini, L. D. (2013). Virtual elements for linear elasticity problems. *Journal on Numerical Analysis, 51*(2), 794–812.

7. Gain, A. L., Talischi, C., & Paulino, G. H. (2014). On the virtual element method for three-dimensional linear elasticity problems on arbitrary polyhedral meshes. *Computer Methods in Applied Mechanics and Engineering, 282,* 132–160.
8. Artioli, E., Beiro da Veiga, L., Lovadina, C., & Sacco, E. (2017). Arbitrary order 2D virtual elements for polygonal meshes: Part I, elastic problem. *Computational Mechanics, 60,* 355–377.
9. Beiro da Veiga, L., Lovadina, C., & Mora, D. (2015). A virtual element method for elastic and inelastic problems on polytope meshes. *Computer Methods in Applied Mechanics and Engineering, 295*, 327–346.
10. Artioli, E., & Taylor, R. L. (2018). VEM for inelastic solids. *Computational Methods in Applied Sciences, 46*, 381–394.
11. Wriggers, P., & Hudobivnik, B. (2017). A low order virtual element formulation for finite elasto-plastic deformations. *Computer Methods in Applied Mechanics and Engineering, 327,* 459–477.
12. De Bellis, M. L., Wriggers, P., Hudobivnik, B., & Zavarise, G. (2018). Virtual element formulation for isotropic damage. *Finite Elements in Analysis and Design, 144,* 38–48.
13. Artioli, E., Beirão da Veiga, L., Lovadina, C., & Sacco, E. (2017). Arbitrary order 2D virtual elements for polygonal meshes: Part II, inelastic problem. *Computational Mechanics, 60,* 643–657.
14. Artioli, E. (2018). Asymptotic homogenization of fibre-reinforced composites: A virtual element method approach. *Meccanica, 53,* 1187–1201.
15. Artioli, E., Marfia, S., & Sacco, E. (2018). High-order virtual element method for the homogenization of long fiber nonlinear composites. *Computer Methods in Applied Mechanics and Engineering, 341*, 571–585.
16. Artioli, E., Beirão Da Veiga, L., & Verani, M. (2020). An adaptive curved virtual element method for the statistical homogenization of random fibre-reinforced composites. *Finite Elements in Analysis and Design, 177*, 103418.
17. Beirão da Veiga, L., Russo, A., & Vacca, G. (2019). The virtual element method with curved edges. *ESAIM: Mathematical Modelling and Numerical Analysis, 53*(2), 375–404.
18. Artioli, E., Beião da Veiga, L., & Dassi, F. (2020). Curvilinear virtual elements for 2D solid mechanics applications. *Computer Methods in Applied Mechanics and Engineering, 359,* 112667.
19. Aldakheel, F., Hudobivnik, B., Artioli, E., Beirão da Veiga, L., & Wriggers, P. (2020). Curvilinear virtual elements for contact mechanics. *Computer Methods in Applied Mechanics and Engineering, 372*, 113394.
20. Wriggers, P., Hudobivnik, B., & Aldakheel F. (2020). A virtual element formulation for general element shapes. *Computational Mechanics, 66*, 963–977.
21. Lene, F., & Leguillon, D. (1982). Homogenized constitutive law for a partially cohesive composite material. *International Journal of Solids and Structures, 18,* 443–458.
22. Hashin, Z. (1991). The spherical inclusion with imperfect interface. *The Journal of Applied Mechanics, 58,* 444–449.
23. Bigoni, D., Serkov, S. K., Valentini, M., & Movchan, A. B. (1998). Asymptotic models of dilute composites with imperfectly bonded inclusions. *International Journal of Solids and Structures, 35*(24), 3239–3258.
24. Artioli, E., Bisegna, P., & Maceri, F. (2010). Effective longitudinal shear moduli of periodic fibre-reinforced composites with radially-graded fibres. *International Journal of Solids and Structures, 47,* 383–397.
25. Artioli, E., & Bisegna, P. (2013). Effective longitudinal shear moduli of periodic fibre-reinforced composites with functionally-graded fibre coatings. *International Journal of Solids and Structures, 50,* 1154–1163.
26. Joyce, D., Parnell, W. J., Assier, R. C., & Abrahams, I. D. (2007). An integral equation method for the homogenization of unidirectional fibre-reinforced media; antiplane elasticity and other potential problems. *Proceedings of the Royal Society A, 473*, 20170080.

Free Bloch Wave Propagation in Periodic Cauchy Materials: Analytical and Computational Strategies

Ferdinando Auricchio, Andrea Bacigalupo, Marco Lepidi, and Simone Morganti

I would like to take this opportunity to wish from the deep of my heart the best to professor Peter Wriggers on occasion of his 70th birthday and every happiness on this special day. Peter has always been a world-renowned, brilliant researcher, an excellent example for our computational mechanics community; however, in the years I have appreciated more and more his friendship, his willingness and easiness in sitting down, coding, exploring new ideas and concepts, as well as enjoying life, traveling, sailing, having a beer and laugh all together. With this completeness I think Peter is a wonderful friend and example for me and for all of us. So, dear Peter, best wishes for many other opportunities to spend time together and for you to enjoy good time with Claudia and your family. Ferdinando

Abstract The mechanical behaviour of periodic three-dimensional non-homogeneous solids can be functionally designed by optimizing their cellular composite microstructure. The present contribution aims at establishing a theoretical and methodological framework to determine the dispersion properties of periodic Cauchy materials, characterized by a generic microstructured repetitive cell. First, two conceptually alternative strategies are formulated to state the continuous complex-valued eigenproblem, governing the free undamped propagation of Bloch waves, in the finite

F. Auricchio
Department of Civil Engineering and Architecture, University of Pavia, Pavia, Italy
e-mail: auricchi@unipv.it

A. Bacigalupo · M. Lepidi
Department of Civil, Chemical and Environmental Engineering, University of Genova, Genova, Italy
e-mail: andrea.bacigalupo@unige.it

M. Lepidi
e-mail: marco.lepidi@unige.it

S. Morganti (✉)
Department of Electrical, Computer, and Biomedical Engineering, University of Pavia, Pavia, Italy
e-mail: simone.morganti@unipv.it

F. Aldakheel et al. (eds.), *Current Trends and Open Problems in Computational Mechanics*, https://doi.org/10.1007/978-3-030-87312-7_5

domain of the periodic cell. Second, a consistent mathematical procedure to convert the governing field equations into an equivalent discrete eigenproblem, suited to be computationally attacked through the finite element method, is outlined.

1 Introduction

The mechanical behaviour of periodic three-dimensional solids can be functionally designed by optimizing their cellular composite architecture. Indeed, the endless possibilities offered by microstructural customization can be profitably conjugated with the rich variety of physical phenomena occurring in periodic media, in order to achieve superior dynamic performances [1, 2]. The recent extraordinary developments in the fields of computational mechanics, intelligent optimization, material micro-engineering and manufacturing technologies are paving the way to systematically transform fascinating conceptual ideas into real-world technical applications [3, 4]. Within this scenario, the present contribution systematizes some analytical and computational approaches to the spectral problem of determining the dispersion properties of periodic Cauchy materials. Different methodological strategies are—first—theoretically established to reduce the field problem in the cell domain (Sect. 2) and—second—adapted to be implemented within a finite element framework (Sect. 3).

2 Wave Propagation in Periodic Cauchy Materials

A three-dimensional non-homogeneous Cauchy material with periodic microstructure is considered (Fig. 1a). The periodic cell is assumed to tessellate the infinite domain $\mathcal{M}$ with the finite subdomain $\mathcal{A} = [-1/2\, d_1, 1/2\, d_1] \times [-1/2\, d_2, 1/2\, d_2] \times [-1/2\, d_3, 1/2\, d_3]$ and boundary $\partial\mathcal{A}$ (Fig. 1b). The partial differential equation of motion governing the free undamped dynamics reads

$$\nabla \cdot \big(\mathfrak{C}(\mathbf{x})\nabla\mathbf{u}(\mathbf{x}, t)\big) = \varrho(\mathbf{x})\ddot{\mathbf{u}}(\mathbf{x}, t), \qquad \forall\mathbf{x} \in \mathcal{M} \tag{1}$$

where $\mathbf{u} = u_i\mathbf{e}_i$ is the displacement of the point at position $\mathbf{x}$ at time t. The elastic tensor $\mathfrak{C} = C_{ijhk}\mathbf{e}_i \otimes \mathbf{e}_j \otimes \mathbf{e}_h \otimes \mathbf{e}_k$ and the mass density ϱ are $\mathcal{A}$-periodic according to the periodicity vectors $\mathbf{v}_p$ $(p = 1, 2, 3)$, yielding

$$\mathfrak{C}(\mathbf{x} + \mathbf{v}_p) = \mathfrak{C}(\mathbf{x}), \qquad \varrho(\mathbf{x} + \mathbf{v}_p) = \varrho(\mathbf{x}), \qquad \forall\mathbf{x} \in \mathcal{M} \tag{2}$$

Employing the time-Fourier transform $\mathcal{F}_t\left[\mathbf{g}(\mathbf{x}, t)\right] = \int_{-\infty}^{\infty} \mathbf{g}(\mathbf{x}, t)\exp(-\imath\omega t)\mathrm{d}t = \hat{\mathbf{g}}(\mathbf{x}, \omega)$ for the generic function $\mathbf{g}(\mathbf{x}, t)$, the transformed equation (1) reads

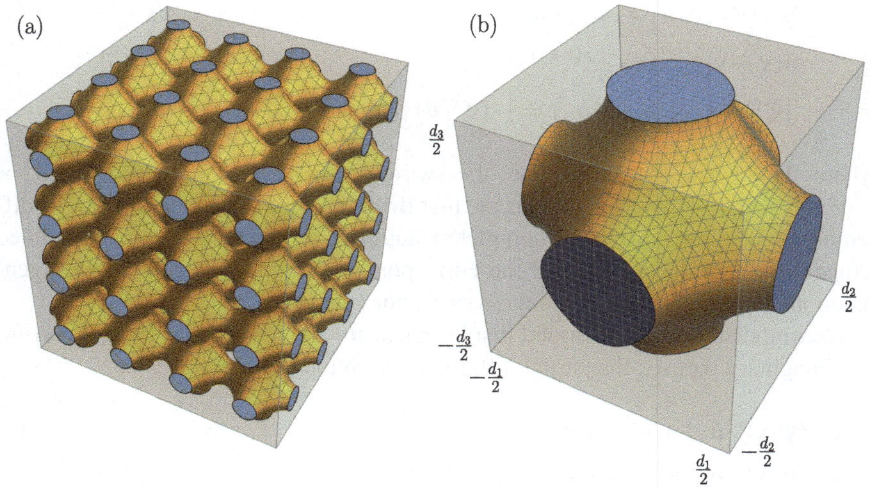

Fig. 1 Non-homogeneous material with periodic microstructure: **a** tessellation of the three-dimensional infinite domain $\mathcal{M}$, **b** periodic cell with finite subdomain $\mathcal{A}$

$$\nabla \cdot \big(\mathfrak{C}(\mathbf{x})\nabla\hat{\mathbf{u}}(\mathbf{x},\omega)\big) + \omega^2\varrho(\mathbf{x})\hat{\mathbf{u}}(\mathbf{x},\omega) = \mathbf{0}, \qquad \forall \mathbf{x} \in \mathcal{M} \tag{3}$$

where $\omega \in \mathbb{R}$ is the angular frequency. Equation (3) is also known as Christoffel equation, whose solution describes the free propagation of elastic waves.

The equation solution is pursued by following two conceptually different strategies. The *direct* strategy is based on tackling the Christoffel equation in the periodic cell domain $\mathcal{A}$ by imposing quasi-periodicity conditions at the cellular boundary $\partial\mathcal{A}$. Differently, the *indirect* strategy is based on tackling the Christoffel equation in the material domain $\mathcal{M}$ by the Floquet-Bloch decomposition and subsequently imposing periodicity conditions at the cellular boundary $\partial\mathcal{A}$. The two alternative strategies, which are detailed in the following, are known to return the same solution in terms of dispersion properties of the freely-propagating elastic waves.

2.1 Direct Strategy

By virtue of the material periodicity, the Christoffel equation can be studied in the periodic cell domain $\mathcal{A}$ by imposing *quasi-periodicity* conditions at the boundary $\partial\mathcal{A}$ on the transformed displacements $\hat{\mathbf{u}}$, as well as on the associated tractions $\hat{\mathbf{T}}\mathbf{n} = \big(\mathfrak{C}\nabla\hat{\mathbf{u}}\big)\,\mathbf{n}$, where $\mathbf{n} = n_i\mathbf{e}_i$ is the outer-pointing normal to the boundary surface. According to this strategy, the partial differential problem governing the free wave propagation reads

$$\begin{aligned}
&\nabla \cdot \left(\mathfrak{C}(\mathbf{x})\nabla\hat{\mathbf{u}}\right) + \omega^2\varrho(\mathbf{x})\hat{\mathbf{u}} = \mathbf{0}, && \forall \mathbf{x} \in \mathcal{A} \\
&\hat{\mathbf{u}}(\mathbf{x}+\mathbf{v}_p) = \hat{\mathbf{u}}(\mathbf{x})\,e^{\imath(\mathbf{k}\cdot\mathbf{v}_p)}, && \forall \mathbf{x} \in \partial\mathcal{A} \\
&\left(\mathfrak{C}\nabla\hat{\mathbf{u}}\right)|_{\mathbf{x}+\mathbf{v}_p}\mathbf{n}(\mathbf{x}+\mathbf{v}_p) = -\left(\mathfrak{C}\nabla\hat{\mathbf{u}}\right)|_{\mathbf{x}}\mathbf{n}(\mathbf{x})\,e^{\imath(\mathbf{k}\cdot\mathbf{v}_p)}, &&
\end{aligned} \tag{4}$$

where $\mathbf{k} = k_i\mathbf{e}_i \in \mathcal{B}$ stands for the wavevector and $\mathcal{B} = [-\pi/d_1, \pi/d_1] \times [-\pi/d_2, \pi/d_2] \times [-\pi/d_3, \pi/d_3]$ is the first Brillouin zone. Equation (4) states a differential eigenproblem, whose complete solution is represented by the real-valued (squared) eigenvalues $\omega(\mathbf{k})$ and the corresponding complex-valued vector eigenfunctions $\hat{\mathbf{u}}(\mathbf{x}, \mathbf{k})$, depending on the wavevector $\mathbf{k} \in \mathcal{B}$.

Decomposing the transformed displacement $\hat{\mathbf{u}} = \hat{\mathbf{u}}_{\Re} + \imath\hat{\mathbf{u}}_{\Im}$ into its real part $\hat{\mathbf{u}}_{\Re}$ and imaginary part $\hat{\mathbf{u}}_{\Im}$, the partial differential problem (4) reads

$$\begin{aligned}
&\nabla \cdot \left(\mathfrak{C}(\mathbf{x})\nabla\hat{\mathbf{u}}_{\Re}\right) + \omega^2\varrho(\mathbf{x})\hat{\mathbf{u}}_{\Re} = \mathbf{0}, && \forall \mathbf{x} \in \mathcal{A} \\
&\nabla \cdot \left(\mathfrak{C}(\mathbf{x})\nabla\hat{\mathbf{u}}_{\Im}\right) + \omega^2\varrho(\mathbf{x})\hat{\mathbf{u}}_{\Im} = \mathbf{0}, && \\
&\hat{\mathbf{u}}_{\Re}(\mathbf{x}+\mathbf{v}_p) = \hat{\mathbf{u}}_{\Re}(\mathbf{x})\cos(\mathbf{k}\cdot\mathbf{v}_p) - \hat{\mathbf{u}}_{\Im}(\mathbf{x})\sin(\mathbf{k}\cdot\mathbf{v}_p), && \forall \mathbf{x} \in \partial\mathcal{A} \\
&\hat{\mathbf{u}}_{\Im}(\mathbf{x}+\mathbf{v}_p) = \hat{\mathbf{u}}_{\Im}(\mathbf{x})\cos(\mathbf{k}\cdot\mathbf{v}_p) + \hat{\mathbf{u}}_{\Re}(\mathbf{x})\sin(\mathbf{k}\cdot\mathbf{v}_p), && \\
&\left(\mathfrak{C}\nabla\hat{\mathbf{u}}_{\Re}\right)|_{\mathbf{x}+\mathbf{v}_p}\mathbf{n}(\mathbf{x}+\mathbf{v}_p) = -\left(\mathfrak{C}\nabla\hat{\mathbf{u}}_{\Re}\right)|_{\mathbf{x}}\cos(\mathbf{k}\cdot\mathbf{v}_p) + \left(\mathfrak{C}\nabla\hat{\mathbf{u}}_{\Im}\right)|_{\mathbf{x}}\sin(\mathbf{k}\cdot\mathbf{v}_p), && \\
&\left(\mathfrak{C}\nabla\hat{\mathbf{u}}_{\Im}\right)|_{\mathbf{x}+\mathbf{v}_p}\mathbf{n}(\mathbf{x}+\mathbf{v}_p) = -\left(\mathfrak{C}\nabla\hat{\mathbf{u}}_{\Im}\right)|_{\mathbf{x}}\cos(\mathbf{k}\cdot\mathbf{v}_p) - \left(\mathfrak{C}\nabla\hat{\mathbf{u}}_{\Re}\right)|_{\mathbf{x}}\sin(\mathbf{k}\cdot\mathbf{v}_p), &&
\end{aligned} \tag{5}$$

where the coupling between the real and imaginary parts of the unknown $\hat{\mathbf{u}}$ is established by the quasi-periodicity boundary conditions.

2.2 Indirect Strategy

As a suited alternative to the direct strategy, the free wave propagation can be studied by an indirect two-step strategy. The first step consists in applying the Floquet-Bloch decomposition to the transformed displacement

$$\hat{\mathbf{u}}(\mathbf{x}) = \tilde{\mathbf{u}}(\mathbf{x})\,e^{\imath(\mathbf{k}\cdot\mathbf{x})} = \tilde{u}_i\mathbf{e}_i\,e^{\imath(k_j x_j)}, \qquad \forall \mathbf{x} \in \mathcal{M} \tag{6}$$

where the complex-valued Bloch amplitude $\tilde{\mathbf{u}}(\mathbf{x}) = \tilde{u}_i(\mathbf{x})\mathbf{e}_i$ is a $\mathcal{A}$-periodic function in the $\mathcal{M}$-domain, which indirectly satisfies the quasi-periodicity conditions (4b). Specifically, the $\tilde{\mathbf{u}}$-components satisfy

$$\tilde{u}_i(\mathbf{x}+\mathbf{v}_p) = \tilde{u}_i(\mathbf{x}), \qquad \forall \mathbf{x} \in \mathcal{M} \tag{7}$$

After decomposition, it is formally convenient to express equation (3) in component form

$$
\begin{aligned}
&\left(C_{ijh\ell}\tilde{u}_{h,\ell}\right)_{,j} - \left(C_{ijh\ell}k_\ell k_j - \omega^2\varrho\delta_{ih}\right)\tilde{u}_h + \\
&\quad + \imath k_j\left[\left(C_{ijh\ell} + C_{i\ell hj}\right)\tilde{u}_{h,\ell} + C_{i\ell hj,\ell}\tilde{u}_h\right] = 0, \qquad \forall \mathbf{x} \in \mathcal{M}
\end{aligned} \tag{8}
$$

where comma indicates partial derivative with respect to the spatial coordinates and δ_{ih} is the Kronecker delta.

Therefore, the second step of the indirect strategy consists in invoking the material periodicity to study equation (8) in the periodic cell domain $\mathcal{A}$, similarly to the direct strategy. The key difference is that the preliminary Floquet-Bloch decomposition of the displacement variable allows the imposition of simple *periodicity* conditions at the boundary $\partial\mathcal{A}$ on the components $\tilde{u}_h$ of the Bloch amplitudes, as well as on the components $\tilde{\sigma}_{hs}n_s = C_{hsrp}\tilde{u}_{r,p}n_s$ of the associated tractions. Accordingly, the partial differential problem governing the free wave propagation reads

$$
\begin{aligned}
&\left(C_{ijh\ell}\tilde{u}_{h,\ell}\right)_{,j} - \left(C_{ijh\ell}k_\ell k_j - \omega^2\varrho\delta_{ih}\right)\tilde{u}_h + \\
&\quad + \imath k_j\left[\left(C_{ijh\ell} + C_{i\ell hj}\right)\tilde{u}_{h,\ell} + C_{i\ell hj,\ell}\tilde{u}_h\right] = 0, \qquad \forall \mathbf{x} \in \mathcal{A} \\
&\tilde{u}_h(\mathbf{x} + \mathbf{v}_p) = \tilde{u}_h(\mathbf{x}), \qquad \forall \mathbf{x} \in \partial\mathcal{A} \\
&\left(C_{hsrp}\tilde{u}_{r,p}\right)\big|_{\mathbf{x}+\mathbf{v}_p} n_s(\mathbf{x} + \mathbf{v}_p) = -\left(C_{hsrp}\tilde{u}_{r,p}\right)\big|_{\mathbf{x}} n_s(\mathbf{x}),
\end{aligned} \tag{9}
$$

Equation (9) states a differential eigenproblem, whose complete solution is given by the real-valued (squared) eigenvalues $\omega(\mathbf{k})$ and the corresponding complex-valued eigenfunctions $\tilde{u}_h(\mathbf{x}, \mathbf{k})$, depending on the wavevector $\mathbf{k} \in \mathcal{B}$.

Decomposing the Bloch components $\tilde{u}_h = \tilde{u}_h^{\Re} + \imath\tilde{u}_h^{\Im}$ into their real part $\tilde{u}_h^{\Re}$ and imaginary part $\tilde{u}_h^{\Im}$, the partial differential problem (9) reads

$$
\begin{aligned}
&\left(C_{ijh\ell}\tilde{u}^{\Re}_{h,\ell}\right)_{,j} - \left(C_{ijh\ell}k_\ell k_j - \omega^2\varrho\delta_{ih}\right)\tilde{u}_h^{\Re} + \\
&\quad - k_j\left[\left(C_{ijh\ell} + C_{i\ell hj}\right)\tilde{u}^{\Im}_{h,\ell} + C_{i\ell hj,\ell}\tilde{u}_h^{\Im}\right] = 0 \qquad \forall \mathbf{x} \in \mathcal{A} \\
&\left(C_{ijh\ell}\tilde{u}^{\Im}_{h,\ell}\right)_{,j} - \left(C_{ijh\ell}k_\ell k_j - \omega^2\varrho\delta_{ih}\right)\tilde{u}_h^{\Im} + \\
&\quad + k_j\left[\left(C_{ijh\ell} + C_{i\ell hj}\right)\tilde{u}^{\Re}_{h,\ell} + C_{i\ell hj,\ell}\tilde{u}_h^{\Re}\right] = 0 \\
&\tilde{u}_h^{\Re}(\mathbf{x} + \mathbf{v}_p) = \tilde{u}_h^{\Re}(\mathbf{x}), \qquad \forall \mathbf{x} \in \partial\mathcal{A} \\
&\tilde{u}_h^{\Im}(\mathbf{x} + \mathbf{v}_p) = \tilde{u}_h^{\Im}(\mathbf{x}), \\
&\left(C_{hsrp}\tilde{u}^{\Re}_{r,p}\right)\big|_{\mathbf{x}+\mathbf{v}_p} n_s(\mathbf{x} + \mathbf{v}_p) = -\left(C_{hsrp}\tilde{u}^{\Re}_{r,p}\right)\big|_{\mathbf{x}} n_s(\mathbf{x}), \\
&\left(C_{hsrp}\tilde{u}^{\Im}_{r,p}\right)\big|_{\mathbf{x}+\mathbf{v}_p} n_s(\mathbf{x} + \mathbf{v}_p) = -\left(C_{hsrp}\tilde{u}^{\Im}_{r,p}\right)\big|_{\mathbf{x}} n_s(\mathbf{x}),
\end{aligned} \tag{10}
$$

where the coupling between the real and imaginary parts of the unknowns $\tilde{u}_h$ is established by the field equations, while the boundary conditions are uncoupled.

3 Dispersion Properties via Finite Element Formulation

In the absence of analytical solutions for complex domain geometries, the differential eigenproblems (4) (or (5)) and (9) (or (10)) can be approached via finite element formulations. Indeed, both strategies of solution can be equally implemented to achieve an equivalent algebraic eigenproblem, although with different methodological approaches. Indeed, the direct strategy can be adopted by employing standard finite elements to discretize a pair of identical domains, each one describing the real and imaginary parts of the mechanical variables, coupled by quasi-periodicity boundary conditions. On the contrary, the indirect strategy requires the formulation of ad-hoc finite elements possessing real and imaginary degrees of freedom at each nodes.

3.1 Direct Strategy

The field equation governing the eigenproblem (4) can be treated according to a weak formulation. Therefore, the domain $\mathcal{A}$ is properly discretized in finite elements. Selecting suited interpolation functions for the transformed displacement $\hat{\mathbf{u}}$, the real-valued stiffness matrix K and mass matrix M associated to the nodal transformed displacements $\hat{\mathsf{q}}$ of the periodic cell are determined. Accordingly, an algebraic field equation is established

$$\left(\mathsf{K} + \omega^2 \mathsf{M}\right) \hat{\mathsf{q}} = \hat{\mathsf{f}} \tag{11}$$

where the frequency ω is real-valued and the nodal displacements $\hat{\mathsf{q}}$ and forces $\hat{\mathsf{f}}$ are complex-valued. The variables can be decomposed in real and imaginary part as $\hat{\mathsf{q}} = \hat{\mathsf{q}}_{\Re} + \imath \hat{\mathsf{q}}_{\Im}$ and $\hat{\mathsf{f}} = \hat{\mathsf{f}}_{\Re} + \imath \hat{\mathsf{f}}_{\Im}$. Generalizing the procedure proposed by Langley for two-dimensional periodic systems [5], the nodal displacements can be condensed in the form

$$\hat{\mathsf{q}} = \mathsf{R}(\mathbf{k}) \hat{\mathsf{p}} \tag{12}$$

where the vector variable $\hat{\mathsf{p}}$ collects a minimal set of nodal displacements at the discretization of the boundary $\partial\mathcal{A}$, and can be decomposed into its real $\hat{\mathsf{p}}_{\Re}$ and imaginary part $\hat{\mathsf{p}}_{\Im}$ in the form $\hat{\mathsf{p}} = \hat{\mathsf{p}}_{\Re} + \imath \hat{\mathsf{p}}_{\Im}$. Therefore, the $\mathbf{k}$-dependent condensation matrix $\mathsf{R}(\mathbf{k})$ collects the quasi-periodicity conditions restraining the nodal displacements at the discretization of cell boundary $\partial\mathcal{A}$ following from the Eq. (4b). Expressing the matrix R through the decomposition $\mathsf{R} = \mathsf{R}_{\Re} + \imath \mathsf{R}_{\Im}$ and substituting equation (12) into Eq. (11) yields

$$\left(\mathsf{K} + \omega^2 \mathsf{M}\right) \left(\mathsf{R}_{\Re} + \imath \mathsf{R}_{\Im}\right) \left(\hat{\mathsf{p}}_{\Re} + \imath \hat{\mathsf{p}}_{\Im}\right) = \left(\hat{\mathsf{f}}_{\Re} + \imath \hat{\mathsf{f}}_{\Im}\right). \tag{13}$$

Therefore, expanding and collecting separately real and imaginary parts, the equation becomes

$$\left(\mathbf{K}+\omega^2\mathbf{M}\right)\left[(\mathbf{R}_{\Re}\hat{\mathbf{p}}_{\Re}-\mathbf{R}_{\Im}\hat{\mathbf{p}}_{\Im})+\imath(\mathbf{R}_{\Im}\hat{\mathbf{p}}_{\Re}+\mathbf{R}_{\Re}\hat{\mathbf{p}}_{\Im})\right]=\left(\hat{\mathbf{f}}_{\Re}+\imath\hat{\mathbf{f}}_{\Im}\right) \tag{14}$$

whose left-hand and right-hand side members can be both pre-multiplied by $\mathbf{R}^{\dagger}=\mathbf{R}_{\Re}^{\top}-\imath\mathbf{R}_{\Im}^{\top}$, yielding

$$\begin{aligned}\left[\mathbf{R}_{\Re}^{\top}\left(\mathbf{K}+\omega^2\mathbf{M}\right)-\imath\mathbf{R}_{\Im}^{\top}\left(\mathbf{K}+\omega^2\mathbf{M}\right)\right]&\left[(\mathbf{R}_{\Re}\hat{\mathbf{p}}_{\Re}-\mathbf{R}_{\Im}\hat{\mathbf{p}}_{\Im})+\imath(\mathbf{R}_{\Im}\hat{\mathbf{p}}_{\Re}+\mathbf{R}_{\Re}\hat{\mathbf{p}}_{\Im})\right]=\\ &=(\mathbf{R}_{\Re}^{\top}\hat{\mathbf{f}}_{\Re}+\mathbf{R}_{\Im}^{\top}\hat{\mathbf{f}}_{\Im})+\imath(\mathbf{R}_{\Re}^{\top}\hat{\mathbf{f}}_{\Im}-\mathbf{R}_{\Im}^{\top}\hat{\mathbf{f}}_{\Re})\end{aligned} \tag{15}$$

where the right-hand term is identically null since body forces are absent in homogeneous equation (4a) and by virtue of the quasi-periodicity conditions of the nodal forces at the discretization of cell boundary $\partial\mathcal{A}$.

Expanding and properly separating real and imaginary part of equation (16), an algebraic eigenproblem is achieved. Specifically, the uncoupled field problems related to the real and imaginary parts $\hat{\mathbf{q}}_{\Re}$ and $\hat{\mathbf{q}}_{\Re}$ of displacement (which might also be related to the nodes of separate identical domains) are governed by condensed real-valued stiffness and mass matrices coupling the real and imaginary parts $\hat{\mathbf{p}}_{\Re}$ and $\hat{\mathbf{p}}_{\Re}$ of the boundary displacements

$$\left(\begin{bmatrix}\mathbf{K}_{\Re\Re} & \mathbf{K}_{\Re\Im}\\ \mathbf{K}_{\Im\Re} & \mathbf{K}_{\Im\Im}\end{bmatrix}+\omega^2\begin{bmatrix}\mathbf{M}_{\Re\Re} & \mathbf{M}_{\Re\Im}\\ \mathbf{M}_{\Im\Re} & \mathbf{M}_{\Im\Im}\end{bmatrix}\right)\begin{pmatrix}\hat{\mathbf{p}}_{\Re}\\ \hat{\mathbf{p}}_{\Im}\end{pmatrix}=\begin{pmatrix}\mathbf{0}\\ \mathbf{0}\end{pmatrix} \tag{16}$$

The algebraic equation (16) is the equivalent counterpart of the differential eigenproblem (6). Specifically, the **k**-dependent matrices are

$$\begin{aligned}\mathbf{K}_{\Re\Re}&=\mathbf{R}_{\Re}^{\top}\mathbf{K}\mathbf{R}_{\Re}+\mathbf{R}_{\Im}^{\top}\mathbf{K}\mathbf{R}_{\Im}, & \mathbf{K}_{\Re\Im}&=\mathbf{R}_{\Im}^{\top}\mathbf{K}\mathbf{R}_{\Re}-\mathbf{R}_{\Re}^{\top}\mathbf{K}\mathbf{R}_{\Im},\\ \mathbf{M}_{\Re\Re}&=\mathbf{R}_{\Re}^{\top}\mathbf{M}\mathbf{R}_{\Re}-\mathbf{R}_{\Im}^{\top}\mathbf{M}\mathbf{R}_{\Im}, & \mathbf{M}_{\Re\Im}&=-\mathbf{R}_{\Im}^{\top}\mathbf{M}\mathbf{R}_{\Re}-\mathbf{R}_{\Re}^{\top}\mathbf{M}\mathbf{R}_{\Im}\end{aligned} \tag{17}$$

with symmetries $\mathbf{K}_{\Im\Im}=\mathbf{K}_{\Re\Re}$, $\mathbf{K}_{\Im\Re}=-\mathbf{K}_{\Re\Im}$, $\mathbf{M}_{\Im\Im}=\mathbf{M}_{\Re\Re}$, $\mathbf{M}_{\Im\Re}=-\mathbf{M}_{\Re\Im}$.

3.2 *Indirect Strategy*

The field equations governing the eigenproblem (9) can be treated according to a weak formulation. Therefore, the domain $\mathcal{A}$ is properly discretized in finite elements. Selecting suited interpolation functions for the Bloch amplitude $\tilde{\mathbf{u}}$, the complex-valued pseudo-stiffness matrix $\mathbf{G}$ and mass matrix $\mathbf{M}$ associated to the nodal Bloch amplitudes $\tilde{\mathbf{q}}$ of the periodic cell are determined. Accordingly, an algebraic field equation is established

$$(\mathbf{G} + \omega^2 \mathbf{M})\, \tilde{\mathbf{q}} = \tilde{\mathbf{f}} \tag{18}$$

where the frequency ω is real-valued and the nodal amplitudes $\tilde{\mathbf{q}}$ and forces $\tilde{\mathbf{f}}$ are complex-valued. Both the variables and the pseudo-stiffness matrix can be decomposed in real and imaginary part as $\tilde{\mathbf{q}} = \tilde{\mathbf{q}}_{\Re} + \imath \tilde{\mathbf{q}}_{\Im}$, $\tilde{\mathbf{f}} = \tilde{\mathbf{f}}_{\Re} + \imath \tilde{\mathbf{f}}_{\Im}$ and $\mathbf{G} = \mathbf{G}_{\Re} + \imath \mathbf{G}_{\Im}$. The nodal amplitudes can be condensed in the form

$$\tilde{\mathbf{q}} = \mathbf{S}\tilde{\mathbf{p}} \tag{19}$$

where $\mathbf{S}$ is a boolean block matrix expressing the periodicity conditions following from the boundary conditions (9b). Remarkably, the real-valued $\mathbf{S}$-matrix can straightforwardly be obtained as $\mathbf{S} = \mathbf{R}(\mathbf{k} = \mathbf{0})$. Substituting equation (19) into Eq. (18) with decomposed $\mathbf{G}$-matrix yields

$$\left(\mathbf{G}_{\Re} + \imath \mathbf{G}_{\Im} + \omega^2 \mathbf{M}\right) \mathbf{S} \left(\tilde{\mathbf{p}}_{\Re} + \imath \tilde{\mathbf{p}}_{\Im}\right) = \left(\tilde{\mathbf{f}}_{\Re} + \imath \tilde{\mathbf{f}}_{\Im}\right) \tag{20}$$

Therefore, expanding, pre-multiplying by $\mathbf{S}^\top$ and collecting separately real and imaginary parts, the equation becomes

$$\begin{aligned} &\left[\mathbf{S}^\top (\mathbf{G}_{\Re} + \omega^2 \mathbf{M}) \mathbf{S} \tilde{\mathbf{p}}_{\Re} - \mathbf{S}^\top \mathbf{G}_{\Im} \mathbf{S} \tilde{\mathbf{p}}_{\Im}\right] + \\ &+ \imath \left[\mathbf{S}^\top (\mathbf{G}_{\Re} + \omega^2 \mathbf{M}) \mathbf{S} \tilde{\mathbf{p}}_{\Im} + \mathbf{S}^\top \mathbf{G}_{\Im} \mathbf{S} \tilde{\mathbf{p}}_{\Re}\right] = \left(\mathbf{S}^\top \tilde{\mathbf{f}}_{\Re} + \imath \mathbf{S}^\top \tilde{\mathbf{f}}_{\Im}\right) \end{aligned} \tag{21}$$

where the right-hand term is identically null since body forces are absent in homogeneous equation (9a) and by virtue of the periodicity conditions (9c) of the nodal forces at the discretization of cell boundary $\partial \mathcal{A}$.

Expanding and properly separating real and imaginary part of equation (21), an algebraic eigenproblem is achieved

$$\left(\begin{bmatrix} \mathbf{G}_{\Re\Re} & \mathbf{G}_{\Re\Im} \\ \mathbf{G}_{\Im\Re} & \mathbf{G}_{\Im\Im} \end{bmatrix} + \omega^2 \begin{bmatrix} \mathbf{N}_{\Re\Re} & \mathbf{O} \\ \mathbf{O} & \mathbf{N}_{\Im\Im} \end{bmatrix} \right) \begin{pmatrix} \mathbf{p}_{\Re} \\ \mathbf{p}_{\Im} \end{pmatrix} = \begin{pmatrix} \mathbf{0} \\ \mathbf{0} \end{pmatrix} \tag{22}$$

which is the equivalent counterpart of the differential eigenproblem (11). Specifically, the $\mathbf{k}$-dependent matrices are

$$\mathbf{G}_{\Re\Re} = \mathbf{S}^\top \mathbf{G}_{\Re} \mathbf{S}, \qquad \mathbf{G}_{\Re\Im} = -\mathbf{S}^\top \mathbf{G}_{\Im} \mathbf{S}, \qquad \mathbf{N}_{\Re\Re} = \mathbf{S}^\top \mathbf{M} \mathbf{S} \tag{23}$$

and the symmetry properties $\mathbf{G}_{\Im\Im} = \mathbf{G}_{\Re\Re}$, $\mathbf{G}_{\Im\Re} = -\mathbf{G}_{\Re\Im}$, $\mathbf{N}_{\Im\Im} = \mathbf{N}_{\Re\Re}$ hold.

4 Conclusions

A theoretical and methodological framework for determining the dispersion properties of periodic Cauchy materials, characterized by a generic microstructured repetitive cell, has been formally established. Two conceptually alternative strategies, based on different mathematical tools, have been synthesized to state the continuous complex-valued eigenproblem, governing the free undamped propagation of Bloch waves. The mathematical procedures required to convert the governing field equations into an equivalent discrete eigenproblem, suited to be computationally solved through the finite element method, have been outlined and comparatively discussed.

Acknowledgements The authors gratefully acknowledge the financial support from National Group of Mathematical Physics (GNFM-INdAM), from Compagnia San Paolo, project MINIERA no. I34I20000380007 and from University of Trento, project UNMASKED 2020.

References

1. Hussein, M. I., Leamy, M. J., & Ruzzene, M. (2014). Dynamics of phononic materials and structures: Historical origins, recent progress, and future outlook. *Applied Mechanics Reviews, 66*(4), 040802.
2. Bertoldi, K., Vitelli, V., Christensen, J., & Van Hecke, M. (2017). Flexible mechanical metamaterials. *Nature Reviews Materials, 2*(11), 1–11.
3. Auricchio, F., Bacigalupo, A., Gambarotta, L., Lepidi, M., Morganti, S., & Vadalà, F. (2019). A novel layered topology of auxetic materials based on the tetrachiral honeycomb microstructure. *Materials & Design, 179*, 107883.
4. Bacigalupo, A., Gnecco, G., Lepidi, M., & Gambarotta, L. (2020). Machine-learning techniques for the optimal design of acoustic metamaterials. *Journal of Optimization Theory and Applications, 187*, 630–653.
5. Langley, R. S. (1993). A note on the force boundary conditions for two-dimensional periodic structures with corner freedoms. *Journal of Sound and Vibration, 167*(2), 377–381.

Divergence Free VEM for the Stokes Problem with No Internal Degrees of Freedom

Lourenco Beirão da Veiga and Giuseppe Vacca

My first direct interaction with Peter was in 2012, where I had the honor of collaborating with him on large deformation problems in elasticity. More recently we are in fruitful contact due to the common interest in the VE method. For a mathematician, it is really a pleasure to collaborate with Peter since, among many other qualities, he possesses a very open mind to new techniques/ideas, a strong understanding of the mathematical aspects behind computational mechanics and, last but not least, he is really a very nice and friendly person to go along with. L. Beirão da Veiga

Abstract In the present contribution we concentrate on fluid mechanics and present a "super-reduced" version of the VEM scheme introduced in previous papers by the same authors for the Navier-Stokes equation, here focusing on the simpler Stokes problem. The basic advantage of the method is, in addition to the capability of handling general polytopal meshes, the property of delivering a truly divergence-free velocity solution. In addition, it exists a reduced version of the method with a smaller pressure and velocity space, but still holding the same accuracy. In this short contribution we briefly review the above mentioned VEM method and introduce a further reduction to the velocity space, leading to a scheme with velocity degrees of freedom only on the element boundaries.

L. Beirão da Veiga (✉) · G. Vacca
Dipartimento di Matematica E Applicazioni, Università Degli Studi di Milano Bicocca, Via Roberto Cozzi 55, 20125 Milano, Italy
e-mail: lourenco.beirao@unimib.it

G. Vacca
e-mail: giuseppe.vacca@unimib.it

F. Aldakheel et al. (eds.), *Current Trends and Open Problems in Computational Mechanics*, https://doi.org/10.1007/978-3-030-87312-7_6

1 Introduction

The Virtual Element Method (VEM) was introduced in [1] as a generalization of the finite element method that is able to cope with general polytopal meshes, even with non-convex and badly-shaped elements. Since its introduction, the VEM enjoyed a large success in the numerical analysis and engineering communities (see for instance [2] and citations therein for some sample paper in solid mechanics). In the present contribution we concentrate on fluid mechanics and present a "super-reduced" version of the VEM scheme introduced in [3, 4] for the Navier-Stokes equation, here focusing on the simpler Stokes problem.

Let $\Omega \subset \mathbb{R}^2$ a polygonal domain. Given the positive "fluid viscosity" constant ν, a field $\boldsymbol{f}$ representing the volumetric forces acting on the fluid and a (compatible) assigned boundary velocity $\boldsymbol{g}$, the Stokes problem in variational form reads

$$
\begin{cases}
\text{Find}(\boldsymbol{u}, p) \in \boldsymbol{V}^{\boldsymbol{g}} \times Q, \text{ such that} & \\
\nu \int_{\Omega} \boldsymbol{\nabla}^S \boldsymbol{u} : \boldsymbol{\nabla}^S \boldsymbol{v} + \int_{\Omega} p \operatorname{div} \boldsymbol{v} = \int_{\Omega} \boldsymbol{f} \cdot \boldsymbol{v} & \text{for all } \boldsymbol{v} \in \boldsymbol{V}^{\boldsymbol{0}}, \\
\int_{\Omega} q \operatorname{div} \boldsymbol{u} = 0 & \text{for all } q \in Q,
\end{cases}
\tag{1}
$$

where $\boldsymbol{u}$, p denote the unknown velocity and pressure field, respectively, $\boldsymbol{\nabla}$ denotes the gradient operator, $\boldsymbol{\nabla}^S$ the symmetric gradient operator, div the divergence operator. Here above, following standard notation in Sobolev spaces, the velocity and pressure spaces are $\boldsymbol{V}^{\boldsymbol{g}} = \left\{ \boldsymbol{v} \in [H^1(\Omega)]^2 \, : \, \boldsymbol{v}|_{\partial\Omega} = \boldsymbol{g} \right\}$ and $Q = L^2_0(\Omega)$. Note that we assume Dirichlet boundary conditions only for simplicity of exposition, the extension to different conditions such as Neumann being trivial.

The basic advantage of the method in [3, 4] is, in addition to the capability of handling general polytopal meshes, the property of delivering a truly divergence-free velocity solution (and not only in a relaxed sense as in standard mixed schemes). In addition, it exists a reduced version of the method with a smaller pressure and velocity space, but still holding the same accuracy. In this short contribution we briefly review the above mentioned VEM method and introduce a further reduction of the velocity space, leading to a scheme with velocity degrees of freedom only on the element boundaries (regardless of the "polynomial" order k). All our construction is described in two dimensions, but the extension of the reduction to the three dimensional case [5] can be easily accomplished along the same lines.

2 Discrete Velocity Spaces

We assume a simple polygonal partition (mesh) of the domain Ω, which we denote by Ω_h. The symbol E will denote a generic element of Ω_h. As in standard finite elements, we build the discrete space for velocities element by element (by introducing the local spaces $\boldsymbol{V}_{h|E}$), and then glue everything continuously.

As is typical in Virtual Elements, the local spaces are defined in an implicit way, through the introduction of an equation (that we do not need to solve!). Let $k \in \mathbb{N}$, $k \geq 2$, denote the "polynomial" degree of the scheme. For any $E \in \Omega_h$, we define

$$\boldsymbol{V}_{h|E} := \Big\{ \boldsymbol{v} \in [H^1(E) \cap C^0(E)]^2 \ \text{ s.t. } \ \boldsymbol{v}_{|e} \in [\mathbb{P}_k(e)]^2 \quad \forall e \in \partial E, \\ \operatorname{div} \boldsymbol{v} \in \mathbb{P}_{k-1}(E), \\ \boldsymbol{\Delta}^S \boldsymbol{v} - \nabla s \in \boldsymbol{x}^\perp \mathbb{P}_{k-3}(E), \quad s \in L^2_0(E) \Big\} \tag{2}$$

where $\mathbb{P}_k$ denotes the polynomials of degree k (with $\mathbb{P}_{-1} = \{0\}$), the operator $\boldsymbol{\Delta}^S := \operatorname{div} \boldsymbol{\nabla}^S$ and $\boldsymbol{x}^\perp = [-y, x]$ with (x, y) local coordinates centered on the element barycenter. The discrete velocity functions solve a local Stokes-like problem, where the field s takes the role of an auxiliary pressure variable. Note that $[\mathbb{P}_k(E)]^2 \subset \boldsymbol{V}_{h|E}$, which is important in relation to the interpolation properties of the space. It can be checked that the dimension of the velocity space $\boldsymbol{V}_{h|E}$ in (2) corresponds to the sum of the dimensions of the data space (boundary and volume) minus 1, such correction following from the Stokes theorem that enforces a data compatibility condition.

Remark 1 It can also be shown that, by applying the rot operator $(-\partial_y, \partial x)$, the first equation in (2) is equivalent to rot $\boldsymbol{\Delta}^S \boldsymbol{v} \in \mathbb{P}_{k-3}(E)$.

We now proceed by reducing the dimension of the space (2). While the first reduction step below was proposed in [3], the second one is novel. The first natural step is to exploit that the exact velocity satisfies $\operatorname{div}\boldsymbol{u} = 0$. Therefore, we would like to directly enforce such condition also in the local discrete space, thus ensuring a reduction of the DoFs without depleting the approximation properties of the scheme. Unfortunately, the condition $\operatorname{div}\boldsymbol{v} = 0$ cannot be enforced locally, since due to the Stokes theorem it would yield a constraint on the boundary values, that is $\int_{\partial E} \boldsymbol{v} \cdot \boldsymbol{n} = 0$. But we can still reduce a lot the local discrete space by requiring that the divergence is constant. We thus obtain the *reduced* space

$$\boldsymbol{V}^{red}_{h|E} := \Big\{ \boldsymbol{v} \in [H^1(E) \cap C^0(E)]^2 \ : \ \boldsymbol{v}|_e \in [\mathbb{P}_k(e)]^2 \ \forall e \in \partial E, \\ \operatorname{rot} \boldsymbol{\Delta}^S \boldsymbol{v} \in \mathbb{P}_{k-3}(E) \ \text{ and } \ \operatorname{div} \boldsymbol{v} \in \mathbb{P}_0(E) \Big\}, \tag{3}$$

where we also made use of Remark 1. It clearly holds $\boldsymbol{V}^{red}_{h|E} \subset \boldsymbol{V}_{h|E}$. Note that $\boldsymbol{V}^{red}_{h|E}$ contains all polynomials $[\mathbb{P}_k(E)]^2$ with constant divergence.

A further space reduction step can be applied, under the following assumption

(A) the loading $\boldsymbol{f}$ in (1) is a gradient.

Note that the above condition includes many cases of interest, as often the volume loading term is either negligible (thus $\boldsymbol{f} = 0$) or representing the gravity force (and thus a gradient). Under condition **(A)**, we immediately have for the exact solution rot $\boldsymbol{\Delta}^S \boldsymbol{u} = 0$ and therefore we can include such condition in the definition of our local space without depleting the approximation properties of the scheme. By doing so we have a reduction also in the first equation of (3), leading to an even smaller space

$$\begin{aligned} \boldsymbol{V}_{h|E}^{RED} := \Big\{ \boldsymbol{v} \in & [H^1(E) \cap C^0(E)]^2 \ : \ \boldsymbol{v}|_e \in [\mathbb{P}_k(e)]^2 \ \forall e \in \partial E, \\ & \mathrm{rot}\, \boldsymbol{\Delta}^S \boldsymbol{v} = 0 \ \text{ and } \ \mathrm{div}\, \boldsymbol{v} \in \mathbb{P}_0(E) \Big\} \,. \end{aligned} \tag{4}$$

Given the boundary data, the integral of the divergence is assigned through the Stokes theorem ($\int_E \mathrm{div} \boldsymbol{v} = \int_{\partial E} \boldsymbol{v} \cdot \boldsymbol{n}$). Therefore the functions in the space $\boldsymbol{V}_{h|E}^{RED}$ are uniquely defined by its boundary values, which in turn will allow to have no "internal to element" degrees of freedom.

We therefore take the following set of degrees of freedom for the space $\boldsymbol{V}_{h|E}^{RED}$:

- two point-wise evaluations (one for each component) at each element vertex;
- two point-wise evaluations (one for each component) at $k-2$ distinct points for each edge of ∂E.

The above operators clearly constitute a set of interpolating DoFs for the continuous and piecewise $[\mathbb{P}_k]^2$ polynomial boundary space, c.f. (4). Therefore, due to the observation above, they also constitute a set of DoFs for the space $\boldsymbol{V}_{h|E}^{RED}$. Note that the original space $\boldsymbol{V}_{h|E}$ would require, in addition to the DoFs above on ∂E, also $k(k-1)$ Dofs inside the element. A similar amount is needed in the corresponding standard finite elements, such as in the Crouzeix-Raviart velocity space.

3 A Projection Operator and the Discrete Problem

Projector operators take a key role in the construction of Virtual Element schemes. We start by introducing a projection $\Pi : \boldsymbol{V}_{h|E}^{RED} \to \mathcal{P}_k(E)$, where

$$\mathcal{P}_k = \Big\{ \mathbf{q} \in [\mathbb{P}_k]^2 \mid : \mathrm{div} \mathbf{q} \in \mathbb{P}_0, \ \mathrm{rot}\, \boldsymbol{\Delta}^S \mathbf{q} = 0 \Big\} \subset \boldsymbol{V}_{h|E}^{RED} \,. \tag{5}$$

The projector Π is defined as the orthogonal projection onto $\mathcal{P}_k(E)$ with respect to the L^2-type scalar product $(\boldsymbol{\nabla}^S \cdot, \boldsymbol{\nabla}^S \cdot)$. More precisely, for any $\boldsymbol{v} \in \boldsymbol{V}_{h|E}^{RED}$, we define $\Pi(\boldsymbol{v}) \in \mathcal{P}_k(E)$ by

$$\begin{cases} \displaystyle\int_E \boldsymbol{\nabla}^S(\boldsymbol{v} - \Pi \boldsymbol{v}) : \boldsymbol{\nabla}^S \boldsymbol{p} = 0 \quad \forall \boldsymbol{p} \in \mathcal{P}_k(E) \,, \\ \displaystyle\int_{\partial E} (\boldsymbol{v} - \Pi \boldsymbol{v}) \cdot \boldsymbol{\psi} = 0 \quad \forall \boldsymbol{\psi} \in \mathrm{span}\{(1,0), (0,1), (-y, x)\}. \end{cases} \tag{6}$$

The second equation above is needed only to fix the kernel of the $\boldsymbol{\nabla}^S$ operator, otherwise the operator Π would not be well defined.

We now show that the operator Π is *computable* only in terms of the DoF values of $\boldsymbol{v}$, in other words without the need to actually calculate the function $\boldsymbol{v}$ inside the element. The second equation in (6) involves an integration of polynomial functions on the boundary, and is therefore computable by standard 1D integration rules. In the first equation, the only non-polynomial term is the product $\int_E \boldsymbol{\nabla}^S\boldsymbol{v} : \boldsymbol{\nabla}^S\boldsymbol{p}$, with $\boldsymbol{v} \in \boldsymbol{V}^{RED}_{h|E}$ and $\boldsymbol{p} \in \mathcal{P}_k(E)$. We thus need to show that such term is computable, in the sense above. It is trivial to check that the second condition in (5) implies $\boldsymbol{\Delta}^S\boldsymbol{p} = \nabla q_{k-1}$ for some $q \in \mathbb{P}_{k-1}(E)$ with $\int_E q_{k-1} = 0$. Thus we get

$$\begin{aligned}\int_E \boldsymbol{\nabla}^S\boldsymbol{v} : \boldsymbol{\nabla}^S\boldsymbol{p} &= -\int_E \boldsymbol{v}\cdot\boldsymbol{\Delta}^S\boldsymbol{p} + \int_{\partial E}\big([\boldsymbol{\nabla}^S\boldsymbol{p}]\boldsymbol{n}\big)\cdot\boldsymbol{v}\\ &= -\int_E \boldsymbol{v}\cdot\nabla q_{k-1} + \int_{\partial E}\big([\boldsymbol{\nabla}^S\boldsymbol{p}]\boldsymbol{n}\big)\cdot\boldsymbol{v}\\ &= \int_E (\mathrm{div}\boldsymbol{v})\, q_{k-1} - \int_{\partial E}(\boldsymbol{v}\cdot\boldsymbol{n})q_{k-1} + \int_{\partial E}\big([\boldsymbol{\nabla}^S\boldsymbol{p}]\boldsymbol{n}\big)\cdot\boldsymbol{v}\\ &= \int_{\partial E}\big([\boldsymbol{\nabla}^S\boldsymbol{p} - q_{k-1}\boldsymbol{I}]\boldsymbol{n}\big)\cdot\boldsymbol{v}\end{aligned} \tag{7}$$

where $\boldsymbol{n}$ denotes the outward unit normal to E and $\boldsymbol{I}$ denotes the identity matrix, and the bulk integral in the third row of (7) vanishes being div$\boldsymbol{v}$ constant and $\int_E q_{k-1} = 0$. The remaining term is computable (by using a Gauss integration rule edge by edge) since they involve only $\boldsymbol{v}$ on the boundary, that is a piecewise polynomial. As an interesting byproduct, the computation of the above projection never involves an integration (not even of polynomials) on the polygon E, but only cheap integrations on its boundary ∂E.

Once the above projection is available, the rest of the construction follows standard steps in the Virtual Element framework [1] and is here reported briefly. We start by introducing a discrete bilinear form $a_h^E(\cdot,\cdot)$ that approximates the local form $\nu\int_E \boldsymbol{\nabla}^S\boldsymbol{v} : \boldsymbol{\nabla}^S\nabla w$. We define

$$a_h^E(\boldsymbol{v},\boldsymbol{w}) = \nu\Big(\int_E \boldsymbol{\nabla}^S\Pi\boldsymbol{v} : \boldsymbol{\nabla}^S\Pi\boldsymbol{w} + s_h^E((I-\Pi))\boldsymbol{v},(I-\Pi))\boldsymbol{w})\Big) \qquad \forall \boldsymbol{v},\boldsymbol{w} \in \boldsymbol{V}^{RED}_{h|E},$$

where the stabilizing form $s_h^E(\cdot,\cdot)$ is simply given by the scalar product of the DoF evaluation vectors of its two entries. Such choice guarantees, under suitable mesh assumptions, that the energy $a_h^E(\boldsymbol{v},\boldsymbol{v}) \sim \int_E ||\boldsymbol{\nabla}^S\boldsymbol{v}||^2$, c.f. [4]. The bilinear form $a_h^E(\cdot,\cdot)$ is exact (i.e. it exactly corresponds to the form $\nu\int_E \boldsymbol{\nabla}^S\boldsymbol{v} : \boldsymbol{\nabla}^S\boldsymbol{w}$) whenever one of the two entries is in $\mathcal{P}_k$, which is a kind of consistency property. The global bilinear form is assembled element by element as usual

$$a_h(\boldsymbol{v},\boldsymbol{w}) = \sum_{E\in\Omega_h} a_h^E(\boldsymbol{v},\boldsymbol{w}) \qquad \forall \boldsymbol{v},\boldsymbol{w} \in \boldsymbol{V}_h^{RED}$$

where clearly

$$\boldsymbol{V}_h^{RED} = \left\{ \boldsymbol{v} \in [H^1(\Omega)]^2 \ : \ \boldsymbol{v}|_E \in \boldsymbol{V}_{h|E}^{RED} \ \forall E \in \Omega_h \right\}.$$

Regarding the loading term, following assumption **(A1)** we have $\boldsymbol{f} = \nabla\phi$ and thus we can compute such term with an integration by parts (see below). Finally, in order to approximate our Stokes problem, we need a pressure space. We simply take

$$Q_h = \left\{ q \in L_0^2(\Omega) \ : \ q|_E \in \mathbb{P}_0(E) \ \forall E \in \Omega_h \right\}. \tag{8}$$

We obtain the following discrete scheme.

$$\begin{cases} \text{Find}(\boldsymbol{u}_h, p_h) \in \boldsymbol{V}_h^{RED,\boldsymbol{g}} \times Q_h, \text{ such that} \\ a_h(\boldsymbol{u}_h, \boldsymbol{v}_h) + \displaystyle\int_\Omega p_h \, \mathrm{div} \boldsymbol{v}_h = -\int_\Omega \phi \, \mathrm{div} \boldsymbol{v}_h & \forall \boldsymbol{v}_h \in \boldsymbol{V}_h^{RED,\boldsymbol{0}}, \\ \displaystyle\int_\Omega q_h \, \mathrm{div} \, \boldsymbol{u}_h = 0 & \forall q_h \in Q_h, \end{cases} \tag{9}$$

where

$$\boldsymbol{V}_h^{RED,\boldsymbol{g}} = \left\{ \boldsymbol{v} \in \boldsymbol{V}_h^{RED} \ : \ \boldsymbol{v}|_{\partial\Omega} = \boldsymbol{g}_I \right\}$$

with $\boldsymbol{g}_I$ a piecewise polynomial interpolation of $\boldsymbol{g}$ preserving the compatibility condition $\int_{\partial\Omega} \boldsymbol{g}_I \cdot \boldsymbol{n} = 0$.

Remark 2 It is trivial to check that $\mathrm{div} \boldsymbol{V}_h^{RED} \subset Q_h$ and as a consequence the discrete solution satisfies $\mathrm{div} \boldsymbol{u}_h = 0$ in a point-wise sense and not only in a relaxed sense as in standard mixed Finite Elements. This leads to a set of advantages, see for instance [4].

We close this section with some theoretical results. The stability and convergence results for the original scheme associated to the space $\boldsymbol{V}_{h|E}$ (and the reduced version $\boldsymbol{V}_{h|E}^{RED}$) where proved in [3]. Such results can be extended also to the novel reduction here proposed (making use of the local spaces $\boldsymbol{V}_{h|E}^{RED}$). The key observation is that the constraints appearing in the definition of $\boldsymbol{V}_{h|E}^{RED}$ and $\mathcal{P}_k(E)$ are also satisfied by the exact solution $\boldsymbol{u}$, and thus do not lead to a loss in the approximation properties of the scheme. The following result holds under standard (in the polygonal community) mesh regularity assumptions.

Theorem 1 *There exists a unique solution* $(\boldsymbol{u}_h, p_h)$ *of the discrete problem* (9). *Let* $(\boldsymbol{u}, p) \in [H^s(\Omega)]^2 \times H^{s-1}(\Omega)$, $1 < s \leq k+1$ *be the solution of* (1). *Then, under the same mesh assumptions of [3], it holds*

$$\|\boldsymbol{u} - \boldsymbol{u}_h\|_{H^1(\Omega)} \leq C h^{s-1} |\boldsymbol{u}|_{H^s(\Omega)},$$
$$\|\overline{p} - p_h\|_{L^2(\Omega)} \leq C h^{s-1} \left(|\boldsymbol{u}|_{H^s(\Omega)} + |p|_{H^{s-1}(\Omega)} \right),$$

where $\overline{p}$ denotes the piecewise constant function that on each element E takes the average value of p over the element.

Note that, as a consequence of the div-free condition in Remark 2, the error on the velocity solution is not polluted by the pressure interpolation error, as it happens in more standard methods. Since the discrete pressure is piecewise constant, see (8), a higher order convergence rate, as the one in Theorem 1, can be obtained for $(\overline{p} - p_h)$ but not for the true error $(p - p_h)$. Nevertheless, one can apply the same element-wise procedure of [3] and obtain a post-processed pressure $p_h^\star$ that is a piecewise polynomial of degree $(k-1)$ satisfying an $O(h^k)$ convergence rate.

4 Numerical Tests

In this section we numerically verify the theoretical convergence rate in Theorem 1. In order to compute the VEM error between the exact solution $\boldsymbol{u}_{\rm ex}$ and the VEM solution $\boldsymbol{u}_h$, we consider the computable H^1-like error quantity

$$\texttt{err}(\boldsymbol{u}_h)^2 := \sum_{E \in \Omega_h} \|\boldsymbol{u}_{\rm ex} - \Pi_{k-1}^{0,E} \boldsymbol{\nabla} \boldsymbol{u}_h\|_{L^2(E)}^2 \,.$$

We consider the Stokes equation on the unit square $\Omega = [0, 1]^2$, the load term and the Dirichlet boundary conditions are chosen in accordance with the analytical solution

$$\boldsymbol{u}_{\rm ex}(x, y) = e^x \begin{pmatrix} \sin y + y \cos y - x \sin y \\ -x \cos y - y \sin y - \cos y \end{pmatrix} \qquad p_{\rm ex}(x, y) = \sin(\pi x) \cos(4\pi y)$$

The domain Ω is partitioned with the following sequences of polygonal meshes: `QUADRILATERAL` distorted meshes, `TRIANGULAR` meshes, `CVT` (Centroidal Voronoi Tessellations) meshes, `RANDOM` meshes (see Fig. 1). For each family of meshes we take the sequence with diameter $h = 2^{-2}, 2^{-3}, 2^{-4}, 2^{-5}, 2^{-6}$. The polynomial degree of accuracy for the numerical test is $k = 3$.

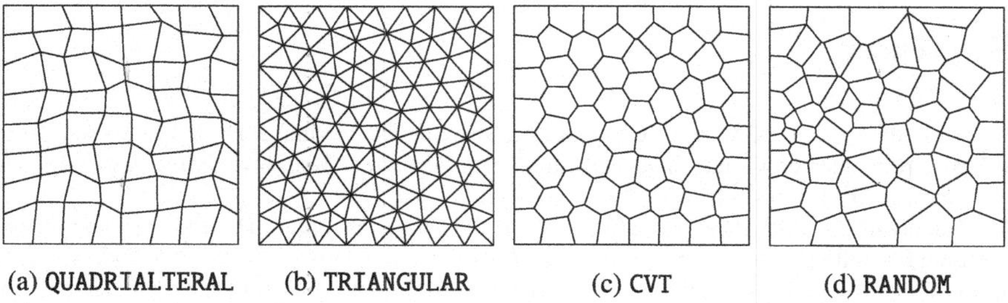

(a) `QUADRIALTERAL` (b) `TRIANGULAR` (c) `CVT` (d) `RANDOM`

Fig. 1 Example of the adopted polygonal meshes

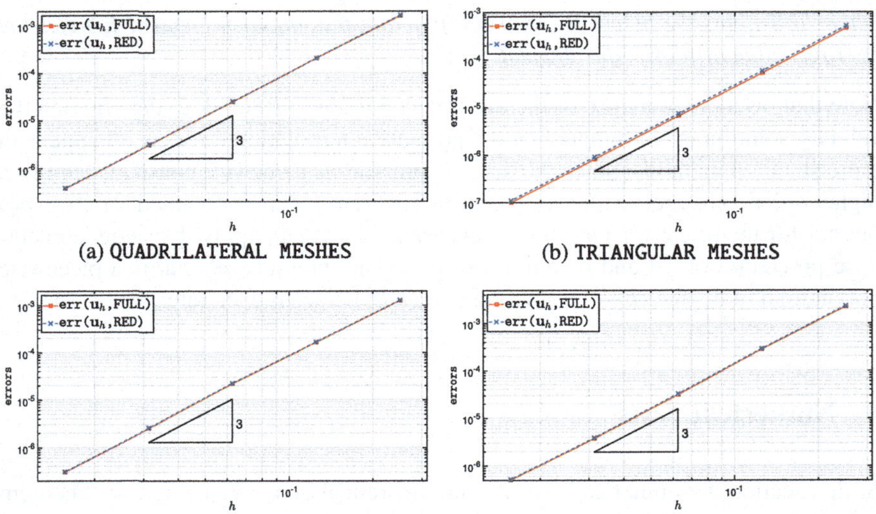

(a) QUADRILATERAL MESHES (b) TRIANGULAR MESHES

(c) CVT MESHES (d) RANDOM MESHES

Fig. 2 Convergence lines for the "FULL" scheme and the "RED" scheme

In Fig. 2 we display the errors $\mathtt{err}(\boldsymbol{u}_h)$ for the the sequences of meshes aforementioned. In order to assess the performance of the proposed reduced space (4) (labeled as "RED") we plot also the results obtained with the complete scheme (2) (labeled as "FULL"). We draw some observations:

- Both methods exhibit the expected rate of convergence.
- The "FULL" scheme and the "RED" scheme produce almost identical results, analogous results where obtained for the pressure error after post-processing (not reported).
- The method is robust with respect to the mesh distortion.

Finally we stress that the "RED" scheme (9) has $11N_P$ less unknowns with respect to its "FULL" counterpart, where N_P is the number of the mesh polygons.

References

1. Beirão da Veiga, L., Brezzi, F., Cangiani, A., Manzini, G., Marini, L. D., & Russo, A. (2013). Basic principles of Virtual Element Methods. *Mathematical Models and Methods in Applied Sciences, 23*(1), 199–214.
2. Hudobivnik, B., Aldakheel, F., & Wriggers, P. (2019). A low order 3D virtual element formulation for finite elasto-plastic deformations. *Computational Mechanics, 63,* 253–269.
3. Beirão da Veiga, L., Lovadina, C., & Vacca, G. (2017). Divergence free virtual elements for the Stokes problem on polygonal meshes. *ESAIM: Mathematical Modelling and Numerical Analysis, 51*(2), 509–535.

4. Beirão da Veiga, L., Lovadina, C., & Vacca, G. (2018). Virtual elements for the Navier-Stokes problem on polygonal meshes. *SIAM Journal on Numerical Analysis, 56*(3), 1210–1242.
5. Beirão da Veiga, L., Dassi, F., & Vacca, G. (2020). The stokes complex for virtual elements in three dimensions. *Mathematical Models and Methods in Applied Sciences, 30*(03), 477–512.

Strategy for Preventing Membrane Locking Through Reparametrization

Simon Bieber, Bastian Oesterle, Manfred Bischoff, and Ekkehard Ramm

The senior authors have a friendly relationship with Peter Wriggers since many years. The influence of our common mentor Erwin Stein as well as our frequent connections to the University of Berkeley have similarly shaped us. We have worked closely with Peter in various activities like the FE im Schnee workshops and various conferences, such as "Particles". The junior authors have always been inspired by Peter Wriggers as one of the leading figures in the field of Computational Mechanics, particularly through his seminal textbooks.

Abstract The contribution takes up the concept of preventing locking a priori in the theory of thin-walled structures instead of curing it during discretization. After briefly summarizing the successful concept for avoiding transverse shear locking through reparametrization of primary variables for beams, plates and shells we concentrate on the same approach for preventing also membrane locking. Here, we describe first steps referring to the plane curved Bernoulli beam as a conceptual proof for the new method. Inspired by the so-called Mixed Displacement variational method we discuss three variants of replacing primary displacement parameters by alternative variables.

S. Bieber (✉) · B. Oesterle · M. Bischoff · E. Ramm
Institute for Structural Mechanics, University of Stuttgart, Stuttgart, Germany
e-mail: bieber@ibb.uni-stuttgart.de

B. Oesterle
e-mail: oesterle@ibb.uni-stuttgart.de

M. Bischoff
e-mail: bischoff@ibb.uni-stuttgart.de

E. Ramm
e-mail: ramm@ibb.uni-stuttgart.de

F. Aldakheel et al. (eds.), *Current Trends and Open Problems in Computational Mechanics*, https://doi.org/10.1007/978-3-030-87312-7_7

1 Introduction

The authors follow the objective of preventing geometrical locking for thin-walled structures a priori on the theory level, i.e. prior to and independent of the chosen discretization. Equal order interpolations for all involved fields are assumed. We successfully derived primal displacement-based formulations for transverse shear deformable beams, plates and shells from the virtual work principle, which are a priori free from transverse shear locking and curvature thickness locking [1, 2]. The main idea follows a reparametrization of primal kinematic parameters in the sense of a hierarchic formulation, introducing either hierarchic rotations or displacements. Let us illustrate this concept of reparametrization using the plane Timoshenko beam as model problem:

standard disp. formulation	reparametrized formulations		
$\gamma = w' + \varphi$ $\kappa = \varphi'$	$\gamma = \gamma$ $\kappa = -w'' + \gamma'$	&	$\gamma = w_s'$ $\kappa = -w_b''$

(1)

It is apparent that the function spaces for the degrees of freedom w and φ are unbalanced for equal order interpolation eventually causing transverse shear locking. If the hierarchic (incremental) rotation γ is used instead the unbalance is shifted from the transverse shear γ to κ allowing a pure bending state with $\gamma = 0$ without parasitic shear strains. In other words the formulation is a priori free from shear locking. If in turn the transverse displacement w is decomposed into bending and shear part, $w = w_b + w_s$ it is clearly shown that the function spaces in both equations are fully balanced avoiding again shear locking. A similar strategy, though limited to beams, was presented in [3], where the authors developed a locking-free single-variable formulation. In [4] we extended the approach of reparametrization also into the geometrically nonlinear regime for thin-walled structures with arbitrarily large rotations. In these studies membrane locking has been prevented by applying either the Hellinger-Reissner variational principle or the Discrete Strain Gap (DSG) scheme [5].

Following the above mentioned main objective, the present study focuses on also preventing membrane locking a priori on the theory level. For membrane locking see Sect. 3.

We proposed a first concept preventing all kinds of geometrical locking a priori including membrane locking utilizing a novel variational method [6]. Starting from a modified Hellinger-Reissner principle which uses displacements and strains as variables the new method replaces the strains by independent displacement-like variables. Consequently, the approach is denoted as Mixed Displacement (MD) method where these additional degrees of freedom (auxiliary displacements) follow a dual concept. They counterbalance (weaken) the too stiff solution of standard primal formulations. The idea was inspired by the DSG method; in fact, the extra variables are directly related to the strain gaps. The equal order interpolation for all involved fields adjusts the function spaces such that all geometrical locking phenomena, including membrane locking, are automatically avoided, in fact for all discretization methods. The approach is briefly elaborated for the Bernoulli beam in Sect. 2.2.

2 Curved Bernoulli Beam

In the present study, we limit ourselves to a plane curved beam formulation as a model problem for the subsequently derived locking-free formulations.

2.1 *Standard Displacement Formulation* (u, w)

Let us consider the in-plane deformation of a geometrically linear planar curved Bernoulli beam equipped with a curvilinear coordinate $s \in [0, L]$, a length L and constant Young's modulus E, thickness t and radius of curvature $R(s)$, which is defined to be positive in the specific geometric configuration shown in Fig. 1.

On the beam's center line we define the displacements (u, w), where $u(s)$ is the tangential displacement and $w(s)$ the inward radial displacement, see e.g. [7, 8]. The axial strain ε and the bending-curvature κ are given as

$$\boxed{\begin{aligned} \varepsilon &= u' - \frac{w}{R} \\ \kappa = \varphi' &= -w'' - \left(\frac{u}{R}\right)' \end{aligned}} \tag{2}$$

with rotation $\varphi = -w' - \frac{u}{R}$. The prime $(\bullet)'$ indicates a derivative with respect to s. The constitutive equations define the axial (membrane) force N and the bending moment M

$$N = EA\varepsilon, \qquad M = EI\kappa, \tag{3}$$

with cross section area A and second moment of inertia I. The governing differential equations are

$$\begin{aligned} N' - \frac{M'}{R} &= 0, \\ \frac{N}{R} + M'' &= 0, \end{aligned} \tag{4}$$

expressing the equilibrium in tangential and radial direction; for simplicity no distributed loads are considered. The boundary value problem is completed with the

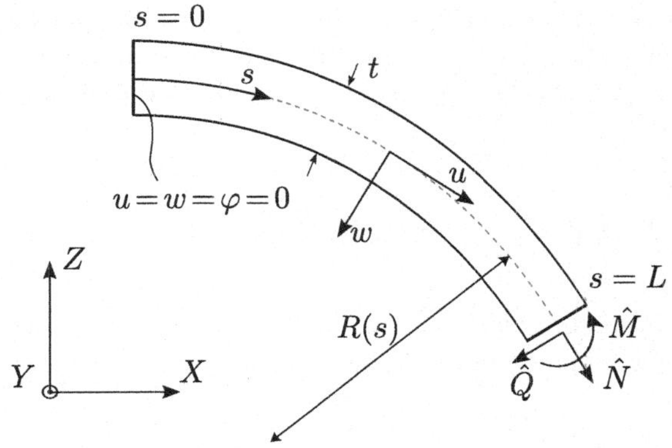

Fig. 1 Schematic description of the model problem

essential and natural boundary conditions at the two ends of the beam (see Fig. 1). The corresponding weak form is obtained via the principle of virtual work

$$\int_0^L (\delta\varepsilon EA\varepsilon + \delta\kappa EI\kappa)\,\mathrm{d}s - \left(\delta u\hat{N} + \delta w\hat{Q} + \delta\varphi\hat{M}\right)\Big|_{s=L} = 0, \tag{5}$$

with virtual displacements δu and δw.

Remark 1 For the special case of a straight center line (which implies $R = \infty$) the terms which contain $1/R$ in Eq. (2) vanish and two decoupled formulations are recovered: a truss with axial degree of freedom u and a straight beam with transverse displacement w.

2.2 *Mixed Displacement Formulation* (u, w, u_ε)

As explained above for the MD formulation the set of primary variables (u, w) is enhanced by an extra independent displacement u_ε leading to a third kinematic equation $\varepsilon = u'_\varepsilon$ additional to $\varepsilon_\mathrm{u} := \varepsilon(u, w)$ and κ defined in Eq. (2) (here index "u" is introduced to distinguish the two axial strains). The constraint equation $\varepsilon - \varepsilon_\mathrm{u} = 0$ is satisfied in a weak sense based on the variational expression

$$\delta\Pi^{\mathrm{int}}_{\mathrm{MD}}(u, w, u_\varepsilon) = \delta\Pi^{\mathrm{b}} + \int_0^L EA\left\{\delta u'_\varepsilon(u'_\varepsilon - \varepsilon_\mathrm{u}) - \delta\varepsilon_\mathrm{u} u'_\varepsilon\right\}\mathrm{d}s. \tag{6}$$

Here the bending part $\delta\Pi^{\mathrm{b}}$ remains unaffected and corresponds to the one of Eq. (5).

3 Membrane Locking

Phenomenologically, locking can be defined by the following symptoms:

1. Displacements are severely underestimated for coarse meshes. In the pre-asymptotic range the order of convergence is below the optimal one.
2. For a given mesh the error depends on a certain model parameter and also the size of the pre-asymptotic range with sub-optimal convergence depends on this parameter. In the case of thin-walled structures, this parameter is the slenderness.
3. Non-physical stresses occur, mostly in an oscillating manner.

Membrane locking results from the inability of a finite element formulation to represent pure bending deformations (inextensional deformations) if the elements are curved , see [9]. The origin of the problem can be explained in an illustrative manner for the simple example of a clamped circular ring (R = const., cf. Fig. 4, below). A pure bending deformation requires the membrane strain to vanish, $\varepsilon = 0$, such that with Eq. (2)

$$u' = \frac{w}{R} \quad \Rightarrow \quad \kappa = -w'' - \frac{w}{R^2}. \tag{7}$$

If polynomials or splines are used as shape functions, a constant curvature $\kappa = \kappa_0$ together with $(7)_2$ implies the condition that w is constant. Due to the applied boundary condition, this constant must be zero and consequently $w = 0$ in the entire beam. From Eq. $(7)_2$ it then follows that also $\kappa_0 = 0$. Therefore, the condition of vanishing membrane strains $\varepsilon = 0$ implies zero curvature and thus zero bending moments. In other words, if curvature and bending moments are supposed to be *non-zero*, this implies $\varepsilon \neq 0$, thus effectively excluding inextensional deformations. It is also obvious from this explanation that for straight beams with $R = \infty$ the problem does not exist.

In practical computations, the curvature for a given load will be underestimated and oscillating membrane strains occur. Possible remedies in the context of curved beams are *selective reduced integration* and a so called *field consistent* approximation, i.e. using different polynomial degrees for the approximation of different degrees of freedom, see [7]. However, a successful application is restricted to beams with constant radius of curvature and critically depends on the discretization scheme.

To demonstrate the effects of membrane locking we consider a thin elliptical cantilever with thickness $t = 0.25$ subject to tip force, see Fig. 2a. The distributions of the axial force N obtained with a standard displacement formulation (u, w) and the MD formulation (u, w, u_ε) are shown in Fig. 2b. While the axial force obtained by the MD formulation captures the analytical solution correctly, the standard displacement solution for N shows severe oscillations, although 3rd order shape functions are used. This is a clear indication for membrane locking.

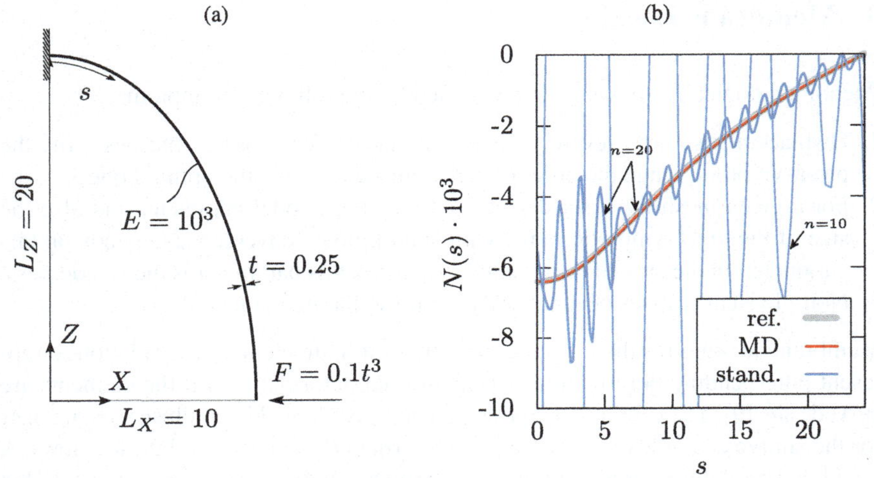

Fig. 2 (**a**) Clamped elliptical beam with slenderness $R/t \in [20, 160]$. (**b**) Axial forces with NURBS of $p = 3$ and number of control points $n = 10$ and $n = 20$

4 Reparametrizations to Avoid Membrane Locking

The main purpose of this paper is to provide a proof of concept for the feasibility of formulations based on suitable reparametrizations, i.e. replacing primary variables by alternative substitutes, to prevent membrane locking *independent of discretization*. In all the three versions shown below, a primal finite element formulation based on the virtual work principle with equal order interpolation of all independent variables of the respective reparametrization is applied. Non-Uniform Rational B-Splines (NURBS) $N_{i,p}$ of polynomial degree p and C^{p-1}-continuity are used as shape functions. For further reading on isogeometric analysis in the context of modeling plane curved beams we refer to [8].

For simplicity of presentation, we restrict the following theoretical analysis to circular beams with constant radius of curvature R.

4.1 *First Reparametrization* (u, u_ε)

Similar to the MD formulation, a new variable u_ε is introduced, replacing w in the set of primary variables by directly satisfying the constraint $\varepsilon - \varepsilon_u = 0$. The modified kinematic equations for $(u,\ u_\varepsilon)$ are

$$\boxed{\begin{aligned} \varepsilon &= u_{\varepsilon}' \\ \kappa &= -R\left(u' - u_{\varepsilon}'\right)'' - \frac{u'}{R} \end{aligned}} \tag{8}$$

where the radial displacement is recovered as $w = R(u' - u_{\varepsilon}')$. A generalization to a varying R is straightforward. Nevertheless, the formulation needs C^2-continuity. Most importantly, it allows non-trivial ($\kappa \neq 0$) inextensional (pure bending) deformations with $\varepsilon = 0$. Corresponding finite element formulations are automatically free of membrane locking. However, the limiting case of a straight beam with $R = \infty$ is not included and needs special attention.

4.2 Second Reparametrization (u_ε, w)

This formulation is based on extracting u from the integrated constraint equation $\varepsilon = 0$

$$u = u_\varepsilon + \frac{\int_s w(s)\, \mathrm{d}s}{R}. \tag{9}$$

Thus, u is eliminated instead of w from the kinematic equations, leaving (u_ε, w) as primary variables. In this version the kinematic relations are

$$\boxed{\begin{aligned} \varepsilon &= u_{\varepsilon}' \\ \kappa &= -w'' - \frac{u_{\varepsilon}'}{R} - \frac{w}{R^2} \end{aligned}} \tag{10}$$

Again, the axial strain ε remains fully decoupled from w, which results in an a priori locking-free formulation. Only C^1-continuity is required. However, the reparametrized variable u in (9) involves an integral operator which complicates practical implementation of this formulation. Furthermore, a similar reparametrization that is also valid for a varying radius of curvature $R(s)$ is rather complex.

4.3 Third Reparametrization (u_ε, A_w)

In order to bypass the integral operator of reparametrization (9) we redefine:

$$w := A_w', \quad \text{with} \quad A_w(s) = \int_s w(s)\, \mathrm{d}s. \tag{11}$$

A_w represents the integrated transverse displacement w. It is only defined up to an additive constant, which has to be eliminated by feasible constraints. Together with

u_ε it replaces both original variables u and w

$$
\begin{aligned}
u &= u_\varepsilon + \frac{A_w}{R}, \\
w &= A_w'.
\end{aligned}
\tag{12}
$$

By substituting (12) into the standard kinematic Eq. (2) we obtain the reparametrized kinematic relation (for the special case R =const.) as

$$
\boxed{
\begin{aligned}
\varepsilon &= u_\varepsilon' \\
\kappa &= -A_w''' - \frac{u_\varepsilon'}{R} - \frac{A_w'}{R^2}
\end{aligned}
}
\tag{13}
$$

which is similar to (8) and (10) in terms of the favorable property of a decoupled axial strain relation. This again allows inextensional deformations without parasitic strains. No integral expression appears in u anymore. However, a disadvantage is the third derivative in (13). Thus, a finite element approximation of A_w within a standard Galerkin approach requires at least C^2-continuity. Again, the formulation becomes complex for a varying radius of curvature $R(s)$.

4.4 Mechanical Interpretation of u_ε

All of the three different versions of reparametrization sketched in the previous section involve a degree of freedom u_ε. In this section, its geometric significance is explained and illustrated by studying a simple deformation of a quarter circular arch, as shown in Fig. 3. The idea is to apply a pure inextensional deformation with a constant change of radius $\hat{w}$. The inextensional constraint $w = u'R$ (cf. Eq. $(7)_1$) along with a prescribed constant curvature $\kappa = \kappa_0$ yields

$$
-u'''R - \frac{u'}{R} = \kappa_0, \quad \text{with boundary conditions} \quad \left.\begin{aligned} u &= 0 \\ w &= -\hat{w} \\ \varphi &= 0 \end{aligned}\right\} \text{ at } s = 0. \tag{14}
$$

To provide the desired insight into the role of u_ε, we divide the deformation process into two consecutive deformation steps:

1. A change of radius $w = -\hat{w}$ that produces a purely *extensional deformation*

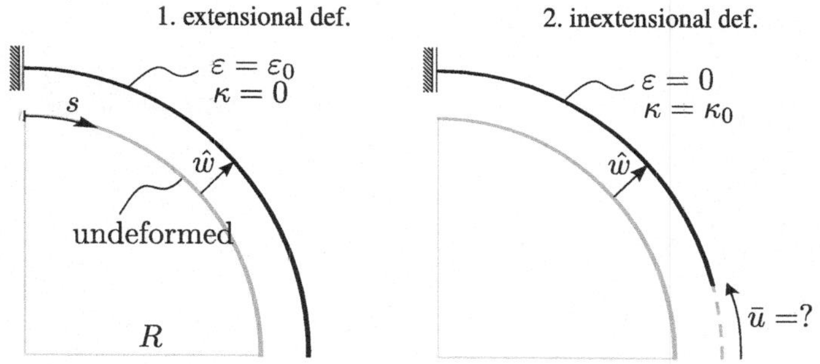

Fig. 3 Two consecutive deformation steps

$$\begin{pmatrix} u = 0 \\ w = -\hat{w} \\ u_\varepsilon = \dfrac{\hat{w}}{R}s \\ A_w = -\hat{w}s \end{pmatrix} \rightarrow \boxed{\begin{matrix} \varepsilon = \dfrac{\hat{w}}{R} =: \varepsilon_0 \\ \kappa = 0 \end{matrix}}$$

By definition u_ε is the integral of the membrane strain along the arc length s. It is equal to an axial displacement $u(s)$ that would lead to the same axial strain ε for the case $w = 0$.

2. A subsequent axial deformation $u(s)$ that eliminates the non-zero ε from the first step and thus yields a purely *inextensional* deformation.

$$\begin{pmatrix} u = -\dfrac{\hat{w}}{R}s \\ w = -\hat{w} \\ u_\varepsilon = 0 \\ A_w = -\hat{w}s \end{pmatrix} \rightarrow \boxed{\begin{matrix} \varepsilon = 0 \\ \kappa = \dfrac{\hat{w}}{R^2} =: \kappa_0 \end{matrix}}$$

Obviously, this axial displacement is exactly equal to the negative u_ε from the first step. Thus, u_ε can be interpreted as the variable that effectively cancels parasitic membrane strains and thus removes membrane locking.

5 Numerical Example

A clamped quarter circular ring subject to a tip load, shown in Fig. 4a, serves as a benchmark to test the performance of the proposed formulations with respect to membrane locking. With $\theta = s/R$ the analytical solution for the radial and tangential displacement is

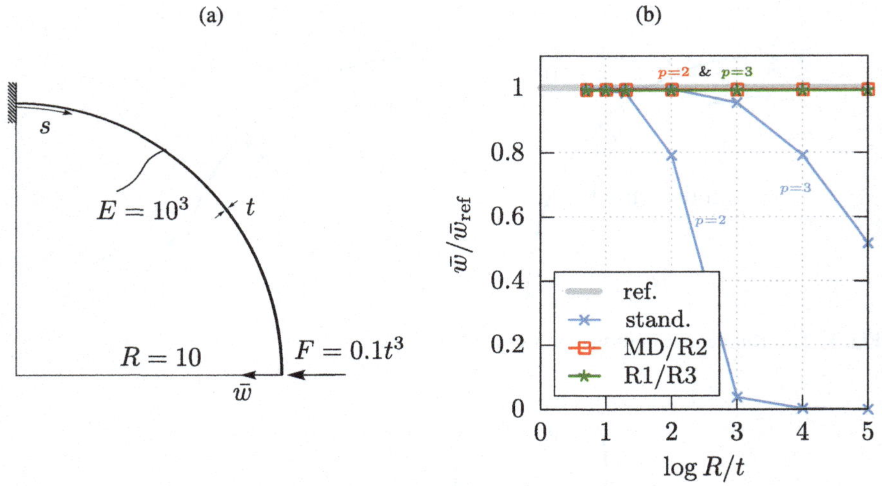

Fig. 4 (**a**) Clamped circular beam. (**b**) Radial tip-displacement $\bar{w}$ versus slenderness for various formulations; NURBS, $n = 10$, $p = 2$ and $p = 3$

$$u_{\mathrm{ref}}(s) = -\frac{F}{2}\left(\frac{s\cos\theta - R\sin\theta}{EI} - \frac{s\cos\theta + R\sin\theta}{EA}\right), \tag{15}$$

$$w_{\mathrm{ref}}(s) = \frac{F}{2}\left(\frac{sR^2\sin\theta}{EI} + \frac{s\sin\theta}{EA}\right), \tag{16}$$

with the tip displacements

$$u_{\mathrm{ref}}(s = L) \approx 0.6 - 0.005t^2, \qquad w_{\mathrm{ref}}(s = L) \approx 0.942478 + 0.000785t^2. \tag{17}$$

The axial forces and bending moments are

$$N_{\mathrm{ref}}(s) = -F\cos(\theta), \qquad M_{\mathrm{ref}}(s) = -FR\cos(\theta). \tag{18}$$

All simulations are performed with NURBS shape functions, as described in Sect. 4. Here, n denotes the number of control points. Boundary conditions are enforced via Lagrange multiplier. The element formulations considered are:

- **stand.**: primal (u, w)-formulation from Sect. 2.1
- **MD**: dual (mixed) (u, w, u_ε)-formulation from Sect. 2.2
- **R1**: primal (u_ε, u)-formulation from Sect. 4.1
- **R2**: primal (u_ε, w)-formulation from Sect. 4.2
- **R3**: primal (u_ε, A_w)-formulation from Sect. 4.3

Fig. 4b shows a typical locking diagram for a discretization with $n = 10$ control points. The normalized tip displacement $\bar{w}$ is plotted versus the slenderness R/t in order to visualize whether or not the quality of the results deteriorates with increasing slenderness. The critical parameter R/t does not influence the numerical solution of the MD formulation and all newly proposed reparametrized primal formulations, confirming that these are free of membrane locking. Here, MD and R2 ($p = 2$), as well as R1 and R3 ($p = 3$), yield numerically identical results. The results from the standard displacement formulation deteriorate with increasing R/t, which indicates locking. Performance can be improved by increasing the polynomial degree, but the principal problem of locking is not solved.

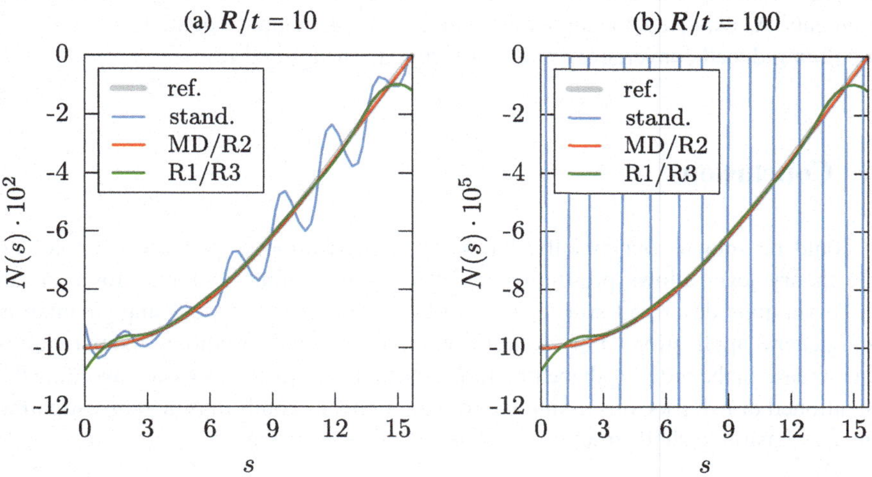

Fig. 5 Axial force versus arc length s for a thick (**a**) and a thin (**b**) beam; NURBS, $n = 10$, $p = 3$

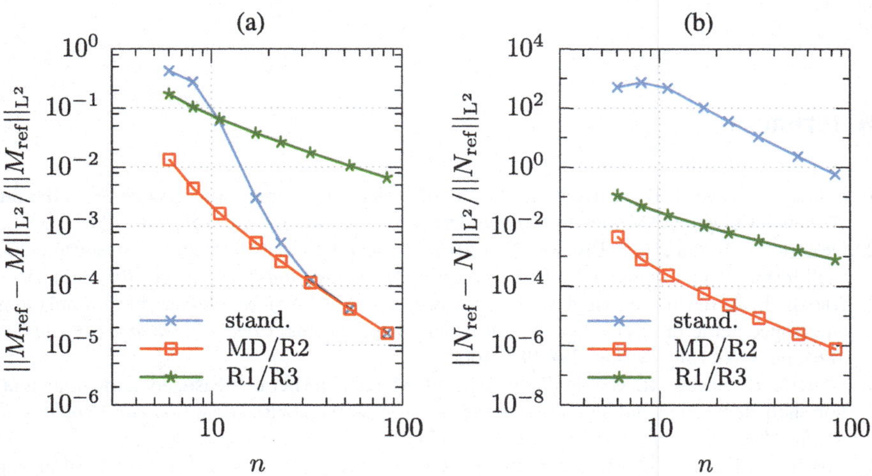

Fig. 6 Relative L^2-norm error of M (**a**) and N (**b**) for a thin beam of slenderness $R/t = 1000$ versus n; NURBS, $p = 3$

Figure 5 shows the axial force diagram for a thick (a) and a moderately thin (b) beam. The standard displacement formulation shows severe oscillations, which are particularly pronounced for higher slenderness. All other formulations are able to capture the reference solution and the quality does not change with slenderness. For R1 and R3 mild oscillations are observed at the boundaries. These, however, vanish with p-or h-refinement (not shown).

These results are confirmed by the relative L^2 error norms for M and N. Figure 6 shows the corresponding convergence diagrams for a thin beam. In contrast to the standard formulation, uniform convergence rates are observed for all locking-free elements. As expected, the higher derivatives in the kinematic equations (8) and (13) result in reduced convergence rates for formulations R1 and R3.

6 Conclusions

It could be shown conceptually that a reparametrization of primary variables on the theory level allows preventing intrinsically membrane locking. However, all three versions described still lead to obstacles for their practical implementation and general applications. For example, an increase of the variational index requires an increase in the order of shape functions. Further studies are necessary avoiding the mentioned drawbacks, discussing the role of boundary conditions, and investigating their extension to shell structures, also in the nonlinear range.

Acknowledgements Support for this research was provided by the Deutsche Forschungsgemeinschaft (DFG) under Grant BI722/11-1 and Grant OE 728/1-1. This support is gratefully acknowledged.

References

1. Echter, R., Oesterle, B., & Bischoff, M. (2013). A hierarchic family of isogeometric shell finite elements. *Computer Methods in Applied Mechanics and Engineering, 254,* 170–180.
2. Oesterle, B., Ramm, E., & Bischoff, M. (2016). A shear deformable, rotation-free isogeometric shell formulation. *Computer Methods in Applied Mechanics and Engineering, 307,* 235–255.
3. Kiendl, J., Auricchio, F., Hughes, T. J. R., & Reali, A. (2015). Single-variable formulations and isogeometric discretizations for shear deformable beams. *Computer Methods in Applied Mechanics and Engineering, 284,* 988–1004.
4. Oesterle, B., Sachse, R., Ramm, E., & Bischoff, M. (2017). Hierarchic isogeometric large rotation shell elements including linearized transverse shear parametrization. *Computer Methods in Applied Mechanics and Engineering, 321,* 383–405.
5. Koschnick, F., Bischoff, M., Camprubi, N., & Bletzinger, K.-U. (2005). The discrete strain gap method and membrane locking. *Computer Methods in Applied Mechanics and Engineering, 194,* 2444–2463.
6. Bieber, S., Oesterle, B., Ramm, E., & Bischoff, M. (2018). A variational method to avoid locking-independent of the discretization scheme. *International Journal for Numerical Methods in Engineering, 114,* 801–827.

7. Prathap, G. (1985). The curved beam/deep arch/finite ring element revisited. *International Journal for Numerical Methods in Engineering, 21,* 389–407.
8. Bouclier, R., Elguedj, T., & Combescure, A. (2012). Locking free isogeometric formulations of curved thick beams. *Computer Methods in Applied Mechanics and Engineering, 245–246,* 144–162.
9. Stolarski, H., & Belytschko, T. (1982). Membrane locking and reduced integration for curved elements. *The Journal of Applied Mechanics, 49,* 172–176.

Model-Free Fracture Mechanics and Fatigue

Pietro Carrara, Michael Ortiz, and Laura De Lorenzis

To Peter, our warmest wishes of a Happy Birthday and of many years of continued health, joy and success.

Abstract We summarize our recent work on data-driven fracture mechanics. The governing equations stemming from variational principles are completed with a set of discrete data points encoding the information about the material behavior, thus the fracture-related modeling assumptions are completely removed. The solution at a given load step is identified as the point within the data set that best satisfies equilibrium and compatibility conditions stemming from global or local minimality of a free energy function. The data-driven approach is successfully tested for rate-independent and rate-dependent fracture including fatigue.

1 Introduction

Recently, the paradigm of model-free data-driven computational mechanics was advocated and first applied to elastic [1] and then to inelastic materials [2]. The basic idea is that, since boundary value problems in mechanics are defined using an

P. Carrara · L. De Lorenzis (✉)
ETH Zürich, D-MAVT, Institute of Mechanical Systems, Computational Mechanics Group, Zürich, Switzerland
e-mail: ldelorenzis@ethz.ch

P. Carrara
e-mail: pcarrara@ethz.ch

M. Ortiz
Division of Engineering and Applied Science, California Institute of Technology, Pasadena, CA, USA
e-mail: ortiz@caltech.edu

F. Aldakheel et al. (eds.), *Current Trends and Open Problems in Computational Mechanics*, https://doi.org/10.1007/978-3-030-87312-7_8

epistemic set of conservation laws and an empirical set of constitutive equations, the uncertainty stemming from the latter can be avoided by replacing the constitutive relationships with a data set of discrete observations termed *material points*. The data-driven solution strategy relies on the definition of a discrete quantity, generally termed *distance*, which attains its minimum in correspondence of the material point that best represents the solution for the imposed load [1].

In this paper, we summarize the main results of our recent work [3, 4] where we investigate model-free data-driven fracture mechanics. The following topics are addressed: i. rate -independent and -dependent fracture; ii. the type of distance adopted within the solution procedure; iii. the type of data sets available; iv. the effect of the scattering within the material data set; v. convergence of the data-driven results to classical solutions obtained adopting analytical constitutive laws.

2 Classical Formulation

We consider first a pre-cracked linear elastic solid where crack propagation happens under mode-I, planar, rate-independent conditions. Hence, the crack size can be characterized by a scalar parameter a and the linear elastic constitutive behavior of the bulk material by a *compliance function* $C(a)$ that is exactly known. The deformation is represented by an effective opening displacement Δ and the conjugate stress by an effective force P.

Assuming a displacement-driven process and given an initial crack length a_0, in the rate-independent case we aim at finding the crack length a^* and the load P corresponding to the imposed displacement Δ, where $a^* \geq a_0$ due to irreversibility. The free energy reads

$$F(\Delta, a) = E(\Delta, a) + F_R(a), \tag{1}$$

where $E(\Delta, a)$ and $F_R(a)$ are the elastic and dissipative energy terms.

The governing equations are obtained through the minimization of (1) following two alternative procedures. The first entails a *global minimization*, namely

$$F\left(\Delta, a^*\right) \leq F\left(\Delta, a\right) \quad \forall a \geq a_0 \quad \text{or} \quad a^*\left(\Delta, a_0\right) = \operatorname{argmin}\left\{F\left(\Delta, a\right) : a \geq a_0\right\}. \tag{2}$$

Another possibility is to look for a *metastable* state or *local minimum* of the free energy, i. e. we look for a^* such that the following condition holds

$$\exists h \geq 0 \;:\; a^*\left(\Delta, a_0\right) = \operatorname{argmin}\left\{F\left(\Delta, a\right) \;:\; a \geq a_0\,,\; \left|a - a^*\right| \leq h\right\}. \tag{3}$$

Expanding (3) in Taylor series up to first order we obtain the Kuhn-Tucker (KT) conditions

$$\begin{gathered} a^* - a_0 \geq 0, \quad G\left(\Delta, a^*\right) - G_R^{QS}\left(a^*\right) \leq 0 \quad \text{and} \\ \left[G\left(\Delta, a^*\right) - G_R^{QS}\left(a^*\right)\right]\left(a^* - a_0\right) = 0 \end{gathered} \tag{4}$$

where $G(\Delta, a) = -\partial E/\partial a$ and $G_R^{QS}(a) = dF_R/da$ are the energy release rate and its quasi-static critical value, respectively.

Global and local minimization are equivalent only if the free energy is a convex function, otherwise results might differ. In particular, the local approach forbids transitions between states separated by energetic barriers. These transitions are allowed for a globally stable process leading to smaller allowed states domain and violating the causality principle [5]. This results in anticipated crack jumps and in the possibility to nucleate cracks in absence of defects in contrast with Griffith theory.

The free energy (1) does not account for the crack-tip kinetics that can be introduced defining the crack tip velocity $v_t = da_t/dt = \dot{a}_t$, where $(\bullet)_t = (\bullet)(t)$, t is the time and $v_t \geq 0$ to impose crack irreversibility. The analysis is restricted to velocities small enough to neglect the inertial effects. The dissipated energy becomes history dependent and is defined by means of a history independent dissipation potential $\mathcal{D}(a, v)$ as

$$F_R(t, a, [0, t]) = \int_0^t \mathcal{D}(a(s), v(s))\, ds \quad \text{and} \quad G_R(a, v) = \frac{dF_R}{da} = \frac{\mathcal{D}}{v}(a, v)\,. \tag{5}$$

where $da = v\,dt$ is used and $G_R(a, v)$ is the critical energy release rate.

The rate-dependent problem thus aims at finding the crack tip velocity v^* that minimizes the free energy function (1) where the dissipation assumes the form (5a). Also, we have that $G_R(a, v \to 0) = G_R^{QS}(a)$ and, in real cases, $G_R(a, v)$ is continuous and non-decreasing in v, making the rate-dependent problem convex [4, 6]. Hence, local or global minimality principles are equivalent and can be expressed as

$$\begin{gathered} v_t^* \geq 0\,, \quad G(\Delta_t, a_t) - G_R(a_t, v_t^*) \leq 0 \quad \text{and} \\ \left[G(\Delta_t, a_t) - G_R(a_t, v_t^*)\right] v_t^* = 0 \end{gathered} \tag{6}$$

Under some regularity conditions (6) can be also written as [6]

$$v_t^* = \frac{da_t^*}{dt} = \tilde{G}_R^{-1}\left(G(\Delta_t, a_t)\right)\,, \tag{7}$$

where $\tilde{G}_R^{-1}\left(G(\Delta_t, a_t)\right) = G_R^{-1}\left(G(\Delta_t, a_t)\right)$ if $G(\Delta_t, a_t) > G_R^{QS}(a_t)$ and 0 otherwise. The structure of (7) is very similar to the one describing the crack growth rate curve in case of fatigue. In the latter case, the role of time is played by the number of cycles N. Thus, the crack tip velocity is $v = da/dN$, which, once substituted in (7), delivers a Paris-like fatigue law in terms of energy release rate.

3 Data-Driven Approach

The classical solution of the fracture propagation problem requires the knowledge of the critical energy (or energy release rate) as a function of the crack length (or, in rate-dependent cases, of the crack length and velocity). This function, postulated to be endowed with some properties [3, 5], is usually a-priori assumed and best fitted to a set of experimental observation $\hat{\xi}_i$, $i = 1, ..., n$. This process introduces a subjective bias and manipulates part of the information carried by the material data set.

Data-driven fracture mechanics aims at characterizing fracture propagation without resorting to analytical material modeling. Instead, we directly make use of the discrete data set $\mathcal{D}_R = \left\{\hat{\xi}_i,\ i = 1, ..., n\right\}$. The global and local problems (2) and (3) entail now a discrete minimization over $\mathcal{D}_R$. This approach considers the inherent presence of scattering—namely, the *noise* affecting $\mathcal{D}_R$—as an inherent property and thus automatically accounts for it.

The numerical examples illustrated in this section are related to a double cantilever beam that is loaded imposing a displacement Δ_T to a connected linear elastic device [7]. As follows, we summarize the most important results. For all details, see [3, 4].

3.1 Rate-Independent Fracture

Although alternatives are possible, in [3] the following distances are proposed

- **Global minimization problem** (2): $\hat{\xi}_i = (\hat{a}_i,\ \hat{F}_{R,i})$
  1. *Free energy (1)*: $d_i = E(\Delta, \hat{a}_i) + \hat{F}_{R,i},\ \hat{a}_i \geq a_0$
- **Local minimization problem** (3): $\hat{\xi}_i = (\hat{a}_i,\ \hat{G}_{R,i})$
  1. *KT condition (4)* : $d_i = \left|\left[G\left(\Delta, \hat{a}_i\right) - \hat{G}^{QS}_{R,i}\left(a_0\right)\right]\left(a_0 - \hat{a}_i\right)\right|$, $\hat{a}_i \geq a_0,\ G\left(\Delta, \hat{a}_i\right) \leq \hat{G}^{QS}_{R,i}$
  2. *Closest-point projection*: $d_i = \min\limits_{a \geq a_0}\left\{\left\|\left(a - \hat{a}_i,\ G(\Delta, a) - \hat{G}_{R,i}\right)\right\|\right\}$, $\hat{a}_i \geq a_0,\ G\left(\Delta, \hat{a}_i\right) \leq \hat{G}_{R,i_0},\ i_0 : \hat{a}_{i_0} = a_0$

Global vs. local minimization. In [3] the aforementioned approaches are compared under rate-independent conditions using in-silico data sets to mimic different analytical resistance models affected or not by noise (Fig. 1). Figure 1a, b compare the data-driven results with those obtained adopting a classical solution scheme and an analytical fracture constitutive law, while Fig. 1c, d show how data-driven search procedures select the solution for an illustrative load step. The results show that

- Local and global data-driven approaches reproduce their analytical counterparts and are able to predict crack jumps without any *ad-hoc* criterion (Fig. 1a, b). In particular, global data-driven solutions suffer of the same drawbacks of their analytical counterparts mentioned in Sect. 2.
- The global minimization approach is very sensitive to the presence of noise resulting in a systematic overestimation of the crack size (Fig. 1a).

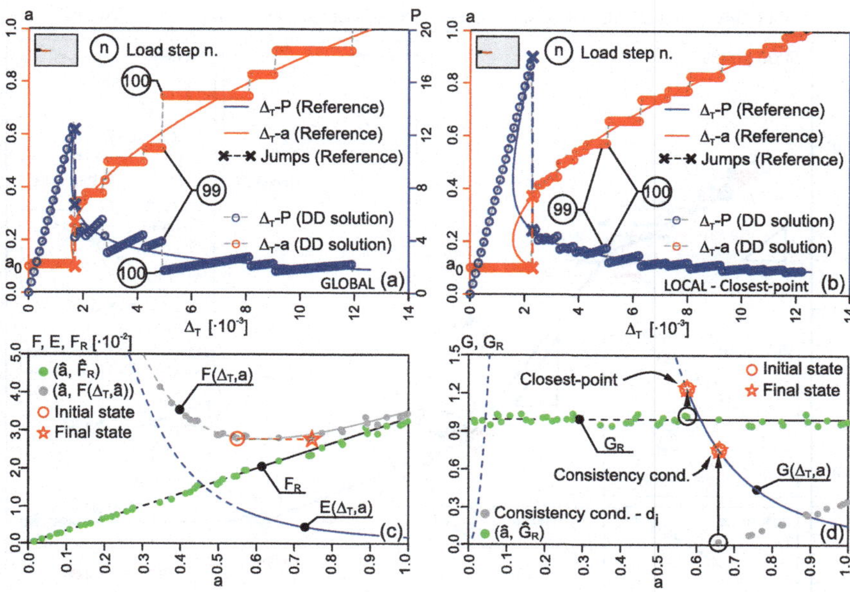

Fig. 1 Data-driven results for Griffith fracture with a noisy material data set: comparison between reference and data-driven results using the global minimization (**a**) and the closest-point projection distance (**b**) (distance based on the KT condition gives similar results). Data-driven solution at load step 100 for global (**c**) and local (**d**) minimization

- In presence of multiple possible meta-stable states, the local minimization approaches are biased toward the one corresponding to the maximum dissipation in spite of the causality principle.
- Local minimization based on the KT condition usually selects crack sizes larger than that expected at equilibrium while closest-point projection is robust with respect to presence of noise and number of data (Fig. 1b).

Convergence. The convergence of the proposed approaches to the reference classical results is also evaluated in [3] and the main findings can be summarized as follows

Convergence with respect to the number of points in $\mathcal{D}_R$

- In absence of noise, all the methods converge linearly to the reference results.
- The error with respect to the reference results for the closest-point projection strategy is always lower than for the alternatives (e. g., noiseless limit in Fig. 2).

Convergence with respect to the noise amplitude

- The approaches based on local minimization converge faster than that based on global minimization (Fig. 2).
- Overestimation of the crack size negatively affects the convergence of the global approach and of the local approach with the KT condition (Fig. 2).

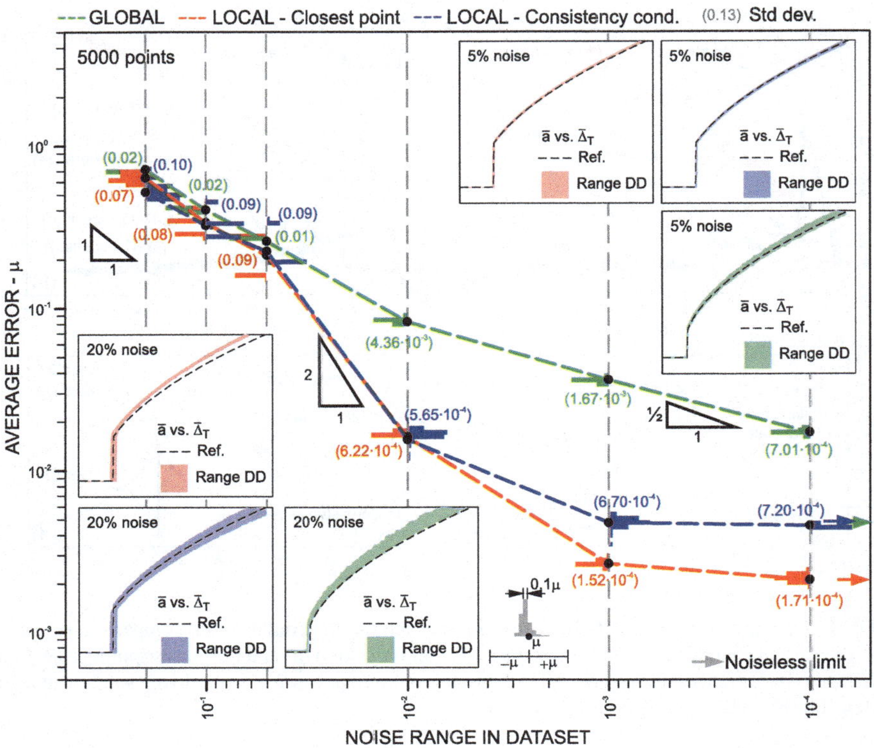

Fig. 2 Convergence with respect to the noise amplitude to the classical solution including frequency histograms and average μ of the error for 100 randomly assigned data set composed of 5000 points. The pop-up graphs indicate the range of the data-driven solutions for 20% and 5% noise amplitude

3.2 Rate-Dependent Fracture and Fatigue

For the rate-dependent case [4] we leverage the results of Sect. 3.1 and rely only on the closest-point projection distance. Upon a time discretization, we define an incremental solution strategy based on a Crank-Nicholson time integration scheme. Depending on the type of database available two different solution strategies can be devised as follows

1. **Complete data set**: if $\hat{\boldsymbol{\xi}}_i = (\hat{a}_i,\ \hat{v}_i,\ \hat{G}_{R,i})$, then the closest-point projection distance is computed assuming an energy release rate function of the specimen parametrized in the plane $\Delta a(v) - G$, where $\Delta a(v)$ is the crack size increment that stems from the selected integration scheme.
2. **Implicit Griffith-like data set**: if $\hat{\boldsymbol{\xi}}_i = (\hat{v}_i,\ \hat{G}_{R,i})$, then the data set is independent on the crack size leading to an *implicit Griffith-like* rate-independent limit for $v \to 0$. In this case, the closest-point projection distance at the generic step $k+1$ is computed rescaling the crack tip velocity axes following the time integration scheme (Fig. 3a).

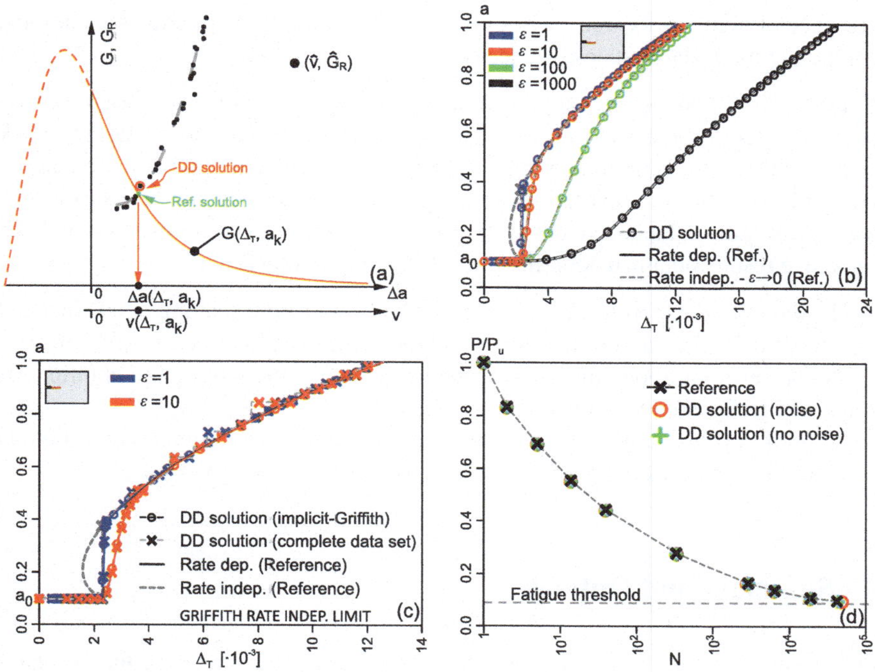

Fig. 3 Schematic example of a closest-point projection data-driven search procedure for rate-dependent fracture and comparison with the classical solution (**a**). Comparison between reference and data-driven results varying the loading rate: for an implicit Griffith-like data set (**b**) and using a complete data set reproducing a Griffith-like quasi-static limit (**c**). Modified Wöhler curve for fatigue tests (**d**)

Complete versus implicit Griffith-like data set. The performance of the proposed approaches is assessed in [4] using data sets artificially generated to reproduce different analytical models and a loading ramp $\Delta_{T,t} = \sqrt{\varepsilon t}$ where ε is the loading rate parameter. The following observation are drawn

- Similarly to the rate-independent case, the rate-dependent data-driven approaches are able to reproduce the crack evolution predicted solving (6) with an analytical critical energy release rate curve (Fig. 3b, c).
- If a complete data set contains observations from rate-independent tests, the quasi-static solutions in the spirit of [3] are automatically selected when relevant (Fig. 3b, c).
- For low loading rates, the rate-independent results are closely approached by the rate-dependent ones (Fig. 3c).
- The closest-point projection approach is reliable and robust with respect to noise for both types of data sets (Fig. 3b, c).

Rate-independent versus—dependent fracture. The rate-dependent approach regularizes some pathological behaviors observed in the rate-independent case in presence of non-convex free energies that include multiple snap-back branches and competing meta-stable states [3] (see Sect. 3.1).

Fatigue. The rate-dependent approach can be adapted to reproduce the fatigue behavior. In this case, the measurements used to calibrate the Paris-like laws relate the crack size increment to the nominal range of energy release rate applied at the crack tip, i. e. neglecting the crack evolution during the cycle [4]. Also, in this case the closest-point projection distance reduces to $d_i = \hat{G}_{r,i} - G(\Delta, a)$. The comparison with the reference results obtained adopting the *NASGRO* law shows that

- The data-driven approach correctly reproduces the curve relating crack size and number of cycles for different loads and the modified Wöhler curve (Fig. 3d).
- The presence of a load threshold value below which the fatigue life is infinite is automatically encoded in the data set (Fig. 3d).
- A simple cycle-jump approach can be used to limit the computational burden retaining a good accuracy of the results also in case of noisy data sets.

4 Summary and Outlook

In this paper, we outlined our recent work on model-free data-driven fracture mechanics. Rate-independent and -dependent fracture as well as fatigue are accounted for and the obtained results highlight that data-driven results correctly reproduce crack propagation as predicted by standard methods. Also, the approach based on the minimization of a closest-point projection distance from the data set is reliable and robust with respect to the presence of noise.

References

1. Kirchdoerfer, T., & Ortiz, M. (2016). Data-driven computational mechanics. *Computer Methods in Applied Mechanics and Engineering, 304*, 81–101.
2. Eggersmann, T., Kirchdoerfer, R., Reese, S., Stainier, L., & Ortiz, M. (2019). Model-Free data-driven inelasticity. *Computer Methods in Applied Mechanics and Engineering, 350*, 81–99.
3. Carrara, P., De Lorenzis, L., Stainier, L., & Ortiz, M. (2020). Data-driven fracture mechanics. *Computer Methods in Applied Mechanics and Engineering, 372*, 113390.
4. Carrara, L., Ortiz M., & De Lorenzis, L. (2021). Data-driven rate-dependent fracture mechanics. *Journal of the Mechanics and Physics of Solids, 155*, 104559.
5. Negri, M., & Ortner, C. (2008). Quasi-static crack propagation by Griffith's criterion. *Mathematical Models and Methods in Applied Sciences, 18*(11), 1895–1925.
6. Negri, M. (2010). From rate-dependent to rate-independent brittle crack propagation. *Journal of Elasticity, 98*(2), 159–187.
7. Hutchinson, J. W. (1979). *A course in Nonlinear Fracture Mechanics*. Lyngby: Technical University of Danmark.

Node Based Non-invasive Form Finding Revisited—The Challenge of Remeshing

Michael Caspari, Michael Stefan Schwarz, and Paul Steinmann

This contribution is dedicated to Peter Wriggers with whom we collaborated during the past 12 years within the TCRC 73 on modeling and simulating sheet-bulk metal forming.

Abstract In the numerical simulation of forming processes, the material configuration describes the semi-finished part, whereas the spatial configuration represents the finished part. In order to keep the subsequent finishing steps of the manufactured component to a minimum, the aim is to produce the finished part within small geometric tolerances. Thereby, a predefined target configuration defines the optimal shape of the part. Here, non-invasive form finding is designed to minimize the difference between the spatial configuration and the target configuration during an iterative procedure. The material configuration, ergo the shape of the semi-finished part, is successively optimized. Starting point of the optimization is the computation of the difference between the computed spatial and the target configuration. The present paper deals with the computation of this difference. Special attention is paid to the variable number of design parameters. The varying number of design parameters is due to the use of re-meshing within the FE-simulation.

M. Caspari (✉) · M. S. Schwarz · P. Steinmann
Institute of Applied Mechanics, Friedrich-Alexander University, Erlangen, Germany
e-mail: michael.caspari@fau.de

M. S. Schwarz
e-mail: michael.stefan.schwarz@fau.de

P. Steinmann
e-mail: paul.steinmann@fau.de

F. Aldakheel et al. (eds.), *Current Trends and Open Problems in Computational Mechanics*, https://doi.org/10.1007/978-3-030-87312-7_9

1 Introduction

Non-invasive form finding addresses the increasing demands in industry regarding the interrelation between ecology and economy. According to this, it is beneficial for both aspects to reduce the components production times and to operate in a material-saving manner. In this context, the collaborative research center TCRC 73 on sheet bulk metal forming (SBMF) [1] investigates, among other things, possibilities to reduce the production steps of a forming process. The output of the forming process is controlled by use of adapted semi-finished parts. This avoids the necessity of subsequent machining of the finished part and also enables the properties of the part to be controlled.

Herein, non-invasive form finding serves as an optimization strategy which determines the optimal shape of tailored blanks by using an iterative approach. The optimized shape of semi-finished part is defined by the reduced geometric deviation between the shape of the finished part and the target shape.

2 A Brief Outline of Non-invasive Form Finding

Within the numerical simulation of the forming process, the decisive shape of the semi-finished part is described by the position of the design nodes $\mathbf{X}^{\mathrm{D}} = \left[\boldsymbol{X}^{1^{\mathrm{T}}}, ..., \boldsymbol{X}^{n_{\mathrm{dsgn}}{}^{\mathrm{T}}}\right]^{\mathrm{T}}$ of the discrete material configuration $\mathscr{B}_0^{\mathrm{h}}$. Likewise, the shape of the finished part is represented by the design nodes $\mathbf{x}^{\mathrm{D}} = \left[\boldsymbol{x}^{1^{\mathrm{T}}}, ..., \boldsymbol{x}^{n_{\mathrm{dsgn}}{}^{\mathrm{T}}}\right]^{\mathrm{T}}$ of the corresponding computed spatial configuration $\mathscr{B}_{\mathrm{t}}^{\mathrm{h}}$. Since the algorithm is node-based, the target shape is also described as a discretized configuration with the nodal positions $\boldsymbol{x}_{\mathrm{tg}}^{D}$ with $D = 1, ...n_{\mathrm{dsgn,tg}}$, representing the design surface.

The nodal difference vector $\boldsymbol{d}^{D}$ is computed between the design nodes of the computed spatial and the target configuration. The computation is performed by projecting the computed spatial design nodes onto the target surface [3]. The nodal difference enters into the objective function as the local least squares error. The resulting scheme is presented in Fig. 1. The nodal positions of the material configuration are iteratively updated with respect to the nodal difference. The optimization problem is summarized in Table 1.

2.1 The Quasi-Newton Update Iteration Step

For the update of the material design coordinates $\mathbf{X}^{\mathrm{D}}$ a Quasi-Newton iteration step is realized. Starting point is the minimization of the global objective function with respect to the material coordinates.

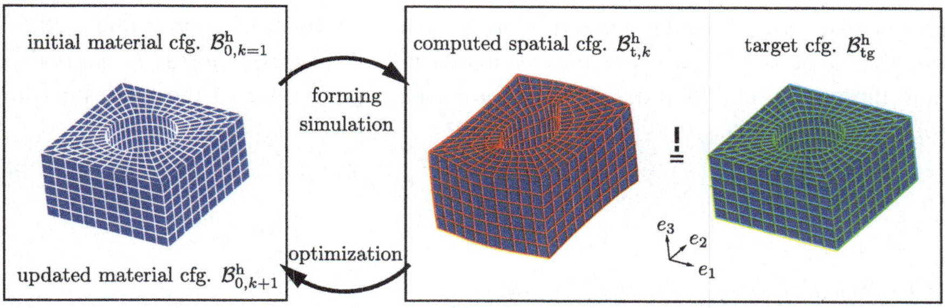

Fig. 1 The non-invasive optimization updates the material configuration by measuring the difference between the computed spatial and target configuration [2]

Table 1 The optimization problem of non-invasive form finding

Objective funct	$\delta(\mathbf{x}^{\rm D}, \mathbf{x}^{\rm D}_{\rm tg}) = \sum_{D=1}^{n_{\rm dsgn}} \frac{1}{2} \Vert \boldsymbol{d}^D \Vert_2^2$ with nodal differences $\boldsymbol{d}^D$ for $D = 1, \ldots, n_{\rm dsgn}$
Design variables	Nodal positions of material design nodes $\boldsymbol{X}^D$ for $D = 1, \ldots, n_{\rm dsgn}$
State equation	Motion $\boldsymbol{\varphi} : \mathbf{X}^{\rm D} \in \mathcal{B}_0^{\rm h} \to \mathbf{x}^{\rm D} \in \mathcal{B}_{\rm t}^{\rm h}$

$$\left.\frac{\partial \delta(\mathbf{x}^{\rm D}, \mathbf{x}^{\rm D}_{\rm tg})}{\partial \mathbf{X}^{\rm D}}\right|_{\mathbf{X}^{\rm D}_{\rm opt}} \overset{!}{=} \mathbf{0}\,. \tag{1}$$

The global objective function is the sum of the local objective functions which are evaluated at the spatial design nodes. This, in turn, is the least squares error of the difference vector:

$$\delta(\mathbf{x}^{\rm D}, \mathbf{x}^{\rm D}_{\rm tg}) = \sum_{D=1}^{n_{\rm dsgn}} \delta^D \left(\boldsymbol{\varphi}(\boldsymbol{X}^D), \boldsymbol{x}^D_{\rm tg}\right) = \frac{1}{2} \sum_{D=1}^{n_{\rm dsgn}} \Vert \boldsymbol{d}^D \Vert_2^2\,. \tag{2}$$

After some further manipulation, which can be found in [4], eventually, a local Quasi-Newton iteration step follows as:

$$\boldsymbol{X}^D \leftarrow \boldsymbol{X}^D - \alpha \frac{\partial^2 \delta^D(\boldsymbol{\varphi}(\boldsymbol{X}^D), \boldsymbol{x}^D_{\rm tg})^{-1}}{\partial \boldsymbol{X}^D \partial \boldsymbol{X}^D} \cdot \frac{\partial \delta^D(\boldsymbol{\varphi}(\boldsymbol{X}^D), \boldsymbol{x}^D_{\rm tg})}{\partial \boldsymbol{X}^D}\,. \tag{3}$$

Therein, the line search parameter α is already included. The line search parameter is redefined in every iteration, within its limits $\alpha \in]0;1]$. It controls the update intensity and ensures a regular mesh. Within Eq. 3 the first and second derivative of the objective function with respect to the material coordinates can be identified. Those are known as the Jacobian $\boldsymbol{J}^D =: -\boldsymbol{F}^{D^{\mathsf{T}}} \cdot \boldsymbol{d}^D$ and the Hessian $\boldsymbol{H}^D =: -\boldsymbol{G}^{D^{\mathsf{T}}} \cdot \boldsymbol{d}^D + \boldsymbol{F}^{D^{\mathsf{T}}} \cdot \boldsymbol{F}^D$. For the approximation of the second deformation gradient $\boldsymbol{G}^{D^{\mathsf{T}}} = \frac{\partial}{\partial \boldsymbol{X}^D}\left[\boldsymbol{F}^{D^{\mathsf{T}}}\right]$ different methods have been compared in [4].

Eventually, a Gauss-Newton method has proven to be the most appropriate method. Finally, the deformation gradient as evaluated at the Gaussian points is L^2-smoothed onto the nodes. The final update step for a local design node of the $k-$th iteration follows as:

$$\boldsymbol{X}_{k+1}^{D} = \boldsymbol{X}_{k}^{D} + \alpha \tilde{\boldsymbol{F}}_{k}^{D^{-1}} \cdot \boldsymbol{d}_{k}^{D} . \tag{4}$$

2.2 Non-invasive Optimization

A special feature of the optimization routine is characterized by the non-invasive approach. This states that the optimization does not interfere with the mechanical formulation of the numerical simulation. Therefore it is possible to use conventional and application-oriented nonlinear FE-solvers. The non-invasive approach is schematically presented in Fig. 2.

Within an iteration loop (Fig. 2-red) a fictitious elastic problem is highlighted, which has not been mentioned before. The updated positions of the material design nodes are computed according to Eq. 4. However, there are two more classes of nodes within the entire configurations $\mathcal{B}_0^{\mathrm{h}}$, $\mathcal{B}_t^{\mathrm{h}}$. The fixed nodes correspond to a Dirichlet boundary condition of the forming simulation. Therefore they stay unchanged during the update procedure. Furthermore, controlled nodes do not contribute to the design surface of the component, they are located for example in the interior of the body. Those kind of nodes are controlled via a fictitious elastic update procedure. The update of the position of the material coordinates renders as Dirichlet boundary condition of the fictitious elastic problem in order to obtain the updated nodal position of the controlled nodes.

The non-invasive approach of coupling the optimization and FE-Solver (indicated with red arrows) is realized by transferring nodal positions between both tools via subroutines and file exchange. Due to this, application-specific solvers may be used. Merely appropriate subroutines for data exchange have to be adapted.

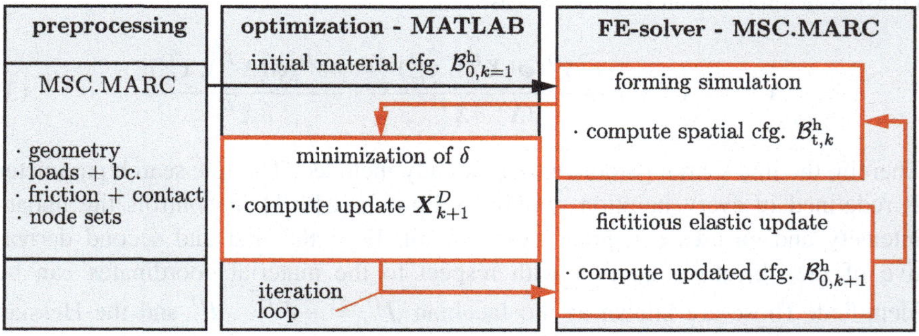

Fig. 2 Workflow of the non-invasive optimization [5]

3 Mesh Transformation for Dealing with Remeshing

A very useful feature of application-oriented commercial solvers is the use of mesh adaptive strategies. Thereby, the mesh is adapted to the forming state. Besides mesh adaption by means of h-/r-/p adaptive strategies, some solvers use remeshing. Error estimators are used to determine the numerical error of the mesh. If a threshold is passed, remeshing takes place. Furthermore, it is also possible to determine user-defined and problem-specific time increments for remeshing. The choice is up to the user. If the mesh is adapted to the current deformation state, high degrees of deformation are possible while maintaining a regular and numerically accurate mesh. Elements are thus prevented from degenerating.

If remeshing takes place during the simulation, the number of nodes changes. This is schematically pictured in Fig. 3. In the upper row, the number of nodes changes during every remeshing step. This useful feature for performing the FE-simulation of a forming process poses a challenge to the node based optimization algorithm. The final update step in Eq. 4 requires an equal number of design nodes for the material $\mathcal{B}_0^h$ and spatial $\mathcal{B}_t^h$ configuration.

In [6] a strategy is presented to use the information of remeshing in order to transform the original material configuration into the shape of the spatial configuration without changing the number of nodes. Using the inverted spatial configuration, the difference vector is computed similar as before. The inversion of the remeshing is done by comparing the spatial and remeshed configuration at the increment when remeshing takes place. To determine the mutual position of new and old nodes in the

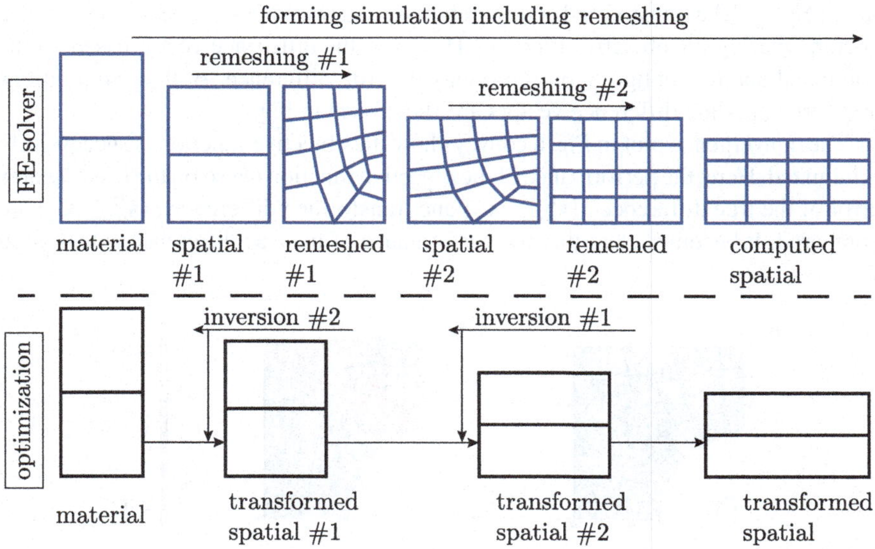

Fig. 3 Schematic representation of the transformation of the material nodes into the computed spatial configuration $\mathcal{B}_{t,comp}^h$ to obtain the transformed spatial configuration $\mathcal{B}_{t,trans}^h$ [6]

increment of remeshing, Parametric Inversion [7] is used. By inverting the remeshing process and transforming the material configuration, the transformed spatial configuration $\mathcal{B}^{\mathrm{h}}_{\mathrm{t,trans}}$ is obtained.

4 Academic Example

In the following, an academic example of a cuboid with hole is used to demonstrate the performance of the approach. Special attention is paid to the computation of the difference vector, which is evaluated at the spatial nodes determined by forward transformation. The example is already depicted in Fig. 1. However, due to the limited feasibility of remeshing within a 3D FE-simulation, the example is reduced to two dimensions.

A quarter of the cuboid is shown in Fig. 4a. Obviously, boundary conditions are used to take advantage of symmetry. The material parameters are taken from [8] for the common steel DC04 including isotropic hardening. In Fig. 4a the original material configuration is highlighted. In Fig. 4b the computed spatial configuration is compared to the target configuration which is indicated by a green dashed line. During the optimization, the deviation between computed spatial configuration and target configuration will be iteratively reduced. Initially, no remeshing was used in this FE-simulation since it is not necessary in order to obtain useful numerical results. However, it is used in the subsequent simulations for comparison purposes.

By using remeshing within the numerical simulation, the optimization deals with two spatial configurations. The computed $\mathcal{B}^{\mathrm{h}}_{\mathrm{t,comp}}$ and the transformed spatial configuration $\mathcal{B}^{\mathrm{h}}_{\mathrm{t,trans}}$. The transformed configuration is used to compute the spatial difference vector, entering the objective function. However, the difference vector based on the computed spatial configuration represents the true difference. Both configurations and corresponding difference vectors are showcased in Fig. 5.

The convergence plot in Fig. 6 conveys how the objective function is successively minimized. Here, the performance of the objective function based on the least squares error of the true difference $\delta(\mathbf{x}^{\mathrm{D}}_{\mathrm{comp}}, \mathbf{x}_{\mathrm{tg}})$ and transformed difference $\delta(\mathbf{x}^{\mathrm{D}}_{\mathrm{trans}}, \mathbf{x}_{\mathrm{tg}})$ are compared. It becomes clear that the performance of the true difference is less good

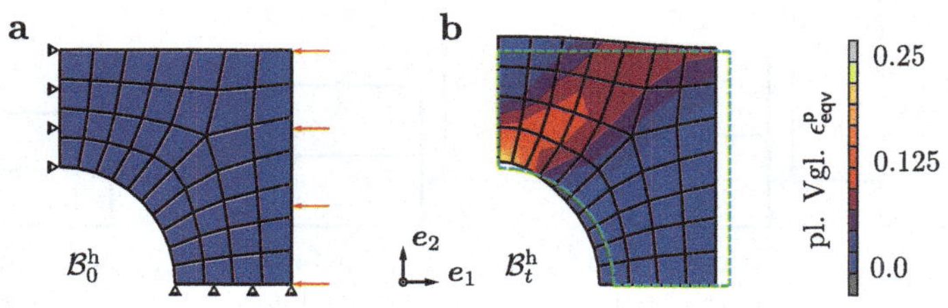

Fig. 4 Spatial and material configuration of the Cuboid with a hole FE-Simulation

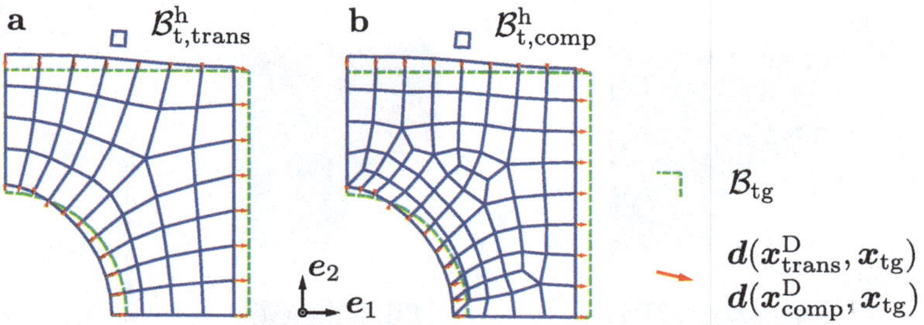

Fig. 5 Computation of the difference vector, based on the transformed $\mathcal{B}^{h}_{t,trans}$ (**a**) and computed $\mathcal{B}^{h}_{t,comp}$ (**b**) spatial configuration

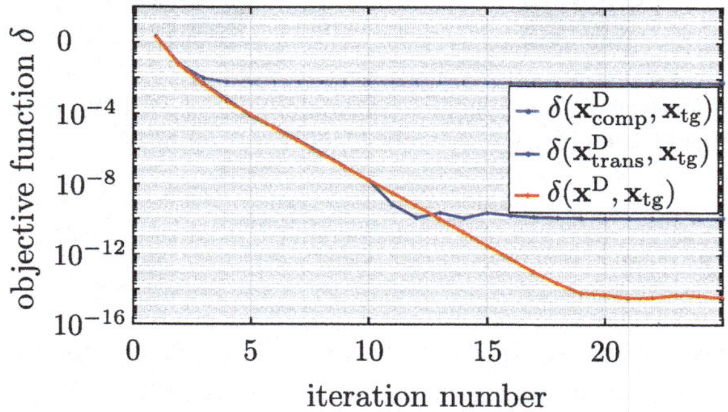

Fig. 6 Convergence plots of the optimization of the example in Fig. 7

than the transformed difference underlying the optimization. This result must be taken into account when using remeshing within the optimization. Especially it gets clear, that a convergence plot based on the true difference already converges at a lower number of iterations. Both convergence plots are compared to the optimization of the example without remeshing. Due to the significantly higher convergence, this curve serves as a benchmark.

The deviation between spatial and target configuration is crucial for the applicability of the optimization algorithm. However, if sufficient convergence is ensured, the resulting shape is of major importance. For this purpose Fig. 7 compares the optimized configurations. The example without remeshing is compared to the result with remeshing. It becomes clear that the spatial configurations of both optimizations show no visible deviations to the target configurations in the eye-ball norm. This was to be expected due to the 10^{-10} digit convergence gap between both convergence plots. However, more important is the optimized material configuration. This represents the semi-finished part shape to be produced. Here again, it is clear that the same configuration is obtained regardless of the chosen method.

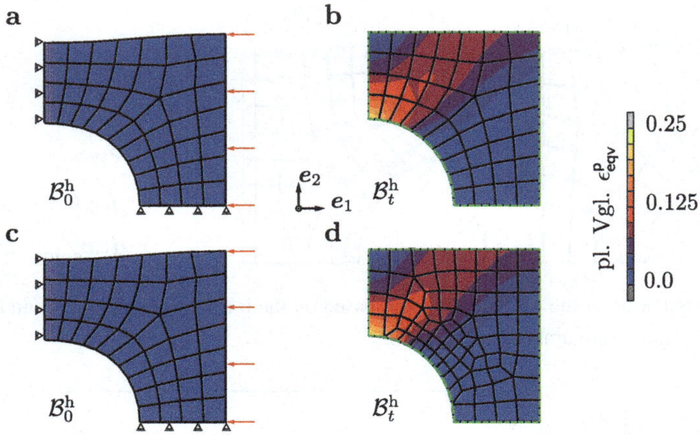

Fig. 7 Optimized material $\mathcal{B}_0^h$ and spatial configurations $\mathcal{B}_{t,comp}^h$ without (**a**, **b**) and with (**c**, **d**) remeshing within the FE-simulation

5 Summary

By considering the objective function based on the true difference, the influence of transforming the remeshing on the convergence behavior becomes clear for the first time. A comparison with the benchmark, however, clearly shows that this resulting deviation has no discernible influence on the optimized shape.

Acknowledgements This study was supported by the German Research Foundation (DFG) within the scope of the Transregional Collaborative Research Centre for sheet-bulk metal forming (TCRC 73, Subproject C3). The authors are in addition grateful to all students who supported the realization of this work.

References

1. Merklein, M., Allwood, J. M., Behrens, B.-A., Brosius, A., Hagenah, H., Kuzman, K., et al. (2012). Bulk forming of sheet metal. *CIRP Annals—Manufacturing Technology, 61*(2), 725–745.
2. Söhngen, B., Caspari, M., Willner, K., & Steinmann, P. (2021). On optimization strategies for inverse problems in metalforming. In M. Merklein & A. E. Tekkaya, & B. A. Behrens (Eds.), *Sheet Bulk Metal Forming*. TCRC73 2020. Lecture Notes in Production Engineering. Cham: Springer.
3. Caspari, M., Landkammer, P., & Steinmann, P. (2018). A non-invasive node-based form finding approach with discretization-independent target configuration. *Advanced Modeling and Simulation in Engineering Sciences, 5*(1), art. no. 11.
4. Landkammer, P., Caspari, M., & Steinmann, P. (2018). Improvements on a non-invasive, parameter-free approach to inverse form finding. *Computational Mechanics, 61*(4), 443–447.

5. Caspari, M., Landkammer, P., & Steinmann, P. (2018). Illustration of an improved non-invasive form finding algorithm. In *AIP Conference Proceedings—21st International ESAFORM Conference on Material Forming, Palermo, Italy* (Vol. 1960, art. no. 110003)
6. Caspari, M., Landkammer, P., & Steinmann, P. (2020). Shape optimization of a backward extrusion process using a non-invasive form finding algorithm. *Procedia Manufacturing, 47*, 873–880.
7. Crawford, R. H., Anderson, D. C., & Waggenspack, W. N. (1989). Mesh rezoning of 2D isoparametric elements by inversion. *International Journal for Numerical Methods in Engineering, 28*(3), 523–531.
8. Schmaltz, S., & Willner, K. (2014). Comparison of different Biaxial tests for the inverse identification of sheet steel material parameters. *Strain, 50*(5), 389–403.

Micropolar Modelling of Periodic Cauchy Materials Based on Asymptotic Homogenization

Maria Laura De Bellis, Andrea Bacigalupo, and Giorgio Zavarise

Academia is a wide world with a lot of colleagues, some coworkers, and few friends. We have had the privilege to have Peter as a great friend.
Hence, thanks Peter, for all we learned from you and for your friendship.

Abstract A micropolar-based asymptotic homogenization approach for the analysis of composite materials with periodic microstructure is proposed. The macro descriptors are directly linked to both suitable perturbation functions, obtained via asymptotic homogenization scheme, and micropolar two-dimensional deformation modes. A properly conceived energy equivalence between the macroscopic point and a microscopic representative portion of the periodic composite material is introduced to derive the overall micropolar constitutive tensors. The resulting constitutive tensors are not affected by the choice of the periodic cell.

1 Introduction

Periodic manufactured composites are widely adopted in many engineering fields. In this context, it is crucial to accurately reproduce the material response in a synthetic, but accurate way. To this aim the study of multi-scale homogenization techniques is a very interesting and debated topic in literature [1, 2]. Homogenization approaches,

M. L. De Bellis (✉)
Department INGEO, University of Chieti-Pescara, Viale Pindaro 42, Pescara, Italy
e-mail: marialaura.debellis@unich.it

A. Bacigalupo
Department DICCA, University of Genoa, via Montallegro 1, Genoa, Italy
e-mail: andrea.bacigalupo@unige.it

G. Zavarise
Department DISEG, Turin Polytechnic, Corso Duca degli Abruzzi 24, Turin, Italy
e-mail: giorgio.zavarise@polito.it

F. Aldakheel et al. (eds.), *Current Trends and Open Problems in Computational Mechanics*, https://doi.org/10.1007/978-3-030-87312-7_10

based on Cauchy continua at both the microscopic and macroscopic scales, may reveal ineffective since are not able to describe size effects as well as the dispersion properties of periodic heterogeneous materials. To overcome these drawbacks, non-local homogenization schemes can be effectively exploited.

In this framework, a micropolar macroscopic modelling of periodic Cauchy materials based on asymptotic homogenization approach is here proposed. In Sect. 2 the governing equations at both microscopic and macroscopic scales are recalled. Section 3 is devoted to define the micro-macro kinematic relations and the asymptotic expansion of the microscopic governing equations. The upscaling relations and a properly defined kinematic map are, then, fully developed in Sect. 4. The generalized macro-homogeneity condition is, then, derived in Sect. 5. An illustrative application is proposed in Sect. 6. Finally conclusions are drawn in Sect. 7.

2 Microscopic and Macroscopic Governing Equations

Focus is on a 2D periodic heterogeneous composite material, as in Fig. 1a, in the framework of linearised kinematics. A Cauchy continuum, subject to stresses induced by periodic body forces, is adopted. The periodic cell $\mathcal{A} = [-\varepsilon/2, \varepsilon/2] \times [-\varepsilon/2, \varepsilon/2]$, whose characteristic size is ε, is shown in Fig. 1b together with the corresponding unit cell Q in Fig. 1c characterized by periodicity vectors $\mathbf{v}_1$, $\mathbf{v}_2$.

The governing equations at the microscopic scale read

$$\nabla \cdot \left(\mathbb{C}^m \left(\frac{\mathbf{x}}{\varepsilon}\right) \nabla \mathbf{u}(\mathbf{x})\right) + \mathbf{b}(\mathbf{x}) = \mathbf{0}, \tag{1}$$

where $\mathbb{C}^m$ is the Q-periodic elasticity tensor, $\mathbf{b}$ are the zero-mean-value $\mathcal{L}$-periodic body force, with $\mathcal{L} = [-L/2; L/2] \times [-L/2; -L/2]$, for $L >> \varepsilon$. It follows that the microscopic displacement field $\mathbf{u}$ explicitly depends on both $\mathbf{x}$ and $\boldsymbol{\xi} = \mathbf{x}/\varepsilon$.

At the macroscopic scale a micropolar continuum [3] is considered with governing equations

$$\begin{aligned}
&\nabla \cdot \left[\mathbb{G}^M \left(\nabla \mathbf{U}(\mathbf{x}) + \epsilon_{3hk} \left(\mathbf{e}_h \otimes \mathbf{e}_k\right) \Phi\left(\mathbf{x}\right)\right)\right] + \nabla \cdot \left(\mathbf{Y}^M \nabla \Phi(\mathbf{x})\right) + \mathbf{b}(\mathbf{x}) = \mathbf{0},\\
&\nabla \cdot \left[\mathbf{Y}^{M^T} \left(\nabla \mathbf{U}(\mathbf{x}) + \epsilon_{3hk}(\mathbf{e}_h \otimes \mathbf{e}_k)\Phi(\mathbf{x})\right)\right] + \nabla \cdot \left(\mathbf{S}^M \nabla \Phi(\mathbf{x})\right) +\\
&- \epsilon_{3ij}(\mathbf{e}_i \otimes \mathbf{e}_j) : \left[\mathbb{G}^M \left(\nabla \mathbf{U}(\mathbf{x}) + \epsilon_{3hk}(\mathbf{e}_h \otimes \mathbf{e}_k)\Phi(\mathbf{x})\right) + \mathbf{Y}^M \nabla \Phi(\mathbf{x})\right] + c_3(\mathbf{x}) = 0,
\end{aligned} \tag{2}$$

where $\mathbb{G}^M$, $\mathbf{Y}^M$ and $\mathbf{S}^M$ are the constitutive tensors, $\mathbf{b}$, c_3 are the generalized body forces, $\mathbf{U}(\mathbf{x})$ and $\Phi(\mathbf{x})$, are the macro-displacement and micropolar rotation field, respectively.

In what follows a procedure aimed at identifying the macroscopic constitutive tensors, characterizing the equivalent micropolar continuum, are derived from the mechanical properties available at the microscopic scale.

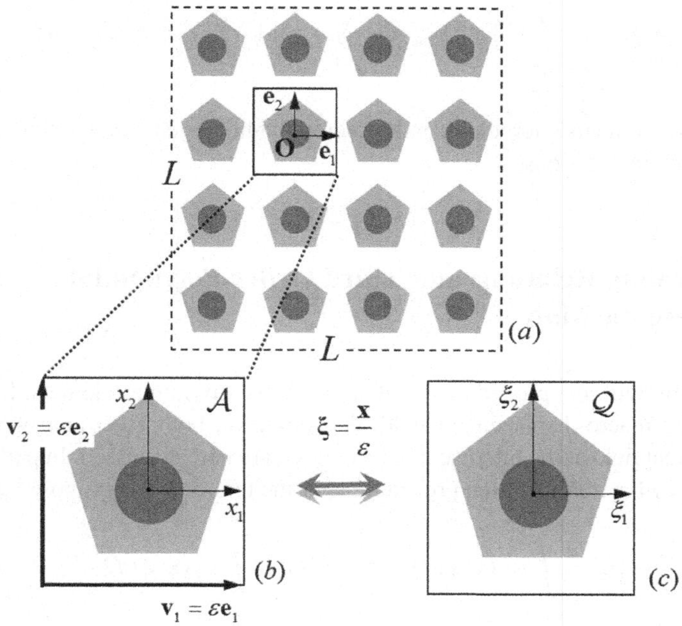

Fig. 1 Heterogeneous material at the microscopic level. **a** Cluster of periodic cells with structural dimensions L; **b** Periodic Cell $\mathcal{A}$ with characteristic size ε; **c** unit cell Q

3 Micro-Macro Kinematic Relations and Asymptotic Expansion of the Microscopic Governing Equations

In line with the asymptotic homogenization scheme (see i.e. [4]), the following asymptotic expansion of the microscopic displacement field is taken

$$u_i\left(\mathbf{x}, \xi = \frac{\mathbf{x}}{\varepsilon}\right) = \left(U_i(\mathbf{x}) + \sum_{l=1}^{+\infty} \varepsilon^l \sum_{|q|=l} N_{ijq}^{(l)}(\boldsymbol{\xi}) \frac{\partial^{|q|}}{\partial x_q} U_j(\mathbf{x})\right)\Bigg|_{\xi=\frac{\mathbf{x}}{\varepsilon}}, \tag{3}$$

where U_i are macro-displacement components, $N_{kpq_1}^{(l)}$ are zero-mean-value Q-periodic perturbation functions and q a multi-index of length l. The Eq. (3) can be plug into the microscopic governing Eq. (1). After suitable manipulations, by collecting the terms with equal power ε, and imposing the so-called solvability condition in the class of Q-periodic functions, a hierarchical sequence of partial differential problems, known as *cell problems*, is obtained as in [5]. The perturbation functions at order ε^0, ε^1 and ε^2, i.e. $N_{hpq_1}^{(1)}$, $N_{hpq_1q_2}^{(2)}$ and $N_{hpq_1q_2q_3}^{(3)}$ are determined, together with the overall first order constitutive tensor components

$$C_{iq_2pq_1} = \int_Q C^m_{rjkl} \left(N^{(1)}_{riq_2,j} + \delta_{ir}\delta_{jq_2}\right)\left(N^{(1)}_{kpq_1,l} + \delta_{pk}\delta_{lq_1}\right) d\boldsymbol{\xi}, \tag{4}$$

that are key elements intervening in the micropolar homogenization scheme, detailed in the following sections.

4 Upscaling Relations and Third Order Polynomial Kinematic Map

Accordingly with the procedure detailed in [6], the upscaling relations, linking the generalized macro-displacement field, depending on both $\mathbf{U}(\mathbf{x})$ and $\Phi(\mathbf{x})$, to the displacement field at the microscopic level $\mathbf{u}(\mathbf{x})$, are here discussed. In particular for $W_{ij} = -\epsilon_{3ij}\Phi$ a minimization procedure over the unit cell Q is proposed as follows

$$U_1(\mathbf{x}) = \int_Q u_1(\mathbf{x}, \boldsymbol{\xi}) d\boldsymbol{\xi}, \qquad U_2(\mathbf{x}) = \int_Q u_2(\mathbf{x}, \boldsymbol{\xi}) d\boldsymbol{\xi},$$

$$\min_{W_{ij}} \mathcal{F}\left[\omega^*_{ij}(W_{ij})\right] = \min_{W_{ij}} \int_Q \left(||\omega_{ij}(\mathbf{x}, \boldsymbol{\xi}) - \omega^*_{ij}(\mathbf{x}, \boldsymbol{\xi})||_2\right)^2 d\boldsymbol{\xi}, \tag{5}$$

where ω_{ij} are the components of the micro infinitesimal rotation tensor and ω^*_{ij} are the components of a properly defined skew-symmetric tensor expressed in the form

$$\omega^*_{ij} = W_{ij} + \sum_{l=1}^{+\infty} \sum_{\substack{|q|=l \\ |r|=l}} \frac{1}{2}\left(N^{(l)}_{ijq,r} - N^{(l)}_{jiq,r}\right) \tilde{W}_{qr}. \tag{6}$$

By plugging (6) truncated at the first order into (5) c, after solving the minimization problem, the components of the micropolar rotation tensor are obtained as

$$W_{ij}(\mathbf{x}) = \frac{\int_Q M_{ijpq_1}(\boldsymbol{\xi})\, \omega_{pq_1}(\mathbf{x}, \boldsymbol{\xi}) d\boldsymbol{\xi}}{\int_Q \delta_{pr}\delta_{q_1s_1} M_{hkpq_1}(\boldsymbol{\xi}) M_{hkrs_1}(\boldsymbol{\xi}) d\boldsymbol{\xi}}, \tag{7}$$

with $M_{ijpq_1} = \delta_{ip}\delta_{jq_1} + \frac{1}{2}\left(N^{(1)}_{ijp,q_1} - N^{(1)}_{jip,q_1}\right)$.
The upscaling relations are now particularized by truncating the microscopic displacement field at the third order, i.e.

$$
\begin{aligned}
u_i^{III}(\mathbf{x},\boldsymbol{\xi}) = {} & U_i(\mathbf{x}) + \varepsilon N^{(1)}_{ijq_1}(\boldsymbol{\xi}) H_{jq_1}(\mathbf{x}) + \\
& + \varepsilon^2 N^{(2)}_{ijq_1q_2}(\boldsymbol{\xi})\,\kappa_{jq_1q_2}(\mathbf{x}) + \varepsilon^3 N^{(3)}_{ijq_1q_2q_3}(\boldsymbol{\xi})\,\kappa_{jq_1q_2q_3}(\mathbf{x}),
\end{aligned} \tag{8}
$$

where $U_i(\mathbf{x})$ is the macroscopic displacement field and $H_{jq_1}(\mathbf{x})$, $\kappa_{jq_1q_2}(\mathbf{x})$ and $\kappa_{jq_1q_2q_3}(\mathbf{x})$ its gradient and higher order gradients.
Concerning the macro-displacement, a third-order Taylor polynomial expansion is chosen

$$
U_i(\mathbf{x}) = \bar{U}_i + \bar{H}_{ip_1}x_{p_1} + \frac{1}{2}\bar{\kappa}_{ip_1p_2}x_{p_1}x_{p_2} + \frac{1}{6}\bar{\kappa}_{ip_1p_2p_3}x_{p_1}x_{p_2}x_{p_3}, \tag{9}
$$

where the coefficients $\bar{U}_i$, $\bar{H}_{ip_1}$, $\bar{\kappa}_{ip_1p_2}$ and $\bar{\kappa}_{ip_1p_2p_3}$ are the macroscopic fields evaluated at point $\mathbf{x} = \mathbf{0}$, respectively. In order to identifying a micropolar continuum, the 20 independent coefficients must be properly reduced to 6 as detailed in [6].

At this point, by replacing the Eq. (9), suitably manipulated, in (8), the polynomial approximation of the microscopic displacement field is

$$
\begin{aligned}
u_1(\mathbf{x},\boldsymbol{\xi}) = {} & \mathcal{B}_1^1(\mathbf{x},\boldsymbol{\xi})\bar{E}_{11} + \mathcal{B}_1^2(\mathbf{x},\boldsymbol{\xi})\bar{E}_{22} + \mathcal{B}_1^3(\mathbf{x},\boldsymbol{\xi})\bar{E}_{12} + \\
& + \mathcal{B}_1^4(\mathbf{x},\boldsymbol{\xi})\bar{\kappa}_{122} + \mathcal{B}_1^5(\mathbf{x},\boldsymbol{\xi})\bar{\kappa}_{211} + \mathcal{B}_1^6(\mathbf{x},\boldsymbol{\xi})\bar{\kappa}_{1222}, \\
u_2(\mathbf{x},\boldsymbol{\xi}) = {} & \mathcal{B}_2^1(\mathbf{x},\boldsymbol{\xi})\bar{E}_{11} + \mathcal{B}_2^2(\mathbf{x},\boldsymbol{\xi})\bar{E}_{22} + \mathcal{B}_2^3(\mathbf{x},\boldsymbol{\xi})\bar{E}_{12} + \\
& + \mathcal{B}_2^4(\mathbf{x},\boldsymbol{\xi})\bar{\kappa}_{122} + \mathcal{B}_2^5(\mathbf{x},\boldsymbol{\xi})\bar{\kappa}_{211} + \mathcal{B}_2^6(\mathbf{x},\boldsymbol{\xi})\bar{\kappa}_{1222},
\end{aligned} \tag{10}
$$

where $\bar{E}_{ip_1} = (\bar{H}_{ip_1} + \bar{H}_{p_1i})/2$ and the localization functions $\mathcal{B}_i^j(\mathbf{x},\boldsymbol{\xi})$ depend on the overall first order elastic tensor components and on the perturbation functions. By plugging equations (10) in Eqs. (7), the micropolar rotation field $\Phi(\mathbf{x})$ and, in turn, the macroscopic curvature tensor components $K_1 = \partial\Phi/\partial x_1$ and $K_2 = \partial\Phi/\partial x_2$ are obtained.

5 Generalized Macro-Homogeneity Condition

The overall micropolar elastic properties are derived by exploiting a generalized macro-homogeneity condition, establishing an energy equivalence between the macroscopic and the microscopic scales. Accordingly with [4, 7], the *microscopic mean strain energy* is defined as

$$
\bar{\mathcal{E}}_m \doteq \frac{1}{2}\int_{\mathcal{A}}\int_{Q} \underline{\boldsymbol{\varepsilon}}(\mathbf{x},\boldsymbol{\xi})^T\, \underline{\underline{\mathbf{C}}}^m(\boldsymbol{\xi})\, \underline{\boldsymbol{\varepsilon}}(\mathbf{x},\boldsymbol{\xi})\, d\boldsymbol{\xi}\, d\mathbf{x}, \tag{11}
$$

where $\underline{\underline{\mathbf{C}}}^m$ is the elasticity matrix and $\underline{\boldsymbol{\varepsilon}}$ is the microscopic strain vector in the standard matrix notation. Under the assumption of scale separation, the $\bar{\mathcal{E}}_m$ related to the strain field $\underline{\boldsymbol{\varepsilon}}_0 = \underline{\boldsymbol{\varepsilon}}(\mathbf{x} = \mathbf{0}, \boldsymbol{\xi})$ can be introduced as

$$\bar{\mathcal{E}}_m^0 \doteq \frac{1}{2} \int_{\mathcal{A}} \int_{Q} \underline{\boldsymbol{\varepsilon}}(\mathbf{x} = \mathbf{0}, \boldsymbol{\xi})^T \underline{\underline{\mathbf{C}}}^m (\boldsymbol{\xi}) \underline{\boldsymbol{\varepsilon}}(\mathbf{x} = \mathbf{0}, \boldsymbol{\xi}) \, d\boldsymbol{\xi} d\mathbf{x} = \frac{|\mathcal{A}|}{2} \int_{Q} \underline{\boldsymbol{\varepsilon}}_0^T \underline{\underline{\mathbf{C}}}^m \underline{\boldsymbol{\varepsilon}}_0 d\boldsymbol{\xi}, \quad (12)$$

being $|\mathcal{A}|$ the area of the periodic cell. More specifically, this strain vector is determined from the microscopic displacement components in Eq. (10), taking the form

$$\underline{\boldsymbol{\varepsilon}}_0 = \underline{\underline{\mathbf{B}}}^{\Xi} \underline{\boldsymbol{\Xi}} + \underline{\underline{\mathbf{B}}}^{\Upsilon} \underline{\boldsymbol{\Upsilon}}, \quad (13)$$

being $\underline{\boldsymbol{\Xi}} = \left\{ \bar{E}_{11} \; \bar{E}_{22} \; \bar{E}_{12} \; \bar{\kappa}_{1222} \right\}^T$, $\underline{\boldsymbol{\Upsilon}} = \left\{ \bar{\kappa}_{122} \; \bar{\kappa}_{211} \right\}^T$ and $\underline{\underline{\mathbf{B}}}^{\Xi}, \underline{\underline{\mathbf{B}}}^{\Upsilon}$ properly defined strain localization matrices. The related *microscopic mean strain energy density* is consequently derived

$$\begin{aligned} \bar{\phi}_m^0 = \frac{\bar{\mathcal{E}}_m^0}{|\mathcal{A}|} = \frac{1}{2} \Bigg(& \underline{\boldsymbol{\Xi}}^T \int_Q \underline{\underline{\mathbf{B}}}^{\Xi^T} \underline{\underline{\mathbf{C}}}^m \underline{\underline{\mathbf{B}}}^{\Xi} d\boldsymbol{\xi} \, \underline{\boldsymbol{\Xi}} + \underline{\boldsymbol{\Upsilon}}^T \int_Q \underline{\underline{\mathbf{B}}}^{\Upsilon^T} \underline{\underline{\mathbf{C}}}^m \underline{\underline{\mathbf{B}}}^{\Upsilon} d\boldsymbol{\xi} \, \underline{\boldsymbol{\Upsilon}} + \\ & + \underline{\boldsymbol{\Xi}}^T \int_Q \underline{\underline{\mathbf{B}}}^{\Xi^T} \underline{\underline{\mathbf{C}}}^m \underline{\underline{\mathbf{B}}}^{\Upsilon} d\boldsymbol{\xi} \, \underline{\boldsymbol{\Upsilon}} + \underline{\boldsymbol{\Upsilon}}^T \int_Q \underline{\underline{\mathbf{B}}}^{\Upsilon^T} \underline{\underline{\mathbf{C}}}^m \underline{\underline{\mathbf{B}}}^{\Xi} d\boldsymbol{\xi} \, \underline{\boldsymbol{\Xi}} \Bigg). \end{aligned} \quad (14)$$

Furthermore, the macroscopic strain energy density evaluated in $\mathbf{x} = \mathbf{0}$ is introduced

$$\phi_M^0 = \frac{1}{2} \left(\underline{\boldsymbol{\Gamma}}_0{}^T \underline{\underline{\mathbf{G}}}^M \underline{\boldsymbol{\Gamma}}_0 + \underline{\mathbf{K}}_0{}^T \underline{\underline{\mathbf{S}}}^M \underline{\mathbf{K}}_0 + \underline{\boldsymbol{\Gamma}}_0{}^T \underline{\underline{\mathbf{Y}}}^M \underline{\mathbf{K}}_0 + \underline{\mathbf{K}}_0{}^T \underline{\underline{\mathbf{Y}}}^{M^T} \underline{\boldsymbol{\Gamma}}_0 \right), \quad (15)$$

where the asymmetric micropolar strain vector and the curvature vector, evaluated in $\mathbf{x} = \mathbf{0}$ are defined as

$$\underline{\boldsymbol{\Gamma}}_0 = \underline{\underline{\mathbf{A}}}_{\Gamma}^{\Xi} \underline{\boldsymbol{\Xi}} + \underline{\underline{\mathbf{A}}}_{\Gamma}^{\Upsilon} \underline{\boldsymbol{\Upsilon}}, \qquad \underline{\mathbf{K}}_0 = \underline{\underline{\mathbf{A}}}_{K}^{\Xi} \underline{\boldsymbol{\Xi}} + \underline{\underline{\mathbf{A}}}_{K}^{\Upsilon} \underline{\boldsymbol{\Upsilon}}, \quad (16)$$

being $\underline{\underline{\mathbf{A}}}_{\Gamma}^{\Xi}$, $\underline{\underline{\mathbf{A}}}_{K}^{\Xi}$, $\underline{\underline{\mathbf{A}}}_{\Gamma}^{\Upsilon}$, $\underline{\underline{\mathbf{A}}}_{K}^{\Upsilon}$ transformation matrices. By exploiting the generalized macro-homogeneity condition, establishing the equivalence between the microscopic and macroscopic strain energy density $\bar{\phi}_m^0 \doteq \phi_M^0$ $(\mathbf{x} = \mathbf{0})$, after some manipulations, the overall elastic micropolar matrices are determined. In the case the periodic cell is characterized by centrosymmetric topology, it results

$$\begin{aligned} \underline{\underline{\mathbf{G}}}^M &= \int_Q \underline{\underline{\mathbf{A}}}_{\Gamma}^{\Xi^{-T}} \underline{\underline{\mathbf{B}}}^{\Xi^T} \underline{\underline{\mathbf{C}}}^m \underline{\underline{\mathbf{B}}}^{\Xi} \underline{\underline{\mathbf{A}}}_{\Gamma}^{\Xi^{-1}} d\boldsymbol{\xi}, \\ \underline{\underline{\mathbf{S}}}^M &= \int_Q \underline{\underline{\mathbf{A}}}_{K}^{\Upsilon^{-T}} \underline{\underline{\mathbf{B}}}^{\Upsilon^T} \underline{\underline{\mathbf{C}}}^m \underline{\underline{\mathbf{B}}}^{\Upsilon} \underline{\underline{\mathbf{A}}}_{K}^{\Upsilon^{-1}} d\boldsymbol{\xi}, \end{aligned} \quad (17)$$

and the coupling matrix is $\underline{\underline{\mathbf{Y}}}^M = \mathbf{0}$.

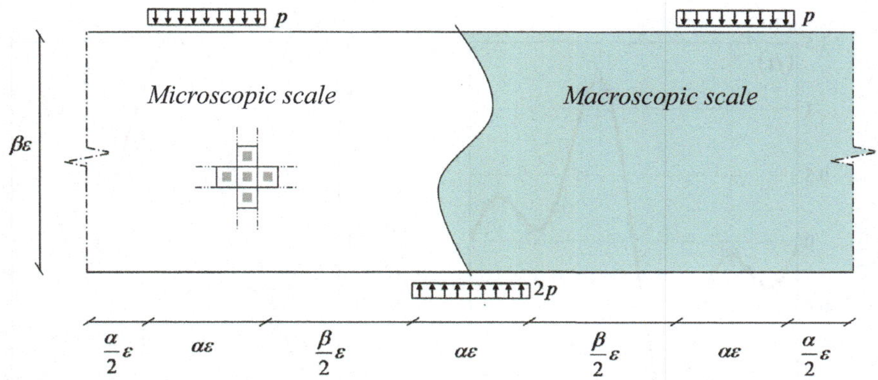

Fig. 2 Strip undergoing discontinuous periodic forces: schematic of the heterogeneous medium versus the homogenized one

6 Benchmark Test

As an example, a strip of two-phase periodic medium, realized by assembling, along horizontal and vertical directions, periodic cells with a stiff matrix (phase 1) embedding soft square inclusions (phase 2), see Fig. 2, is considered. The size of the inclusion is $\varepsilon/2$ and the material is characterized by $\eta_E = E_2/E_1 = 3/50$ and $\eta_\nu = \nu_1/\nu_2 = 1$ (with $E_1 = 500$ GPa, $\nu_1 = 0.1$). The specimen undergoes a system of discontinuous periodic forces, with period $L_1 = (4\alpha + \beta)\varepsilon$, located on both top and bottom sides of the strip characterized by width equal to $\beta\varepsilon$, with $\alpha = 4$, $\beta = 10$, and ε being the size of the periodic cell. Plane strain conditions are assumed. The numerical results of the micro-mechanical model, in terms of displacement components, along a horizontal line located at a distance $9/2\varepsilon$ from the top side of the strip, are compared with the respective ones obtained considering the micropolar homogenized model. In Fig. 3a the dimensionless micro-mechanical displacement component u_1/L_1 (blue dotted line), and the respective macro-mechanical one U_1/L_1 (red solid line) are plotted versus the dimensionless coordinate x_1/L_1. Analogously, in Fig. 3b the dimensionless components u_2/L_1 and U_2/L_1 are reported with the same line styles. For both microscopic and macroscopic displacement components a very good agreement is shown.

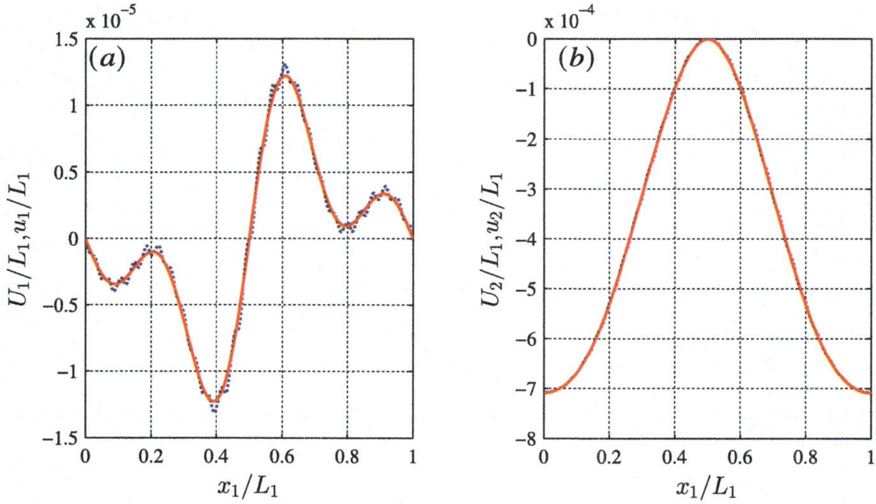

Fig. 3 Comparison between dimensionless micro-mechanical displacement components, in blue dotted lines, and the respective macro-mechanical ones versus x_1/L_1. **a** Components $u_1/L_1, U_1/L_1$; **b** components u_2/L_1, U_2/L_1

7 Conclusions

A micropolar modelling of periodic Cauchy materials based on asymptotic homogenization approach is proposed. Consistent upscaling relations, able to define the generalized macro-displacement components of the micropolar continuum in terms of the micro-displacement components and the perturbation functions, are proposed. Overall micropolar elastic properties are derived via a generalized macro-homogeneity condition. It results that overall elastic tensors do not depend on the choice of the periodic cell and in the limit case of locally homogeneous material the characteristic lengths tend to zero.

Acknowledgements The authors gratefully acknowledge financial support from National Group of Mathematical Physics (GNFM-INdAM), from the Compagnia San Paolo, project MINIERA no. I34I20000380007, from University of Trento, project UNMASKED 2020, project Search for Excellence Ud'A 2019 and project PRIN 2017 - 20173C478N - XFAST-SIMS: Extra fast and accurate simulation of complex structural systems.

References

1. Bakhvalov, N., & Panasenko, G. (1984). *Homogenization: Averaging processes in periodic media*. Dordrecht, Boston, London: Kluwer Academic Publishers.
2. Zohdi, T. I., & Wriggers, P. (2008). *An introduction to computational micromechanics*. Springer Science & Business Media.

3. Mindlin, R. D. (1964). Micro-structure in linear elasticity. *Archive for Rational Mechanics and Analysis, 16*(1), 51–78.
4. Bacigalupo, A. (2014). Second-order homogenization of periodic materials based on asymptotic approximation of the strain energy: Formulation and validity limits. *Meccanica, 49*(6), 1407–1425.
5. De Bellis, M. L., Bacigalupo, A., & Zavarise G. (2019). Characterization of hybrid piezoelectric nanogenerators through asymptotic homogenization. *Computer Methods in Applied Mechanics and Engineering, 355*, 1148–1186.
6. Bacigalupo, A., De Bellis, M. L., & Zavarise, G. (2021). Asymptotic homogenization approach for the micropolar modelling of periodic materials. *Submitted 2021*.
7. Smyshlyaev, V., & Cherednichenko, K. (2000). On rigorous derivation of strain gradient effects in the overall behaviour of periodic heterogeneous media. *Journal of the Mechanics and Physics of Solids, 48*(6), 1325–1357.

Experimental and Numerical Investigation of Granules as Crash-Absorber in Ship Building

Alexander Düster and Christian Woitzik

This contribution is dedicated to the occasion of Professor Wriggers' 70th birthday. We summarize part of the results of a joint project with him that we completed recently. Our regular meetings during this cooperation were scientifically very inspiring. We had the chance to learn from his very impressive knowledge and expertise in the broad field of mechanics. In addition, it is always a great pleasure to spend time with him because of his very pleasant personality. We wish Professor Wriggers all the best for the future.

Abstract We summarize results related to the experimental and numerical investigation of granular materials used as crash-absorber in ships with a double hull construction. The granular material is filled into the cavity between the two hull layers in order to increase the safety in the event of a collision of the ship with another vessel. The benefit of the filling material is twofold. First of all it helps to carry the load from the outer hull to the inner hull such that both hulls can absorb kinetic energy. Furthermore, the crushing of the granules contributes to the dissipation of kinetic energy. In order to quantify the crash-absorbing effect of the granular material, experimental as well as numerical investigations of a simplified double hull construction are carried out.

1 Introduction

The increasing usage of the North Sea and the Baltic Sea for wind energy and the growing traffic raises the risk of collisions between ships as well as between ships and offshore structures. To improve the safety of ships, the International Maritime

A. Düster (✉) · C. Woitzik
Numerical Structural Analysis with Application in Ship Technology, Institute for Ship Structural Design and Analysis, Hamburg University of Technology, Hamburg, Germany
e-mail: alexander.duester@tuhh.de

C. Woitzik
e-mail: christian.woitzik@tuhh.de

F. Aldakheel et al. (eds.), *Current Trends and Open Problems in Computational Mechanics*, https://doi.org/10.1007/978-3-030-87312-7_11

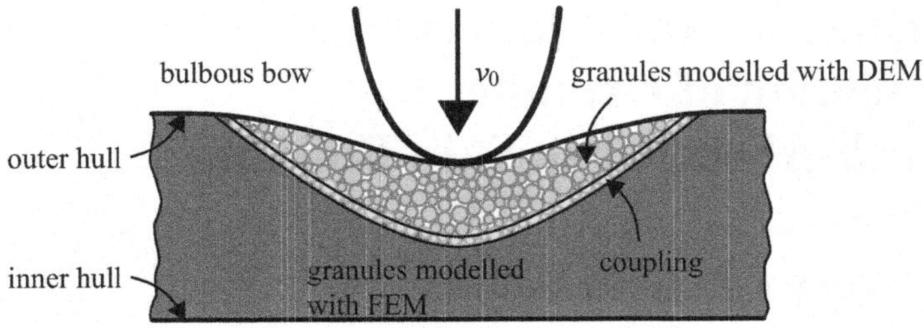

Fig. 1 Collision of a bulbous bow with a double hull filled with particles. The sketch illustrates the idea of a two-scale modelling approach in which the DEM and FEM are coupled

Organization (IMO) [3] requires that oil tankers must have double hulls. To further increase the collision safety of ships, [12, 13] proposed to fill the cavity between the double hulls with a lightweight granular material. Figure 1 sketches a situation in which the bulbous bow of a vessel collides with a double hull ship. The motivation to use filling material is based on two main effects. First of all, the granules help to transfer the load from the outer to the inner hull. In this way, the inner hull is deformed as well—and the role of the filling material can be thought of as similar to the core of a sandwich plate. The second effect is related to the dissipation of kinetic energy as the granules start to crush. The decision to use granules to achieve these effects is based on practical considerations. This kind of filling material can be easily pumped into the cavities and also removed for inspection purposes as required by classification societies.

In a joint project with Professor Wriggers, a two-scale simulation method was developed and validated by experimental investigations in order to investigate these effects in more detail. To this end, the granules were modelled with two different approaches. In the near field of the collision, the Discrete Element Method (DEM) was used to account for the large and inelastic deformations of the granular material, also including the crushing of particles. Since the computational effort of the DEM would be too high to treat the whole domain with this approach, the far field was modelled as a continuum using the Finite Element Method (FEM) for its discretization. The coupling of the DEM and FEM approach is described in [8, 9]. The mechanical investigation and modelling of the particles can be found in [14]. Crushing of particles in the context of the DEM is addressed in [10]. In this contribution, we want to address the experimental investigation of particles used as crash-absorber in a simplified double hull. Furthermore, we will present a FEM approach to simulate the experiment. To this end, we will summarize the results reported in [15].

2 Experimental Testing

In this section, we describe the experimental testing of particles as crash-absorber in double hull ships. Figure 2 shows the three different types of granules considered in our studies. Based on investigations presented in [14], we decided to use Poraver [2] as

Fig. 2 Three different types of granules: Poraver 2–4 [2], Omega-CXL [4], and Leca 2–4 [5]

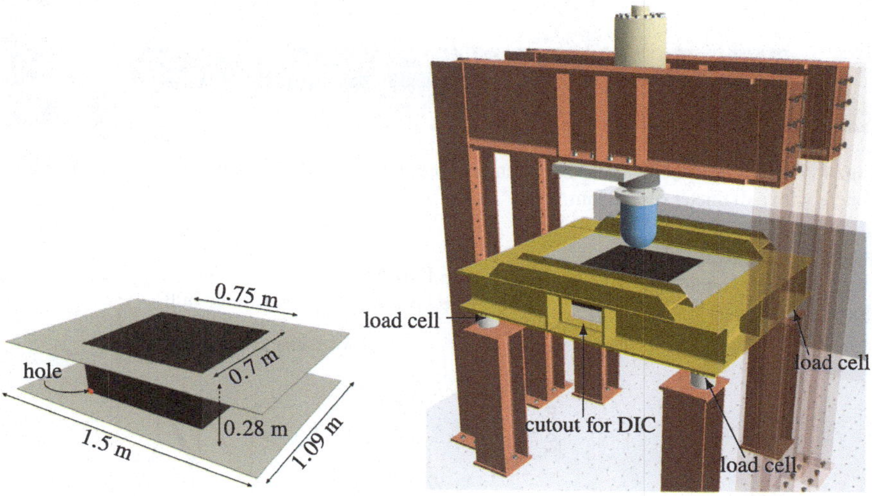

Fig. 3 Simplified double hull structure (left) and experimental setup (right) for the indentation/penetration test (gray/black: test specimen; gold: rigid frame; blue indenter)

crash-absorbing material. The multicellular expanded glass granules have a diameter of 2–4 mm and a bulk density of 190 kg/m^3. A detailed study of the material—determining further properties such as Young's modulus and crushing strength—is presented in [14]. The granular material is filled into a simplified double hull, as depicted in Fig. 3 (left). The double hull structure is fixed in the test rig shown in Fig. 3 (right). Similar tests were carried out in [6, 7, 11]. In a quasi-static experiment, the indenter is driven (with 0.2 mm/s) into the double hull, which is welded to a rigid frame, see Fig. 4. The experiment is carried out both with an empty double hull and with one that is filled with Poraver. The force-displacement curves for both the empty and the filled double hull structure are presented in Fig. 5. The unloading at a displacement of approximately 350 mm is due to the limited stroke length. In order to continue the indentation process, an extension of the indenter was installed. The peaks in the curves represent the rupture of the outer and inner hulls. It is interesting to note that the filled double hull structure yields a higher load level, especially

Fig. 4 Double hull subjected to an indenter representing a bulbous bow. The indenter is a half-sphere with a diameter of 270 mm

Fig. 5 Force-displacement curve for empty and filled double hull

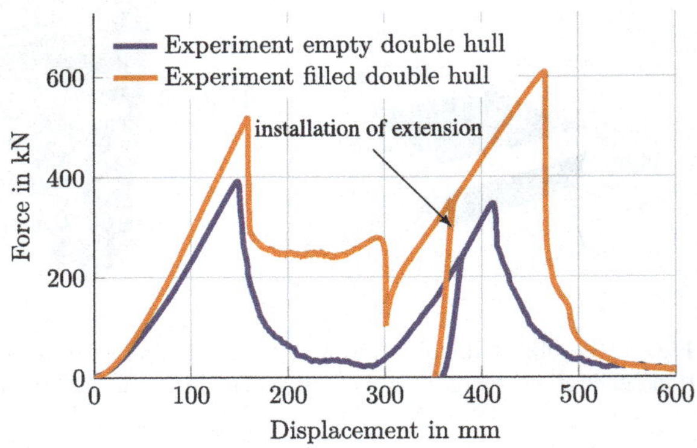

also for the part of the penetration where the outer hull is ruptured. During the penetration of the outer hull, the presence of the granules increases the stiffness of the whole structure, comparable to a sandwich core structure. The energy absorption is displayed in Fig. 6. The gain in energy due to the usage of granules is measured with respect to the rupture of the inner hull, as indicated with the vertical lines at a displacement of about 410 and 460 mm, respectively. In this setting, it amounts to approximately 146%. From this, it is obvious that filling the double hull with particles can increase the energy absorption quite significantly. It is interesting to note that the filled structure ruptures later, which could be due to the granules acting as a layer between the steel parts, hindering abrasive contact.

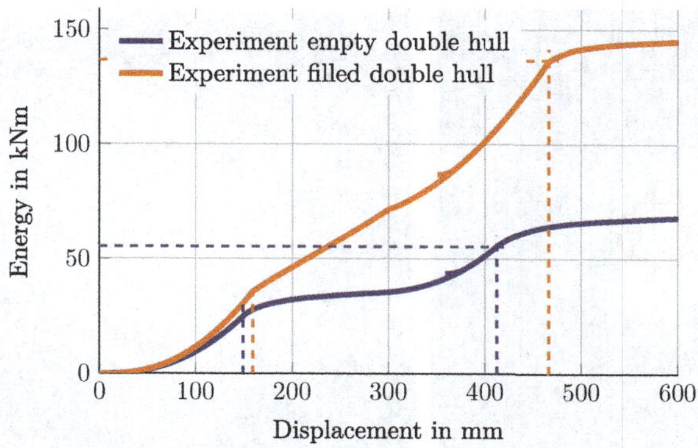

Fig. 6 Comparison of energy absorption for the empty and the filled double hull

3 Numerical Simulation

As mentioned in the introduction, numerical methods were developed and used to simulate the collision process. Here, we will briefly describe the FEM approach applied to simulate the process of the experiment discussed in the previous section. The structure is discretized taking advantage of the commercial finite element software package Abaqus [1]. The spatial discretization of the granules utilizes eight-noded hexahedral elements with reduced integration. The steel structure is meshed into Reissner-Mindlin shell elements using five quadrature points over the thickness. The constitutive behavior of the granules is based on the Mohr-Coulomb model, whereas the steel is described with an elastoplastic material model including damage. The temporal discretization features an explicit time-stepping scheme.

A comparison of the computed and measured von Mises strains on the surface of the inner hull is depicted in Fig. 7. The strains were measured using an optical digital image correlation (DIC) system. From the figure, a good agreement between simulation and experiment is evident.

Figures 8 and 9 show a comparison between the experiment and the simulation of the empty and, respectively, the filled double hull structure. Both figures demonstrate a good agreement between simulation and experiment. For the simulation with granules, the proposed method is not able to capture the entire indentation process, due to large element distortions. Additionally, the continuum model lacks effects such as grain crushing.

A comparison between the measured and the computed absorbed energy is depicted in Fig. 10. Again, a good agreement between the simulation and the experiment can be observed.

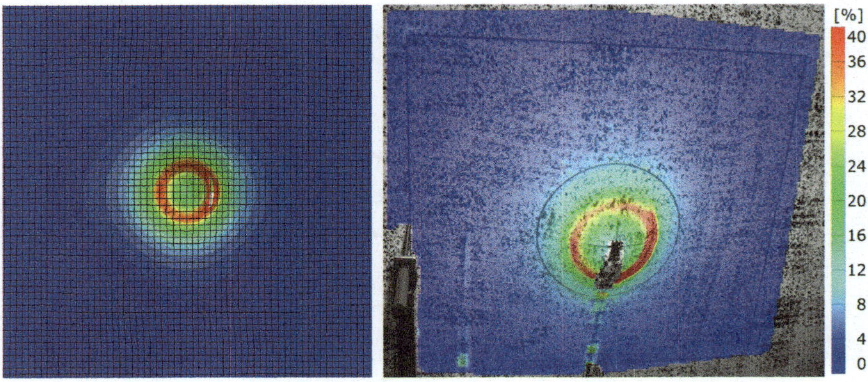

Fig. 7 The results of the FEM computation of the strains (left) are compared to the experimental results (right) obtained applying a DIC-system at the onset of rupture

Fig. 8 Force-displacement curve for the empty double hull. Comparison between simulation and experiment

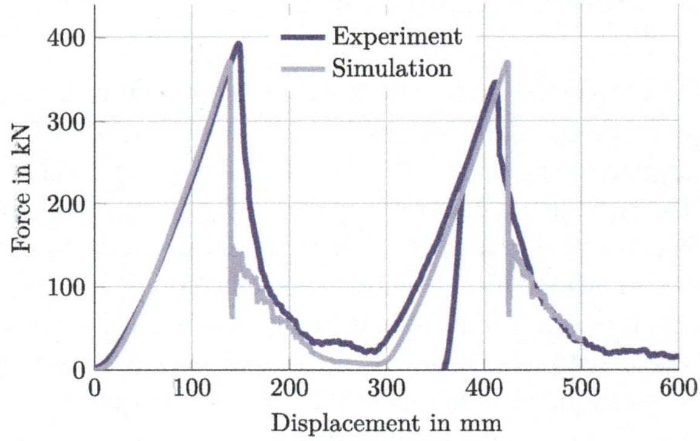

Fig. 9 Force-displacement curve for the filled double hull. Comparison between simulation and experiment

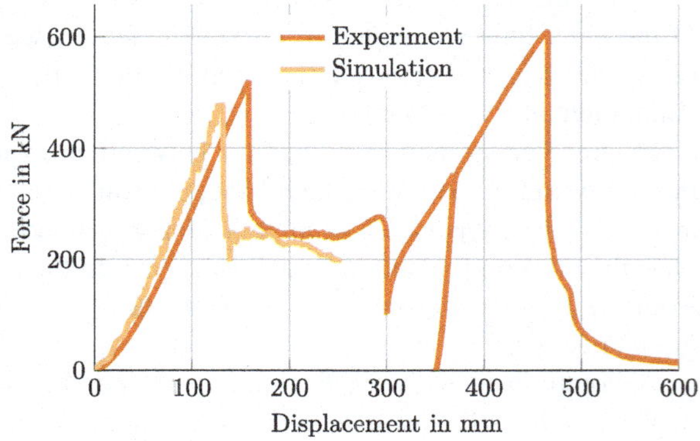

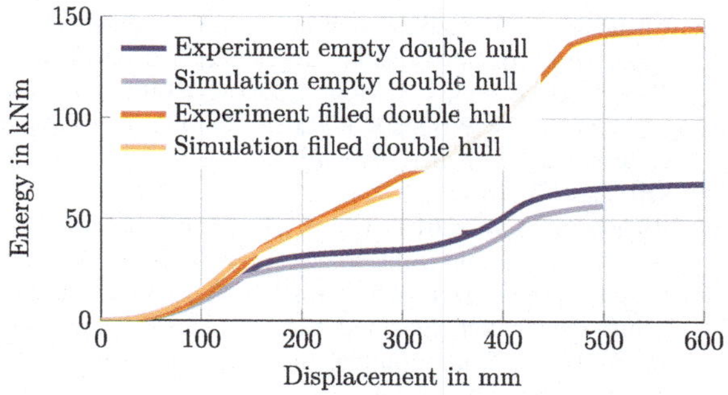

Fig. 10 Energy absorption for the filled double hull. Comparison of simulation and experiment

4 Conclusions

We have presented an experimental and numerical investigation of a simplified double hull structure filled with granular materials. The usage of granular material helps to increase the crashworthiness of ships—as was demonstrated by the experiments. Furthermore, it was shown that a finite element approach allows to simulate the collision process quite accurately. However, it should be mentioned that the modelling of the granular material undergoing finite deformations is problematic when using the classical Mohr-Coulomb model. This can be overcome by applying the two-scale approach mentioned in the introduction, which is based on coupling the DEM with the FEM. A detailed description of the method can be found in the literature we referred to.

Acknowledgements The support of the Deutsche Forschungsgemeinschaft (DFG) under grant number DU 405/9/project number 268649611 is gratefully acknowledged.

References

1. http://www.simulia.com/.
2. Dennert Poraver GmbH. Poraver 2–4 expanded glass. Retrieved October 11, 2020 from https://www.poraver.com/en/poraver/.
3. International Maritime Organization. Retrieved October 15, 2020 from http://www.imo.org/.
4. OSTHOFF OMEGA GROUP. Omega C-XL expanded glass. Retrieved October 11, 2020 from https://www.lehvoss.de/en/products/other-raw-materials-additives-and-functional-fillers/.
5. Saint-Gobain Weber. Leca 2–4 expanded clay. Retrieved October 11, 2020 from https://www.leca.com.
6. Alsos, H. S., & Amdahl, J. (2009). On the resistance to penetration of stiffened plates, Part I—Experiments. *International Journal of Impact Engineering, 36*, 799–807.
7. Alsos, H. S., Amdahl, J., & Hopperstad, O. S. (2009). On the resistance to penetration of stiffened plates, Part II: Numerical analysis. *International Journal of Impact Engineering, 36*, 875–887.
8. Chaudry, M., Woitzik, C., Düster, A., & Wriggers, P. (2018). Experimental and numerical characterization of expanded glass granules. *Computational Particle Mechanics, 5*, 297–312. https://doi.org/10.1007/s40571-017-0169-0

9. Chaudry, M., Woitzik, C., Düster, A., & Wriggers, P.: A multi-scale DEM-FEM coupled approach for the investigation fo granules as crash-absorber in ship building. *Computational Particle Mechanics,* 1–16. https://doi.org/10.1007/s40571-021-00401-5
10. Chaudry, M., & Wriggers, P. (2018). On the computational aspects of comminution in discrete element method. *Computational Particle Mechanics, 5*, 175–189. https://doi.org/10.1007/s40571-017-0161-8
11. Ringsberg, J. W. (2010). Characteristics of material, ship side structure response and ship survivability in ship collisions. *Ships and Offshore Structures, 5*, 51–66.
12. Schöttelndreyer, M. (2015). Füllstoffe in der Konstruktion: Ein Konzept zur Verstärkung von Schiffsseitenhüllen. Doctoral Thesis, Hamburg University of Technology. https://doi.org/10.15480/882.1258. http://tubdok.tub.tuhh.de/handle/11420/1261.
13. Schöttelndreyer, M., & Lehmann, E. (2015). A concept about strengthening of ship side structures verified by quasi-static collision experiments. In *Proceedings of the 12th International Conference on the Stability of Ships and Ocean Vehicles*, Glasgow, UK (pp. 363–372).
14. Woitzik, C., & Düster, A. (2017). Modelling the material parameter distribution of expanded granules. *Granular Matter, 19*(52), 1–12. https://doi.org/10.1007/s10035-017-0735-4
15. Woitzik, C., & Düster, A. (2020). Experimental investigation of granules as crash-absorber in ship building. *Ships and Offshore Structures,* 1–12. https://doi.org/10.1080/17445302.2020.1727179.

On Hydraulic Fracturing in Fully and Partially Saturated Brittle Porous Material

Wolfgang Ehlers, Alixa Sonntag, and Arndt Wagner

After long years of successfully covering the broad field of continuum and computational mechanics, Professor Peter Wriggers is recommended to look to both sides of time. In the backward view, Prof. Wriggers has achieved merits in all areas of academia, research and education, university management as vice president, book author and journal editor, especially as editor-in-chief of Computational Mechanics. He has been honoured by plenty of awards and additionally by three honorary doctorates. The forward view is simple and promising. As emeritus professor, Peter Wriggers has a lot of choices. Rid of occupational ties, he has the freedom to do what he wants. This is a unique chance. We sincerely wish Professor Peter Wriggers to make use of this chance and to further live a successful and content life.

Abstract Hydraulic fracturing or in short fracking is a technique, where a fracking fluid is pressed in the underground with the goal to open fractures for the exploitation of oil and natural gas, for geothermal purposes and further applications. Although fracking techniques are widely used in practice, they are scientifically not well established. This is due to the fact that the underground is a complicated matter made of rock or soil filled with fluids, such as water and air. In a continuum- and computational-mechanical environment, fluid- and gas-filled porous media are treated on the basis of the Theory of Porous Media, thus building the foundation of a successful description of both saturated and partially saturated porous media. In the present article, this is enhanced by the phase-field strategy applied to fracturing processes.

W. Ehlers (✉) · A. Sonntag · A. Wagner
Institute of Applied Mechanics, University of Stuttgart, Stuttgart, Germany
e-mail: Wolfgang.Ehlers@mechbau.uni-stuttgart.de

A. Sonntag
e-mail: Alixa.Sonntag@mechbau.uni-stuttgart.de

A. Wagner
e-mail: Arndt.Wagner@mechbau.uni-stuttgart.de

F. Aldakheel et al. (eds.), *Current Trends and Open Problems in Computational Mechanics*, https://doi.org/10.1007/978-3-030-87312-7_12

1 Introduction

The description of porous media with pore content is easily achieved on the basis of the Theory of Porous Media (TPM), compare Ehlers [1] for a general overview. Treating saturated and unsaturated media means that a coupled system of balance equations for two or three components, solid, liquid and gas, has to be considered. This system leads to strongly coupled formulations that have to be solved by the finite-element analysis (FEA), for details, see Wriggers [2]. When hydraulic fracturing comes into play, the phase-field approach to fracture is a convenient tool, as it regularises a sharp crack-surface topology in the porous solid by diffuse crack zones governed by a single scalar auxiliary variable, the phase field. Here, the reader is referred, for example, to Miehe et al. [3] or to Kuhn and Müller [4]. Applying the phase-field approach to fracturing processes in porous media leads to an additional equation that has to be coupled to the governing system of equations, compare Ehlers and Luo [5] for dynamic hydraulic fracturing of fully saturated material or Heider and Markert [6] for the quasi-static case.

In hydraulic fracturing, a fluid is pressed in the underground with the goal to either yield a sharp fracture or a diffuse fracture domain that basically can be compared to a damaged zone. To set an example for this behaviour, consider a sandstone as porous medium, where the sand grains with usual grain sizes between 0.063 and 0.2 mm have been glued together in a long-lasting period by cap-rock pressure and cementation processes. After the fracking process, the bonds between the grains are broken, while the fractures can either be open or closed depending of the local deformation governed by the fluid pressure, compare Ehlers and Luo [7]. Following Ehlers and Luo [5, 7], where only fully saturated material of solid and liquid has been considered, the present article also includes partially saturated material of solid, liquid and gas.

2 Governing Equations

Based on isothermal conditions, the components φ^α of the overall model φ, where $\alpha = \{S,\ L,\ G\}$ stands for solid, liquid and gas, are governed by their mass and momentum balances yielding

$$\begin{aligned} (\rho^\alpha)'_\alpha \,+\, \rho^\alpha \operatorname{div} \overset{\prime}{\mathbf{x}}_\alpha &= 0\,, \\ \rho^\alpha \overset{\prime\prime}{\mathbf{x}}_\alpha &= \operatorname{div} \mathbf{T}^\alpha \,+\, \rho^\alpha\, \mathbf{g} \,+\, \hat{\mathbf{p}}^\alpha\,. \end{aligned} \tag{1}$$

Therein, $\rho^\alpha = n^\alpha \rho^{\alpha R}$ is the partial density of φ^α built from the product of the volume fraction n^α and the intrinsic (effective or real) density $\rho^{\alpha R}$. In case of material incompressible components, such as the solid φ^S and the pore liquid φ^L, the intrinsic density remains constant, while the partial density can vary through variations of the volume fractions. Furthermore $\overset{\prime}{\mathbf{x}}_\alpha$ and $\overset{\prime\prime}{\mathbf{x}}_\alpha$ are the velocity and acceleration of φ^α at the current position $\mathbf{x}$, while $(\,\cdot\,)'_\alpha$ defines the material time derivative of $(\,\cdot\,)$ following the motion of φ^α. In addition, div $(\,\cdot\,)$ is the divergence operator corresponding to grad $(\,\cdot\,) = \partial(\,\cdot\,)/\partial\mathbf{x}$. Finally, $\mathbf{T}^\alpha$ and $\hat{\mathbf{p}}^\alpha$ are the Cauchy stress and the direct momentum production of φ^α, and $\mathbf{g}$ is the gravitation vector. Note in passing that $\hat{\mathbf{p}}^\alpha$ is the local volumetric mean of the contact forces acting upon φ^α through the other components of the overall model.

The above set of equations has to be closed by convenient constitutive relations for all components.

2.1 Fracturing of Brittle Porous Solids

Considering brittle porous solids, $\mathbf{T}^S$ depends on the pore pressure p^{FR} and the effective stress $\mathbf{T}^S_{\mathit{eff}}$ via

$$\mathbf{T}^S = -n^S p^{FR}\,\mathbf{I} + \mathbf{T}^S_{\mathit{eff}} \quad \text{with} \quad \mathbf{T}^S_{\mathit{eff}} = \rho^S \frac{\partial \psi^S}{\partial \mathbf{F}_S} \mathbf{F}^T_S\,, \tag{2}$$

where ψ^S is the Helmholtz free energy per unit solid mass and $\mathbf{F}_S$ the solid deformation gradient. Based on earlier work on non-porous solids [3, 4] and on fully liquid-saturated porous media [5, 7], the Helmholtz free energy has been shown to depend on three terms, a tensile, a compressive and a fracture energy. For details, compare Ehlers and Luo [5]. Thus, in the framework of small deformations, the effective stress yields

$$\begin{aligned}\mathbf{T}^S_{\mathit{eff}} = \left[(1-\phi^S)^2 + \eta^S_r\right] &\left[2\,\mu^S\,\boldsymbol{\varepsilon}^+_S + \lambda^S \left(\frac{\operatorname{tr}\boldsymbol{\varepsilon}_S + |\operatorname{tr}\boldsymbol{\varepsilon}_S|}{2}\right)\,\mathbf{I}\right] + \\ &+2\,\mu^S\,\boldsymbol{\varepsilon}^-_S + \lambda^S \left(\frac{\operatorname{tr}\boldsymbol{\varepsilon}_S - |\operatorname{tr}\boldsymbol{\varepsilon}_S|}{2}\right)\,\mathbf{I}\,.\end{aligned} \tag{3}$$

Therein, $\boldsymbol{\varepsilon}^+_S$ and $\boldsymbol{\varepsilon}^-_S$ contain the positive and negative eigenvalues of the linearised Lagrangean strain $\boldsymbol{\varepsilon}_S$, while μ^S and λ^S are the Lamé constants. Note that the first summand of (3) is multiplied by a function depending on the phase-field variable ϕ^S varying between zero for fully intact material and one for fully broken material, while the viscous stress resistance η^S_r guarantees a minimum of tensile forces even in the fully broken case.

From thermodynamical considerations, one furthermore concludes to the evolution equation for the phase-field variable reading

$$(\phi^S)'_S = \frac{1}{M}\left[2(1-\phi^S)\,\rho^S_0\,\psi^{S+} - G_c\left(\frac{\phi^S}{\epsilon} - \epsilon\,\operatorname{div}\operatorname{grad}\phi^S\right)\right] \tag{4}$$

with the mobility parameter M and the tensile energy $\rho_0^S \psi^{S+}$ depending on the partial solid density ρ_0^S at the solid reference configuration at time $t = t_0$, cf. [5].

2.2 Fluid Components

From thermodynamical considerations, one recovers Dalton's law [8, 9]

$$p^{FR} = s^L p^{LR} + (1 - s^L) p^{GR} \tag{5}$$

stating that the pore pressure p^{FR} is composed of the effective liquid and gas pressures p^{LR} and p^{GR} by the liquid saturation $s^L = n^L / n^P$ with $n^P = n^L + n^G$ as the porosity. While $n^S = n_0^S (\det \mathbf{F}_S)^{-1}$ and $n^P = 1 - n^S$ can be found from the solid deformation and the initial solidity n_0^S, the liquid saturation s^L has to be found by a constitutive assumption as a function of the capillary pressure $p^C = p^{GR} - p^{LR}$, where φ^G is the non-wetting and φ^L the wetting phase. As the liquid is materially incompressible, there is no constitutive equation for p^{LR}. However, as the effective gas pressure p^{GR} can be found from the ideal gas law, p^{LR} can be computed from p^C. In contrast to the standard p^C-over-s^L functions of van-Genuchten or Brooks-and-Corey type valid for spontaneous imbibition and drainage problems, fracking processes proceed from forcing the fluid into the porous medium, such that forced curves are needed, compare Blunt [10].

When the fluid pressures are known, the fluid stresses and the direct momentum productions can be found as

$$\begin{aligned}
\mathbf{T}^L &= -n^L p^{LR}\, \mathbf{I} + \mathbf{T}^L_{fric}\,, \\
\mathbf{T}^G &= -n^G p^{GR}\, \mathbf{I}\,, \\
\hat{\mathbf{p}}^L &= p^{FR} \operatorname{grad} n^L - s^L p^C \operatorname{grad} n^G + \hat{\mathbf{p}}^L_{fric}\,, \\
\hat{\mathbf{p}}^G &= p^{GR} \operatorname{grad} n^G + \hat{\mathbf{p}}^G_{fric}\,,
\end{aligned} \tag{6}$$

where frictional stresses of the pore gas have been dropped as they are negligible compared to the frictional liquid stresses, compare Ehlers [11, 12]. Based on thermodynamical considerations [7], the frictional liquid stresses and momentum productions read

$$\begin{aligned}
\mathbf{T}^L_{fric} &= 2\,\mathcal{I}\,(\phi^S)^2 n^L \mu^{LR} \mathbf{D}_L\,, \\
\hat{\mathbf{p}}^\beta_{fric} &= -[1 - \mathcal{I} + \mathcal{I}(1 - \phi^S)^2](n^\beta)^2 \frac{\gamma^{\beta R}}{k_r^\beta} \mathbf{w}_\beta
\end{aligned} \tag{7}$$

with $\beta = \{L, G\}$. In these equations, $\mathcal{I}$ is the crack-opening indicator (COI), compare Ehlers and Luo [7], μ^{LR} is the intrinsic or effective shear viscosity of the pore liquid, and $\mathbf{D}_L$ is the symmetric part of the liquid velocity gradient $\mathbf{L}_L = \operatorname{grad} \mathbf{v}_L$ with

$\mathbf{v}_L = \overset{\prime}{\mathbf{x}}_L$. Furthermore, $\gamma^{\beta R} = \rho^{\beta R}|\mathbf{g}|$ is the specific weight of φ^β, while $k_r^\beta = \kappa_r^\beta k^\beta$ is the relative hydraulic conductivity with κ_r^β as the relative permeability factor that transforms the hydraulic conductivity k^β valid for a single pore fluid φ^β towards k_r^β valid for multiple pore fluids, such as φ^L and φ^G. Finally, $\mathbf{w}_\beta = \overset{\prime}{\mathbf{x}}_\beta - \overset{\prime}{\mathbf{x}}_S$ defines the seepage velocity. For further details, compare Ehlers [11, 12].

Equation (7) need some additional explanations. For example, $(7)_2$ reduces to the usual extra part of the direct momentum production, when the term in bracket equals one. This is the case, when the COI is zero, while ϕ^S can be arbitrary between zero and one, or when the COI is one and ϕ^S vanishes. Note that the COI acts as a switch governing two possibilities. On the one hand, $\phi^S = 0$ characterises the fact that there is no fracture, such that only the drag force governed by $(7)_2$ exists. On the other hand, $\phi^S = 1$ either indicates a closed fracture, where, as before, only the drag force exists, or it indicates an open fracture, such that the drag force shrinks with growing values of ϕ^S, while the Navier-Stokes term governed by $(7)_1$ increases.

2.3 *Equations Governing the Numerical Computations*

Proceeding from the basic balance equations (1) and the constitutive setting given through (2)–(7), the strong form of equations governing the dynamic computational process of triphasic media of solid, pore liquid and pore gas are found as

$$
\begin{aligned}
0 &= (n^L)'_S + n^L \operatorname{div} \overset{\prime}{\mathbf{x}}_S + \operatorname{div}(n^L \mathbf{w}_L)\,,\\
0 &= (n^G \rho^{GR})'_S + n^G \rho^{GR} \operatorname{div} \overset{\prime}{\mathbf{x}}_S + \operatorname{div}(n^G \rho^{GR} \mathbf{w}_G)\,,\\
n^S \rho^{SR} \overset{\prime\prime}{\mathbf{x}}_S + \textstyle\sum_\beta n^\beta \rho^{\beta R} (\mathbf{v}_\beta)'_\beta &= -\operatorname{grad} p^{FR} + \operatorname{div}(\mathbf{T}^S_{\mathit{eff}} + \mathbf{T}^L_{\mathit{fric}}) + \rho\, \mathbf{g}\,,\\
n^L \rho^{LR} (\mathbf{v}_L)'_L &= -n^L \operatorname{grad} p^{LR} + \operatorname{div} \mathbf{T}^L_{\mathit{fric}} + n^L \rho^{LR} \mathbf{g} + \hat{\mathbf{p}}^L_{\mathit{fric}} +\\
&\quad + p^C [\,(1 - s^L) \operatorname{grad} n^L - s^L \operatorname{grad} n^G\,]\,,\\
n^G \rho^{GR} (\mathbf{v}_G)'_G &= -n^G \operatorname{grad} p^{GR} + n^G \rho^{GR} \mathbf{g} + \hat{\mathbf{p}}^G_{\mathit{fric}}\,,\\
(\phi^S)'_S &= \frac{1}{M} [2(1 - \phi^S) \mathcal{H} - G_c (\frac{\phi^S}{\epsilon} - \epsilon \operatorname{div} \operatorname{grad} \phi^S)]
\end{aligned}
\tag{8}
$$

with $\rho = \sum_\alpha n^\alpha \rho^{\alpha R}$ as the density of the overall medium and $\mathcal{H}$ as history variable based on $\rho_0^S \psi^{S+}$, cf. [3, 5]. These equations together with the initial and boundary conditions of the problems at hand yield a strongly coupled system of partial differential equations that can be solved numerically by the tool PANDAS.[1] In the framework of the Bubnov-Galerkin method, the volume and mass balances $(8)_{1,2}$ are tested by δp^{LR} and δp^{GR}, the overall momentum balance $(8)_3$ by $\delta \mathbf{u}_S$, the liquid and gas momentum balances $(8)_{4,5}$ by $\delta \mathbf{v}_L$ and $\delta \mathbf{v}_G$, and the phase-field evolution equation $(8)_6$ by $\delta \phi^S$. Depending on the boundary conditions, δp^{LR} can alternatively be substituted by δs^L.

---

[1] **P**orous Media **a**daptive **n**onlinear finite-element solver based on **d**ifferential **a**lgebraic **s**ystems, cf. http://www.get-pandas.com.

While the above set of equations describes partially saturated porous media as a triphasic model, this set can easily be reduced towards biphasic media that are sufficient for the solution of fully saturated problems.

3 Numerical Example

The following example, compare Fig. 1, has been designed to illustrate the different fracturing behaviour of fully saturated and partially saturated material. Based on PANDAS, the problem has been computed by use of Taylor-Hood elements with quadratic approximation functions for $\mathbf{u}_S$, $\mathbf{v}_L$ and $\mathbf{v}_G$, and linear approximation functions for the remainder of primary variables, s^L, p^{GR} and ϕ^S. As can be seen from Fig. 1, a notched block of brittle porous material is investigated under symmetry conditions, such that only the upper half has to be taken into consideration. To guarantee symmetry, the problem is computed without gravitational forces but under consideration of inertia terms. The notch represents an initial crack, where a fluid is pressed into the material with a horizontal velocity of $\bar{v}_L = 5 \cdot 10^{-3}\, t\, \text{m}^3/(\text{m}^2\text{s})$ increasing with time t. Before the fracking starts, the initial conditions represent the case, where everything is at rest under atmospheric pressure. The mesh is symmetric with 23,432 elements including a refinement at the initial crack and and along the symmetry line.

Figures 2 and 3 show the results, especially at five different time steps. Figure 2 displays the fracking process by the phase-field variable ϕ^S. As can be seen from this figure, the fracture evolution is fast in fully saturated material compared to the partially saturated case. This is due to the fact, that the pore gas is compressible, such that it takes a longer time until the pore pressure p^{FR} reaches a value that is

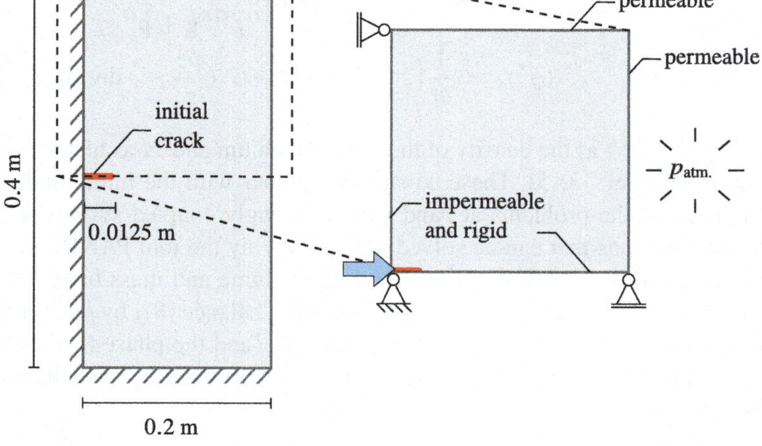

Fig. 1 Sketch of the initial-boundary-value problem of the hydraulic fracturing problem

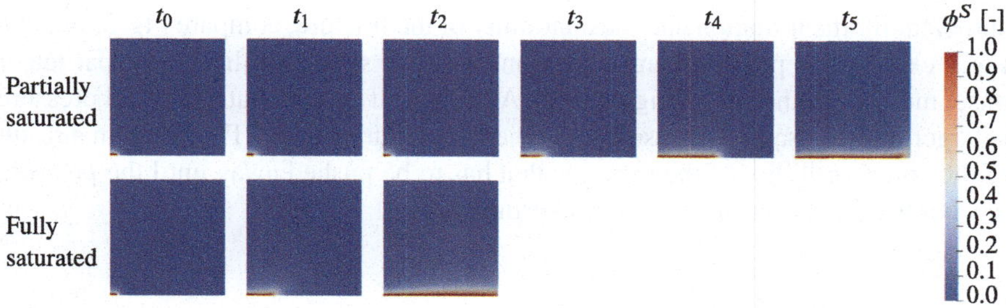

Fig. 2 Evolution of the phase-field variable ϕ^S during the fracturing problem for partially and fully saturated porous material

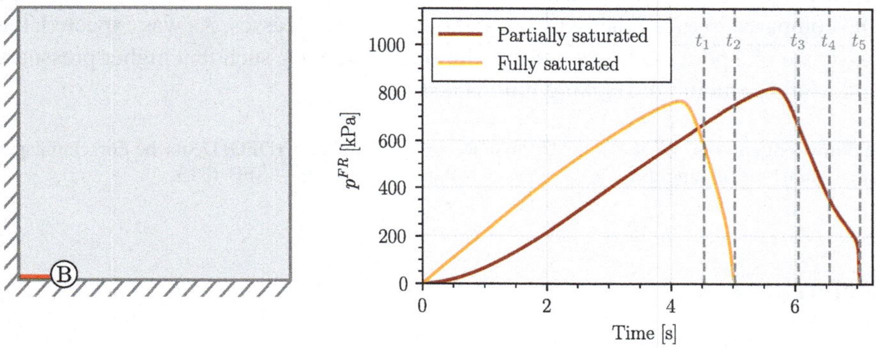

Fig. 3 Depiction of the temporal development of the effective pore pressure p^{FR} at the initial crack tip (point B) for the partially and fully saturated model

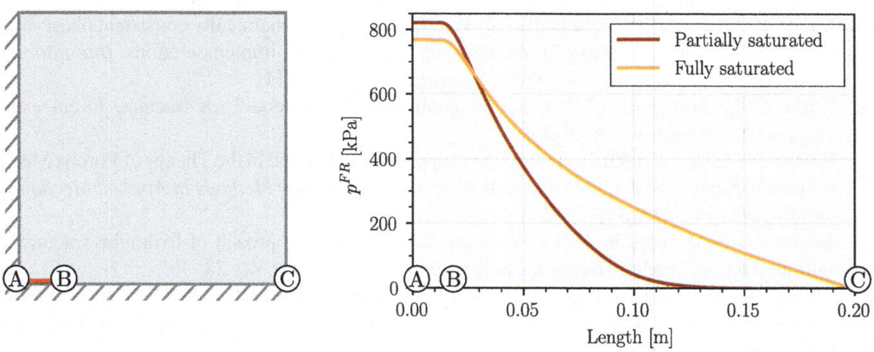

Fig. 4 Effective pore pressure field over the length A – C of the specimen for the partially and fully saturated model at the particular onset time of the crack propagation

sufficient for the initiation and the continuity of the fracture, compare Fig. 3. Finally, Fig. 4 exhibits the pore pressure evolution along the symmetry line at the particular onset moment of the fracturing process. As before, it is seen that the initial pressure is larger in the unsaturated case than in the fully saturated one. This is again a result of the compressibility of the pore gas that has to be pushed away until the pressure reaches a value necessary for the fracturing process.

4 Conclusion

In the present article, fully and partially saturated brittle material have been discussed and compared to each other with respect to fracking processes. As was expected, the existence of a pore gas hinders the evolution of fractures, such that higher pressures and a larger amount of fracking fluid is necessary.

Acknowledgements We thank the German Research Foundation (DFG, Deutsche Forschungsgemeinschaft) for funding this work – Project Number 327154368 – SFB 1313.

References

1. Ehlers, W. (2002). Foundations of multiphasic and porous materials. In W. Ehlers & J. Bluhm (Eds.), *Porous Media: Theory, Experiments and Numerical Applications* (pp. 3–86). Berlin: Springer.
2. Wriggers, P. (2008). *Nonlinear Finite Element Methods*. Berlin: Springer.
3. Miehe, C., Welschinger, F., & Hofacker, M. (2010). Thermodynamically consistent phase-field models of fracture: Variational principles and multi-field FE implementations. *International Journal for Numerical Methods in Engineering, 83*, 1273–1311.
4. Kuhn, C., & Müller, R. (2010). A continuum phase field model for fracture. *Engineering Fracture Mechanics, 77*, 3625–3634.
5. Ehlers, W., & Luo, C. (2017). A phase-field approach embedded in the Theory of Porous Media for the description of dynamic hydraulic fracturing. *Computer Methods in Applied Mechanics and Engineering, 315*, 348–368.
6. Heider, Y., & Markert, B. (2017). A phase-field modeling approach of hydraulic fracture in saturated porous media. *Mechanics Research Communications, 80*, 38–46.
7. Ehlers, W., & Luo, C. (2018). A phase-field approach embedded in the Theory of Porous Media for the description of dynamic hydraulic fracturing, Part II: The crack-opening indicator. *Computer Methods in Applied Mechanics and Engineering, 341*, 429–442.
8. Dalton, J. (1802). On the expansion of elastic fluids by heat. *Essay IV of Memoirs of the Literary and Philosophical Society of Manchester, 5*, 595–602.
9. Ehlers, W. (2018). Effective Stresses in multiphasic porous media: A thermodynamic investigation of a fully non-linear model with compressible and incompressible constituents. *Geomechanics for Energy and the Environment, 15*, 35–46.
10. Blunt, J. (2017). *Multiphase Flow in Permeable Media: A Pore-Scale Perspective*. Cambridge: Cambridge University Press. Chapter 5.

11. Ehlers, W. (2009). Challenges of porous media models in geo- and biomechanical engineering including electro-chemically active polymers and gels. *International Journal for Advances in Engineering Sciences and Applied Mathematics, 1*, 1–24.
12. Ehlers, W. (2020). Darcy, Forchheimer, Brinkman and Richards: Classical hydromechanical equations and their significance in the light of the TPM. *Archive of Applied Mechanics*. https://doi.org/10.1007/s00419-020-01802-3.

Efficient Two-Scale Modeling of Porous Media Using Numerical Model Reduction with Fully Computable Error Bounds

Fredrik Ekre, Ralf Jänicke, Fredrik Larsson, and Kenneth Runesson

This contribution is dedicated to Professor Peter Wriggers. Besides being an outstanding scientist, he is always friendly and willing to generously share his deep knowledge.

Abstract The microscale problem arizing from computational homogenization of porous media problems is solved by adopting the concept of Numerical Model Reduction. Thereby, the displacement and pore pressure are the unknown fields. A suitable reduced basis is constructed for the pore pressure approximation using Proper Orthogonal Decomposition (POD), whereby it is possible to compute the appropriate basis for the displacement field in the spirit of Nonlinear Transformation Field Analysis (NTFA). Inexpensive fully computable error bounds are obtainable, and the performance of the error estimates is illustrated in this paper.

1 Introduction

This paper is devoted to the computationally efficient multi-scale simulation of hydro-mechanical coupling and Darcy-type fluid transport in heterogeneous pororelastic materials such as, for example, clayey soil, rock or concrete. In such materials,

F. Ekre (✉) · R. Jänicke · F. Larsson · K. Runesson
Division of Material and Computational Mechanics, Chalmers University of Technology, Gothenborg, Sweden
e-mail: fredrik.ekre@chalmers.se

R. Jänicke
e-mail: ralf.janicke@chalmers.se

F. Larsson
e-mail: fredrik.larsson@chalmers.se

K. Runesson
e-mail: kenneth.runesson@chalmers.se

F. Aldakheel et al. (eds.), *Current Trends and Open Problems in Computational Mechanics*, https://doi.org/10.1007/978-3-030-87312-7_13

heterogeneity might become manifest in domains with different saturating fluids (compressibility contrast) or in domains with fluctuating mechanical properties of the solid skeleton (stiffness and permeability contrast). Moreover, heterogeneity may occur on multiple length scales, ranging from the pore scale to much larger length scales (e.g. in layered soil or rock formations).

Hence, it is a challenge to model hydro-mechanical coupling and fluid transport in a computationally efficient way, and it is therefore imperative to develop and implement suitable simulation strategies. This paper aims at providing a review of our recent and ongoing research in this field: In Sect. 2, we demonstrate how to treat multi-scale fluid transport associated with hydro-mechanical coupling in poroelastic media by means of Variationally Consistent Computational Homogenization (VCH). To ensure computationally efficiency, we enrich VCH with Numerical Model Reduction (NMR). In Sect. 3, we present a strategy to estimate the approximation error associated with NMR in an inexpensive way.

2 Computational Homogenization with Model Reduction

In this section, we present the VCH concept for multi-scale poroelastic media combined with NMR. For a more detailed introduction to the overall concept of VCH, we refer to Larsson et al. [1]. For the related field of thermo-elasticity, Temizer and Wriggers [2] might be a useful reference. The NMR concept for poroelastic media has recently been discussed in detail in Jänicke et al. [3]. The two-scale space-variational formulation of linear poroelasticity can be expressed as the problem of finding the displacement $\boldsymbol{u}(\bullet, t)$ and the fluid pressure $p(\bullet, t)$ in the appropriately defined two-scale trial set that solve

$$\int_\Omega \left[a_\Box^{(u)}(\boldsymbol{u}; \delta\boldsymbol{u}) - b_\Box(p; \delta\boldsymbol{u})\right] \mathrm{d}\Omega = \int_{\Gamma_\mathrm{N}^{(u)}} \boldsymbol{t}^\mathrm{p} \cdot \delta\boldsymbol{u} \, \mathrm{d}\Gamma, \tag{1a}$$

$$\int_\Omega \left[b_\Box(\delta p; \dot{\boldsymbol{u}}) + m_\Box(\dot{p}; \delta p) + a_\Box^{(p)}(p; \delta p)\right] \mathrm{d}\Omega = \int_{\Gamma_\mathrm{N}^{(p)}} w^\mathrm{p} \, \delta p \, \mathrm{d}\Gamma, \tag{1b}$$

for suitable (dual-scale) test functions $\delta\boldsymbol{u}$ and δp. The pertinent space-variational forms in (1) are defined as

$$a_\Box^{(u)}(\boldsymbol{u}; \boldsymbol{v}) := \langle \boldsymbol{\varepsilon}[\boldsymbol{u}] : \mathbb{E} : \boldsymbol{\varepsilon}[\boldsymbol{v}] \rangle_\Box, \quad a_\Box^{(p)}(p; q) := \langle \boldsymbol{\zeta}[p] \cdot \boldsymbol{K} \cdot \boldsymbol{\zeta}[q] \rangle_\Box, \tag{2a}$$

$$b_\Box(p; \boldsymbol{v}) := \langle \alpha \, p \, \boldsymbol{I} : \boldsymbol{\varepsilon}[\boldsymbol{v}] \rangle_\Box = \langle \alpha \, p \, (\boldsymbol{v} \cdot \boldsymbol{\nabla}) \rangle_\Box, \quad m_\Box(p; q) := \langle \beta \, p \, q \rangle_\Box. \tag{2b}$$

Here, we use the linear symmetric strain $\boldsymbol{\varepsilon}$ and the pressure gradient $\boldsymbol{\zeta}$ with the material properties $\mathbb{E}$ (fourth order elasticity modulus expressed via shear modulus G and bulk modulus K), $\boldsymbol{K}$ (second order permeability tensor), α (Biot coefficient) and β (intrinsic compression compliance of the pore fluid). Moreover, angle brackets represent volume averaging over the square/cubic RVE domain $\Omega_\Box$ located at each macro-scale spatial point $\bar{x} \in \Omega$. Following the standard approach of so-called first-order homogenization, we decompose the sub-scale fields $\boldsymbol{u}$ and p into macro-scale parts, $\boldsymbol{u}^{\mathrm{M}}$ and p^{M}, and micro-scale or fluctuating parts, $\boldsymbol{u}^{\mu}$ and p^{μ}, within each RVE such that

$$\boldsymbol{u} = \boldsymbol{u}^{\mathrm{M}}[\bar{\boldsymbol{u}}] + \boldsymbol{u}^{\mu}, \quad \text{with} \quad \boldsymbol{u}^{\mathrm{M}}[\bar{\boldsymbol{u}}](\boldsymbol{x}, t) := \bar{\boldsymbol{u}} + \bar{\boldsymbol{h}}(t) \cdot [\boldsymbol{x} - \bar{\boldsymbol{x}}], \quad \bar{\boldsymbol{h}} := \bar{\boldsymbol{u}} \otimes \boldsymbol{\nabla}, \tag{3a}$$

$$p = p^{\mathrm{M}}[\bar{p}] + p^{\mu}, \quad \text{with} \quad p^{\mathrm{M}}[\bar{p}](\boldsymbol{x}, t) := \bar{p} + \bar{\boldsymbol{\zeta}}(t) \cdot [\boldsymbol{x} - \bar{\boldsymbol{x}}], \quad \bar{\boldsymbol{\zeta}} := \boldsymbol{\nabla} \bar{p}. \tag{3b}$$

Choosing purely macroscopic test functions $\delta \boldsymbol{u} = \boldsymbol{u}^{\mathrm{M}}[\delta \bar{\boldsymbol{u}}]$ and $\delta p = p^{\mathrm{M}}[\bar{p}]$ leads to the homogenized macro-scale problem

$$\int_{\Omega} \bar{\boldsymbol{\sigma}} : [\delta \bar{\boldsymbol{u}} \otimes \boldsymbol{\nabla}] \mathrm{d}\Omega = \int_{\Gamma_{\mathrm{N}}^{(u)}} \bar{\boldsymbol{t}}^{\mathrm{p}} \cdot \delta \bar{\boldsymbol{u}} \, \mathrm{d}\Gamma, \tag{4a}$$

$$\int_{\Omega} \dot{\bar{\Phi}} \delta \bar{p} \, \mathrm{d}\Omega + \int_{\Omega} \left[\dot{\bar{\boldsymbol{Q}}} - \bar{\boldsymbol{q}} \right] \cdot \boldsymbol{\nabla} \delta \bar{p} \, \mathrm{d}\Omega = \int_{\Gamma_{\mathrm{N}}^{(u)}} \bar{w}^{\mathrm{p}} \delta \bar{p} \, \mathrm{d}\Gamma, \tag{4b}$$

with the macro-scale fields

$$\bar{\boldsymbol{\sigma}} = \langle \boldsymbol{\sigma} \rangle_{\Box} = \langle \mathbb{E} : \boldsymbol{\varepsilon}[\boldsymbol{u}] - \alpha \, p \boldsymbol{I} \rangle_{\Box}, \tag{5a}$$

$$\bar{\boldsymbol{q}} = \langle \boldsymbol{w} \rangle_{\Box} = -\langle \boldsymbol{K} \cdot \boldsymbol{\zeta}[p] \rangle_{\Box}, \tag{5b}$$

$$\bar{\Phi} = \langle \Phi \rangle_{\Box} = \langle \phi + \alpha \, \boldsymbol{I} : \boldsymbol{\varepsilon}[\boldsymbol{u}] + \beta \, p \rangle_{\Box}, \tag{5c}$$

$$\bar{\boldsymbol{Q}} = \langle \Phi \, [\boldsymbol{x} - \bar{\boldsymbol{x}}] \rangle_{\Box} = \langle [\phi + \alpha \, \boldsymbol{I} : \boldsymbol{\varepsilon}[\boldsymbol{u}] + \beta \, p] \, [\boldsymbol{x} - \bar{\boldsymbol{x}}] \rangle_{\Box}. \tag{5d}$$

In addition, $\bar{\boldsymbol{t}}^{\mathrm{p}}$ and $\bar{w}^{\mathrm{p}}$ are defined as the suitably homogenized quantities on the Neumann boundary parts. Choosing purely micro-scale test functions instead leads to

$$a_{\Box}^{(u)}(\boldsymbol{u}; \delta \boldsymbol{u}) - b_{\Box}(p; \delta \boldsymbol{u}) - d_{\Box}^{(u)}(\boldsymbol{\lambda}; \delta \boldsymbol{u}) = 0, \tag{6a}$$

$$b_{\Box}(\delta p; \dot{\boldsymbol{u}}) + m_{\Box}(\dot{p}; \delta p) + a_{\Box}^{(p)}(p; \delta p) + d_{\Box}^{(p)}(\mu; \delta p) + \bar{\mu} \, \langle\langle \delta p \rangle\rangle_{\Box} = 0, \tag{6b}$$

$$-d_{\Box}^{(u)}(\delta \boldsymbol{\lambda}; \boldsymbol{u}) = -d_{\Box}^{(u)}(\delta \boldsymbol{\lambda}; \boldsymbol{u}^{\mathrm{M}}[\bar{\boldsymbol{u}}, \bar{\boldsymbol{h}}]), \tag{6c}$$

$$d_{\Box}^{(p)}(\delta \mu; p) = d_{\Box}^{(p)}(\delta \mu; p^{\mathrm{M}}[\bar{p}, \bar{\boldsymbol{\zeta}}]), \tag{6d}$$

$$\delta \bar{\mu} \, \langle\langle p \rangle\rangle_{\Box} = \delta \bar{\mu} \, \bar{p}. \tag{6e}$$

Here, the weak constraints (6c, 6d) ensure micro-periodicity of the fluctuation fields $\boldsymbol{u}^{\mu}$ and p^{μ}. Hence, we define the variational operators

$$d_{\Box}^{(u)}(\boldsymbol{\lambda}, \boldsymbol{u}) := \frac{1}{|\Omega_{\Box}|} \int_{\Gamma_{\Box}^{+}} \boldsymbol{\lambda} \cdot [\![\boldsymbol{u}]\!]_{\Box} \, \mathrm{d}\Gamma, \quad d_{\Box}^{(p)}(\mu, p) := \frac{1}{|\Omega_{\Box}|} \int_{\Gamma_{\Box}^{+}} \mu \, [\![p]\!]_{\Box} \, \mathrm{d}\Gamma, \tag{7}$$

where we introduced the "difference operator" $[\![\bullet]\!]_\square(\boldsymbol{x}) := \bullet(\boldsymbol{x}) - \bullet(\boldsymbol{x}^-(\boldsymbol{x}))$; $\boldsymbol{x} \in \Gamma_\square^+$ is an "image point" whereas $\boldsymbol{x}^-(\boldsymbol{x}) \in \Gamma^- = \Gamma_\square \setminus \Gamma_\square^+$ is the corresponding "mirror point". Moreover, (6e) imposes, in a weak fashion, that the macroscopic pressure $\bar{p}$ can be computed as the surface average $\langle\langle p \rangle\rangle_\square$ of the pore pressure of the RVE boundary. Finally, due to rigid body invariance, we set $\bar{\boldsymbol{u}} = \boldsymbol{0}$.

It is important to remark that the macro-scale formulation (4) together with the RVE formulation (6) and the averaging rules (5) enable us to execute two-scale simulations in FE^2 fashion. However, we aim at making use of the transient but linear properties of the equation system. Therefore, we introduce an alternative decomposition of the solution fields into stationary and transient parts according to $\boldsymbol{u} = \boldsymbol{u}_{\text{stat}}\{\bar{\boldsymbol{\varepsilon}}, \bar{p}, \bar{\boldsymbol{\zeta}}\} + \tilde{\boldsymbol{u}}^\mu$ and $p = p_{\text{stat}}\{\bar{\boldsymbol{\varepsilon}}, \bar{p}, \bar{\boldsymbol{\zeta}}\} + \tilde{p}^\mu$. The variables $\boldsymbol{\lambda}$, $\boldsymbol{\mu}$ and $\bar{\mu}$ can be decomposed accordingly. The stationary fields can be computed from the stationary problem

$$a_\square^{(u)}(\boldsymbol{u}_{\text{stat}}; \delta\boldsymbol{u}) - b_\square(p_{\text{stat}}; \delta\boldsymbol{u}) - d_\square^{(u)}(\boldsymbol{\lambda}_{\text{stat}}; \delta\boldsymbol{u}) = 0, \tag{8a}$$

$$a_\square^{(p)}(p_{\text{stat}}; \delta p) + d_\square^{(p)}(\mu_{\text{stat}}; \delta p) + \bar{\mu}_{\text{stat}}\langle\langle \delta p \rangle\rangle_\square = 0, \tag{8b}$$

$$-d_\square^{(u)}(\delta\boldsymbol{\lambda}; \boldsymbol{u}_{\text{stat}}) = -d_\square^{(u)}(\delta\boldsymbol{\lambda}; \boldsymbol{u}^{\text{M}}[\bar{\boldsymbol{u}}, \bar{\boldsymbol{\varepsilon}}]), \tag{8c}$$

$$d_\square^{(p)}(\delta\mu; p_{\text{stat}}) = d_\square^{(p)}(\delta\mu; p^{\text{M}}[\bar{p}, \bar{\boldsymbol{\zeta}}]), \tag{8d}$$

$$\delta\bar{\mu}\,\langle\langle p_{\text{stat}} \rangle\rangle_\square = \delta\bar{\mu}\,\bar{p}. \tag{8e}$$

Since the problem in (8) is linear and local in time, we may compute the solution in terms of "sensitivities" for the macroscopic loading cases $\bar{\boldsymbol{\varepsilon}}$, $\bar{p}$ and $\bar{\boldsymbol{\zeta}}$ such that

$$\boldsymbol{u}_{\text{stat}}(\boldsymbol{x}, t) = \sum_{i,j} \hat{\boldsymbol{u}}_{ij}^{(\bar{\varepsilon})}(\boldsymbol{x})(\bar{\boldsymbol{\varepsilon}}(t))_{ij} + \hat{\boldsymbol{u}}^{(\bar{p})}(\boldsymbol{x})\bar{p}(t) + \sum_i \hat{\boldsymbol{u}}_i^{(\bar{\zeta})}(\boldsymbol{x})(\bar{\boldsymbol{\zeta}}(t))_i, \tag{9a}$$

$$p_{\text{stat}}(\boldsymbol{x}, t) = \sum_{i,j} \hat{p}_{ij}^{(\bar{\varepsilon})}(\boldsymbol{x})(\bar{\boldsymbol{\varepsilon}}(t))_{ij} + \hat{p}^{(\bar{p})}(\boldsymbol{x})\bar{p}(t) + \sum_i \hat{p}_i^{(\bar{\zeta})}(\boldsymbol{x})(\bar{\boldsymbol{\zeta}}(t))_i. \tag{9b}$$

The truly transient RVE problem can now be defined as

$$a_\square^{(u)}(\tilde{\boldsymbol{u}}^\mu; \delta\boldsymbol{u}) - b_\square(\tilde{p}^\mu; \delta\boldsymbol{u}) - d_\square^{(u)}(\tilde{\boldsymbol{\lambda}}^\mu; \delta\boldsymbol{u}) = 0, \tag{10a}$$

$$\begin{aligned} b_\square(\delta p; \dot{\tilde{\boldsymbol{u}}}^\mu) + m_\square(\dot{\tilde{p}}^\mu; \delta p) + a_\square^{(p)}(\tilde{p}^\mu; \delta p) \\ + d_\square^{(p)}(\tilde{\mu}^\mu; \delta p) + \tilde{\bar{\mu}}^\mu\,\langle\langle \delta p \rangle\rangle_\square = -b_\square(\delta p; \dot{\boldsymbol{u}}_{\text{stat}}) - m_\square(\dot{p}_{\text{stat}}; \delta p), \end{aligned} \tag{10b}$$

$$-d_\square^{(u)}(\delta\boldsymbol{\lambda}; \tilde{\boldsymbol{u}}^\mu) = 0, \tag{10c}$$

$$d_\square^{(p)}(\delta\mu; \tilde{p}^\mu) = 0, \tag{10d}$$

$$\delta\bar{\mu}\,\langle\langle \tilde{p}^\mu \rangle\rangle_\square = 0. \tag{10e}$$

We establish NMR in the spatial domain by introducing a set of linearly independent global basis functions $\{\hat{p}_a(\boldsymbol{x})\}_{a=1}^{M_{\text{R}}}$ called "pressure modes". The pressure modes are identified in a training phase by solving (10) with different loadings $\dot{\boldsymbol{u}}_{\text{stat}}$

and $\dot{p}_{\text{stat}}$ with a POD of extracted snapshots of $\tilde{p}^{\mu}$. The "exact" solution $\tilde{p}^{\mu}$ is then approximated by $\tilde{p}^{\mu}_{\text{R}}$ as

$$\tilde{p}^{\mu}(\boldsymbol{x},t) \approx \tilde{p}^{\mu}_{\text{R}}(\boldsymbol{x},t) := \sum_{a=1}^{M_{\text{R}}} \hat{p}_a(\boldsymbol{x})\xi_a(t), \tag{11}$$

where $\{\xi_a(t)\}_{a=1}^{M_{\text{R}}}$ are "mode activity" parameters. Furthermore, since (10a) is local in time, we can compute the pertinent displacement basis fields $\hat{\boldsymbol{u}}_a$ such that $\tilde{\boldsymbol{u}}_{\text{R}} = \sum_{a=1}^{M_{\text{R}}} \hat{\boldsymbol{u}}_a(\boldsymbol{x})\boldsymbol{\xi}_a(t)$. Inserting these approximations into (10b) allows to extract an evolution equation for the mode activity parameters

$$\left[\underline{\hat{S}} + \underline{\hat{M}}\right]\underline{\dot{\xi}} + \underline{\hat{K}}\underline{\xi} = \underline{\hat{f}}\left[\dot{\bar{\boldsymbol{\varepsilon}}}, \dot{\bar{p}}, \dot{\bar{\boldsymbol{\zeta}}}\right], \tag{12}$$

where the matrix entries are defined as

$$\hat{S}_{ab} := a_{\square}^{(u)}(\hat{\boldsymbol{u}}_a; \hat{\boldsymbol{u}}_b) = \langle \boldsymbol{\varepsilon}[\hat{\boldsymbol{u}}_a] : \mathbb{E} : \boldsymbol{\varepsilon}[\hat{\boldsymbol{u}}_b]\rangle_{\square}, \tag{13a}$$

$$\hat{M}_{ab} := m_{\square}(\hat{p}_a; \hat{p}_b) = \langle \beta\, \hat{p}_a \hat{p}_b\rangle_{\square}, \tag{13b}$$

$$\hat{K}_{ab} := a_{\square}^{(p)}(\hat{p}_a; \hat{p}_b) = \langle \boldsymbol{\zeta}[\hat{p}_a] \cdot \boldsymbol{K} \cdot \boldsymbol{\zeta}[\hat{p}_b]\rangle_{\square}. \tag{13c}$$

Finally, the RHS of (12) is defined as

$$\hat{f}_a\left[\dot{\bar{\boldsymbol{\varepsilon}}}, \dot{\bar{p}}, \dot{\bar{\boldsymbol{\zeta}}}\right] = -\sum_{i,j}\left[\hat{b}^{(\bar{\varepsilon})}_{ij,a} + \hat{m}^{(\bar{\varepsilon})}_{ij,a}\right](\dot{\bar{\boldsymbol{\varepsilon}}})_{ij} - \left[\hat{b}^{(\bar{p})}_a + \hat{m}^{(\bar{p})}_a\right]\dot{\bar{p}} - \sum_i\left[\hat{b}^{(\bar{\zeta})}_{i,a} + \hat{m}^{(\bar{\zeta})}_{i,a}\right](\dot{\bar{\boldsymbol{\zeta}}})_i, \tag{14}$$

where $\hat{b}^{(\bar{\bullet})}_{\bullet,a} := b_{\square}(\hat{p}_a; \hat{\boldsymbol{u}}^{(\bar{\bullet})}_{\bullet})$ and $\hat{m}^{(\bar{\bullet})}_{\bullet,a} := m_{\square}(\hat{p}_a; \hat{p}^{(\bar{\bullet})}_{\bullet})$.

Altogether, the FE^2 problem of solving the macro-scale problem (4) and the RVE problem (6) in a nested fashion is reduced to solving the macro-scale problem (4) together with the evolution equation (12) "online" whilst the linear transient RVE problem (10) can be pre-computed in an "offline" phase.

We conclude this section with a numerical simulation of a 3D consolidation test. We assume a microstructure consisting of spherical inclusions embedded in a matrix material. Both, matrix and inclusions, are chosen poroelastic. An example RVE realization is displayed in Fig. 1. The poroelastic material parameters of the two phases (matrix/inclusions) are chosen as: $G = 3.1/6.1\,\text{GPa}$, $K = 3.7/7.4\,\text{GPa}$, $\alpha = 0.9075/0.8150$, $\beta = 0.06185/4.563\,\text{GPa}^{-1}$, $k = 10/100\,\text{mD(Pa\,s)}^{-1}$, where $1\,\text{mD} \approx 1\text{e-}12\text{m}^2$. After having pre-computed the linear transient RVE problem "offline" and, thereby, after having identified the relevant system matrices (13), we are able to solve the (reduced) FE^2 problem of a macro-scale consolidation experiment where the macro-scale computation domain is defined as a vertical column (length $\bar{L} = 20\,\text{m}$) under a vertical load. Solving (4) together with (12) "online" results in the transient settlement of the surface shown in Fig. 2. The computation times on a standard desktop computer are 1000 s for the RVE pre-computations ("offline") and 400 s for solving the reduced FE^2 macro-scale problem ("online") with 16 pressure modes.

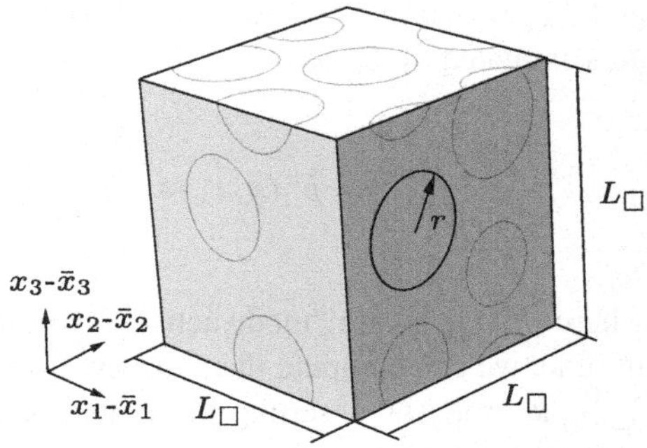

Fig. 1 Poroelastic RVE containing 8 spherical inclusions. $L_\square = 0.2\,\text{m}$, $r = 4.15\text{e-}2\,\text{m}$, $n = 0.3$

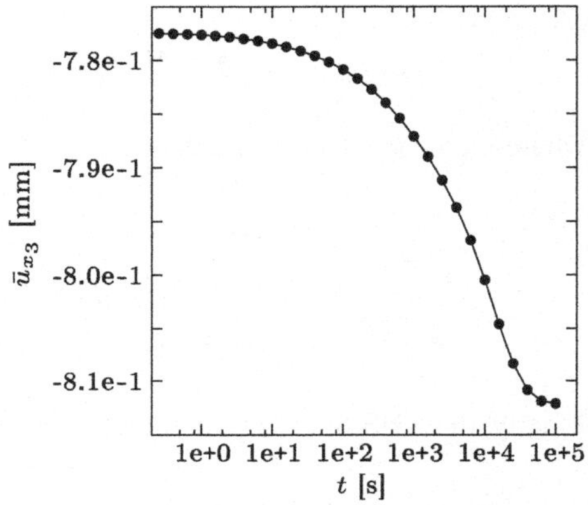

Fig. 2 Settlement (surface displacement) of a macro-scale consolidation test of a column with length $\bar{L} = 20\,\text{m}$

3 Estimation of the NMR Error for the RVE Problem

It is clear that the reduced solution is an approximation of the finite element solution, and the reduced basis is yet another source or error in addition to e.g. space- and time-discretization errors and modeling errors from the computational homogenization framework. In a two-scale setting this error will propagate between the scales and also affect the macroscopic solution. In order to control the accuracy of the solution it is of interest to estimate the error. A numerically efficient a posteriori error estimator for the pertinent RVE problem was presented in Ekre et al. [4], as a step towards a full two-scale error estimation algorithm. For simplicity the estimator was developed for the case of selective homogenization[1] and with homogeneous Dirichlet boundary conditions for the fluctuation fields. In addition, the estimator considers only the error stemming from NMR by assuming that the discrete, fully resolved, finite element solution is exact, i.e. $\tilde{p}^\mu = \tilde{p}^\mu_\text{h}$ where $\tilde{p}^\mu_\text{h}$ is the finite element solution. In what follows we summarize the core concepts of the estimator and refer to [4] for a thorough description.

[1] The pressure field lives solely on the microscopic scale, whereas the displacement field has both a macroscopic and a microscopic part, as usual.

The error equation for the error in pressure, $g(\boldsymbol{x},t) := \tilde{p}^{\mu}(\boldsymbol{x},t) - \tilde{p}^{\mu}_{\mathrm{R}}(\boldsymbol{x},t)$, is defined as

$$A_{\square}(g,q) = L(q) - A_{\square}(\tilde{p}^{\mu}_{\mathrm{R}},q) =: R_{\square}(q), \tag{15}$$

where $A_{\square}(\bullet,\star) = L_{\square}(\star)$ is (10b) integrated in time. Hence, the residual $R_{\square}(\bullet)$ is explicit in terms of the reduced approximation and the macroscale data. Ignoring space and time errors results in the Galerkin-type property $R_{\square}(\Pi_{\mathrm{R}}q) = 0$, where Π_{R} is the projection onto the reduced space.

In order to estimate the error in terms of predefined Quantities of Interest (QoI) we define a linear output functional. As example, consider time-averaged homogenized stress as the QoI:

$$Q_{\square}(q) = \frac{1}{|I|\,|\Omega_{\square}|}\int_I\int_{\Omega_{\square}} [\boldsymbol{e}_1 \otimes \boldsymbol{e}_1 : \mathbb{E} : \boldsymbol{\varepsilon}[\boldsymbol{u}\{q\}] - \alpha q]\,\mathrm{d}\Omega\,\mathrm{d}t. \tag{16}$$

The output functional is used to define the dual problem, i.e.

$$A^{\star}_{\square}(\tilde{p}^{\mu\star},q) = Q_{\square}(q), \tag{17}$$

where $A^{\star}_{\square}(\bullet,\bullet)$ is the dual of $A_{\square}(\bullet,\bullet)$ and $\tilde{p}^{\mu\star}$ is the dual solution. We can now introduce a reduced approximation and define an error equation with a corresponding residual, in a similar fashion as for the primal problem. Once again the residual is explicit in terms of macroscale data and the reduced solution.

To proceed with the estimate we define an auxiliary bilinear form, $\hat{A}_{\square}(\bullet,\bullet)$, with the key feature that it is symmetric and coercive, and it satisfies the condition $\hat{A}_{\square}(q,q) \le A_{\square}(q,q)$, for any field q. We may now introduce the auxiliary error equations,

$$\hat{A}_{\square}(\hat{g},q) = R_{\square}(q), \qquad \hat{A}_{\square}(\hat{g}^{\star},q) = R^{\star}_{\square}(q), \tag{18}$$

where a key feature is that $\hat{g}$ and $\hat{g}^{\star}$ can be solved for locally in time.

We note that $\hat{A}_{\square}(\bullet,\bullet)$ naturally defines a norm $\|\bullet\| = \sqrt{\hat{A}_{\square}(\bullet,\bullet)}$, and we can show that $\|g\| \le \|\hat{g}\|$ and $\|g^{\star}\| \le \|\hat{g}^{\star}\|$. Furthermore, the quantities $\|\hat{g}\|$ and $\|\hat{g}^{\star}\|$ can be bounded explicitly by the discrete residuals.

Finally, the explicit fully computable bound on the Quantity of Interest is obtained by combining the error representations $\hat{g}$ and $\hat{g}^{\star}$ using the bounding properties of the auxiliary form to obtain upper and lower estimates for the error in the Quantity of Interest, i.e. $E^{-}_{\mathrm{Q,est}} \le Q_{\square}(g) \le E^{+}_{\mathrm{Q,est}}$, where

$$E^{\pm}_{\mathrm{Q,est}} = \pm\frac{1}{4}\left[\int_I \frac{\|A(t)\|^2_{\mathrm{m}}}{\lambda}\,\mathrm{d}t + 2\,\|B_0\|^2_{\mathrm{m}} + 2\,\|C_{\mathrm{T}}\|^2_{\mathrm{m}}\right], \tag{19}$$

where $\|\bullet\|_{\mathrm{m}} := \sqrt{m_{\square}(\bullet,\bullet)}$, cf. (2b), and where $A(t)$, B_0, and C_{T} are explicit expressions in terms of macroscale data and the reduced solutions. λ is a computable eigenvalue from the spatially discretized problem.

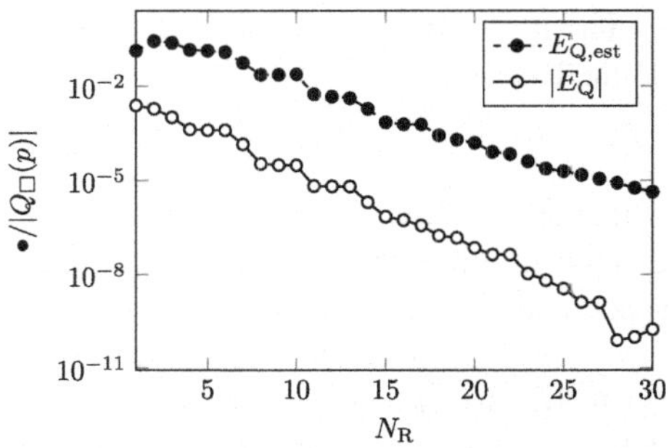

Fig. 3 Exact and estimated relative error in time averaged homogenized stress for the RVE problem using the first 30 POD modes

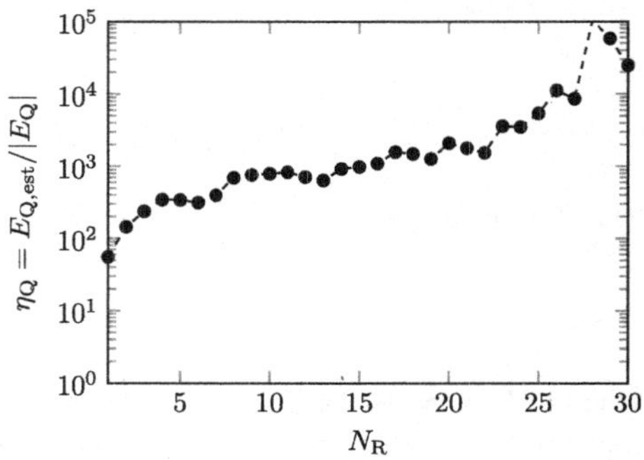

Fig. 4 Effectivity index for the estimator relating the estimated error to the exact error

To demonstrate the estimator, consider the RVE from Sect. 2. The macroscopic data is prescribed in terms of predefined macroscopic strain histories for $\boldsymbol{\varepsilon}_{11}$, $\boldsymbol{\varepsilon}_{22}$, and $\boldsymbol{\varepsilon}_{33}$, defined as in Fig. 10 in [4]. Time-averaged homogenized stress, $\boldsymbol{\sigma}_{11}$, is chosen for the Quantity of Interest, as exemplified in (16). In Fig. 3 the estimated relative error, $E_{\mathrm{Q,est}}$, and exact relative error, E_Q, in time-averaged homogenized stress is plotted for solutions with $N_\mathrm{R} = 1, 2, \ldots, 30$ POD modes. In Fig. 4 the corresponding effectivity index $\eta = E_{\mathrm{Q,est}}/\left|E_\mathrm{Q}\right|$ is plotted. We note that, while the estimate bounds the exact error from above, it overestimates the true error by orders of magnitude. However, what is of importance for a real application is the actual error which is indeed quite small even for moderate number of modes.

4 Conclusions

We have presented a method for reducing the RVE problem arising from computational homogenization of heterogeneous poro-elastic materials. The reduced basis, constructed using POD modes, transforms the RVE problem(s) to a system of linear evolution equations. Using a reduced basis introduces an extra source of errors to the model. As an initial step to a full two-scale error estimator we presented an estimator for the RVE problem with guaranteed bounds on the NMR error. Potential future developments include application to non-linear problems, and, for the estimator, a scheme for adaptively selecting the appropriate (number of) modes for each individual RVE problem.

References

1. Larsson, F., Runesson, K., Saroukhani, S., & Vafadari, R. (2011). Computational homogenization based on a weak format of micro-periodicity for RVE-problems. *Computer Methods in Applied Mechanics and Engineering*. https://doi.org/10.1016/j.cma.2010.06.023.
2. Temizer, I., & Wriggers, P. (2011). Homogenization in finite thermoelasticity. *Journal of the Mechanics and Physics of Solids*. https://doi.org/10.1016/j.jmps.2010.10.004.
3. Jänicke, R., Larsson, F., & Runesson, K. (2020). A poro-viscoelastic substitute model of fine-scale poroelasticity obtained from homogenization and numerical model reduction. *Computational Mechanics*. https://doi.org/10.1007/s00466-019-01808-x.
4. Ekre, F., Larsson, F., Runesson, K., & Jänicke, R. (2020). A posteriori error estimation for numerical model reduction in computational homogenization of porous media. *International Journal for Numerical Methods in Engineering*. https://doi.org/10.1002/nme.6504.

Perspectives on the Master-Master Contact Formulation

Alfredo Gay Neto

I know Professor Wriggers since 2013 when he supervised me during a post-doc period. Since that, we have started collaborating in academic works. His incredible competence and encouraging skills always inspired me to go further. Personally, he is a good friend and is extremely kind. I would like to express my congratulations for his 70th birthday and thank him for the numerous times he hosted me in Hannover, for fruitful periods of nice and fun research on computational contact mechanics.

Abstract This chapter provides an overview and perspectives of the master-master contact formulation, as a strategy to model pointwise contact between bodies. The formulation may be applied together with the Finite Element Method (FEM), the Discrete Element Method (DEM) and Multibody Dynamics (MBD). The Local Contact Problem (LCP) is presented and discussed, as well as its degeneration concept. An example of application is provided, such as current challenges and possible directions of research on the subject.

1 Introduction

Mathematical models involving contact between rigid or flexible bodies are of great interest in mechanical sciences and engineering. In literature, it is possible to identify distinct techniques to address such kind of problem, particularly by engineers, mathematicians and physicians. However, all share a common aspect that has to be included in the model, which is the main mechanical characteristic of the contact phenomenon: a constraint that does not permit penetration between the involved bodies.

A. Gay Neto (✉)
Polytechnic School at the University of São Paulo, São Paulo, Brazil
e-mail: alfredo.gay@usp.br

F. Aldakheel et al. (eds.), *Current Trends and Open Problems in Computational Mechanics*, https://doi.org/10.1007/978-3-030-87312-7_14

Let two bodies B_A and B_B be candidate to contact. In a given configuration, their external surfaces are Γ_A and Γ_B. In a scenario of contact occurrence, one may name Γ_c as the contact region (or contact patch), shared by subsets of material points from Γ_A and Γ_B. In Γ_c, one may find pairs of material points in Γ_A and Γ_B that define local contact interactions. The overall contact action is distributed on Γ_c, being an integral quantity.

1.1 Master-Slave Scheme

As usually Γ_c is unknown a priori, in order to handle contact occurrence in a numerical model one needs first to establish which are the pairs of material points on both surfaces Γ_A and Γ_B, which are candidate to contact. This can be done by the classical procedure called "master-slave", in which slave points $\boldsymbol{x}_{As}$ are elected in Γ_A. Their counterparts candidates in the so-called "master" surface Γ_B are here denoted by $\boldsymbol{x}_B$. One has to find, for each slave point, which is its counterpart in Γ_B. This is done by the solution of a Local Contact Problem (LCP), interpreted in this context as an orthogonal projection: a geometrical problem. The solution of the LCP for a given slave point is $\bar{\boldsymbol{x}}_B$. Let $\boldsymbol{n}_B$ be the external normal direction of the master surface, evaluated at $\bar{\boldsymbol{x}}_B$. With that, a gap quantity may be defined as:

$$g_n = (\boldsymbol{x}_{As} - \bar{\boldsymbol{x}}_B) \cdot \boldsymbol{n}_B. \tag{1}$$

The sign of g_n holds an important information: penetration occurrence (or not) of a given slave point onto the master surface. To detect the contact occurrence, one has to solve, for each configuration of interest, the LCP for a collection of slave points and, after, for each one, verify the condition given by $g_n \geq 0$. When contact takes place, one can represent the contact normal pressure on the interface by p_n. Finally, it is possible to write the Karush-Kuhn Tucker (KKT) conditions to characterize the frictionless contact mechanical constraint:

$$g_n \geq 0,\ p_n \leq 0,\ g_n p_n = 0. \tag{2}$$

Equation (2) may be interpreted as a condition of non-penetration. When the bodies are not in contact, KKT conditions are fulfilled because $g_n > 0$ and $p_n = 0$. When contact takes place, a contact traction $p_n \neq 0$ appears, but in this case $g_n = 0$. Thus, contact pressure may be interpreted as a contribution for each body's Neumann boundary condition, thus introducing a distributed load on Γ_c in order to fulfill the required mechanical constraint of non-penetrability.

1.2 Contact Contributions in a Numerical Model

In a numerical model, to address contact one has to (i) identify if contact occurs and (ii) include proper mechanical constraints in the model, in case of contact detection. For sake of simplicity we here discuss only the contact normal contributions. The following contact potential W_c may be defined:

$$W_c = \int_{\Gamma_c} \lambda_n g_n d\Gamma_c, \tag{3}$$

where λ_n is a Lagrange multiplier function defined on Γ_c. The first variation of W_c gives

$$\delta W_c = \int_{\Gamma_c} (\delta\lambda_n g_n + \lambda_n \delta g_n) \, d\Gamma_c, \tag{4}$$

which can be interpreted as a contact contribution, to be included together with other terms on the weak-form of a set of differential equations of motion of a multibody system. The assumptions regarding the flexibility of the bodies may introduce distinct strategies to obtain the equations of motion. One may find in [1, 2] a view from the FEM perspectives to model flexible bodies, while in [3, 4] the focus is given to rigid bodies.

In the context, $p_n = \lambda_n$ gives a natural interpretation of the Lagrange multiplier, from the physical point of view. A key point to describe the contact non-penetration constraint and its forthcoming contributions to the model lies on the definition of the gap quantity g_n, which is a result of a geometrical description, such as the master-slave scheme, and the solution of the corresponding LCP.

To obtain the integral terms in Eq. (4) one may find distinct techniques. A possibility is to sample slave points on Γ_A and attribute to each one an area of influence. Then, the integral term is approximated as a collection of pointwise contacts, whereas each one has a single LCP problem. This is referred in literature as the "node-to-segment" approach, since it is a natural way to handle the problem when modeling the bodies using the FEM. Each node or integration point, therefore, is a natural candidate to be a slave point. As examples of references, the reader may find in [5] some initial discussions on this strategy, while in [6] there are some discussions on the patch-test.

Another alternative is to keep the integral aspect on Eq. (4) and compute it choosing integration points as slave points. Then, the so-called surface-to-surface approaches are developed, and the Mortar Method schemes are of great interest (see e.g.: [7–10]).

Here we will not enter in the discussion of the methods that may be employed to fulfill contact constraints. The presentation is made with the Lagrange multiplier method, only. The reader finds alternatives, such as the penalty method, the augmented Lagrange method, among others, which are well described in [11].

2 Pointwise Contact

There are scenarios in which the surface patch Γ_c is small, when compared to geometric scales of interest in the model. Then, one may simplify Eq. (4) as follows:

$$\delta W_c = \delta\lambda_n g_n A + \lambda_n \delta g_n A, \tag{5}$$

where A is the area of the contact patch. This represents an homogenization of the contact traction, interpreted as a pointwise action. In our overview here presented we are restricted to the normal direction. The normal force magnitude is given by $|\lambda_n A_n|$. One may enhance Eq. (5) to include tangential (friction) terms by a similar procedure (see [11]).

2.1 Master-Master Scheme

In particular scenarios of pointwise contact, the mechanical representation of contact can be done by a single pair of action-reaction forces. Based on that, in these cases the idea of not choosing a priori a set of slave points on a surface and performing the master-slave procedure may be sought. Indeed, one may include on the LCP additional information, thus encompassing both surfaces nature as "masters" and seek for a pair of material points candidate to contact. With that, we abandon the intrinsic biasing of surfaces present on the master-slave scheme. The location of a pointwise mechanical action-reaction would be found in both Γ_A and Γ_B. This idea is the basis of the so-called master-master contact scheme.

This kind of strategy may be employed in a natural way when addressing the contact between beams in a FEM context, as proposed in [12, 13] and further enhanced by [14]. The topic is also addressed in [15, 16], among other contributions in literature. Beam self-contact with this kind of technique is done in [17].

The main drawback is the possibility of non-uniqueness of the LCP when one has parallel beams (or more generally, conformal contact conditions). When this is the case, the pointwise mechanical representation of contact loads would be better suited as a distributed load, as long as the action is broadly distributed. This may represent scenarios of a distributed contact action along a line (or a curve in space) or along an area (flat or curved).

When this is the case, alternative approaches based on the master-slave schemes appear as a way to replace the master-master approach. A prior choice of slave points would change the LCP and eliminate its non-uniqueness. The difficulty on that, however, lies on the fact that a given problem of interest may present a transition from pointwise to non-pointwise contact actions (or the opposite) along the model evolution. For such cases [18] introduced a model to provide a smooth change between both approaches and [19] created a scheme to add more points to a master-master-based contact description, when almost parallel beams appear. We may note that all

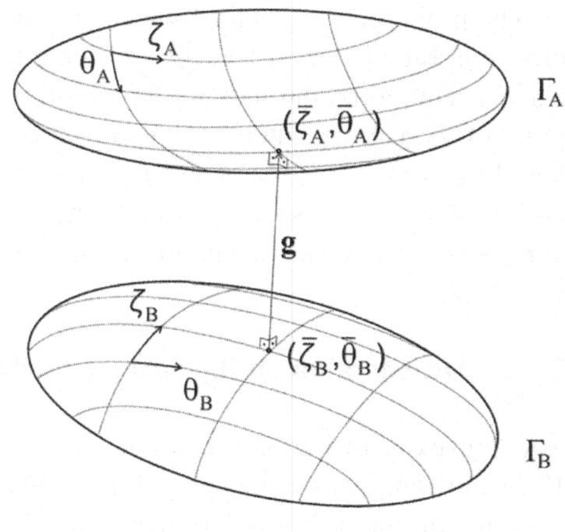

Fig. 1 Surfaces and the vector gap quantity definition. Adapted from [22]

the complexity of this kind of model lies on the LCP and in a proper definition for a gap quantity. Therefore, it is related to geometry. The concept of master-master contact degeneration is a way to circumvent this geometric problem (see Sect. 2.1.2).

2.1.1 Surface-Surface Approach

Based on seminal ideas of beam-to-beam contact, [20, 21] have proposed a surface-surface approach to handle the contact between beams, formulating the LCP as a 4-variable problem. Figure 1 shows two parameterized surfaces in a given configuration. One may define convective coordinates for both, organized as $\boldsymbol{c}_A = \left[\zeta_A \ \theta_A\right]$ and $\boldsymbol{c}_B = \left[\zeta_B \ \theta_B\right]$. It is interesting to consider that surfaces may deform. For that, one may introduce two vectors of general degrees of freedom (DOF) $\boldsymbol{d}_A$ and $\boldsymbol{d}_B$, which rule surfaces Γ_A and Γ_B, respectively. Next, one may define the vector gap quantity $\boldsymbol{g}$:

$$\boldsymbol{g} = \Gamma_A\left(\boldsymbol{c}_A, \boldsymbol{d}_A\right) - \Gamma_B\left(\boldsymbol{c}_B, \boldsymbol{d}_B\right) \tag{6}$$

It is convenient to introduce the vectors $\boldsymbol{c} = \left[\zeta_A \ \theta_A \ \zeta_B \ \theta_B\right]^T$ and $\boldsymbol{d} = \left[\boldsymbol{d}_A^T \ \boldsymbol{d}_B^T\right]^T$. With that, the LCP in this context is defined as: given a fixed $\boldsymbol{d}$, find the particular set of convective coordinates $\bar{\boldsymbol{c}} = \left[\bar{\zeta}_A \ \bar{\theta}_A \ \bar{\zeta}_B \ \bar{\theta}_B\right]^T$ such that the following relations hold:

$$\begin{bmatrix} \Gamma_{A,\zeta_A} \cdot \boldsymbol{g} \\ \Gamma_{A,\theta_A} \cdot \boldsymbol{g} \\ -\Gamma_{B,\zeta_B} \cdot \boldsymbol{g} \\ -\Gamma_{B,\theta_B} \cdot \boldsymbol{g} \end{bmatrix} = \boldsymbol{o}. \tag{7}$$

One may interpret Eq. (7) as a set of orthogonality relations. When all are fulfilled, one may illustrate the gap vector as in Fig. 1.

The proposed LCP can be viewed as an optimization problem. If one defines an objective function $f_1(\boldsymbol{c}) = \frac{1}{2}\boldsymbol{g} \cdot \boldsymbol{g}$ (see e.g.: [22]), the condition $\nabla f_1 = \boldsymbol{o}$ leads to Eq. (7). Therefore, the seek for this solution may be seen as a search for a stationary point of f_1. Unfortunately, not always the solution of interest is a minimum. When handling contact in practical problems, we have to identify overlapping surfaces and, in these cases, seek for a maximum penetration solution. This creates certain complexity when addressing this kind of problem numerically (see [23] for details).

2.1.2 Degeneration of the Surface-Surface Approach

The convexity of the optimization problem given by the LCP depends on Γ_A and Γ_B. Conformal surfaces approaching contact interaction represent an example that leads to non-uniqueness of the solution $\bar{\boldsymbol{c}}$. In these cases, one finds null eigenvalue(s) on the Hessian $\nabla^2 f_1$. This leads to a limitation in master-master surface-surface scheme. However, the possibility of degeneration of the LCP presented in [22, 24] is a possibility to remedy this drawback.

The idea of degeneration of the LCP is quite simple: instead of keeping the LCP as a 4-variable optimization problem, in certain cases of interest one may choose a priori fixed value(s) for selected convective coordinate(s), thus reducing the dimension of the LCP. As a simple case, the reader may think of a particular case, in which a prior choice is made fixing values to known $\bar{\zeta}_A$ and $\bar{\theta}_A$. With that, the LCP is re-defined, now keeping as variables only ζ_B and θ_B. Note that, geometrically, this proposed case recovers the master-slave procedure described in Sect. 1.1. However, this is only one of the main possibilities of degeneration. In [22], a degenerative operator $\boldsymbol{P}_s$ was proposed as follows:

$$\boldsymbol{c}_s = \boldsymbol{P}_s^T \boldsymbol{c}, \tag{8}$$

where $\boldsymbol{c}_s$ is the vector with degenerated convective coordinates, with dimension $s \in \mathbb{N} | 1 \leq s \leq 4$. The operator $\boldsymbol{P}_s$ not only includes a projection onto a possibly smaller space of convective coordinates, but a basis change. With a proper choice of degenerative operator, one may recover the curve-curve master-master scheme, here discussed in the context of beam-to-beam contact (Sect. 2.1). One obtains, in this case, a 2-variable LCP, in which a single convective coordinate is kept free for each contacting surface. It is also possible to create distinct contact geometrical descriptions, such as curve-surface interaction, leading to a 3-variable LCP, or point-curve interaction, which leads to a single-variable LCP.

We see LCP degeneration applicability mainly for two scenarios: (i) when non-uniqueness is expected and (ii) when a known singularity exists in a contact surface. The first represents a way to avoid facing the null eigenvalues of the LCP Hessian. When conformal contact is expected, a collection of material points may be chosen a priori as contact-candidate in a conformal contact locus. This naturally creates a distributed action composed of multiple pointwise loads, leading to a collection of master-slave schemes, recovering ideas of the node-to-segment approach for FEM. The second application permits usage of master-master surface-surface-based approaches in non-smooth surfaces. When a sharp edge or a non-smooth transition of normal direction occurs in a surface, one may abandon certain orthogonality relations in Eq. 7 by using the degenerative operator. With that, it is possible to handle cases of contact involving tips of surfaces with singularities, for example.

As an example to illustrate how LCP degeneration may be useful, we here show a simple scenario of contact involving domino blocks. Let two domino blocks be resting in the ground, as shown in Fig. 2a. Each block is represented by a single beam finite element, which external surface is plot in Fig. 2. Each lateral surface of a beam is parameterized as in [20]: an extruded super-elliptical cross-section surface, which is the region assumed as candidate to contact with the ground or with the other domino block. The ground is considered as flat and rigid.

The initial scenario is already problematic: the bottom surfaces of domino blocks are not included in the contact model and they touch the ground. The tip of each beam lateral surface, however, also touches the ground. To include this effect in the model and use only the lateral surfaces of the beams to define contacts, one can introduce the degeneration concept. The lateral surfaces of the beams are obviously non-orthogonal to the ground. Then, one may establish the contact between the ground and each beam lateral surface with degenerations on each surface by a pre-choice of material points exactly on each domino block bottom location, along the perimeter of that cross-section. This leads to a collection of master-slave LCP's, straightforwardly solved and leading to the set of pointwise contact actions as shown in Fig. 2a. Figure 2e shows a representation of friction in this scenario, which has actually null forces.

In a second moment, an external force pulls the right block towards the left one. The right block pivotes in the ground with a normal and friction re-distribution, as shown in Fig. 2b and f, respectively. In the sequence, a lateral impact occurs between the upper cross-section of the right block and the lateral of the left one. Figure 2c depicts the normal forces representation in a given time-frame during this interaction, and Fig. 2g shows the corresponding friction forces, much more pronounced. For this interaction we employed a degeneration on the right block beam lateral surface, thus fixing a single convective coordinate, associated with the top of the block.

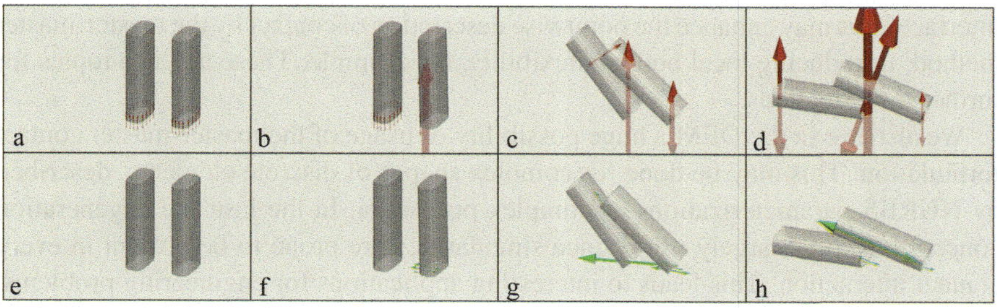

Fig. 2 Sequence of frames for the contact interaction between two domino blocks. Figure **a**, **b**, **c** and **d** shows arrows representing normal forces. Figure **e**, **f**, **g** and **h** shows arrows representing friction forces, where the size of each arrow is proportional to the force magnitude

However, the position along the perimeter of the cross section of an expected pointwise interaction was set free, which is naturally chosen along the model evolution by a sequence of 3-variable LCP's, solved as long as the time-integration and contact detection evolve.

Finally, at the end, the bottom of the left block touches the lateral surface of the right one, leading to final configuration (static), whose normal forces are represented in Fig. 2d and friction in Fig. 2h. The degeneration treatment for this new contact leads to a similar 3-variable LCP, as previously commented.

This simple example is full of singularities and non-uniqueness scenarios, and shows ideas on how the LCP degeneration in the context of surface-surface master-master formulation may be useful. All the contacts herein described were addressed by a single numerical implementation, only employing distinct degenerative operators, in each case. This is quite convenient and avoids separate treatments for distinct geometrical scenarios: many cases are embedded in the same contact formulation.

3 Challenges for Future Research

We see many interesting perspectives for future research on the master-master contact formulation. As we already discussed in this chapter, the main applications are found in cases where a contact pointwise description is natural. This includes many scenarios of beam-to-beam contact. However, there are practical cases of conformal contact, which may be addressed by the degeneration of the LCP. Here the main challenge is to automatically identify these situations and degenerate the LCP, accordingly. This would lead to a general beam-to-beam formulation based on a surface-surface approach, which also may be used in the context of rigid-bodies.

In [25] some applications of the master-master techniques were done in multibody rigid-flexible dynamics context. Particularly, recent applications involving wheel-rail contact may be found in [26]. The well-known algorithms in multibody-dynamics field for representing rolling with creepage scenarios may be tested together with the master-master approach (see e.g.: [3] for a broad view of these creep constitutive laws in a computational framework). Other local constitutive tangential or normal interface laws may enhance the pointwise description of contact by the master-master method, introducing local bodies' flexibility, for example. These are also topics for further investigations.

We also see in the DEM a huge possibility of usage of the master-master contact formulation. This may be done for complex shapes of discrete elements, described by NURBS parameterizations or complex polyhedra. In the last, the degeneration concept has to be largely used, since singularities are prone to be present in every contact interaction. This leads to interesting applications for engineering problems, such as granular materials with complex grain shapes, such as sand or railway ballast, to give some examples. The detailed shape description for granular particles is a need when aiming at reproducing realistic macroscale properties of some materials, such as sand. In this context, the surface-surface approach of the master-master formulation together with its degenerations may be extremely helpful as proposed in the work [27].

Furthermore, the solution and characterization of the LCP and its degenerations still claim an in-depth mathematical discussion. Therefore, we see here also space for developments in the basis of the mathematical formulation, which can give interesting insights also for further novel applications.

Acknowledgements This study was financed by Alexander von Humboldt Foundation and in part by the Coordenação de Aperfeiçoamento de Pessoal de Nível Superior - Brasil (CAPES) - Finance Code 001.

References

1. Wriggers, P. (2008). *Nonlinear Finite Element Methods*. Berlin Heidelberg: Springer.
2. Bonet, J., & Wood, R. (2008). *Nonlinear Continuum Mechanics for Finite Element Analysis* (2nd ed.). Cambridge: Cambridge University Press.
3. Shabana, A. (1998). *Dynamics of Multibody Systems*. Cambridge: Cambridge University Press.
4. Bauchau, O. A. (2011). *Flexible Multibody Dynamics*. Berlin: Springer.
5. Wriggers, P., Van, T. V., & Stein, E. (1990). Finite-element-formulation of large deformation impact-contact-problems with friction. *Computers and Structures, 37*, 319–333.
6. Zavarise, G., & De Lorenzis, L. (2009). A modified node-to-segment algorithm passing the contact patch test. *International Journal for Numerical Methods in Engineering, 79*, 379–416.
7. Puso, M. A. (2004). A 3D mortar method for solid mechanics. *International Journal for Numerical Methods in Engineering, 59*(3), 315–336.
8. Puso, M. A., & Laursen, T. A. (2004). A mortar segment-to-segment contact method for large deformation solid mechanics. *Computer Methods in Applied Mechanics and Engineering, 193*, 601–629.
9. Fischer, K. A., & Wriggers, P. (2005). Frictionless 2D contact formulations for finite deformations based on the mortar method. *Computational Mechanics, 36*, 226–244.
10. Popp, A., Seitz, A., Gee, M., & Wall, W. (2013). A dual mortar approach for improved robustness and consistency of 3D contact algorithms. *Computer Methods in Applied Mechanics and Engineering, 264*, 67–80.
11. Wriggers, P. (2002). *Computational Contact Mechanics*. West Sussex: Wiley.
12. Wriggers, P., & Zavarise, G. (1997). On contact between three-dimensional beams undergoing large deflections. *Communications in Numerical Methods in Engineering, 13*, 429–438.
13. Zavarise, G., & Wriggers, P. (2000). Contact with friction between beams in 3-d space. *International Journal for Numerical Methods in Engineering, 49*, 977–1006.
14. Litewka, P., & Wriggers, P. (2002). Friction contact between 3D beams. *Computational Mechanics, 28*, 26–39.
15. Konyukhov, A., & Schweizerhof, K. (2010). Geometrically exact covariant approach for contact between curves. *Computer Methods in Applied Mechanics and Engineering, 199*, 2510–2531.
16. Meier, C., Popp, A., & Wall, W. A. (2016). A finite element approach for the line-to-line contact interaction of thin beams with arbitrary orientation. *Computer Methods in Applied Mechanics and Engineering, 308*, 377–413.
17. Gay Neto, A., Pimenta, P. M., & Wriggers, P. (2014). Self-contact modeling on beams experiencing loop formation. *Computational Mechanics, 55*, 193–208.
18. Meier, C., Wall, W. A., & Popp, A. (2017). A unified approach for beam-to-beam contact. *Computer Methods in Applied Mechanics and Engineering, 315*(1), 972–1010.
19. Litewka, P. (2013). Enhanced multiple-point beam-to-beam frictionless contact finite element. *Computational Mechanics, 52*, 1365–1380.

20. Gay Neto, A., Pimenta, P., & Wriggers, P. (2016). A master-surface to master-surface formulation for beam to beam contact. part i: frictionless interaction. *Computer Methods in Applied Mechanics and Engineering, 303*, 400–429.
21. Gay Neto, A., Pimenta, P., & Wriggers, P. (2017). A master-surface to master-surface formulation for beam to beam contact. part ii: frictional interaction. *Computer Methods in Applied Mechanics and Engineering, 319*, 146–174.
22. Gay Neto, A., & Wriggers, P. (2019). Computing pointwise contact between bodies: a class of formulations based on master master approach. *Computational Mechanics, 64*(3), 585–609.
23. Gay Neto, A., & Wriggers, P. (2020). Numerical method for solution of pointwise contact between surfaces. *Computer Methods in Applied Mechanics and Engineering, 365*(15), 112971.
24. Gay Neto, A., & Wriggers, P. (2020). Master-master frictional contact and applications for beam-shell interaction. *Computational Mechanics*.
25. de Campos, P., & Gay Neto, A. (2018). Rigid Body formulation in a finite element context with contact interaction. *Computational Mechanics*.
26. Higa, D.N., Kina, E.J., & Gay Neto, A. (2020). Wheelset-rail mechanical model for a steady-state dynamic condition and prediction of rolling contact fatigue locci. Vehicle System Dynamics.
27. Gay Neto, A., Wriggers, P. (2021). Discrete element model for general Polyhedra. *Computational Particle Mechanics*. Available at: https://link.springer.com/article/10.1007%2Fs40571-021-00415-z

Remarks on the History of Glacier Research and the Flow Law of Ice

Dietmar Gross

I've known Peter Wriggers personally since the late 1980s. Our relationship became particularly intense after he was appointed to the TU Darmstadt in 1990. At the Institute of Mechanics, Wriggers and I formed a joint working group, in which there was a climate, both scientifically and personally, that can hardly be imagined better. The joint activities as textbook authors go back to this time. Our relationships as colleagues, authors and friends survived his move to Hannover in 1998 uninfluenced and they will undoubtedly continue in the future. The special relationship with the Mechanics in Darmstadt is underlined by the honorary doctorate that my university awarded him in 2015.

Abstract The paper provides a short review on the behavior of glacier ice from the beginnings in the 17th century via the formulation of a flow law in the fifties of the forgoing century up to the present. It is based in part on the recent extended review by Hutter and Gross [1], but has been supplemented by additional aspects. It focuses on the macroscopic constitutive description in the framework of classical continuum mechanics while micromechanical aspects are only touched in passing.

1 Beginnings of Glacier Ice Research

The science of the physics of ice crystals started almost 500 years ago with Cardano, followed by Kepler, Hooke, Dalton, the two Braggs and Pauling. Here, we are concerned with naturally formed ice of glaciers, ice sheets and ice shelves which cover more than 10% of earth's land mass. Glaciers as natural phenomena at high altitudes of mountainous regions were attractive, if simply for their mystic appearance. Prior to the 17th century, not even the deformability of glacier ice was recognized, certainly not admitted. For example, Johann Moraltus (1645–1733) postulated in 1669

D. Gross (✉)
Division of Solid Mechanics, TU Darmstadt, Darmstadt, Germany
e-mail: gross@mechanik.tu-darmstadt.de

F. Aldakheel et al. (eds.), *Current Trends and Open Problems in Computational Mechanics*, https://doi.org/10.1007/978-3-030-87312-7_15

Glaciers to be rigid objects. Only valley inhabitants close to them knew it better and accepted some deformability.

With this background in the year 1705, Johann Jakob Scheuchzer (1672–1733) visited Swiss Glaciers and proposed a theory on their motion. He knew from physics that water is expanding in the freezing process to ice, and that the force of expansion is so large that cartridge rounds which are filled with water that freezes, are blasted into pieces. Scheuchzer assumed that the water in glacier fissures and crevasses that freezes would extend with such excessive power that its force will unquestionably push the glacier downward. This concept, often called *Dilatation Theory*, was later adopted by Jean de Charpentier, Louis Agassiz and others. Basic thought of this concept was the belief that glaciers are permanent storehouses of coldness, capable to freeze all water that percolates through them.

About in the year 1760, Altmann and Grüner brought forward their opinion that glaciers would move by means of sliding along their beds. Almost 40 years later, Horace Bénédict de Saussure (1740–1799) revived this concept, which became the so-called *de Saussure Theory*.

The 18th century also brought the first ideas about the deformability of ice. Simply the concept of sliding should have brought the science mountaineers to the postulation of the deformability of glacier ice. Yet, none of the above mentioned scientists attributes the terms viscosity and kneadability to characterize the deformability of the ice, even though the appearances of many glaciers suggest these terminologies, if it were not so contradictory to any daily experience with brittle ice. In 1773 André Bordier from Geneva compared glacier ice with mollified wax and mentioned for the first time its flexibility and extensibility. However, Bordiers concepts were unheard in the natural scientific community in the1770 s. They were reborn more than 60 years later by Louis Rendu, the later bishop of Annecy (Fig. 1). In 1840 he submitted his Théorie des Glaciers de la Savoie to the Royal Academy of Savoy where he states among others, [2]:

– Between Mer de Glace and a river, there exists such a perfect similarity that it is impossible to find a circumstance in a glacier which would not equally occur in a

Fig. 1 **a** Louis Rendu, **b** James David Forbes, **c** John Tyndall

river.
– There exists a large set of facts, which seem to force us to believe that glacier ice possesses some sort of extensibility, which allows it to adjust to the local circumstances, to thin, to swell and to contract as if it were a soft dough.

Of course, at this time, a characterization of viscous or plastic behavior of glacier ice could not yet be phrased in terms of a constitutive relation.

2 First Measurements and Link with Young Thermodynamics

First rough measurements of the flow velocity of a glacier were done by Franz Josef Hugi (1791–1855) who spent in 1830–32 a lot of time on the Unteraar-Glacier. The measurements were continued by Louis Agassiz (1807–1873) who periodically undertook expeditions to this glacier since 1838. Here he set flow markers and drilled holes to get some information about the velocity and the temperature distribution. In 1841, as a visitor of Agassiz, the physicist James Forbes (1809–1868) from Edinburgh observed these measurements and heard the first time about Rendu's theory. One year later, in 1842, Forbes started his own, precise measurements with theodolites at the Mer de Glace near Chamonix. Already in the same year, on basis of his observations and data, he set the cornerstones of his *viscous or plastic theory of glaciers* which he complemented by additional measurements in the following years until 1845 [3]. In his theory he stated that a glacier essentially is a viscous (plastic) 'semirigid' flowing mass containing a crack and vein-structure whose fluidity depends on parameters like temperature, water contents and so on, [4]. In his numerous publications on glaciers since 1843 he gave credit also to Louis Rendu who, on request, had sent him his treatise.

Approximately at the same time the basic laws of thermodynamics have been found. Though Carnot's principle of 1824 initially was ignored, its importance was recognized by William Thomson (the later Lord Kelvin) in conjunction with James Prescott Joule's work on the mechanical equivalent of heat of 1850. In 1851 W. Thomson reconciled both principles, postulating the 1st and 2nd law of thermodynamics. It this context, a number of other scientists have to be mentioned, among them Robert Mayer, Rudolf Clausius and Hermann von Helmholtz. The behaviour of ice was an essential issue in this context. In fact, Thomson's seminal paper starts with the phenomenon that two ice blocks consolidate into one after inducing surface melting by rubbing them together. From then on, melting or freezing of ice and the young thermodynamics were closely linked. Melting of ice under pressure and refreezing when pressure is reduced was later named by John Tyndall (1820–1893) as *regelation*; it was unknown to James Forbes until mid-1850.

In 1855, John Tyndall started working on ice and glaciers. In the laboratory he observed the deformability of ice under pressure, crack formation, fracture and regelation. In addition, during a short visit to the Alps in 1856, which in subsequent years was followed by longer stays, he made his own impression of glaciers. This was enough for him to attack Forbes in early 1857 and to completely put in question his viscous glacier theory [5]. A main issue of the attack was the term 'viscous' which never was clearly defined by Forbes. For Tyndall this term described the property of a fluid that could permanently be stretched under tension without fracture what was obviously not the case for glacier ice. Contrary, he explained the 'apparent quasi-viscous' behavior through continuous cleavage and refreezing, i.e. through regelation.

From today's perspective, both properties, viscous flow, micro and macro-cracking combined with regelation contribute to the movement of a glacier, where viscous flow is clearly the dominating mechanism. At that time, however, the harsh attack did not lead to an objective scientific debate, but to the so-called 'Great Glacier Controversy' between two parties which lasted until 1875. It was partly led by inexpert means and aroused public interest due to its heat.

3 Creep Law for Ice

At the mid 19-hundreds further progress in developing a physically based theory for the motion of large ice masses was hampered by the nonexistence of a material theory, coupled with the basic physical laws, and by the primitive state of experimental techniques for the determination of the constitutive relations. Nevertheless, the behavior of ice has been further investigated primarily on the laboratory scale. Towards the end of the century, it was well known that ice is a polycrystalline material and that the single crystal has a hexagonal structure. McConnel and Kidd showed around 1890 that ice under tension or bending can be severely plastically deformed if this is done sufficiently slowly, see e.g. [6]. They also showed that a single crystal deforms by slip on the basal planes. Further works in the first half of the 20th century, such as Höppler [7] or Perutz and Seligman [8] dealt in particular with the viscosity (plasticity) of ice and its dependence on temperature, crystal orientation, etc. However, the final step to a macroscopic flow law did not take place; probably because, unlike as in engineering for metallic materials. no need was seen for it yet.

The situation changed in the mid-50s of the previous century. In 1947 the all-rounder Max Perutz, 1962 Nobel Prize Winner in Chemistry, wrote a report on the state of research on the flow of glaciers, in which he called for the experimental determination of the creep rate of ice near 0 °C as a function of shear stress, crystal size, crystal orientation and hydrostatic pressure [9]. Perutz was also the driving scientist who, in 1949, brought together the Cambridge glaciologists, rheologists and metallurgists, including Egon Orowan as prominent scientist, for a joint meeting on *The Flow of Ice and of other Solids*; it can be regarded as the starting point of intensified glacier ice research. Independently of the requested laboratory experiments, Perutz, J. Gerrard and A. Roch, using a 130 m long vertical steel tube, carried out since 1948 for over 3 years direct measurements of the velocity profile of the glacier at the Jungfraujoch, see e.g. [10]. Their results in 1950 suggested a nonlinear relationship between shear strain rate $\dot{\gamma}$ and shear stress τ in the form

$$\dot{\gamma} = k\,\tau^{2.3} \tag{1}$$

where the stress exponent in the final publication of 1952 was corrected to $n \approx 1.5 \ldots 4$. To the author's best knowledge, this was the first time that a power law was proposed to describe the macroscopic behavior of ice. Furthermore, the results of these measurements did not support the 'extrusion theory' according to Streiff-Becker, which assumed faster moving domains in the interior of the glacier. Also the results of rigid-plastic ice models, which were still favored by Orowan and Nye at that time, became questionable. In this context it should be mentioned that Norton's power law of 1929 for creep of metals in the higher temperature regime was well known in the metal community but, though Orowan certainly knew this law, it had not yet found its application to ice.

Laboratory creep tests on isotropic polycrystalline ice were done first under supervision of Perutz and Orowan in Cambridge by John Glen for uniaxial compression (1952, 1953, 1955, 1974) and fully independently in Switzerland by Samuel Steinemann for uniaxial compression/tension and torsional shear (1954, 1956, 1958), see e.g. [11, 12]. As a result of their tests both scientists proposed for the strain rate $\dot{\varepsilon}$ in a certain stress range the (uniaxial) power law

$$\dot{\varepsilon} = \operatorname{sign}(\sigma)\, f(T)\, |\sigma|^n \qquad \text{with} \qquad n = 2 \ldots 4\,. \tag{2}$$

Here the function $f(T)$ describes the significant dependence on the temperature T, which for temperatures distant from the melting point was approximated by the Arrhenius-type relation $f(T) = K \exp(-Q/RT)$. For a fixed Temperature T and a restricted stress or deformation range the exponent n is regarded as constant, but in general it depends on these quantities: $n = n(T, \sigma)$. Note that for the strain rate (velocity gradient) the notation $\dot{\varepsilon}$ is used, which should not be confused with the time derivative of the infinitesimal strain ε.

To make the power law (2) applicable to more general two- or three-dimensional stress states in ice, John Nye [13] generalized it, by assuming that the material behavior is isotropic, incompressible, independent of the 3rd invariant of the stress deviator and that the stresses and strain rates are co-axial:

$$\dot{\varepsilon}_{ij} = \dot{e}_{ij} = \frac{3}{2}\frac{\dot{\varepsilon}_{\mathrm{e}}}{\sigma_{\mathrm{e}}}\, s_{ij} = \frac{3}{2}\, f(T)\, \sigma_{\mathrm{e}}^{n-1}\, s_{ij}\,. \tag{3}$$

In (3) the stresses σ_{ij} and strain rates $\dot{\varepsilon}_{ij}$ are linked with the respective deviatoric quantities s_{ij} and $\dot{e}_{ij}$ by $s_{ij} = \sigma_{ij} - (\sigma_{kk}/3)\,\delta_{ij}$ and $e_{ij} = \varepsilon_{ij} - (\varepsilon_{kk}/3)\,\delta_{ij}$ while the equivalent stress σ_{e} and strain rate $\dot{\varepsilon}_{\mathrm{e}}$ are defined as $\sigma_{\mathrm{e}}^2 = \frac{3}{2}\, s_{ij}s_{ij} = 3\, II_s$ and $\dot{\varepsilon}_{\mathrm{e}}^2 = \frac{2}{3}\, \dot{e}_{ij}\dot{e}_{ij} = \frac{4}{3}\, II_{\dot{e}}$ with II_s and $II_{\dot{e}}$ being the 2nd invariants of the stress and strain rate deviators. For the sake of completeness, we give the 3rd invariant of the stress deviator as well as the 'effective' and 'octahedral' quantities, which frequently are used as alternatives: $III_s = \det(\mathbf{s})$, $\tau_{eff}^2 = II_s = \frac{3}{2}\,\tau_{oct}^2$, $\dot{\varepsilon}_{eff}^2 = II_{\dot{e}} = \frac{3}{2}\,(\dot{\gamma}_{oct}/2)^2$.

Glen's and Steinemann's investigations were supplemented in the following decades by various other tests under different loading conditions, most of which, however, adhered to the structure of the material law (3), for references see e.g. [1]. For example, to improve the constitutive law, the power-function was replaced by a sinh-function or by polynomials better matching the experiments. Furthermore, to avoid the singularity for the viscosity, associated with the power law, a constant residual stress has been incorporated.

One important issue was recognized early. While Glen's experiments essentially covered only the strain range $\varepsilon < 10\,\%$, where secondary creep with minimum strain rates prevails, Steinemann investigated also the strain range $\varepsilon > 10\,\%$, i.e. tertiary creep with accelerated strain rates, and he emphasized its relevance. The increased strain rates can be explained by recrystallisation and texture formation or, in other words, by deformation induced anisotropy. The flow data for secondary creep are relatively easy to determine. To apply these data also for tertiary creep, the experimentalists, ignoring the anisotropy, simply introduced in the late 1970th a multiplicative strain rate 'enhancement factor' E into the power law (3), see e.g. [14]. It is defined as $E = \dot{\varepsilon}_{ter}/\dot{\varepsilon}_{min}$ where the strain rates are measured at constant temperature and stress at an initially isotropic ice, i.e. in general E depends (indirectly) on T and σ. In the context of this rough simplification it is worth to be mentioned that large polar ice masses are in the state of tertiary creep. Introducing E, the power law (3) nowadays is often written in the form

$$\dot{\varepsilon}_{ij} = E\,A\,II_s^{(n-1)/2}\,s_{ij} \qquad \text{with} \qquad A = A_0 \exp(-Q/RT')\,, \tag{4}$$

where A_0, R and T' are a constant, the universal gas constant and the temperature difference to the pressure melting point, respectively. The activation energy Q depends on the temperature (cold and warm ice) and also n depends on the temperature and the stress state but is usually set to $n = 3$ by default.

Due to the dependence of the numerous parameters on temperature, stress or deformation state and rates, it is hardly surprising that the flow data used by different groups vary widely and thus lead to uncertainties, especially in long-term forecasts, see [1, 15].

4 Beyond the Power Law

In the research of large polar ice sheets or glaciers Glen's (or Norton's) flow law (4) is extensively used, sometimes in slightly modified form, and it is implemented in nearly all numerical simulation tools. Reasons for this are its simplicity and the assumption or belief that this law captures the essential macroscopic properties of the material. Nevertheless, there is evidence that this constitutive law has obvious weaknesses and should be improved.

One weak point concerns the representation of the non-linear flow behavior. It is well known from the material theory of incompressible, isotropic creeping fluids that the general form of the temperature dependent constitutive law can alternatively be written as

$$\dot{\mathbf{e}} = \psi_1\,\mathbf{s} + \psi_2\,[\mathbf{s}^2 - \tfrac{2}{3}\,II_s\,\mathbf{1}] \quad \text{or} \quad \dot{\mathbf{s}} = \phi_1\,\mathbf{e} + \phi_2\,[\mathbf{e}^2 - \tfrac{2}{3}\,II_e\,\mathbf{1}]$$
$$\text{with} \quad \psi_{1,2} = \psi_{1,2}(II_s, III_s, T)\,, \quad \phi_{1,2} = \phi_{1,2}(II_e, III_e, T)\,, \tag{5}$$

where $\dot{\mathbf{e}}$ and $\mathbf{s}$ are the strain rate and stress deviators and $II_{s,e}$ and $III_{s,e}$ the 2nd and 3rd invariants of the stress and strain rate deviators; $\mathbf{1}$ is the unit tensor. Such a fluid is called a *Reiner-Riwlin*[1] *fluid*. It can be shown, that (5_1) consistently reduces to

$$\dot{\mathbf{e}} = \psi\,\mathbf{s} \quad \text{with} \quad \psi = \psi(II_s, T) \tag{6}$$

if independence of the 3rd invariant III_s is postulated.

If ψ in (6) is chosen as $\psi = EA[k + II_s^{(n-1)/2}] = EA[k + \tau_{eff}^{(n-1)}]$, we obtain as a special case Glen's law (4), supplemented by a constant k to assure a finite viscosity at zero stress. Smith and Morland (1981) have proposed the choice $\psi = EA[k + \sum_0^j c_i\,\tau_{eff}^{2i}]$ (with $j = 2$ or $j = 3$) which enables better adaption to the experimental data.

Independent of the choice of ψ in (6) remains the question whether the assumed independence from III_s and the co-axiality of deviatoric stresses and strain rates, indicated by (6), are justified. These questions were asked very early on, as e.g. by Glen (1958); Hutter. Morland or Baker, but they have not been sufficiently investigated and mostly pushed aside by applied glaciologists, see e.g. [1, 16]. For example, an assumed validity of (6) or (4) implies that, independent of the specific loading, all results (datapoints) of $\dot{\gamma}_{oct}$ versus τ_{oct} tests at same temperature should lie in a log-log plot on one and the same straight line, whose slope is given by the exponent n. That Steinemann's test results for combined pressure-shear test do not meet this condition was already shown by Glen 1958. But this discrepancy can also be found in most recent works as e.g. in Treverrov et al. [14] where for isotropic ice at −2 °C the exponents for uniaxial compression and for simple shear were determined as $n_c = 3.3 \pm 0.3$ and $n_s = 2.9 \pm 0.3$, respectively. Note that for uniaxial compression we have $III_s = -2p^3/27 = \frac{2}{\sqrt{27}}\,(II_s)^{3/2}$ while for simple shear $III_s = 0$ holds. In any case, a review of the available experimental results shows that the co-axiality and/or III_s-independence assumption does not hold in general [1].

The experimental determination of the two material functions $\psi_{1,2}$ or $\phi_{1,2}$ in (5) is not possible from single tests as uniaxial compression or simple shear, since in these cases III_s is not independent of II_s or III_s is zero. However, in principle it is possible to determine e.g. $\phi_{1,2}$ from combined pressure-shear tests with an independently prescribed pressure $\sigma_{33} = -\sigma$ and a shear stress $\sigma_{13} = \tau$. For given test conditions ('Melbourne experiments') $\dot{\varepsilon}_{11} = 0$ (confined deformation) and $\sigma_{22} = 0$ (unconfined deformation), the rates $\dot{\varepsilon}_{33} = -\dot{\varepsilon}$ and $\dot{\varepsilon}_{13} = \dot{\gamma} = \dot{\kappa}/2\lambda_3$ should be measured; additionally, if desired, also σ_{11} can be measured. Here κ is the tangent of the shear angle and λ_3 the stretch in 3-direction. Without going in the details of the

[1] Regarding Riwlin see remark in [1].

calculation, the material functions under these circumstances follow from, [1, 17]

$$\phi_1 = \frac{\dot{\bar{\gamma}}\,\tau - \dot{\bar{\varepsilon}}\,\sigma}{\dot{\bar{\gamma}}^2 - 2\dot{\bar{\varepsilon}}^2}, \qquad \phi_2 = \frac{2\dot{\bar{\varepsilon}}\,\tau - \dot{\bar{\gamma}}\,\sigma}{\dot{\bar{\gamma}}(\dot{\bar{\gamma}}^2 - 2\dot{\bar{\varepsilon}}^2)}. \tag{7}$$

In addition, the universal relationship $\dot{\bar{\gamma}}(\sigma_{11} - \sigma) = \dot{\bar{\varepsilon}}\,\tau$ must apply, which can serve as a check. The bars over the symbols indicate melt point normalised quantities. Unfortunately, there are currently no corresponding test results available which allow a complete determination of the two material functions. Existing test results allow also no clear statement with respect to the dependence on III_s but they indicate a non-coaxiality of deviatoric stresses and strain rates [17].

The statements made here are limited to isotropic ice, i.e. to secondary creep. A corresponding restricted extension to tertiary creep with anisotropy evolution has recently been proposed by Morland and Staroszczyk (2020) [18] but will not discussed here. It should only be emphasized again that the characterization of tertiary creep using a physically unclear enhancement factor is unsatisfactory from the continuum-mechanical perspective.

In this brief overview, only the macroscopic flow of glacier ice has been addressed. Short-time processes like calving or the formation and propagation of cracks or rifts are also important on this scale. For this purpose visco-elastic constitutive laws are needed that combine the brittle-elastic with the viscous behavior of ice. Finally, it should be mentioned that ice is currently mostly only modeled on one scale, either the macro or the microscale. However, a scale transition and/or a two-scale modeling would be advantageous for processes in which significant changes in the microstructure take place.

Acknowledgements The author is grateful to Professor Angelika Humbert for helpful comments.

References

1. Hutter, K., & Gross, D. (2019). A historical tour of glacier ice on Earth and its role in climate dynamics. *Journal of Earth and Environmental Sciences, 7*, 174.
2. le Chanoine Rendu, M. (1840). Theorie des Glaciers de la Savoie, Chambery.
3. Forbes, J. D. (1843). An attempt to explain the leading phenomena of glaciers. *Edinburgh New Philosophical Journal, 35*, 221–252.
4. Rowlinson, J. S. (1971). The theory of glaciers. *Notes and Records of the Royal Society of London, 26*, 189–202.
5. Tyndall, J., & Huxley, T. H. (1857). On the structure and motion of glaciers. *Philosophical Transactions of the Royal Society London, 147*, 327–346.
6. McConnel, Jc., & Kidd, D. A. (1888). On the plasticity of glacier and other ice. *Proceedings of the Royal Society London, 44*, 331–367.
7. Höppler, F. (1941). Die Plastizität des Eises. *Kolloid-Zeitschrift, 97*, 154–160.
8. Perutz, M. F., & Seligman, G. (1939). A crystallographic investigation of glacier structure and the mechanism of glacie flow. *Proceedings of the Royal Society London, A172*, 335–360.
9. Perutz, M. F. (1947). Report on problems relating to the flow of glaciers. *Journal of Glaciology, 1*, 47–51.

10. Perutz, M. F. (1950). Glaciology-the flow of glaciers. *The Observatory, 70,* 63–69.
11. Glen, J. W. (1953). The creep of polycrystalline ice. *Proceedings of the Royal Society London, A228,* 519–538.
12. Steinemann, S. (1958). Experimentelle Untersuchungen zur Plastitzität von Eis, Beiträge zur Geologie der Schweiz, Hydrologie, No. 10.
13. Nye, J. F. (1953). The flow law of ice from experiments in glacier tunnels, laboratory experiments and the Jungfraufirn borehole experiment. *Proceedings of the Royal Society London, 219,* 477–489.
14. Treverrov, A., et al. (2012). The tertiary creep of polycrystalline ice: experimental evidence for stress-dependent levels of strain-rate enhancement. *Journal of Glaciology, 58,* 301–314.
15. Zeitz, M. et al. Sensitivity of ice flow to uncertainty in flow law parameters in an idealized one-dimensional geometry, The Cryosphere, Discuss. https://doi.org/10.5194/tc-2020-79
16. Baker, R. Is creep of ice really independent of the third deviatoric stress invariant? The Physical Basis of Ice Sheet Modelling (Proceedings of the Vancouver Symposium, August 1987), 7-16, IAHS Publ. no. 170.
17. Morland, J. W., & Staroszczyk, R. (2019). The viscous relation for the initial isotropic response of ice. *Cold Regions Science and Technology, 162,* 11–18.
18. Morland, J. W., & Staroszczyk, R. (2020). A constitutive law for the viscous and tertiary creep responses of ice to applied stress. *Cold Regions Science and Technology*. https://doi.org/10.1016/j.coldregions.2020.103034.

Anisotropic Failure Criteria in Relation to Crack Phase-Field Modeling at Finite Strains

Osman Gültekin and Gerhard A. Holzapfel

I know Peter since about three decades. At that time, Peter had just been appointed as a full professor at the Institute for Mechanics at TH Darmstadt in Germany, and I was a young postdoctoral researcher. I still remember a meeting in Darmstadt and a discussion about contact and impact problems, and, of course, about FEAP and a new compiler that promised to make the software faster. From 1993 I was then a visiting scholar at Stanford joining the group of JC Simo, and Peter was a frequent visitor during that time. Our two joint papers also originate from this time – I still remember the discussions in the European-style Cafe Borrone (which, by the way, is still there), an important meeting point at El Camino Real in Menlo Park. Over the years we have met again and again, in Hanover, at various conferences (such as COMPLAS) and also at his ECCOMAS (Thematic) Conference on "Biomedical Technology", which he started in 2013. I respect Peter very much not also as an outstanding scientist but also as a person. The last time we met was in Graz just before the corona pandemic, on the occasion of a dissertation defense on anisotropic crack phase-field modeling, which was also an opportunity for various updates. The 70th birthday is a round and important one, and you, Peter, should celebrate it properly. Then we should maybe meet again at Cafe Borrone to think about a new paper.

Gerhard

O. Gültekin
Department of Mechanical Engineering, Middle East Technical University (METU), Çankaya, Ankara, Turkey
e-mail: osmang@metu.edu.tr

G. A. Holzapfel (✉)
Institute of Biomechanics, Graz University of Technology, Graz, Austria
e-mail: holzapfel@tugraz.at

Department of Structural Engineering, Norwegian University of Science and Technology (NTNU), Trondheim, Norway

F. Aldakheel et al. (eds.), *Current Trends and Open Problems in Computational Mechanics*, https://doi.org/10.1007/978-3-030-87312-7_16

Abstract The estimation of rupture in fibrous soft biological tissues has emerged as a central task in medical monitoring and risk assessment of diseases such as aortic dissection and aneurysms. In an attempt to address the rupture phenomenon in fibrous soft tissues, a computational framework featuring an anisotropic crack phase-field approach is reviewed which involves a structure tensor aligning the crack along the collagen fibers, and the crack driving source term informed by several anisotropic failure criteria proposed so far. Subsequently, we focus on failure surfaces determined by the failure criteria. Finally, we provide a brief summary and discussion in the numerical rupture modeling of fibrous soft tissues.

1 Introduction

Prediction of rupture in fibrous soft tissues remains an elusive topic as it involves tangled series of coupled biomechanical processes which can hardly be identified during mechanical experiments and imaging. In this respect, computational models can help physicians to better assess the risk of rupture involved in diseases such as atherosclerosis, aneurysms and aortic dissection, see, e.g., Humphrey and Holzapfel [1] and Gültekin et al. [2].

Fibrous soft tissues demonstrate anisotropic mechanical responses which originate from the collagenous texture blended in the isotropic matrix material. In an attempt to address this characteristics of fibrous soft tissues and to describe a physically relevant failure, the following sections review a brief mathematical framework within the context of anisotropic crack phase-field, bringing the local features of the strength of materials approach and the non–local features of fracture mechanics on the same platform. On the geometrical side, anisotropic rupture is characterized by a structure tensor describing the evolution of the phase-field in the direction of the fibers, whereas, on the material side, anisotropic failure criteria, as introduced in [3–6], account for the anisotropic constitutive and failure behavior that originate from the tissue structure by means of the crack driving source term. Finally, an analysis on their numerical performance, i.e. their capability to describe an admissible failure surface, is presented by means of homogeneous numerical tests capturing uniaxial and biaxial extensions.

However, most of the research papers on phase-field modeling deal with brittle materials, whereby Wriggers and co-workers contributed substantially to the advanced modeling and simulations of brittle fracture. In this regard, it is worth to mention the very recent study [7] on an adaptive global-local approach applied to anisotropic phase-filed brittle fracture, which also contains a substantial list of references.

2 Anisotropic Crack Phase-Field Modeling

This section deals with phase-field modeling of anisotropic fracture in soft biological tissues with relevant anisotropic failure criteria describing the onset of macroscopic cracking in the tissue.

2.1 *Geometrical Aspects of Anisotropic Phase-Field Modeling*

The coupled problem of fracture is described by the deformation map φ and the auxiliary crack phase-field d, i.e.

$$\varphi_t(\mathbf{X}) : \begin{cases} \mathcal{B} \times \mathcal{T} & \rightarrow & \mathcal{S}, \\ (\mathbf{X}, t) & \mapsto & \mathbf{x} = \varphi(\mathbf{X}, t), \end{cases} \qquad d : \begin{cases} \mathcal{B} \times \mathcal{T} & \rightarrow & [0, 1], \\ (\mathbf{X}, t) & \mapsto & d(\mathbf{X}, t), \end{cases} \tag{1}$$

where φ maps a material point $\mathbf{X} \in \mathcal{B} \subset \mathbb{R}^3$ at time $t_0 \in \mathcal{T} \subset \mathbb{R}^+$ in the reference configuration onto $\mathbf{x} \in \mathcal{S} \subset \mathbb{R}^3$ at current time $t \in \mathcal{T} \subset \mathbb{R}^+$ located in the spatial configuration, while the crack phase-field d interpolates between the intact ($d = 0$) and the ruptured ($d = 1$) state of the material. A key aspect is to achieve a diffusive crack topology that smears out over a solid domain unlike sharp crack topology.

According to Miehe et al. [8, 9] a length-scale parameter l is introduced. The sharp crack surface topology at time t can be denoted by $\Gamma(t) \subset \mathbb{R}^2$ in the solid $\mathcal{B}$, with the definition $\Gamma(t) = \int_\Gamma \mathrm{d}A$. In contrast, a diffusive crack simply approximates the sharp crack surface by a volume integral in the form of the regularized crack surface functional

$$\Gamma_l(d) = \int_{\mathcal{B}} \gamma(d, \nabla d; \mathcal{L})\mathrm{d}V, \qquad \gamma(d, \nabla d; \mathcal{L}) = \frac{1}{2l}(d^2 + \nabla d \cdot \mathcal{L}\nabla d), \tag{2}$$

where γ refers to a generally anisotropic volume-specific crack surface.

Now, restricting the modeling framework to arterial walls, with two families of collagen fibers characterized by the unit vectors $\mathbf{M}$ and $\mathbf{M}'$, anisotropy is accounted by a second-order structure tensor $\mathcal{L}$, i.e.

$$\mathcal{L} = l^2(\mathbf{I} + \omega_{\mathrm{M}}\mathbf{M} \otimes \mathbf{M} + \omega_{\mathrm{M}'}\mathbf{M}' \otimes \mathbf{M}'), \tag{3}$$

where $\mathbf{I}$ denotes the second-order identity tensor. The structure tensor (3) aligns the evolution of the crack in the direction of the collagen fibers through the two parameters ω_{M} and $\omega_{\mathrm{M}'}$ that regulate the transition from weak to strong anisotropy. For isotropic solids, the parameters $\omega_{\mathrm{M}} = \omega_{\mathrm{M}}' \equiv 0$. Generally, these parameters must lie in an open domain, i.e. $-1 < \omega_i < \infty$ where $i \in \{\mathrm{M}, \mathrm{M}'\}$ in order to satisfy the ellipticity condition for the regularized crack surface $\Gamma_l(d)$.

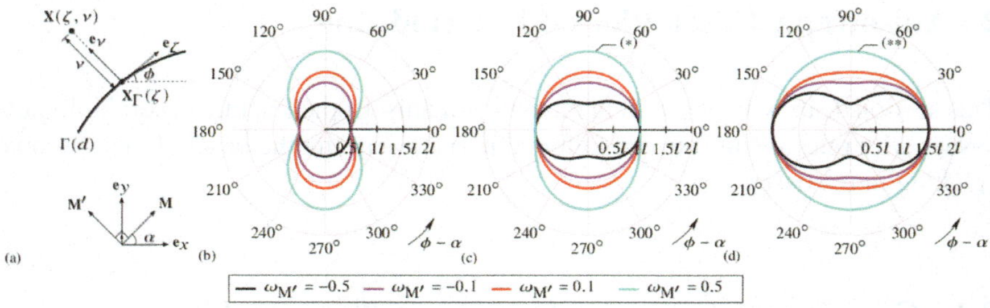

Fig. 1 **a** Sharp crack $\Gamma \in \mathcal{B}$ at point $\mathbf{X}_\Gamma(\zeta)$ together with the global and local coordinates systems (x, y) and (ζ, ν). Polar plots of the effective length scale parameter $l_e(\phi, \alpha)$ shown for an orthotropic case ($\mathbf{M} \perp \mathbf{M}'$) where $\alpha = 0°$ in (5), with **b** $\omega_M = -0.5$; **c** $\omega_M = 0.0$; **d** $\omega_M = 0.5$. The values of $\omega_{M'}$ are shown in different color, see the legend (adopted from [12])

The second-order structure tensor $\boldsymbol{\mathcal{L}}$, as defined in (3), motivates the concept of effective length scale parameter, which can be represented through polar plots. To this end, suppose that the sharp crack topology $\Gamma \in \mathcal{B}$ as a parametric curve $\mathbf{X}_\Gamma(\zeta)$ is traced out by ζ, i.e. $\zeta \rightarrow \mathbf{X}_\Gamma(\zeta)$, so that the position $\forall \mathbf{X} \in \mathcal{B}$ can be uniquely determined, i.e.

$$\mathbf{X}(\zeta, \nu) = \mathbf{X}_\Gamma(\zeta) + \nu \mathbf{e}_\nu. \tag{4}$$

The base vectors $\mathbf{e}_\zeta$ and $\mathbf{e}_\nu$ stand for the respective unit tangent and the normal at point $\mathbf{X}_\Gamma(\zeta)$, and furnish the local coordinate system (ζ, ν) along the global Cartesian system (x, y), see Fig. 1a.

Now, let the angle between the x-axis and the tangent of the crack at the position $\mathbf{X}_\Gamma(\zeta)$ be $\phi = \angle(\mathbf{e}_x, \mathbf{e}_\zeta)$, while the angle between the fiber orientation $\mathbf{M}$ and the x-axis is represented by $\alpha = \angle(\mathbf{e}_x, \mathbf{M})$. By assuming that the effective length scale parameter l_e is sufficiently small compared with the length of the sharp crack $|\Gamma|$, we get

$$l_e(\phi, \alpha) = l^2[1 + \omega_M \cos(\phi - \alpha) + \omega_{M'} \sin(\phi - \alpha)], \tag{5}$$

for a generally orthotropic case, where the second family of fibers $\mathbf{M}'$ is aligned perpendicular to the first fiber family $\mathbf{M}$, see Fig. 1a.

Figure 1b–d depict the polar plots of $l_e(\phi, \alpha)$, as given in (5), for specific choices of ω_M and $\omega_{M'}$. Specifically, the plot denoted by (∗), see Fig. 1c, recovers the transversely isotropic distribution of the effective length scale parameter for $\omega_{M'} = 0.5$ and vanishing ω_M. Moreover, the plot (∗∗), see Fig. 1d, retrieves isotropy with a wider range ($l_e = 1.5l$) for $\omega_M = \omega_{M'} \equiv 0.5$. This feature is the natural consequence of the second-order phase-field models providing a two-fold symmetry unlike the fourth-order phase-field models conferring a four-fold symmetry resembling a trefoil, a case seen for materials having cubic symmetry, see, e.g., Li et al. [10] and Teichtmeister et al. [11] for more details.

2.2 Balance Equations of Phase-Field Modeling of Rupture

The balance equations, namely the Euler-Lagrange equations, describing the multi-field problem of fracture result from the minimization principle of the global power balance under the quasi-static process of loading. The power balance comprises the rate of the elastic energy storage and the external power functional as well as the rate-dependent dissipation functional due to cracking of the solid. For an elaborate formulation of the balance equations we refer to Gültekin et al. [3, 12]. Consequently, the balance of linear momentum and the non-local evolution of the crack phase-field are the following two resultant local equations, i.e.

$$\boxed{\begin{aligned} &1:\ \ J\mathrm{div}(J^{-1}\boldsymbol{\tau}) + \rho_0\tilde{\boldsymbol{\gamma}} = \mathbf{0}, \\ &2:\ \ \eta\dot{d} + d - \nabla\cdot(\boldsymbol{\mathcal{L}}\nabla d) = 2(1-d)\mathcal{H}, \end{aligned}} \tag{6}$$

where J, ρ_0 and $\tilde{\boldsymbol{\gamma}}$ denote the determinant of the deformation gradient ($J = \det \mathbf{F}$), the material density and the prescribed body force, respectively. In $(6)_2$, η and $\mathcal{H}$ indicate the viscosity determining the viscous over-force, and the dimensionless crack driving source term, respectively. Particularly, the source term is characterized by a ramp-type function as

$$\mathcal{H}(t) = \max_{s\in[0,t]} \left[\langle \overline{\mathcal{H}}(s) - 1\rangle\right]. \tag{7}$$

The Macaulay brackets filter out the positive values for $\overline{\mathcal{H}}(s) - 1$ and keeps the solid intact until a threshold value is reached, i.e. the failure criterion. Therefore, the crack phase-field does not evolve for $\overline{\mathcal{H}}(s) < 1$. It should also be highlighted that (7) takes always into account the maximum value of $\overline{\mathcal{H}}(s) - 1$ in the deformation history ensuring the irreversibility of cracking ($\dot{d} \geq 0$). The dimensionless characteristics of $\overline{\mathcal{H}}$ allows for different choices in regard to failure criteria which will be covered in the proceeding subsection.

2.3 Constitutive Aspects of Anisotropic Phase-Field Modeling

Fibrous soft tissues exhibit an anisotropic morphology, thereby an anisotropic mechanical response is expected which needs to inform the source term in (7). Subsequently, we provide a short description of anisotropic failure criteria. For simplicity, the ensuing formulations are established according to the assumption that the principal axes of anisotropy lie on the axes of reference. Nonetheless, transformation of stress components can be achieved without much effort [12].

2.3.1 Energy-Based Anisotropic Failure Criterion

Two distinct failure processes are assumed to govern rupture of the isotropic matrix material and the anisotropic collagen fibers, as suggested by Gültekin et al. [3]. This leads to two different evolution equations for the phase-field, i.e. $(6)_2$, as to isotropic and anisotropic material responses. Accordingly, one can describe distinct isotropic and anisotropic forms of the structure tensor $\boldsymbol{\mathcal{L}}$ in (3) as

$$\boldsymbol{\mathcal{L}}^{\text{iso}} = l^2\mathbf{I}, \qquad \boldsymbol{\mathcal{L}}^{\text{ani}} = l^2(\omega_{\text{M}}\mathbf{M} \otimes \mathbf{M} + \omega_{\text{M}'}\mathbf{M}' \otimes \mathbf{M}'), \tag{8}$$

in conjunction with the dimensionless crack driving functions given as

$$\overline{\mathcal{H}}^{\text{iso}} = \frac{\hat{\Psi}_0^{\text{iso}}}{g_{\text{c}}^{\text{iso}}/l}, \qquad \overline{\mathcal{H}}^{\text{ani}} = \frac{\hat{\Psi}_0^{\text{ani}}}{g_{\text{c}}^{\text{ani}}/l}. \tag{9}$$

Therein, $g_{\text{c}}^{\text{iso}}/l$ and $g_{\text{c}}^{\text{ani}}/l$ are the distinct critical Griffith-type fracture energies over the length scale for the matrix material and for the fibers, respectively, while $\hat{\Psi}_0^{\text{iso}}$ and $\hat{\Psi}_0^{\text{ani}}$ refer to the effective isotropic and anisotropic free-energy functions related to the intact (undamaged) response, see, e.g., Holzapfel et al. [13]. Upon the superposition of the isotropic and anisotropic evolution equations, one can obtain a modified form of the crack evolution $(6)_2$, i.e. [12]

$$\eta\dot{d} + d - \frac{1}{2}\nabla \cdot (\boldsymbol{\mathcal{L}}\nabla d) = (1-d)\mathcal{H}, \tag{10}$$

for which $\overline{\mathcal{H}} = \overline{\mathcal{H}}^{\text{iso}} + \overline{\mathcal{H}}^{\text{ani}}$ enters in (7), where $\overline{\mathcal{H}}^{\text{iso}}$ and $\overline{\mathcal{H}}^{\text{ani}}$ define the dimensionless crack driving forces for the isotropic and the anisotropic part, respectively.

2.3.2 Stress-Based Anisotropic Tsai-Wu Failure Criterion

The Tsai-Wu criterion is based on the strength of the material at which the stress space intercepts the assumed failure surface, see Tsai and Wu [6]. Accordingly, the dimensionless crack driving function with respect to the effective Cauchy stress tensor $\boldsymbol{\sigma}_0$ assumes a composition of two scalar functions, i.e.

$$\overline{\mathcal{H}} = \boldsymbol{T} : \boldsymbol{\sigma}_0 + \boldsymbol{\sigma}_0 : \mathbb{T} : \boldsymbol{\sigma}_0, \tag{11}$$

where $\boldsymbol{T}$ and $\mathbb{T}$ denote the second and fourth-order strength tensors, respectively. Through assumptions and simplifications introduced by symmetry relations we end up with $T_{ii} = 1/(\sigma_i^{\text{u}})^2$ for the diagonal terms of the fourth-order strength tensor, which relate to the ultimate normal and shear stresses σ_i^{u}, with $i \in \{1, \dots, 6\}$.

2.3.3 Stress-Based Anisotropic Hill Failure Criterion

Considered as the anisotropic extension of the von Mises–Huber criterion, the Hill criterion uses a quadratic form of $\overline{\mathcal{H}}$ such that [4]

$$\overline{\mathcal{H}} = \boldsymbol{\sigma}_0^{\text{vm}} : \mathbb{T} : \boldsymbol{\sigma}_0^{\text{vm}}, \tag{12}$$

where $\boldsymbol{\sigma}_0^{\text{vm}}$ represents the effective von Mises stress tensor. The components of $\boldsymbol{\sigma}_0^{\text{vm}}$ can be defined in terms of the general stress components, i.e. $\sigma_{0_1}^{\text{vm}} = \sigma_{0_1} - \sigma_{0_2}$, $\sigma_{0_2}^{\text{vm}} = \sigma_{0_2} - \sigma_{0_3}$, $\sigma_{0_3}^{\text{vm}} = \sigma_{0_3} - \sigma_{0_1}$, $\sigma_{0_4}^{\text{vm}} = \sigma_{0_4}$, $\sigma_{0_5}^{\text{vm}} = \sigma_{0_5}$, $\sigma_{0_6}^{\text{vm}} = \sigma_{0_6}$. The fourth-order strength tensor $\mathbb{T}$ pertains to the effective normal and shear stresses, see [12] for more details.

2.3.4 Principal Stress Criterion

Developed on the basis of principal stresses the criterion by Raina and Miehe [5] reports on the spectral decomposition of the effective Cauchy stress tensor and takes the positive principal stresses into account, i.e.

$$\boldsymbol{\sigma}_0^+ = \sum_{i=1}^{3} \langle \sigma_{0_i} \rangle \boldsymbol{n}_i \otimes \boldsymbol{n}_i, \tag{13}$$

where σ_{0_i} denote the effective principal stresses, and $\boldsymbol{n}_i$ are the corresponding eigenvectors for $i \in \{1, 2, 3\}$. Accordingly, $\overline{\mathcal{H}}$ may be rewritten as

$$\overline{\mathcal{H}} = \boldsymbol{\sigma}_0^+ : \mathbb{T} : \boldsymbol{\sigma}_0^+, \tag{14}$$

where the fourth-order strength tensor $\mathbb{T}$ reads in the index notation $(\mathbb{T})_{ijkl} = (A_{ik}A_{jl} + A_{il}A_{jk})/4\sigma_{\text{crit}}^2$, where σ_{crit} denotes the reference critical stress associated with uniaxial loading in a certain axis that can be conceptually replaced by an ultimate stress. Therein, $\boldsymbol{A}$ is expressed in index notation with $i, j, k, l \in \{1, 2, 3\}$, see [5] for more details.

2.3.5 Homogeneous Tests on the Respective Failure Surfaces

This example demonstrates the 3D failure surfaces associated with the criteria provided in the Sects. 2.3.1–2.3.4, thereby pinpointing to the onset of macro-cracking in the solid. The problem setup involves a homogeneous case with a unit cube discretized by a single hexahedral element resolving the analytical solution for the deformation and stress. The sample, regarded as transversely isotropic, undergoes a series of uniaxial and biaxial deformations [12].

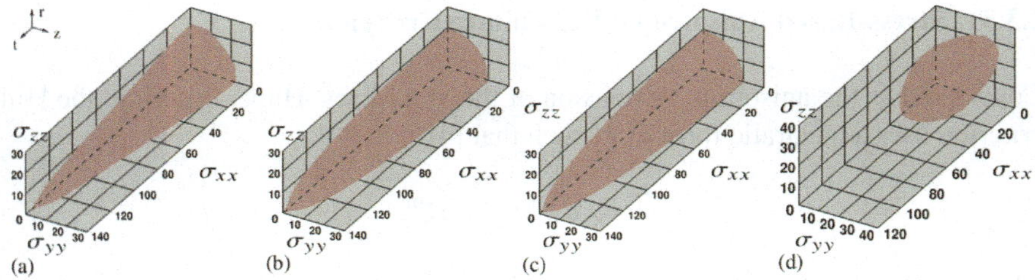

Fig. 2 Failure surfaces in regard to the Cauchy stresses σ_{xx}, σ_{yy} and σ_{zz} (in kPa) at which the failure conditions are satisfied, leading to $d > 0$ for **a** the energy-based; **b** the Tsai-Wu; **c** the maximum principal stress; **d** the Hill failure criterion (adopted from [12])

Figure 2a–c illustrate the resulting failure surfaces at the instance when $d \neq 0$ for the energy-based, the Tsai-Wu and the principal stress criterion, respectively. Ellipsoidal failure surfaces are evident. The Hill criterion in Fig. 2d, however, induces a surface that diverges from being elliptic. In particular, the isotropic failure envelope on the yz-plane eventually becomes discernible, which recovers the von Mises–Huber criterion, as expected.

3 Discussion

For several materials such as fibrous soft tissues, anisotropic rupture is not only a geometrical phenomenon but also inherent in the micro-structure due to the collagen fibers embedded in an isotropic matrix material, which undoubtedly entails the use of an anisotropic crack driving source term. Although the majority of the failure criteria treated here are capable to provide admissible anisotropic failure surfaces, see Fig. 2a–c, for homogeneous cases, the modeling of crack propagation demands non-homogeneous numerical tests whereby the numerical stability plays a crucial role, see [12] for a comparative study. Meso–scale and micro–scale structural information along with statistical aspects in hyperelastic and rupture responses of fibrous soft tissues (through, e.g., data-driven modeling or machine learning) has to communicate with the extant macro–scale approaches discussed here if clinical support is to be a desired goal of such models in the future.

References

1. Humphrey, J. D., & Holzapfel, G. A. (2012). Mechanics, mechanobiology, and modeling of human abdominal aorta and aneurysms. *Journal of Biomechanics, 45,* 805–814.
2. Gültekin, O., Hager, S. P., Dal, H., & Holzapfel, G. A. (2019). Computational modeling of progressive damage and rupture in fibrous biological tissues: application to aortic dissection. *Biomechanics and Modeling in Mechanobiology, 18,* 1607–1628.

3. Gültekin, O., Dal, H., & Holzapfel, G. A. (2016). A phase-field approach to model fracture of arterial walls: theory and finite element analysis. *Computer Methods in Applied Mechanics and Engineering, 312,* 542–566.
4. Hill, R. (1948). A theory of the yielding and plastic flow of anisotropic metals. *Proceedings of the Royal Society London A, 193,* 281–297.
5. Raina, R., & Miehe, C. (2016). A phase-field model for fracture in biological tissues. *Biomechanics and Modeling in Mechanobiology, 15,* 479–496.
6. Tsai, S. W., & Wu, E. M. (1971). A general theory of strength of anisotropic materials. *Journal of Composite Materials, 5,* 58–80.
7. Noii, N., Aldakheel, F., Wick, T., & Wriggers, P. (2020). An adaptive global-local approach for phase-field modeling of anisotropic brittle fracture. *Computer Methods in Applied Mechanics and Engineering, 361,* 112744.
8. Miehe, C., Welschinger, F., & Hofacker, M. (2010). Thermodynamically consistent phase-field models of fracture: Variational principles and multi-field FE implementations. *International Journal for Numerical Methods in Engineering, 83,* 1273–1311.
9. Miehe, C., Welschinger, F., & Hofacker, M. (2010). A phase field model for rate-independent crack propagation: Robust algorithmic implementation based on operator splits. *Computer Methods in Applied Mechanics and Engineering, 199,* 2765–2778.
10. Li, B., Peco, C., Millán, D., Arias, I., & Arroyo, M. (2015). Phase-field modeling and simulation of fracture in brittle materials with strongly anisotropic surface energy. *International Journal for Numerical Methods in Engineering, 102,* 711–727.
11. Teichtmeister, S., Kienle, D., Aldakheel, F., & Keip, M.-A. (2017). Phase-field modeling of fracture in anisotropic brittle solids. *International Journal of Non-Linear Mechanics, 97,* 1–21.
12. Gültekin, O., Dal, H., & Holzapfel, G. A. (2018). Numerical aspects of anisotropic failure in soft biological tissues favor energy-based criteria: A rate-dependent anisotropic phase-field model. *Computer Methods in Applied Mechanics and Engineering, 331,* 23–52.
13. Holzapfel, G. A., Gasser, T. C., & Ogden, R. W. (2000). A new constitutive framework for arterial wall mechanics and a comparative study of material models. *Journal of Elasticity, 61,* 1–48.

A Poroelastic Element for FEAP Using AceGen

Ajay B. Harish, Robert L. Taylor, and Sanjay Govindjee

We dedicate this article to Peter Wriggers on the occasion of his 70th birthday, acknowledging with appreciation his longstanding connections to the University of California, Berkeley.

Abstract The error free generation of elements for complex field equations is generally challenging. The process however can be helped via the judicious use of automated systems that allow for the specification of weak forms and interpolation choices and subsequently produce source code that can be compiled into general purpose finite element programs. In this contribution, we illustrate the use of one such system, AceGen, coupled to FEAP and applied to transient linear poroelasticity.

1 Introduction to Poroelasticity

Poroelastic problems arise in many interesting areas of engineering and science, ranging from classical civil engineering to modern biological sciences. The dominant presentation of poroelasticity is generally attributed to the efforts of Biot [1, 2] and countless subsequent studies. As a computational problem poroelasticity, even in the linear setting, presents interesting challenges due to the unique nature of the coupling between the variation in fluid content and the deformation of the media. In particular, it is well-known that in a finite element setting the governing field equations are most effectively interpolated with continuous but unequal orders for

A. B. Harish · R. L. Taylor (✉) · S. Govindjee
University of California, Berkeley, CA, USA
e-mail: rlt@ce.berkeley.edu

A. B. Harish
e-mail: ajaybh@berkeley.edu

S. Govindjee
e-mail: s_g@berkeley.edu

F. Aldakheel et al. (eds.), *Current Trends and Open Problems in Computational Mechanics*, https://doi.org/10.1007/978-3-030-87312-7_17

the pore pressure and the displacements. The first apparent recognition of this point appears in Sandhu and Wilson [3], where a T6/T3 element was proposed (continuous quadratic displacements with continuous linear pressures) for seepage problems. Later Hood and Taylor [4] introduced a Q8/Q4 varient within the context of solving the Navier-Stokes equations; see also Huyakorn et al. [5] where the Q9/Q4 extension of this element appears. The challenge of the numerical problem to this date still attracts continued attention [6].

The programming of such elements while straightforward can be tricky and tedious to get correct and error free. One option for assisting in this effort is to use automated systems that take as input variational statements and directly produce elements, including features to perform automated differentiation. The DOLFIN [7] library presents one popular choice with C++ and Python interfaces. An alternate system that is additionally capable of generating elements for several major general purpose finite element codes is AceGen [8]. In this work we illustrate its use on the problem of linear poroelasticity in conjunction with FEAP [9].

2 Linear Poroelasticity

2.1 Theory

The constitutive equations for the stress, $\boldsymbol{\sigma}$, and variation in fluid content, ζ, in a linear isotropic poroelastic material are given by [10]

$$\begin{aligned} \boldsymbol{\sigma} &= 2\,G\,\boldsymbol{\epsilon} + (K^{(d)} - \tfrac{2}{3}G)(\mathrm{tr}\boldsymbol{\epsilon})\mathbf{1} - \alpha\,p\,\mathbf{1} \\ \zeta &= \alpha\,(\mathrm{tr}\boldsymbol{\epsilon}) + \Upsilon\,p\,, \end{aligned} \tag{1}$$

where $\boldsymbol{\epsilon}$ is the linear strain and p is the pore pressure. The volumetric fluid flux, $\mathbf{q}$, is given by

$$\mathbf{q} = -k\,\nabla p\,. \tag{2}$$

The required material parameters to define the model are given by the elastic shear moduli G and drained bulk modulus $K^{(d)}$, the Biot modulus $M = 1/\Upsilon$, the Biot coefficient α and the permeability k. The relationship between the undrained and drained bulk modulus is given by

$$K^{(u)} - K^{(d)} = \alpha^2 M\,. \tag{3}$$

We also note that the Skempton compressibility index is defined for isotropic materials by

$$B = \frac{1}{\alpha}\left(1 - \frac{K^{(d)}}{K^{(u)}}\right)\,. \tag{4}$$

The governing differential equations are given by linear momentum balance

$$\text{div}\boldsymbol{\sigma} + \mathbf{b} = \rho\,\ddot{\mathbf{u}} \tag{5}$$

and, using the second of (1), fluid balance

$$\dot{\zeta} = \Upsilon\,\dot{p} + \alpha\,\text{div}\dot{\mathbf{u}} = -\text{div}\mathbf{q}\,. \tag{6}$$

2.2 *Variational Equations*

The governing equations may be expressed in terms of the pair of weak forms

$$\begin{aligned} G_u(\mathbf{u}, p; \delta\mathbf{u}) &= \int_V \left[\delta\mathbf{u}^T\,(\rho\,\ddot{\mathbf{u}} - \mathbf{b}) + \delta\boldsymbol{\epsilon}^T\boldsymbol{\sigma}\right]\mathrm{d}V - \int_{\partial V_t} \delta\mathbf{u}^T\bar{\mathbf{t}}\,\mathrm{d}S = 0 \\ G_p(\mathbf{u}, p; \delta p) &= \int_V \left[\delta p\,(\Upsilon\,\dot{p} + \alpha\,\text{div}\dot{\mathbf{u}}) - (\nabla\delta p)^T\mathbf{q}\right]\mathrm{d}V + \int_{\partial V_p} \delta p\,\bar{q}\,dS = 0\,, \end{aligned} \tag{7}$$

where $\mathbf{t} = \mathbf{n}^T\boldsymbol{\sigma}$ is the boundary traction and $q = \mathbf{n}^T\mathbf{q}$ is the normal boundary flux; an over-bar denotes a specified value.

2.3 *Finite Element Solution*

A finite element approximation for the theory presented above may be given as

$$\mathbf{x} = N_a(\boldsymbol{\xi})\,\tilde{\mathbf{x}}_a \;\;;\;\; \mathbf{u} = N_a(\boldsymbol{\xi})\,\tilde{\mathbf{u}}_a \;\text{ and }\; p = N_a^p(\boldsymbol{\xi})\,\tilde{p}_a\,, \tag{8}$$

where N_a and N_a^p are shape functions expressed in terms of parent coordinates $\boldsymbol{\xi}$ and $\tilde{\mathbf{u}}_a$, $\tilde{\mathbf{x}}_a$ and $\tilde{p}_a$ are nodal values of the displacements, coordinates and pressure, respectively. The possibility of using different interpolations for the dependent variables is motivated by a need to reduce the possibility of spurious oscillations in solutions [11]. The strains and gradient of the pressure may be expressed, assuming summation convention, by

$$\boldsymbol{\epsilon} = \mathbf{B}_a\,\tilde{\mathbf{u}}_a \;\text{ and }\; \nabla p = \mathbf{b}_a\,\tilde{p}_a\,, \tag{9}$$

where for the two-dimensional plane case considered here, the $\mathbf{B}_a$ and $\mathbf{b}_a$ matrices are defined by

$$\mathbf{B}_a = \begin{bmatrix} N_{a,x} & 0 \\ 0 & N_{a,y} \\ 0 & 0 \\ N_{a,y} & N_{a,x} \end{bmatrix} \quad \text{and} \quad \mathbf{b}_a = \begin{bmatrix} N^p_{a,x} \\ N^p_{a,y} \end{bmatrix} . \tag{10}$$

Inserting the approximations into the weak forms (7) yields the semi-discrete form

$$\begin{aligned} G_u &= \delta \tilde{\mathbf{u}}_a^T \left[\mathbf{M}_{ab}\, \ddot{\tilde{\mathbf{u}}}_b + \mathbf{P}_a - \mathbf{f}_a \right] = 0 \\ G_p &= \delta \tilde{p}_a \left[C_{ab}\, \dot{\tilde{p}}_b + \mathbf{G}_{ab}\, \dot{\tilde{\mathbf{u}}}_b - J_a \right] = 0 , \end{aligned} \tag{11}$$

where $\mathbf{M}_{ab}$, $\mathbf{P}_a$, $\mathbf{f}_a$, C_{ab}, $\mathbf{G}_{ab}$ and J_a are defined by

$$\begin{array}{ll} \mathbf{M}_{ab} = \int_V N_a\, \rho\, N_b \, \mathrm{d}V\, \mathbf{I}; & \mathbf{P}_a = \int_V \mathbf{B}_a^T \boldsymbol{\sigma}\, \mathrm{d}V \\ C_{ab} = \int_V N_a \Upsilon\, N_b \, \mathrm{d}V; & \mathbf{G}_{ab} = \int_V N_a^p\, \alpha\, \mathbf{b}_b^T \, \mathrm{d}V \\ \mathbf{f}_a = \int_V N_a \mathbf{b}\, \mathrm{d}V + \int_{\partial V_t} N_a \bar{\mathbf{t}}\, \mathrm{d}S; & J_a = \int_V \mathbf{b}_a^T \mathbf{q}\, \mathrm{d}V - \int_{\partial V_p} N_a^p\, \bar{q}\, \mathrm{d}S . \end{array} \tag{12}$$

A Newton solution then may be given as

$$\begin{bmatrix} \mathbf{M}_{ab} & \mathbf{0} \\ \mathbf{0} & 0 \end{bmatrix} \begin{Bmatrix} d\ddot{\tilde{\mathbf{u}}}_b \\ d\ddot{\tilde{p}}_b \end{Bmatrix} + \begin{bmatrix} \mathbf{0} & \mathbf{0} \\ \mathbf{G}_{ab} & C_{ab} \end{bmatrix} \begin{Bmatrix} d\dot{\tilde{\mathbf{u}}}_b \\ d\dot{\tilde{p}}_b \end{Bmatrix} + \begin{bmatrix} \mathbf{K}_{ab} & -\mathbf{Q}_{ab} \\ \mathbf{0} & -H_{ab} \end{bmatrix} \begin{Bmatrix} d\tilde{\mathbf{u}}_b \\ d\tilde{p}_b \end{Bmatrix} = \begin{Bmatrix} \mathbf{R}_a \\ r_a \end{Bmatrix} , \tag{13}$$

where, in addition to quantities defined above,

$$\begin{aligned} &\mathbf{K}_{ab} = \int_V \mathbf{B}_a^T \mathbf{D} \mathbf{B}_b \, \mathrm{d}V \;; \qquad H_{ab} = \int_V k\, \mathbf{b}_a^T \mathbf{b}_b \, \mathrm{d}V \\ &\mathbf{Q}_{ab} = \int_V \mathbf{b}_a\, \alpha\, N_b^p \, \mathrm{d}V = \mathbf{G}_{ba}^T \;; \; r_a = J_a - C_{ab}\, \dot{\tilde{p}}_b - \mathbf{G}_{ab}\, \dot{\tilde{\mathbf{u}}}_b \\ &\mathbf{R}_a = \mathbf{f}_a - \mathbf{P}_a - \mathbf{M}_{ab}\, \ddot{\tilde{\mathbf{u}}}_b . \end{aligned} \tag{14}$$

If a time discretization is introduce such that [9, 12]

$$d\dot{\tilde{\mathbf{u}}}_b = c_2\, d\tilde{\mathbf{u}}_b \;\; ; \;\; d\ddot{\tilde{\mathbf{u}}}_b = c_3\, d\tilde{\mathbf{u}}_b \;\text{ and }\; d\dot{\tilde{p}}_b = c_2\, d\tilde{p}_b , \tag{15}$$

then (14) becomes an algebraic equation given by

$$\begin{bmatrix} \mathbf{K}_{ab} + c_3\, \mathbf{M}_{ab} & -\mathbf{G}_{ba}^T \\ c_2 \mathbf{G}_{ab} & H_{ab} + c_2\, C_{ab} \end{bmatrix} \begin{Bmatrix} d\tilde{\mathbf{u}}_b \\ d\tilde{p}_b \end{Bmatrix} = \begin{Bmatrix} \mathbf{R}_a \\ r_a \end{Bmatrix} , \tag{16}$$

with updates given by

$$\tilde{\mathbf{u}}_a^{i+1} = \tilde{\mathbf{u}}_a^i + d\tilde{\mathbf{u}}_a \;\text{ and }\; \tilde{p}_a^{i+1} = \tilde{p}_a^i + d\tilde{p}_a \tag{17}$$

in which i denotes the iteration number. For linear systems proper discretization should lead to convergence in one iteration.

3 Automated Computational Modeling Using AceGen

The above theory is implemented for use in the finite element program *FEAP* [13] using an element module developed using the automated computational program AceGen [8]. The element module developed utilizes a Q9/Q4 Taylor-Hood type interpolation for the $\mathbf{u}/p$ dependent variables, respectively. In addition, the tangent matrix is separated into three parts for the terms in (13) related to the different time derivative orders and combined as described in Taylor and Govindjee [9]. This permits the use of the standard *FEAP* time integration routines directly. In the interest of brevity, we do not detail these steps here. However, the AceGen modules used in this work, along with documentation can be accessed at https://github.com/bhajay/Coupled-Problems; see also [22].

4 Results and Discussion

In this section, the element is validated using two test problems.

4.1 Mandel Problem

In the theory proposed by Biot [1], Mandel showed that when a soil is loaded by a constant load, the pore-pressure can initially increase before decreasing to a final value of zero. This effect, later came to be known as the Mandel-Cryer effect and was also confirmed experimentally [14, 15]. Mandel [16] and others [17, 18] have presented a set of three problem to discuss this non-monotonic variation of pore-pressure in porous media, one of these problems is considered here.

As shown in the Fig. 1, an infinitely long (out of the plane) rectangular plate of width $2a$ is sandwiched between two rigid and frictionless plates. Additionally, drainage is allowed on the two lateral sides, which are stress free. A generalized plane strain condition is considered by preventing any deformation or flux in the direction perpendicular to the plane. At time $t = 0$, a vertical force F is applied and remains constant. At the instant of loading, the pore pressure is homogeneous while instantaneously it drops to zero at the two sides, i.e. at $|x| = a$. Since the pressure is constant in the vertical direction, a single row of 25 Q9/Q4 elements is used for the analysis (employing symmetry).

The material properties are: Shear modulus: ($G = 1$); drained Poisson ratio: ($\nu^{(d)} = 0.2$); permeability: ($k = 1$); density: ($\rho = 0$); and Skempton coefficient: ($B = 1$). Two cases are considered: Compressible (with $\nu^{(u)} = 0.4$) and incompressible (with $\nu^{(u)} = 0.5$).

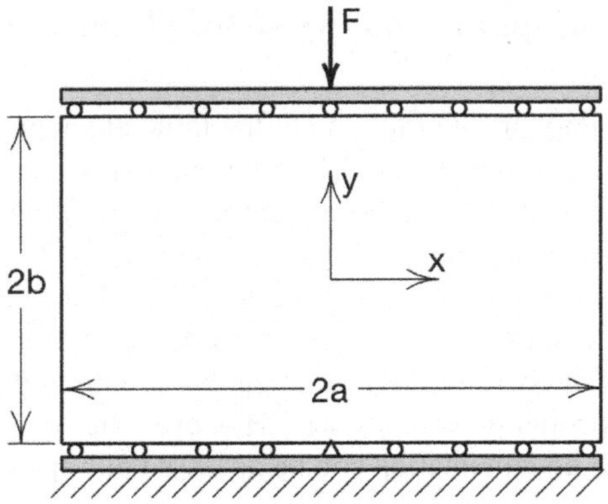

Fig. 1 Geometry for the Mandel problem

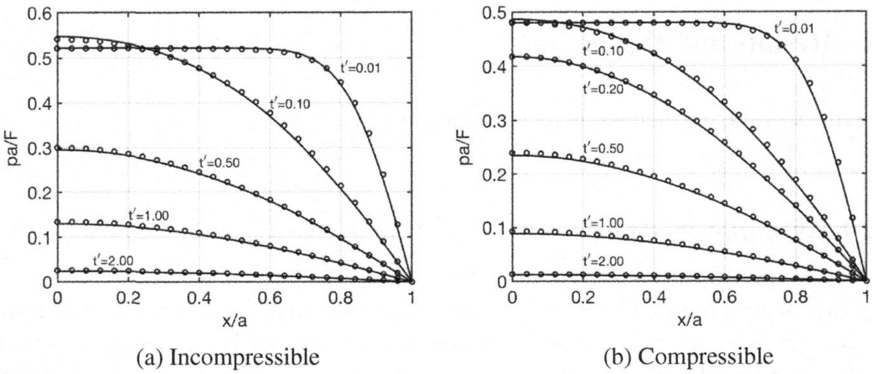

Fig. 2 Comparison of analytical solution (solid lines) with FEA solution from FEAP using the Q9/Q4 element generated by AceGen (circles). Time labels t' correspond to non-dimensional time $t' = t\frac{kG}{a^2}\left[\frac{2}{3}\frac{B}{\alpha}\frac{1+\nu^{(u)}}{1-\nu^{(u)}}\frac{1-\nu^{(d)}}{1-2\nu^{(d)}}\right]$

In Fig. 2, the non-dimensional pressure is compared to the analytical solution given by Cheng and Detournay [19, 20] as a function of the distance from the center of the plate for the two cases.

4.2 Consolidation Problem

As a second verification example we look at the time dependent consolidation problem proposed by Booker and Small [21] for modeling surface footings on horizontally layered soils on a rigid base. The example considered here is an infinite finite depth strip subjected to periodic uniform loading as shown in Fig. 3a. Due to symmetry the region modeled is shown in Fig. 3b. The imposed boundary conditions

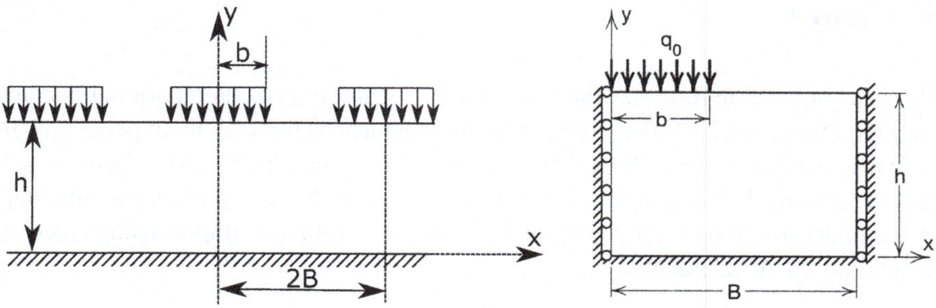

(a) Actual geometry (b) Geometry with boundary conditions

Fig. 3 Geometry and boundary conditions considered for the Booker problem

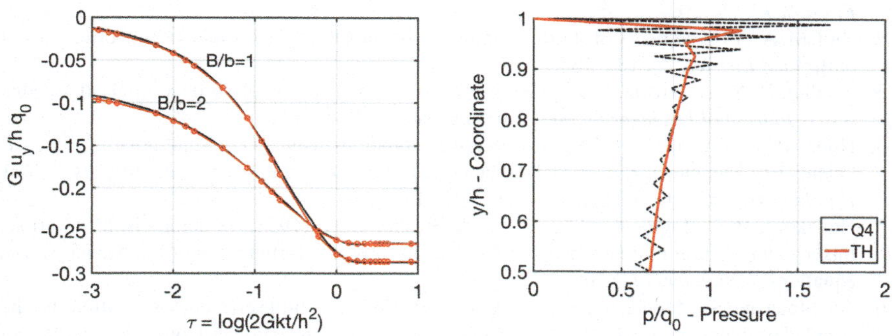

(a) Non-dimensional displacement (b) Pressure variation

Fig. 4 Consolidation problem. Left: Dimensionless settlement versus time. Red markers FEAP result with AceGen Q9/Q4 element. Solid lines from Booker and Small [21]. Right: Comparison of pore pressure at the center line for a Q4 element with equal order interpolation versus Q9/Q4 Taylor-Hood interpolation

are: $u_x(0, y) = u_x(\mathsf{B}, y) = 0$ and $\tau_{xy}(0, y) = \tau_{xy}(\mathsf{B}, y) = 0$. At the bottom a rough condition is assumed with $u_x(x, 0) = u_y(x, 0) = 0$. The top surface is considered to be permeable with $p(x, \mathsf{h}) = 0$.

The material parameters considered for the analysis are $G = 1$, $\nu^{(d)} = 0.3$, $k = 1$, $\alpha = B = 1$ and $\rho = 0$. The geometry is meshed using 40×40 Q9/Q4 elements. The geometric ratio for B:b is considered for two values: 1:1 and 2:1. The applied load is $q_o = 1$. The resulting dimensionless time settlement behavior at $(x, y) = (0, \mathsf{h})$ is shown in Fig. 4a.

Figure 4a shows that the time settlement behavior changes dramatically with the B:b ratio. Figure 4b shows the excess pore pressure at the center line of the domain and its variation with depth. The Q4 element shows significant oscillations and instability as compared to the solutions from the Q9/Q4 element.

5 Closure

This brief note has highlighted the utility and use of automated computational systems for generating (Fortran) source code for general purpose finite element programs. In the particular case shown, the modestly complex transient Taylor-Hood element was generated using AceGen for the finite element program FEAP, and subsequently verified against solutions available in the literature; for further examples using AceGen to generate FEAP elements, see [22].

References

1. Biot, M. A. (1941). General theory of three-dimensional consolidation. *Journal of Applied Physics, 12*, 155–164.
2. Biot, M. A. (1955). Theory of elasticity and consolidation for a porous anisotropic solid. *Journal of Applied Physics, 26*, 182–185.
3. Sandhu, R. S., & Wilson, E. L. (1969). Finite-element analysis of seepage in elastic media. *ASCE Journal of Engineering Mechanics Division, 95*, 641–652.
4. Hood, P., & Taylor, C. (1974). Navier-stokes equations using mixed-interpolation. In J. T. Oden, O. C. Zienkiewicz, R. H. Gallagher, & C. Taylor (Eds.), *Finite Element Methods in Flow Problems* (pp. 121–132). Huntsville AL: UAH Press.
5. Huyakorn, P. S., Taylor, C., Lee, R. L., & Gresho, P. M. (1978). A comparison of various mixed-interpolation finite elements in the velocity-pressure formulation of the Navier-Stokes equations. *Computers and Fluids, 6*, 25–35.
6. Teichtmeister, S., Mauthe, S., & Miehe, C. (2019). Aspects of finite element formulations for the coupled problem of poroelasticity based on a canonical minimization principle. *Computational Mechanics, 64*, 685–716.
7. Logg, A., & Wells, G. N. (2010). DOLFIN: Automated finite element computing. *ACM Transactions on Mathematical Software, 37*, (article 20, 28 pages).
8. Korelc, J., & Wriggers, P. (2016). *Automation of Finite Element Methods*. Switzerland: Springer International Publishing.
9. Taylor, R. L., & Govindjee, S. *FEAP - A Finite Element Analysis Program, Programmer Manual*. University of California, Berkeley. http://projects.ce.berkeley.edu/feap.
10. Anand, L., & Govindjee, S. (2020). *Continuum Mechanics of Solids*. Oxford: Oxford University Press.
11. Pastor, M., Mira, P., & Fernandez Merodo, J. A. (2004). Practical aspects of the finite element method. In M. Pastor & C. Tamagnini (Eds.), *Numerical Modelling in Geomechanics* (pp. 256–276). London: Kogan Page.
12. Zienkiewicz, O. C., Taylor, R. L., & Zhu, J. Z. (2013). *The Finite Element Method: Its Basis and Fundamentals* (7th ed.). Oxford: Elsevier.
13. Taylor, R. L., & Govindjee, S. *FEAP - A Finite Element Analysis Program, User Manual*. University of California, Berkeley. http://projects.ce.berkeley.edu/feap.
14. Gibson, R. E., Knight, K., & Taylor, P. W. (1963). A critical experiment to examine theories of three-dimensional consolidation. In *Proceedings of European Conference on Soil Mechanics, Wiesbaden*, pp. 69–76
15. Verruijt, A. (1965). Discussion on consolidation of a massive sphere. In *Proceedings of the 6th International Conference on Soil Mechanics, Montreal*, pp. 401–402.
16. Mandel, J. (1953). Consolidation des sols (étude mathématique). *Géotechnique, 3*, 287–299.
17. Cryer, C. (1963). A comparison of the three-dimensional consolidation theories and Terzaghí. *Quarterly Journal of Mechanics and Applied Mathematics, 16*, 401–412.

18. De Leeuw, E. H. (1965). The theory of three-dimensional consolidation applied to cylindrical bodies. In *Proceedings of the 6th International Conference on Soil Mechanics and Foundation Engineering*, pp. 287–290.
19. Cheng, A.H.-D., & Detournay, E. (1988). A direct boundary element method for plane strain poroelasticity. *International Journal for Numerical and Analytical Methods in Geomechanics, 12,* 661–572.
20. Wang, H. F. (2000). *Theory of Linear Poroelasticity: With Applications to Geomechanics and Hydrogeology*. Oxford: Princeton University Press.
21. Booker, J. R., & Small, J. C. (1982). Finite layer analysis of consolidation. I. *International Journal for Numerical and Analytical Methods in Geomechanics, 6*, 151–171.
22. Harish, A. B., Taylor, R. L., & Govindjee, S. (2021). Automated generation of user elements (UEL) for FEAP, SEMM Report UCB/SEMM-2021/01, University of California, Berkeley (2021).

Contact Formulation for Second Gradient Materials

Christian Hesch and Stefan Schuß

This paper is dedicated to Professor Peter Wriggers, who was the co-reviewer of my phd-thesis in 2007. I had the great opportunity to meet him nearly every year on conferences for the last 15 years. I enjoyed every debate on mechanics and university politics. Congratulations and many more healthy and happy years to come. (C. Hesch)

Abstract A first investigation on higher-order contact boundary conditions is presented. A short summary on second gradient materials is followed by a general higher-order domain decomposition formulation for optimal boundary conditions. Afterwards, the application on contact problems is discussed.

1 Introduction

General higher gradient theories have been investigated in the pioneering and most fundamental work of Mindlin [1], see also the work of Toupin [2] and Eringen [3]. A series of publications establish a link between the microstructure and the macroscopic formulation as generalized continua, see [4, 5] among others, see also [6, 7] for more details on the mathematical structure. Specific mechanical problems like elastic nets have been addressed by Steigmann et al. [8, 9], for application on panthographic structures see dell'Isola et al. [10].

C. Hesch (✉) · S. Schuß
Chair of Computational Mechanics, University of Siegen, Paul-Bonatz-Str. 9-11, 57076 Siegen, Germany
e-mail: christian.hesch@uni-siegen.de

S. Schuß
e-mail: stefan.schuss@uni-siegen.de

F. Aldakheel et al. (eds.), *Current Trends and Open Problems in Computational Mechanics*, https://doi.org/10.1007/978-3-030-87312-7_18

In this contribution, we investigate possible contact formulations for second gradient models. Therefore, we make use of previous developments for higher-order domain decomposition problems as presented in Schuß et. al. [11] and Dittmann et al. [12, 13], and extend them towards contact mechanics as presented in Dittmann et al. [14], where we investigated porous-ductile fracture in non-linear thermo-elasto-plastic solids.

2 Finite Strain Second Gradient Material

We start with a short summary of non-linear elasticity. Therefore, we consider a continuum body with reference configuration $\Omega_0 \subset \mathbb{R}^3$ undergoing a deformation characterised by its deformation map $\boldsymbol{\varphi} : \bar{\Omega}_0 \rightarrow \mathbb{R}^3$, such that the current configuration is given by $\Omega = \boldsymbol{\varphi}(\Omega_0)$. As usual, material points are labelled with uppercase letters, $\mathbf{X} \in \Omega_0$, whereas spatial points are denoted by lowercase letters, $\mathbf{x} \in \Omega$. The deformation gradient $\mathbf{F} : \Omega_0 \rightarrow \mathbb{R}^{3\times 3}$ and the second gradient of the deformation map $\mathbf{G} : \Omega_0 \rightarrow \mathbb{R}^{3\times 3\times 3}$ are defined by $\mathbf{F} = \nabla\boldsymbol{\varphi}$ and $\mathbf{G} = \nabla\mathbf{F}$, respectively. Here, $\nabla(\bullet)$ is the gradient with respect to the material coordinates, such that a vector field $\mathbf{a}$ and a second-order tensor field $\mathbf{A}$ give rise to

$$[\nabla\mathbf{a}]_{iJ} = \frac{\partial[\mathbf{a}]_i}{\partial[\mathbf{X}]_J} \quad \text{and} \quad [\nabla\mathbf{A}]_{iJK} = \frac{\partial[\mathbf{A}]_{iJ}}{\partial[\mathbf{X}]_K}. \tag{1}$$

We assume that the material behaviour is governed by a sufficiently smooth free energy function $\Psi : \Omega_0 \rightarrow \mathbb{R}$, $\Psi = \Psi(\mathbf{F}, \mathbf{G})$. This allows to write the internal stored energy of the continuum body as

$$\Pi^{int} = \int_{\Omega_0} \Psi(\mathbf{F}, \mathbf{G})\, \mathrm{d}V. \tag{2}$$

To establish the principle of virtual work, we introduce the spaces of admissible solutions

$$\mathcal{V}^{\varphi} := \left\{\boldsymbol{\varphi} \in \mathcal{H}^2(\Omega_0)^3 \mid \det(\mathbf{F}) = J > 0 \,\text{in}\, \Omega_0,\ \boldsymbol{\varphi} = \bar{\boldsymbol{\varphi}} \,\text{on}\, \Gamma^{\varphi}\right\} \tag{3}$$

and virtual (or test) functions

$$\mathcal{V}^{\delta\varphi} := \left\{\delta\boldsymbol{\varphi} \in \mathcal{H}^2(\Omega_0)^3 \mid \delta\boldsymbol{\varphi} = 0 \,\text{on}\, \Gamma^{\varphi}\right\}, \tag{4}$$

where $\mathcal{H}^2(\Omega_0)$ denotes the Sobolev space of square integrable functions with square integrable weak derivatives up to the order 2 and $\bar{\varphi}$ is a prescribed deformation. The boundary $\Gamma := \partial\Omega_0$ of the reference domain is assumed to be divided into two open subsets $\Gamma^{\varphi}, \Gamma^{\sigma} \subset \Gamma$, the Dirichlet and Neumann boundaries, which satisfy

$$\bar{\Gamma}^{\varphi} \cup \bar{\Gamma}^{\sigma} = \Gamma \text{ and } \Gamma^{\varphi} \cap \Gamma^{\sigma} = \emptyset, \tag{5}$$

inducing a decomposition of the boundary $\gamma := \partial\Omega$ of the actual configuration such that

$$\bar{\gamma}^{\varphi} \cup \bar{\gamma}^{\sigma} = \gamma \text{ and } \gamma^{\varphi} \cap \gamma^{\sigma} = \emptyset, \tag{6}$$

where $\gamma^{\varphi} = \boldsymbol{\varphi}(\Gamma^{\varphi})$ and $\gamma^{\sigma} = \boldsymbol{\varphi}(\Gamma^{\sigma})$. Thus, we obtain for the virtual work of the internal stored energy

$$\delta\Pi^{int} = \int_{\Omega} \underbrace{J^{-1}\frac{\partial\Psi}{\partial\mathbf{F}}\mathbf{F}^{\mathrm{T}}}_{=\boldsymbol{\sigma}} : \nabla_{\mathbf{x}}\delta\boldsymbol{\varphi} + \underbrace{J^{-1}\frac{\partial\Psi}{\partial\mathbf{G}} : \left(\mathbf{F}^{\mathrm{T}} \otimes \mathbf{F}^{\mathrm{T}}\right)}_{=\mathfrak{p}} \vdots \nabla_{\mathbf{x}}^{2}\delta\boldsymbol{\varphi}\, \mathrm{d}v, \tag{7}$$

where $\boldsymbol{\sigma}$ is the Cauchy stress tensors and $\mathfrak{p}$ the corresponding third order stress tensor. To identify and collect the corresponding boundary conditions for later use in contact formulations, we apply twice integration by parts

$$\begin{aligned}\delta\Pi^{int} = \int_{\Omega} & (\nabla_{\mathbf{x}} \cdot (\nabla_{\mathbf{x}} \cdot \mathfrak{p} - \boldsymbol{\sigma}) \cdot \delta\boldsymbol{\varphi} + \\ & \nabla_{\mathbf{x}} \cdot [(\boldsymbol{\sigma} - \nabla_{\mathbf{x}} \cdot \mathfrak{p})^{\mathrm{T}}\, \delta\boldsymbol{\varphi}] + \nabla_{\mathbf{x}} \cdot (\mathfrak{p} : \nabla_{\mathbf{x}}\delta\boldsymbol{\varphi})\big)\, \mathrm{d}v.\end{aligned} \tag{8}$$

Here, $\nabla_{\mathbf{x}} \cdot (\bullet)$ represents the divergence operator w.r.t. the spatial coordinates, such that $[\nabla_{\mathbf{x}} \cdot \boldsymbol{\sigma}]_i = [\boldsymbol{\sigma}]_{ij,j}$ and $[\nabla_{\mathbf{x}} \cdot \mathfrak{p}]_{ij} = [\mathfrak{p}]_{ijk,k}$. Moreover, $\mathfrak{p} : \nabla_{\mathbf{x}}\delta\boldsymbol{\varphi} = [\mathfrak{p}]_{ijk}[\nabla_{\mathbf{x}}\delta\boldsymbol{\varphi}]_{ij}$, where we have made use of the Einstein summation convention. Application of the divergence theorem on the second and third term yields

$$\delta\Pi^{int} = \int_{\Omega} \nabla_{\mathbf{x}} \cdot (\nabla_{\mathbf{x}} \cdot \mathfrak{p} - \boldsymbol{\sigma}) \cdot \delta\boldsymbol{\varphi} \mathrm{d}v + \int_{\gamma} \big(\delta\boldsymbol{\varphi} \cdot [(\boldsymbol{\sigma} - \nabla_{\mathbf{x}} \cdot \mathfrak{p})\, \mathbf{n}] + \nabla\delta\boldsymbol{\varphi} : (\mathfrak{p}\, \mathbf{n})\big) \mathrm{d}a, \tag{9}$$

where $\mathbf{n}$ denotes the outer normal vector field on γ. Note that the last term can be decomposed using an orthogonal decomposition $\nabla_{\perp} \cdot (\bullet) = \nabla_{\mathbf{x}}(\bullet) : (\mathbf{n} \otimes \mathbf{n})$ and $\nabla_{\parallel} \cdot (\bullet) = \nabla_{\mathbf{x}}(\bullet) : (\mathbf{I} - \mathbf{n} \otimes \mathbf{n})$, such that we can apply a surface divergence theorem to obtain additional information on γ.

Next, following Mindlin and Tiersten [15], we postulate that the conservation of momentum, moment of momentum and the mechanical energy holds in the current configuration (again assuming the absence of inertia terms)

$$\mathbf{0} = \int_{\gamma^{\sigma}} \mathbf{t}_n \, \mathrm{d}a + \int_{\Omega} \mathbf{f} \, \mathrm{d}v, \tag{10}$$

$$\mathbf{0} = \int_{\gamma^{\sigma}} \mathbf{r} \times \mathbf{t}_n + \mathbf{m}_n \, \mathrm{d}a + \int_{\Omega} \mathbf{r} \times \mathbf{f} + \mathbf{c} \, \mathrm{d}v, \tag{11}$$

$$\int_{\Omega} \dot{\Psi} \mathrm{d}v = \int_{\gamma^\sigma} \mathbf{t}_n \cdot \mathbf{v} + \mathbf{m}_n \cdot \mathbf{l} \, \mathrm{d}a + \int_{\Omega} \mathbf{f} \cdot \mathbf{v} + \mathbf{c} \cdot \mathbf{l} \, \mathrm{d}v. \tag{12}$$

Here, $\mathbf{t}_n$ is the force-stress vector, $\mathbf{f}$ a body force, $\mathbf{m}_n$ the couple-stress vector, $\mathbf{c}$ the distributed body couple, $\mathbf{v}$ the material velocity, $\mathbf{r}$ the spatial position vector and $\mathbf{l} = \frac{1}{2}\nabla_{\mathbf{x}} \times \mathbf{v}$ denotes the spin vector using the spatial gradient $\nabla_{\mathbf{x}}(\bullet)$, such that

$$\nabla_{\mathbf{x}}(\mathbf{v}) = \underbrace{\nabla_{\mathbf{x}}^{sym}(\mathbf{v})}_{=:\mathbf{d}} + \underbrace{\nabla_{\mathbf{x}}^{skw}(\mathbf{v})}_{=:\boldsymbol{\omega}}, \quad \mathrm{axl}(\boldsymbol{\omega}) = \mathbf{l}, \tag{13}$$

where $\mathrm{axl}(\boldsymbol{\omega})$ is the axial vector, i.e. $[\mathrm{axl}(\boldsymbol{\omega})]_i = -\frac{1}{2}\epsilon_{ijk}[\boldsymbol{\omega}]_{jk}$ using the Levi-Civita permutation tensor ϵ_{ijk}. Moreover, we define $[[\mathbf{l}]_\times]_{ij} = -\epsilon_{ijk}[\mathbf{l}]_k$, such that $\mathrm{axl}([\mathbf{l}]_\times) = \mathbf{l}$. Noting that $\mathbf{t}_n = \boldsymbol{\sigma}\mathbf{n}$, we get from (10), after application of the divergence theorem, the usual force-stress equation $\nabla_{\mathbf{x}} \cdot \boldsymbol{\sigma} + \mathbf{f} = \mathbf{0}$. Furthermore, for the first term in (11), we can write

$$\int_{\gamma^\sigma} \mathbf{r} \times \mathbf{t}_n \, \mathrm{d}a = -\int_{\gamma^\sigma} \mathbf{n} \cdot (\boldsymbol{\sigma} \times \mathbf{r}) \, \mathrm{d}a = -\int_{\Omega} \nabla_{\mathbf{x}} \cdot (\boldsymbol{\sigma} \times \mathbf{r}) \, \mathrm{d}v, \tag{14}$$

using again the divergence theorem. Hence, we obtain

$$-\int_{\Omega} \nabla_{\mathbf{x}} \cdot (\boldsymbol{\sigma} \times \mathbf{r}) \, \mathrm{d}v = \int_{\Omega} \mathbf{r} \times (\nabla_{\mathbf{x}} \cdot \boldsymbol{\sigma}) + \boldsymbol{\sigma} \overset{\times}{:} \mathbf{I} \, \mathrm{d}v, \tag{15}$$

where $[\boldsymbol{\sigma} \overset{\times}{:} \mathbf{I}]_l = \epsilon_{ljk}[\boldsymbol{\sigma}_{ij}]\delta_{ik}$, which is equal to $-2\,\mathrm{axl}(\boldsymbol{\tau})$. Eventually, introducing the usual couple-stress dyadic $\boldsymbol{\mu}$ as $\mathbf{m}_n = \boldsymbol{\mu}\mathbf{n}$ and application of the divergence theorem on the second term in (11) gives rise to

$$\int_{\gamma^\sigma} \mathbf{m}_n \mathrm{d}a = \int_{\Omega} \nabla_{\mathbf{x}} \cdot \boldsymbol{\mu} \mathrm{d}v. \tag{16}$$

Assuming that the force-stress equation is valid, we obtain the couple-stress equation $\nabla_{\mathbf{x}} \cdot \boldsymbol{\mu} + \mathbf{c} - 2\,\mathrm{axl}(\boldsymbol{\sigma}) = \mathbf{0}$. Noting that $\boldsymbol{\sigma}^{skw} = [\mathrm{axl}(\boldsymbol{\sigma})]_\times$, we further get

$$\boldsymbol{\sigma}^{skw} = \frac{1}{2}[\nabla_{\mathbf{x}} \cdot \boldsymbol{\mu}]_\times + \frac{1}{2}[\mathbf{c}]_\times, \tag{17}$$

such that the force-stress equation becomes

$$\nabla_{\mathbf{x}} \cdot \boldsymbol{\sigma}^{sym} + \frac{1}{2}\nabla_{\mathbf{x}} \cdot [\nabla_{\mathbf{x}} \cdot \boldsymbol{\mu}]_\times + \mathbf{f} + \frac{1}{2}\nabla_{\mathbf{x}} \cdot [\mathbf{c}]_\times = \mathbf{0}. \tag{18}$$

Note that the divergence $\frac{\partial}{\partial \mathbf{x}_j}[\mathbf{A}]_{ij} = [\mathbf{A}]_{ij,j}$ can be used to write $\frac{1}{2}\nabla_{\mathbf{x}} \cdot [\nabla_{\mathbf{x}} \cdot \boldsymbol{\mu}]_\times = \nabla_{\mathbf{x}} \times \nabla_{\mathbf{x}} \cdot \boldsymbol{\mu}$. Moreover, since $\frac{1}{2}\nabla_{\mathbf{x}} \cdot [\nabla_{\mathbf{x}} \cdot (\boldsymbol{\mu} : \mathbf{I})]_\times = 0$, only the deviator $\boldsymbol{\mu}^{dev} = \boldsymbol{\mu} - \frac{1}{3}\,\mathrm{tr}(\boldsymbol{\mu})\mathbf{I}$ is required, i.e. one constitutive equation less has to be considered.

In this final step, we use the energy balance equation in (12) to establish the link between the virtual work in (7), the structure of the boundaries as derived in (9) and the force-stress and couple-stress equations in (18). For the force-stress vector follows

$$\int_{\gamma^\sigma} \mathbf{t}_n \cdot \mathbf{v}\,\mathrm{d}a = \int_\Omega \boldsymbol{\sigma} : \nabla_{\mathbf{x}}(\mathbf{v}) + (\nabla_{\mathbf{x}} \cdot \boldsymbol{\sigma}) \cdot \mathbf{v}\mathrm{d}v, \tag{19}$$

and for the couple-stress vector we get

$$\int_{\gamma^\sigma} \mathbf{m}_n \cdot \mathbf{l}\,\mathrm{d}a = \int_\Omega \boldsymbol{\mu} : \nabla_{\mathbf{x}}(\mathbf{l}) + (\nabla_{\mathbf{x}} \cdot \boldsymbol{\mu}) \cdot \mathbf{l}\mathrm{d}v. \tag{20}$$

Assuming that the force-stress and couple-stress equations are valid, we obtain

$$\int_\Omega \dot{\Psi}\mathrm{d}v = \int_\Omega \underbrace{\boldsymbol{\sigma}^{sym} : \nabla_{\mathbf{x}}(\mathbf{v})}_{=\boldsymbol{\sigma}:\mathbf{d}} + \boldsymbol{\mu} : \nabla_{\mathbf{x}}(\mathrm{axl}(\boldsymbol{\omega}))\mathrm{d}v. \tag{21}$$

Note that $\mathrm{tr}(\nabla_{\mathbf{x}}(\mathrm{axl}(\boldsymbol{\omega}))) = 0$, such that the trace-term does not contribute to the internal energy and we can use again the deviator $\boldsymbol{\mu}^{dev}$ instead of $\boldsymbol{\mu}$.

With regard to (7), we can now state that

$$\left(\boldsymbol{\sigma} - J^{-1}\frac{\partial \Psi}{\partial \mathbf{F}}\mathbf{F}^{\mathrm{T}}\right) : \mathbf{d} - J^{-1}\frac{\partial \Psi}{\partial \mathbf{F}}\mathbf{F}^{\mathrm{T}} : \boldsymbol{\omega} + \boldsymbol{\mu} : \nabla_{\mathbf{x}}(\mathrm{axl}(\boldsymbol{\omega})) - \mathfrak{p} \,\vdots\, \nabla^2_{\mathbf{x}}(\mathbf{v}) = 0. \tag{22}$$

In consequence, we can write for the last term

$$\left(-\frac{1}{2}[\boldsymbol{\mu}]_{mj}\epsilon_{jin} - [\mathfrak{p}]_{inm}\right)[\nabla^2_{\mathbf{x}}(\mathbf{v})]_{inm} = 0. \tag{23}$$

Following Spencer and Soldatos [16] with respect to the work of Toupin [2], we demand that the symmetric part with respect to the n and m components of the terms in the brackets of (23) have to be zero, as the anti-symmetric parts are always zero due to the second term. Thus, we obtain

$$-\frac{1}{2}\left([\boldsymbol{\mu}]_{mj}\epsilon_{jin} + [\boldsymbol{\mu}]_{nj}\epsilon_{jim}\right) = \left[\frac{\partial \Psi}{\partial \mathbf{G}}\right]_{iJK}\left([\mathbf{F}]_{mJ}[\mathbf{F}]_{nK} + [\mathbf{F}]_{nJ}[\mathbf{F}]_{mK}\right), \tag{24}$$

and pre-multiplication with ϵ_{rin} yields eventually

$$3[\boldsymbol{\mu}]_{mr} - [\boldsymbol{\mu}]_{nn}\delta_{rm} = -2\epsilon_{rin}\left[\frac{\partial\Psi}{\partial\mathbf{G}}\right]_{iJK}\big([\mathbf{F}]_{mJ}[\mathbf{F}]_{nK} + [\mathbf{F}]_{nJ}[\mathbf{F}]_{mK}\big), \tag{25}$$

which is the constitutive law for the deviatoric part of the couple-stress $\boldsymbol{\mu}$. Noting that we can rewrite the couple-stress vector in (20) as $\mathbf{m}_n \cdot \mathbf{l} = 2[\mathbf{m}]_\times : \nabla_\mathbf{x}(\mathbf{v})$ and replacing the velocity by the virtual displacements, we obtain

$$\int_\Omega \boldsymbol{\sigma} : \nabla_\mathbf{x}^{sym}(\delta\boldsymbol{\varphi}) + \boldsymbol{\mu}^{dev} : \nabla_\mathbf{x}(\mathrm{axl}(\nabla_\mathbf{x}^{skw}(\delta\boldsymbol{\varphi})))\mathrm{d}v = \int_{\gamma^\sigma} \mathbf{t}_n \cdot \delta\boldsymbol{\varphi} + 2[\mathbf{m}]_\times : \nabla_\mathbf{x}(\delta\boldsymbol{\varphi})\mathrm{d}a + \int_\Omega \mathbf{f}\cdot\delta\boldsymbol{\varphi} + \frac{1}{2}\mathbf{c}\nabla_\mathbf{x} \times \delta\boldsymbol{\varphi}\mathrm{d}v, \tag{26}$$

and if Ψ is constructed such that $[\mathfrak{p}]_{inm} = -\frac{1}{2}[\boldsymbol{\mu}]_{mj}\epsilon_{jin}$ (see Toupin [2] for possible formulations), we end with

$$\int_\Omega \boldsymbol{\sigma} : \nabla_\mathbf{x}^{sym}(\delta\boldsymbol{\varphi}) + \mathfrak{p} \,\vdots\, \nabla_\mathbf{x}^2(\mathbf{v})\mathrm{d}v = \int_{\gamma^\sigma} \mathbf{t}_n \cdot \delta\boldsymbol{\varphi} + 2[\mathbf{m}]_\times : \nabla_\mathbf{x}(\delta\boldsymbol{\varphi})\mathrm{d}a + \int_\Omega \mathbf{f}\cdot\delta\boldsymbol{\varphi} + \frac{1}{2}\mathbf{c}\nabla_\mathbf{x} \times \delta\boldsymbol{\varphi}\mathrm{d}v. \tag{27}$$

3 Contact Constraints

We start here with a short summary in higher-order domain decomposition problems, before we consider contact and impact formulations. Therefore let $\Omega^{(1)}$, $\Omega^{(2)}$ be two domains with an open common boundary $\gamma^c = \gamma^c_{(1)} = \gamma^c_{(2)}$, such that

$$\gamma_{(i)} = \bar{\gamma}^{\varphi}_{(i)} \cup \bar{\gamma}^{\sigma}_{(i)} \cup \bar{\gamma}^{c}_{(i)} \tag{28}$$

and

$$\gamma^{\varphi}_{(i)} \cap \gamma^{\sigma}_{(i)} = \emptyset, \quad \gamma^{\varphi}_{(i)} \cap \gamma^{c}_{(i)} = \emptyset, \quad \gamma^{\sigma}_{(i)} \cap \gamma^{c}_{(i)} = \emptyset, \quad i = 1, 2. \tag{29}$$

To preserve in general C^n-continuity across the interface in a weak sense, the corresponding deformation mappings $\boldsymbol{\varphi}^{(i)}$ have to fulfil the optimality condition

$$\sum_{|\mathbf{j}|\leq n} h^{2|\mathbf{j}|}\|\partial^\mathbf{j}\boldsymbol{\varphi}^{(1)} - \partial^\mathbf{j}\boldsymbol{\varphi}^{(2)}\|^2_{L^2(\gamma^c)^3} = \inf_{\mathbf{w}\in\mathcal{V}^{\varphi}_{(1)}} \sum_{|\mathbf{j}|\leq n} h^{2|\mathbf{j}|}\|\partial^\mathbf{j}\mathbf{w} - \partial^\mathbf{j}\boldsymbol{\varphi}^{(2)}\|^2_{L^2(\gamma^c)^3}, \tag{30}$$

cf. Dittmann et al. [12], where $\mathbf{j} \in \mathbb{N}_0^3$ is a multi-index denoting the respective derivative and h is a length scale dependent parameter ensuring physically correct dimensions. In Dittmann et al. [13], the typical length of an element is used, other choices are possible. Introducing a suitable space of Lagrange multiplier $\mathcal{M}$, we obtain from (30) the constraints

$$\Phi = \sum_{|\mathbf{j}|\leq n} h^{2|\mathbf{j}|} \int_{\gamma^c} \partial^{\mathbf{j}}\delta\boldsymbol{\lambda} \cdot (\partial^{\mathbf{j}}\boldsymbol{\varphi}^{(1)} - \partial^{\mathbf{j}}\boldsymbol{\varphi}^{(2)}) \underset{\cdot}{a} = 0, \quad \forall \delta\boldsymbol{\lambda} \in \mathcal{M}, \tag{31}$$

which reads for first gradient materials (weak C^0-coupling)

$$\Phi = \int_{\gamma^c} \delta\boldsymbol{\lambda} \cdot \left(\boldsymbol{\varphi}^{(1)} - \boldsymbol{\varphi}^{(2)}\right) \underset{\cdot}{a} = 0, \quad \forall \delta\boldsymbol{\lambda} \in \mathcal{M} \tag{32}$$

and for second gradient materials (weak C^1-coupling)

$$\Phi = \int_{\gamma^c} \delta\boldsymbol{\lambda} \cdot \left(\boldsymbol{\varphi}^{(1)} - \boldsymbol{\varphi}^{(2)}\right) + h^2 \nabla_{\mathbf{x}}(\delta\boldsymbol{\lambda}) : \left(\nabla_{\mathbf{x}}(\boldsymbol{\varphi}^{(1)}) - \nabla_{\mathbf{x}}(\boldsymbol{\varphi}^{(2)})\right) \underset{\cdot}{a} = 0, \tag{33}$$

for all $\delta\boldsymbol{\lambda} \in \mathcal{M}$.

Concerning contact in a first gradient material, we have to decompose the constraints in normal and tangential directions, since we have to obey different physical constitutive laws. In normal direction, the non-penetration condition have to be valid, in tangential direction, possible friction laws are required. Introducing the projector $\mathbf{I} = (\mathbf{I} - \mathbf{n} \otimes \mathbf{n}) + \mathbf{n} \otimes \mathbf{n}$, where $\mathbf{n}$ denotes the outer normal vector field on $\gamma_{(1)}^{(c)}$, we obtain

$$\Phi = \int_{\gamma^c} \delta\lambda_n \underbrace{\mathbf{n} \cdot \left(\boldsymbol{\varphi}^{(1)} - \boldsymbol{\varphi}^{(2)}\right)}_{=:g_n} + \delta\boldsymbol{\lambda}_t \cdot \left(\boldsymbol{\varphi}^{(1)} - \boldsymbol{\varphi}^{(2)}\right) \underset{\cdot}{a}, \tag{34}$$

where $\delta\lambda_n = \delta\boldsymbol{\lambda} \cdot \mathbf{n}$ and $\delta\boldsymbol{\lambda}_t = \delta\boldsymbol{\lambda}\,(\mathbf{I} - \mathbf{n} \otimes \mathbf{n})$. The constitutive relations require that

$$\lambda_n \geq 0, \quad g_n \leq 0 \text{ and } \lambda_n g_n = 0, \tag{35}$$

representing the classical Karush–Kuhn Tucker conditions. For the tangential part $\boldsymbol{\lambda}_t$, classical formulations like Coulomb's friction law can be taken into account. For the gradient constraints follows, that they have to obey the constitutive relations as well: If λ_n is equal zero due to the positive gap, the constraints in all directions are released, including the gradient constraints. If the gap is closed, $\boldsymbol{\lambda}_t$ is restricted due to the used constitutive relation for the friction. Consequently, the constraints for second gradient material read

$$\Phi = \int_{\gamma^c} \delta\lambda_n \mathbf{n} \cdot \left(\boldsymbol{\varphi}^{(1)} - \boldsymbol{\varphi}^{(2)}\right) + \delta\boldsymbol{\lambda}_t \cdot \left(\boldsymbol{\varphi}^{(1)} - \boldsymbol{\varphi}^{(2)}\right) \underset{\cdot}{a} + \int_{\gamma^c} h^2 \nabla_{\mathbf{x}}(\delta\lambda_n \mathbf{n}) : \left(\nabla_{\mathbf{x}}(\boldsymbol{\varphi}^{(1)}) - \nabla_{\mathbf{x}}(\boldsymbol{\varphi}^{(2)})\right) \underset{\cdot}{a} + \int_{\gamma^c} h^2 \nabla_{\mathbf{x}}(\delta\boldsymbol{\lambda}_t) : \left(\nabla_{\mathbf{x}}(\boldsymbol{\varphi}^{(1)}) - \nabla_{\mathbf{x}}(\boldsymbol{\varphi}^{(2)})\right) \underset{\cdot}{a} \tag{36}$$

This has to be discretised using suitable Mortar methods. We refer again to Dittmann et al. [12] for a detailed discussion on Mortar constraints and their application on higher-order domain decomposition methods.

4 Conclusions

In this paper, we presented a second gradient formulation for non-linear materials in a nutshell. Afterwards, suitable domain decomposition methods are used to derive a possible contact formulation for a second gradient material. Several unresolved issues have to be addressed in future work: First, is a moment transfer in the contact area physically reasonable? Second, if so, the normal constraints can be implemented with regard to the Karush–Kuhn Tucker conditions. However, the application of the tangential constraints in terms of, e.g., Coulombs friction law remains a challenge, since the action of the gradient of the constitutively restricted Lagrange multipliers is unclear. Especially the interpretation as couple-stress vector remains a challenge.

References

1. Mindlin, R. A. (1964). Micro-structure in linear elasticity. *Archive for Rational Mechanics and Analysis, 16*, 51–78.
2. Toupin, R. A. (1962). Elastic materials with couple-stresses. *Archive for Rational Mechanics and Analysis, 11*, 385–414.
3. Eringen, A. C. (1999). *Microcontinuum field theories I: Foundations and solids*. Berlin: Springer.
4. Eringen, A. C. (1983). Theories of nonlocal plasticity. *International Journal of Engineering Science, 21*, 741–751.
5. Maugin, G. A., & Metrikine, A. V. (2010). *Mechanics of generalized continua*. Berlin: Springer.
6. Javili, A., dell'Isola, F., & Steinmann, P. (2013). Geometrically nonlinear higher-gradient elasticity with energetic boundaries. *Journal of the Mechanics and Physics of Solids, 61*(12), 2381–2401.
7. Kirchner, N., & Steinmann, P. (2005). A unifying treatise on variational principles for gradient and micromorphic continua. *Philosophical Magazine, 85*, 3875–3895.
8. Steigmann, D. J. (2018). Equilibrium of elastic lattice shells. *Journal of Engineering Mathematics, 109*, 47–61.

9. Steigmann, D. J., & dell'Isola, F. (2015). Mechanical response of fabric sheets to three-dimensional bending, twisting, and stretching. *Acta Mechanica Sinica*.
10. dell'Isola, F., Giorgio, I., Pawlikowski, M., & Rizzi, N. L. (2016). Large deformations of planar extensible beams and pantographic lattices: heuristic homogenization, experimental and numerical examples of equilibrium. *Proceedings of the Royal Society of London A: Mathematical, Physical and Engineering Sciences, 472*(2185).
11. Schuß, S., Dittmann, M., Klinkel, S., Wohlmuth, B., & Hesch, C. (2019). Multi-patch isogeometric analysis for Kirchhoff-Love shell elements. *Computer Methods in Applied Mechanics and Engineering, 349*, 91–116.
12. Dittmann, M., Schuß, S., Wohlmuth, B., & Hesch, C. (2019). Weak C^n coupling for multi-patch isogeometric analysis in solid mechanics. *International Journal for Numerical Methods in Engineering, 118*(3), 678–699.
13. Dittmann, M., Schuß, S., Wohlmuth, B., & Hesch, C. (2020). Crosspoint modification for multi-patch isogeometric analysis. *Computer Methods in Applied Mechanics and Engineering, 360*, 112768.
14. Dittmann, M., Aldakheel, F., Schulte, J., Schmidt, F., Krüger, M., Wriggers, P., & Hesch, C. (2020). Phase-field modeling of porous-ductile fracture in non-linear thermo-elasto-plastic solids. *Computer Methods in Applied Mechanics and Engineering, 361*, 112730.
15. Mindlin, R. D., & Tiersten, H. F. (1962). Effects of couple-stresses in linear elasticity. *Archive for Rational Mechanics and Analysis, 11*, 415–448.
16. Spencer, A. J. M., & Soldatos, K. P. (2007). Finite deformations of fibre-reinforced elastic solids with fibre bending stiffness. *International Journal of Non-Linear Mechanics, 42*, 355–368.

Locking-Free Mixed Finite Element Methods and Their Spurious Hourglassing Patterns

Moritz Hille, Robin Pfefferkorn, and Peter Betsch

This article is dedicated to Professor Peter Wriggers who fostered the successful development of Computational Mechanics and who has made significant scientific contributions to the field. This includes mixed finite element methods for large deformation problems, a topic also addressed in the present work.

Abstract We present the five field mixed finite element formulation introduced by Armero and extend it to 3D problems. It combines the nonlinear mixed pressure element with an enhanced assumed strain (EAS) method employing the transposed Wilson modes. The well-known mixed pressure element arises from a Hu-Washizu-like variational principle, where dilatation and pressure are independent variables. This functional is further modified using the EAS framework to get the mixed formulation presented in this work. The element is compared to several mixed pressure and EAS element formulations showing its great performance in alleviating volumetric and shear locking in large deformation problems. The main focus of the present work is spurious hourglassing of mixed finite elements that arise in hyperelastic and elasto-plastic simulations.

Keywords Mixed finite elements · Enhanced assumed strain · Hyperelasticity · Elasto-plasticity · Robustness · Numerical instabilities

M. Hille · R. Pfefferkorn · P. Betsch (✉)
Karlsruhe Institute of Technology, Institute of Mechanics, Karlsruhe, Germany
e-mail: peter.betsch@kit.edu

M. Hille
e-mail: moritz.hille@kit.edu

R. Pfefferkorn
e-mail: robin.pfefferkorn@kit.edu

F. Aldakheel et al. (eds.), *Current Trends and Open Problems in Computational Mechanics*, https://doi.org/10.1007/978-3-030-87312-7_19

1 Introduction

The finite element method (FEM) (see e.g. Wriggers [19]) is one of the most significant methods to numerically solve partial differential equations (PDE) in solid continuum mechanics. Its key idea is to subdivide a body in the name giving finite elements (FE) and to approximate the deformation of this body using the weak form of a PDE.

One of the major phenomenons that limits the applicability of low-order displacement elements is locking, which denotes numerical stiffening of the body. It causes too small deformations and slow convergence with mesh refinement. A distinction is made between shear and volumetric locking occurring in bending dominated problems and in the incompressible limit, respectively. In 1985 Simo et al. [14] presented a mixed element formulation based on a Hu-Washizu [17] three field functional introducing extra variables for dilatation and pressure such that the deviatoric and volumetric part of the strain tensor can be treated separately. This alleviates the effects of volumetric locking for nearly incompressible materials.

In 1990 and 1992 Simo and Rifai [13] and Simo and Armero [11] presented another widely used mixed finite element method called enhanced assumed strain method (EAS) for linear and nonlinear deformations, respectively. This method introduced enhanced strains in addition to the compatible strains with ansatz functions based on the previously introduced popular incompatible displacement modes (see Wilson [18]). While EAS elements work extremely well for small deformation problems, they unfortunately suffer from spurious hourglassing instabilities under large deformation compression as first pointed out by Wriggers and Reese [20]. A first remedy to this problem was proposed by Korelc and Wriggers [6], who introduced the method of transposed Wilson modes in 1996 solving this problem for polyconvex elastic materials. However, the method proposed in Korelc and Wriggers [6] is not frame-invariant if it is subject to large rotations, which was corrected by Glaser and Armero [3]. For more details on the long history of EAS elements, including various ansatz functions and transformations, see e.g. Pfefferkorn and Betsch [7]. Even though the transposed Wilson-modes eliminated hourglassing in hyperelastic problems, the element still suffers from hourglassing in elasto-plastic simulations. To the best knowledge of the authors, there exists no EAS element that overcomes this problem without introducing unphysical expressions with parameters to be set by the user.

In this paper we discuss an element formulation presented by Armero [1] in 2000, that combines the mixed pressure formulation of Simo et al. [14] with the EAS method of Glaser and Armero [3]. The key idea is to enhance only the non-diagonal elements of the deformation gradient yielding a specific enhancement of the deviatoric part of the strain tensor, since the locking of the volumetric part is taken care of by the mixed pressure ansatz. This yields a five field variational functional which is the basis for a stable and hourglassing-free finite element method for nonlinear elastic material behavior with great results in computing deformations of hyperelastic as well as elasto-plastic problems. Most remarkably, the element presented by

Armero [1] shows no hourglassing in 2D elasto-plastic simulations. Hence Armero [1] remedies the shear locking of the mixed pressure element of Simo et al. [14] whilst circumventing the spurious hourglassing effects of the EAS formulation by Simo and Armero [11] in 2D. The 3D case, however, has not been covered so far. In this paper we present a 3D version of this element formulation and provide numerical tests for 3D elastic and plastic material behavior. We show most importantly that, unfortunately, the 3D formulation of this element is not hourglassing-free in elasto-plastic simulations.

The paper is structured into four chapters. Section 2 covers the mixed finite element method of Armero [1]. Section 3 contains numerical tests assessing the performance and stability of the mixed element presented in Sect. 2 by comparing it to mixed EAS and mixed pressure element formulations. It is shown that the mixed pressure EAS element has extremely high robustness (concerning number of load steps and number of Newton-Raphson (NR) iterations) and is not susceptible to locking. However, in 3D it is not hourglassing-free in standard tests. Finally, Sect. 4 concludes the paper and its results with a short summary.

2 The Five-Field Finite Element Formulation

In this section we give an overview of the five-field finite element formulation, which was first presented by Armero [1] for 2D problems. In the present work we focus on the 3D formulation.

2.1 *Deformation Gradient*

We first introduce the deformation gradient $\mathbf{F}$ since its structure is the key to both mixed methods on which the present element relies. It combines the volumetric and deviatoric split used for the mixed pressure element in Simo et al. [14] with an additive enhancement. The purely displacement based deformation gradient

$$\mathbf{F}_{\varphi}(\boldsymbol{\varphi}) = \text{Grad}\,(\boldsymbol{\varphi}) = \frac{\partial \boldsymbol{\varphi}(\mathbf{X})}{\partial \mathbf{X}} \tag{1}$$

is modified in the form

$$\mathbf{F}(\boldsymbol{\varphi}, \theta, \boldsymbol{\Gamma}) = \theta^{1/3} \widehat{J}^{-1/3} \widehat{\boldsymbol{F}}. \tag{2}$$

Therein, θ is the (independent) dilatation and $\widehat{J} = \det(\widehat{\mathbf{F}})$. Moreover, the enhancement is denoted by $\widehat{(\bullet)}$ and the enhanced deformation gradient $\widehat{\mathbf{F}}$ is given by

$$\widehat{\mathbf{F}}(\boldsymbol{\varphi}, \boldsymbol{\Gamma}) = \mathbf{F}_{\varphi}(\boldsymbol{\varphi}) + \widetilde{\mathbf{F}}(\boldsymbol{\varphi}, \boldsymbol{\Gamma}), \tag{3}$$

where

$$\widetilde{\mathbf{F}}(\boldsymbol{\varphi}, \boldsymbol{\Gamma}) = \mathbf{F}_{\varphi,0}(\boldsymbol{\varphi})\mathbf{F}_{\Gamma}(\boldsymbol{\Gamma}) \tag{4}$$

is the enhanced part of the deformation gradient, which depends apart from the enhanced degrees of freedom $\boldsymbol{\Gamma}$ on the deformations $\boldsymbol{\varphi}$ via $\mathbf{F}_{\varphi,0}$. The latter denotes the compatible deformation gradient (1) evaluated at the element centroid and ensures frame invariance (see Simo et al. [12] and Glaser and Armero [3]). Finally, $\mathbf{F}_{\Gamma}$ includes the enhanced degrees of freedom Γ_i arranged in vector $\boldsymbol{\Gamma}$ and is specified in Sect. 2.3.

2.2 Variational Framework

The classic variational potential for the pure displacement element formulation is extended by two constraints. The first condition on kinematic level, $J = \theta$ with $J = \det(\mathbf{F}_{\varphi})$, is enforced by the Lagrange multiplier p in analogy to Simo et al. [14]. The second constraint ensures that the enhanced part of the deformation gradient $\widetilde{\mathbf{F}}$ vanishes on the continuum level and is enforced by the second Lagrange multiplier $\mathbf{P}$. This yields the five-field functional

$$\Pi(\boldsymbol{\varphi}, \theta, p, \boldsymbol{\Gamma}, \mathbf{P}) = \int_{\mathcal{B}_0} \left[W(\mathbf{C}) + p\,(J - \theta) - \mathbf{P} : \widetilde{\mathbf{F}}\right] \mathrm{d}V + \Pi_{\text{ext}}(\boldsymbol{\varphi})\,, \tag{5}$$

where W is an arbitrary hyperelastic strain-energy function characterizing the behavior of body $\mathcal{B}$, which is mapped from its reference configuration $\mathcal{B}_0$ to its current configuration by deformation map $\boldsymbol{\varphi}$. Furthermore, Π_{ext} denotes the potential of external forces not further specified here and $\mathbf{C} = \mathbf{F}^{\mathrm{T}}\mathbf{F}$ is the modified Cauchy–Green tensor computed using $\mathbf{F}(\boldsymbol{\varphi}, \boldsymbol{\Gamma})$.

The five independent fields in (5) require the five stationary conditions

$$\delta_{\varphi}\Pi = \int_{\mathcal{B}_0} \operatorname{dev}(\boldsymbol{\tau}) : \widehat{\nabla}^{\mathrm{s}}(\delta\boldsymbol{\varphi}) + pJ\mathbf{I} : \nabla^{\mathrm{s}}(\delta\boldsymbol{\varphi}) - \delta_{\varphi}\widetilde{\mathbf{F}} : \mathbf{P}\,\mathrm{d}V + \delta_{\varphi}\Pi_{\text{ext}} = 0, \tag{6a}$$

$$\delta_{\theta}\Pi = \int_{\mathcal{B}_0} \left(\frac{1}{3}\operatorname{tr}(\boldsymbol{\tau})\theta^{-1} - p\right) \delta\theta \mathrm{d}V = 0, \tag{6b}$$

$$\delta_{p}\Pi = \int_{\mathcal{B}_0} \delta p\,(J - \theta)\,\mathrm{d}V = 0, \tag{6c}$$

$$\delta_{\Gamma}\Pi = \int_{\mathcal{B}_0} \operatorname{dev}(\boldsymbol{\tau}) : \operatorname{sym}\left(\mathbf{F}_{\varphi,0}\delta_{\Gamma}\mathbf{F}_{\Gamma}\widehat{\mathbf{F}}^{-1}\right) - \delta_{\Gamma}\widetilde{\mathbf{F}} : \mathbf{P}\,\mathrm{d}V = 0, \tag{6d}$$

$$\delta_{P}\Pi = -\int_{\mathcal{B}_0} \delta\mathbf{P} : \widetilde{\mathbf{F}}\,\mathrm{d}V = 0, \tag{6e}$$

where $\boldsymbol{\tau}$ is the constitutive Kirchhoff stress tensor and assumes the usual form

$$\boldsymbol{\tau} = \mathbf{FSF}^{\mathrm{T}} \quad \text{with} \quad \mathbf{S} = 2\frac{\partial W(\mathbf{C})}{\partial \mathbf{C}} \tag{7}$$

for a hyperelastic material model.[1] Furthermore, $\widehat{\nabla}^{\mathrm{s}}(\bullet)$ is a modified symmetric gradient in the spatial configuration that arises from the derivative of the Cauchy–Green tensor $\mathbf{C}$ with respect to $\boldsymbol{\varphi}$. The modified gradient $\widehat{\nabla}(\bullet)$ is given by

$$\widehat{\nabla}(\bullet) := \widehat{\nabla}_{\mathrm{x}}(\bullet) = \left[\nabla_{\mathrm{X}}(\bullet) + \mathbf{F}_{\Gamma}^{\mathrm{T}}\nabla_{0}(\bullet)\right]\widehat{\mathbf{F}}^{-1}, \tag{8}$$

where $\nabla_0(\bullet) = \nabla_{\mathrm{X}}(\bullet)\big|_{\xi=0}$ and $\nabla_{\mathrm{X}}(\bullet) = \mathrm{Grad}(\bullet)$.

2.3 Discretization

For all simulations we consider the common isoparametric concept and the Bubnov–Galerkin method. Thus, the discretized $(\bullet)^{\mathrm{h}}$ and elementwise $(\bullet)^{e}$ position $\mathbf{X}^{\mathrm{h},e}$ and displacement $\boldsymbol{\varphi}^{\mathrm{h},e}$ as well as its variation $\delta\boldsymbol{\varphi}^{\mathrm{h},e}$ are given by

$$\mathbf{X}^{\mathrm{h},e} = \sum_{I\in\mathcal{I}} N_I^e \mathbf{X}_I^e, \quad \boldsymbol{\varphi}^{\mathrm{h},e} = \sum_{I\in\mathcal{I}} N_I^e \boldsymbol{\varphi}_I^e \quad \text{and} \quad \delta\boldsymbol{\varphi}^{\mathrm{h},e} = \sum_{I\in\mathcal{I}} N_I^e \delta\boldsymbol{\varphi}_I^e \tag{9}$$

for all elements $e \in \{1, 2, \ldots, n_e\}$ in the subdivision $\mathcal{B}_0^{\mathrm{h}} = \bigcup_{e=1}^{n_e} \Omega^e$ of body $\mathcal{B}_0$. Furthermore, $\mathcal{I}$ is the set of nodes in the reference element $\Omega_{\square}$. The standard trilinear Lagrangian shape functions for the eight noded cube are employed.

Constant ansatz functions are used for the dilatation θ and Lagrange multiplier p since they match well with linear shape functions on the deformation $\boldsymbol{\varphi}$ if θ and p are restricted to each element Ω^e as discussed e.g. in Hughes [4]. This yields

$$\theta^{\mathrm{h},e} = \text{const} \quad \text{and} \quad p^{\mathrm{h},e} = \text{const} \tag{10}$$

in every element.

This element does not fulfill the LBB-condition (cf. Boffi et al. [2]). This means, that for some boundary conditions checkerboard solutions for the approximated pressure arises. Fortunately, this can easily be fixed by various workarounds e.g. $\mathcal{L}^2$-smoothing (cf. Wriggers [19]).

Finally, we discuss the approximation of $\mathbf{F}_{\Gamma}(\boldsymbol{\Gamma})$ introduced in (4). Armero [1] uses transposed Wilson modes which are first described in the work of Korelc and Wriggers [6]. The basic idea of the element proposed by Armero [1] is to only use the non-diagonal modes.[2] We extend this formulation to the three-dimensional space which yields an enhancement field of the form

[1] Note that all terms in (6) refer to the spatial configuration except for terms containing $\mathbf{P}$ or $\delta\mathbf{P}$ since these terms vanish on discrete level due to the orthogonality condition given below.

[2] We also tested the same element with all nine transposed Wilson Modes for the enhanced field. That element is even softer than the element of Armero, which is too soft in pure bending problems

$$\mathbf{F}_\Gamma(\boldsymbol{\Gamma}^e) = \frac{j_0^{h,e}}{j^{h,e}(\boldsymbol{\xi})} \left(\mathbf{J}_0^{h,e}\right)^{-T} \left(\begin{bmatrix} 0 & \Gamma_1^e \xi & \Gamma_2^e \xi \\ \Gamma_3^e \eta & 0 & \Gamma_4^e \eta \\ \Gamma_5^e \zeta & \Gamma_6^e \zeta & 0 \end{bmatrix} \right) \left(\mathbf{J}_0^{h,e}\right)^{-1}, \tag{11}$$

where $\mathbf{J}_0^{h,e} = \mathbf{J}^{h,e}\big|_{\xi=\mathbf{0}}$ and $\mathbf{J}^{h,e}$ is the Jacobian matrix of the isoparametric map. Analogously, their determinants are given by $j^{h,e} = \det(\mathbf{J}^{h,e})$ and $j_0^{h,e} = \det(\mathbf{J}^{h,e})\big|_{\xi=\mathbf{0}}$, respectively.

In analogy to standard EAS elements the present five-field element formulation fulfills the patch test provided that $\int_{\Omega_\square} \mathbf{F}_\Gamma \, d\Omega_\square = 0$ is fulfilled (see Armero [1]).

The usual $\mathcal{L}^2$-orthogonality condition assumed between discrete stress and strain field yields

$$\int_{\Omega^e} \widetilde{\mathbf{F}}^{h,e} : \delta \mathbf{P}^{h,e} \, dV = \int_{\Omega^e} \delta \widetilde{\mathbf{F}}^{h,e} : \mathbf{P}^{h,e} \, dV = 0 \tag{12}$$

on element level. This allows to eliminate $\mathbf{P}$ in the discrete case such that only four fields remain in the discretized version of functional (5).

3 Numerical Investigations

This chapter covers benchmarks testing the performance of the finite element formulation presented in Sect. 2. The element is compared to the standard isoparametric displacement element and popular (well-working) mixed elements. Armero [1] already described and evaluated the mixed pressure EAS element using various 2D tests. Thus, focus of this work lies on the 3D formulation.

The element proposed by Armero is denoted Q1/P0ET2. Its 3D extension covered in this work is named H1/P0ET6. The 3D elements used for comparison in the following numerical examples are:

- H1, the standard isoparametric 8-node displacement element,
- H1/E9, the classic EAS element formulation (see Simo and Armero [11]) and H1/ET9, the EAS formulation with transposed Wilson modes (see Glaser and Armero [3]),
- HA1/ET12, an extension of H1/ET9 with additional volumetric modes using a 9-point integration (see Pfefferkorn and Betsch [7]),
- H1/E9-MIP and HA1/E12-MIP the EAS elements with improved robustness by use of the mixed integration point (MIP) method (see Pfefferkorn et al. [9]),
- H1/P0, the three field mixed pressure element (see Simo et al. [14]),
- H1/S18, the well known assumed stress element by Pian and Sumihara [10], which has 18 stress modes in 3D.

---

with undistorted meshes. Thus, the element with all 9 enhanced modes is not taken into account in subsequent investigations.

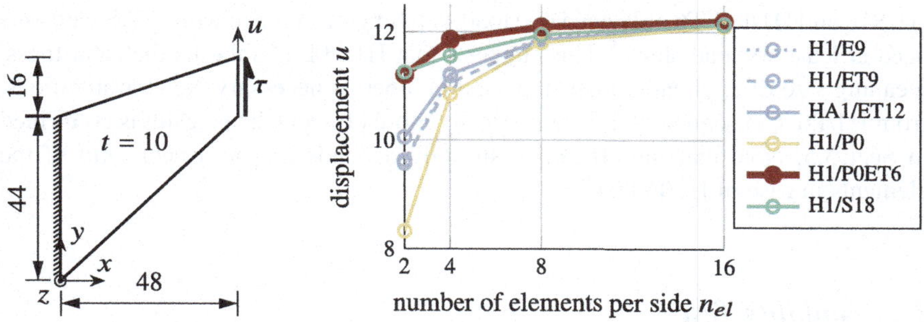

Fig. 1 Cook's membrane: system (left) and displacement u of the hindmost node in the right upper corner (right)

We consider two material models for all simulations. First, the nonlinear neo-Hookean material model with strain-energy function

$$W = \frac{\mu}{2}(\text{tr}(\mathbf{C}) - 3) + \frac{\lambda}{2}(\ln J)^2 - \mu \ln J \tag{13}$$

for elastic behavior.[3,4] Second, the multiplicative J_2-plasticity model with nonlinear isotropic hardening and the Hencky elastic law as proposed by e.g. Simo [15]. The material parameters of that model are set to their usual values (see e.g. [1], [3] or [8]).

3.1 Cook's Membrane

The first test described in the present work is the Cook's membrane test. Figure 1 shows the system, which is clamped on the left. We employ the elastic model given in (13) with material parameters $\mu = 756.00$ and $\lambda = 8.2669 \cdot 10^4$.

The focus of this test is on shear and volumetric locking. To determine the elements performance, we evaluate the displacement u in y-direction of the top right corner caused by equally distributed shear stress $\tau = 100$. The right image in Fig. 1 shows this deformation in dependence of the number of elements per side for the various mixed elements (there are always 2 elements in direction of the thickness).

The worst element is H1/P0 which suffers from shear locking since only the volumetric parts are modified in this formulation. On the other end, H1/P0ET6 exhibits the best results with fastest convergence with mesh refinement.

Another interesting result of this investigation is the number of load steps required to achieve convergence of Newton's method. Regardless of the mesh size, H1/P0,

---

[3] Note that no volumetric deviatoric split in W (13) is considered in the present work in contrast to many works on mixed elements with approximation of the pressure (see e.g. Simo et al. [14]).

[4] An inverse stress strain relation of (13) for H1/S18 is given in [9].

H1/S18 and H1/P0ET6 only need one load step for convergence while EAS elements need at least six load steps.[5] This suggests that H1/P0ET6 inherits the robustness, meaning size of applicable load steps and number of necessary Newton iterations, from H1/P0. Consequently, it is numerically more efficient. This result is confirmed in Sect. 3.5. Note that the efficiency strongly depends on the aspect ratio of the elements in case of H1/P0ET6.

3.2 Stability Test

This section describes the modal analysis presented in Armero [1], where the eigenvalues of the stiffness matrix are evaluated to examine possible instabilities from so called hourglass modes. Armero [1] covers the 2D version of this test while this paper focuses on 3D testing as mentioned above.

Every eigenpair linked to an eigenvalue ω_i of the element stiffness matrix denotes a specific mode that describes one kind of motion or deformation that the element can perform. Besides the six rigid body and the six constant strain modes, there are twelve hourglass modes in 3D with corresponding eigenvalue ω_i^{hour}. These eigenvalues can cause locking and numerical instabilities and sometimes yield non-physical deformations in numerical computations. Negative hourglass eigenvalues can cause typical hourglass patterns, thus, the construction of element formulations that exhibit only positive eigenvalues should be the goal.

For the 3D version of this test we consider a single finite element $\Omega^e = [-1, 1]^3$. It is deformed in a state of uniaxial stress which allows to analytically compute the principal stretch $\lambda_2 = \lambda_3$ for any given λ_1 using conditions $\tau_2 = \tau_3 = 0$.[6] Armero [1] showed that for the plane strain case of the problem depicted in Fig. 2 the hourglass-eigenvectors are always constant and thus let us readily compute the corresponding eigenvalues. In the uniaxial 3D case the hourglass-eigenvectors depend on the stiffness matrix and thus on the material properties and the element formulation. However, they can still be computed analytically as shown in Pfefferkorn and Betsch [8].

Since the eigenmodes are well known, a matrix of eigenvectors can be described and a product of the stiffness matrix and the hourglass matrix yields the eigenvalues. For the elastic case we use the Lamé parameters $\lambda = 10^5$ and $\mu = 20$.

Before we list the 3D results, we give a brief summary of previous results for 2D elements. The first thorough work on the instabilities under compression of Q1/E4 is Wriggers and Reese [20]. Viehbahn et al. [16] showed for Q1/S5 and Wriggers and Reese [20] for Q1/E4 that these two elements show instabilities in uniaxial pressure

[5] Elements with MIP method need even slightly more iterations in this example, which shows that there are cases where this method is not advantageous. Nevertheless, it improves convergence in many cases (see Pfefferkorn et al. [9]).

[6] Due to the non-linearity of this expression a solver e.g. Newton's method might be required to solve for λ_2. However, for the considered material models, an analytic solution is possible.

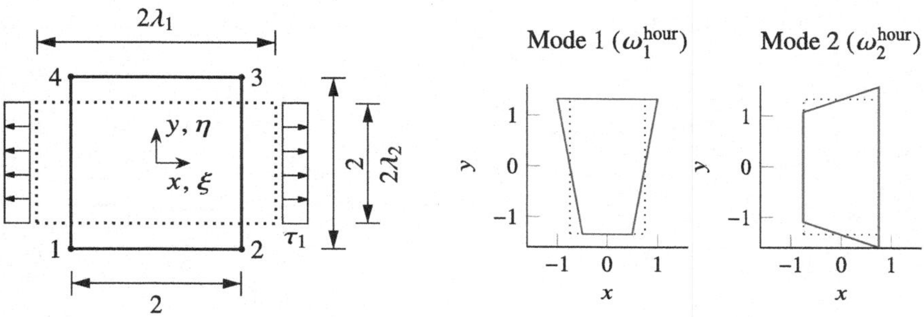

Fig. 2 Setup for stability test 2D (left) and hourglass modes 2D (right)

($\omega_2^{\text{hour}} < 0$ for $\lambda_1 < 1$).[7] All other elements tested exhibit no negative hourglass eigenvalues as shown by Armero [1]. In the case of the elasto-plastic material model only Q1/E4 shows instabilities in uniaxial compression while all other EAS elements suffer from instabilities under tension (see also Sect. 3.3).

There are twelve hourglass eigenvalues in the three-dimensional case. Due to symmetry of the uniaxial stress problem, some of these twelve modes are identical except for a rotation around the x-axis (see Pfefferkorn and Betsch [8]). Thus, only the unique eigenvalues are shown in Fig. 3 for the neo-Hookean and for the elasto-plastic material model, respectively.

For the elastic material behavior mode 6 of Q1/P0ET6 is slightly negative under compression although this seems not to influence its stability in practical simulations.

For the plastic material behavior Q1/P0 shows the best results since all eigenvalues are strictly positive. H1/E9, HA1/ET12 and H1/P0ET6 all have instabilities in the plastic case since various eigenvalues are below zero. In particular, modes 7 and 8 have different shapes and much lower values for H1/P0ET6 than for standard EAS elements. This ultimately leads to the hourglassing patterns observed in Sect. 3.4. The EAS elements exhibit negative modes as well. In particular, ω_6^{hour} is negative. However, this mode is not activated in distorted meshes as shown in Sect. 3.4, whereas it leads to hourglassing situations in plane strain necking problems (see Sect. 3.3).

3.3 Necking Plane Strain

In this and the following example we consider a necking simulation using the elasto-plastic material and focus on hourglassing patterns. A rectangular bar with a length of $2L = 53.334$ and a width of $2R = 12.826$ is subject to prescribed displacements $\bar{v}$. Necking is initiated by a geometric imperfection in form of a linear reduction of R to $\bar{R}$ by ΔR. In order to capture the hourglassing patterns, a total of 200 load steps is considered to reach $\bar{v} = 7$. Note that only one fourth of the specimen needs to be

[7] Q1/S5 has a different behavior in tension than all other elements which might emerge from problems with the Legendre transformation.

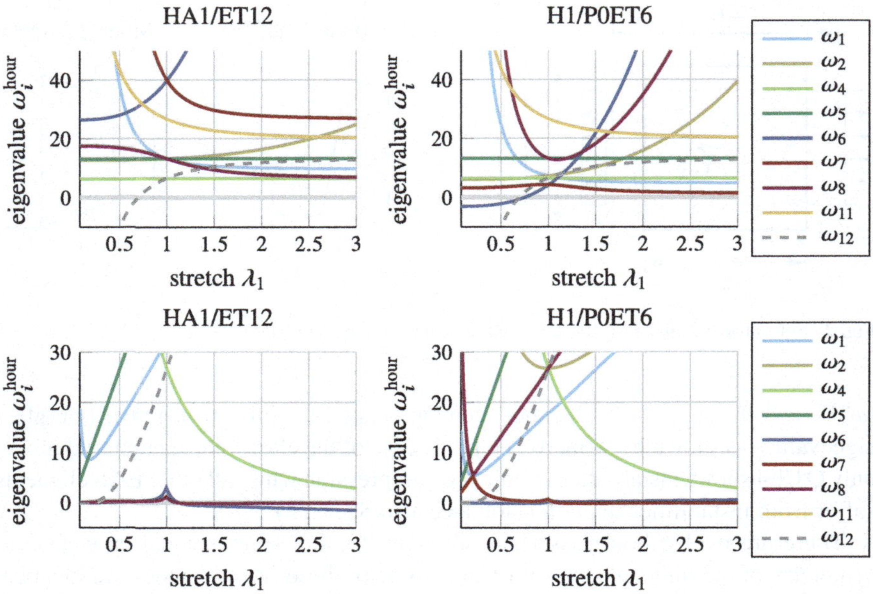

Fig. 3 Hourglass eigenvalues of 3D elements for the neo-Hookean (top) and elasto-plastic (bottom) material model

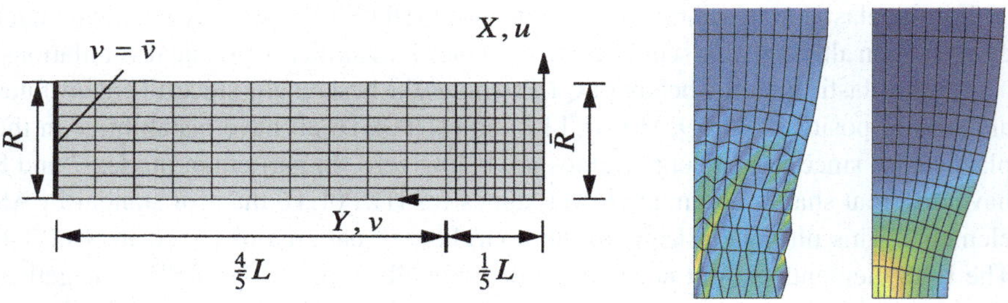

Fig. 4 Thin bar: setup (left) and deformed mesh for $\bar{v} = 5.6$ of Q1/ET4 (middle) and Q1/P0ET2 (right)

simulated due to symmetry. For more details see e.g. Armero [1] or Pfefferkorn and Betsch [8].

As shown in Fig. 4, exemplary for Q1/ET4, all EAS based element formulations suffer from hourglassing patterns while Q1/P0 and Q1/P0ET2 are hourglassing-free. Here, the negative eigenvalues of Mode 6 (or rather its 2D equivalent) observed in Sect. 3.2 cause the hourglassing patterns.

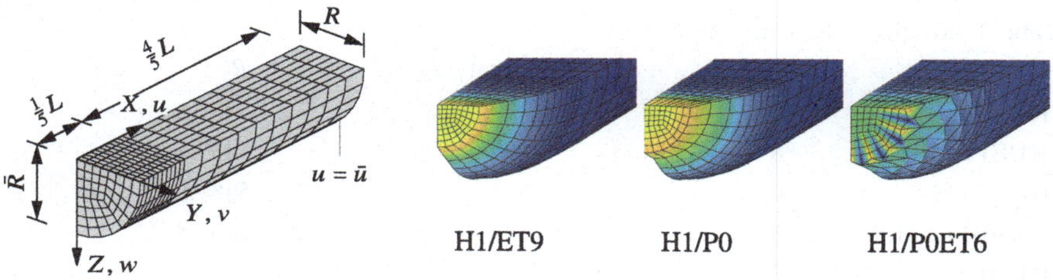

Fig. 5 Circular bar: setup (left) and from left to right the deformed meshes of H1/ET9, H1/P0 and H1/P0ET6

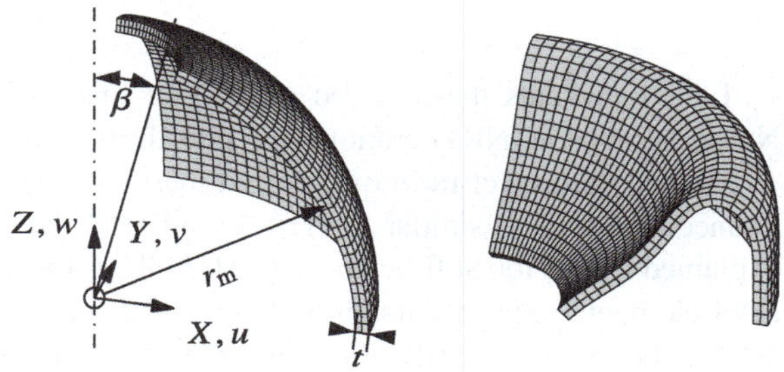

Fig. 6 Spherical shell with opening: setup (left), deformed mesh (right)

3.4 Necking Circular Bar

In analogy to Sect. 3.3, the 3D version of the necking problem is conducted on one eighth of a circular bar of length $2L = 53.334$ and Radius $R = 6.413$. After a total of 50 load steps opposite results to Sect. 3.3 are obtained as shown in Fig. 5. All EAS based elements are hourglassing-free, but the element of Armero, H1/P0ET6, shows hourglassing patterns. In dissent to the results of Simo and Armero [11] H1/P0 is hourglassing-free. Thus, for the current problem with its distorted elements, the negative Mode 6 does not seem to induce hourglassing while the negative modes 7 and 8 observed for H1/P0ET6 induce hourglassing.

3.5 Spherical Shell with Opening

The final test presented in the present paper is the spherical shell with opening problem modeled using the neo-Hookean nonlinear elastic material model with $\lambda = 1.2115 \cdot 10^5$ and $\mu = 8.0769 \cdot 10^4$. The left image in Fig. 6 shows the setup with middle radius $r = 10$, thickness 0.5 and opening angle $\beta = 18°$ while the right image shows the deformed mesh computed with element formulation H1/E9. The shell is deformed by prescribed displacements applied at the inner edge of the top opening. For more details on setup and procedure see e.g. Korelc et al. [5].

Table 1 Results of the spherical shell test

element type	req. n_{steps}	total n_{NR}	R_z
H1/ET9	9	77	5487.5
H1/ET9-MIP	8	58	5487.5
HA1/ET12	9	77	5489.8
HA1/ET12-MIP	8	58	5489.8
H1/P0	3	27	5773.5
H1/P0ET6	3	29	4920.6
H1/S18	3	27	5454.1

Table 1 presents the required number of load steps n_{steps} and total number of Newton-Raphson (NR) iterations n_{NR} plus the resulting reaction force in z-direction R_z.[8] Since the aspect ratio of the used mesh is around 1, H1/P0ET6 exhibits pronounced robustness similar to H1/P0 (and H1/S18). The low reaction force R_z is explained by the too soft behavior of H1/P0ET6 for nearly undistorted meshes. All EAS elements require a much higher number of load steps as well as more than double the amount of NR iterations. We emphasize that this behavior can not be observed for H1/P0ET6 if poor aspect ratios occur. In that case H1/P0ET6 performs extremely poor and even worse than EAS elements.[9]

4 Conclusion

The present paper covers the mixed pressure EAS element formulation of Armero [1] with an extension to 3D as well as thorough evaluation of the performance of the element regarding numerical instabilities such as spurious hourglassing patterns as well as locking and robustness in NR-iterations.

Although the element formulation of Armero [1] is stable and hourglassing-free for regular meshes with nonlinear elastic material behavior, some problems occur in case of distorted meshes and for elasto-plastic material behavior.

In particular, while Q1/P0ET2 solves the hourglass problems, that EAS element formulations such as Q1/ET4 suffer from in case of 2D elasto-plastic simulations (see Armero [1]), converse results are observed in 3D computations. Here, H1/P0ET6 shows hourglassing and EAS elements show no spurious results.

In conclusion, there is yet an EAS based element formulation to be found that is free of spurious hourglassing effects with nonlinear elastic and elasto-plastic material behavior in 2D and 3D and that provides the required robustness for different problem

[8] We use the Newton tolerance $||R|| < 1 \cdot 10^{-8}$ as well as a maximum of 20 numerical iterations per load step.

[9] The convergence test of a circular ring (see e.g. Pfefferkorn et al. [9]) provides a good example of the poor behavior of H1/P0ET6 in shell-like problems.

formulations. Finding such an element remains the task for element development. Nevertheless, the element formulation of Armero [1] provides an interesting mixed element with many desirable properties and works especially well in 2D.

References

1. Armero, F. (2000). On the locking and stability of finite elements in finite deformation plane strain problems. *Computers and Structures, 75*(3), 261–290. https://doi.org/10.1016/S0045-7949(99)00136-4.
2. Boffi, D., Brezzi, F., & Fortin, M. (2013). Mixed finite element methods and applications. Heidelberg: Springer. https://doi.org/10.1007/978-3-642-36519-5
3. Glaser, S., & Armero, F. (1997). On the formulation of enhanced strain finite elements in finite deformations. *Engineering Computations, 14*(7), 759–791. https://doi.org/10.1108/02644409710188664.
4. Hughes, T. (2000). *The finite element method: Linear static and dynamic finite element analysis*. Mineola: Dover Publication.
5. Korelc, J., Šolinc, U., & Wriggers, P. (2010). An improved EAS brick element for finite deformation. *Computational Mechanics, 46*(4), 641–659. https://doi.org/10.1007/s00466-010-0506-0
6. Korelc, J., & Wriggers, P. (1996). Consistent gradient formulation for a stable enhanced strain method for large deformations. *Engineering Computations, 13*(1), 103–123. https://doi.org/10.1108/02644409610111001
7. Pfefferkorn, R., & Betsch, P. (2019). On transformations and shape functions for enhanced assumed strain elements. *International Journal for Numerical Methods in Engineering, 120*(2), 231–261. https://doi.org/10.1002/nme.6133
8. Pfefferkorn, R., & Betsch, P. (2020). Extension of the enhanced assumed strain method based on the structure of polyconvex strain-energy functions. *International Journal for Numerical Methods in Engineering, 121*(8), 1695–1737. https://doi.org/10.1002/nme.6284
9. Pfefferkorn, R., Bieber, S., Oesterle, B., Bischoff, M., & Betsch, P. (2020). Improving efficiency and robustness of EAS elements for nonlinear problems. International Journal for Numerical Methods in Engineering (submitted).
10. Pian, T. H. H., & Sumihara, K. (1984). Rational approach for assumed stress finite elements. *International Journal for Numerical Methods in Engineering, 20*(9), 1685–1695. https://doi.org/10.1002/nme.1620200911.
11. Simo, J., & Armero, F. (1992). Geometrically non-linear enhanced strain mixed methods and the method of incompatible modes. *International Journal for Numerical Methods in Engineering, 33*(7), 1413–1449. https://doi.org/10.1002/nme.1620330705
12. Simo, J., Armero, F., & Taylor, R. (1993). Improved versions of assumed enhanced strain tri-linear elements for 3D finite deformation problems. *Computer Methods in Applied Mechanics and Engineering, 110*(3), 359–386. https://doi.org/10.1016/0045-7825(93)90215-J
13. Simo, J., & Rifai, M. (1990). A class of mixed assumed strain methods and the method of incompatible modes. *International Journal for Numerical Methods in Engineering, 29*, 1595–1638. https://doi.org/10.1002/nme.1620290802
14. Simo, J., Taylor, R., & Pister, K. (1985). variational and projection methods for the volume constraint in finite deformation elasto-plasticity. *Computer Methods in Applied Mechanics and Engineering, 51*, 177–208. https://doi.org/10.1016/0045-7825(85)90033-7
15. Simo, J. C. (1992). Algorithms for static and dynamic multiplicative plasticity that preserve the classical return mapping schemes of the infinitesimal theory. *Computer Methods in Applied Mechanics and Engineering, 99*(1), 61–112. https://doi.org/10.1016/0045-7825(92)90123-2.

16. Viebahn, N., Schröder, J., & Wriggers, P. (2019). An extension of assumed stress finite elements to a general hyperelastic framework. *Advanced Modeling and Simulations in Engineering Science, 6*(9). https://doi.org/10.1186/s40323-019-0133-z
17. Washizu, K. (1975). *Variational methods in elasticity and plasticity*. Oxford: Pergamon Press.
18. Wilson, E., Taylor, R., Doherty, W., & Ghaboussi, J. (1973) Incompatible displacement models. In: Fenves, S., Perrone, N., Robinson, A., & Schnobrich, W. (eds.) Numerical and Computer Methods in Structural Mechanics (pp. 43–57). New York. https://doi.org/10.1016/B978-0-12-253250-4.50008-7
19. Wriggers, P. (2008). *Nonlinear finite element methods*. Berlin: Springer. https://doi.org/10.1007/978-3-540-71001-1.
20. Wriggers, P., & Reese, S. (1996). A note on enhanced strain methods for large deformations. *Computer Methods in Applied Mechanics and Engineering, 135*(3), 201–209. https://doi.org/10.1016/0045-7825(96)01037-7

Adaptive Virtual Element Method for Large-Strain Phase-Field Fracture

Blaž Hudobivnik, Fadi Aldakheel, and Peter Wriggers

A dedication to Professor Peter Wriggers. I got to know you, a great man, just six months into my Doctorate study in 2012, in a great city of Vienna, which we Slovenians call Dunaj and Germans Wien. We walked around the town with your wife and Professor Korelc and we got some Sachertorte. And just 4 years later after my Promotion, I was fortunate to be invited to work as a post-doctoral student at your Institute. It was my honour working close with you for the last 5 years, discussing problems and to be your right hand in coding, it brought me great joy. I thank you for this chance you gave me, confidence in my work and that I was able work at your side all this years. I wish you health and happiness and that we can continue to closely cooperate in the future for many years to come!

—Blaž Hudobivnik

Abstract In this contribution, the phase-field (PF) approach to brittle fracture is extended to adaptively refined meshes at finite strains. Such mesh refinement produces regular structured elements with hanging nodes at edges. These hanging nodes can be included to the mesh by employing the Virtual Element Method (VEM). The model performance is demonstrated by two representative numerical examples.

B. Hudobivnik (✉) · F. Aldakheel · P. Wriggers
Institute of Continuum Mechanics, Leibniz University Hannover, Garbsen, Germany
e-mail: hudobivnik@ikm.uni-hannover.de

F. Aldakheel
e-mail: aldakheel@ikm.uni-hannover.de

P. Wriggers
e-mail: wriggers@ikm.uni-hannover.de

F. Aldakheel et al. (eds.), *Current Trends and Open Problems in Computational Mechanics*, https://doi.org/10.1007/978-3-030-87312-7_20

1 Introduction

With the ongoing development of modern element technologies, promising numerical simulation tools were created for failure analysis of solids. In this regard, discretization schemes like the Finite Difference Method, the Finite Element Method and the Boundary Element Method are well established numerical tools for solving various science and engineering problems. Additionally new approaches like Meshless methods for arbitrary deformations, IsoGeometric Analysis and the eXtended Finite Element Method for fracture mechanics problems are continuously evolved and can be efficiently used within a specific problem range. Hence, the art of modeling means here to pick the right numerical solution method that provides accurate results in the most time efficient way. Each of the methods described above has its own specifications and thus needs experts for a correct and efficient application.

In this work a relatively new discretization technique the Virtual Element Method (VEM) will be presented that introduces some new features to the numerical solution of problems in solid mechanics [1, 2]. VEM has proven to be a competitive discretization scheme for meshes with highly irregular shaped elements that can even be non convex. Moreover, VEM allows the usage of an arbitrary polygonal (2D) and polyhedral (3D) element shapes with arbitrary number of nodes. So far applications of virtual elements have been devoted to linear elastic deformations in [3], contact problems in [4, 5], second order approximation (serendipity elements) in [6, 7], elasto-plastic deformations in [8] and fracture mechanics in [9–11].

This work extends the virtual element method further towards adaptive fracture mechanics at large deformations. The modeling of crack formation is achieved in a convenient way by continuum phase-field approaches to fracture, which are based on the regularization of sharp crack discontinuities [12]. Phase-field modeling of fracture has been attracting considerable attention in recent years due to its capability of capturing complex crack patterns in various problems in solid mechanics. In the current contribution, a robust and efficient adaptive virtual element method is proposed within the phase-field fracture. This reduce the computation cost remarkable and allows for complex democratization scheme.

2 Phase-Field Modeling of Brittle Fracture

In this section, we outline a variational approach to brittle fracture in elastic solids at finite strains.

2.1 Basic Kinematics

Let $\Omega \in \mathcal{R}^{\delta}$ with $\delta = 2, 3$ be a solid body. The response of fracturing solid at material points $\boldsymbol{x} \in \Omega$ and time t is described by the displacement field $\boldsymbol{u}(\boldsymbol{x}, t)$ and the crack phase-field $d(\boldsymbol{x}, t)$ with $\dot{d} \geq 0$. Here, $d(\boldsymbol{x}, t) = 0$ and $d(\boldsymbol{x}, t) = 1$ represent the unbroken and fully broken state of the material, respectively. The deformation gradient $\boldsymbol{F}$ can be defined from displacement gradient as

$$\boldsymbol{F} = \frac{\partial \boldsymbol{x}}{\partial \boldsymbol{X}} = \frac{\partial (\boldsymbol{X} + \boldsymbol{u})}{\partial \boldsymbol{X}} = \boldsymbol{I} + \nabla \boldsymbol{u} \,. \tag{1}$$

For the phase-field problem, a sharp-crack surface topology $\Gamma \to \Gamma_l$ is regularized by the crack surface functional

$$\Gamma_l(d) = \int_{\Omega} \gamma_l(d, \nabla d)\, dV \quad \text{with} \quad \gamma_l(d, \nabla d) = \frac{d^2}{2l_f} + \frac{l_f}{2}|\nabla d|^2 \tag{2}$$

based on the crack surface density function γ_l per unit volume of the solid and the fracture length scale parameter l_f that governs the regularization. Hence, the combination of elasticity with the first-order gradient damage modeling focuses on the set

$$\text{Constitutive State Variables: } C := \{\boldsymbol{F}, d, \nabla d\} \,. \tag{3}$$

2.2 Constitutive Work Density Function

The constitutive work density function W is assumed to depend on the constitutive state variables C introduced in Eq. (3). It consists of the sum

$$W(C) = W_{\text{bulk}}(\boldsymbol{F}, d) + W_{\text{frac}}(d, \nabla d) \tag{4}$$

of a degrading elastic bulk energy W_{bulk} and a contribution due to fracture W_{frac}, which contains the accumulated dissipative energy. The hyper-elastic bulk contribution in Eq. (4) is assumed to be a non-linear function as

$$W_{\text{bulk}}(\boldsymbol{F}, d) = g(d)\ (\psi_{\text{iso}}(\boldsymbol{F}) + \psi_{\text{vol}}(J_{F+})) + \psi_{\text{vol}}(J_{F-}) \qquad \text{with} \tag{5}$$

$$\psi_{\text{iso}}(\boldsymbol{F}) = \frac{\mu}{2}\left(J_F^{-\frac{2}{3}}\,\text{tr}(\boldsymbol{F}\boldsymbol{F}^{\text{T}}) - 3\right) \qquad \text{isohoric part and} \tag{6}$$

$$\psi_{\text{vol}}(J_F) = \frac{\kappa}{4}\,(J_F - 1 - 2\log(J_F)) \qquad \text{volumetric part and} \tag{7}$$

$$J_{F\pm} = \langle J_F \rangle_{\pm} = (J_F \pm |J_F|)/2, \quad \langle \Box \rangle_{\pm} \qquad \text{is the Macaulay bracket}$$

where, κ and μ are the material bulk and shear moduli and $J_F = \det(\boldsymbol{F})$ is the determinant of deformation gradient $\boldsymbol{F}$. The function $g(d) = (1-d)^2$ models the degradation of the stored elastic energy of the solid due to fracture and it degrades the positive volumetric part and full shear part. The fracture contribution in Eq. (4) is defined as

$$W_{\text{frac}}(d, \nabla d) = \mathcal{G}_c \; \gamma_l(d, \nabla d) \tag{8}$$

where, $\mathcal{G}_c > 0$ is the Griffith's critical energy release rate.

2.3 Governing Equations

2.3.1 Balance of Linear Momentum

The first equation is the stress equilibrium or the quasi-static form of the balance of linear momentum defined as

$$\text{Div}\,(\boldsymbol{P}) + \overline{\boldsymbol{f}} = \boldsymbol{0}, \quad \text{with} \quad \boldsymbol{P} = \partial_{\boldsymbol{F}} W \;, \tag{9}$$

where $\overline{\boldsymbol{f}}$ are volume forces and the first Piola–Kirchhoff stress tensor, which can be obtained from hyper elastic strain energy W in (4).

2.3.2 The Fracture Phase-Field Equation

Evolution of the regularized crack surface functional (2) can be driven by the constitutive functions as outlined in [9], postulating a global evolution equation of regularized crack surface as

$$\frac{d}{dt}\Gamma_l(d) = \frac{1}{l_f}\int_{\Omega} [\,(1-d)\mathcal{H} - \eta\dot{d}]\;\dot{d}\;\text{dV} \geq 0\;, \tag{10}$$

where $\eta \geq 0$ is a material parameter that characterizes the artificial/numerical viscosity of the crack propagation. Various criteria can be chosen here (see [12]), in this case the crack driving force was chosen to be only the volumetric part:

$$\mathcal{H} = \max_{s\in[0,t]} D(\boldsymbol{x}, s) \geq 0 \quad \text{with} \quad D := \psi_{\text{vol}}(J_{F+}) + \psi_{\text{iso}}(\boldsymbol{F}) \tag{11}$$

is introduced as a local history variable that accounts for the irreversibility of the phase-field evolution by filtering out a maximum value of what is known as the crack driving state function D. Then the evolution statement (10) provides the local equation for the evolution of the crack phase-field in the domain Ω along with its homogeneous Neumann boundary condition as

$$(d - l_f^2 \Delta d) + \eta \dot{d} + (d - 1)\mathcal{H} = 0 \tag{12}$$

with $\nabla d \cdot \boldsymbol{n} = 0$ on $\partial\Omega$. Here, $\boldsymbol{n}$ represents the outward normal on $\partial\Omega$. The fracture pseudo potential can then be defined as:

$$\widetilde{W}_{\text{frac}}(d, \nabla d) = d\dot{d}\eta - g(d)\mathcal{H} + \mathcal{G}_c \gamma_l(d, \nabla d) \tag{13}$$

The specific potential can then be defined as:

$$W(C) = W_{\text{bulk}}(\boldsymbol{F}, d) + \widetilde{W}_{\text{frac}}(d, \nabla d) \tag{14}$$

3 The VEM

In this chapter a standard VEM is introduced. The basics of VEM is explained in [2]. The first step is to rewrite element unknowns $\mathfrak{p}_h = (u_h \; v_h \; d_h)$ [13, 14], known only at element nodes and edges as:

$$\mathfrak{p}_h = \mathfrak{p}_\Pi + (\mathfrak{p}_h - \mathfrak{p}_\Pi)\,. \tag{15}$$

Here Π is a projection operator $\Pi(\mathfrak{p}) = \mathfrak{p}_\Pi$, that projects quantity $\mathfrak{p}$ on first order polynomial space.

3.1 The VEM Projection

The projected variables are defined as:

$$\mathfrak{p}_\Pi = \begin{pmatrix} u_\pi \\ v_\pi \\ d_\pi \end{pmatrix} = \mathbf{a}\mathbf{N}_\Pi = \begin{bmatrix} a_{1\,1} & a_{1\,2} & a_{1\,3} \\ a_{2\,1} & a_{2\,2} & a_{2\,3} \\ a_{3\,1} & a_{3\,2} & a_{3\,3} \end{bmatrix} \begin{pmatrix} 1 \\ X \\ Y \end{pmatrix} \tag{16}$$

for every 2D element with the set of unknown parameters $\mathbf{a}$, 3 per nodal degree of freedom (DOF), in this case total 9 parameters. The goal now is to compute $\mathbf{a}$ as a map of the nodal degrees of freedom of a virtual element $\mathbf{p}_e$.

The computation of this polynomial function is based on the requirement that the remainder $(\mathfrak{p}_h - \mathfrak{p}_\Pi)$ and its gradient are orthogonal to any first order polynomial $\mathfrak{p}_p$ and its gradient. This yields

$$\int_\Omega \left[\mathfrak{p}_p \cdot (\mathfrak{p}_h - \mathfrak{p}_\Pi)\right] \Omega = 0 \text{ and } \int_\Omega \left[\nabla\mathfrak{p}_p \cdot (\nabla\mathfrak{p}_h - \nabla\mathfrak{p}_\Pi)\right] \mathrm{d}\Omega = 0\,. \tag{17}$$

Since both $\nabla \mathfrak{p}_p$ and $\nabla \mathfrak{p}_\Pi$ are constant at element level, (17) reduces to the condition that the gradients computed from the projected and real displacements must be equal. Considering latter, the gradient of projection can be directly defined from (17) as:

$$\nabla \mathfrak{p}_\Pi |\Omega| = \int_\Omega \nabla \mathfrak{p}_h d\Omega = \int_{\partial\Omega} \mathfrak{p}_h \otimes \boldsymbol{N}\, \mathrm{d}s. \tag{18}$$

Because $\mathfrak{p}_h$ and its gradient are unknown on element Ω the integral on the right-hand-side was also transformed to the boundary integral by using the divergence theorem. The $\boldsymbol{N} = (N_X\ N_Y)$ is the outward unit normal to the element surface. For the two-dimensional case this results in

$$\nabla \mathfrak{p}_\Pi = \begin{bmatrix} \nabla \mathfrak{p}_{\Pi 1} \\ \nabla \mathfrak{p}_{\Pi 2} \\ \nabla \mathfrak{p}_{\Pi 3} \end{bmatrix} = \begin{bmatrix} a_{12}\ a_{13} \\ a_{22}\ a_{23} \\ a_{32}\ a_{33} \end{bmatrix} = \frac{1}{|\Omega|} \sum_{\mathcal{E}=1}^{n_v} \int_{\mathcal{E}} \begin{bmatrix} u(s)N_X(s)\ u(s)N_Y(s) \\ v(s)N_X(s)\ v(s)N_Y(s) \\ d(s)N_X(s)\ d(s)N_Y(s) \end{bmatrix} ds\,, \tag{19}$$

right hand side represents the edge by edge integration of $\mathfrak{p}_h \otimes \boldsymbol{N}$. The $s \in [0, L_\mathcal{E}]$ represents a curvilinear coordinate along the edge $\mathcal{E}$ of length $L_\mathcal{E}$, defining edge coordinate $\boldsymbol{X}(s) = \left((1 - \frac{s}{L_\mathcal{E}})\boldsymbol{X}_\mathcal{E} + \frac{s}{L_\mathcal{E}}\boldsymbol{X}_{\mathcal{E}+1}\right)$. Similarly the components of $\mathfrak{p}_h = \left((1 - \frac{s}{L_\mathcal{E}})\mathfrak{p}_{h\,\mathcal{E}} + \frac{s}{L_\mathcal{E}}\mathfrak{p}_{h\,\mathcal{E}+1}\right)$, normal $\boldsymbol{N}(s) = \left(Y(s)_{,s}\ - Y(s)_{,s}\right)/L_\mathcal{E}$ and length $L_\mathcal{E} = |\boldsymbol{X}_{,s}|$ are defined in terms of s at the boundary of the virtual element.

For completeness of the polynomial in (16) the rest of parameters (a_{11}, a_{21} and a_{31}) have to be determined. They follow from the condition (17), this simplifies for linear VEM (see [2]) to the contrition that the average vertex value of nodal degrees of freedom $\mathfrak{p}_h(\boldsymbol{X}_i)$ equals to the average of their projections $\mathfrak{p}_\Pi(\boldsymbol{X}_i)$ the projection $\mathfrak{p}_\Pi$

$$\frac{1}{n_v}\sum_{i=1}^{n_v} \mathfrak{p}_\Pi(\boldsymbol{X}_i) = \frac{1}{n_v}\sum_{i=1}^{n_v} \mathfrak{p}_h(\boldsymbol{X}_i) \rightarrow a_{I1} = \frac{1}{n_v}\left(\sum_i^{n_v} \mathfrak{p}_{h\,I}(\boldsymbol{X}_i) + \nabla \mathfrak{p}_{\Pi\,I} \cdot \sum_i^{n_v} \boldsymbol{X}_i\right) \tag{20}$$

$$\mathfrak{p}_{h\,I} \subset \mathfrak{p}_h(\boldsymbol{X}_i) = (u_h(\boldsymbol{X}_i)\ v_h(\boldsymbol{X}_i)\ d_h(\boldsymbol{X}_i)) = (\mathbf{p}_e)_{i \in (3i-2,\,3i-1,\,3i)} \tag{21}$$

Now all kinematic quantities in C can be expressed in terms of the projection function (16) as:

$$\boldsymbol{u}_\Pi = (\mathfrak{p}_{\Pi 1}\ \mathfrak{p}_{\Pi 2})\,,\ \ \nabla \boldsymbol{u}_\Pi = (\nabla \mathfrak{p}_{\Pi 1}\ \nabla \mathfrak{p}_{\Pi 2})^{\mathrm{T}}\,,\ d_\Pi = \mathfrak{p}_{\Pi 3} \text{ and } \nabla d_\Pi = \nabla \mathfrak{p}_{\Pi 3}\,. \tag{22}$$

3.2 Construction of the Virtual Element

Knowing the potential (14) and the kinematical quantities (22) a virtual element formulation can be derived as described in [14]. With the split in (15) the energy can be also split in a part compatibility by the deformation gradient of the projection and a stabilization part i.e. $U = U_c + U_{stab}$. Here the same potential has been chosen for both parts (i.e. $U_{stab} := \beta \Delta U_c$), interpolating between with factor β [9].

$$U(\mathbf{p}) = \mathsf{A}_{e=1}^{n_e} U_e(\mathbf{p}_e), \text{ where } U_e(\mathbf{p}_e) = [(1-\beta)U_c(C_\Pi|_e) + \beta U_c(C_{h\mathcal{T}}|_e)] \quad (23)$$

where A is the assembly operator. The energy contribution of each virtual element resulting from the displacement is defined as:

$$U_c(\mathfrak{p}_\square) = \int_\Omega [W(C_\square) - \bar{f} \cdot u_\square]\, d\Omega - \int_{\partial\Omega} \bar{t} \cdot u_\square\, ds\,. \quad (24)$$

The projection parameters from (16) and consequently $\mathfrak{p}_\Pi$ depend on the nodal degrees of freedom $\mathbf{p}_e$ through Eqs. (19) and (20) and therefore also energy function of projection (14). The residual vector $\mathbf{R}_e$ and the stiffness matrix $\mathbf{K}_e$ can be computed at element level from U_e. This yields

$$\mathbf{R}_e = \frac{\partial U_e(\mathbf{p}_e)}{\partial \mathbf{p}_e} \quad \text{and} \quad \mathbf{K}_e = \frac{\partial \mathbf{R}_e(\mathbf{p}_e)}{\partial \mathbf{p}_e} \quad (25)$$

that are explicitly computable through $\mathbf{p}_e$. The matrices (25) are evaluated directly from potential by using *AceGen* software tool, which has automatic differentiation capabilities [15].

4 Mesh Refinement

A simple refining algorithm was implemented. The algorithm is based on a refinement function $f(e)$ which loops over each element e. If $f(e)$ yields $true$, refinement is triggered on element e. In that case, a new node is inserted on a position of element centroid $X_C|_e = 1/|\Omega_e| \int_{\Omega_e} X \mathrm{d}\Omega$ and additional new nodes are inserted at each element edge $\mathcal{E}$ $X_C|_\mathcal{E} = (X|_\mathcal{E} + X|_{\mathcal{E}+1})/2$ if needed (the algorithm checks if the edge has already been previously refined from neighbouring element, thus avoiding the new insertion). After, the $n_\mathcal{E} = n_v$ new elements are created, connecting points $X_C|_\mathcal{E}$, $X|_\mathcal{E}$, $X_C|_{\mathcal{E}+\infty}$ and $X_C|_e$ into a new element for each edge $\mathcal{E}$. Figure 1 demonstrates the mesh refinement algorithm. Starting with 2×2 mesh in Fig. 1a and refining around a line plotted in Fig. 1a for refinement depths in Fig. 1b–e.

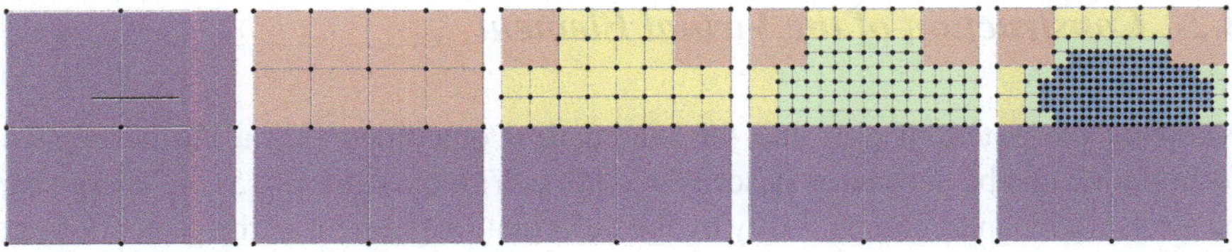

Fig. 1 The mesh Refinement algorithm, step-by-step. **a** Starting mesh with line to refine around **b–c** refinement steps

Table 1 Material parameters

Id	Description	Symbol	Value	Unit
1	Elastic modulus	E	2.1×10^4	$\frac{\mathrm{kN}}{\mathrm{m}^2}$
2	Poisson ratio	ν	0.2	
3	fracture length scale	l_f	0.015	mm
4	Griffith's critical energy release rate	$\mathcal{G}_c$	10	$\frac{\mathrm{kN}}{\mathrm{mm}}$
5	Viscosity of crack	η	0.01	$\frac{\mathrm{kNs}}{\mathrm{mm}^2}$

5 Numerical Examples

The performance of the shown refinement and phase-field model will be demonstrated in this chapter. On the computational side, a robust and efficient *staggered scheme* is employed using the software tool AceFEM in the numerical implementation to compute the unknowns (displacement $\boldsymbol{u}$ and crack phase-field d). Within each load step a quadratic convergence is achieved, due to the fact that all formulations are linearized in a consistent manner using AceGen. The benchmark tests considers a square block ($L = H = 1$ mm) with a horizontal notch placed at the middle height. With material parameters summarized in (Table 1).

5.1 Single-Edge Notched Tension Test

The standard tension example has displacements at top and bottom fully clamped and the top is vertically displaced at rate $\bar{\boldsymbol{u}}_X = 1$ mm/s until final failure at time $t = 0.1$ s, as outlined in Fig. 2. The results show good match between fine and adaptively refined mesh in Fig. 3b–c and a good match between VEM and FEM. In fact FEM exhibits some locking behaviour, but the major point of the contribution was to show that the

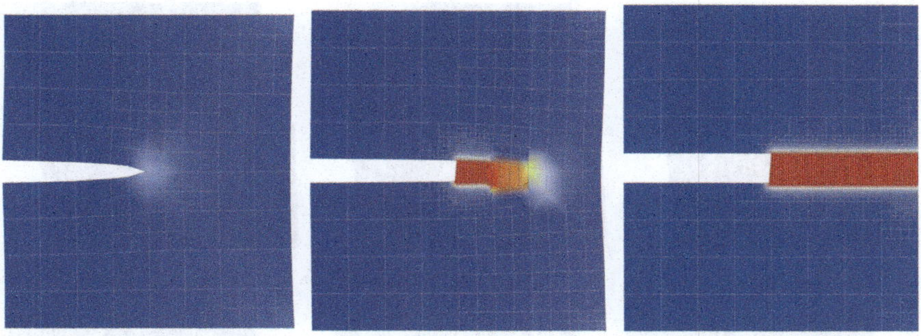

(a) Initiation (b) Propagation (c) Final state

Fig. 2 Single-edge notched tension test. VEM-Adaptive crack phase-field evolution for different deformation stages until final failure

(a) FEM fine Q1 (b) VEM fine Q1 (c) VEM refined

Fig. 3 The FEM (**a**) and VEM (**b**) on fine mesh with $2^7 \times 2^7$ and VEM final refined state (**c**)

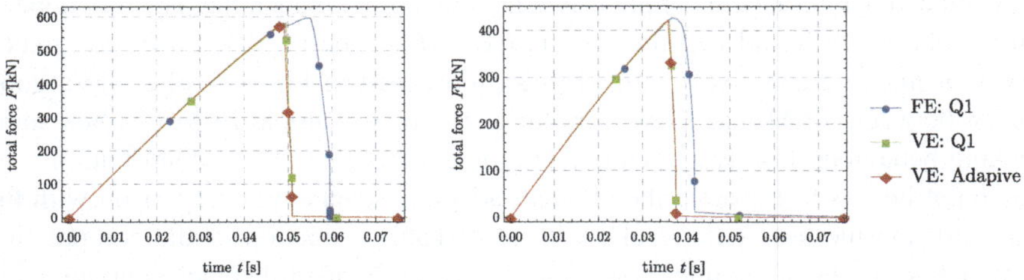

(a) $\mathcal{H} = \max_{s\in[0,t]}(\psi_{\text{vol}}(J_{F+}))$ (b) $\mathcal{H} = \max_{s\in[0,t]}(\psi_{\text{vol}}(J_{F+}) + \psi_{\text{iso}}(\boldsymbol{F}))$

Fig. 4 The load displacement curve of FEM, VEM on Q1 fine mesh and VEM with adaptive re-meshing for the tensile example

adaptive and fine meshes produce comparable results, and as outlined in Fig. 4 are in good agreement.

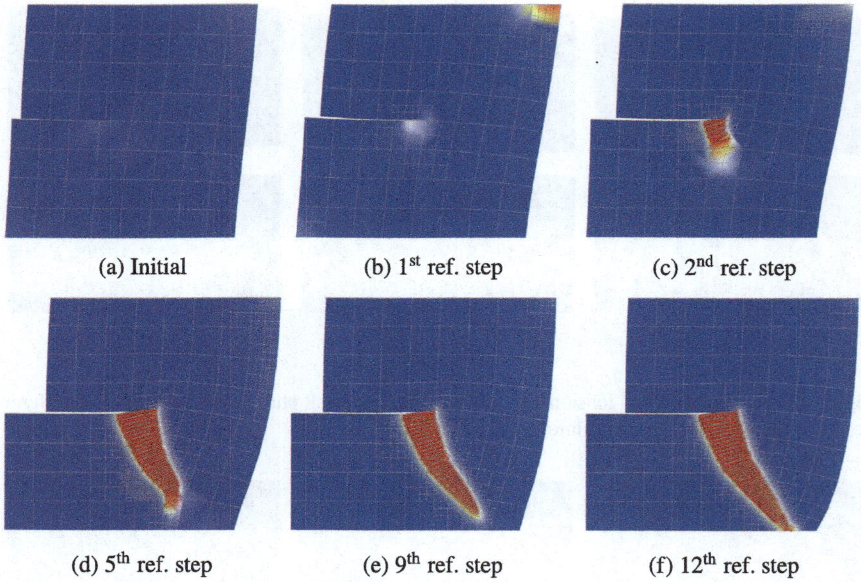

Fig. 5 Single-edge notched shear test. VEM-Adaptive crack phase-field evolution for different deformation stages until final failure

5.2 Single-Edge Notched Shear Test

The shear example has displacements at top and bottom fully clamped and the top is vertically displaced at rate $\bar{\boldsymbol{u}}_X = 1$ mm/s until final failure at time $t = 0.2$ s, as outlined in Fig. 5.

Unlike Tension, shear example is much more unpredictable, the path will depend on the chosen split and fracture criteria (see [16]). The results show good match between fine and adaptively refined mesh illustrated in Fig. 6b and c and a good match between VEM and FEM. Similar to the first example FEM exhibits some locking behaviour, however such outcome is not the goal of this contribution. The key point here is to illustrate the efficient adaptive scheme in comparison with the standard fine meshes. This results with remarkable reduction of the computation time, while producing similar results, as shown in the load-displacement curves in Fig. 7.

Furthermore, we investigated two different fracture deriving forces $\mathcal{H}$, namely only with positive volumetric part (Fig. 7a) or with the full deviatoric part (Fig. 7b). In both cases, the developed mesh adaptive scheme illustrates good results in comparison with the classical fine meshes of VEM/FEM.

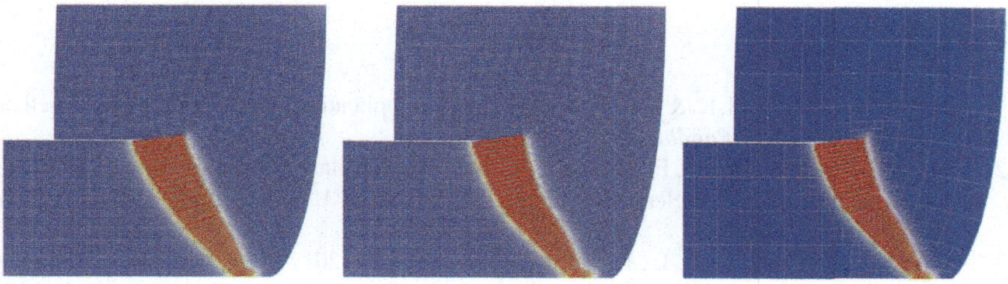

(a) FEM fine Q1 (b) VEM fine Q1 (c) VEM refined

Fig. 6 The FEM (**a**) and VEM (**b**) on fine mesh with $2^7 \times 2^7$ and VEM final refined state (**c**)

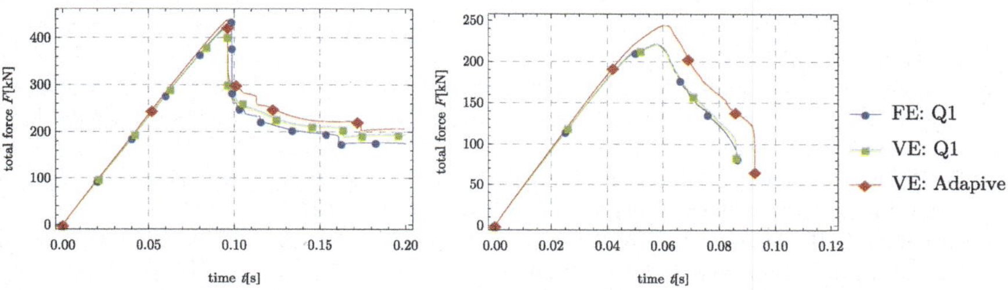

(a) $\mathcal{H} = \max_{s\in[0,t]}(\psi_{vol}(J_{F+}))$ (b) $\mathcal{H} = \max_{s\in[0,t]}(\psi_{vol}(J_{F+}) + \psi_{iso}(\boldsymbol{F}))$

Fig. 7 The load displacement curve of FEM, VEM on Q1 fine mesh and VEM with adaptive re-meshing for Shear example

6 Conclusion

In this contribution a robust algorithm was shown that allows adaptive refinement of meshes based on the gradient and value of phase-field variable on large deformations. The algorithm requires introduction of hanging nodes, thus the use of Virtual Element Method is beneficial here. The results show good agreement between fine and adaptively refined results. Due to the fact the standard mesh was relatively coarse, the response is slightly stiffer. However a remarkable computational time reduction, yet good results close to the standard formulation was achieved.

Acknowledgements BH gratefully acknowledge financial support to this work by the German Research Foundation (DFG) with the cluster of excellence PhoenixD (EXC 2122). FA gratefully acknowledges support for this research by the "German Research Foundation" (**DFG**) within **SPP** 2020 under the project WR 19/58-2.

References

1. Wriggers, P., Aldakheel, F., & Hudobivnik, B. (2019). Application of the virtual element method in mechanics. *GAMM-Rundbriefe, 1*(2019), 4–10.
2. Beirão Da Veiga, L., Brezzi, F., Cangiani, A., Manzini, G., Marini, L. D., & Russo, A. (2013). Basic principles of virtual element methods. *Mathematical Models and Methods in Applied Sciences, 23*(1), 199–214.
3. Artioli, E., Beirão da Veiga, L., Lovadina, C., & Sacco, E. (2017). Arbitrary order 2D virtual elements for polygonal meshes: Part I, elastic problem. *Computational Mechanics, 60*(3), 355–377.
4. Wriggers, P., Rust, W. T., & Reddy, B. D. (2016). A virtual element method for contact. *Computational Mechanics, 58*(6), 1039–1050.
5. Aldakheel, F., Hudobivnik, B., Artioli, E., da Veiga, L. B., & Wriggers, P. (2020). Curvilinear virtual elements for contact mechanics. *Computer Methods in Applied Mechanics and Engineering,372*, 113394.
6. Wriggers, P., Hudobivnik, B., & Aldakheel, F. (2021). Nurbs-based geometries: A mapping approach for virtual serendipity elements. *Computer Methods in Applied Mechanics and Engineering,378*, 113732.
7. De Bellis, M. L., Wriggers, P., & Hudobivnik, B. (2019). Serendipity virtual element formulation for nonlinear elasticity. *Computers & Structures,223*, 106094.
8. Hudobivnik, B., Aldakheel, F., & Wriggers, P. (2018). A low order 3d virtual element formulation for finite elasto-plastic deformations. *Computational Mechanics, 63*(2), 253–269.
9. Aldakheel, F., Hudobivnik, B., Hussein, A., & Wriggers, P. (2018). Phase-field modeling of brittle fracture using an efficient virtual element scheme. *Computer Methods in Applied Mechanics and Engineering, 341,* 443–466.
10. Hussein, A., Aldakheel, F., Hudobivnik, B., Wriggers, P., Guidault, P.-A., & Allix, O. (2019). A computational framework for brittle crack-propagation based on efficient virtual element method. *Finite Elements in Analysis and Design, 159,* 15–32.
11. Aldakheel, F., Hudobivnik, B., & Wriggers, P. (2019). Virtual element formulation for phase-field modeling of ductile fracture. *International Journal for Multiscale Computational Engineering, 17*(2), 181–200.
12. Miehe, C., Hofacker, M., Schänzel, L.-M., & Aldakheel, F. (2015). Phase field modeling of fracture in multi-physics problems. part ii. Coupled brittle-to-ductile failure criteria and crack propagation in thermo-elastic-plastic solids. *Computer Methods in Applied Mechanics and Engineering, 294,* 486–522.
13. Aldakheel, F., Hudobivnik, B., & Wriggers, P. (2019). Virtual elements for finite thermo-plasticity problems. *Computational Mechanics, 64*(5), 1347–1360.
14. Wriggers, P., & Hudobivnik, B. (2017). A low order virtual element formulation for finite elasto-plastic deformations. *Computer Methods in Applied Mechanics and Engineering, 327,* 459–477.
15. Korelc, J., & Wriggers, P. (2016). *Automation of finite element methods*. Berlin: Springer International Publishing.
16. Aldakheel, F. (2016). Mechanics of nonlocal dissipative solids: Gradient plasticity and phase field modeling of ductile fracture. Stuttgart: Institut für Mechanik (Bauwesen), Lehrstuhl I, Universität Stuttgart. https://doi.org/10.18419/opus-8803.

Galerkin Formulations with Greville Quadrature Rules for Isogeometric Shell Analysis: Higher Order Elements and Locking

Thomas J. R. Hughes, Zhihui Zou, Michael A. Scott, Roger A. Sauer, and Eshwar J. Savitha

I am not sure when I first met Peter Wriggers, but it was no later than 1983/84 when he was a postdoc at UC Berkeley. During that time, he and Juan Simo jointly developed the first consistent tangent operator for contact, a great step forward in Computational Contact Mechanics. Peter returned to Berkeley in 1988 and came to Stanford quite often to work with Juan. Over the years, I interacted with Peter in a multitude of professional capacities and enjoyed and valued his friendship. I visited him in Hannover in 2009 and, with his assistant, İlker Temizer, we initiated the first applications of IGA to contact problems. (Tom Hughes)

Abstract We propose new Greville quadrature schemes that asymptotically require only four in-plane points for Reissner–Mindlin (RM) shell elements and nine in-plane points for Kirchhoff–Love (KL) shell elements in B-spline and NURBS-based iso-

T. J. R. Hughes (✉) · Z. Zou
Oden Institute for Computational Engineering and Sciences, The University of Texas at Austin, Austin, TX, USA
e-mail: hughes@oden.utexas.edu

Z. Zou
e-mail: zhihui@utexas.edu

M. A. Scott
Coreform LLC, Orem, UT, USA
e-mail: mike@coreform.com

R. A. Sauer · E. J. Savitha
Aachen Institute for Advanced Study in Computational Engineering Science, RWTH Aachen University, Aachen, Germany
e-mail: sauer@aices.rwth-aachen.de; roger.sauer@pg.edu.pl

E. J. Savitha
e-mail: savitha@aices.rwth-aachen.de

R. A. Sauer
Faculty of Civil and Environmental Engineering, Gdansk University of Technology, ul. Narutowicza 11/12, 80-233 Gdansk, Poland

F. Aldakheel et al. (eds.), *Current Trends and Open Problems in Computational Mechanics*, https://doi.org/10.1007/978-3-030-87312-7_21

geometric shell analysis, independent of the polynomial degree of the elements. For polynomial degrees 5 and 6, the approach delivers high accuracy, low computational cost, and alleviates membrane and transverse shear locking.

1 Introduction

In this paper we address the problem of creating shell finite elements within the Isogeometric Analysis (IGA) paradigm, which, in structural mechanics, amounts to employing the same kinematic description (i.e., specification of the displacement field) as that being utilized in the definition of geometry emanating from a Computer Aided Design (CAD) representation [1]. We focus on B-splines and NURBS, as these CAD technologies dominate industrial usage. IGA offers a fundamental advantage in shell modeling, namely, precise, or even exact, geometric representation, and this is no doubt important as it is well known that even small geometric imperfections can significantly affect results in thin shell buckling, indicating numerical approximations of geometry may also be a primary source of error. Nevertheless, there are still major barriers to creating effective IGA shell elements, and these are shared by traditional finite element methods.

A primary concern are "locking phenomena", specifically, transverse shear locking and membrane-bending locking. Transverse shear locking is not a consequence of curved shell geometry; it is present as well for flat plate and straight beam models. In the development of shell finite elements, the main challenge to overcome, and a remaining open problem, is membrane-bending locking. It is apparent that curved, higher-order, traditional shell elements have not distinguished themselves heretofore because curvature is the root cause of membrane-bending coupling, hence locking. It is no wonder that in industrial software there is a heavy reliance on the lowest-order, four-node, quadrilateral shell elements, despite their inherently low accuracy, because they are typically flat, or almost flat, and minimize membrane-bending coupling within elements thereby. What we would like to have are simple, straightforward, IGA shell elements that would be candidates for inclusion in industrial scale, commercial general-purpose computer programs.

We have pursued a study that starts with the most direct "primal" formulations of shell finite elements, and adheres to the finite element analysis orthodoxy of using high-enough accurate Gauss quadrature rules to ensure stability of the stiffness and mass matrices. Just as in the case of traditional finite elements, there are no exact quadrature rules for non-affine element geometries. So, sufficiently accurate Gauss rules are generally accepted as about the best one can do. We have investigated Reissner–Mindlin (RM) shell theory [2, 3] and Kirchhoff–Love (KL) shell theory [4], which precludes transverse shear deformation and is "rotation free," only requiring displacement degrees of freedom, unlike RM elements, which additionally require rotation or director fields. Our study focused on maximally smooth B-splines and NURBS elements of polynomial order $p = 2, 3, 4, 5$, and 6 for RM theory, and $p = 3, 4, 5$, and 6 for KL theory. The in-plane Gauss point patterns used involve $(p + 1)^2$

points per Bézier element. Based on previous studies, we anticipated severe locking to occur for lower orders of p and mitigation of locking for higher orders of p, and indeed this was the case [5]. For orders $p = 5$ and 6, we found promising results for all tests considered. It seems higher-order elements cure a multitude of ills, but, of course, the obvious drawback is the computational cost associated with the very large number of Gauss quadrature points per element. These orders of p may seem high, but that is probably due to lingering perceptions emanating from experience with classical finite element analysis. With one control point per element, the order of smooth spline elements is asymptotically the same as $p = 1$ in traditional finite element analysis. Given these observations, it seems that the cases $p = 5$ and 6 might provide robust capabilities of the type desired if, and only if, the cost of quadrature could be reduced to an acceptable level, independent of p.

We endeavored to reduce the number of quadrature points to be substantially less than full Gauss quadrature. Greville abscissae, which are in one-to-one correspondence with the control points (i.e., nodes), represent a "one-point" quadrature rule in the sense that there is only one quadrature point per control point. This was our first attempt, but in Galerkin formulations of shell theories it was not effective. However, we found that Greville abscissae were effective, if we redefined the space that determined the Greville abscissae to include, in addition to the basis functions, all the derivatives appearing in the weak form of the problem. To be specific, in the case of maximally smooth RM elements, to determine the Greville abscissae, we used the larger space of pth-order splines that are C^{p-2} continuous. Note that this is one order less continuity than for maximally smooth pth-order splines, which are C^{p-1} continuous. For maximally smooth KL elements, we used the still larger space of pth-order splines that are C^{p-3} continuous. In both cases, we then solve linear, moment fitting equations in each parametric direction to obtain the weights, and then the two-dimensional quadrature points and weights are generated by a simple tensor product of the one-dimensional quantities. This results in, asymptotically, four in-plane quadrature points per RM shell element and nine in-plane quadrature points per KL shell element, which are fewer than those required by full Gauss quadrature for all the cases considered, and substantially fewer in the higher-order cases, with concomitant reductions in computational cost. The accuracy of the Greville rules is found to be commensurate with full Gauss quadrature.

2 Greville Quadrature

2.1 Definition of Greville Quadrature

Numerical integration of a univariate function, $f(x)$, can be written as

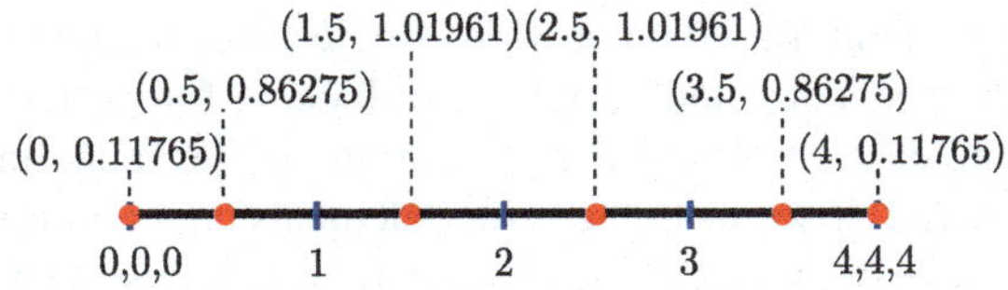

Fig. 1 Greville quadrature points and weights for a quadratic B-spline basis with knot vector $\Xi = \{0, 0, 0, 1, 2, 3, 4, 4, 4\}$. Red dots denote the locations of the quadrature points and $(\cdot, \cdot)$ indicates $(x_i, w_i), i = 1, 2, \cdots, 6$

$$\int_{\hat{I}} f \mathrm{d}\hat{I} \approx \sum_{I=1}^{n} f(x_I) w_I, \tag{1}$$

where f is the integrand, $\hat{I}$ is the integral domain, $\{x_I\}_{I=1}^n$ are the n quadrature points, and $\{w_I\}_{I=1}^n$ are the corresponding weights. Given a univariate p-degree ($p \geq 2$) B-spline basis $\{N_I\}_{I=1}^n$ with an open knot vector $\Xi = \left\{\xi_1, \xi_2, \ldots, \xi_{n+p+1}\right\}$, we propose a way to determine the quadrature points and weights as follows: the Greville abscissae $\{x_I\}_{I=1}^n$, where $x_I = \frac{1}{p}(\xi_{I+1} + \xi_{I+2} + \cdots + \xi_{I+p})$, are chosen to be the quadrature points, and the weights $\{w_I\}_{I=1}^n$ are determined so that the quadrature rule can exactly integrate all linear combinations of the univariate B-spline basis $\{N_I\}_{I=1}^n$. This can be accomplished by solving the following moment fitting system of equations

$$\begin{bmatrix} \int_{\hat{I}} N_1(\xi) \mathrm{d}\xi \\ \int_{\hat{I}} N_2(\xi) \mathrm{d}\xi \\ \vdots \\ \int_{\hat{I}} N_n(\xi) \mathrm{d}\xi \end{bmatrix} = \begin{bmatrix} N_1(x_1) & N_1(x_2) & \cdots & N_1(x_n) \\ N_2(x_1) & \ddots & \cdots & N_2(x_n) \\ \vdots & & \vdots & \vdots \\ N_n(x_1) & \cdots & \cdots & N_n(x_n) \end{bmatrix} \begin{bmatrix} w_1 \\ w_2 \\ \vdots \\ w_n \end{bmatrix}, \tag{2}$$

where the left-hand side contains the moments, which are computed exactly using full Gauss quadrature. As the Greville abscissae are taken as quadrature points, we refer to this quadrature rule as the Greville quadrature. Figure 1 shows the Greville quadrature points and weights for a univariate quadratic B-spline basis associated with the knot vector $\Xi = \{0, 0, 0, 1, 2, 3, 4, 4, 4\}$. Note that the Greville quadrature points and weights are calculated with respect to the global parametric domain of the patch. To utilize the method in existing FEA routines we can easily map these quadrature points into a parent element coordinate system through an affine mapping.

For a bivariate B-spline or NURBS basis, the quadrature points and weights are efficiently obtained through a simple tensor product of the corresponding univariate quantities.

2.2 Greville Quadrature for Shells

The Greville quadrature proposed in Sect. 2.1 lays down a general framework for determining quadrature points and weights, i.e., preselecting the Greville points as the quadrature points and then generating the quadrature weights by solving a moment fitting equation system. However, for a specific isogeometric Galerkin formulation, a proper integration accuracy is necessary to ensure that the resulting linear equation system is stable and accurate. By construction, the Greville quadrature rule can exactly integrate all B-spline basis functions $\{N_I\}_{I=1}^n$ adopted in (2). Therefore, one can easily control the quadrature accuracy by using specific B-spline bases to build the quadrature rule. In this section, we propose different B-spline bases to build quadrature rules for KL and RM shells.

Assuming the highest order of derivatives in the Galerkin formulation is k and the univariate B-spline basis along one of the parametric directions in the Galerkin formulation is $\{N_I^p\}_{I=1}^n$ with knot vector $\Xi = \{\xi_1, \xi_2, \ldots, \xi_{n+p+1}\}$, the Greville quadrature rules for analysis should be constructed in a way such that all basis functions $\{N_I^p\}_{I=1}^n$ and their derivatives of order less than or equal to k are integrated exactly. In other words, (2) should be satisfied for all functions in $\{N_{I,m}^p \mid 1 \leq I \leq n, 0 \leq m \leq k\}$. Notice that these functions are included in a set of new B-spline basis functions $\left\{\tilde{N}_I^p\right\}_{I=1}^{\tilde{n}}$, with knot vector $\tilde{\Xi}$ obtained by increasing the multiplicity of each interior knot of Ξ by k. It is preferable to use $\left\{\tilde{N}_I^p\right\}_{I=1}^{\tilde{n}}$ to build the Greville quadrature rules, because, in this way, we can avoid calculating the derivatives of the B-spline basis functions $\{N_I^p\}_{I=1}^n$ and the quadrature points are naturally the Greville quadrature points calculated from the knot vector $\tilde{\Xi}$. In what follows, we will use the notations

$$\mathcal{S}_0^p = \{N_I^p\}_{I=1}^n \quad \text{and} \quad \mathcal{S}_k^p = \left\{\tilde{N}_I^p\right\}_{I=1}^{\tilde{n}}, \quad k \in \{1, 2\}, \tag{3}$$

to indicate different B-spline bases.

According to the rules given above, for KL shells, the quadrature rule along one direction will be constructed with $\mathcal{S}_2^p$, and for RM shells, it will be constructed with $\mathcal{S}_1^p$. A two-dimensional quadrature rule is simply the tensor product of two one-dimensional quadrature rules as mentioned in Sect. 2.1. To distinguish these two quadrature rules for KL and RM shells, we will refer to them as GREVI-K and GREVI-R, respectively, hereafter. For a cubic B-spline basis with knot vector $\Xi = \{0, 0, 0, 0, 1, 2, 3, 4, 4, 4, 4\}$, the one-dimensional quadrature points and weights for GREVI-K and GREVI-R are illustrated in Fig. 2. It is clear that the GREVI-R and GREVI-K rules, asymptotically, only involve two and three quadrature points in each parametric direction per element, respectively, regardless of the basis degrees. Consequently, only four and nine in-plane quadrature points are required for RM and KL shell elements.

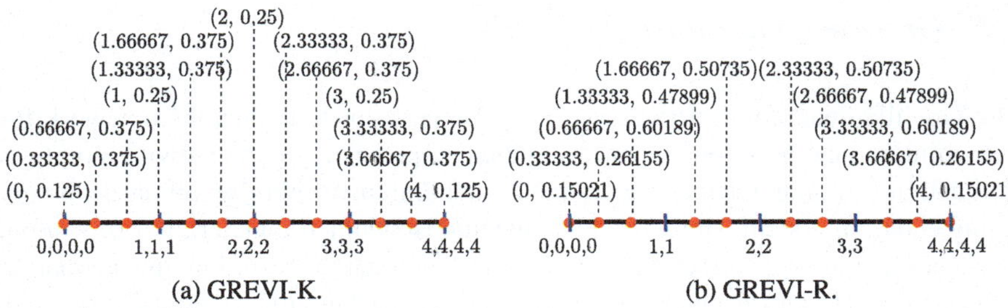

Fig. 2 Quadrature points and weights of GREVI-K and GREVI-R for a cubic B-spline basis with knot vector $\Xi = \{0, 0, 0, 0, 1, 2, 3, 4, 4, 4, 4\}$. Red dots denote the locations of the quadrature points, and $(\cdot, \cdot)$ indicates the global quadrature point and weight pair (x_i, w_i)

Remark 1. The Greville quadrature weights are not always positive for an arbitrary knot vector. For example, if a knot interval of Ξ is extremely small compared to adjacent intervals, it is possible for the GREVI-K and GREVI-R quadrature rules to exhibit negative weights locally. Quadrature rules with negative weights are prone to instability and not preferred in engineering analysis. In this work we confine ourselves to uniform knot vectors. With uniform knot vectors we only see negative weights for the GREVI-K rule with $p = 4$. How to effectively remove the negative weights for arbitrary knot vectors is non-trivial and will be addressed in future work.

2. For $p = 2$, the multiplicities of the interior knots of the resulting knot vector $\tilde{\Xi}$ will be three for GREVI-K. Therefore, each element is an independent Bézier patch and the quadrature rule needs to be determined on the element level through (2). The resulting quadrature points will be distributed by the Simpson's rule, and unfortunately the two coincident quadrature points at the element interface can not be combined into one point due to the discontinuous second order derivatives [6]. As a result, the number of quadrature points will be the same as for full Gauss quadrature and thus we will not explore the case of $p = 2$ for GREVI-K further. □

2.3 *Scordelis–Lo Roof*

The Scordelis–Lo roof problem is part of the so-called shell obstacle course [7] and tests a shell element's ability to handle both membrane and bending modes. An 80° arc of a cylinder with radius $R = 25$, length $L = 50$, and thickness $t = 0.25$ or 0.025 is supported on each end by a rigid diaphragm. It is loaded with its own weight $q_z = 90$. The material has Young's Modulus, $E = 4.32 \times 10^8$, and Poisson's ratio $\nu = 0$. Figure 3 shows the problem setup. The initial mesh of the *whole* model consists of 4×4 maximally smooth elements.

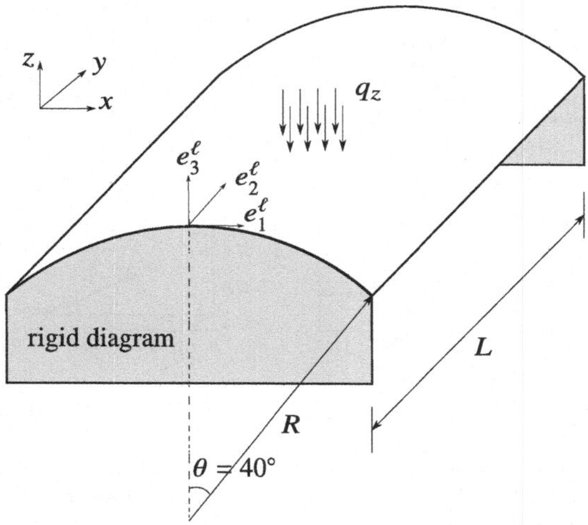

Fig. 3 Schematic for the Scordelis–Lo roof problem

The maximum displacement occurs on the free edge at $\frac{L}{2}$. For $t = 0.25$, the usual FEA solution converges to 0.3006 for KL shells [4] and 0.3024 for RM shells [2, 7]. For $t = 0.025$, the reference solution given in [8] is 32.0 for KL shells and we also take it as the reference solution for RM shells in this work. The maximum displacement on the free edge at $\frac{L}{2}$ is monitored and results for the KL shell are shown in Fig. 4. For $p = 3$ and 4, $t = 0.25$, as shown in Fig. 4a, the GREVI-K rule obtains slightly worse results than the GAUSS rule with the initial mesh, but with one refinement they almost achieve identical results that are close to the reference solution. As the degrees increase to $p = 5$ and 6, GREVI-K and GAUSS obtain nearly coincident results and, since locking is alleviated largely by higher-order bases, good results are achieved with even the initial mesh. As the shell thins, i.e., $t = 0.025$, membrane locking becomes more severe and the results converge more slowly for both quadrature rules as shown in Fig. 4b. In this case, the GREVI-K rule achieves superior results for $p = 4$ while, for $p = 5$ and 6, the results obtained by GREVI-K and GAUSS hardly differ from each other.

Since this problem is membrane dominated, transverse shear locking is not significant. Figure 5 demonstrates that for $p = 3$ to 6 the RM shell with GAUSS and GREVI-R converges in a similar way as the KL shell with GAUSS and GREVI-K shown in Fig. 4. For $p = 2$, the GREVI-R rule underperforms the GAUSS rule but otherwise both rules perform about the same.

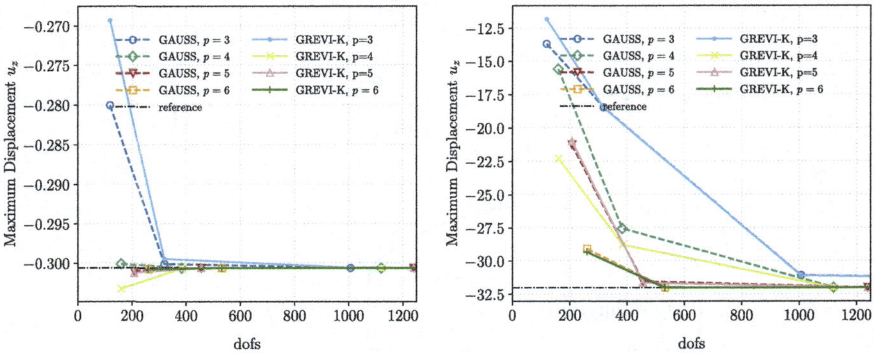

(a) t = 0.25. (b) t = 0.025.

Fig. 4 Scordelis–Lo roof modeled as a KL shell: Convergence of the maximum displacement u_z with GAUSS and GREVI-K, degrees $p = 3$ to 6, and maximally smooth elements. The whole roof is modeled with an initial 4×4 mesh

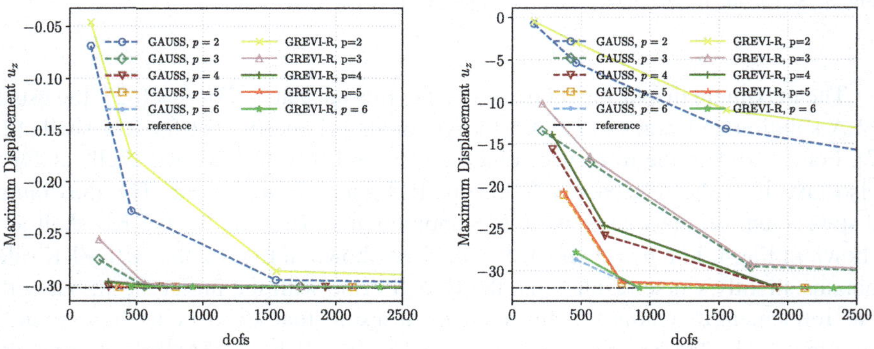

(a) t = 0.25. (b) t = 0.025.

Fig. 5 Scordelis–Lo roof modeled as an RM shell: Convergence of the maximum displacement u_z with GAUSS and GREVI-R, degrees $p = 2$ to 6, and maximally smooth elements. The whole roof is modeled with an initial 4×4 mesh

3 Conclusions

We proposed Greville quadrature schemes for isogeometric shell analysis. The quadrature points are chosen to be Greville abscissae for B-splines and NURBS, but the spaces from which the rules emanate are unusual. The proposed method for Reissner–Mindlin (RM) shells, referred to as GREVI-R, is a Greville quadrature scheme based on pth-order basis functions, but with one-order lower continuity across element interfaces than the pth-order, maximally smooth, basis functions used for analysis. The proposed scheme for Kirchhoff–Love (KL) shells, referred to as GREVI-K, constructs the Greville quadrature points based on pth-order basis functions, but with continuity two orders lower than the maximally smooth basis used for analysis. The quadrature weights are determined by solving linear moment fitting equations. These methods are free of rank deficiency and spurious modes. At the same time, they achieve similar accuracy as full Gauss quadrature. The proposed methods for RM and KL shell elements only involve four and nine in-plane quadrature points, respectively, per Bézier element, compared to the usual $(p + 1)^2$ in-plane quadrature points for elements with standard "full" Gauss integration. As increasing the basis order does not asymptotically increase degrees-of-freedom or the number of quadrature points for higher-order basis functions, the proposed methods are efficient, robust, accurate and alleviate locking. For further details of the methodology and a comprehensive evaluation, please see our paper [9].

Acknowledgements Thomas J. R. Hughes and Zhihui Zou acknowledge support from the Office of Naval Research grant N00014-17-1-2039 and through the Department of Defense, Navy, Contract N6833518C0014, respectively, with Coreform, LLC.

References

1. Hughes, T. J. R., Cottrell, J. A., & Bazilevs, Y. (2005). Isogeometric analysis: CAD, finite elements, NURBS, exact geometry and mesh refinement. *Computer Methods in Applied Mechanics and Engineering, 194*(39), 4135–4195.
2. Dornisch, W., Müller, R., & Klinkel, S. (2016). An efficient and robust rotational formulation for isogeometric Reissner-Mindlin shell elements. *Computer Methods in Applied Mechanics and Engineering, 303*, 1–34.
3. Zou, Z., Scott, M. A., Miao, D., Bischoff, M., Oesterle, B., & Dornisch, W. (2020). An isogeometric Reissner-Mindlin shell element based on Bézier dual basis functions: Overcoming locking and improved coarse mesh accuracy. *Computer Methods in Applied Mechanics and Engineering, 370*, 113283.
4. Kiendl, J., Bletzinger, K.-U., Linhard, J., & Wüchner, R. (2009). Isogeometric shell analysis with Kirchhoff-Love elements. *Computer Methods in Applied Mechanics and Engineering, 198*(49–52), 3902–3914.
5. Echter, R., & Bischoff, M. (2010). Numerical efficiency, locking and unlocking of NURBS finite elements. *Computer Methods in Applied Mechanics and Engineering, 199*(5–8), 374–382.
6. Schillinger, D., Hossain, S. J., & Hughes, T. J. R. (2014). Reduced Bézier element quadrature rules for quadratic and cubic splines in isogeometric analysis. *Computer Methods in Applied Mechanics and Engineering, 277*, 1–45.
7. Macneal, R. H., & Harder, R. L. (1985). A proposed standard set of problems to test finite element accuracy. *Finite Elements in Analysis and Design, 1*(1), 3–20.
8. Greco, L., Cuomo, M., & Contrafatto, L. (2018). A reconstructed local $\overline{\mathrm{B}}$ formulation for isogeometric Kirchhoff-Love shells. *Computer Methods in Applied Mechanics and Engineering, 332*, 462–487.
9. Zou, Z., Hughes, T. J. R., Scott, M. A., Sauer, R. A., & Savitha, E. J. (2020). Galerkin formulations of isogeometric shell analysis: Alleviating locking with Greville quadratures and higher-order elements. *Computer Methods in Applied Mechanics and Engineering, 380*(2021), 113757.

Thermodynamic Topology Optimization of Layered Anisotropic Materials

Dustin R. Jantos and Philipp Junker

To Professor Peter Wriggers on occasion of his 70th birthday: I thank you for your warm welcome at the Institute of Continuum Mechanics, your manifold support, and our trustful cooperation. I am looking forward to our common time at IKM.

—P. Junker

Abstract Anisotropic materials are often used for high-performance components and thus the optimization of structures produced with those materials is of major interest. To optimize such structures, the topology as well as the material orientation should be considered as design variables for maximum performance. Most common production processes of anisotropic materials consider additive manufacturing either by layering laminates of fiber reinforced composites or 3D-printing. To this end, we present a variational and thermodynamic optimization model for the topology based on a density approach in combination with a local continuous fiber angle optimization in the three-dimensional space. To tackle the mentioned production restrictions, the fibers are restricted to be parallel to a globally defined layer plane, which accounts for the layered production process. The layer plane is either optimized as well or can be prescribed by the user. In addition, a filtering technique for the fibers is presented to constrain the maximum fiber curvature within the layers.

1 Introduction

Over the last decades topology optimization (TO) became a more and more important tool to save resources. Overviews and reviews over different topology optimization methods can be found in the literature [1, 2]. Anisotropic materials are often used for

D. R. Jantos (✉) · P. Junker
Institute of Continuum Mechanics, Leibniz University Hannover, Hannover, Germany
e-mail: jantos@ikm.uni-hannover.de

P. Junker
e-mail: junker@ikm.uni-hannover.de

F. Aldakheel et al. (eds.), *Current Trends and Open Problems in Computational Mechanics*, https://doi.org/10.1007/978-3-030-87312-7_22

high-performance components and thus not only the optimization of the topology but also the optimization of the material orientation is important. However, production of anisotropic materials is often constrained to layer-wise additive manufacturing (AM), as for example for laminates of layered fiber reinforced composites [3, 4] or 3D-printing techniques [5–8], e.g. components produced with Selective Laser Melting (SLM) or Fused Deposition Modeling (FDM) (also called Fused Filament Fabrication) posses inherent anisotropic mechanical properties, especially for the latter, if reinforced with carbon or metal fibers [9]. Components produced with AM may have nearly arbitrary geometries but the fibers are aligned layer-wise and all fibers are parallel to a given plane due to the nature of the manufacturing process.

Besides general works of TO with anisotropic materials including the optimization of the local fiber orientation [10–13], research on direct combinations of AM and TO can be found in literature: in [14], a Discrete Material Optimization (DMO)[1] model was applied to optimize the material orientation within each layer of beams with given cross section profile.

The authors in [15, 16] investigated components optimized with Continuous Fiber Angle Optimization (CFAO)[2] and produced by FDM/FFF but—as most studies—only for the two-dimensional case. In [17, 18], the authors used a CFAO combined with topology optimization for 2.5-dimensional boundary value problems, i.e. three-dimensional problems with prescribed (curved) fiber planes, and produced the optimization results with voxel-based multimaterial jetting.

Special constraints have been incorporated to TO approaches, e.g. overhang constraint introducing an allowable self-supporting angle as additional constraint for the TO to reduce the requirement of support structures within 3D-printing processes [19]. An overview over the challenges and requirements on TO models regarding AM can be found in [16, 20, 21] for 3D-Printing and in [22] for laminates.

Most works in literature assume the layer plane to be given, i.e. the problem becomes a stacking of two-dimensional problem solutions. In contrast, within this publication, we derive a thermodynamic optimization model for layered anisotropic materials based on a fully three-dimensional setting, i.e. the design space and boundary value problem are defined in a three-dimensional setting without further restrictions on the geometry of the component. We combine a density based topology optimization with a three-dimensional CFAO approach which yields locally optimized fiber directions which are all restricted to be parallel to a single layer plane. This layer plane is also determined by a continuous angle optimization, or can be prescribed by the user. For the optimization of the material orientation, i.e. fiber direction, we present a novel approach based on Euler angles as design variables for the rotation parametrization based on [23]: two globally defined Euler angles parametrize the

[1] DMO for material orientation optimization is used by applying the same anisotropic material, but with different orientation as set of discrete number of varying materials. Thus, the local material orientation is not optimized freely, but determined by the material choice and the possible orientations are limited to the chosen set of materials.

[2] In CFAO, the local material orientation is parametrized by continuous rotation parameters. Thus, it is possible to optimize the material orientation freely.

orientation of the layer plane in three-dimensional space for the whole design. A third Euler angle is defined locally which yields fiber directions parallel to the layer plane for each point in the design space. In addition, we present a filtering technique to constrain the maximum curvature of the fibers within the layer plane based on the filtering technique presented in [23].

The remainder of the article is structured as follows: the equations for the thermodynamic optimization are derived in Sect. 2 and their numerical solution including the material orientation filter for the fiber curvature constraint are explained in Sect. 3. Numerical results for two different materials, varying maximum fiber curvature and prescribed layer planes are presented in Sect. 4. Finally, the article is concluded in Sect. 5.

2 Thermodynamic Optimization

2.1 Design Variables

The design variables $\boldsymbol{d} = \{\chi, \boldsymbol{\alpha}\}$ for the topology optimization of anisotropic materials for the three-dimensional case are given by

$$\boldsymbol{d} = \begin{cases} \chi : & \text{density} \rightarrow \text{topology} \\ \boldsymbol{\alpha} = \{\phi, \nu, \omega\} : & \text{Euler angles} \rightarrow \text{orientation}\,. \end{cases} \tag{1}$$

The continuous density variable $\chi \in [0, 1]$ describes the presence ($\chi = 1$) and absence ($\chi = 0$) of material.

The material orientation is described by the set of continuous rotation parameters $\boldsymbol{\alpha}$ describing the rotation of an anisotropic base material (c.f. Sect. 2.2), i.e. a continuous fiber angle optimization (CFAO). To this end, we define the set of Euler angles $\boldsymbol{\alpha} = \{\phi, \nu, \omega\}$ to describe the rotation matrix $\boldsymbol{Q}$ for the rotation defined by a sequential z-x'-z'' rotation (see Fig. 1), i.e. a rotation around the z-axis followed by the rotation around the rotated x-axis (indicated by x') followed by the rotation around the rotated z-axis (indicated by z''). Thus, the first two rotations define the final orientation of the z-axis given by z'', i.e. the x''-y''-plane. Then, the third rotation describes an additional rotation of around the z''-axis and thus the final orientation within the x''-y''-plane. Therefore, if the first two rotations are defined as global, i.e. are identical for all discretized points $\boldsymbol{x} \in \Omega$, the z''-axis describes the normal of the layers of a layered (anisotropic) material, i.e. for example the plane normal of laminates or the print plane of a 3D-printer. The third and final rotation describes the fiber direction within the layers, i.e. within the x''-y''-plane.

The rotation is parametrized by the three Euler angles as follows

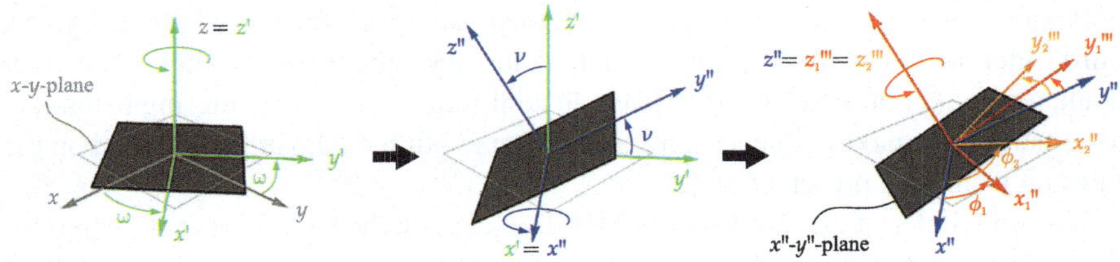

Fig. 1 Illustration of a z-x'-z'' rotation with the Euler angles ω and ν defining the x''-y''-plane within the first two rotations and ϕ_i defining the final orientations of the x_i''- and y_i''-axes within the x''-y''-plane

$$\boldsymbol{\alpha} = \begin{pmatrix} \phi(\boldsymbol{x}) \\ \nu \\ \omega \end{pmatrix} \begin{array}{l} \rightarrow \quad \text{rotation around } z''\text{-axis} \\ \rightarrow \quad \text{rotation around } x'\text{-axis} \\ \rightarrow \quad \text{rotation around } z\text{-axis} \end{array}, \tag{2}$$

where $\phi = \phi(\boldsymbol{x})$ is a local design variable and $\nu \neq \nu(\boldsymbol{x})$ and $\omega \neq \omega(\boldsymbol{x})$ are global design variables. With the abbreviation $s_k = \sin(\alpha_k)$ and $c_k = \cos(\alpha_k)$ the rotation matrix is defined by

$$\boldsymbol{Q}(\boldsymbol{\alpha}) = \begin{pmatrix} c_1 c_3 - s_1 c_2 s_3 & -s_1 c_2 c_3 - c_1 s_3 & s_1 s_2 \\ s_1 c_3 + c_1 c_2 s_3 & c_1 c_2 c_3 - s_1 s_3 & -c_1 s_2 \\ s_2 s_3 & s_2 c_3 & c_2 \end{pmatrix} \tag{3}$$

2.2 *Material Definition*

For the remainder of the publication, we assume all higher order tensors to be given in their respective Mehrabadi–Cowin notation [24]. Thus, the stresses $\boldsymbol{\sigma}$ and strains $\boldsymbol{\varepsilon}$ are given as the corresponding six-dimensional first order tensors and the material stiffness tensors and the Rotation-matrix $\boldsymbol{Q}$ are given as the corresponding six-dimensional second order tensors.

Within the respective Mehrabadi–Cowin notation the rotation of a elasticity stiffness tensor $\mathbb{E}^0$ can be written as [24]

$$\mathbb{E}(\boldsymbol{d}) = \mathbb{E}(\chi, \boldsymbol{\alpha}) = \rho(\chi)\left[\boldsymbol{Q}(\boldsymbol{\alpha})\right]^{\mathrm{T}} \cdot \mathbb{E}^0 \cdot \boldsymbol{Q}(\boldsymbol{\alpha}) \tag{4}$$

with the Power-Law material interpolation

$$\rho(\chi) = (1 - \kappa)\, \chi^3 + \kappa \quad \text{with } \rho(\chi) \in [\kappa, 1] \tag{5}$$

known from SIMP approaches [25] which penalizes intermediate densities $0 < \chi < 1$. Herein, $\kappa = 10^{-9}$ is a small but non-zero numerical value to prevent singularities within the finite element method.

With the given rotation defined in Sect. 2.1, the z''-axis corresponds to the layer normal of the plane to which all fibers are oriented in parallel. Thus, the third spacial direction of the base materials $\mathbb{E}^0$ must correspond to the inter-plane stiffness, i.e. between layers. Consequently, the first spacial direction of the base materials correspond to the stiffness in fiber direction and the second spacial direction correspond to the stiffness between two parallel fibers within the z''-axis-plane.

2.3 Optimization Model

The thermodynamic optimization is based on the stationary condition of the Hamilton functional [26]

$$\mathcal{H}[\boldsymbol{u}, \boldsymbol{d}, \boldsymbol{p}] := \mathcal{G}[\boldsymbol{u}, \boldsymbol{d}, \boldsymbol{p}] + \mathcal{D}_p[\boldsymbol{p}] - \mathcal{R}[\boldsymbol{d}] + C[\boldsymbol{d}] \rightarrow \underset{\boldsymbol{u},\boldsymbol{d},\boldsymbol{p}}{\text{stat}} , \tag{6}$$

which postulates that the Gibbs energy $\mathcal{G}$, the dissipative work $\mathcal{D}_p$, and the rearrangement functional $\mathcal{R}$ become stationary under given constraints C. The unknowns of the problems are the displacements $\boldsymbol{u}$, the design variables $\boldsymbol{d}$, and the internal variables $\boldsymbol{p}$ describing the (physical) microstructure.

The dissipative work due to physical changes of the microstructure described by the internal variables $\boldsymbol{p}$ is given by $\mathcal{D}_p$. Typical examples for such internal variables might be damage, plastic strain, and volume fractions of material phases. However, in the presented model, only linear elastic material behavior is applied so that no internal variables $\boldsymbol{p}$ are included and $\mathcal{D}_p$ can be omitted. The energetic contribution due to design changes, i.e. rearrangement of the microstructure in terms of the design variables $\boldsymbol{d}$ is given by

$$\mathcal{R}(\boldsymbol{d}) = \mathcal{D}_d(\boldsymbol{d}) + \mathcal{F}(\boldsymbol{d}) , \tag{7}$$

where $\mathcal{D}_d$ describes the energy due to local rearrangement and $\mathcal{F}$ the energy due to convective rearrangement. The rearrangement functional $\mathcal{R}$ describes the energy consumed to enforce design changes, or in other words: external work is required to change the design since it is not an intrinsic or natural material behavior, which motivates the negative sign of $\mathcal{R}$ in the energy functional.

The individual functionals are specified as follows: the Gibbs energy for a linear elastic material is given by

$$\mathcal{G} = \int_{\Omega} \Psi \, \mathrm{d}V - \int_{\Omega} \boldsymbol{b} \cdot \boldsymbol{u} \, \mathrm{d}V - \int_{\partial\Omega} \boldsymbol{t} \cdot \boldsymbol{u} \, \mathrm{d}A \tag{8}$$

for a given design space Ω with the Helmholtz free energy

$$\Psi(\boldsymbol{u}, \boldsymbol{d}) = \frac{1}{2} \, \boldsymbol{\varepsilon}(\boldsymbol{u}) \cdot \mathbb{E}(\boldsymbol{d}) \cdot \boldsymbol{\varepsilon}(\boldsymbol{u}) . \tag{9}$$

The body forces $\boldsymbol{b}$ and traction forces $\boldsymbol{t}$ are given as constant, i.e. they are neither a function of the displacements $\boldsymbol{u}$ nor the design $\boldsymbol{d}$.

Additional constraints are required for the density variable χ to prevent unphysical designs or trivial solutions. The total volume of the component is constrained by

$$c(\boldsymbol{d}) = \int_\Omega \chi \,\mathrm{d}V - \varrho_\Omega\, \Omega = 0 \tag{10}$$

with the prescribed (relative) total volume ϱ_Ω. Accounting for the interval constraint $\chi \in [0, 1]$, the functional C reads

$$C = \lambda\, c(\boldsymbol{d}) + \int_\Omega \gamma_\chi\, \chi \,\mathrm{d}V \tag{11}$$

with the Lagrange multiplier λ and the Karush–Kuhn–Tucker parameter γ_χ which are determined by a bisection algorithm within each iteration step [27].

The functional for convective rearrangement

$$\mathcal{F} = \frac{1}{2} \int_\Omega r_\chi^2\, w_\chi \|\nabla \chi\|^2 \tag{12}$$

is used to enforce a gradient enhanced regularization which suppresses checkerboarding, grants mesh independent results, and allows to control the minimum member size given by r_χ. Herein, w_χ is added as weight factor, so that the numerical value of r_χ corresponds to a physical length, i.e. the prescribed minimum width of geometric features within the topology optimization. No gradient enhanced regularization or constraints are applied for the material orientation parameters $\boldsymbol{\alpha}$. However, a modified version of the material orientation filter from [23] is presented in Sect. 3.3 which provides control over the maximum fiber curvature and within layers.

The functional for local arrangement is given by [26]

$$\mathcal{D}_d = \int_\Omega \frac{\partial \Phi}{\partial \dot{\boldsymbol{d}}} \cdot \boldsymbol{d} \,\mathrm{d}V \tag{13}$$

with a function Φ similar to dissipation functions, see [26],

$$\Phi = \frac{1}{2}\eta_\chi w_\chi \dot{\chi}^2 + \frac{1}{2}\eta_\alpha \psi \left\| \dot{\boldsymbol{Q}} \cdot \boldsymbol{Q}^{-1} \right\|^2 , \tag{14}$$

where the application of the tensor of angular velocity $\dot{\boldsymbol{Q}} \cdot \boldsymbol{Q}^{-1}$ grants frame indifference. The so-called viscosities η_χ and η_α as well as the weights w_χ and $\psi := \boldsymbol{\sigma} \cdot \boldsymbol{\varepsilon}$ contribute to numerical damping to provide a stable convergence of the solution.

2.4 Stationarity Conditions and Evolution Equations

The stationary condition (omitting internal variables $\boldsymbol{p}$) of Eq. (6) yields

$$\delta\mathcal{G}[\boldsymbol{u},\boldsymbol{d}](\delta\boldsymbol{u}) = 0 \quad \forall\, \delta\boldsymbol{u} \tag{15}$$

$$\delta\mathcal{G}[\boldsymbol{u},\boldsymbol{d}](\delta\boldsymbol{d}) - \delta\mathcal{D}_d[\boldsymbol{d}](\delta\boldsymbol{d}) - \delta\mathcal{F}[\boldsymbol{d}](\delta\boldsymbol{d}) + \delta\mathcal{C}[\boldsymbol{d}](\delta\boldsymbol{d}) = 0 \quad \forall\, \delta\boldsymbol{d}\,. \tag{16}$$

The evaluation of the stationary condition given in Eq. (16) allows for the derivation of the evolution equations which are differential equations whose (iterative) solution yields the optimal design. The variation with respect to the density variable χ in its strong form yields the evolution equation

$$\dot{\chi} = \frac{1}{\eta_\chi w_\chi}\Big(-\tau_\chi + r_\chi^2\, w_\chi\, \nabla\cdot\nabla\chi - \lambda - \gamma_\chi \Big) \quad \forall\, \boldsymbol{x} \in \Omega \tag{17}$$

including the Neumann boundary condition

$$\nabla\chi \cdot \boldsymbol{n} = 0 \quad \forall\, \boldsymbol{x} \in \partial\Omega\,. \tag{18}$$

The evaluation of the stationary condition in Eq. (16) with respect to the Euler angles $\boldsymbol{\alpha}$ yields

$$\int_\Omega \left(\frac{\partial\Psi}{\partial\boldsymbol{\alpha}} - \frac{\partial\Phi}{\partial\dot{\boldsymbol{\alpha}}} \right) \cdot \delta\boldsymbol{\alpha}\, \mathrm{d}V = 0 \quad \forall\, \delta\boldsymbol{\alpha}\,. \tag{19}$$

Replacing the Helmholtz energy Ψ with the definition of the driving forces according to Eq. (26), and the dissipation function Φ from Eq. (14) with the relation

$$\left\| \dot{\boldsymbol{Q}} \cdot \boldsymbol{Q}^{-1} \right\|^2 = \dot{\phi}^2 + \dot{\nu}^2 + \dot{\omega}^2 + 2\,\dot{\phi}\,\dot{\omega}\,\cos(\nu) \tag{20}$$

yields

$$\dot{\phi} = -\frac{1}{2\eta_\alpha}\left(\frac{\tau_\phi}{\psi} - \mu\cos(\nu) \right) \quad \forall\, \boldsymbol{x} \in \Omega,\ \nu \neq n\pi,\ n \in \mathbb{Z} \tag{21}$$

$$\dot{\nu} = -\frac{1}{2\eta_\alpha}\,\frac{\langle\tau_\nu\rangle}{\langle\psi\rangle} \tag{22}$$

$$\dot{\omega} = -\frac{1}{2\eta_\alpha}\,\mu \quad \nu \neq n\pi,\ n \in \mathbb{Z} \tag{23}$$

with the abbreviations

$$\mu = \frac{\langle\tau_\omega\rangle - \cos(\nu)\,\langle\tau_\phi\rangle}{\big(1-\cos^2(\nu)\big)\,\langle\psi\rangle} \quad \text{and} \quad \langle\ldots\rangle = \int_\Omega \ldots\,\mathrm{d}V\,. \tag{24}$$

The evolution Eq. (21) of the first Euler angle ϕ, which is supposed to define the local fiber orientation within the material layers, contains a local part with $\frac{\tau_\phi}{\psi}$ and a global part with $\mu\cos(\nu)$, whereas the two remaining evolution equations contain only global quantities. A special case, the so-called gimbal lock is given when the second Euler angle ν is a natural multiplier of π. In this case, the first and third Euler angle, ϕ and ω, respectively, describe a rotation around the same axis and singularities in $\dot{\phi}$ and $\dot{\omega}$ appear due to $\cos(\nu)^2 = 1$. To circumvent the singularity we set $\mu = 0$, such that

$$\dot{\omega} = 0 \quad \text{and} \quad \dot{\phi} = -\frac{1}{2\eta_\alpha}\frac{\tau_\phi}{\psi} \quad \text{for } \nu = n\pi,\ n \in \mathbb{Z}\,. \tag{25}$$

This is reasonable since $\dot{\phi}$ and $\dot{\omega}$ describe the same rotation and are redundant in this special case.

The driving forces τ_d serve as sensitivities and emerge naturally by evaluating the stationary condition of the Hamilton principle. The driving force of a (scalar) design variable d is defined by

$$\tau_d := -\frac{\partial \Psi}{\partial d} = -\frac{1}{2}\,\boldsymbol{\varepsilon}\cdot\frac{\partial \mathbb{E}}{\partial d}\cdot\boldsymbol{\varepsilon}\,. \tag{26}$$

Here, the driving forces for the design variables $\boldsymbol{d} = \{\chi, \boldsymbol{\alpha}\}$ are computed to be

$$\tau_\chi = -\frac{\rho'}{2}\,\boldsymbol{\varepsilon}\cdot\left[\boldsymbol{Q}\right]^{\mathrm{T}}\cdot\mathbb{E}^0\cdot\boldsymbol{Q}\cdot\boldsymbol{\varepsilon} \tag{27}$$

$$\tau_{\alpha_i} = -\frac{\rho}{2}\,\boldsymbol{\varepsilon}\cdot\left(\left[\frac{\partial \boldsymbol{Q}}{\partial \alpha_i}\right]^{\mathrm{T}}\cdot\mathbb{E}^0\cdot\boldsymbol{Q} + \boldsymbol{Q}^{\mathrm{T}}\cdot\mathbb{E}^0\cdot\frac{\partial \boldsymbol{Q}}{\partial \alpha_i}\right)\cdot\boldsymbol{\varepsilon} \tag{28}$$

with $\rho' = 3\,(1-\kappa)\,\chi^2$ (see Eq. (5)) and $\alpha_i \in \{\phi, \nu, \omega\}$.

3 Numerical Solution

3.1 Program Structure

The optimal design is determined by an iterative process over the iteration steps i consisting of an alternating finite element analysis to solve the mechanical equilibrium (15) for the displacements $\boldsymbol{u}^i$ with a given design $\boldsymbol{d}^i$ and a design update by solving the evolution equations. The design variables $\boldsymbol{d}$ are discretized element-wise. Thus, the driving forces τ_d and energy ψ for the evolution equations must be evaluated within the elements center of mass [28, 29].

$$\boldsymbol{d}^{i+1} = \boldsymbol{d}^i + \Delta t_d\,\dot{\boldsymbol{d}}(\boldsymbol{u}^i, \boldsymbol{d}^i)\,. \tag{29}$$

Thus, after discretization, only element-wise algebraic equations but no equation systems need to be solved for the design update (except for the calculation of the displacements field $\boldsymbol{u}$ via FEM). The optimization is considered to be converged if the relative change of the structural compliance energy becomes less than a the tolerance

$$\frac{\langle \Psi^{i+1} \rangle - \langle \Psi^{i} \rangle}{\langle \Psi^{i} \rangle} < 10^{-6} \, . \tag{30}$$

The weight function for the regularization is given by

$$w_\chi = \left[\int_\Omega \chi\,(1-\chi)\,\mathrm{d}V \right]^{-1} \int_\Omega \chi\,(1-\chi)\,|\tau_\chi|\,\mathrm{d}V \tag{31}$$

The Laplace-operator $\nabla \cdot \nabla\chi$ in the evolution equation for the density variable χ including the Neumann boundary conditions (18) are calculated for element-wise discretized design variables via the neighbored element method introduced in [27]. Herein, an algorithm with internal loop and appropriate choice of the time increment Δt_χ w.r.t. the mesh width h are provided to grant numerical stability of the conditionally stable explicit time discretization. For more details and explanations of the implementation on the method, we refer to the original publication. The evaluation of the evolution equation for the material orientation does not require any internal loops and can be evaluated once parallel to the update-loop of the topology. We chose $\Delta t = t^{i+1} - t^{i} = 1$, so that the time steps i are equivalent to the iteration steps of the optimization, i.e. number of required FEM solutions (Fig. 2).

3.2 *Initial Conditions*

It is well-known that the numerical solutions of the non-convex compliance minimization problem depend on the initial conditions for the design variables [30, 31]. Thus, the initial values for the design variables should be chosen carefully to circumvent undesired local minima. Therefore, we apply objective and homogenous initial conditions, i.e. without user specific "guesses", which are unbiased and do not require any additional input besides the obligatory boundary value problem and material parameters.

For the initial density variable, we apply a homogenous distribution that satisfies the volume constraint (10) which yields $\chi^0 = \varrho_\Omega \neq \chi^0(\boldsymbol{x})$. For the initial material orientation $\boldsymbol{\alpha}^0 \neq \boldsymbol{\alpha}^0(\boldsymbol{x})$, we introduce a weighted homogenization based on the principal stress directions, which are the optimal solution for shear-weak anisotropic materials [10, 32] and are therefore a reasonable consideration:

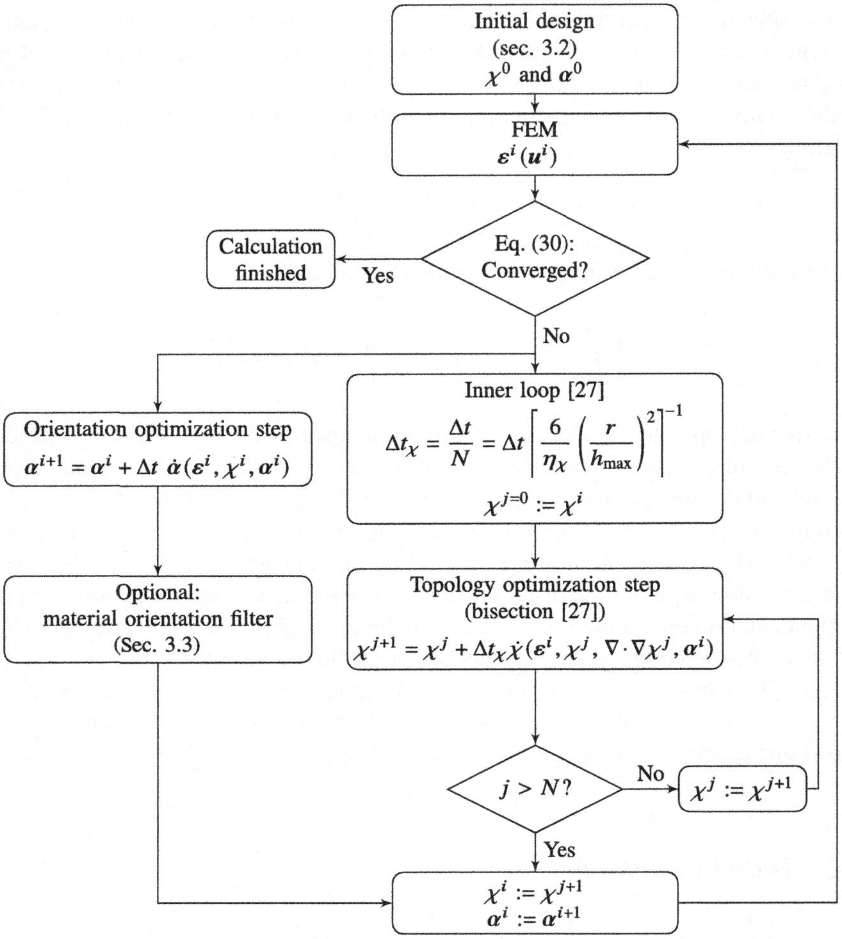

Fig. 2 Flowchart of the numerical solution of the proposed model

1. The boundary value problem is evaluated once via a single FEM step with an isotropic material with $E^{\text{iso}} = 1$, the Poisson's ratio $\nu^{\text{iso}} = 0.25$, and the homogenous density distribution χ^0 and the stresses $\sigma^{\text{iso}}(x)$ are calculated.[3]
2. With the relations from [24, 33], we determine a material distribution $\mathbb{E}^{\text{aniso}}(x)$, whose principal directions coincide with the directions of the principal stresses $\sigma^{\text{iso}}(x)$.
3. We homogenize $\mathbb{E}^{\text{aniso}}(x)$, so that we find an initial material configuration $\mathbb{E}^0 \neq \mathbb{E}^0(x)$ in which all fibers are aligned in one direction and are therefore parallel

[3] A more appropriate isotropic material would be the Haar measure of the applied anisotropic material $\mathbb{E}(\chi^0, \alpha)$ [23, 33]. However, due to the linear theory, the magnitude of the Young's modulus does not influence the result and the influence of the Possion's ratio to the optimization is also negligible.

to an (initial) layer plane. The homogenization is weighted by the norm of the stresses

$$\mathbb{E}^0 = \left[\int_\Omega \|\boldsymbol{\sigma}^{\text{iso}}\| \, \mathrm{d}V\right]^{-1} \int_\Omega \|\boldsymbol{\sigma}^{\text{iso}}\| \, \mathbb{E}^{\text{aniso}} \, \mathrm{d}V \tag{32}$$

4. Finally, the algorithm given in [23] is used to determine the initial values for the Euler angles $\boldsymbol{\alpha}^0$ which correspond to the orientation of the initial material configuration $\mathbb{E}^0$.

3.3 Material Orientation Filter

We present now a filtering technique that constrains the maximum fiber curvature within a given layer with normal $\boldsymbol{n}$, i.e. preserves the layer-wise orientation of the fibers. The layer normal $\boldsymbol{n}$ results either from the global Euler angles ν^i and ω^i of the current iteration step or is prescribed by the user. The filter is based on the technique presented in [23] which applies a distance-weighted homogenization on the material stiffness tensor to circumvent errors due to the symmetry and periodicity of rotations. For explanations on the individual algorithm steps, we refer to [23] and for detailed explanations on material symmetries, material principal directions, and the correspondence to eigensystems, we refer to [34].

The following steps are executed after each update step i of the Euler angles (i.e. evaluation of the evolution equations) for each finite element $e \in N_e$:

1. Calculate the material stiffness tensor of the unfiltered design $\mathbb{E}_e = \mathbb{E}(\chi_e^i, \boldsymbol{\alpha}_e^i)$
2. Apply the convolution operator

$$\mathbb{E}_e^s = \left[\sum_f^{N_e} \Omega_f \, w_\alpha(\Delta x_{ef})\right]^{-1} \sum_f^{N_e} \Omega_f \, w_\alpha(\Delta x_{ef}) \, \mathbb{E}_f \tag{33}$$

with the points f surrounding e to find the material stiffness tensor $\mathbb{E}_e^s$ corresponding to a smoothed design. The respective volume of the elements is indicated by Ω_f. The distance weight is given by

$$w_\alpha(\Delta x_{ef}) = \max\left(0,\ 1 - r^{-1}\|\boldsymbol{x}_e - \boldsymbol{x}_f\|\right) \tag{34}$$

with the midpoint coordinates of the elements $\boldsymbol{x}_e$ and $\boldsymbol{x}_f$. The filter radius r

$$r = r_\alpha + \left(\frac{h}{2} - r_\alpha\right) \boldsymbol{n} \cdot \frac{\boldsymbol{x}_e - \boldsymbol{x}_f}{\|\boldsymbol{x}_e - \boldsymbol{x}_f\|} \tag{35}$$

constrains the fiber curvature to the maximum value of $r_\alpha^{-1}/\sqrt{2}$ within the layer plane given by its normal $\boldsymbol{n}$. The parameter h describes the average distance between two finite elements. Thus, the effective filter radius becomes the user prescribed value r_α for all surrounding points f within the layer plane around the point e and is gradually reduced to $\frac{h}{2}$ for elements which are not part of the layer plane.

3. Calculate the three principal directions $\boldsymbol{e}_k^s$ of the smoothed design $\mathbb{E}_e^s$ and assign them to the fiber direction $\boldsymbol{e}_\mathrm{f}^s$, layer normal $\boldsymbol{e}_\mathrm{n}^s$, and their orthogonal direction $\boldsymbol{e}_\mathrm{o}^s$ of $\mathbb{E}_e^s$, i.e. its basis.[4] An efficient eigenvalue analysis for elasticity tensors of order four is presented in [34].
4. To ensure that the fibers are not aligned out of the layer plane by the filtering, the filtered material directions $\boldsymbol{e}_k^*$ are defined as follows:

   4.1 The layer normal does not change, so that $\boldsymbol{e}_\mathrm{n}^* = \boldsymbol{n}$.
   4.2 The fiber direction is the projection of $\boldsymbol{e}_\mathrm{f}^s$ to the layer plane $\boldsymbol{e}_\mathrm{f}^* = (\boldsymbol{n} \times \boldsymbol{e}_\mathrm{f}^s) \times \boldsymbol{n}$.
   4.3 The remaining direction is given by $\boldsymbol{e}_\mathrm{o}^* = \boldsymbol{n} \times \boldsymbol{e}_\mathrm{f}^*$ to ensure an orthogonal basis.

5. With the algorithm described in [23], a set of Euler angles $\boldsymbol{\alpha}_e^*$ can be extracted from the basis described by $\{\boldsymbol{e}_\mathrm{f}^*, \boldsymbol{e}_\mathrm{o}^*, \boldsymbol{n}\}$.
6. The original values of the Euler angles $\boldsymbol{\alpha}_e^i$ are replaced by the filtered angles $\boldsymbol{\alpha}_e^*$

4 Numerical Results

4.1 Material Parameters and Boundary Conditions

For the simulations, we apply the orthotropic material (in Voigt notation)

$$\mathbb{E}^0 = \begin{pmatrix} \frac{1}{E_1} & -\frac{\nu_{12}}{E_1} & -\frac{\nu_{13}}{E_1} & 0 & 0 & 0 \\ & \frac{1}{E_2} & -\frac{\nu_{23}}{E_2} & 0 & 0 & 0 \\ & & \frac{1}{E_3} & 0 & 0 & 0 \\ & & & \frac{1}{2G_{23}} & 0 & 0 \\ & \text{sym} & & & \frac{1}{2G_{13}} & 0 \\ & & & & & \frac{1}{2G_{12}} \end{pmatrix}^{-1} \tag{36}$$

with the numerical values for an ABS material processed with FDM/FFF according to [5] given in Table 1.

The parameter E_1 corresponds to the first principal material direction with highest stiffness, i.e. the printing direction (fiber direction). The parameter E_3 corresponds

[4] For example, if the base material $\mathbb{E}^0$ possesses the highest stiffness in fiber direction, then the smoothed fiber direction $\boldsymbol{e}_\mathrm{f}^s$ is equal to the eigenvector $\boldsymbol{e}_k^s$ corresponding to the largest eigenvalue.

Table 1 Material parameters for a component made of ABS and fabricated by FDM/FFF according to [5] with $E_1 > E_2 > E_3$

$E_1 = 1653.0$	$G_{23} = 540.5$	$\nu_{23} = 0.2707$
$E_2 = 1391.7$	$G_{13} = 369.6$	$\nu_{13} = 0.3209$
$E_3 = 1072.9$	$G_{12} = 554.0$	$\nu_{12} = 0.4391$

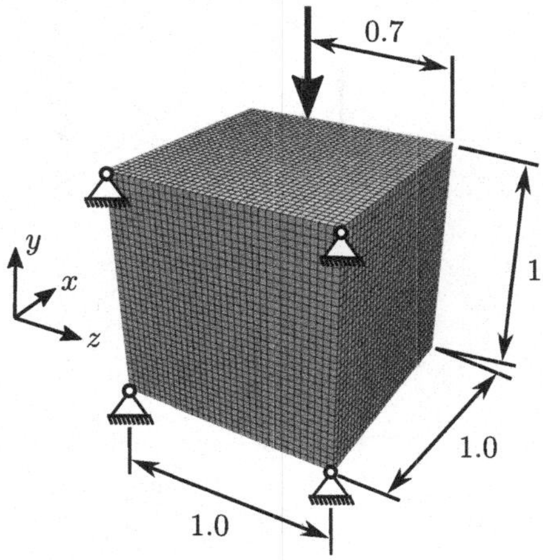

Fig. 3 Boundary conditions for the bending problem. The displacements are fixed in all direction in the four supports. The force is applied at the top edge and is pointing in (negative) y-direction. The design space is discretized by $35 \times 35 \times 35$ elements with $h = \frac{1}{35}$. Minimum member size for the topology is $r_\chi = 1.5h \approx 0.043$

to the third principal material direction with lowest stiffness, i.e. the printing plane normal (layer normal). The parameters for numerical damping within the evolution equation, i.e. the viscosity, for this material are given by $\frac{\Delta t}{\eta_\alpha} = 10.0$. The viscosity for the density variable is chosen independently of the material with $\eta_\chi = 15$. The values for the viscosities are independent of any boundary value problem.

The boundary conditions for the numerical examples presented in the following sections are given in Figs. 3 and 4. The according minimum member sizes for the topology r_χ are given in the respective captions.

4.2 Results Without Material Orientation Filter

Figures 5 and 7 show the evolution within the optimization of the layer normal and the structure stiffness $(\langle\Psi^i\rangle)^{-1}$ for the bending and torsion problem given in Figs. 3 and 4, respectively. The angle between the initial layer normal $\boldsymbol{n}^0$ and the layer normal within the respective iteration step $\boldsymbol{n}^i$ is plotted in degree. The orientation of the layer normal in each iteration step is marked on a sphere with the initial layer plotted as dashed line and the final normal as solid line. Figures 6 and 8 present the final results

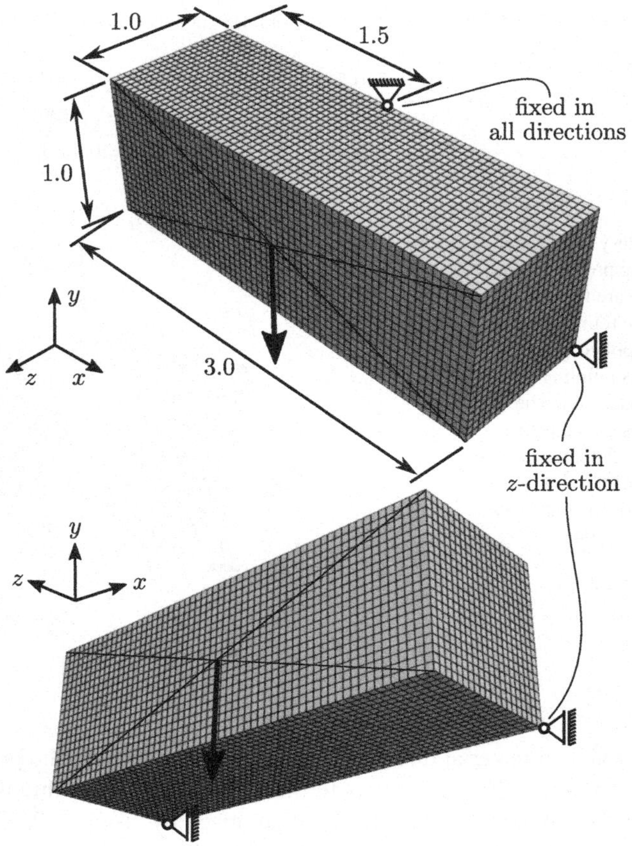

Fig. 4 Boundary conditions for the torsion problem. The displacements are fixed in all direction in the support on the top in the middle of the edge and only in z-direction in the supports at the bottom corners at the back. The force is applied in the middle of the front face and is pointing in (negative) y-direction. The design space is discretized by $60 \times 20 \times 20$ elements with $h = \frac{1}{20}$. Minimum member size for the topology is $r_\chi = 1.5h = 0.075$

with the fiber paths denoted by black dashes[5] from different perspectives and slices through the layer plane for different positions.[6]

For both examples, the method yields smooth convergence of the stiffness. For the torsion problem (Fig. 7), the layer normal converges rather quickly. However, for the bending problem (Fig. 5), the layer normal rotates around the x-axis in positive direction and changes its rotation direction at around iteration step 40 due to changes

[5] The element-wise discretized values of the density are interpolated and the iso-surface with threshold value $\chi = 0.5$ is visualized.

[6] Some fibers in the three-dimensional pictures seem to be outside of the structure but they are actually floating in front of the sliced surface: the fibers are plotted within the element mid-points and are not projected onto the surface to avoid interpolation and possible smoothing effects.

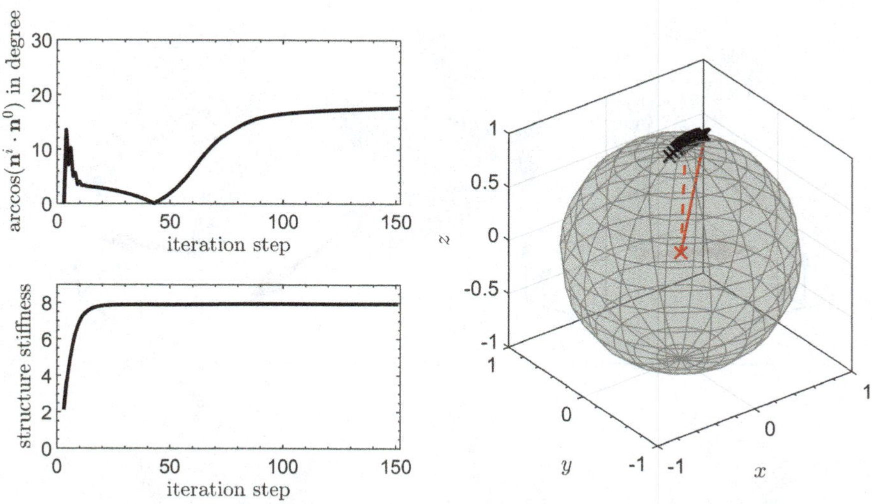

Fig. 5 Evolution of the layer normal (angle to initial vector and orientation) and stiffness over iterations steps for the bending problem

in the topology, which does not change after iteration 60 (only the shape changes negligibly). The influence of the changed layer on the stiffness seems negligible, at least for the material with rather weak anisotropy.

Usually, the optimal fiber direction would coincide with the truss direction, i.e. direction with largest normal stresses. Due to the globally restricted layer normal, the optimal layer should allow the majority of fibers to align to the trusses. For the bending problem in Fig. 6, the longest truss, whose longitudinal section is shown in the first slice, dictates the layer normal. Fibers in differently oriented trusses cannot follow its direction and follow instead the stress trajectories projected to the layer plane.

4.3 Material Orientation Filter

We will now examine the functionality of the material orientation filter exemplary for the bending problem. The evolutions of the layer normal for two different filter radii $r_\alpha = \{2h, 4h\}$ are given in Fig. 9. An overview over the values of the final structure stiffness and layer normal are given in Table 2 and the final topologies and fiber orientations from different perspectives and slices are given in Fig. 10.

The layer normal converges to its final value after small oscillations within the first 20 iteration steps. The structure stiffness converges without oscillation in a smooth manner and the filter has only negligible influence on the final structure stiffness, i.e. the difference to the results without material orientation filter is about 2%. The final layer normal is nearly identical for both values of the filter radius. However,

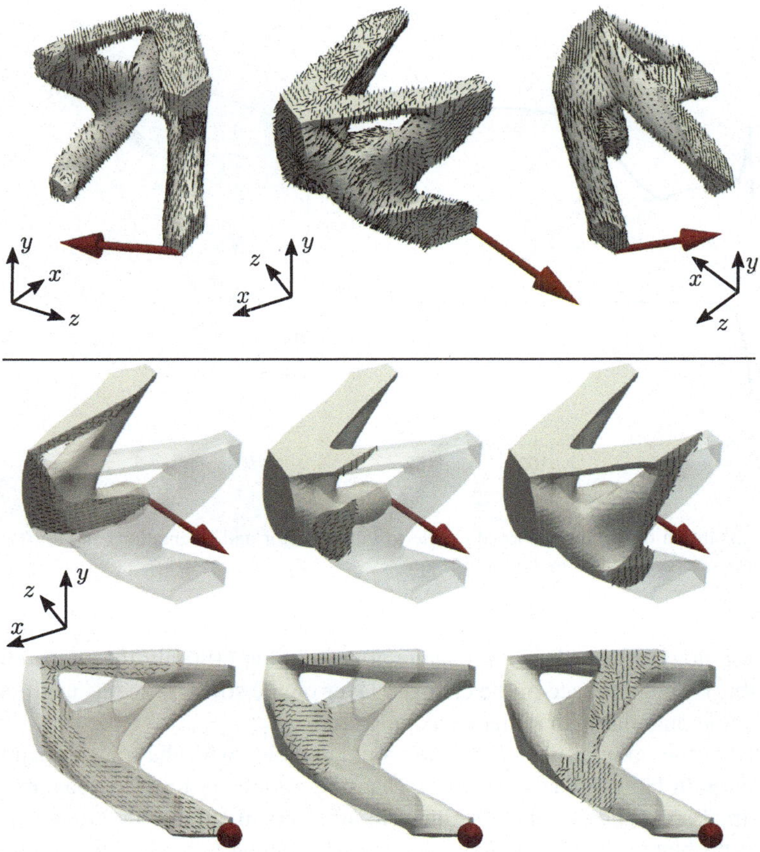

Fig. 6 Final result for the bending problem in Fig. 3 with fiber path at the surface from different perspectives (top) and different slices according to the layer normal including the perspective directly onto the layer (bottom). The large arrow denotes the optimized layer normal computed to be $\boldsymbol{n} = (0.343 \;\; 0.125 \;\; 0.931)^{\mathrm{T}}$

Table 2 Final values of structure stiffness and layer normal for the bending problem for varying material orientation filter radii r_α

Final result:	Stiffness	Layer normal $\boldsymbol{n}$
$r_\alpha = 0$	7.943 (102.07%)	$(0.343 \;\; 0.125 \;\; 0.931)^{\mathrm{T}}$
$r_\alpha = 2h$	7.806 (100.31%)	$(-0.162 \;\; -0.059 \;\; 0.985)^{\mathrm{T}}$
$r_\alpha = 4h$	7.782 (100.0%)	$(-0.159 \;\; -0.057 \;\; 0.985)^{\mathrm{T}}$

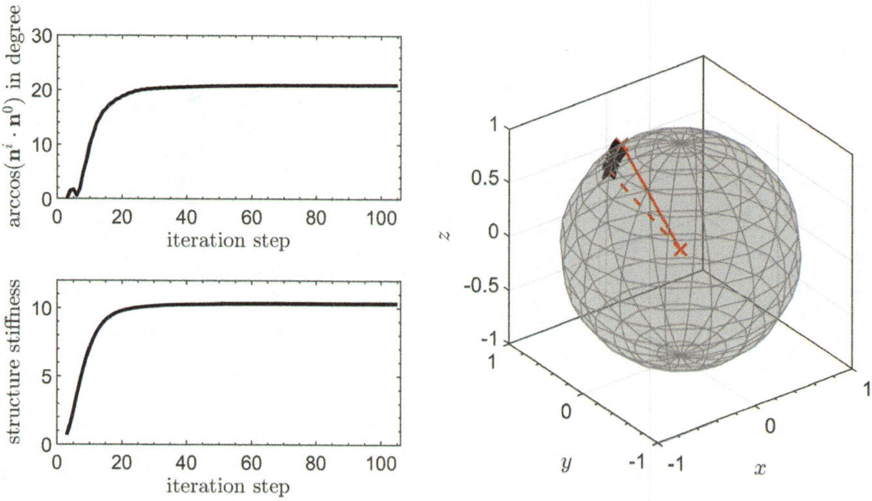

Fig. 7 Evolution of the layer normal (angle between initial and current orientation) and stiffness over iterations steps for the torsion problem

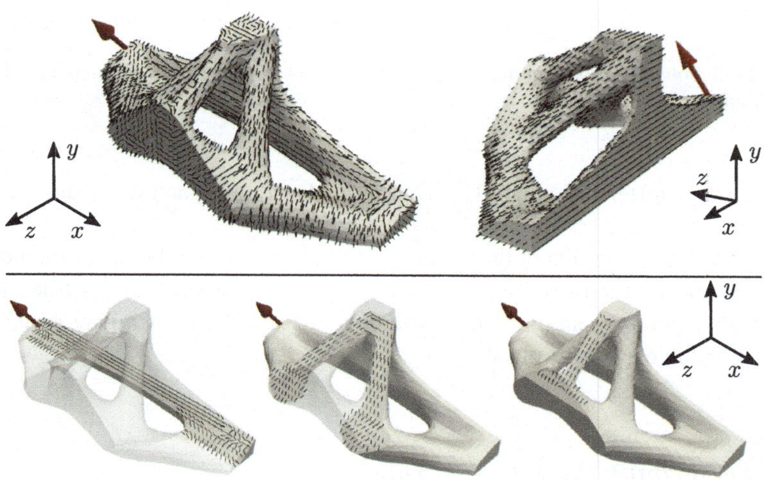

Fig. 8 Final result for the torsion problem in Fig. 4 with fiber path at the surface from different perspectives (top) and different slices according to the layer normal (bottom). The large arrow denotes the optimized layer normal computed to be $\boldsymbol{n} = (0.000 \;\; 0.787 \;\; 0.617)^{\mathrm{T}}$

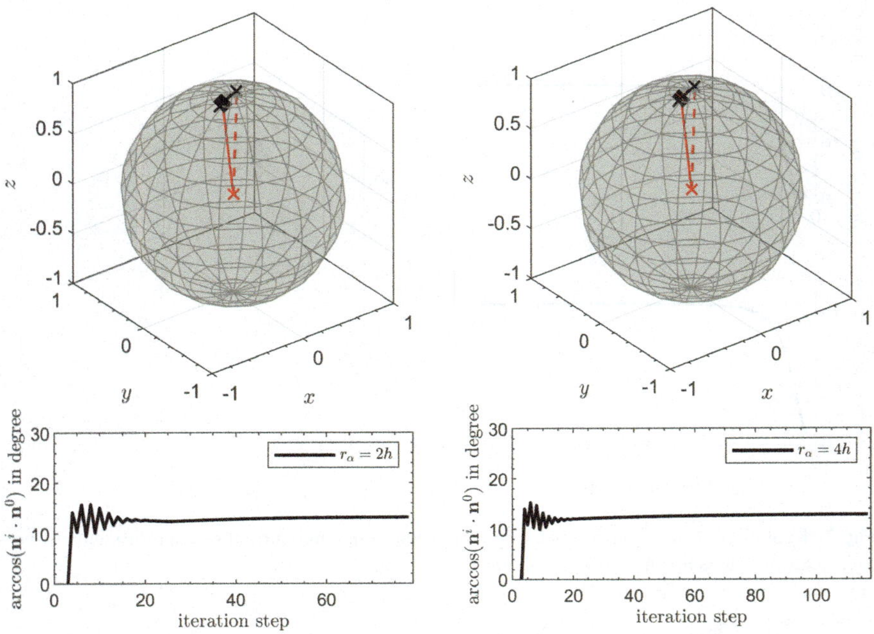

Fig. 9 Evolution of the layer normal for the bending problem with material orientation filter and filter radius $r_\alpha = 2h$ (left) and $r_\alpha = 4h$ (right)

those normals differs by about 31° from normal without material orientation filter (Table 2).

Nevertheless, the effect of the material orientation filter on the fiber orientation is rather significant: the overall fiber path ways are much smoother depending on the chosen filter radius r_α and the maximum fiber curvature is constrained to the value $r_\alpha^{-1}/\sqrt{2}$ within each layer plane.

4.4 Prescribing the Layer Normal

Due to fabrication constraints or reduction of overhangs, it may be reasonable to prescribe the layer normal. This can be achieved with the presented model by simply prescribing the second and third angle as constant. However, prescribing the layer normal can decrease the structure stiffness significantly. Table 3 shows an overview over the final structure stiffness for the varying layer normals and filter radii for the torsion problem: the design with optimized layer normal is always superior to the designs with prescribed layer normal by approximately 5 and 10%. The structure stiffness is decreased with larger filter radii r_α which is reasonable since the constraint on the fiber curvature decreases the possible design space. In other words: if a smooth

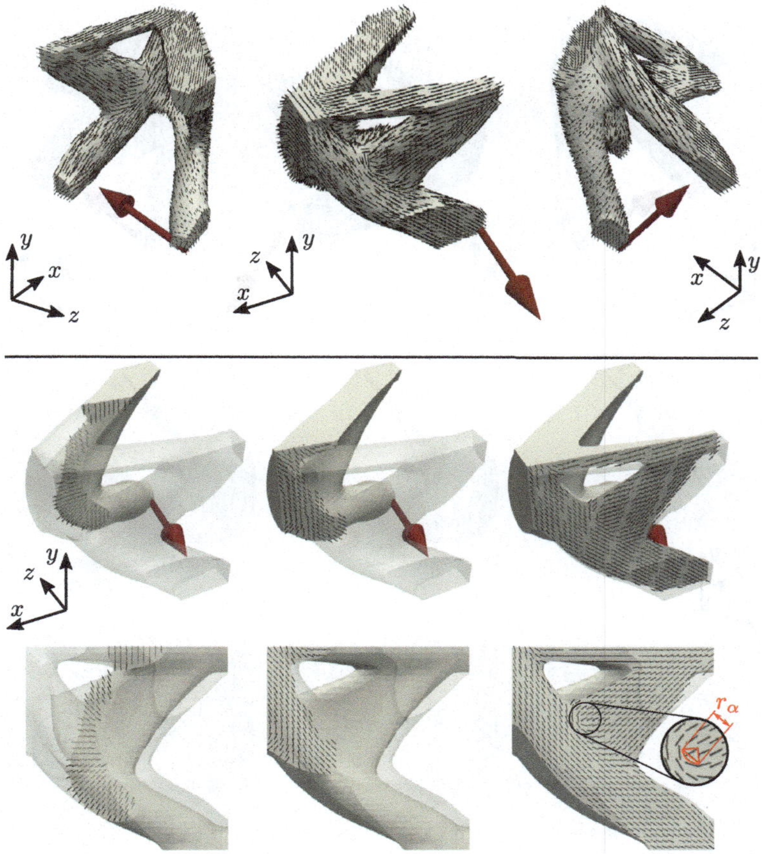

Fig. 10 Final result for the bending problem in Fig. 3 with material orientation filter with filter radius $r_\alpha = 2h \approx 0.057$. The large arrow denotes the layer normal $\boldsymbol{n} = (-0.162 \;\; -0.059 \;\; 0.985)^{\mathrm{T}}$

fiber path would yield a better design, then (a correct) optimization should also yield such a smooth result without the constraint.

5 Conclusions and Outlook

We presented a thermodynamic topology optimization model for three-dimensional layered anisotropic materials as special case for a continuous fiber angle optimization: the globally defined normal of the layers as well as the local fiber orientation within each layer are optimized by a set of three continuous Euler angles. Herein, two of the three Euler angles are defined globally and define the layer normal for additive manufacturing to which all fibers must be parallel. The third Euler angle defines the local fiber orientation within the layer plane.

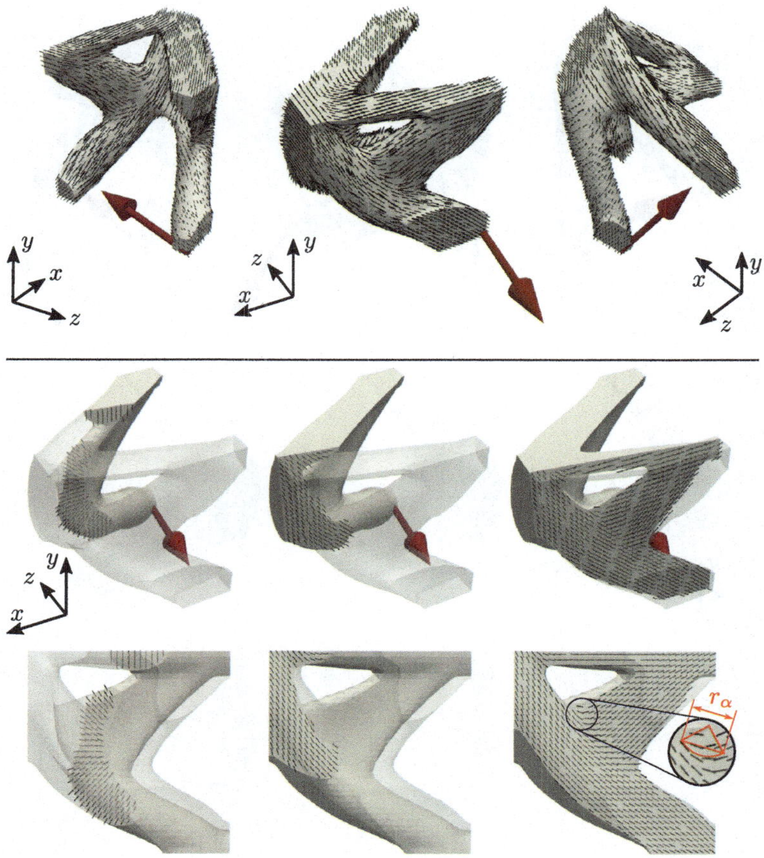

Fig. 11 Final result for the bending problem in Fig. 3 with material orientation filter with filter radius $r_\alpha = 4h \approx 0.114$. The large arrow denotes the layer normal $\boldsymbol{n} = (-0.159\ \ -0.057\ \ 0.985)^{\mathrm{T}}$. The quarter circle shows the maximum fiber curvature which is constrained by the filter radius r_α [23]

Table 3 Final values of structure stiffness for the torsion problem with optimized and prescribed layer normals with relative stiffness decrease corresponding to the variation of the filter radius

Layer normal:	Optimized	y-direction	z-direction
$r_\alpha = 0$	10.327 (100.0%)	9.981 (100.0%)	9.519 (100.0%)
$r_\alpha = 2h$	9.742 (94.33%)	9.606 (96.24%)	9.139 (96.01%)
$r_\alpha = 4h$	9.701 (93.94%)	9.079 (90.96%)	9.006 (94.61%)

The topology and layer plane are optimized freely, i.e. no geometry features or orientations are required as input by the user, although the layer plane(s) can be prescribed by the user. However, the stiffness of the final structure is always higher with the optimized layer normal.

An optional material orientation filter is presented to constrain the maximum fiber curvature within the layer plane. The filter results in much smoother fiber pathways but reduces the stiffness of the structure due to restrictions on the possible design.

References

1. Sigmund, O., & Maute, K. (2013). Topology optimization approaches. *Structural and Multidisciplinary Optimization, 48*(6), 1031–1055.
2. Deaton, J. D., & Grandhi, R. V. (2014). A survey of structural and multidisciplinary continuum topology optimization: Post 2000. *Structural and Multidisciplinary Optimization, 49*(1), 1–38.
3. Mazumdar, S. (2001). *Composites manufacturing: Materials, product, and process engineering*. Boca Raton: CRC Press.
4. Mallick, P. K. (2007). *Fiber-reinforced composites: Materials, manufacturing, and design.* Boca Raton: CRC Press.
5. Bellini, A., & Güçeri, S. (2003). Mechanical characterization of parts fabricated using fused deposition modeling. *Rapid Prototyping Journal.*
6. Tekinalp, H. L., Kunc, V., Velez-Garcia, G. M., Duty, C. E., Love, L. J., Naskar, A. K., et al. (2014). Highly oriented carbon fiber-polymer composites via additive manufacturing. *Composites Science and Technology, 105,* 144–150.
7. Bikas, H., Stavropoulos, P., & Chryssolouris, G. (2016). Additive manufacturing methods and modelling approaches: A critical review. *The International Journal of Advanced Manufacturing Technology, 83*(1–4), 389–405.
8. Wessels, H., Gieseke, M., Weißenfels, C., Kaierle, S., Wriggers, P., & Overmeyer, L. (2017). Simulation von selective laser melting prozessen. *Additive manufacturing quantifiziert* (pp. 145–162). Berlin: Springer.
9. Parandoush, P., & Lin, D. (2017). A review on additive manufacturing of polymer-fiber composites. *Composite Structures, 182,* 36–53.
10. Hörnlein, H. R. E. M., Kočvara, M., & Werner, R. (2001). Material optimization: Bridging the gap between conceptual and preliminary design. *Aerospace Science and Technology, 5*(8), 541–554.
11. Hvejsel, C. F., & Lund, E. (2011). Material interpolation schemes for unified topology and multi-material optimization. *Structural and Multidisciplinary Optimization, 43*(6), 811–825.
12. Nomura, T., Dede, E. M., Lee, J., Yamasaki, S., Matsumori, T., Kawamoto, A., & Kikuchi, N. (2015). General topology optimization method with continuous and discrete orientation design using isoparametric projection. *International Journal for Numerical Methods in Engineering, 101*(8), 571–605.
13. Petrovic, M., Nomura, T., Yamada, T., Izui, K., & Nishiwaki, S. (2018). Orthotropic material orientation optimization method in composite laminates. *Structural and Multidisciplinary Optimization, 57*(2), 815–828.
14. Blasques, J. P., & Stolpe, M. (2012). Multi-material topology optimization of laminated composite beam cross sections. *Composite Structures, 94*(11), 3278–3289.
15. Hoglund, R., & Smith,D. E. (2015). Non-isotropic material distribution topology optimization for fused deposition modeling products. In *Proceeding of the Solid Freeform Fabrication Symposium, Austin, TX* (p. 888).
16. Liu, J., Gaynor, A. T., Chen, S., Kang, Z., Suresh, K., Takezawa, A., et al. (2018). Current and future trends in topology optimization for additive manufacturing. *Structural and Multidisciplinary Optimization, 57*(6), 2457–2483.
17. Boddeti, N., Ding, Z., Kaijima, S., Maute, K., & Dunn, M. L. (2018). Simultaneous digital design and additive manufacture of structures and materials. *Scientific reports, 8*(1), 1–10.

18. Boddeti, N., Tang, Y., Maute, K., Rosen, D. W., & Dunn, M. L. (2020). Optimal design and manufacture of variable stiffness laminated continuous fiber reinforced composites. *Scientific Reports, 10*(1), 1–15.
19. Gaynor, A. T., & Guest, J. K. (2016). Topology optimization considering overhang constraints: Eliminating sacrificial support material in additive manufacturing through design. *Structural and Multidisciplinary Optimization, 54*(5), 1157–1172.
20. Zegard, T., & Paulino, G. H. (2016). Bridging topology optimization and additive manufacturing.*Structural and Multidisciplinary Optimization, 53*(1), 175–192.
21. Meng, L., Zhang, W., Quan, D., Shi, G., Tang, L., Hou, Y., et al. (2020). From topology optimization design to additive manufacturing: Today's success and tomorrow's roadmap. *Archives of Computational Methods in Engineering, 27*(3), 805–830.
22. Yingjie, X., Zhu, J., Zhen, W., Cao, Y., Zhao, Y., & Zhang, W. (2018). A review on the design of laminated composite structures: Constant and variable stiffness design and topology optimization. *Advanced Composites and Hybrid Materials, 1*(3), 460–477.
23. Jantos, D. R., Hackl, K., & Junker, P. (2020). Topology optimization with anisotropic materials, including a filter to smooth fiber pathways. *Structural and Multidisciplinary Optimization* (pp. 1–20).
24. Mehrabadi, M. M., & Cowin, S. C. (1990). Eigentensors of linear anisotropic elastic materials. *The Quarterly Journal of Mechanics and Applied Mathematics, 43*(1), 15–41.
25. Bendsøe, M. P., & Sigmund, O. (1999). Material interpolation schemes in topology optimization. *Archive of Applied Mechanics, 69*(9–10), 635–654.
26. Junker, P., & Balzani, D. (2021). A new variational approach for the thermodynamic topology optimization of hyperelastic structures. *Computational Mechanics, 67*(2), 455–480.
27. Jantos, D. R., Hackl, K., & Junker, P. (2019). An accurate and fast regularization approach to thermodynamic topology optimization. *International Journal for Numerical Methods in Engineering, 117*(9), 991–1017.
28. Wriggers, P. (2009). Mixed finite element methods-theory and discretization. *Mixed finite element technologies* (pp. 131–177). Berlin: Springer.
29. Schröder, J., & Wriggers, P. (2016). *Advanced finite element technologies* (Vol. 566). Berlin: Springer.
30. Sigmund, O., & Petersson, J. (1998). Numerical instabilities in topology optimization: A survey on procedures dealing with checkerboards, mesh-dependencies and local minima. *Structural Optimization, 16*(1), 68–75.
31. Rion, V., & Bruyneel, M. (2006). Topology optimization of membranes made of orthotropic material. *Collection of Papers from Prof. Nguyen Dang Hung's former students* (pp. 107–120).
32. Bendsøe, M. P., & Sigmund, O. (2003). *Topology optimization: Theory, methods and applications*. Berlin: Springer.
33. Hackl, K. (1999). On the representation of anisotropic elastic materials by symmetric irreducible tensors. *Continuum Mechanics and Thermodynamics, 11*(6), 353–369.
34. Cowin, S. C., & Mehrabadi, M. M. (1987). On the identification of material symmetry for anisotropic elastic materials. *Quarterly Journal of Mechanics and Applied Mathematics, 40,* 451–476.

A Review of Nonlocality in Computational Contact Mechanics

David Kamensky, Mert D. Alaydin, and Yuri Bazilevs

This article summarizes early and recent work on nonlocal approaches in computational contact mechanics and is dedicated to Prof. Peter Wriggers on the occasion of his 70th birthday. Peter's impactful research has shaped the field of computational contact mechanics and has enabled us to tackle many computational challenges with confidence. The senior author would also like to thank Peter for his continued mentorship and support.

Abstract This chapter reviews a class of methods for computational contact mechanics, where the contact problem is regularized using nonlocal interaction, to simplify discretization. This discussion is guided by an analogy to computational fracture mechanics, where nonlocal regularizations are widely employed to obtain robust computational models. Particular emphasis is given to a class of regularizations based on nonlocal integral operators. Such regularizations are analogous to the peridynamic models gaining popularity in the fracture mechanics community. The use of ideas from peridynamics to include physically-consistent friction in such regularizations illustrates the potential value of exploring nonlocal contact modeling through the contact–fracture analogy.

1 Introduction

Computational contact mechanics [1] is notoriously difficult. The primary practical challenge is identification and discretization of the solution-dependent contact region, on which an inequality constraint must be imposed. As such, numerous research

D. Kamensky
Department of Mechanical and Aerospace Engineering, University of California, San Diego, La Jolla, CA, USA
e-mail: dmkamensky@eng.ucsd.edu

M. D. Alaydin · Y. Bazilevs (✉)
School of Engineering, Brown University, Providence, RI, USA
e-mail: yuri_bazilevs@brown.edu

F. Aldakheel et al. (eds.), *Current Trends and Open Problems in Computational Mechanics*, https://doi.org/10.1007/978-3-030-87312-7_23

efforts have sought to regularize the problem in such a way as to simplify the necessary computational geometry tasks. The present chapter reviews a class of regularizations that reformulate the problem in terms of nonlocal interaction between the contacting bodies. While we shall identify many instances of such contact methods, going back several decades in the literature, they have not been grouped according to this shared characteristic until quite recently. Our hope is that the present review will inspire productive cross-pollination between these related ideas.

To that end, we make an analogy to the field of computational fracture mechanics, where the role of nonlocality in a variety of practical methods has led to increasing exchange of ideas, exemplified by a recent series of workshops organized by Oak Ridge National Laboratory [2–4]. The computational fracture mechanics community now recognizes that the emerging methods of phase-field fracture [5] and peridynamics [6] both rely crucially on spreading interaction over a volume encompassing both sides of a crack, which makes the problem much easier to discretize. From that perspective, it is not surprising that a similar spreading of contact interactions might likewise simplify discretization—the key challenge in simulating both fracture and contact mechanics without any regularization is to parameterize and couple fields across a solution-dependent interface between solid bodies.

We begin, in Sect. 2, with the nonlocal contact methods that came first historically: those added directly to (or implicit in) discrete models. We then consider two classes of regularized continuous problems. Section 3 reviews methods based on introducing an auxiliary problem governed by a local partial differential equation (PDE) which mediates interaction between physically-separated objects, while Sect. 4 discusses nonlocal contact regularizations based on integral operators. We give particular emphasis to the way in which some methods from Sect. 2 emerge naturally as discretizations of nonlocal integral operators. Section 5 then goes into greater depth on how the fracture–contact analogy can inspire novel ideas, by summarizing the frictional extension of methods from Sect. 4. Section 6 draws conclusions and discusses future directions for the topic.

2 Nonlocal Interactions in Discretized Models

Historically, the use of action at a distance to model contact began with methods formulated directly in the discrete setting, rather than as discretizations of a continuous, regularized problem. The spontaneous emergence of such discrete methods speaks for their practical advantages in complex problems. However, much like the ad hoc use of element sizes as length scales to limit damage localization, these ad hoc contact schemes may not converge to any mesh-independent solution.

In the setting of Lagrangian finite element analysis, the so-called "pinball method" developed by Belytschko and collaborators in the early 1990s [7–9] associates a sphere (or "pinball") with each element of a mesh and applies concentrated contact forces at the centers of colliding spheres. The radii of these spheres—tied to the size of mesh elements—introduce a length scale for nonlocal interaction. The resulting geometric approximation introduces a texture to surfaces, which may interfere with sliding contact, even under refinement. However, the simplicity and robustness of the approach have given it an enduring appeal for researchers approaching complex contact and impact problems [10].

Contact between spherical elements is especially appealing in meshfree methods [11], where solid bodies are represented as point clouds, without any explicit surface parameterizations. An example of such a meshfree contact treatment would be kernel contact used in conjunction with the reproducing kernel particle method (RKPM) [12, Sect. 4]. Similarly, some meshfree peridynamic simulations [13] include small spherical elements around quadrature points. Collisions between these spheres are then penalized by "short-range forces" [14, Sect. 3.9] to deter interpenetration of solid bodies.

Contact-like interactions between separated material points may also be implicit in certain discretizations. An example of such implicit contact would be the basic material point method (MPM) [15], which allows Lagrangian quadrature points to move through a background mesh used to construct a function space for material velocity. Thus points from nominally-distinct bodies may influence each other when passing within a length scale proportional to the background mesh element size.

3 Nonlocal Interaction Through a Fictitious Medium

An approach loosely analogous to phase field fracture is to introduce nonlocal interactions through an auxiliary problem, which is itself a local PDE problem. In phase field fracture, this auxiliary problem is the PDE governing the phase field. In contact mechanics, it is equilibrium of an artificial material filling the void region between contacting bodies. This artificial material is given material properties which are carefully designed to emulate the effects of contact.

This idea was developed in the setting of Lagrangian finite elements by Wriggers et al. [16]. The method [16] uses a conforming mesh of the void region to discretize the fictitious material. A similar approach was applied in the finite cell method [17] shortly thereafter [18]. In the finite cell method, solid bodies are immersed in a mesh of their union with the void region. Normally, the void region is assigned vanishing stiffness, and the displacement solution extending into it has no physical interpretation. However, [18] introduces a fictitious contact material governing the extension of the displacement field.

By posing a PDE problem over an unfitted subset of a larger background mesh, the finite cell method is loosely similar to the MPM. Contact through a fictitious medium has been considered for MPM discretizations by the computer graphics community [19]. In particular, the fictitious material is given a modified plastic constitutive model, to simulate friction. This idea had been developed earlier to simulate frictional contact in RKPM discretizations [20, Sect. 3.1].

4 Nonlcal Interaction Through Integral Operators

An approach sharing close formal ties to peridynamics is to formulate contact mechanics in terms of integral operators inducing finite-range repulsive interactions between solid bodies. A particularly simple discretization of such contact forces is obtained by analogy to meshfree quadrature of peridynamics.

Work by Sauer motivated nonlocal repulsive forces between continuum bodies as a coarse-graining of molecular dynamics [21]. The idea of modeling contact through such force fields was further explored by Sauer and de Lorenzis, with an emphasis on nonlocal interactions between surfaces [22]. The idea of repulsive forces between volumes was later revisited by Kamensky et al. [23], with the interpretation of an artificial regularization of contact mechanics. Reference [23] models frictionless contact by introducing a potential energy E_c of the form

$$E_c = \frac{1}{2}\int_{\Omega_0 \setminus B_{R_{\text{self}}}(\mathbf{X}_1)} \int_{\Omega_0} \phi\left(|\mathbf{x}_2 - \mathbf{x}_1|\right)\, d\mathbf{X}_1 d\mathbf{X}_2\,, \tag{1}$$

where Ω_0 is the initial reference configuration of a solid body, $\mathbf{X}_1$ and $\mathbf{X}_2$ are material points in Ω_0 with deformed positions $\mathbf{x}_1$ and $\mathbf{x}_2$, and $\phi : \mathbb{R}_{\geq 0} \to \mathbb{R}_{\geq 0}$ is a kernel function representing the energy contribution of a pair of points as a function of their distance from each other. The radius $R_{\text{self}} > 0$ is the minimum distance between points in the reference configuration which may interact via nonlocal contact forces. This naturally incorporates self contact, but without material points repelling their immediate neighbors, under the obvious assumption that $\phi(x) = 0$ for $x \geq R_{\text{self}}$. As for the functional form of ϕ, [21] considers physically-motivated choices (e.g., a Lennard-Jones potential), while [23] derives conditions on ϕ that are sufficient to prevent interpenetration under some geometrical simplifications.

If both integrals are discretized using a single numerical quadrature rule, one obtains

$$E_c \approx \frac{1}{2}\sum_{i=1}^{N}\left(\sum_{j=1}^{N}\begin{Bmatrix} w_i w_j \phi\left(|\mathbf{x}_j - \mathbf{x}_i|\right) & |\mathbf{X}_j - \mathbf{X}_i| > R_{\text{self}} \\ 0 & \text{otherwise} \end{Bmatrix}\right), \tag{2}$$

where $\{\mathbf{X}_i\}_{i=1}^{N}$, are quadrature points in Ω_0, $\{\mathbf{x}_i\}$ are their deformed positions, and $\{w_i\}$ are quadrature weights. Notably, this is almost formally identical to both the pinball method and short-range forces used in meshfree peridynamics (cf. Sect. 2). It is essentially an N-body problem with finite range of interactions. All nonzero terms in the double sum (2) can be efficiently identified using a standard k–d tree or the simpler data structure proposed in [23, Section 3.1]. This method has already been applied in a variety of challenging contact problems, including atrioventricular heart valve contact [23], knot tightening [23, 24] (Fig. 1), stent–leaflet and stent–artery interactions in transcatheter aortic valve replacements [25], leaflet–leaflet interaction in aortic heart valves [26], and plastic post-buckling with self-contact of shell

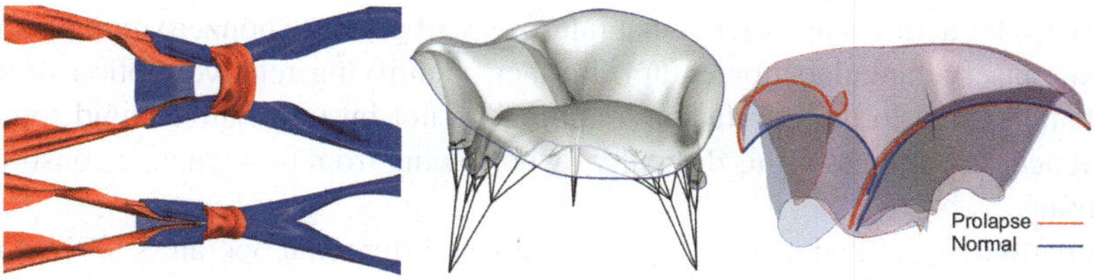

Fig. 1 Examples of complex (self-)contact from [23]. *Left:* Tying a reef knot. *Middle, right:* Simulation of a right atrioventricular (tricuspid) heart valve closure, including normal and prolapsed conditions

Fig. 2 Example of complex (self-)contact from [27]. Crushing of an aluminum soda can using compression and twisting. Contours of the effective plastic strain with blue color indicating low values and red color indicating high values

structures [27] (Fig. 2). The method's ease of implementation has also facilitated its integration with open-source codes, e.g., [24, 26]. Unlike the pinball method or short range interactions, though, the model (1) defines a mesh-independent solution, which one can hope discretizations like (2) converge toward. However, neither the convergence of (2) to (1) nor the convergence of (1) to some local contact problem have been investigated with any rigor beyond qualitative convergence of quantities of interest.

5 Borrowing from Peridynamics: Frictional Nonlocal Contact

We now illustrate the potential productivity of collaboration between fracture mechanics and contact mechanics communities, by looking at how ideas from peridynamics were used to extend the nonlocal contact formulation of [23] to frictional contact [28]. Friction in nonlocal contact presents a fundamental challenge. If contact forces are applied at points that are some distance from the contact interface, balanced force couples acting tangential to that interface will generate nonzero moments, failing to conserve angular momentum. Further, identifying relative motion of these points tangential to the interface as "sliding" is not invariant under rigid rotation. Reference [28] addresses this difficulty by borrowing from peridynamic constitutive modeling.

In particular, [28] adapts the idea of nonlocal differential operators [29], as used to define a nonlocal deformation gradient in peridynamic correspondence models [30, (2.6)]. Reference [28] defines the operator

$$\nabla_{\mathbf{x}}\mathbf{f}(\mathbf{X}_1) = \mathbf{K}_{\mathbf{f}}(\mathbf{X}_1) \cdot (\mathbf{K}_{\mathbf{x}}(\mathbf{X}_1))^{-1} \tag{3}$$

of some vector field $\mathbf{f} : \Omega_0 \rightarrow \mathbb{R}^n$, with

$$\mathbf{K}_{\mathbf{f}}(\mathbf{X}_1) = \int_{B_\delta(\mathbf{X}_1)\cap\Omega_0} \omega\left(|\mathbf{X}_2 - \mathbf{X}_1|\right)\left(\mathbf{f}(\mathbf{X}_2) - \mathbf{f}(\mathbf{X}_1)\right) \otimes (\mathbf{X}_2 - \mathbf{X}_1)\, d\mathbf{X}_2\,, \tag{4}$$

where $\delta > 0$ is a nonlocality length scale and $\omega : [0, \delta) \rightarrow \mathbb{R}_{\geq 0}$ is an influence function. As suggested by notation, this operator is used like a gradient with respect to deformed position $\mathbf{x}$, in a first-order Taylor extrapolation of quantities from interacting points to each other:

$$E_{1\rightarrow 2}\mathbf{f} = \mathbf{f}(\mathbf{X}_1) + (\nabla_{\mathbf{x}}\mathbf{f}(\mathbf{X}_1)) \cdot (\mathbf{x}_2 - \mathbf{x}_1)\,, \tag{5}$$

represents the extrapolation of some field $\mathbf{f}$ from $\mathbf{x}_1$ to $\mathbf{x}_2$. If such an extrapolation is applied to both kinematic variables and their corresponding test functions, it leads to objective and angular momentum-conserving force couples acting tangential to contact interfaces, despite nonzero separation. This is analogous to non-ordinary state-based constitutive models in peridynamics, following the terminology of [31]. Reference [28] applied this new frictional formulation in benchmark problems using finite element discretizations of local PDE-based models of solid mechanics and complex impact problems using peridynamics with damage.

6 Conclusions and Future Directions

This chapter has reviewed the use of nonlocal regularizations of contact mechanics to simplify its discretization. We guided the presentation of work on this topic by analogy to the role of nonlocality in computational fracture mechanics. In Sect. 4, the extension of contact regularizations based on nonlocal integral operators to frictional contact illustrates the potential for productive exchange of ideas between contact and fracture mechanics communities.

A relatively unexplored topic is the potential connection between contact based on nonlocal integral operators and contact based on local formulations of fictitious media. In the analogy to fracture mechanics, this would be similar to the question of correspondence between phase field and peridynamic fracture formulations. It is possible that quick progress in this direction could be made in the meshfree setting, where recent work has already elucidated the close ties between RKPM discretizations of local problems and numerical quadrature of nonlocal integral operators [32, 33]. In the continuous setting, one might explore the limit of interaction range going to zero, as has recently been done for peridynamic modeling of fracture [34]. If nonlocal integral regularizations of contact do indeed converge to classical contact conditions in some rigorous way, there is also the open question of whether asymptotically-compatible [35] discretizations are possible, such as those developed for peridynamics [36].

References

1. Wriggers, P. (2006). *Computational Contact Mechanics* (2nd ed.). Berlin: Springer.
2. *Nonlocal Models in Mathematics, Computation, Science, and Engineering*. (2015). Oak Ridge, TN, United States.
3. *Nonlocal Methods in Fracture*. (2018). Austin, TX, United States.
4. *Experimental and Computational Fracture Mechanics*. (2020). Baton Rouge, LA, United States.
5. Ambati, M., Gerasimov, T., & De Lorenzis, L. (2015). A review on phase-field models of brittle fracture and a new fast hybrid formulation. *Computational Mechanics, 55*(2), 383–405.
6. Silling, S. A. (2000). Reformulation of elasticity theory for discontinuities and long-range forces. *Journal of the Mechanics and Physics of Solids, 48*(1), 175–209.
7. Belytschko, T., & Neal, M. O. (1991). Contact-impact by the pinball algorithm with penalty and Lagrangian methods. *International Journal for Numerical Methods in Engineering, 31*(3), 547–572.
8. Belytschko, T., & Yeh, I. S. (1991). The splitting pinball method for general contact. In *Proceedings of the 10th International Conference on Computing Methods in Applied Sciences and Engineering on Computing Methods in Applied Sciences and Engineering* (pp. 73–87). Commack, NY, USA: Nova Science Publishers, Inc.
9. Belytschko, T., & Yeh, I. S. (1993). The splitting pinball method for contact-impact problems. *Computer Methods in Applied Mechanics and Engineering, 105*(3), 375–393.
10. Casadei, F., Aune, V., Valsamos, G., & Larcher, M. (2016). Generalization of the pinball contact/impact model for use with mesh adaptivity and element erosion in EUROPLEXUS. Technical Report JRC101013, European Commission: Joint Research Centre.
11. Chen, J.-S., Hillman, M., & Chi, S.-W. (2017). Meshfree methods: Progress made after 20 years. *Journal of Engineering Mechanics, 143*(4).
12. Guan, P. C., Chi, S. W., Chen, J. S., Slawson, T. R., & Roth, M. J. (2011). Semi-Lagrangian reproducing kernel particle method for fragment-impact problems. *International Journal of Impact Engineering, 38*(12), 1033–1047.
13. Silling, S. A., & Askari, E. (2005). A meshfree method based on the peridynamic model of solid mechanics. *Computers & Structures, 83*(17), 1526–1535. Advances in Meshfree Methods.
14. Parks, M. L., Littlewood, D. J., Mitchell, J. A., & Silling, S. A. (2012). Peridigm users' guide. Technical Report SAND2012-7800, Sandia National Laboratories.
15. Sulsky, D., Chen, Z., & Schreyer, H. L. (1994). A particle method for history-dependent materials. *Computer Methods in Applied Mechanics and Engineering, 118*(1), 179–196.
16. Wriggers, P., Schröder, J., & Schwarz, A. (2013). A finite element method for contact using a third medium. *Computational Mechanics, 52*(4), 837–847.

17. Schillinger, D., & Ruess, M. (2015). The finite cell method: A review in the context of higher-order structural analysis of CAD and image-based geometric models. *Archives of Computational Methods in Engineering, 22*(3), 391–455.
18. Bog, T., Zander, N., Kollmannsberger, S., & Rank, E. (2015). Normal contact with high order finite elements and a fictitious contact material. *Computers & Mathematics with Applications, 70*(7), 1370–1390.
19. Han, X., Gast, T. F., Guo, Q., Wang, S., Jiang, C., & Teran, J. (2019). A hybrid material point method for frictional contact with diverse materials. *Proceedings of the ACM in Computer Graphics and Interactive Techniques, 2*(2).
20. Chi, S.-W., Lee, C.-H., Chen, J.-S., & Guan, P.-C. (2014). A level set enhanced natural kernel contact algorithm for impact and penetration modeling. *International Journal for Numerical Methods in Engineering, 102*(3–4), 839–866.
21. Sauer, R. A. (2006). *An atomic interaction based continuum model for computational multiscale contact mechanics*. Ph.D. thesis, University of California, Berkeley, Berkeley, California.
22. Sauer, R. A., & De Lorenzis, L. (2013). A computational contact formulation based on surface potentials. *Computer Methods in Applied Mechanics and Engineering, 253*, 369–395.
23. Kamensky, D., Xu, F., Lee, C.-H., Yan, J., Bazilevs, Y., & Hsu, M.-C. (2018). A contact formulation based on a volumetric potential: Application to isogeometric simulations of atrioventricular valves. *Computer Methods in Applied Mechanics and Engineering, 330*, 522–546.
24. Kamensky, D., & Bazilevs, Y. (2018). tIGAr: Automating isogeometric analysis with FEniCS. *Computer Methods in Applied Mechanics and Engineering*.
25. Wu, M. C. H., Muchowski, H. M., Johnson, E. L., Rajanna, M. R., & Hsu, M.-C. (2019). Immersogeometric fluid-structure interaction modeling and simulation of transcatheter aortic valve replacement. *Computer Methods in Applied Mechanics and Engineering, 357*, 112556.
26. Kamensky, D. (2020). Open-source immersogeometric analysis of fluid–structure interaction using FEniCS and tIGAr. *Computers & Mathematics with Applications*. In press.
27. Alaydin, M. D., & Bazilevs, Y. (2021). An updated lagrangian framework for isogeometric kirchhoff–love thin-shell analysis. *Computer Methods in Applied Mechanics and Engineering*. In review.
28. Kamensky, D., Behzadinasab, M., Foster, J. T., & Bazilevs, Y. (2019). Peridynamic modeling of frictional contact. *Journal of Peridynamics and Nonlocal Modeling, 1*(2), 107–121.
29. Madenci, E., Barut, A., & Futch, M. (2016). Peridynamic differential operator and its applications. *Computer Methods in Applied Mechanics and Engineering, 304*, 408–451.
30. Tupek, M. R. (2014). *Extension of the peridynamic theory of solids for the simulation of materials under extreme loadings*. Ph.D. thesis, Massachusetts Institute of Technology, Cambridge, Massachusetts.
31. Silling, S. A., Epton, M., Weckner, O., Xu, J., & Askari, E. (2007). Peridynamic states and constitutive modeling. *Journal of Elasticity, 88*(2), 151–184.
32. Bessa, M. A., Foster, J. T., Belytschko, T., & Liu, W. K. (2014). A meshfree unification: Reproducing kernel peridynamics. *Computational Mechanics, 53*(6), 1251–1264.
33. Hillman, M., Pasetto, M., & Zhou, G. (2020). Generalized reproducing kernel peridynamics: Unification of local and non-local meshfree methods, non-local derivative operations, and an arbitrary-order state-based peridynamic formulation. *Computational Particle Mechanics, 7*(2), 435–469.
34. Lipton, R. (2014). Dynamic brittle fracture as a small horizon limit of peridynamics. *Journal of Elasticity, 117*(1), 21–50.
35. Tian, X., & Du, Q. (2014). Asymptotically compatible schemes and applications to robust discretization of nonlocal models. *SIAM Journal on Numerical Analysis, 52*(4), 1641–1665.
36. Trask, N., You, H., Yu, Y., & Parks, M. L. (2019). An asymptotically compatible meshfree quadrature rule for nonlocal problems with applications to peridynamics. *Computer Methods in Applied Mechanics and Engineering, 343*, 151–165.

Optimal Control for Phase-Field Fracture: Algorithmic Concepts and Computations

Denis Khimin, Marc C. Steinbach, and Thomas Wick

This work is dedicated to Professor Peter Wriggers. We appreciate the past and ongoing collaboration with his institute in the research fields of computational mechanics. All the best and good health for his coming years.

Abstract In this work, we present an algorithmic realization for computing optimal control problems with quasi-static phase-field fracture as a PDE constraint. The phase-field fracture problem is formulated in a quasi-monolithic approach resulting in a nonlinear forward problem. The optimization problem is formulated within a reduced approach, where the state variable is eliminated. To this end, a globalized reduced Newton algorithm is employed. Our algorithmic developments are substantiated with a numerical example.

1 Introduction

This work is devoted to algorithmic concepts of phase-field fracture based optimization. Overviews of current trends in phase-field fracture are provided in [3, 12] and the monograph [11]. The basic mathematical model of phase-field fracture optimization was developed and analyzed in [9, 10]. Based on known algorithms for PDE-constrained optimization (e.g., [2, 5, 7]), numerical computations were carried out in [6] using the open-source package DOpElib [4]. In this contribution, we sum-

D. Khimin (✉) · M. C. Steinbach · T. Wick
Institute of Applied Mathematics, Leibniz University Hannover, Welfengarten 1, 30167 Hannover, Germany
e-mail: khimin@ifam.uni-hannover.de

M. C. Steinbach
e-mail: mcs@ifam.uni-hannover.de

T. Wick
e-mail: thomas.wick@ifam.uni-hannover.de

F. Aldakheel et al. (eds.), *Current Trends and Open Problems in Computational Mechanics*, https://doi.org/10.1007/978-3-030-87312-7_24

marize algorithmic concepts and substantiate our developments with a numerical example.

2 Problem Statements

In this section, we first state the forward problem and then the optimization problem. The former is given by a regularized discrete time fracture problem which uses a penalty ansatz for the growth irreversibility constraint.

2.1 Phase-Field Fracture Forward Problem

To formulate the problem, we need to introduce some notation. We consider a bounded domain $\Omega \subset \mathbb{R}^2$. Its boundary is partitioned as $\partial\Omega = \Gamma_N \dot{\cup} \Gamma_D$ where Γ_D and Γ_N have nonzero Hausdorff measure. Next we define two function spaces, $V := H^1_D(\Omega; \mathbb{R}^2) \times H^1(\Omega)$ and $Q := L^2(\Gamma_N)$, where V denotes the space for the displacement-field and the phase-field and Q is the control space. In the next step we introduce a function $E^\gamma_\varepsilon(q; u, \varphi)$ from which we derive our forward problem. $E^\gamma_\varepsilon(q; u, \varphi)$ is defined as the sum of the regularized total energy of a crack plus a penalty term for our time dependent irreversibility constraint. The regularized total energy of a crack is given by

$$E_\varepsilon(q; u, \varphi) := \frac{1}{2}\left(g(\varphi)\mathbb{C}e(u), e(u)\right) - (q, u)_{\Gamma_N} + G_c\Gamma_\varepsilon(\varphi), \tag{1}$$

where q denotes a force that is applied in orthogonal direction to $\Gamma_N \subset \partial\Omega$, u is the vector-valued displacement field, φ is the phase-field, $\mathbb{C}$ is the elasticity tensor and $e(u)$ the symmetric gradient. The so-called degradation function $g(\varphi) := (1-\kappa)\varphi^2 + \kappa$ helps to extend the displacements to the entire domain Ω. The term $G_c\Gamma_\varepsilon(\varphi) := \frac{1}{2\varepsilon}\|1-\varphi\| + \frac{\varepsilon}{2}\|\nabla\varphi\|^2$ is a regularized form of the Hausdorff measure [1].

So far the problem consists in finding a function $\mathbf{u} := (u, \varphi)$ that minimizes the regularized total energy (1) subject to the irreversibility constraint $\varphi(t_2) \leq \varphi(t_1)$ for $t_1 \leq t_2$. In the sequel, the constraint is being replaced by a penalty term in the cost functional. To ensure differentiability up to second order [8] we define the penalty function as $R(\varphi^{i-1}; \varphi^i) := \frac{1}{4}\left\|(\varphi^i - \varphi^{i-1})^+\right\|^4_{\mathcal{L}^4}$. Finally the regularized discrete forward problem consists in finding a vector

$$\mathbf{u} := (\mathbf{u}^i)_{i=1}^M = (u, \varphi) = (u^i, \varphi^i)_{i=1}^M = (u(t_i), \varphi(t_i))_{i=1}^M \in V^M$$

that solves the following optimization problem for given initial data $(u^0, \varphi^0) \in V$ and given control $q := (q^i)_{i=1}^M \in Q^M$:

$$\min_{\mathbf{u}} \; E_\varepsilon^\gamma(q^i, \varphi^{i-1}; u^i, \varphi^i) := E_\varepsilon(q^i; u^i, \varphi^i) + \gamma R(\varphi^{i-1}; \varphi^i), \tag{C^γ}$$

for $i = 1, \ldots, M$ and $\gamma > 0$.

2.2 *Optimization Problem*

We formulate the following regularized NLP. For given $(u^0, \varphi^0) \in V$ we seek a solution $(q, \mathbf{u}) \in (Q, V)^M$ to (NLP$^\gamma$):

$$\begin{aligned} \min_{q,\mathbf{u}} \quad & J(q,\mathbf{u}) = \frac{1}{2}\sum_{i=1}^{M} \|\varphi^i - \varphi_d^i\|^2 + \frac{\alpha}{2}\sum_{i=1}^{M} \|q^i\|_{\Gamma_N}^2 \\ \text{s.t.} \quad & (q^i, \mathbf{u}^i) \text{ solves (EL}^\gamma\text{) for each } i = 1, \ldots, M, \end{aligned} \tag{NLP$^\gamma$}$$

where φ_d is some desired phase-field function. The existence of a global solution of (NLP$^\gamma$) has been shown in [9, Theorem 4.3] for functions that are non-negative and weakly semi-continuous. Here, $\mathbf{u}^i = (u^i, \varphi^i) \in V^M$ of (C^γ) solves the Euler-Lagrange equations for describing the phase-field fracture propagation, see e.g., [9],

$$\begin{aligned} &\big(g(\varphi^i)\mathbb{C}e(u^i), e(v)\big) - \big(q^i, v\big)_{\Gamma_N} = 0, \\ &G_c\varepsilon\big(\nabla\varphi^i, \nabla\psi\big) - \frac{G_c}{\varepsilon}\big(1-\varphi^i, \psi\big) + (1-\kappa)\big(\varphi^i\mathbb{C}e(u^i) : e(u^i), \psi\big) \\ &+ \gamma\big(((\varphi^i - \varphi^{i-1})^+)^3, \psi\big) = 0, \end{aligned} \tag{EL$^\gamma$}$$

for every pair of test functions $(v, \psi) \in V$ and $i = 1, \ldots, M$.

3 Reduced Optimization Problem

To formulate a reduced optimization problem, we assume the existence of a solution operator $S\colon Q^M \to V^M$ via equation (EL$^\gamma$). With this solution operator we can reduce the cost functional $J(q, \mathbf{u})$ to $j\colon Q^M \to \mathbb{R}$, $j(q) := J(q, S(q))$. This yields the unconstrained reduced optimization problem

$$\min_q \; j(q), \quad q \in Q^M. \tag{NLP$^\gamma_{\text{red}}$}$$

In the next step we formulate the Lagrangian function for (NLP$^\gamma$) by means of a functional $a\colon (Q \times V)^M \to \mathbb{R}$ that combines the constraints (EL$^\gamma$). Given a Lagrange multiplier $\mathbf{z} = (\mathbf{z}^i)_{i=1}^M = (z, \varphi_z) = (z^i, \varphi_z^i)_{i=1}^M \in V^M$, that functional reads

$$a(q, \mathbf{u})(\mathbf{z}) := \sum_{i=1}^{M} \Big[\big(g(\varphi^i)\mathbb{C}e(u^i)), e(z^i)\big) - (q^i, z^i)_{|\Gamma_N} + G_c\varepsilon\big(\nabla\varphi^i, \nabla\varphi_z^i\big) \\ - \frac{G_c}{\varepsilon}\big(1 - \varphi^i, \varphi_z^i\big) + (1-\kappa)\big(\varphi^i\mathbb{C}e(u^i) : e(u^i), \varphi_z^i\big) + \gamma\big(((\varphi^i - \varphi^{i-1})^+)^3, \varphi_z^i\big)\Big]. \tag{2}$$

The Lagrangian $\mathcal{L}\colon (Q \times V \times V)^M \to \mathbb{R}$ for (NLP$^\gamma$) is then defined as

$$\mathcal{L}(q, \mathbf{u}, \mathbf{z}) := J(q, \mathbf{u}) - a(q, \mathbf{u})(\mathbf{z}). \tag{3}$$

We will use this Lagrangian to obtain computable representations of the first and second order derivatives of j. To find a respresentation for the first derivative $j'(q)(\tau q)$ in the direction τq, we have to solve two equations. First we need to find a solution $\boldsymbol{u} = S(q) \in V$ of the state equation $\mathcal{L}'_{\mathbf{z}}(q, \mathbf{u}, \mathbf{z})(\boldsymbol{\phi}) = 0 \Leftrightarrow a(q, \mathbf{u})(\boldsymbol{\phi}) = 0\, \forall \boldsymbol{\phi} \in V^M$. Then we have to solve the adjoint equation $\mathcal{L}'_{\mathbf{u}}(q, \mathbf{u}, \mathbf{z})(\boldsymbol{\phi}) = 0$ for $\mathbf{z} \in V$. Finally we obtain the representation

$$j'(q)(\tau q) = \mathcal{L}'_q(q, \mathbf{u}, \mathbf{z})(\tau q). \tag{4}$$

For an analogous representation of $j''(q)(\delta q, \tau q)$ we first need to find a solution $\boldsymbol{\delta u} \in V$ of the tangent equation $\mathcal{L}''_{q\mathbf{z}}(q, \mathbf{u}, \mathbf{z})(\delta q, \boldsymbol{\phi}) + \mathcal{L}''_{\mathbf{uz}}(q, \mathbf{u}, \mathbf{z})(\boldsymbol{\delta u}, \boldsymbol{\phi}) = 0$. After that we solve the adjoint Hessian equation $\mathcal{L}''_{q\mathbf{u}}(q, \mathbf{u}, \mathbf{z})(\delta q, \boldsymbol{\phi}) + \mathcal{L}''_{\mathbf{uu}}(q, \mathbf{u}, \mathbf{z})(\boldsymbol{\delta u}, \boldsymbol{\phi}) + \mathcal{L}''_{\mathbf{zu}}(q, \mathbf{u}, \mathbf{z})(\boldsymbol{\delta z}, \boldsymbol{\phi}) = 0$ for $\boldsymbol{\delta z} \in V$. This results in the representation

$$j''(q)(\delta q, \tau q) = \mathcal{L}''_{qq}(q, \mathbf{u}, \mathbf{z})(\delta q, \tau q) + \mathcal{L}''_{\mathbf{u}q}(q, \mathbf{u}, \mathbf{z})(\boldsymbol{\delta u}, \tau q) \\ + \mathcal{L}''_{\mathbf{z}q}(q, \mathbf{u}, \mathbf{z})(\boldsymbol{\delta z}, \tau q). \tag{5}$$

Additional theory for these representations can be found in [2, Theorem 2.1].

For the spatial discretization, we employ a classical Galerkin finite element method on a quadrilateral mesh. To this end, we use conforming finite element spaces $V_{h_1} \subset H^1_D(\Omega; \mathbb{R}^2)$, $V_{h_2} \subset H^1(\Omega)$, and Q_h for the control.

Now we are ready to formulate our discrete problem, where we seek a solution $(q_h, \boldsymbol{u}_h) := (q_h(t_i), \boldsymbol{u}_h(t_i))_{i=1}^M$ with $\varphi_h(0) = \varphi_h^0$ of the following NLP:

$$\begin{aligned} \min_{q_h, \mathbf{u}_h} \quad & J(q_h, \mathbf{u}_h) \\ \text{s.t.} \quad & \big(g(\varphi_h^i)\mathbb{C}e(u_h^i), e(v_h)\big) - \big(q_h^i, v_h\big)_{\Gamma_N} = 0\, \forall (v_h, \psi_h) \in V_{h_1} \times V_{h_2}, i = 1, \ldots, M, \\ & G_c\varepsilon\big(\nabla\varphi_h^i, \nabla\psi_h\big) - \frac{G_c}{\varepsilon}\big(1 - \varphi_h^i, \psi_h\big) + (1-\kappa)\big(\varphi_h^i\mathbb{C}e(u_h^i) : e(u_h^i), \psi_h\big) \\ & + \gamma\big(((\varphi_h^i - \varphi_h^{i-1})^+)^3, \psi_h\big) = 0\, \forall (v_h, \psi_h) \in V_{h_1} \times V_{h_2}, i = 1, \ldots, M. \end{aligned}$$

The discrete reduced cost functional $j_h\colon Q_h^M \to \mathbb{R}$ is defined in terms of the discrete solution operator S_h as $j_h(q_h) := J(q_h, S_h(q_h))$. This yields the (unconstrained) discrete reduced NLP

$$\min_{q_h} j_h(q_h), \quad q_h \in Q_h^M. \tag{6}$$

Next, we define a basis $\{\tau q_j \mid j = 1, \dots, n_q\}$ of our n_q-dimensional control space Q_h^M in order to solve (6) with Newton's method. In the unconstrained problems ($\text{NLP}^{\gamma}_{\text{red}}$) and (6) we seek an optimal control $q \in Q_h^M$ that satisfies the first order necessary optimality condition $j_h'(q)(\tau q) = 0 \; \forall \tau q \in Q_h^M$. The Newton increment δq of this condition is determined from the linear system

$$j_h''(q)(\delta q, \tau q) = -j_h'(q)(\tau q) \quad \forall \tau q \in Q_h^M. \tag{7}$$

Provided that $j_h''(q)$ is positive definite, (7) is also a necessary *and sufficient* condition for the strongly convex quadratic problem

$$\min_{\delta q} m(q, \delta q) := j_h(q) + j_h'(q)(\delta q) + \frac{1}{2} j_h''(q)(\delta q, \delta q) \quad \forall \delta q \in Q_h^M. \tag{8}$$

Thus, any solution δq of (8) solves (7) and vice versa. Before we proceed with system (7), we define vectors and matrices that represent the derivatives of the discrete reduced cost functional j_h:

$$(\nabla j_h(q), \tau q)_{Q^M} := j_h'(q)(\tau q) \quad \forall \tau q \in Q_h^M,$$
$$\left(\nabla^2 j_h(q)\delta q, \tau q\right)_{Q^M} := j_h''(q)(\delta q, \tau q) \quad \forall \delta q, \tau q \in Q_h^M.$$

With this notation (7) becomes:

$$\left(\nabla^2 j_h(q)\delta q, \tau q_j\right)_{Q^M} = -\left(\nabla j_h(q), \tau q_j\right)_{Q^M}, \quad j = 1, \dots, n_q. \tag{9}$$

We can reformulate (9) in vector notation by replacing $\nabla j_h(q) \in Q_h^M$ with the corresponding coefficient vector $\boldsymbol{f} \in \mathbb{R}^{n_q}$:

$$\left(\nabla j_h(q), \tau q_j\right)_{Q^M} = \sum_{i=1}^{n_q} \left(\boldsymbol{f}_i \tau q_i, \tau q_j\right)_{Q^M} = \sum_{i=1}^{n_q} \boldsymbol{f}_i (\tau q_i, \tau q_j)_{Q^M}.$$

Since that equation has to hold for every element τq_j, $\boldsymbol{f}$ can be computed as a solution of the linear system

$$\boldsymbol{G}\boldsymbol{f} = \left((\nabla j_h(q), \tau q_j)_{Q^M}\right)_{j=1}^{n_q}, \tag{10}$$

where $\boldsymbol{G}_{ji} := \left(\tau q_i, \tau q_j\right)_{Q^M}$ denotes the Gramian matrix of the basis of Q_h^M. Having constructed the right hand side of (9), we do the same for the left hand side and replace δq by its coefficient vector $\boldsymbol{d}$:

$$\left(\nabla^2 j_h(q)\delta q, \tau q_j\right)_{Q^M} = \sum_{i=1}^{n_q} \boldsymbol{d}_i \left(\nabla^2 j_h(q)\tau q_i, \tau q_j\right)_{Q^M}.$$

Thus we compute $\boldsymbol{d}$ as solution of the linear system

$$\boldsymbol{K}\boldsymbol{d} = \left(\left(\nabla^2 j_h(q)\delta q, \tau q_j \right)_Q \right)_{j=1}^{n_q}, \tag{11}$$

where $\boldsymbol{K}_{ji} := (\nabla^2 j_h(q)\tau q_i, \tau q_j)_{Q^M}$. Using (10) and (11), we finally replace (9) by the equivalent linear system $\boldsymbol{H}\boldsymbol{d} = -\boldsymbol{f}$ with coefficient matrix $\boldsymbol{H} := \boldsymbol{G}^{-1}\boldsymbol{K}$. If the dimension n_q is large, constructing the entire matrix $\boldsymbol{H}$ becomes expensive. Instead we can just compute a coefficient vector for the product $\nabla^2 j_h(q)\delta q \in Q_h^M$ by solving $\left(\nabla^2 j_h(q)\delta q, \tau q_j\right)_{Q^M} = \sum_{i=1}^{n_q} \boldsymbol{h}_i \left(\tau q_i, \tau q_j\right)_{Q^M}$. To conclude this section, we introduce the notation $\langle \boldsymbol{a}, \boldsymbol{b}\rangle := \boldsymbol{a}^T \boldsymbol{G}\boldsymbol{b}$ and $|\boldsymbol{a}| := \langle \boldsymbol{a}, \boldsymbol{a}\rangle^{\frac{1}{2}}$. Now $(\mathbb{R}^{n_q}, |\cdot|)$ is isometrically isomorphic to $(Q_h^M, \|\cdot\|_{Q^M})$, and the functional m from (8) can be written as:

$$m(q, \boldsymbol{d}) = j_h(q) + \langle \boldsymbol{f}, \boldsymbol{d}\rangle + \frac{1}{2}\langle \boldsymbol{H}\boldsymbol{d}, \boldsymbol{d}\rangle. \tag{12}$$

The conjugate gradient (CG) method can be used for minimizing m in step 4 of Algorithm 4.1 below.

4 Algorithmic Realization

Algorithm 4.1 For a given control $q^0 = (q^0(t_i))_{i=1}^M \in Q_h^M$, iterate over $k = 0, 1, \ldots$

1. Calculate $\mathbf{u}^k = (\mathbf{u}^k(t_i))_{i=1}^M \in (V_{h_1} \times V_{h_2})^M$ from the discrete state equation: $\mathcal{L}'_{\mathbf{z}}(q^k, \mathbf{u}^k, \mathbf{z}^k)(\boldsymbol{\phi}) = 0 \Leftrightarrow a(q^k, \mathbf{u}^k)(\boldsymbol{\phi}) = 0$, for all $\boldsymbol{\phi} \in (V_{h_1} \times V_{h_2})^M$.
2. Calculate $\mathbf{z}^k = (\mathbf{z}^k(t_i))_{i=1}^M \in (V_{h_1} \times V_{h_2})^M$ from the discrete adjoint equation: $\mathcal{L}'_{\mathbf{u}}(q^k, \mathbf{u}^k, \mathbf{z}^k)(\boldsymbol{\phi}) = 0$ for all $\boldsymbol{\phi} \in (V_{h_1} \times V_{h_2})^M$.
3. Build the coefficient vector $\boldsymbol{f}$ of $\nabla j_h(q^k)$ by applying the representation (4) to $\tau q = \tau q_j$ for $j = 1, \ldots, n_q$: $(\nabla j_h(q^k))_j = \hat{\mathcal{L}}'_q(q^k, \mathbf{u}^k, \mathbf{z}^k)(\tau q_j)$. Then solve the following system:

$$\boldsymbol{G}\boldsymbol{f} = \left(j'_h(q^k)(\tau q_j)\right)_{j=1}^{n_q} \quad \text{where} \quad \boldsymbol{G}_{ij} = (\tau q_j, \tau q_i)_{Q^M}.$$

4. Calculate a solution δq of $j''_h(q^k)(\delta q, \tau q_j) = -j'_h(q^k)(\tau q_j)$ with $\tau q_j \in Q_h^M$ via

$$\min_{\boldsymbol{d} \in \mathbb{R}^{n_q}} m(q^k, \boldsymbol{d}), \tag{13}$$

   using an iterative algorithm that only needs matrix-vector products with the Hessian. These products will be calculated in Algorithm 4.2.
5. Choose a step length ν_k by an Armijo backtracking method.
6. Set $q^{k+1} = q^k + \nu_k \delta q$.
7. Increase $k = k + 1$.

Repeat until $|\boldsymbol{f}| = \left\|\nabla j_h(q^k)\right\|_{Q^M} < \mathrm{TOL}$. □

Algorithm 4.2 (2) Let $\mathbf{u}^k$ and z^k be given with respect to the current control q^k.

1. Calculate $\delta\mathbf{u}^k$ by solving the tangent equation.
2. Calculate δz^k by solving the adjoint Hessian equation.
3. Construct the coefficient vector for $\nabla^2 j_h(q)\delta q$ by using representation (5) with $\tau q = \tau q_j$ for $j = 1, \ldots, n_q$. Then solve the linear system

$$\boldsymbol{G}\boldsymbol{h} = \left(j_h''(q^k)(\delta q, \tau q_j)\right)_{j=1}^{n_q}. \quad \square$$

Remark 1 A documentation and implementation of Newton's method for solving the inner nonlinear phase-field fracture problem can be found in [4].

5 Numerical Example

We wish to find an optimal control that produces some given phase-field φ_d. Here we apply our force q in orthogonal direction to the top side $\Gamma_N \subset \partial\Omega$ of the domain $\Omega := (-1, 1)^2$. Furthermore we split $\partial\Omega$ in three subsets, $\partial\Omega = \Gamma_N \cup \Gamma_D \cup \Gamma_{\text{free}}$, where $\Gamma_N := [-1, 1] \times \{1\}, \Gamma_D := [-1, 1] \times \{-1\}, \Gamma_{\text{free}} := \{\pm 1\} \times [-1, 1]$. On Γ_N we apply our force q and on Γ_D we enforce Dirichlet boundary conditions $u = 0$.

In this example the control space Q_h is one-dimensional. The applied force q is constant in time, $q^i = q$, and the initial data is given by $\boldsymbol{u}^0 = (u(t_0), \varphi(t_0)) := (0; \varphi^0)$, where the initial phase-field φ^0 is defined as follows:

$$\varphi^0(x, y) := \begin{cases} 0, & x \in (-0.1 - h, 0.1 + h) \text{ and } y \in (-h, h), \\ 1, & \text{else.} \end{cases} \tag{14}$$

This initial condition describes a horizontal crack between the points $(-0.1 - h, 0)$ and $(0.1 + h, 0)$ with a thickness of $2h$. Before we can solve (NLP$^\gamma$) we have to define some constants. The regularization parameter for the phase-field is $\varepsilon := 2h = 0.088$, where h denotes the diameter of the mesh elements. The parameter κ in the coefficient function $g(\varphi)$ is $\kappa := 10^{-10}$. The main regularization parameter α is 10^{-6}, and the penalty parameter γ is 10^8. The given phase-field φ_d continues the initial phase-field φ to the left side of the domain,

$$\varphi_d(x, y) := \begin{cases} 0, & x \in (-1, 0.1 + h) \text{ and } y \in (-h, h), \\ 1, & \text{else.} \end{cases} \tag{15}$$

Our results are presented in Table 1. Every row consists of 8 entries. The first value (Iter.) is the current Newton iteration, the second value (CG) is the number of CG iterations that were necessary for minimizing m. The remaining values are the relative and absolute Newton residuals (Rel. residual, Abs. residual), the cost functional J and its tracking part $\frac{1}{2}\sum_{i=1}^{5} \|\varphi(t_i) - \varphi_d(t_i)\|_2^2$ (Cost, Tracking), the maximal force $|q_{\max}|$ on Γ_N (Force), and the norm-square of the gradient of the reduced cost functional,

Table 1 Results of the numerical example

Iter.	CG	Rel. residual	Abs. residual	Cost	Tracking	Force	Gradient
0	—	1.0	0.005 66	4.68	0.251	1000	1.00×10^{-6}
1	2	0.00673	$3.82 \times 10^{-}5$	0.165	0.165	26.2	4.15×10^{-11}
2	3	1.05×10^{-5}	5.93×10^{-8}	0.165	0.165	0.882	3.52×10^{-15}
3	18	7.36×10^{-9}	4.17×10^{-11}	0.165	0.165	0.000215	1.74×10^{-21}
4	9	3.76×10^{-11}	2.13×10^{-13}	0.165	0.165	1.36×10^{-6}	4.52×10^{-26}

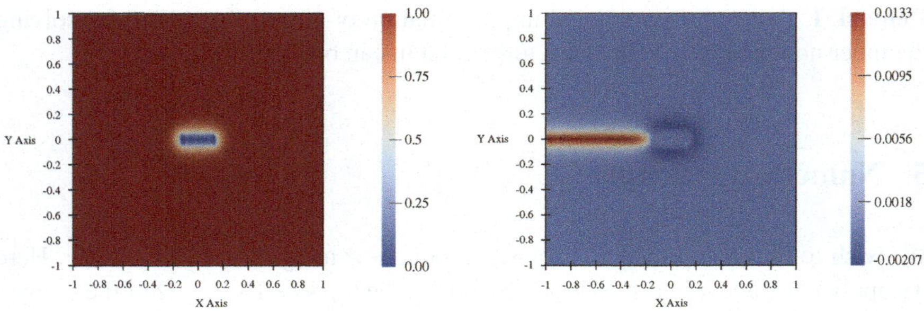

Fig. 1 Left: phase-field, right: adjoint phase-field

$\|\nabla j(q)\|_2^2$ (Gradient). All values are rounded to three significant digits. Finally, solution plots are provided in Fig. 1.

Acknowledgements This work has been supported by the German Research Foundation, Priority Program 1962 (DFG SPP 1962) with in the subproject *Optimizing Fracture Propagation using a Phase-Field Approach* with the project number 314067056.

References

1. Ambrosio, L., & Tortorelli, V. (1992). On the approximation of free discontinuity problems. *Bollettino dell'Unione Matematica Italiana B, 6*, 105–123.
2. Becker, R., Meidner, D., & Vexler, B. (2007). Efficient numerical solution of parabolic optimization problems by finite element methods. *Optimization Methods and Software, 22*(5), 813–833.
3. Bourdin, B., & Francfort, G. A. (2019). Past and present of variational fracture. *SIAM News, 52*(9).
4. Goll, C., Wick, T., & Wollner, W. (2017). DOpElib: Differential equations and optimization environment; A goal oriented software library for solving PDEs and optimization problems with PDEs. *Archive of Numerical Software, 5*(2), 1–14.
5. Hinze, M., Pinnau, R., Ulbrich, M., & Ulbrich, S. (2009). *Optimization with PDE constraints*. Number 23 in Mathematical modelling: theory and applications. Dordrecht u.a.: Springer.
6. Khimin, D. (2020). Numerische Modellierung von Optimalsteuerungsproblemen für Phasenfeld-Riss-Ausbreitung. Master's thesis, Leibniz University Hannover.
7. Meidner, D. (2008). *Adaptive Space-Time Finite Element Methods for Optimization Problems Governed by Nonlinear Parabolic Systems*. Ph.D. thesis, University of Heidelberg.

8. Meyer, C., Rademacher, A., & Wollner, W. (2015). Adaptive optimal control of the obstacle problem. *SIAM Journal on Scientific Computing, 37*(2), A918–A945.
9. Neitzel, I., Wick, T., & Wollner, W. (2017). An optimal control problem governed by a regularized phase-field fracture propagation model. *SIAM Journal on Control and Optimization, 55*(4), 2271–2288.
10. Neitzel, I., Wick, T., & Wollner, W. (2019). An optimal control problem governed by a regularized phase-field fracture propagation model. Part II: The regularization limit. *SIAM Journal on Control and Optimization, 57*(3), 1672–1690.
11. Wick, T. (2020). *Multiphysics Phase-Field Fracture: Modeling, Adaptive Discretizations, and Solvers*. De Gruyter, Berlin, Boston, 12 Oct.
12. Wu, J.-Y., Nguyen, V. P., Thanh Nguyen, C., Sutula, D., Bordas, S., & Sinaie, S. (2019). Phase field modelling of fracture. *Advances in Applied Mechanics, 53*, 09.

A Strong Form Meshfree Collocation Method: Engineering Applications Including Frictional Contact

Tae-Yeon Kim, Jeong-Hoon Song, and Tod A. Laursen

This article is dedicated to my friend and longtime colleague Peter Wriggers on his 70th birthday. We were first introduced by my doctoral advisor Juan Simo on the Stanford campus in 1990, and I have been so pleased and privileged to collaborate with Peter on so many things, including organization of conferences and minisymposia, short courses, and various publication projects throughout the years. Thank you Peter for your friendship, your generosity, and your mentorship over the course of my career, and my very best wishes for health, happiness, and continued productivity in the years to come (T. Laursen).

Abstract This article provides the brief review of the recently developed strong form meshfree collocation method. The method directly discretizes a strong form with approximated derivatives from the moving least-squares approximation using the Taylor polynomial of the unknown variable. The approximations of derivatives of any order can be generated in the process of computing the shape function without further cost. The method does not require mesh structure and numerical integration, and adaptivity can be easily achieved by locally refining collocation points. The discretization of the strong form using the derivative approximation is briefly described based on a frictional contact problem. Moreover, recent applications and developments of the method for various engineering problems are briefly presented.

T.-Y. Kim
Department of Civil Infrastructure and Environmental Engineering, Khalifa University of Science and Technology, Abu Dhabi, UAE
e-mail: taeyeon.kim@ku.ac.ae

J.-H. Song
Department of Civil, Environmental and Architectural Engineering, University of Colorado at Boulder, Boulder, CO, USA
e-mail: jh.song@colorado.edu

T. A. Laursen (✉)
Department of Civil, Structural and Environmental Engineering, University at Buffalo, Buffalo, NY, USA
e-mail: Tod.Laursen@suny.edu

F. Aldakheel et al. (eds.), *Current Trends and Open Problems in Computational Mechanics*, https://doi.org/10.1007/978-3-030-87312-7_25

1 A Strong Form Meshfree Collocation Method

The strong form meshfree collocation method is based on the idea of Kim and Kim [1] and Yoon and Song [2–4]. The method discretizes directly a strong form with the derivative approximation generated from the moving least-squares approximation using a Taylor expansion of the unknown variable. This section provides the brief procedure of constructing such derivative approximation. Further details can be found in [1–4].

1.1 *Approximation of Derivative Operators*

Let $\mathbf{x} = (x_1, \ldots, x_n)$ be an n-dimensional real vector and $\boldsymbol{\alpha} = (\alpha_1, \ldots, \alpha_n)$ be an n-tuple of non-negative integers. The $\boldsymbol{\alpha}$th-power of $\mathbf{x}$ is defined by $\mathbf{x}^{\boldsymbol{\alpha}} = x_1^{\alpha_1} x_2^{\alpha_2} \cdots x_n^{\alpha_n}$. The $\boldsymbol{\alpha}$th-order derivative of a smooth function $f(\mathbf{x})$ with respect to $\mathbf{x}$ is defined by

$$D_{\mathbf{x}}^{\boldsymbol{\alpha}} f(\mathbf{x}) = \frac{\partial^{|\boldsymbol{\alpha}|} f(\mathbf{x})}{\partial x_1^{\alpha_1} \partial x_2^{\alpha_2} \cdots \partial x_n^{\alpha_n}} \tag{1}$$

where $|\boldsymbol{\alpha}| = \sum_{i=1}^{n} \alpha_i$.

The mth-order Taylor polynomial for approximating a function $u(\mathbf{x})$ at the local center $\bar{\mathbf{x}}$ can be expressed as

$$u(\mathbf{x}; \bar{\mathbf{x}}) = \sum_{|\boldsymbol{\alpha}| \leq m} \frac{(\mathbf{x} - \bar{\mathbf{x}})^{\boldsymbol{\alpha}}}{\boldsymbol{\alpha}!} D_{\mathbf{x}}^{\boldsymbol{\alpha}} u(\bar{\mathbf{x}}) = \mathbf{p}_m^{\mathrm{T}}(\mathbf{x}; \bar{\mathbf{x}}) \mathbf{M}(\bar{\mathbf{x}}) \tag{2}$$

where $\boldsymbol{\alpha}!$ is the factorial of $\boldsymbol{\alpha}$ and the polynomial vector $\mathbf{p}_m^{\mathrm{T}}(\mathbf{x}; \bar{\mathbf{x}})$ and the derivative coefficient vector $\mathbf{M}(\bar{\mathbf{x}})$ are defined by

$$\mathbf{p}_m^{\mathrm{T}}(\mathbf{x}; \bar{\mathbf{x}}) = \left[\frac{(\mathbf{x} - \bar{\mathbf{x}})^{\boldsymbol{\alpha}_1}}{\boldsymbol{\alpha}_1!}, \ldots, \frac{(\mathbf{x} - \bar{\mathbf{x}})^{\boldsymbol{\alpha}_L}}{\boldsymbol{\alpha}_L!} \right], \quad \mathbf{M}^{\mathrm{T}}(\bar{\mathbf{x}}) = \left[D_{\mathbf{x}}^{\boldsymbol{\alpha}_1} u(\bar{\mathbf{x}}), \ldots, D_{\mathbf{x}}^{\boldsymbol{\alpha}_L} u(\bar{\mathbf{x}}) \right], \tag{3}$$

in which $\boldsymbol{\alpha}_i$'s are an n-tuple of non-negative integers and $L = (n + m)!/n!m!$ is the number of the components of $\mathbf{p}_m^{\mathrm{T}}$.

Bearing in mind of the idea of the moving least-squares approximation, minimizing the discrete form of the weighted, discrete L_2-norm

$$\mathbf{J}(\mathbf{a}) = \sum_{I=1}^{N} w \left(\frac{\mathbf{x}_I - \bar{\mathbf{x}}}{\rho} \right) [\mathbf{p}^{\mathrm{T}}(\mathbf{x}_I) \mathbf{M}(\bar{\mathbf{x}}) - u_I]^2 \tag{4}$$

with respect to $\mathbf{M}(\bar{\mathbf{x}})$ yields

$$\mathbf{M}(\bar{\mathbf{x}}) = \mathbf{M}^{-1}(\bar{\mathbf{x}}) \mathbf{B}(\bar{\mathbf{x}}) \mathbf{u}^{\mathrm{T}}. \tag{5}$$

The matrices **M** and **B** can be defined by

$$\mathbf{M}(\bar{\mathbf{x}}) = \sum_{I=1}^{N} w\left(\frac{\mathbf{x}_I - \bar{\mathbf{x}}}{\rho}\right) \mathbf{p}_m(\mathbf{x}_I; \bar{\mathbf{x}})\mathbf{p}_m^{\mathrm{T}}(\mathbf{x}_I; \bar{\mathbf{x}}), \tag{6}$$

$$\mathbf{B}(\bar{\mathbf{x}}) = \left[w\left(\tfrac{\mathbf{x}_1-\bar{\mathbf{x}}}{\rho}\right)\mathbf{p}_m(\mathbf{x}_1; \bar{\mathbf{x}}), \cdots, w\left(\tfrac{\mathbf{x}_N-\bar{\mathbf{x}}}{\rho_{\bar{\mathbf{x}}}}\right)\mathbf{p}_m(\mathbf{x}_N; \bar{\mathbf{x}})\right] \tag{7}$$

where $w\left(\frac{\mathbf{x}_I-\bar{\mathbf{x}}}{\rho}\right)$ is the weight function with a compact support (or domain of influence) in which its size is determined by the dilation parameter ρ, N is the number of nodes included in the support of the weight function, and u_I is the nodal solution for neighbor node I.

Substituting **x** for $\bar{\mathbf{x}}$ and replacing $\mathbf{M}(\mathbf{x})$ with $\mathbf{D}_{\mathbf{x}}^{\alpha}u(\mathbf{x})$ in (5) give rise to the $\boldsymbol{\alpha}$th derivative approximation of $u(\mathbf{x})$

$$\mathbf{D}_{\mathbf{x}}^{\alpha}u(\mathbf{x}) = \sum_{I=1}^{N} \Phi_I^{\alpha}(\mathbf{x})u_I \tag{8}$$

where $\boldsymbol{\alpha} = (\alpha_1, \alpha_2)$ be a 2-tuple of non-negative integers and $\boldsymbol{\Phi}_I^{\alpha}(\mathbf{x})$ is the $\boldsymbol{\alpha}$th-order derivative of the shape function at node I defined as

$$\boldsymbol{\Phi}_I^{\alpha}(\mathbf{x}) = \mathbf{e}_{\alpha}^{\mathrm{T}}\mathbf{M}^{-1}(\mathbf{x})\mathbf{p}(\mathbf{x}_I; \mathbf{x})w\left(\frac{\mathbf{x}_I - \mathbf{x}}{\rho}\right) \tag{9}$$

where $\mathbf{e}_{\alpha}^{\mathrm{T}} = [e_0, \ldots, e_m]$ with its component defined as $e_k = 1$ if $k = \alpha$ and $e_k = 0$ otherwise for $k = 0, \ldots, m$.

It is worthwhile to mention that (8) does not require actual differentiation as shown in (9). As a consequence, this method doesn't require the regularity of the weight function to ensure the regularity of the shape functions.

1.2 Discretization of a Strong Form for Frictional Contact

The derivate approximation (8) allows for straightforward computation of derivative operators that can directly discretize a strong form of any order as well as their boundary conditions. We briefly provide the application of (8) to frictional contact problems by directly imposing contact constraints as part of Neumann boundary conditions on the elastostatic equation $\nabla \cdot \boldsymbol{\sigma} + \mathbf{b} = 0$ with $\boldsymbol{\sigma}$ being the Cauchy stress and **b** the body force. Details are referred to as Almansi et al. [5].

Upon substituting the constitutive relation of $\boldsymbol{\sigma}$ for linear elastic isotropic material, the elastostatic equation can be written in terms of the displacement **u** as

$$\mu\Delta\mathbf{u} + (\lambda + \mu)\nabla(\nabla \cdot \mathbf{u}) + \mathbf{b} = 0 \quad \text{in } \Omega \tag{10}$$

where λ and μ are Lamé constants. Similarly, Dirichlet and Neumann boundary conditions and contact constraints can be written as

$$\begin{aligned} \mathbf{u} &= \bar{\mathbf{u}} \quad \text{on } \Gamma_u, \\ 2\mu\mathbf{n}\cdot\boldsymbol{\epsilon} + \lambda\mathbf{n}\cdot\mathbf{1}(\nabla\cdot\mathbf{u}) &= \bar{\mathbf{t}} \quad \text{on } \Gamma_t, \\ 2\mu\mathbf{n}\cdot\boldsymbol{\epsilon} + \lambda\mathbf{n}\cdot\mathbf{1}(\nabla\cdot\mathbf{u}) &= \mathbf{t}_c \quad \text{on } \Gamma_c \end{aligned} \tag{11}$$

where $\boldsymbol{\epsilon} = (\nabla\mathbf{u} + (\nabla\mathbf{u})^{\mathrm{T}})$ is the strain tensor, $\mathbf{n}$ is the unit outward normal vector to domain Ω, $\bar{\mathbf{u}}$ is the prescribed displacement on Γ_u, $\bar{\mathbf{t}}$ is the prescribed traction on Γ_t, and $\mathbf{t}_c$ is the unknown contact traction on Γ_c.

By substituting (8) into (10), the discrete form of (10) can be obtained as

$$\begin{aligned} &\sum_{I=1}^{N}\{[(\lambda+2\mu)\boldsymbol{\Phi}_{IJ}^{(2,0)} + \mu\boldsymbol{\Phi}_{IJ}^{(0,2)}]u_{1I} + (\lambda+\mu)\boldsymbol{\Phi}_{IJ}^{(1,1)}u_{2I}\} + b_{1J} = 0, \\ &\sum_{I=1}^{N}\{(\lambda+\mu)\boldsymbol{\Phi}_{IJ}^{(1,1)}u_{1I} + [\mu\boldsymbol{\Phi}_{IJ}^{(2,0)} + (\lambda+2\mu)\boldsymbol{\Phi}_{IJ}^{(0,2)}]u_{2I}\} + b_{2J} = 0 \end{aligned} \tag{12}$$

where $\mathbf{x}_J$ indicates interior nodes within Ω and $(\cdot)_J$ indicates the value at $\mathbf{x}_J$, e.g., $\boldsymbol{\Phi}_{IJ}^{(2,0)} = \boldsymbol{\Phi}_{I}^{(2,0)}(\mathbf{x}_J)$. Similarly, the discrete forms of Dirichlet and Neumann boundary conditions at boundary nodes can be obtained by substituting (8) into $(11)_1$ and $(11)_2$ as shown in (4.23) and (4.26) in [5].

The contact traction can be decomposed into the normal and tangential components, i.e., $\mathbf{t}_c = \mathbf{t}_N - \mathbf{t}_T = t_N\mathbf{n}u - t_T\boldsymbol{\tau}$ where t_N and t_T are subject to the classical Kuhn-Tucker constraints governing contact interaction as in (3.4) and (3.7) of [5] and $\boldsymbol{\nu}$ and $\boldsymbol{\tau}$ are the unit normal and tangential vectors on Γ_c. For the normal contact constraint, the penalty regularization is achieved by $t_N = \epsilon_N < g >$ where $< g >$ is the Macaulay bracket of a gap function g and ϵ_N is the normal penalty parameter. With the tangential penalty parameter ϵ_T, the penalty regularization for frictional contact constraint can be constructed by (3.7) and (3.8) in [5] along with a trial state/return mapping algorithm to determine the Columb frictional traction as in (3.9) and (3.10) of [5]. Substituting (8) into $(11)_3$ results in the discrete forms of the contact constraints for both stick and slip as follows. For the stick case,

$$\begin{aligned} &\sum_{I=1}^{N}\{[(\lambda+2\mu)\boldsymbol{\Phi}_{IJ}^{(1,0)}n_1 + \mu\boldsymbol{\Phi}_{IJ}^{(0,1)}n_2]u_{1I} + [\lambda\boldsymbol{\Phi}_{IJ}^{(0,1)}n_1 + \mu\boldsymbol{\Phi}_{IJ}^{(1,0)}n_2]u_{2I}\} \\ &\qquad -\epsilon_N < g(\mathbf{x}_J) > \nu_1 + \epsilon_T H(g(\mathbf{x}_J))(\mathbf{u}\cdot\boldsymbol{\tau})\tau_1 = 0, \\ &\sum_{I=1}^{N}\{[\mu\boldsymbol{\Phi}_{IJ}^{(0,1)}n_1 + \lambda\boldsymbol{\Phi}_{IJ}^{(1,0)}n_2]u_{1I} + [(\lambda+2\mu)\boldsymbol{\Phi}_{IJ}^{(0,1)}n_2 + \mu\boldsymbol{\Phi}_{IJ}^{(1,0)}n_1]u_{2I}\} \\ &\qquad -\epsilon_N < g(\mathbf{x}_J) > \nu_2 + \epsilon_T H(g(\mathbf{x}_J))(\mathbf{u}\cdot\boldsymbol{\tau})\tau_2 = 0, \end{aligned} \tag{13}$$

and for the slip case,

$$
\begin{aligned}
&\sum_{I=1}^{N}\{[(\lambda+2\mu)\boldsymbol{\Phi}_{IJ}^{(1,0)}n_1+\mu\boldsymbol{\Phi}_{IJ}^{(0,1)}n_2]u_{1I}+[\lambda\boldsymbol{\Phi}_{IJ}^{(0,1)}n_1+\mu\boldsymbol{\Phi}_{IJ}^{(1,0)}n_2]u_{2I}\}\\
&\qquad -\epsilon_N < g(\mathbf{x}_J) > \nu_1+\mu\epsilon_N < g(\mathbf{x}_J) > \mathrm{sign}(\mathbf{u}\cdot\boldsymbol{\tau})\tau_1=0,\\
&\sum_{I=1}^{N}\{[\mu\boldsymbol{\Phi}_{IJ}^{(0,1)}n_1+\lambda\boldsymbol{\Phi}_{IJ}^{(1,0)}n_2]u_{1I}+[(\lambda+2\mu)\boldsymbol{\Phi}_{IJ}^{(0,1)}n_2+\mu\boldsymbol{\Phi}_{IJ}^{(1,0)}n_1]u_{2I}\}\\
&\qquad -\epsilon_N < g(\mathbf{x}_J) > \nu_2+\mu\epsilon_N < g(\mathbf{x}_J) > \mathrm{sign}(\mathbf{u}\cdot\boldsymbol{\tau})\tau_2=0,
\end{aligned}
\tag{14}
$$

where $\mathbf{x}_J$ belongs to contact nodes on Γ_c and $H(g)$ is the Heaviside function. Notice that the system of the equation is nonlinear due to the presence of contact constraints in (13) and (14). To use a full Newton-Raphson iteration scheme as a nonlinear solver, the residual of the system can be obtained by assembling the discrete forms of Dirichlet and Neumann boundary conditions, (12), and (13) for stick and (14) for slip. The tangent stiffness matrix can be computed by linearizing (13) for stick and (14) for slip. Notice that (13) and (14) can be used for both one- and two-body frictional contact with the proper definition of the gap function $g(\mathbf{x})$. The detailed procedure for the nonlinear solver can be found in Sect. 4 of [5].

2 Applications Including Frictional Contact

The method described in the previous section has several advantages due to the nature of the meshfree collocation method. The distinct feature of the method is an easy treatment of adaptive refinement because it does not require grid or mesh structure and numerical integration along with mesh connectivity. As a result, the method can simplify modeling of engineering problems requiring high accuracy locally. Another advantage of the method is the computation of higher-order derivative approximations without further cost. Using such derivative approximations, the method can easily evaluate a strong form of any order at spatially distributed collocation points. Upon making use of such advantages of the method, it has been further developed and applied to various engineering problems. The brief overview of the recent progress is provided in this section.

We first begin by reviewing the accuracy and efficiency of the method which has been performed by few researchers. Kim and Kim [1] provided L^2-convergence studies for Poisson and Stokes problems on both uniformly and non-uniformly distributed collocation points. The accuracy was tested for three values of the Taylor polynomial order, i.e., $m = 2, 3, 4$, with varying the domain of influence ρ. For the Poisson problem, while no significant difference with $m = 2$ and $m = 3$ was observed, relatively higher accuracy was obtained for $m = 4$. For the Stokes problem, the same accuracy was observed in both velocity and pressure with $m = 4$ as shown in Fig. 1a. Another

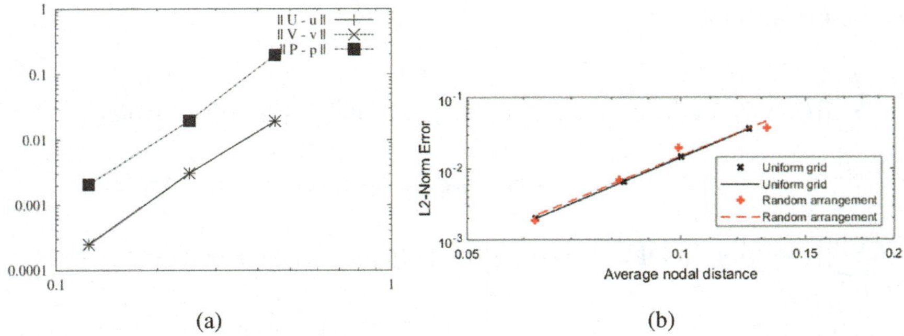

Fig. 1 Convergence studies: **a** Stokes flow [1] and **b** Large-scale wind driven ocean circulation [6]

convergence study was performed by Beel et al. [6] for the large-scale wind driven ocean circulation problems in both the discrete L^2- and L^∞-norm. They observed that the order of the convergence rate is approximately close to the order of the Taylor expansion without the boundary layer for both uniform and non-uniformly distributed collocation points as shown in Fig. 1b. On the other hand, with the presence of the boundary layer, low convergence rates and largely unpredictable error behavior were observed for randomly distributed collocation points (see Figs. 12 and 13 in [6]).

The computational efficiency of the meshfree collocation method was studied by Song et al. [7]. They used PETSc for parallel computing and performed numerical study of the polycrystalline solidification process. The computational time of the method was compared with that of the finite-difference method. They observed that the meshfree collocation method is more computationally demanding at each time step when compared with the finite-difference method. This is because the method requires sufficient amount of nodes within the compact support to ensure the invertibility of the moment matrix **M**. However, relatively larger time step size with the method than the finite-difference method can be chosen, indicating that the difference between two methods becomes not significant.

An easy adaptive refinement of the method has been employed to various engineering problems to improve the accuracy of the solution. Almansi et al. [5] took advantage of the adaptivity to model frictional contact problems. The accuracy of the contact algorithm was examined for the frictional Hertzian contact by modeling a half cylinder compressed by both horizontal and vertical uniform displacement as shown in Fig. 15 of [5]. Collocation points were non-uniformly refined in the vicinity of the contact area where the stress is highly concentrated as shown in Fig. 2. The accuracy of the method was verified by comparing the normal and tangential tractions on the contact surface with the results from the finite-element method using ABAQUS (see Fig. 17 in [5]). Both tractions from the proposed collocation method are quantitatively close to those of the finite-element method.

Beel et al. [6] applied the advantage of the adaptive refinement to the large-scale wind driven ocean circulation problem. Both uniformly and non-uniformly

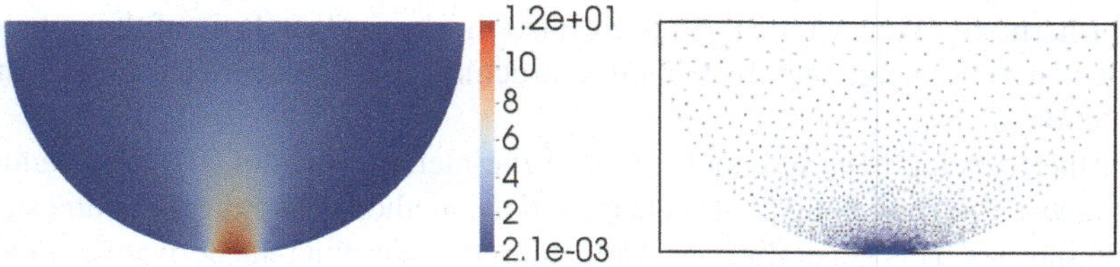

(a) (b)

Fig. 2 Adaptive refinement for frictional contact [5]: **a** Contour plot of the von misses stress and **b** Non-uniformly refined collocation points near contact area

collocation points were refined near the strong western boundary layer as shown in Fig. 14 of [6]. As summarized in Table 6 of [6], higher accuracy with uniform and non-uniform local refinements of collocation points was obtained than uniformly and non-uniformly distributed collocation points without local refinement. In addition, the strong form used in this study was the fourth-order partial differential equation with the strong western boundary layer. They showed that a sixth-order Taylor polynomial instead of fourth- and fifth-order polynomials is necessary to obtain reliable solutions due to the presence of the western boundary layer.

Another application of the adaptive refinement was done by Lee et al. [8]. They used the uniformly refined collocation points near the crack tip for dynamic crack propagation as shown in Figs. 8 and 13 of [8]. Moreover, the topology change due to crack extension was modeled by simple addition and deletion of collocation points. Good agreement with the analytical solution in terms of dynamic energy release rate and the direction of crack growth was observed with a relatively small number of collocation points.

The meshfree collocation method has been also applied to a phase-field model. Fu et al. [7] employed the method to the three-dimensional time dependent phase-field model for modeling the solidification process of a polycrystalline material along with the temperature change due to the latent heat. The solidification of polycrystalline nickel (Ni) from undercooled melt was selected to demonstrate the robustness and flexibility of the method. In Fig. 3, we display the polycrystalline structure of Ni obtained using non-uniformly distributed collocation points in three-dimension.

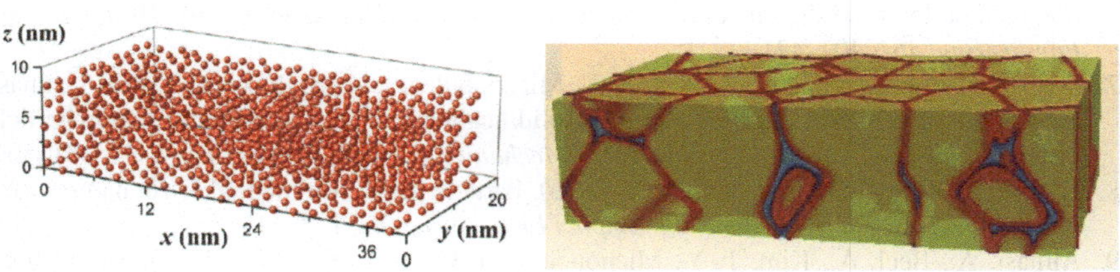

(a) (b)

Fig. 3 Polycrystalline solidification in three-dimension using a phase-field model [7]: **a** Non-uniformly distributed collocation points **b** Solidification of polycrystalline Ni

Furthermore, Almansi et al. [9] performed mechanical analysis of the polycrystalline materials using the polycrystalline structures obtained from the solidification simulation.

Further development and application of the method was done by Yoon and his colleagues. Yoon and Song [2–4] generalized the method to capture weak and strong discontinuities. This generalization was achieved by deriving the derivative approximation including enrichment terms to capture weak and strong discontinuities along the interface or moving boundary. The original derivative approximation (8) can be recovered without enrichment terms. Moreover, Yoon et al. [10] modeled simple material nonlinear problems by directly discretizing the force balance equation using the double derivative approximation without using the second-order derivative approximation. Yoon et al. [11] also applied the method for the simulation of a proportionally damped system subjected to dynamic load and the fracture simulation of cracked concrete beam.

Acknowledgements T.-Y. Kim gratefully appreciates the support from Khalifa University Internal Research Funding (CIRA-2019-009, No. 8474000170).

References

1. Kim, D. W., & Kim, Y. S. (2003). Point collocation methods using the fast moving least square reproducing kernel approximation. *International Journal for Numerical Methods in Engineering, 56,* 1445–1464.
2. Yoon, Y. C., & Song, J. H. (2014). Extended particle difference method for weak and strong discontinuity problems: Part I. Derivation of the extended particle derivative approximation for the representation of weak and strong discontinuities. *Computational Mechanics, 53,* 1087–1103.
3. Yoon, Y. C., & Song, J. H. (2014). Extended particle difference method for weak and strong discontinuity problems: Part II. Formulations and applications for various interfacial singularity problems. *Computational Mechanics, 53,* 1105–1128.
4. Yoon, Y. C., & Song, J. H. (2014). Extended particle difference method for moving boundary problems. *Computational Mechanics, 54,* 723–743.
5. Almasi, A., Kim, T. Y., Laursen, T. A., & Song, J. H. (2019). A strong form meshfree collocation method for frictional contact on a rigid obstacle. *Computer Methods in Applied Mechanics and Engineering, 357,* 112597.
6. Beel, A., Kim, T. Y., Jiang, W., & Song, J. H. (2019). Strong form-based meshfree collocation method for wind-driven ocean circulation. *Computer Methods in Applied Mechanics and Engineering, 351,* 404–421.
7. Song, J. H., Fu, Y., Kim, T. Y., Yoon, Y. C., Michopoulos, J. G., & Rabczuk, T. (2018). Phase field simulations of coupled microstructure solidification problems via the strong form particle difference method. *International Journal of Mechanics and Materials in Design, 14,* 491–509.
8. Lee, S. H., Kim, K. H., & Yoon, Y. C. (2016). Particle difference method for dynamic crack propagation. *International Journal of Impact Engineering, 87,* 132–145.
9. Almasi, A., Beel, A., Kim, T. Y., Michopoulos, J. G., & Song, J. H. (2019). Strong-form collocation method for solidification and mechanical analysis of polycrystalline materials. *Journal of Engineering Mechanics, 145,* 04019082.

10. Yoon, Y. C., Schaefferkoetter, P., Rabczuk, T., & Song, J. H. (2019). New strong formulation for material nonlinear problems based on the particle difference method. *Engineering Analysis with Boundary Elements, 98,* 310–327.
11. Yoon, Y. C., Kim, K. H., & Lee, S. H. (2016). Dynamic particle difference method for the analysis of proportionally damped system and cracked concrete beam. *International Journal of Fracture, 203*(1), 237–262.

A Mixed XFEM Formulation to Simulate Dynamic Wave Propagation in Nearly Incompressible Materials

Verena Klempt and Stefan Löhnert

In 1994 I got to know Peter Wriggers. He used to be my engineering mechanics and computational mechanics teacher, PhD supervisor and mentor during my post-doc time. It has always been a pleasure to work together with him, and I am really grateful that he gave me the opportunity to build up my own research group and to follow my scientific goals. Thanks to his role model function I have chosen my career path. I wish him a Happy Birthday and many more healthy, happy and successful years to come. (Stefan Löhnert).

Abstract A mixed XFEM formulation including inertia terms for the efficient, locking free simulation of cracks and heterogeneities in nearly incompressible materials under dynamic loads is presented. It is shown, that this formulation can accurately predict wave speeds and reflection patterns at heterogeneities in nearly incompressible materials for coarser meshes, than the standard XFEM element with first order shape functions.

1 Introduction

Using lower order finite element formulations to simulate nearly incompressible materials leads to severe locking behaviour. For dynamic problems this results in an overestimation of the wave propagation speed and in incorrect reflection patterns at discontinuities. This can be observed especially for quasi-longitudinal waves, where the main displacement of a material point is parallel to the wave propagation direction, but the material is also able to expand transversally.

V. Klempt (✉) · S. Löhnert
Technische Universität Dresden, Institute of Mechanics and Shell Structures, Dresden, Germany
e-mail: verena.klempt@tu-dresden.de

S. Löhnert
e-mail: stefan.loehnert@tu-dresden.de

F. Aldakheel et al. (eds.), *Current Trends and Open Problems in Computational Mechanics*, https://doi.org/10.1007/978-3-030-87312-7_26

One remedy is the utilisation of mixed methods, based on the Hellinger-Reissner principle, like the so-called Q1P0 element [1]. Despite being a first order element, it shows excellent behaviour due to a mixed approach including an additional element-wise constant volumetric stress approximation.

The eXtended Finite Element Method (XFEM) is well established to simulate heterogeneities and cracks. By means of so called enrichment functions the extended approximation space is equipped to handle discontinuities in the solution field that do not align with the mesh. An overview is given e.g. in [2].

To accurately and efficiently simulate arbitrary discontinuities in nearly incompressible materials, a mixed enriched approach is useful. Such a formulation has been proposed in [3] for two dimensional applications using triangular elements and in [4] for two and three dimensional applications using quadrilateral and hexahedral elements, respectively. As these element formulations employ an enriched ansatz for the displacement field based on bilinear or trilinear shape functions and an enriched ansatz for the volumetric stress field based on piecewise constant shape functions, it is called XQ1XP0 element following the naming of the Q1P0 element. Here, the degrees of freedom for the volumetric stress can be reduced via static condensation just like for the original Q1P0 element. This also includes the degrees of freedom of the volumetric stress enrichment.

In this contribution we present the extension of the XQ1XP0 formulation by inertia terms in order to apply it to dynamic problems. We show that it is better suited to accurately calculate the wave propagation speed and reflection patterns in nearly incompressible materials with heterogeneities than a standard XFEM element based on first order shape functions.

2 The XQ1XP0 Formulation for Small Deformations and Dynamic Problems

Consider a domain $\Omega \subset \mathbb{R}^3$ and its external boundary $\partial\Omega$. The boundary is subdivided into the Neumann boundary $\partial\Omega_\mathrm{t}$ and the Dirichlet boundary $\partial\Omega_\mathrm{u}$ where surface tractions $\bar{\boldsymbol{t}}$ and prescribed displacements $\bar{\boldsymbol{u}}$ are imposed, respectively. Therefore $\partial\Omega = \partial\Omega_\mathrm{t} \cup \partial\Omega_\mathrm{u}$ applies. The initial boundary value problem that needs to be solved fot the time dependent displacement field $\boldsymbol{u} = \boldsymbol{u}(\boldsymbol{x}, t)$ is given by

$$\mathrm{div}[\boldsymbol{\sigma}] + \boldsymbol{f} = \rho\ddot{\boldsymbol{u}} \qquad \text{in } \Omega \tag{1}$$

$$\boldsymbol{\sigma} \cdot \boldsymbol{n} = \bar{\boldsymbol{t}} \qquad \text{on } \partial\Omega_t \tag{2}$$

$$\boldsymbol{u} = \bar{\boldsymbol{u}} \qquad \text{on} \partial\Omega_u \tag{3}$$

$$\boldsymbol{u} = \boldsymbol{u}_0 \qquad \text{in } \Omega \text{ at } t = t_0 \tag{4}$$

$$\dot{\boldsymbol{u}} = \dot{\boldsymbol{u}}_0 \qquad \text{in } \Omega \text{ at } t = t_0. \tag{5}$$

Here $\boldsymbol{\sigma}$ describes the Cauchy stress tensor field, $\boldsymbol{f}$ is the body force density vector, ρ is the mass density of the body, $\dot{\boldsymbol{u}} = \frac{\mathrm{d}\boldsymbol{u}}{\mathrm{d}t}$ and $\ddot{\boldsymbol{u}} = \frac{\mathrm{d}^2\boldsymbol{u}}{\mathrm{d}t^2}$ are the vectors of velocity and acceleration, respectively, and $\boldsymbol{n}$ is the unit outward normal on the boundary $\partial\Omega_\mathrm{t}$. Assuming small deformations the strain tensor is given by

$$\boldsymbol{\varepsilon} = \boldsymbol{\varepsilon}(\boldsymbol{u}) = \mathrm{grad}^{\mathrm{sym}}[\boldsymbol{u}] \tag{6}$$

along with the constitutive equation

$$\boldsymbol{\sigma} = K\,\mathrm{tr}[\boldsymbol{\varepsilon}]\mathbf{1} + 2\mu\boldsymbol{\varepsilon}^D \quad \text{in } \Omega \tag{7}$$

for linear isotropic elasticity, where K und μ are the bulk modulus and the shear modulus, respectively, and $\mathbf{1}$ is the second order unit tensor. The deviator of the strain tensor $\boldsymbol{\varepsilon}^D$ is defined as

$$\boldsymbol{\varepsilon}^D = \boldsymbol{\varepsilon} - \frac{1}{3}\mathrm{tr}[\boldsymbol{\varepsilon}]\mathbf{1} \quad . \tag{8}$$

Introducing a second variable p, namely the volumetic stress,

$$p = K\,\mathrm{tr}[\boldsymbol{\varepsilon}] \tag{9}$$

the total potential can be expressed as a two field Hellinger-Reissner type formulation. Through variation with respect to the two independent variables $\boldsymbol{u}$ and p, we obtain the equations

$$\int_\Omega \left(\boldsymbol{\varepsilon}^D(\delta\boldsymbol{u}) : \boldsymbol{\sigma}^D(\boldsymbol{u}) + p\,\mathrm{div}[\delta\boldsymbol{u}] + \delta\boldsymbol{u}\cdot\rho\ddot{\boldsymbol{u}}\right)\mathrm{d}\Omega = \int_\Omega \delta\boldsymbol{u}\cdot\boldsymbol{f}\,\mathrm{d}\Omega + \int_{\partial\Omega_\mathrm{t}} \delta\boldsymbol{u}\cdot\bar{\boldsymbol{t}}\,\mathrm{d}\partial\Omega_\mathrm{t} \tag{10}$$

and

$$\int_\Omega \delta p\left(K^{-1}p - \mathrm{tr}[\boldsymbol{\varepsilon}(\boldsymbol{u})]\right)\mathrm{d}\Omega = 0 \quad . \tag{11}$$

The deviatoric stresses are calculated with $\boldsymbol{\sigma}^D = 2\mu\boldsymbol{\varepsilon}^D$. To account for discontinuities in the displacement field the XFEM is used for discretisation. Due to geometrically dependent enrichment functions, the extended ansatz is able to reproduce discontinuities virtually independently of the mesh. It reads

$$\boldsymbol{u}^h = \sum_{i\in I} N_i(\boldsymbol{\xi})\,\boldsymbol{u}_i + \sum_{j\in J} N_j(\boldsymbol{\xi})\left(\sum_{k=1}^{n_x^u} f_k(\boldsymbol{\xi})\,\boldsymbol{a}_{jk}\right) \tag{12}$$

where N_i are the shape functions, $\boldsymbol{\xi}$ are the reference coordinates, I is the set of all nodes in the domain, J is the set of enriched nodes, n_x^u is the number of displacement enrichments used, f_k are the enrichment functions, chosen for the particular discontinuity and $\boldsymbol{u}_i$ and $\boldsymbol{a}_{jk}$ are the standard and the enriched degrees of freedom corresponding to the respective enrichment function. The virtual displacement $\delta \boldsymbol{u}^h$ and the acceleration $\ddot{\boldsymbol{u}}^h$ are discretised accordingly.

In general, not only the displacement field may have mesh independent strong or weak discontinuities. The volumetric stress may also be discontinuous within an element. Therefore, the volumetric stress is enriched as well:

$$p^h = p^e + \sum_{m=1}^{n_x^p} g_m(\boldsymbol{\xi})\, q_m^e \quad . \tag{13}$$

Here, p^e is the same element-wise constant volumetric stress as used for the Q1P0 element. Additionally, the enrichment functions g_m together with their corresponding degrees of freedom q_m^e account for the discontinuity. n_x^p is the number of used enrichment functions for the volumetric stress ansatz. Again, the virtual volumetric stress δp^h is discretised accordingly.

Defining the vectors

$$\hat{\boldsymbol{u}}_i = \begin{pmatrix} \boldsymbol{u}_i \\ \boldsymbol{a}_{j1} \\ \boldsymbol{a}_{j2} \\ \vdots \\ \boldsymbol{a}_{jn_x^u} \end{pmatrix}, \; \ddot{\hat{\boldsymbol{u}}}_i = \begin{pmatrix} \ddot{\boldsymbol{u}}_i \\ \ddot{\boldsymbol{a}}_{j1} \\ \ddot{\boldsymbol{a}}_{j2} \\ \vdots \\ \ddot{\boldsymbol{a}}_{jn_x^u} \end{pmatrix}, \; \delta\hat{\boldsymbol{u}}_i = \begin{pmatrix} \delta\boldsymbol{u}_i \\ \delta\boldsymbol{a}_{j1} \\ \delta\boldsymbol{a}_{j2} \\ \vdots \\ \delta\boldsymbol{a}_{jn_x^u} \end{pmatrix}, \; \hat{\boldsymbol{N}}_i = \begin{pmatrix} N_i \\ N_i f_1 \\ N_i f_2 \\ \vdots \\ N_i f_{n_x^u} \end{pmatrix} \tag{14}$$

and

$$\hat{\boldsymbol{p}}^e = \begin{pmatrix} p^e \\ q_1^e \\ q_2^e \\ \vdots \\ q_{n_x^p}^e \end{pmatrix}, \; \delta\hat{\boldsymbol{p}}^e = \begin{pmatrix} \delta p^e \\ \delta q_1^e \\ \delta q_2^e \\ \vdots \\ \delta q_{n_x^p}^e \end{pmatrix}, \; \hat{\boldsymbol{g}} = \begin{pmatrix} 1 \\ g_1 \\ g_2 \\ \vdots \\ g_{n_x^p} \end{pmatrix} \tag{15}$$

facilitates writing the ansatz functions in shorter notation

$$\boldsymbol{u}^h = \sum_{i\in I} \hat{\boldsymbol{N}}_i \cdot \hat{\boldsymbol{u}}_i\,, \quad \ddot{\boldsymbol{u}}^h = \sum_{i\in I} \hat{\boldsymbol{N}}_i \cdot \ddot{\hat{\boldsymbol{u}}}_i \quad \text{and} \quad \delta\boldsymbol{u}^h = \sum_{i\in I} \hat{\boldsymbol{N}}_i \cdot \delta\hat{\boldsymbol{u}}_i \tag{16}$$

as well as

$$\boldsymbol{p}^h = \hat{\boldsymbol{g}}_i \cdot \hat{\boldsymbol{p}}^e \quad \text{and} \quad \delta\boldsymbol{p}^h = \hat{\boldsymbol{g}}_i \cdot \delta\hat{\boldsymbol{p}}^e \quad . \tag{17}$$

The strains can be computed by $\boldsymbol{\varepsilon}^h = \sum_{i\in I} \hat{\boldsymbol{B}}_i \cdot \hat{\boldsymbol{u}}_i$ with the definition of the matrix $\hat{\boldsymbol{B}}_i = \text{grad}^{\text{sym}}[\hat{\boldsymbol{N}}_{\text{i}}]$.

Let n_n be the number of nodes per element. The discretised form of Eq. (11) for one element with its volume Ω^e is then

$$\delta \hat{\boldsymbol{p}}^e \cdot \int_{\Omega^e} \left(K^{-1} \hat{\boldsymbol{g}} \otimes \hat{\boldsymbol{g}} \cdot \hat{\boldsymbol{p}}^e - \hat{\boldsymbol{g}}\, \mathrm{tr}\left[\sum_{i=1}^{n_n} \hat{\boldsymbol{B}}_i \cdot \hat{\boldsymbol{u}}_i \right] \right) \mathrm{d}\Omega^e = 0 \tag{18}$$

and can therefore be rewritten as

$$\hat{\boldsymbol{p}}^e = \boldsymbol{M}^{-1} \cdot \int_{\Omega^e} \hat{\boldsymbol{g}}\, \mathrm{tr}\left[\sum_{i=1}^{n_n} \hat{\boldsymbol{B}}_i \cdot \hat{\boldsymbol{u}}_i \right] \mathrm{d}\Omega^e \tag{19}$$

using the coefficient matrix

$$\boldsymbol{M} = \int_{\Omega^e} K^{-1} \hat{\boldsymbol{g}} \otimes \hat{\boldsymbol{g}}\, \mathrm{d}\Omega^e \ . \tag{20}$$

Equation (19) shows that it is possible to solve for the variables $\hat{\boldsymbol{p}}^e$ on element level. This static condensation procedure is similar to the one used within the Q1P0 element formulation and has the advantage that no additional degrees of freedom for the volumetric stress need to be introduced in the global equation system. As a consequence the discretised weak form can be written independent of the variables defining the volumetric stress:

$$\begin{aligned}
&\sum_{i,j=1}^{n_n} \delta \hat{\boldsymbol{u}}_i \cdot \int_{\Omega^e} \hat{\boldsymbol{B}}_i : \mu\left(\mathcal{I} - \frac{1}{3} \mathbf{1} \otimes \mathbf{1} \right) : \hat{\boldsymbol{B}}_j\, \mathrm{d}\Omega^e \cdot \hat{\boldsymbol{u}}_j \\
&+ \int_{\Omega^e} \hat{\boldsymbol{g}}\, \mathrm{tr}\left[\sum_{i=1}^{n_n} \hat{\boldsymbol{B}}_i \cdot \delta \hat{\boldsymbol{u}}_i \right] \mathrm{d}\Omega^e \cdot \boldsymbol{M}^{-1} \cdot \int_{\Omega^e} \hat{\boldsymbol{g}}\, \mathrm{tr}\left[\sum_{j=1}^{n_n} \hat{\boldsymbol{B}}_j \cdot \hat{\boldsymbol{u}}_j \right] \mathrm{d}\Omega^e \\
&+ \sum_{i,j=1}^{n_n} \delta \hat{\boldsymbol{u}}_i \cdot \int_{\Omega^e} \hat{\boldsymbol{N}}_i \otimes \hat{\boldsymbol{N}}_j\, \mathrm{d}\Omega^e \cdot \ddot{\hat{\boldsymbol{u}}}_j \\
&= \sum_{i=1}^{n_n} \delta \hat{\boldsymbol{u}}_i \cdot \int_{\Omega^e} \hat{\boldsymbol{N}}_i \cdot \boldsymbol{f}\, \mathrm{d}\Omega^e + \sum_{i=1}^{n_n} \delta \hat{\boldsymbol{u}}_i \cdot \int_{\partial \Omega^e_t} \hat{\boldsymbol{N}}_i \cdot \bar{\boldsymbol{t}}\, \mathrm{d}\partial\Omega^e_t \ .
\end{aligned} \tag{21}$$

Here, $\mathcal{I}$ is the symmetric fourth order unit tensor.

3 Quasi-longitudinal Wave Reflection Pattern of a Cylindrical Heterogeneity

In [4] the XQ1XP0 element has been tested with various quasi-static examples. It has been shown to exhibit superior convergence properties compared to the standard XFEM formulation. In the following chapter, however, we will show, that the XQ1XP0 element is also capable of reproducing accurate wave propagation and accurate wave reflection patterns at a discontinuity in nearly incompressible materi-

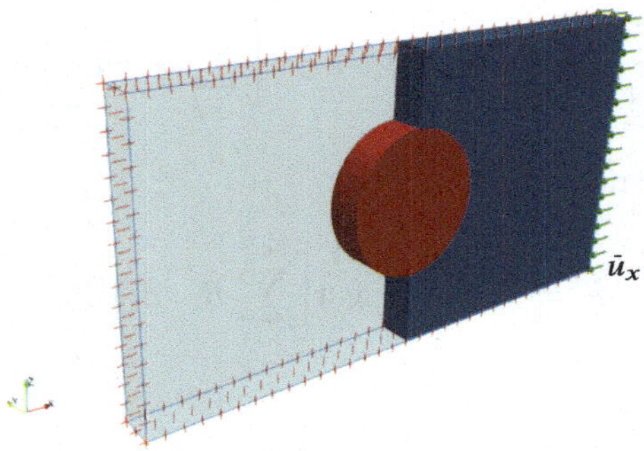

Fig. 1 Problem setup for the example with a cylindrical heterogeneity in a nearly incompressible matrix material

als for dynamic problems. If the displacement of a material point is parallel to the propagation direction of the wave, the wave is called a longitudinal wave. In nearly incompressible materials they propagate with extreme speed due to the limitation of volumetric change. However, if the body is able to expand transversally and the wavelenght is chosen big enough compared to the transverse body dimension, the wave slows down considerably. It is now called a quasi-longitudinal wave.

A body with the dimensions 100 mm × 6.25 mm × 50 mm and a cylindrical inclusion having a diameter of $d = 25$ mm is considered, see Fig. 1 also for the boundary conditions. To avoid multiple reflection patterns which distract from the reflections due to the heterogeneity, the z-displacement at the upper and lower boundary is impeded. Due to the small dimension in y-direction, the body is able to easily expand transversally. The matrix material is chosen to be nearly incompressible rubber and the inclusion is steel. The material properties can be found in Table 1.

To reproduce the weak discontinuity in the displacement field and in the volumetric stress field at the interface between matrix and inclusion, the modified absolute value enrichment is used, see [5] and [4] for further information.

The displacements $\bar{u}_x = 1$ mm are linearly imposed in ten time steps on the right boundary and then reduced to zero in the next ten time steps. After that, the applied displacements on the boundary are kept zero. This ensures that the wave lenght is larger than the dimension in y-direction. The time step size is chosen to be $\Delta t = 1.9765 \cdot 10^{-5}\,s$. For the time integration the Newmark scheme is employed with the parameters $\beta = 0.25$ and $\gamma = 0.5$.

Table 1 Material parameters for the nearly incompressible rubber matrix and the steel inclusion

Parameter	Unit	Rubber (blue)	Steel (red)
μ	$\frac{\text{kN}}{\text{mm}^2}$	0.0003	79.3
K	$\frac{\text{kN}}{\text{mm}^2}$	0.1499	171.817
ρ	$\frac{\text{kg}}{\text{mm}^3}$	$0.9 \cdot 10^{-6}$	$7.85 \cdot 10^{-6}$
c	$\frac{\text{mm}}{\text{s}}$	31 619	5 124 936

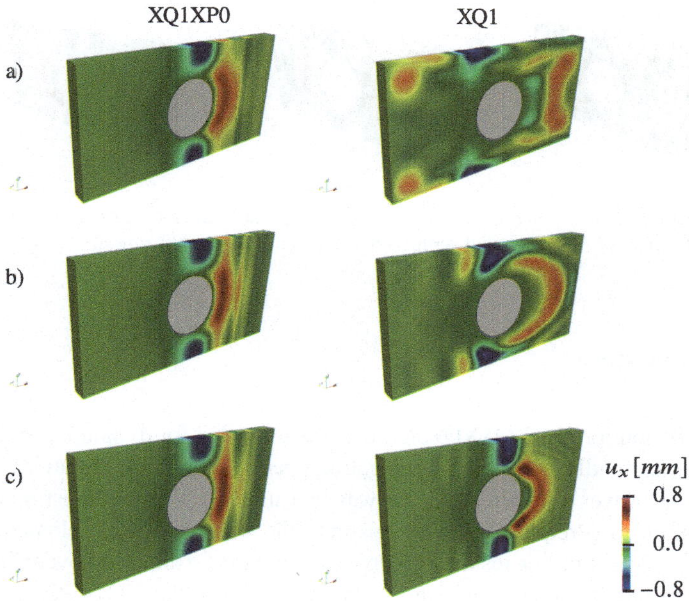

Fig. 2 Displacement in x-direction after 80 time steps (left: XQ1XP0, right: XQ1) with **a** 8 192 elements **b** 65 536 elements **c** 524 288 elements

Table 2 Approximate wave propagation speed for the XQ1XP0 element and the XQ1 element for three different discretisations

Element	(a)	(b)	(c)
XQ1XP0	31 622 $\frac{\mathrm{mm}}{\mathrm{s}}$	31 620 $\frac{\mathrm{mm}}{\mathrm{s}}$	31 620 $\frac{\mathrm{mm}}{\mathrm{s}}$
XQ1	55 654 $\frac{\mathrm{mm}}{\mathrm{s}}$	43 005 $\frac{\mathrm{mm}}{\mathrm{s}}$	34 784 $\frac{\mathrm{mm}}{\mathrm{s}}$

In Fig. 2 the displacement in x-direction after 80 timesteps is shown for the XQ1XP0 formulation and the standard XFEM formulation (XQ1) for different mesh resolutions. One can see that the XQ1 element leads to a too fast wave propagation for coarse meshes. The wave has already reached the left boundary and has reflected back. However, with the finest discretisation it has just reached the inclusion. In Table 2 the measured wave propagation speed is shown for the two different XFEM formulations and three different mesh resolutions. One can see that the mixed XQ1XP0 formulation leads to a much faster convergence than the XQ1 formulation.

In Fig. 3 the results for the XQ1 element for the different discretisations at different points in time are shown, such that the wave has approximately travelled equally far in each simulation. Still, the reflexion patterns due to the heterogeneity change significantly with increasing refinement and converge to the results of the XQ1XP0 element, for which the reflexion patterns are relatively similar for every refinement step, cf. Fig. 2 on the left.

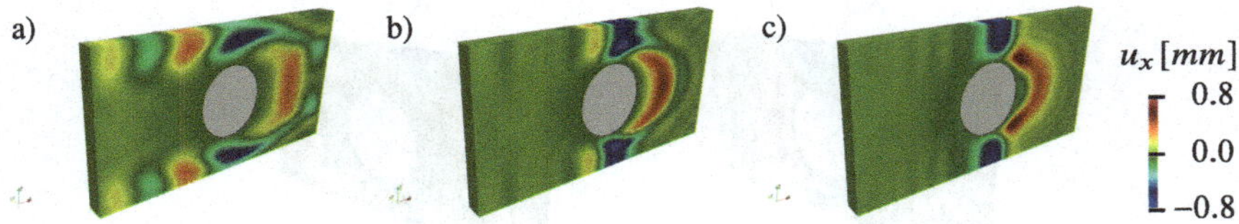

Fig. 3 Displacement in x-direction for the XQ1 element **a** after 61 timesteps and with 8 192 elements **b** after 68 timesteps and with 65 536 elements **c** after 80 timesteps and with 524 288 elements

4 Conclusions

In this contribution the XQ1XP0 element was extended for dynamic problems. Furthermore, its handling of wave propagation speed and reflection patterns of quasi-longitudinal waves at heterogeneities was investigated. While almost no additional numerical effort is required in comparison to the standard XFEM element, this formulation shows superior results. Where the standard formulation overestimates the wave propagation speed even using far more elements, the XQ1XP0 element does not and is therefore also capable of accurately depicting reflection patterns even for coarse meshes. In the present contribution only results for the simulation of heterogeneities are shown. However, similarly good behaviour can be observed for the application of the mixed XFEM formulation to dynamic wave propagation in nearly incompressible fractured media.

References

1. Simo, J. C., Taylor, R. L., & Pister, K. S. (1985). Variational and projection methods for the volume constraint in finite deformation elasto-plasticity. *Computer Methods in Applied Mechanics and Engineering*. https://doi.org/10.1016/0045-7825(85)90033-7
2. Fries, T.-P., & Belytschko, T. (2010). The extended/generalized finite element method: An overview of the method and its applications. *International Journal for Numerical Methods in Engineering*. https://doi.org/10.1002/nme.2914
3. Legrain, G., Moës, N., & Huerta, A. (2008). Stability of incompressible formulations enriched with X-FEM. *Computer Methods in Applied Mechanics and Engineering*. https://doi.org/10.1016/j.cma.2007.08.032
4. Loehnert, S., & Munk, L. (2020). A mixed extended finite element for the simulation of cracks and heterogeneities in nearly incompressible materials and metal plasticity. *Engineering Fracture Mechanics*. https://doi.org/10.1016/j.engfracmech.2020.107217
5. Moës, N., Cloirec, M., Cartraud, P., & Remacle, J.-F. (2003). A computational approach to handle complex microstructure geometries. *Computer Methods in Applied Mechanics and Engineering*. https://doi.org/10.1016/S0045-7825(03)00346-3

What Machine Learning Can Do for Computational Solid Mechanics

Siddhant Kumar and Dennis M. Kochmann

"Most people say that it is the intellect which makes a great scientist. They are wrong: it is character." (A. Einstein) Here's to a notable scientist and great character on the occasion of his 70th birthday – happy birthday, Peter! (D. M. Kochmann).

Abstract Machine learning has found its way into almost every area of science and engineering, and we are only at the beginning of its exploration across fields. Being a popular, versatile and powerful framework, machine learning has proven most useful where classical techniques are computationally inefficient, which applies particularly to computational solid mechanics. Here, we dare to give a non-exhaustive overview of potential avenues for machine learning in the numerical modeling of solids and structures and offer our (subjective) perspective on what is yet to come.

1 Introduction

We are at the beginning of an exciting era in which the numerical solution of complex problems as well as the design and discovery of (meta-)materials and the solution of inverse problems are accelerated by big data and advances in machine learning (ML) strategies. Contrary to traditional approaches based on intricate theoretical insight, exhaustive simulations and/or experiments, data-driven learning methods exploit large data sets to identify otherwise unknown relations, thus creating efficient and invertible maps between input and output parameters – such as simulation parameters

S. Kumar (✉)
Department of Materials Science and Engineering, Delft University of Technology, 2628CD Delft, The Netherlands
e-mail: Sid.Kumar@tudelft.nl

D. M. Kochmann
Mechanics & Materials Lab, Department of Mechanical and Process Engineering, ETH Zürich, 8092, Zürich, Switzerland
e-mail: dmk@ethz.ch

F. Aldakheel et al. (eds.), *Current Trends and Open Problems in Computational Mechanics*, https://doi.org/10.1007/978-3-030-87312-7_27

and resulting outcomes, or (meta-)material structure and resulting properties. The latter particularly offers a quick identification and reduction of the design space, eventually leading to a faster turnaround of material design by demand.

2 Material Modeling

The computational modeling of materials and structures has relied on solving century-old partial differential equations (PDEs) with ever improved numerical techniques. Especially when it comes to multiscale problems, the computational expense of simulations – involving coupled problems on various length and/or time scales – has remained a prohibitive challenge. ML strategies are promising for significantly accelerating not only multiscale simulations but also for identifying new constitutive models as well as for solving systems of PDEs (see Fig. 1).

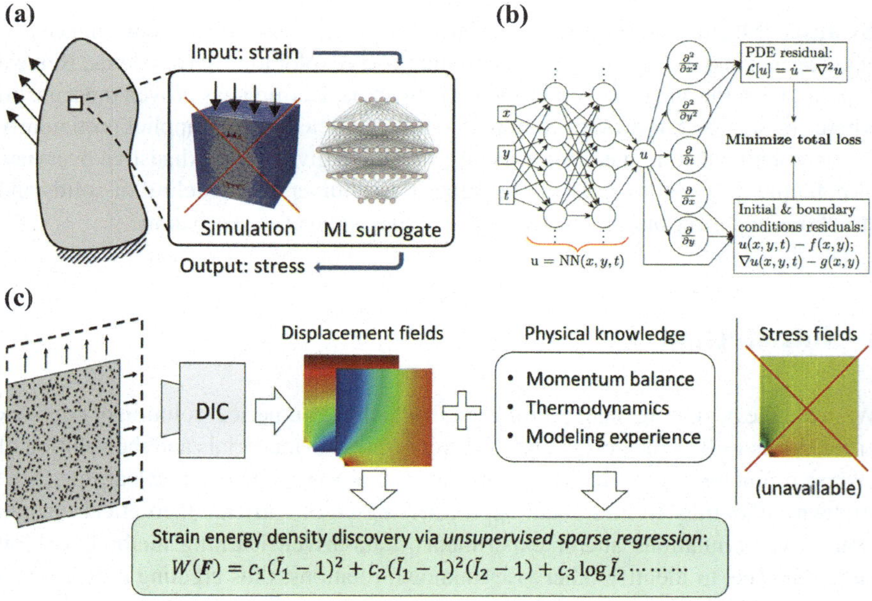

Fig. 1 Representative examples of ML applications in material modeling. **a** ML-based surrogate models for bypassing microscale simulations in multiscale systems. **b** PINNs [1, 2] for solving PDEs (e.g., two-dimensional heat equation shown here). **c** Unsupervised discovery of interpretable and parsimonious constitutive laws using only displacement data and physical knowledge [3]

2.1 Accelerating Multiscale Simulations

Data-driven and ML-based methods have been most effective in overcoming the computational limitations of multiscale simulations by bypassing expensive lower-scale calculations and thereby accelerating macroscale simulations (Fig. 1a). Assuming a separation of scales between the microstructure and the macroscale boundary value problem (BVP), the effective constitutive response from the microscale is approximated by a pre-trained ML model to replace costly representative volume element (RVE) simulations. In computational mechanics, the focus has mostly been on *surrogate models* for the homogenized microscale stress-strain response in FE^2-type settings. Commonly, this has relied on supervised ML models which approximate the constitutive manifold based on training examples of stress-strain pairs. Recent work [4–6] has developed surrogate constitutive models that capture strong nonlinearity as well as path dependence (e.g., learning plastic yield surfaces for complex composites and polycrystalline metals). They were also shown to generalize well to arbitrary strain paths and are easily integrated into finite-element (FE) frameworks. Towards accelerated macroscale simulations, Capuano and Rimoli [7] developed a general ML-driven *smart finite element* method, which bypasses the need for explicitly solving internal displacement fields within an element while enforcing physical constraints. All such approaches rely on learning a model a-priori (offline) from homogenized RVE data and, consequently, require a large number of computationally expensive RVE-level simulations to obtain sufficient data for training the model. Alternatively, surrogate modeling methods can also leverage the spatial structure of the RVEs to reduce the training data requirements. Recent techniques such as graph neural networks and nonlocal pooling may take advantage of the spatial proximity of grains in polycrystalline systems to learn their constitutive response [8] and texture evolution [9] in a data-efficient manner.

2.2 Data-Driven Constitutive Models: Beyond Simulation-Based Training

While the above approaches rely heavily on RVE simulation data to learn the constitutive response, recent work on *model-free data-driven approaches* [10, 11] take a different path by using experimental data directly instead of building surrogate models. Aiming at avoiding physical-modeling biases, this results in a purely data-driven framework that searches for the closest stress-strain pair (within the dataset) which satisfies the conservation laws and compatibility conditions. These approaches have even been extended to the modeling of inelasticity [12] and fracture [13]. Since interpolation-based modeling is generally avoided, current challenges include generalization and extrapolation capabilities to data not available in the (training) dataset, especially when dealing with high-dimensional and noisy data. In contrast to the purely model-free philosophy is the approach of physics-informed discovery of *inter-*

pretable constitutive models. The strategy is to use experience in physical modeling to reduce the dependence on data and improve generalizability. The most popular approach has relied on sparse regression to discover a mathematically and physically interpretable form of the underlying constitutive equation [3, 14] (in contrast to, e.g., black-box neural networks). The method involves creating an exhaustive library of mathematical functions and then sparsely selecting the combination of those which best explains the data. The motivation behind sparsity originates from the principle of Occam's razor: most physical models in nature are parsimonious in description. One of the biggest challenges in realizing data-driven constitutive models for practical application lies in material characterization; e.g., using digital image correlation (DIC) full-field strain maps become available but stress data is hard to obtain. To this end, recent work [3] (Fig. 1c) discovered interpretable constitutive models in an unsupervised manner – i.e., without stress data – which presents a promising departure from curve-fitting and supervised learning based on stress data.

2.3 Learning to Solve PDEs

ML has also been successfully deployed to bypass the expensive numerical solution of highly nonlinear and high-dimensional BVPs. The goal is to achieve real-time solutions for a given family of BVPs with applications ranging from surgical robots to time-critical control systems. Initial attempts in this direction relied on supervised training based on thousands of BVP solutions with different boundary conditions and/or material inhomogeneity. Those ML approaches treat the inputs (boundary conditions and material distribution) and outputs (solution fields) as images and reduce the problem to image-to-image regression, using, e.g., convolutional neural networks [15]. However, such methods are data-intensive and showed limited generalization to unseen data. Additionally, the learnt model only applies to the mesh resolution of the training data. To address those limitations, a new class of ML methods, *physics-informed neural networks* (PINNs) [1, 2] (Fig. 1b), has been proposed to solved BVPs in an unsupervised manner, i.e., using data without labels. Analogous to, e.g., the conventional Rayleigh-Ritz method, a neural network as a function of time and position is chosen as the ansatz for the solution field. Its trainable parameters are tuned to minimize the loss in the strong or weak form of the PDE and satisfaction of the boundary conditions. The major advantage of this approach is that it is unsupervised, i.e., it does not require any training dataset generated by expensive simulations. Instead, physical constrains (via the PDE-based loss) are exploited to replace the training data. Leveraging the differentiability of neural networks, this approach can also been used to calibrate parametric models of PDEs based on the observed solutions [16]. Though PINNs are successful in learning to solve BVPs, they suffer from mesh dependence of the discretization used to evaluate the PDE loss during training [17]. To this end, new methods [17, 18] are being developed for operator learning, which aims to learn the mapping in the functional space and avoid mesh dependence. In computational solid mechanics, currently both PINNs and operator learning methods have been applied only to simple PDEs in regular domains but are promising for future application to complex material systems such as those with sharp discontinuities and localization, multiscale metallic systems, or loss of uniqueness in the solutions (e.g., see [19–21]).

3 Design of (meta-)materials

Tailoring a material's microstructure to achieve targeted properties is a major challenge in natural material systems due to physical limitations. Mechanical metamaterials [22] based on random or periodic truss, plate or shell architectures overcome those limitations by letting an engineered microscale produce novel and exciting effective macroscale material properties. The natural challenge that follows is the inverse design: how can one efficiently reverse-engineer a metamaterial's design to achieve a certain properties? Traditionally, material design has heavily relied on trial and error and intuition, which is expensive and explores only a small subset of the design space. This calls for new ML-assisted computational design approaches.

3.1 Accelerating Topology Optimization

Topology optimization [23] (e.g., based on the Solid Isotropic Material with Penalization method, SIMP) has become a standard for designing structures that optimize design objectives such as minimum compliance or tailored anisotropy. This is achieved by optimizing the material distribution (solid or void) across all voxels in a discretized domain. Analogously, spatially-variant metamaterials can be designed by optimizing the design parameter(s) for each voxel. Since most metamaterials do not admit (semi-)analytical expressions of their homogenized responses, the latter requires expensive nested simulations based on lower-scale RVE calculations. The computational expenses are further compounded by sensitivity calculations, which rely on computing numerical derivatives (i.e., perturbing the design parameters and re-running costly RVE simulations). Similar to accelerating multiscale simulations, ML-based *models for structure-property relations* have proven beneficial for the topology optimization of metamaterials [24, 25]. Deep neural networks are particularly advantageous in learning high-dimensional and highly nonlinear structure-property maps. They also naturally provide the exact sensitivities using automatic differentiation (which forms the backbone of neural network training by back-propagation) and avoid the computational costs and precision issues associated with numerical differentiation. Sobolev training [26], wherein sensitivity information is used for learning, can improve the accuracy, as shown, e.g., for the elastic stiffness of trusses [24]. A key open challenge arises from the topology optimization of spatially-variant structures, where prior approaches have optimized microstructural design parameters but within a limited design space since small-scale unit cells must

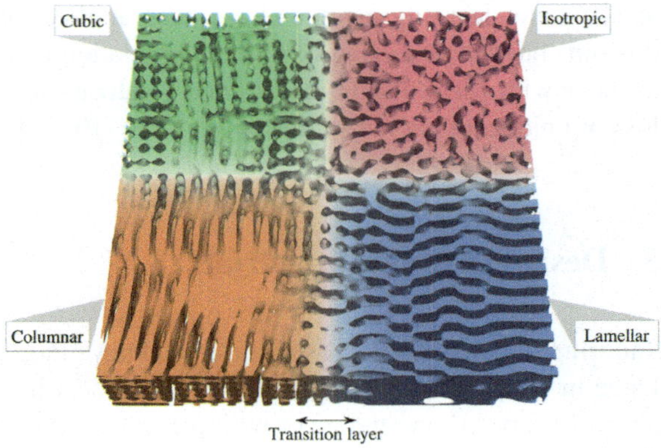

Fig. 2 Spindoid metamaterials, based on an approximation of structures observed during anisotropic spinodal decomposition, offer seamlessly tunable elastic anisotropy (including, e.g., cubic-, columnar-, isotropic-, and lamellar-type topologies) and functional grading [27]

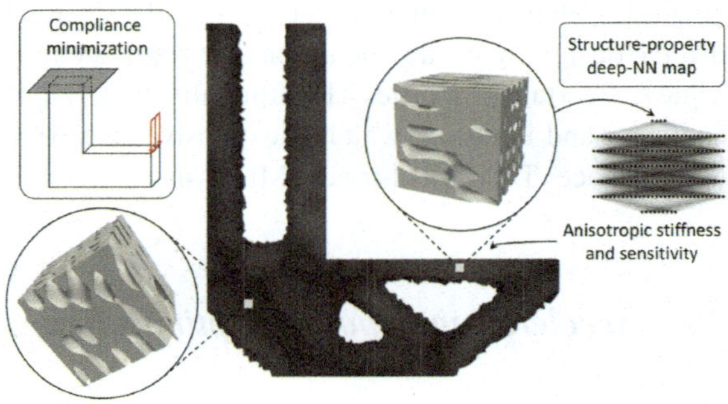

Fig. 3 Topology optimization of spinodoid metamaterials for minimum compliance [28]: a deep neural network provides the map from the topology to anisotropic stiffness and sensitivity

be compatible for practical purposes to form a macroscale structure (e.g., knowing the optimal truss unit cell at each point within a macroscale body does not ensure that a smooth grating between the distinct unit cells exists). As a remedy, we recently introduced *spinodoid metamaterials* [27] (Fig. 2), which are composed of smooth bi-continuous topologies inspired by those observed in the naturally-occurring process of spinodal decomposition. Spinodoids possess a tremendous design space of anisotropic topological and mechanical properties. Their non-periodicity is a departure from periodic unit cells and avoids tessellation-related issues while admitting seamless transitions in between different anisotropic designs. This allows for combined topology optimization [28] (Fig. 3) of local microstructural design parameters (and in turn, the local anisotropy), material distribution (solid vs. void), and material orientation, yielding better designs than SIMP. This is made possible by an efficient deep neural network-based surrogate model for the anisotropic design-stiffness map, which is particularly advantageous in strongly nonlinear and multiply-connected design space of spinodoids.

3.2 *Efficiently Exploring Design Spaces*

Exploring or searching a design space to identify optimal designs is a topical challenge both at the material level (optimizing, e.g., molecular structures for target properties) and at the structural level of metamaterials (optimizing, e.g., stiffness and strength by unit cell design). Searching a design space faces two challenges. First, owing to its high dimensionality, the size of a design space increases exponentially with the number of design parameters, which presents computational limitations on the amount of data that can be generated for learning surrogate models of the structure-property relations. Feature selection techniques (e.g., principal component analysis and autoencoders [29]) can be used to identify the important features and reduce the dimensionality. Alternatively, the regions of the design space relevant to the design objective can be adaptively identified using, e.g., Bayesian and active learning techniques [30, 31]. Second, most design spaces do not admit an explicit design parameterization. E.g., a composite can be represented as a binary image, whose the pixels describe the spatial arrangement of two materials. To this end, generative ML methods such as variational autoencoders (VAE) and generative adversarial networks (GANs) have been used to learn an abstract parameterization of the design space [32, 33]. These are unsupervised ML methods, i.e., they do not require labeled training data and usually involve jointly training a pair of neural networks against each other. E.g., an *encoder* in a VAE abstracts features from the designs and maps them to a lower-dimensional latent space where similar designs are located close to each other. The features must be sufficiently informative to allow reconstructing the design from the latent space parameterization using another neural network called the *decoder*. The latent space, now representing an abstract design parameterization, can be further used as input to a regressor for surrogating structure-property relations. Alternatively, genetic algorithms and gradient-based optimization methods can be used to explore the latent space and search for the optimal design. Generative modeling is a promising direction for designing complex material systems. However, training of such models is usually unstable and remains as an active ML research area. From the perspective of application to (meta-)materials, recent works (e.g. [32, 33]) rely on datasets with millions of samples and are limited to two-dimensional designs. Therefore, future improvements in the scalability of these methods are needed for application to a data-efficient design of material systems with higher-dimensional and more complex topologies.

3.3 *Inverting Structure-Property Maps*

The process of developing (meta-)materials with tailored properties has traditionally followed a *forward design* paradigm: computationally homogenizing (or experimentally testing) a wide range of possible structures results in an effective *structure-to-property map*. Capitalizing on this strategy, methods such as topology optimization

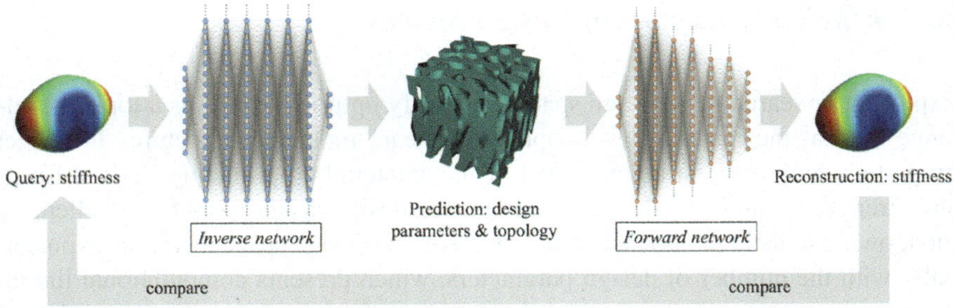

Fig. 4 Schematic of the inverse design framework [27], which has been applied to the design of spinodoid metamaterials with targeted anisotropic elastic stiffness. The stiffness is visualized via the elasticity surface, wherein each point on the surface denotes Young's modulus in that direction

and genetic algorithms are unfortunately computationally expensive and sensitive to the initial design guess. While ML-based generative methods such as GANs and VAEs (Sect. 3.2) reduce computational costs, they still require solving an optimization problem based on the surrogate model to achieve a targeted property, and they may converge to local optima in the design space. A beneficial alternative is the on-demand *inverse design* of optimal architectures, wherein an appropriate design is identified directly which achieves a targeted effective property. Unfortunately, identifying or learning an inverse design map is inherently ill-posed, because multiple topologies may have similar effective properties (e.g., the same elastic stiffness can be achieved by different spinodoid designs [27]). Conventional ML strategies require training with a distance metric in the output (structural design) space – defining how "close" the structures are to each other. Yet, the one-to-many nature of property-to-structure maps is prohibitive in defining such a metric and, in turn, precludes the applicability of related ML strategies.

To render the inverse design problem well-posed, we introduced an ML approach based on the integration of two neural networks for both the forward and inverse problems [27], which is illustrated in Fig. 4 for the example of spinodoid metamaterials with inverse-designed anisotropic elastic stiffness. The *inverse network* takes the target stiffness as an input and predicts a design topology. Of course, this predicted topology can be different from the one that was used to generate the training example. Therefore, we do not compare topologies but their effective properties. Since the forward problem is well-posed (each design has unique properties), we use a *forward network* to serve as a surrogate to RVE-based homogenization (pre-trained using simulation data) and reconstructs the stiffness of any predicted design. The inverse network is then trained to minimize the reconstruction error of the target stiffness, and together with the forward network provides a two-way structure-property map.

This on-demand inverse design paradigm is sufficiently general for application to a wide range of (meta-)materials and effective properties. It can, e.g., benefit the design of patient-specific bone-mimetic scaffolds and implants [27] – bone being highly anisotropic and heterogeneous in its topological and mechanical properties, so that inverse designing bone implants which locally match the native bone properties holds promise for improving the long-term compatibility of bone implants and avoiding atrophy due to property mismatch.

4 Conclusions and Outlook

In this brief contribution, we could only provide a glimpse of the powerful opportunities provided by introducing ML into computational mechanics – from accelerating multiscale modeling and topology optimization to bypassing constitutive modeling or solving complex systems of PDEs all the way to identifying and inverting structure-property maps. To add a personal perspective, ML in the mechanics community is both underrated and overrated at the same time. ML-based models are best viewed as powerful and scalable approximators. However, using ML for just about everything is certainly not advised. We are already seeing the limitations of black-box ML methods when applied to physical problems, particularly related to extrapolation, generalization, computational expense, and amount of training data. The future lies in the integration of classical physics-based methods with ML methods to address these challenges. By endowing interpretability and insights to data-driven models, we can begin to uncover and understand previously unknown physical phenomena, which have been our pursuits for over a century in mechanics.

References

1. Raissi, M., Perdikaris, P., & Karniadakis, G. (2017). Physics informed deep learning (part i): Data-driven solutions of nonlinear partial differential equations. arXiv:1711.10561.
2. Raissi, M., Perdikaris, P., Karniadakis, G. E. (2019). Physics-informed neural networks: A deep learning framework for solving forward and inverse problems involving nonlinear partial differential equations. *Journal of Computational Physics, 378*, 686–707.
3. Flaschel, M., Kumar, S., & De Lorenzis, L. (2021). Unsupervised discovery of interpretable hyperelastic constitutive laws. *Computer Methods in Applied Mechanics and Engineering, 381*, 113852. https://doi.org/10.1016/j.cma.2021.113852
4. Huang, D., Niklas Fuhg, J. Weißenfels, & Wriggers, P. (2020). A machine learning based plasticity model using proper orthogonal decomposition. *Computer Methods in Applied Mechanics and Engineering, 365*, 113008.
5. Mozaffar, M., Bostanabad, R., Chen, W., Ehmann, K., Cao, J., & Bessa, M. A. (2019). Deep learning predicts path-dependent plasticity. *Proceedings of the National Academy of Sciences, 116*(52), 26414–26420.
6. Reimann, D., Nidadavolu, K., ul Hassan, H., Vajragupta, N., Glasmachers, T., Junker, P. et al. (2019). Modeling macroscopic material behavior with machine learning algorithms trained by micromechanical simulations. *Frontiers in Materials, 6*, 181.
7. Capuano, German, & Rimoli, Julian J. (2019). Smart finite elements: A novel machine learning application. *Computer Methods in Applied Mechanics and Engineering, 345*, 363–381.
8. Vlassis, Nikolaos N., Ma, Ran, & Sun, WaiChing. (2020). Geometric deep learning for computational mechanics Part I: anisotropic hyperelasticity. *Computer Methods in Applied Mechanics and Engineering, 371*, 113299.

9. Pandey, Anup, & Pokharel, Reeju. (2021). Machine learning based surrogate modeling approach for mapping crystal deformation in three dimensions. *Scripta Materialia, 193*, 1–5.
10. Ibañez, R., Abisset-Chavanne, E., Vicente Aguado, J., Gonzalez, D., Cueto, E. Chinesta, F. et al. (2018). A manifold learning approach to data-driven computational elasticity and inelasticity. *Archives of Computational Methods in Engineering, 25*(1), 47–57.
11. Kirchdoerfer, T., & Ortiz, M. (2016). Data-driven computational mechanics. *Computer Methods in Applied Mechanics and Engineering, 304*, 81–101.
12. Eggersmann, R., Kirchdoerfer, T., Reese, S., Stainier, L., & Ortiz, M. (2019). Model-free data-driven inelasticity. *Computer Methods in Applied Mechanics and Engineering, 350*, 81–99.
13. Carrara, P., De Lorenzis, L., Stainier, L., & Ortiz, M. (2020). Data-driven fracture mechanics. *Computer Methods in Applied Mechanics and Engineering, 372*, 113390.
14. Brunton, S. L., Proctor, J. L., & Nathan Kutz, J. (2016). Discovering governing equations from data by sparse identification of nonlinear dynamical systems. *Proceedings of the National Academy of Sciences, 113*(15), 3932–3937.
15. Zhu, Yinhao, & Zabaras, Nicholas. (2018). Bayesian deep convolutional encoder-decoder networks for surrogate modeling and uncertainty quantification. *Journal of Computational Physics, 366*, 415–447.
16. Raissi, M., Perdikaris, P., & Karniadakis, G. (2017). Physics informed deep learning (part ii): Data-driven discovery of nonlinear partial differential equations. arXiv:1711.10566.
17. Bhattacharya, K., Hosseini, B., Kovachki, N. B., & Stuart, A. (2020). Model reduction and neural networks for parametric pdes. arXiv:2005.03180.
18. Li, Z-Y., Kovachki, N. B., Azizzadenesheli, K., Liu, B., Bhattacharya, K., Stuart, A. et al. (2020). Fourier neural operator for parametric partial differential equations. arXiv:2010.08895.
19. Kumar, S., Tutcuoglu, A. D., Hollenweger, Y., & Kochmann, D. M. (2020). A meshless multi-scale approach to modeling severe plastic deformation of metals: Application to ECAE of pure copper. *Computational Materials Science, 173*, 109329.
20. Kumar, S., Vidyasagar, A., & Kochmann, D. M. (2020). An assessment of numerical techniques to find energy-minimizing microstructures associated with nonconvex potentials. *International Journal for Numerical Methods in Engineering, 121*(7), 1595–1628.
21. Tutcuoglu, A. D., Hollenweger, Y., Stoy, A., & Kochmann, D. M. (2019). High- vs. low-fidelity models for dynamic recrystallization in copper. *Materialia, 7*, 100411.
22. Kochmann, Dennis M., Hopkins, Jonathan B., & Valdevit, Lorenzo. (2019). Multiscale modeling and optimization of the mechanics of hierarchical metamaterials. *MRS Bulletin, 44*(10), 773–781.
23. Sigmund, Ole, & Maute, Kurt. (2013). Topology optimization approaches. *Structural and Multidisciplinary Optimization, 48*(6), 1031–1055.
24. White, Daniel A., Arrighi, William J., Kudo, Jun, & Watts, Seth E. (2019). Multiscale topology optimization using neural network surrogate models. *Computer Methods in Applied Mechanics and Engineering, 346*, 1118–1135.
25. Zhang, Yan, Li, Hao, Xiao, Mi., Gao, Liang, Chu, Sheng, & Zhang, Jinhao. (2019). Concurrent topology optimization for cellular structures with nonuniform microstructures based on the kriging metamodel. *Structural and Multidisciplinary Optimization, 59*(4), 1273–1299.
26. Marian Czarnecki, W., Osindero, S., Jaderberg, M., Swirszcz, G., & Pascanu, R. (2017). Sobolev training for neural networks. arXiv:1706.04859.
27. Kumar, Siddhant, Tan, Stephanie, Zheng, Li., & Kochmann, Dennis M. (2020). Inverse-designed spinodoid metamaterials. *npj. Computational Materials, 6*(1), 73.
28. Zheng, L., Kumar, S., & Kochmann, D. M. (2020). Data-driven topology optimization of spinodoid metamaterials. *Computer Methods in Applied Mechanics and Engineering, 383*, 113894. https://doi.org/10.1016/j.cma.2021.113894
29. Kramer, M. A. (1991). Nonlinear principal component analysis using autoassociative neural networks. *AIChE Journal, 37*(2), 233–243, 2020/11/15.
30. Bessa, Miguel A., Glowacki, Piotr, & Houlder, Michael. (2019). Bayesian machine learning in metamaterial design: Fragile becomes supercompressible. *Advanced Materials, 31*(48), 1904845.

31. Iyer, A., Zhang, Y., Prasad, A., Gupta, P., Tao, S., Wang, Y. et al. (2020). Data centric nanocomposites design via mixed-variable bayesian optimization. *Molecular Systems Design & Engineering, 5*, 1376–1390.
32. Mao, Y., He, Q., & Zhao, X. (2020). Designing complex architectured materials with generative adversarial networks. *Science Advances, 6*(17), eaaz4169.
33. Wang, L., Chan, Y-C., Ahmed, F., Liu, Z., Zhu, P., & Chen, W. (2020). Deep generative modeling for mechanistic-based learning and design of metamaterial systems. *Computer Methods in Applied Mechanics and Engineering, 372*, 113377.

On a Physics-Compatible Approach for Data-Driven Computational Mechanics

Pierre Ladevèze, Paul-William Gerbaud, and David Néron

This paper is dedicated to Peter Wriggers for his 70th birthday. I first met Peter a long time ago and, frankly, I don't remember when our friendship began. Peter is not only an internationally renowned scientist, but also a warm-hearted person who respects and cares for others. We have participated together in many conferences, committees and a number of courses and, most recently, in the IRTG doctoral student exchange program between the University of Hanover and ENS Paris-Saclay directed by him and Olivier Allix. We had a great time together, not only at work. I remember in particular a wine tasting in Udine with also Erwin Stein and Rolf Rannacher and also exceptional evenings with our wives in great restaurants in Paris.

Pierre Ladevèze.

Abstract Within the framework of the thermodynamics of irreversible processes, one introduces a general vision of data-driven computational mechanics, adapted to history-dependent materials, through the concept of the "Experimental Constitutive Manifold" (ECM). The mathematical structure of the ECM, which involves internal state variables, constitutes the material model associated with the available experimental data. The hidden variables are not known a priori but are calculated from the experimental data thanks to the so-called "ECM-Central Problem". This paper also tries to present recent advances in the data-driven computation approach. The potential applications are illustrated through the proposed new way to describe mathematically the material, where a priori assumptions on modelling are absent.

P. Ladevèze (✉) · P.-W. Gerbaud · D. Néron
LMT (ENS Paris-Saclay, CNRS, Université Paris-Saclay, Gif-sur-Yvette, France
e-mail: ladeveze@ens-paris-saclay.fr

F. Aldakheel et al. (eds.), *Current Trends and Open Problems in Computational Mechanics*, https://doi.org/10.1007/978-3-030-87312-7_28

1 Introduction

Data-driven structure calculation has recently become one of the issues impacted by the area of research linked to big data. The pioneering works were due to Ortiz and Chinesta and their teams [1–3]. In these works, Material Science was either annihilated or minimized. Numerous developments followed with a more or less strong presence of physics [4–12]. These works provided original and very effective responses in some situations but are limited to really specific material behaviours or specific problems such as 1D. It follows that the treatment of complex material behaviours, such as viscoplasticity with several internal variables in 2D or 3D, remains a real challenge for which an answer has been given in [8].

This paper aims at showing the recent developments of the data-driven computation approach that we have proposed and which works for history-dependent materials. [8, 13–15]. (Visco)-plastic materials under small perturbation hypothesis are taken as a paradigm for the most complex materials. Our goal is to propose solutions for the description of the material behavior.

The proposed data-driven computation approach is based, first of all, as almost all the data-driven computation approaches, on the separation of the governing equations of the mechanical problem, defined over the space-time domain: equilibrium and compatibility equations in one hand and constitutive equations in the other hand. This separation is the basis of the LATIN solver, which will be the well-suited calculation method [16–18]. Moreover, this separation is also the basis of the Constitutive Relation Error method, developed for both Validation and Verification [19–21]. The first group of equations can be qualified as "exact", in contrast to the constitutive equations that strongly depends on experimental data. These experimental points define what we call the "Experimental Constitutive Manifold" (ECM) whose construction is not a triviality when one considers nonlinear and time-dependent behaviours.

The proposed data-driven computation approach is based on the classical internal variable approach that is used today to describe the material state. However, here the internal hidden variables are not a priori known. There are, however, two particular cases which do not pose any difficulties for us. The first case is that of (visco)plastic materials with isotropic hardening which does not require any hidden internal variable; it is also the case of (visco)plastic materials with kinematic hardening for which the inelastic strain is the hidden internal variable. The hidden internal variables are calculated only from the raw experimental data thanks to the so-called "Central Problem" , which is at the core of this data-driven approach. No other information is used except for the constraints related to the verification of the two principles of Thermodynamics of Irreversible Processes.

Finally, one of the potential applications is illustrated; it is a new way of mathematically describing material from raw experimental data, where a priori assumptions about modelling used in Materials Science are completely absent.

2 Basic Ideas

2.1 The Computation Problem

Let us consider the quasi-static evolution of a structure Ω whose state is defined at any time $t \in [0, T]$ by its displacement field $\underline{u}$, strain field $\boldsymbol{\varepsilon}$ and stress field $\boldsymbol{\sigma}$. This structure is subjected at any time $t \in [0, T]$ to prescribed body forces $\underline{f}_d$, traction force $\underline{F}_d$ over a part $\partial_2\Omega$ of the boundary and displacements $\underline{U}_d$ over the complementary part $\partial_1\Omega$. The class of materials considered is made up of stable materials for which several families can be distinguished:

- elasto-(visco)plasticity where the hidden variables evolve at the same time as the inelastic strain,
- viscoelasticity characterized by viscous phenomena a priori reversible,
- damage related to the evolution of stiffness.

Of course, all these phenomena can be present together. However, as a paradigm, we will take elasto-(visco)plastic materials. The problem to solve can be written: *Find the pair*|, $\boldsymbol{s} = (\dot{\boldsymbol{\varepsilon}}_p, \boldsymbol{\sigma}) \in \boldsymbol{S}^{(0,T)}$ *which satisfies:*

$$(\boldsymbol{A_d}) : \begin{cases} \text{compatibility equation: } \boldsymbol{\varepsilon}_p = \boldsymbol{\varepsilon}(\underline{u}) - \boldsymbol{K}^{-1}\boldsymbol{\sigma},\ \underline{u} \in \mathscr{U}_d^{(0,T)}, \\ \text{equilibrium equation: } \boldsymbol{\sigma} \in \mathscr{S}_d^{(0,T)},\ \text{ where } \boldsymbol{K} \text{ is the elasticity tensor,} \end{cases}$$
$$(\boldsymbol{\Gamma}) : \ \left\{ \text{constitutive relation: } \dot{\boldsymbol{\varepsilon}}_{p|_t} = \mathscr{A}(\boldsymbol{\sigma}_\tau,\ \tau \leq t). \right.$$

2.2 Principle P1: Separation of Equations

Here again, as we did in most of our previous work, we separate the equations that can be seen as "exact" $(\boldsymbol{A_d})$ from the equations that depend very strongly on the experimental data i.e. the constitutive relation $(\boldsymbol{\Gamma})$. This formulation is schematized in Fig. 1. It follows:

$$\textit{Find } \boldsymbol{s} \in (\boldsymbol{A_d}) \textit{ such that: } \quad s_{\text{exact}} = \underset{\boldsymbol{s} \in (\boldsymbol{A_d})}{\operatorname{argmin}}\, d(\boldsymbol{s}, (\boldsymbol{\Gamma})) \tag{1}$$

where $d(\bullet, (\boldsymbol{\Gamma}))$ is a distance to the manifold $(\boldsymbol{\Gamma})$.

2.3 Principle P2: The Experimental Constitutive Manifold

Classically today, Material Science is exploited to construct the $(\boldsymbol{\Gamma})$ manifold which is built from an analytical mathematical model. Here we consider a new approach

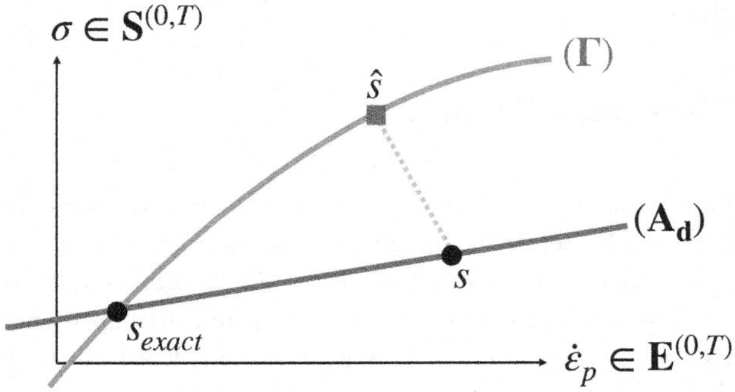

Fig. 1 The data-driven computation approach

in which the manifold $(\boldsymbol{\Gamma})$ is constructed solely from data while respecting the two "exact" principles of the Thermodynamics of Irreversible Processes.

In this paper, we develop the approach completely driven by data. Let us consider that data are a set of histories $\mathscr{H}^{(0,T)}$ of the pair $(\dot{\boldsymbol{\varepsilon}}_p, \boldsymbol{\sigma})$ defined over $[0, T_f]$ with $T_f \leq T$. The elastic properties are supposed known. Instead of the time, we use the material time $p \in [0, P]$ where p is the cumulated inelastic strain, leading to the set of histories $\mathscr{H}^{(0,P)}$. For damage or viscoelastic materials, we propose to use the cumulated strain.

In practice, the interval $[0, P]$ is discretized and consequently, data are the set of points $(\boldsymbol{\varepsilon}_{p,p}, \boldsymbol{\sigma}, \dot{p})_n$ with $n \in \boldsymbol{N}_p$, which defines the raw Experimental Constitutive Manifold (ECM). To go further, we suppose that it exists a hidden internal variable $\boldsymbol{X}$ with $\boldsymbol{X} \in \boldsymbol{R}^q$ such that all the histories belonging to $\mathscr{H}^{(0,P)}$ are described with a good approximation by the system of differential equations:

$$\begin{aligned} \boldsymbol{\varepsilon}_{p,p} &= \boldsymbol{g}(\boldsymbol{\sigma}, \boldsymbol{X}, p) \quad \dot{p} > 0\,, \\ \boldsymbol{X}_{,p} &= \boldsymbol{h}(\boldsymbol{\sigma}, \boldsymbol{X}, p) \quad \dot{p} > 0, \qquad \text{with } \boldsymbol{X} = 0 \text{ at } p = 0\,, \\ \dot{p} &= f(\boldsymbol{\sigma}, \boldsymbol{X}, p) \quad \dot{p} \geq 0\,. \end{aligned} \tag{2}$$

For the plasticity model, the third relation can be replaced by:

$$1 = f(\boldsymbol{\sigma}, \boldsymbol{\sigma}_{,p}, \boldsymbol{X}, p) \quad \dot{p} > 0. \tag{3}$$

To satisfy the two principles of Thermodynamics of Irreversible Processes, additional constraints should be introduced depending if one knows or not the dissipation [14, 15]. In the case where the dissipation is not known, we have proven the additional constraint:

$$\mathrm{Tr}[\boldsymbol{\sigma}\boldsymbol{\varepsilon}_{p,p}] - \boldsymbol{X} \cdot \boldsymbol{X}_{,p} \geq 0 \text{ for } \dot{p} > 0 \text{ and } p \in [0, P], \tag{4}$$

and it follows that the free Helmotz energy is unique.

To compute $\boldsymbol{X} \in \boldsymbol{R}^q$ and its dimension q is what we call the "Central Problem", which is at the core of the proposed data-driven computation approach [8]. Its numerical treatment is given in [13]. $\boldsymbol{U}$ being the functional to minimize, one has formerly:

$$\{\boldsymbol{X}(p)|p \in [0, P]\ \boldsymbol{X} \in \boldsymbol{R}^q\} = \underset{q,\ \boldsymbol{X}'}{\text{argmin}}\ \boldsymbol{U}\big(q, \boldsymbol{X}'(p)\ ;\ p \in [0, P]\quad \boldsymbol{X}' \in \boldsymbol{R}^q\big) \tag{5}$$

and a satisfactory solution is obtained when

$$\frac{\boldsymbol{U}\big(q, \boldsymbol{X}(p)\ ;\ p \in [0, P]\quad \boldsymbol{X} \in \boldsymbol{R}^q\big)}{\underline{\varepsilon}} = O(1)\ , \quad \text{where } \underline{\varepsilon} \text{ is a given threshold.}$$

Finally, the Experimental Constitutive Manifold is defined by the set of points $s_n = \big(\boldsymbol{\varepsilon}_{p,p}, \boldsymbol{X}_{,p}, \dot{p}, \boldsymbol{\sigma}, \boldsymbol{X}\big)_n(p)\, n \in \boldsymbol{N}_p$ which for the calculation should be extended in the best cases by an interpolation process, but in general by an extrapolation procedure. In any case, the domain

$$\Sigma(p) = \Big\{\sum_{i=1}^{N_p} \lambda_i(\boldsymbol{\sigma}, \boldsymbol{X})_i\ ,\ i \in \boldsymbol{N}_p\ ;\ \lambda_i \geq 0 \sum_{i=1}^{N_p} \lambda_i = 1\Big\}$$

must cover the field of use of the material even roughly. Here we use KPCA which computes the $\overline{k}$-coordinates of $\boldsymbol{s}_n\ ,\ n \in \boldsymbol{N}_p$ defining the discrete ECM

$$V_n^k\ ,\ 1 \leq k \leq \overline{k}\quad : \boldsymbol{s}_n(p)\ ,\ p \in [0, P]\ ,\ n \in \boldsymbol{N}_p$$

where $\underline{V}^k\ , k \leq \overline{k}$ are the first eigenvectors of

$$\boldsymbol{K}\ \underline{V}^k = \lambda^k\ \underline{V}^k\ , \quad \text{with} \quad \lambda^{\overline{k}+1} \leq \varepsilon\lambda^1 \qquad (\varepsilon \ll 1),$$
$$\boldsymbol{K}_{ij} = \boldsymbol{K}(\boldsymbol{s}_i, \boldsymbol{s}_j) \quad \text{(kernel)}$$

The extended ECM (ExtECM) is defined for $p \in [0, P]$ by

$$\boldsymbol{s}(\underline{V}) \text{ with } \underline{V} \in \boldsymbol{D}(p) \subset \boldsymbol{R}^{\overline{k}}$$

which means that all points of the ExtECM are defined by the same $\overline{k}$-coordinate system. To get $\boldsymbol{s}(\underline{V})$ from $\underline{V}$ is a classical problem for which several solutions have been already given [22, 23].

2.4 Structure Computation

Using the LATIN method which is very natural here, one has to solve the constitutive relation knowing for example $\boldsymbol{\sigma}(t)\ ,\ t \in [0, T]$. One proposes the following incremental scheme which uses the univocity property of $\boldsymbol{g}$, $\boldsymbol{h}$ and f.

— initialisation at $t = 0$ $\quad \boldsymbol{\sigma}(0) = 0 \ , \ \boldsymbol{X} = 0 \ , \ p = 0$

— $\quad t \rightarrow t + \Delta t \quad$ at t, known $\big(\boldsymbol{\sigma}(t), \boldsymbol{X}(p), \boldsymbol{\varepsilon}_p(t), p(t)\big)$

The ECM (or ExtECM) gives: $\big(\dot{p}, \dot{\boldsymbol{\varepsilon}}_p, \dot{\boldsymbol{X}}\big)_t$

and consequently: $\boldsymbol{\sigma}(t + \Delta t) = \boldsymbol{\sigma}(t) + \Delta\boldsymbol{\sigma}$

$$\big(p, \boldsymbol{\varepsilon}_p, \boldsymbol{X}\big)_{t+\Delta t} = \big(p, \boldsymbol{\varepsilon}_p, \boldsymbol{X}\big)_t + \big(\dot{p}, \dot{\boldsymbol{\varepsilon}}_p, \dot{\boldsymbol{X}}\big)\Delta t$$

3 The Experimental Constitutive Manifold—Illustration

This application is about the construction of the ECM which results here from a given set of experimental histories $\mathcal{H}^{[0,T]}$ through a process of interpolation or extrapolation. This differs from the material identification procedure used in Materials Science, which is based on an analytical mathematical model. All the "experimental" histories will be performed here by classical tests that are being replaced today by much more complex tests where displacements or stress fields are measured [24, 25]. For the sake of simplicity, the experimental points are generated under the hypothesis of plane stress using classical plasticity models.

Data generated by a plasticity model with isotropic hardening—The set of experimental histories $\mathcal{H}_{PM}^{[0,T]}$ is defined thanks to a series of proportional monotonous biaxial tests for which the maximal cumulated strain is P:

— direction $\underline{m}(\alpha)$ with $\alpha \in \left\{m\frac{1}{4\underline{m}} | m \in (-\underline{m}, ...0, 1, ..., \underline{m})\right\}$

— ratio between the two stresses $\underline{n}(\alpha)$: $n \times \frac{1}{\underline{n}}$ with $n \in \left\{ \in (-\underline{n}, ...0, 1, ..., \underline{n})\right\}$

Regularly, cycling loadings are performed without increasing the plastic strain. The ECM does not need any hidden variable: its potential $\boldsymbol{U}$ satisfies to:

$$\boldsymbol{U}(0) = 10^{-12} \ll O(1), \tag{6}$$

The elasticity domain $\boldsymbol{\Sigma}(p)$ is correctly defined and consequently, the use of the ECM (or ExtECM) is equivalent to the use of a plasticity model with isotropic hardening. Moreover, if one assumes that the elastic domain is convex, then it is defined by its border, and therefore is already known from the monotonous loadings. Consequently, the additional cycling loadings are not necessary.

Data generated by a plasticity model with kinematic hardening—To cover the elasticity domain which depends on $\boldsymbol{\varepsilon}_p$, it should be added to the set $\mathcal{H}_P^{[0,T]}$ of proportional monotonous and cyclic loadings, a set of non-proportional loadings $\mathcal{H}_{NP}^{[0,T]}$. We have

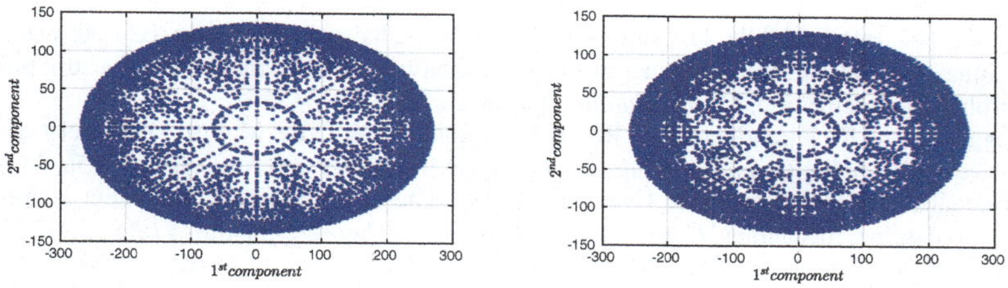

Fig. 2 KPCA images of the ECM for $p = 0.0062$, data generated with a plastic model (only two coordinates are needed): **a** Kinematic hardening; **b** Chaboche-Marquis

performed two-steps tests, starting with a biaxial proportional test $(\tilde{\alpha}, \tilde{n})$ until $p = \frac{P}{4}$ and followed by a new similar test but for $(\tilde{\alpha}', \tilde{n}')$ until $p = P$.

The ECM potential are:

$$U(0) = 124 \gg O(1), \tag{7}$$

$$U(\boldsymbol{\varepsilon}_p) = 0.2 \sim O(1), \tag{8}$$

This set of experimental information allows the construction of the ECM for more complex behaviors including the Bauschinger effect, the distortion of the elasticity domain or the plastic anisotropy, behaviors that are difficult to apprehend from the analytical models of Materials Science. Of course, the ECM brings into play hidden internal variables that are determined by the Central Problem. Figure 2 shows the KPCA image of the ECM generated by two plastic models.

4 Conclusion

The applications of the data-based approach that we propose are envisaged in synergy with Materials Science:

- In situations in which the classical models of Materials Science do not work well or fail to describe the behavior of the material; complex materials (anisotropy, distorsion, ...), correction or extension of a first material model,
- To transform the FE^2 multiscale calculation method into a simple FE method thanks to the calculation of an ECM describing the homogenized macro behavior.

References

1. Kirchdoerfer, T., & Ortiz, M. (2016). Data-driven computational mechanics. *Computer Methods in Applied Mechanics and Engineering, 304*, 81–101.
2. Chinesta, F., Ladeveze, P., Ibanez, R., Aguado, J. V., Abisset-Chavanne, E., & Cueto, E. (2017). Data-driven computational plasticity. *Procedia Engineering, 207*, 209–214.

3. Ibañez, R., Borzacchiello, D., Aguado, J. V., Abisset-Chavanne, E., Cueto, E., Ladeveze, P., & Chinesta, F. (2017). Data-driven non-linear elasticity: Constitutive manifold construction and problem discretization. *Computational Mechanics, 60*(5), 813–826.
4. Eggersmann, R., Kirchdoerfer, T., Reese, S., Stainier, L., & Ortiz, M. (2019). Model-free data-driven inelasticity. *Computer Methods in Applied Mechanics and Engineering, 350,* 81–99.
5. González, D., Chinesta, F., & Cueto, E. (2019). Thermodynamically consistent data-driven computational mechanics. *Continuum Mechanics and Thermodynamics, 31*(1), 239–253.
6. Guo, M., & Hesthaven, J. S. (2019). Data-driven reduced order modeling for time-dependent problems. *Computer Methods in Applied Mechanics and Engineering, 345*, 75–99.
7. Ibáñez, R., Abisset-Chavanne, E., González, D., Duval, J. L., Cueto, E., & Chinesta, F. (2019). Hybrid constitutive modeling: Data-driven learning of corrections to plasticity models. *International Journal of Material Forming, 12*(4), 717–725.
8. Ladevèze, P., Néron, D., & Gerbaud, P. W. (2019). Data-driven computation for history-dependent materials. *Comptes Rendus de l'Academie des Sciences. Mécanique, 347*(11):831–844.
9. Lopez, E., Gonzalez, D., Aguado, J. V., Abisset-Chavanne, E., Cueto, E., Binetruy, C., & Chinesta, F. (2018). A manifold learning approach for integrated computational materials engineering. *Archives of Computational Methods in Engineering, 25*(1), 59–68.
10. Leygue, A., Coret, M., Réthoré, J., Stainier, L., & Verron, E. (2018). Data-based derivation of material response. *Computer Methods in Applied Mechanics and Engineering, 331*, 184–196.
11. Liu, Z., Bessa, M. A., & Liu, W. K. (2016). Self-consistent clustering analysis: An efficient multi-scale scheme for inelastic heterogeneous materials. *Computer Methods in Applied Mechanics and Engineering, 306*, 319–341.
12. Versino, D., Tonda, A., & Bronkhorst, C. A. (2017). Data driven modeling of plastic deformation. *Computer Methods in Applied Mechanics and Engineering, 318*, 981–1004.
13. Gerbaud, P. -W., Ladevèze, P., & Néron, D. (to appear) Solving a fundamental problem at the core of a data-driven approach for history-dependent materials.
14. Ladevèze, P. (2020). *Le calcul piloté par les données pour les matériaux à mémoire?: théorie, pratique, application*. LMT Paris-Saclay: Technical Report.
15. Ladevèze, P., Gerbaud, P. -W., & Néron, D. (to appear). A thermodynamics-compatible approach to data-driven computation.
16. Ladevèze, P. (1989). The large time increment method for the analyse of structures with non-linear constitutive relation described by internal variables. *Comptes rendus de l'Academie des sciences. Série 2, 309*(2):1095–1099.
17. Ladevèze, P. (1999). *Nonlinear computational structural mechanics: New approaches and non-incremental methods of calculation*. New York: Springer.
18. Néron, D., & Ladevèze, P. (2010). Proper generalized decomposition for multiscale and multiphysics problems. *Archives of Computational Methods in Engineering, 17*(4), 351–372.
19. Ladeveze, P., & Leguillon, D. (1983). Error estimate procedure in the finite element method and applications. *SIAM Journal on Numerical Analysis, 20*(3), 485–509.
20. Ladevèze, P., & Chouaki, A. (1999). Application of a posteriori error estimation for structural model updating. *Inverse Problems, 15*(1), 49–58.
21. Ladevèze, P., & Pelle, J.-P. (2005). *Mastering calculations in linear and nonlinear mechanics*. New York: Springer.
22. A. García-González, A. Huerta, S. Zlotnik, and P. Díez. A kernel Principal Component Analysis (kPCA) digest with a new backward mapping (pre-image reconstruction) strategy. *arxiv.org*, (2008):1–16, 2020.
23. M. H. Nguyen and F. D. la Torre Frade. Robust Kernel Principal Component Analysis. In *Proceedings of (NeurIPS) Neural Information Processing Systems*, dec 2008.
24. F. Hild, A. Bouterf, L. Chamoin, H. Leclerc, F. Mathieu, J. Neggers, F. Pled, Z. Tomičević, and S. Roux. Toward 4D mechanical correlation. *Advanced Modeling and Simulation in Engineering Sciences*, 3(1), 2016.
25. Neggers, J., Allix, O., Hild, F., & Roux, S. (2018). Big Data in Experimental Mechanics and Model Order Reduction: Today's Challenges and Tomorrow's Opportunities. *Archives of Computational Methods in Engineering, 25*(1), 143–164.

Wave Propagation in Layered Fiber Composites with Nonlinear Material Laws

Rolf Lammering, Ngoc Nguyen Vu, and Natalie Rauter

Peter and I have known each other since my student days in Hannover. After his habilitation he took over his first co-lecture in my doctoral project. Since then we have never lost sight of each other, both academically and privately. We are also currently united by the fact that we are both vice presidents for research at our universities. I would like to thank Peter for many inspiring discussions and look forward to hopefully many more encounters, in Hamburg, Hannover or wherever.

—Rolf.

Abstract This work deals with numerical investigations of nonlinear wave propagation in cross-ply laminates of carbon fiber reinforced plastics. For this purpose, a nonlinear hyperelastic model for transversely isotropic materials is presented which considers the compressibility of the material. Then, the case is investigated where the wave propagation is in the fiber direction or transversely to it, so that no coupling between the symmetric and asymmetric modes and the horizontal shear waves occurs. Subsequently, the wave propagation at 45° is considered at which coupling takes place. It is shown that the nonlinearity leads to second harmonic modes in both cases. Furthermore a power flux from the primary mode to the second harmonic mode can be achieved in case of excitation at an appropriate frequency resulting in increasing or oscillating amplitudes of the higher harmonic waves. Since the occurrence of higher harmonics can also be seen as an indicator of micro-damage and material deterioration, nonlinear material models may be used to describe composite materials with micro-damage.

R. Lammering (✉) · N. N. Vu · N. Rauter
Institute of Mechanics, Helmut-Schmidt-University/University of the Federal Armed Forces Hamburg, Hamburg, Germany
e-mail: rolf.lammering@hsu-hh.de

N. N. Vu
e-mail: natalie.rauter@hsu-hh.de

N. Rauter
e-mail: n.vn@hsu-hh.de

F. Aldakheel et al. (eds.), *Current Trends and Open Problems in Computational Mechanics*, https://doi.org/10.1007/978-3-030-87312-7_29

1 Introduction

In carbon fiber reinforced polymers (CFRP) fatigue damage starts at an early stage of lifetime [1]. Microstructural damage like microscopic cracks in the matrix material and fiber/matrix interface debonding is observed, which increases during operation and can eventually lead to a catastrophic structural failure [2]. Therefore, the monitoring of composite material structures is essential even at an early stage.

Techniques based on guided waves are very well suited for monitoring tasks. However, it has been shown that linear techniques are not suitable for detecting microstructural damage. Therefore, methods based on nonlinear elastic waves come into play. These methods take advantage of the fact that a damaged structure behaves non-linearly, so that a mono-frequency structural excitation causes not only waves at the excitation frequency, but also higher harmonic wave modes at a multiple of the excitation frequency.

The detection of microstructural damage was first experimentally investigated for isotropic material in [3–5] and was extended to the class of composite structures later, see [6, 7, 9, 10]. To get a better insight into the physics behind this monitoring method and to analyze the further potential numerical simulations are essential beside experimental investigations. In [11] it was shown that numerical investigations on the propagation of elastic waves in structures with microstructural damage can in principle be performed in two different ways. On the one hand, the microstructural damage can be modeled geometrically by cracks whose surfaces come into contact when a wave passes through them. On the other hand, the nonlinear material behavior can be modeled by a nonlinear hyperelastic material model. Both types of modeling lead to comparable results, see [11].

If one wants to take advantage of nonlinear phenomena in the propagation of elastic waves, however, one must keep in mind that the amplitudes of the higher harmonic wave modes are very small. Therefore, it is appropriate to use the so-called cumulative effect, which increases the amplitudes of the higher harmonic wave modes with increasing propagation distance and thus allows an improved amplitude determination, see [8].

This work deals with the simulation of nonlinear wave propagation in CFRP plates. It is based on [12] and extends previous investigations, see e.g. [11], to symmetric cross ply laminates. After the presentation of a nonlinear hyperelastic material model, wave propagation in CFRP is under consideration. The special feature of layered fiber reinforced composites is that symmetric and antisymmetric LAMB-waves as well as horizontal shear waves are only decoupled when they propagate in 0° or 90° directions und thus in fiber direction or perpendicular to it. When propagating in any other direction they are coupled. This fact is considered in the modelling approach by choosing a plane strain state and thus a 2D modelling on the one hand and a 2.5D modelling on the other hand, which allows displacements perpendicular to the plane under consideration. It will be shown, that previous findings can be transferred to stacking sequences in which coupling of various modes take place.

2 Nonlinear Plate Structure

In the case of uncoupled LAMB- and shear horizontal waves, a 2D model of a plate is used, of which a section is investigated in a plane strain state, see Fig. 1. The plate consists of four layers of equal thickness at h= 0.25 mm. At the top side there is a reflection-free area of a length ℓ_e= 0,7 m with virtual sensor points. The total length ℓ of the plate and the total simulation time t_e are determined depending on the excitation frequency. The spatial discretization of this 2D model is achieved by means of 9-node plane elements. In order to simulate the wave propagation with sufficient accuracy, at least 8 elements are selected per smallest wavelength and 20 time steps for the shortest time period.

In the case of coupled LAMB- and shear horizontal waves, displacements in X_2-direction arise. Brick elements with 27 nodes are chosen and symmetrical periodic boundary conditions are defined to ensure a plane wavefront perpendicular to the X_2-axis.

The transversely isotropic material is described by a strain energy function ψ, which is additively split into a linear incompressible part ψ^{lin} and a non-linear compressible part ψ^{nl}

$$\psi = \psi^{lin} + \psi^{nl} \tag{1}$$

The linear part ψ^{lin} can be formulated by using the three invariants I_1, I_2, I_3 and the two pseudo-invariants I_4, I_5 of the right CAUCHY- GREEN tensor $\mathbf{C}$, see [13]. In [14] the following expression for ψ^{lin} based on experimental investigations was presented

$$\begin{aligned}\psi^{lin} =& c_1(I_1-3)^2 - c_2(I_2-3-2(I_1-3)) + c_3(I_4-1)^2 \\ &+ c_4(I_5-1-2(I_4-1)) + c_5(I_4-1)(I_1-3)\end{aligned} \tag{2}$$

The c_i parameters are used to describe the transversely isotropic material. In accordance to [15] the non-linear part is defined in the following way

$$\psi^{nl} = K\left(J^2 + \frac{1}{J^2} - 2\right)^{3/2}, \quad K > 1 \tag{3}$$

Here, J is the Jacobian and K describes the volumetric nonlinearity.

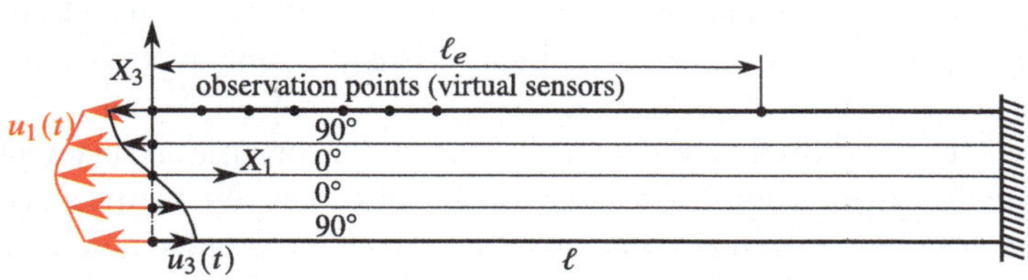

Fig. 1 2D model of the symmetric cross-ply, boundary conditions, and displacement controlled dynamic loading in X_1- and X_3-directions

At the left end of the plate, displacement boundary conditions are applied according to $u_1(t) = u_{01}(X_3)F_1(t)$ and $u_3(t) = u_{03}(X_3)F_2(t)$ where $u_{01}(X_3)$ and $u_{03}(X_3)$ correspond to the displacement field of the selected primary wave mode, see Fig. 1. Additionally, $u_2(t) = u_{02}(X_3)F_1(t)$ is used with the 2.5D model.

3 Cumulative Higher Harmonic Wave Modes

In a nonlinear medium a mono-frequent excitation causes a structural response not only at the excitation frequency but also at higher harmonic frequencies. The higher harmonic wave modes usually have very small amplitudes. So, they subside very quickly due to damping and are difficult to detect in experiments. To overcome this problem, a cumulative effect is used, see [8], for which the following conditions have to be satisfied: (i) a non-zero power flux from the primary to the second harmonic wave mode has to exist and, (ii) the phase velocity of the primary and second harmonic wave modes have to be equal. In this case, the modes are called internally resonant and the amplitude of the higher harmonic wave grows linearly with the propagation distance. If the phase velocities are different it shows an oscillating behavior with a wavelength depending on the difference of phase velocities.

A third condition, the matching of the group velocities, was controversially discussed in the past. This condition is not necessary for the cumulative effect. However, if the group velocities match, both wave modes do not separate and thus, the higher harmonic wave does not extend over the covered propagation distance. So, in order to detect the amplitudes of the fundamentally excited and the higher harmonic Lamb wave mode simultaneously this condition is taken into account as a third one for the selection of suitable mode pairs and the determination of the excitation frequency in this study.

This procedure is shown exemplarily with the help of the dispersion diagram for the phase velocity in Fig. 2. The following material parameters were used: E_1=121.9 GPa, E_2=8.0 GPa, G_{12}=4.0 GPa, G_{23}=3.75 GPa, ν_{12}=0.33. It becomes visible that the S0-mode and the $S1$-mode have the same phase velocities (and group velocities, see [12]) if the primary S0-mode is excited at fh=0.585 MHz mm so that second harmonic $S1$-mode is generated at $2fh$=1.171 MHz mm, see related marks. Figure 2 further shows that the phase velocity does not change significantly in the low frequency range. For this reason several mode pairs S0-S0 can be selected in this frequency range for quasi-cumulative higher harmonic generation, e.g. at fh=0.2 MHz mm.

Figure 3 shows the dispersion diagram for the same geometric and material parameters but in the case of a propagating wave in 45°-direction. Symmetric and antisymmetric LAMB-waves as well as horizontal shear waves are coupled now and denoted by LSHn. The phase velocity of the LSH0- and LSH2- modes do not change significantly in the low frequency range so that corresponding mode pairs can be selected here for quasi-cumulative higher harmonic generation.

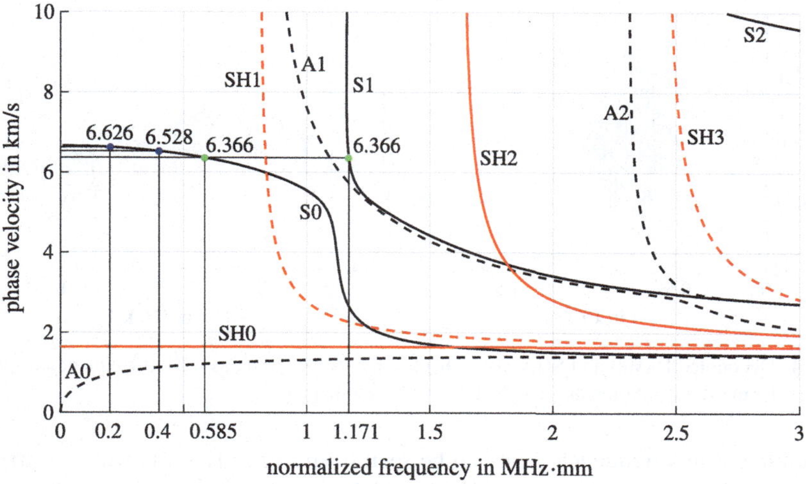

Fig. 2 Phase velocity dispersion diagram for a symmetric cross-ply laminate: S: symmetric, A: antisymmetric LAMB-waves, SH0, 2, ...: symmetric, SH1, 3, ...: antisymmetric shear horizontal waves

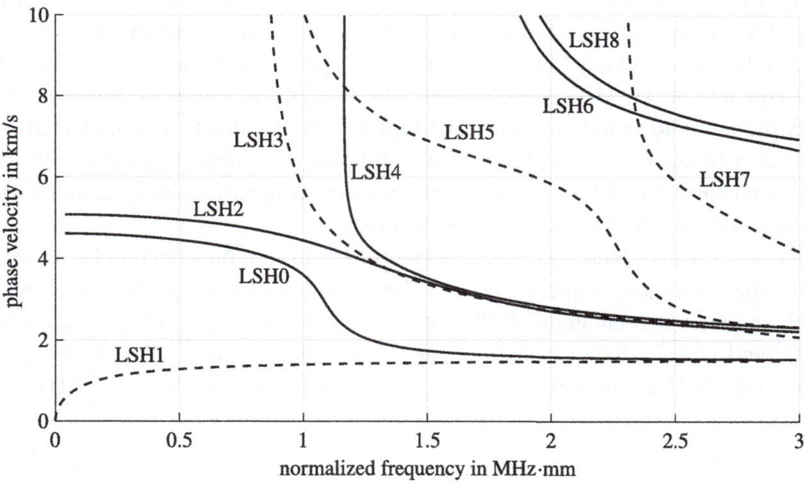

Fig. 3 Phase velocity dispersion diagram for a symmetric cross-ply laminate [45°, −45°, −45°, 45°]: LSH0, 2, ...: symmetric, LSH1, 3, ...: antisymmetric coupled LAMB- and SH-waves

4 Analysis and Results

In the case of the non-coupled waves Fig. 4 shows some results obtained after displacement controlled excitation (see Fig. 1) of the structure at 0.3 MHz by a windowed sine burst signal of 20 cycles with an amplitude of 0.01 μm. The parameter K in Eq. (2) was set to 10 GPa. Generally, the fundamental modes S0 and A0 are

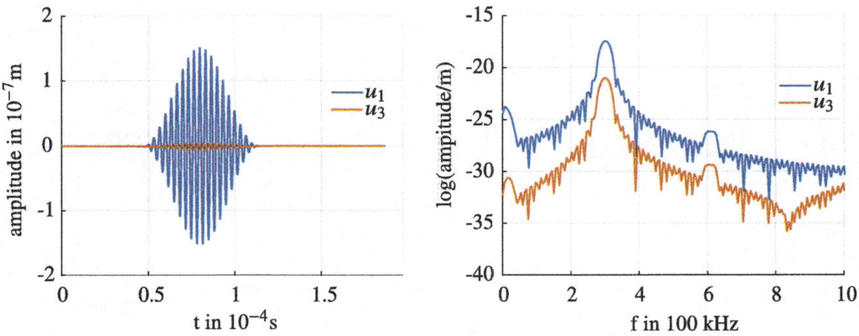

Fig. 4 Decoupled LAMB- and SH-waves. Displacements u_1 and u_3 as well as frequency response at X_1=0.3 m after excitation at f = 0.3 MHz. S0-S0-mode pair

generated at this frequency, as it can be seen from Fig. 2. Here, however, a special type of excitation was chosen, see Fig. 1, which is mode selective and therefore only generates an S0-mode. So, only this mode is visible in Fig. 4, left, where the displacements u_1 and u_3 at the location X_1=0.3 m are shown with respect to time. At the selected excitation frequency, the conditions for cumulative higher harmonic wave modes are met and, therefore, a secondary wave field is generated. This is confirmed by a FFT analysis of the time signals. The results are shown in Fig. 4, right, from which it becomes obvious that the excitation frequency at 0.3 MHz as well as twice the frequency at 0.6 MHz are included. The cumulative effect is shown in Fig. 5. Here, the amplitude ratio A_2/A_1^2 of the inplane displacement u_1 is shown over the covered distance of the propagating wave. It becomes visible, that the amplitude of the second harmonic S0-mode oscillates with increasing propagation distance which is in accordance with theoretical considerations.

In the case of coupled waves corresponding results are presented in Figs. 6 and 7. Again, the excitation frequency is 0.3 MHz. Figure 3 shows that the guided waves, which are excited at an angle of 45° with respect to the fiber orientation, are LSH0, LSH1, and LSH2. However, theoretical considerations show, that only the modes LSH0 and LSH2 are internally resonant. In the following, mode pairs LSH2-LSH2

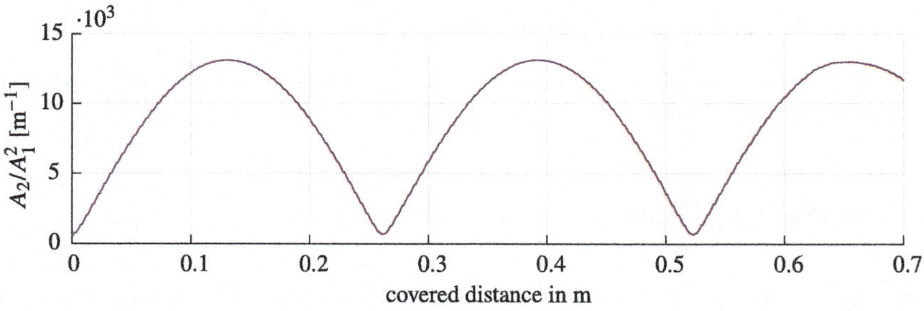

Fig. 5 Decoupled LAMB- and SH-waves. Behavior of the u_1-amplitude of second harmonic S0-mode after excitation at f = 0.3 MHz

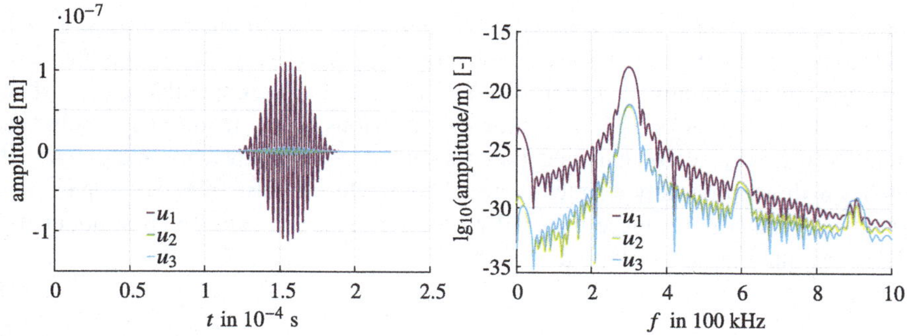

Fig. 6 Coupled LAMB- and SH-waves. Displacements u_1, u_2, and u_3 as well as frequency response at X_1=0.6 m after excitation at f= 0.3 MHz. LSH2-LSH2-mode pair

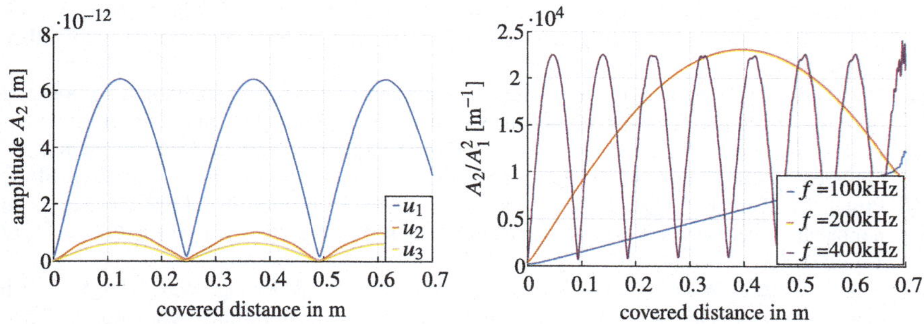

Fig. 7 Coupled LAMB- and SH-waves. Amplitude behavior of second harmonic LSH2-mode after excitation at f= 0.3 MHz

are under consideration. Figure 6, left, shows the three displacement components u_1, u_2, and u_3 at the location X_1=0.6 m as functions of time. Due to the consideration of nonlinearities and the power flux from the primary to the second harmonic mode, twice the excitation frequency is clearly visible in the frequency response. The cumulative effect becomes visible in Fig. 7. Here, the displacement amplitudes of the second harmonic LSH2-mode show an oscillating behavior. Moreover, Fig. 7, right, shows that the period strongly depends on the excitation frequency: the higher the frequency, the lower the period.

5 Concluding Remarks

This study shows that a nonlinear hyperelastic material model allows to simulate the cumulative effect of higher harmonic guided waves in cross-ply laminates made of CFRP. This effect has been shown for wave propagation in 45° and 0° directions and, hence, for situations at which the symmetric and antisymmetric modes as well as the shear horizontal modes are coupled or not, respectively. Regarding the wave propagation, the nonlinear material model thus leads to comparable effects than micro-structural cracks as it was shown in previous work. It may be concluded, that micro-mechanically damaged material can be modelled by a nonlinear material model with appropriately adapted material parameters in case of wave propagation analysis. Furthermore, these material parameters may give information about the degree of material deterioration.

References

1. Kaminski, M., Laurin, F., Maire, J.-F., Rakotoarisoa, C., & Hémon, E. (2015). Fatigue damage modeling of composite structures. *Journal Aerospace Lab*. https://doi.org/10.12762/2015.AL09-06.
2. Harris, B. (2003). Fatigue in composites: Science and technology of the fatigue response of fibre-reinforced plastics. CRC Press and Woodhead, Boca Raton, Fla., and Cambridge, U.K.
3. Bermes, C., Kim, J.-Y., Qu, J., & Jacobs, L. J. (2007). Experimental characterization of material nonlinearity using Lamb waves. *Applied Physics Letters*. https://doi.org/10.1063/1.2431467.
4. Pruell, C., Kim, J.-Y., Qu, J., & Jacobs, L. J. (2009). Evaluation of fatigue damage using nonlinear guided waves. *Smart Materials and Structures*. https://doi.org/10.1088/0964-1726/18/3/035003
5. Xiang, Y., Zhu, W., Liu, C.-J., Xuan, F.-Z., Wang, Y.-N., & Kuang, W.-C. (2015). Creep degradation characterization of titanium alloy using nonlinear ultrasonic technique. *NDT & E International*. https://doi.org/10.1016/j.ndteint.2015.02.001
6. Li, W., Cho, Y., & Achenbach, J. D. (2012). Detection of thermal fatigue in composites by second harmonic Lamb waves. *Smart Materials and Structures*. https://doi.org/10.1088/0964-1726/21/8/085019
7. Rauter, N., & Lammering, R. (2015). Impact damage detection in composite structures considering non-linear Lamb wave propagation. *Mechanics of Advanced Materials and Structures, 15376494*(2014), 907950. https://doi.org/10.1080/15376494.2014.907950.
8. Chillara, V. K., & Lissenden, C. J. (2014). Nonlinear guided waves in plates: A numerical perspective. *Ultrasonics*. https://doi.org/10.1016/j.ultras.2014.04.009.
9. Rauter, N., Lammering, R., & Kühnrich, T. (2016). On the detection of fatigue damage in composites by use of second harmonic guided waves. *Composite Structures*. https://doi.org/10.1016/j.compstruct.2016.05.049
10. Rauter, N., & Lammering, R. (2018). A constitutive model for the analysis of second harmonic Lamb waves in unidirectional composites. *International Journal of Solids and Structures*. https://doi.org/10.1016/j.ijsolstr.2017.11.019
11. Rauter, N. (2017). *Analyse des Einflusses der Werkstoffdegradation auf die nichtlineare Wellenausbreitung in unidirektionalen Compositen - Experimentelle Untersuchung mit numerischer Analyse und Modellbildung -*. Ph.D.-Thesis: Helmut-Schmidt-Universität / Universität der Bundeswehr Hamburg.
12. Ngoc Nguyen, V. (2020). *Zur Wellenausbreitung in geschichteten Faserverbundstrukturen unter Verwendung nichtlinearer Stoffgesetze*. Ph.D.-Thesis: Helmut-Schmidt-Universität / Universität der Bundeswehr Hamburg. https://doi.org/10.24405/9640.

13. Holzapfel, G. A. (2000). *Nonlinear solid mechanics: A continuum approach for engineering*. Chichester, U.K.: Wiley.
14. Reese, S., Raible, T., & Wriggers, P. (2001). Finite element modelling of orthotropic material behaviour in pneumatic membranes. *International Journal of Solids and Structures*. https://doi.org/10.1016/S0020-7683(01)00137-8.
15. Simo, J. C., & Miehe, C. (1992). Associative coupled thermoplasticity at finite strains: Formulation, numerical analysis and implementation. *Computer Methods in Applied Mechanics and Engineering*. https://doi.org/10.1016/0045-7825(92)90170-O.

Finite Element Modelling of In-Stent Restenosis

Kiran Manjunatha, Marek Behr, Felix Vogt, and Stefanie Reese

Dedicated to Peter Wriggers for his 70th birthday (Stefanie Reese). Peter Wriggers already impressed me in 1984, the time of my first semester of civil engineering studies at the Technical University of Hannover. Over the following years I got to know him in various respects, as supervisor of my doctoral thesis, as a daredevil skier, as a scientifically outstanding and exceptional colleague, as a very nice companion in academies and finally also in his function as vice president of the university. I wish him all the best for the years to come and look forward to further inspiring discussions.

Abstract From the perspective of coronary heart disease, the development of stents has come significantly far in reducing the associated mortality rate, drug-eluting stents being the epitome of innovative and effective solutions. Within this work, the intricate process of in-stent restenosis is modelled considering one of the significant growth factors and its effect on constituents of the arterial wall. A multiphysical modelling approach is adopted in this regard. Experimental investigations from the literature have been used to hypothesize the governing equations and the corresponding parameters. A staggered solution strategy is utilised to capture the transport phenomena as

K. Manjunatha (✉) · S. Reese
Institute of Applied Mechanics, RWTH Aachen University, Aachen, Germany
e-mail: kiran.manjunatha@ifam.rwth-aachen.de

S. Reese
e-mail: stefanie.reese@ifam.rwth-aachen.de

M. Behr
Chair for Computational Analysis of Technical Systems, RWTH Aachen University, Aachen, Germany
e-mail: behr@cats.rwth-aachen.de

F. Vogt
Department of Cardiology, Pulmonology, Intensive Care and Vascular Medicine, RWTH Aachen University, Aachen, Germany
e-mail: fvogt@ukaachen.de

F. Aldakheel et al. (eds.), *Current Trends and Open Problems in Computational Mechanics*, https://doi.org/10.1007/978-3-030-87312-7_30

well as the growth and remodeling that follows stent implantation. The model herein developed serves as a tool to predict in-stent restenosis depending on the endothelial injury sustained and the protuberance of stents into the lumen of the arteries.

1 Introduction

Coronary heart disease (CHD) is one of the most prominent causes of mortality and affected 126.5 million lives worldwide as of 2017. CHD is characterised by constriction of the coronary artery arising due to the build-up of plaque which consists of lipids, calcifications, and macrophages. This condition, termed atherosclerosis, leads to the narrowing of pathways for blood flow as well as the loss of elasticity of the arterial wall. Percutaneous coronary intervention (PCI) is the process of placing reinforcement structures called stents within these constricted blood vessels to normalize the blood flow. Unfortunately, this interventional procedure is associated with the risk of in-stent restenosis and stent thrombosis. Drug-eluting-stents have emerged in recent years as the most viable option for revascularisation, significantly reducing the risks associated with PCI. But the mechanisms involved in the restenotic process remain incompletely understood. Quantifying the probability and level of restenosis via simulation of the underlying mechanisms will help in addressing the risks more precisely.

Several computational approaches have been developed in this regard. A multiscale framework involving finite element models and agent-based models has been conceptualised in [17], where the restenotic process was unidirectionally coupled to stresses on the arterial walls. A more recent development on similar grounds has been made in [12], incorporating bidirectional coupling between finite element and agent-based models. A general continuum-based model describing growth and remodelling of soft tissues [7], considering the evolution of constituents in the arterial wall using the constrained-mixture theory, also serves as a computational tool for restenosis prediction. With temporal averaging, a homogenised constrained-mixture model has been developed in [2] as an extension to the constrained-mixture theory. Alternatively, highly resolved transport phenomena occurring in the arterial wall have also been utilised to model the pathophysiology of vascular diseases. Venous neointimal hyperplasia has been quantified considering the inherent pathogenic mechanisms involving growth factors in [1]. A coupled multi-physical approach for quantifying atherosclerosis has been described in [16], wherein the mechanics of bloodflow, mass transport and arterial wall mechanics have been unified under a single framework. An in-stent restenosis predictive model developed on a similar foundation incorporating damage on the arterial wall can be found in [3]. An analogous attempt is made hereof for the prediction and quantification of the after-effects of stent implantation.

1.1 Structure of the Arterial Wall

The arterial wall is composed of three layers: intima, media, and adventitia. The region where the blood flow occurs is called the lumen. Intima is the innermost layer of the arterial wall, immediately adjacent to the lumen. It is consisted of a monolayer of endothelial cells accompanied by a few layers of smooth muscle cells and extends up to the internal elastic lamina. The media occupies the space between the internal and external laminae and is mainly composed of collagen, elastin and smooth muscle cells. Adventitia, the outermost layer, contains loose connective tissue composed of elastin, collagen and fibroblasts.

1.2 In-Stent Restenosis and Platelet-Derived Growth Factor

In-stent restenosis refers to the accumulation of new tissue within the intima leading to a diminished cross-section of the lumen post stent implantation. The underlying mechanism is called neointimal hyperplasia. It is a collaborative effect of migration and proliferation of smooth muscle cells (SMC) in the arterial wall, regulated by intricate signalling pathways that are triggered by certain stimuli, either internal or external to the arterial wall.

The platelet-derived growth factor (PDGF) is a dimer of two peptides linked by a disulfide bond. It has been implicated in vascular remodelling processes, including neointimal hyperplasia, that follow an injury to arterial wall [10]. This can be attributed to its mitogenic and chemoattractant properties. PDGF is secreted by an array of cellular species namely the endothelial cells, SMC, fibroblasts, macrophages and platelets.

The stent implantation procedure damages the endothelial monolayer. Also, depending on the arterial overstretch achieved during the implantation, stent struts partially obstruct the blood flow creating vortices in their wake regions. This causes oscillatory wall shear stresses and hence further damages to the endothelium [9]. PDGF, which is stored in the alpha-granules of the aggregated platelets at endothelial injury sites, is released into the arterial wall. The presence of PDGF upregulates matrix metalloproteinases (MMP) production in the arterial wall. Extracellular matrix (ECM) is a network of collagen and glycoproteins surrounding the SMC and are degraded in the presence of MMP. SMC, which are usually held stationary by the ECM, are rendered free for migration under the action of MMP. Also, a degraded ECM encourages the proliferation of SMC under the presence of growth factors. The focal adhesion sites created due to cleaved proteins in the ECM assist in the migration of SMC. The direction of migration is influenced by the number of adjacent focal adhesion sites available and the local concentration gradient in PDGF. This directional movement of SMC is termed chemotaxis, and results in the accumulation of the proliferated and migrated SMC in the intima of the arterial wall. A positive feedback loop might occur wherein the migrated SMC create a further obstruction in the blood flow and subsequent upregulation of PDGF. The uncontrolled growth of vascular tissue that follows will eventually lead to a complete blockage of the lumen.

2 Mathematical Modelling

In-stent restenosis is understood to be a convoluted phenomenon involving multiple constituents. Only the behaviours of key constituents identified from the perspective of the processes described in Sect. 1.2 are modelled. Blood flow in the lumen is not included in the model and the arterial wall is considered to be made of only the intima and media layers.

2.1 Transport Phenomena

The transport of constituents within the arterial wall is governed by a set of advection-reaction-diffusion equations, the general structure of which for a scalar field ϕ is

$$\frac{\partial \phi}{\partial t} = \underbrace{-\nabla \cdot (\phi\, \boldsymbol{v})}_{\text{advection}} + \underbrace{\nabla \cdot (k\, \nabla \phi)}_{\text{diffusion}} + \underbrace{T_s}_{\text{source}} - \underbrace{T_r}_{\text{reaction}}\,, \tag{1}$$

where $\boldsymbol{v}$ denotes the velocity of the medium of transport and k, the diffusivity of ϕ in the medium. To mathematically model in-stent restenosis, PDGF, ECM and SMC are considered to be the key ingredients. Hence the general structure above is adapted to reflect their respective behaviours. The arterial wall is considered to be quasi-static and hence advective terms arising from the movement of the wall are ignored ($\boldsymbol{v} \approx \boldsymbol{0}$). The symbols and units associated with the transport quantities are declared here for clarity (Table 1).

Platelet-derived growth factor

PDGF is assumed to diffuse throughout the arterial wall. Also, it is internalised by the SMCs during their migration and proliferation. The evolution of the concentration of PDGF is thus modelled by the following equation:

Table 1 Transport variables

Variable	Symbol	Units
PDGF concentration	c_P	mol/mm^3
ECM density	ρ_E	mol/mm^3
SMC density	ρ_S	cells/mm^3

$$\frac{\partial c_P}{\partial t} = \underbrace{\nabla \cdot \left(D_P \, \nabla c_P \right)}_{\text{diffusion}} \quad \underbrace{- \, \alpha \, \rho_S \, c_P}_{\text{consumption by SMC}} \quad , \tag{2}$$

where α refers to an internalisation coefficient and ρ_S, the density of SMCs. In comparison to Eq. (1), we see that only the terms associated with the diffusive and the reactive phenomena appear here. Although PDGF is produced by other cellular species, the net effect observed is degradation. The diffusion coefficient D_P can theoretically vary between positions in the arterial wall but is considered to be constant here.

Extracellular matrix

The major constituent of ECM is collagen, which is considered to be non-diffusive. SMCs recognize a degraded ECM and synthesize collagen. A source term, in the form of a logistic function, is introduced in this regard and an asymptotic threshold for collagen density $\rho_{E,th}$ prescribed. Collagen is degraded by MMP, which in turn is regulated by PDGF. A reactive term is introduced to take care of this degradation. The evolution of ECM density hence reads as follows:

$$\frac{\partial \rho_E}{\partial t} = \underbrace{\beta \, \rho_S \left(1 - \frac{\rho_E}{\rho_{E,th}} \right)}_{\text{synthesis by SMC}} \quad \underbrace{- \, \gamma \, c_P \, \rho_E}_{\text{degradation due to MMP}} \quad . \tag{3}$$

Here, β and γ refer to synthesis and degradation rate coefficients respectively.

Smooth muscle cells

The migration of SMC under the influence of PDGF is modelled using a chemotaxis term [8], which can be interpreted as a pseudo-advective term wherein the velocity of the medium is replaced by the gradient in the ECM density. This corresponds to the fact that higher degradation in the ECM results in higher focal adhesion sites for SMC migration. Additionally, a source term is introduced to model the proliferation of SMC under the presence of PDGF. Hence the evolution of the SMC density is prescribed using the equation:

$$\frac{\partial \rho_S}{\partial t} = \underbrace{\nabla \cdot \left(\chi \, c_P \left(1 - \frac{\rho_E}{\rho_{E,th}} \right) \rho_S \, \nabla \rho_E \right)}_{\text{chemotaxis due to ECM degradation}} + \underbrace{\kappa \, c_P \left(1 - \frac{\rho_E}{\rho_{E,th}} \right) \rho_S}_{\text{proliferation due to PDGF}} . \tag{4}$$

The chemotactic sensitivity χ and the proliferation constant κ are scaled by the concentration of PDGF as well as the logistic coefficient dependent on the ECM density. This takes care of the fact that the migration and proliferation effects increase with an increase in ECM degradation, in the presence of PDGF.

2.2 Arterial Wall Mechanics

The continuum mechanical description of in-stent restenosis requires first the definition of the kinematics of growth and remodelling of the arterial wall, incorporating relevant quantities from the transport phenomena previously described. Constitutive equations hence based on the kinematics, in combination with the balance of linear momentum, will complete the mathematical model description.

2.3 Kinematics

A deformation map $\boldsymbol{\varphi}$ of a particle at position $\boldsymbol{X}$ in the reference configuration Ω_0 at time t_0 to its position $\boldsymbol{x}$ in the current configuration Ω at time t is provided by the deformation gradient $\boldsymbol{F} = \nabla_X \boldsymbol{\varphi}(\boldsymbol{X}, t)$. The right Cauchy-Green tensor is further defined as $\boldsymbol{C} = \boldsymbol{F}^T \boldsymbol{F}$.

For a description of growth, a multiplicative decomposition of the deformation gradient is adopted. Assumption of an intermediate incompatible configuration which achieves a locally stress-free state forms the basis of this split [5, 14]. An additional elastic deformation is needed to ensure the compatibility of the total deformation. Hence the total deformation gradient will be $\boldsymbol{F} = \boldsymbol{F}_e \, \boldsymbol{F}_g$ and the elastic right Cauchy-Green tensor reads $\boldsymbol{C}_e = \boldsymbol{F}_e^T \, \boldsymbol{F}_e$.

The growth deformation gradient is specified as $\boldsymbol{F}_g = \vartheta \, \boldsymbol{I}$, under the assumption of isotropic growth. Hence the compatible elastic deformation gradient will be $\boldsymbol{F}_e = \vartheta^{-1} \boldsymbol{F}$.

To calculate the growth stretch ϑ, it is hypothesised that an increase in volume at a point due to growth at any time t is proportional to the additional SMC transported to that point. A similar idea is postulated for aggregation of foam cells in [16]. If v_s is the volume occupied by a single SMC and N_s the number of additional SMC at a point, then $\Delta V_g = v_s \, N_s$. v_s can be prescribed via the definition of the SMC density of a healthy artery $\rho_{S,h}$, using $v_s = (\rho_{S,h})^{-1}$. If V_g is the volume in the intermediate growth configuration Ω_g at time t and V_0 is that in the reference configuration Ω_0, then we can arrive at an expression for ϑ through the following exercise.

$$\Delta V_g = V_g - V_0 = (\rho_{S,h})^{-1} \, N_s \tag{5}$$

$$\int_{\Omega_g} 1 \, dv_g - \int_{\Omega_0} 1 \, dV = (\rho_{S,h})^{-1} \int_{\Omega} (\rho_S - \rho_{S,h}) \, dv \tag{6}$$

Pulling back the quantities to the initial configuration,

$$\int_{\Omega_0} J_g \, dV - \int_{\Omega_0} 1 \, dV = (\rho_{S,h})^{-1} \int_{\Omega_0} J \, (\rho_S - \rho_{S,h}) \, dV \tag{7}$$

Since Eq. (7) has to hold true locally,

$$J_g - 1 = (\rho_{S,h})^{-1} \, J \, (\rho_S - \rho_{S,h}) \tag{8}$$

$$\Longrightarrow \; J_g = 1 + J \left(\frac{\rho_S}{\rho_{S,h}} - 1 \right). \tag{9}$$

If d is the dimensionality of the problem, then $J_g = \det(\boldsymbol{F}_g) = \vartheta^d$. Therefore,

$$\vartheta = \left(1 + J \left(\frac{\rho_S}{\rho_{S,h}} - 1 \right) \right)^{1/d}. \tag{10}$$

2.4 Hyperelastic Constitutive Model

The constitutive model is derived from the hyperelastic models presented in [4, 6, 13], wherein the arterial wall is assumed to be composed of two helices of collagen fibres, with respective helix angles θ_i^a, embedded in an isotropic ground substance. The associated Helmholtz free energy per unit volume is hence split into an isotropic and an anisotropic part, both of which are dependent on the elastic right Cauchy-Green tensor.

$$\psi(\boldsymbol{C}_e, \boldsymbol{H}_1, \boldsymbol{H}_2) = \psi_{iso}(\boldsymbol{C}_e) + \psi_{ani}(\boldsymbol{C}_e, \boldsymbol{H}_1, \boldsymbol{H}_2) \tag{11}$$

$$\psi_{iso}(\boldsymbol{C}_e) = \frac{\mu^a}{2} \left(\operatorname{tr} \boldsymbol{C}_e - 3\right) - \mu^a \ln J_e + \frac{\Lambda^a}{4} \left(J_e^2 - 1 - 2 \ln J_e\right) \tag{12}$$

$$\psi_{ani}(\boldsymbol{C}_e, \boldsymbol{H}_1, \boldsymbol{H}_2) = \frac{k_1}{2k_2} \sum_{i=1,2} \left(\exp\left[k_2 \langle E_i \rangle^2\right] - 1\right) \tag{13}$$

This stems from the hypothesis that in the stress-free incompatible growth configuration, even the collagen fibres grow due to the synthesis by SMC and hence no residual stresses arise in the incompatible state. The dependence of ψ on $\boldsymbol{C}_e$ embeds its dependence on $\boldsymbol{F}_g$ via

$$\boldsymbol{C}_e = \boldsymbol{F}_g^{-T} \, \boldsymbol{C} \, \boldsymbol{F}_g^{-1} = \vartheta^{-2} \boldsymbol{C}. \tag{14}$$

The generalised structure tensors $\boldsymbol{H}_i$ are constructed based on the local collagen orientations in the reference configuration $\boldsymbol{a}_{0i}$ using

$$\boldsymbol{H}_i = \kappa^a \, \boldsymbol{I} + \left(1 - 3\,\kappa^a\right) \boldsymbol{a}_{0i} \otimes \boldsymbol{a}_{0i} \;, \tag{15}$$

where κ^a is a dispersion parameter. The Green-Lagrange strain like parameter E_i is calculated utilising the relationship $E_i = \boldsymbol{H}_i : \boldsymbol{C}_e - 1$, where the definition of scalar product of second order tensors $\boldsymbol{A} : \boldsymbol{B} = A_i \, B_i$ (Einstein summation convention) is applied.

The first Piola-Kirchhoff stress tensor is deduced from the Helmholtz free energy function using

$$\boldsymbol{P} = \frac{\partial \psi}{\partial \boldsymbol{F}} = \frac{\partial \psi}{\partial \boldsymbol{C}_e} : \frac{\partial \boldsymbol{C}_e}{\partial \boldsymbol{F}}. \tag{16}$$

Using Eq. (14), it can be deduced that

$$\frac{\partial \boldsymbol{C}_e}{\partial \boldsymbol{F}} = \frac{1}{\vartheta^2} \frac{\partial \boldsymbol{C}}{\partial \boldsymbol{F}}. \tag{17}$$

Finally, the balance of linear momentum governs the quasi-static equilibrium of the arterial wall structure.

$$\nabla \cdot \boldsymbol{P} + \boldsymbol{b} = \boldsymbol{0} \tag{18}$$

3 Numerical Methods

The Galerkin weak forms of the governing differential equations are arrived at.

$$\int_\Omega \frac{\partial c_P}{\partial t} \, \delta c_P \, dv = \int_\Gamma (J_P \cdot \boldsymbol{n}) \, \delta c_P \, da - \int_\Omega (\nabla \delta c_P \cdot D_P \nabla c_P) \, dv - \int_\Omega \alpha \, \rho_S \, c_P \, \delta c_P \, dv \tag{19}$$

$$\int_\Omega \frac{\partial \rho_E}{\partial t} \, \delta \rho_E \, dv = \int_\Omega \beta \, \rho_S \left(1 - \frac{\rho_E}{\rho_{E,th}}\right) \delta \rho_E \, dv - \int_\Omega \gamma c_P \rho_E \, \delta \rho_E \, dv \tag{20}$$

$$\int_\Omega \frac{\partial \rho_S}{\partial t} \, \delta \rho_S \, dv = \int_\Gamma (J_S \cdot \boldsymbol{n}) \, \delta \rho_S \, da - \int_\Omega \left(\nabla \delta \rho_S \cdot \chi \, c_P \left(1 - \frac{\rho_E}{\rho_{E,th}}\right) \rho_S \nabla \rho_E\right) dv + \int_\Omega \kappa \, c_P \, \rho_S \left(1 - \frac{\rho_E}{\rho_{E,th}}\right) \delta \rho_S \, dv \tag{21}$$

$$\int_{\Omega_0} \boldsymbol{P} : \delta \boldsymbol{F} \, dV = \int_{\Gamma_0} (\boldsymbol{T} \cdot \delta \boldsymbol{u}) \, dA \tag{22}$$

Here, J_P and J_S denote the PDGF and SMC fluxes across the boundary, $\boldsymbol{n}$ the normal on the boundary and $\boldsymbol{T}$ the traction on the boundary. Body forces are considered absent.

The transport equations are temporally discretised using the semi-implicit Backward-Euler method. Semi-implicitness is attributed to the fact that in each of the weak forms, variables other than the ones discretised in time are carried over from the previous time step. Spatial discretisation is performed using linear finite elements. A decoupled set of linear equations for the time-step updation of transport variables are hence obtained.

$$[\boldsymbol{M} + \Delta t\, \boldsymbol{L} + \Delta t\, \boldsymbol{P}]\, \boldsymbol{c}_{P}^{n+1} = \boldsymbol{M} \boldsymbol{c}_{P}^{n} \tag{23}$$

$$[\boldsymbol{M} + \Delta t\, \boldsymbol{T}]\, \boldsymbol{\rho}_{E}^{n+1} = \boldsymbol{M} \boldsymbol{\rho}_{E}^{n} + \boldsymbol{R} \tag{24}$$

$$\left[\boldsymbol{M} + \Delta t\, \boldsymbol{K} - \Delta t\, \boldsymbol{Q}\right] \boldsymbol{\rho}_{S}^{n+1} = \boldsymbol{M} \boldsymbol{\rho}_{S}^{n} \tag{25}$$

Definitions of matrices in the equations above can be found in the appendix. The diffusion and advection operators within the finite element setting, when discretised with linear finite elements, display instabilities across steep gradients in the transport quantities. This leads to unphysical negative concentrations and densities at the diffusion and advection fronts. To treat this, artificial discrete diffusive fluxes are introduced, which result in over smoothened solution profiles. Hence antidiffusive fluxes, limited using the Zalesak limiter, are utilised to preserve local extrema [11, 15].

The weak form of the balance of linear momentum in Eq. (22) is spatially discretised using the same linear finite elements as defined for the transport problems, and is solved iteratively using the Newton-Raphson method. The residual and its linearization, prior to discretisation, required for the iterative updates are determined as follows.

$$g_u = \int_{\Omega_0} \boldsymbol{P} : \delta \boldsymbol{F}\, dV - \int_{\Gamma_0} (\boldsymbol{T} \cdot \delta \boldsymbol{u})\, dA \tag{26}$$

$$\Delta g_u = \int_{\Omega_0} \delta \boldsymbol{F} : (\mathcal{A} : \Delta \boldsymbol{F})\, dV \tag{27}$$

The fourth order tensor $\mathcal{A}$ is computed using

$$\mathcal{A} = \frac{\partial \boldsymbol{P}}{\partial \boldsymbol{F}} + \frac{\partial \boldsymbol{P}}{\partial \vartheta} \otimes \frac{\partial \vartheta}{\partial \boldsymbol{F}}. \tag{28}$$

Coupling

The transport problems are solved in the current configuration Ω while the balance of linear momentum is solved in the reference configuration Ω_0. A staggered solution strategy is carried out to achieve coupling. The transport problem provides the SMC density to the structural problem while the structural problem updates the current configuration Ω based on the SMC density. An incremental form of Eq. (10) is applied in this regard as follows:

$$\vartheta = \vartheta^{-} \left(1 + \frac{J}{J^{-}} \left(\frac{\rho_S}{\rho_{S,h}} - 1\right)\right)^{1/d}, \tag{29}$$

wherein the quantities $(\ )^{-}$ refer to those from the previous time step. At the end of solution of the structural problem, the transport quantities ϕ are updated at the Gauss-points using $\phi^* = (J^{-}/J)\, \phi$ and extrapolated to the nodes. The staggered solution strategy provides flexibility in case different spatial discretisations are absolutely necessary for the systems being coupled.

4 Numerical Investigation

A longitudinal axisymmetric section of an artery is considered fixed on either ends. A Gaussian distribution of PDGF concentration consisting of three distinct peaks is prescribed in the domain to mimic a PCI injury in the arterial wall (Fig. 1). ECM and SMC densities ($\rho_{E,0}$ and $\rho_{S,0}$) are uniformly prescribed to be those of a healthy artery. Zero flux boundary conditions are prescribed for all the transport quantities on all boundaries. Some of the parameters are deduced from literature and the others tuned to predict the physics qualitatively (Table 2).

Transport simulation is run for a time period of 1 day to perform the mesh and time step convergence analyses since this dictates the spatial and temporal discretisations. The SMC density plots along section A-A are used for the analyses (Figs. 2, 3).

Using the converged discretisations, the coupled simulation is run for a period of 1 day and the results examined. It is expected that the growth ceases when the ECM is healed and PDGF is completely internalized. The evolutions of quantities at a Gauss-point of interest reflect this as shown in Fig. 4. Contraction is observed in some regions of the arterial wall which can be attributed to the migration of SMC away from these regions.

Table 2 Parameter set

Parameter	Value [units]
D_P	0.01 [mm^2/day]
α	1.0×10^{-13} [mm^3/cell/day]
β	5.0×10^{-8} [mol/cell/day]
γ	5.0×10^{17} [mm^3/mol/day]
χ	1.0×10^{19} [mm^5/cell/day]
κ	1.0×10^{-2} [mm^3/mol/cell/day]
$\rho_{E,0}$	7.0×10^{-9} [mol/mm^3]
$\rho_{E,th}$	$1.1 \times \rho_{E,0}$
$\rho_{S,0}$	3.16×10^6 [cells/mm^3]
μ^a	0.02 [MPa]
λ^a	10 [MPa]
k_1	0.112 [MPa]
k_2	20.61 [–]
κ^a	0.1 [–]
θ_i^a	41 [°]

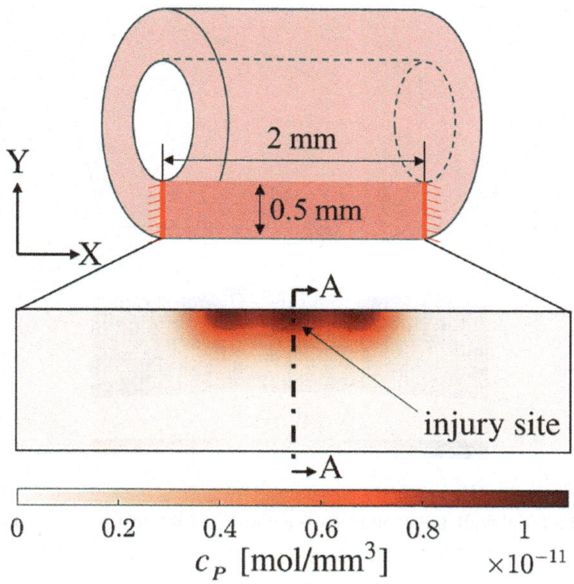

Fig. 1 Problem setup

Fig. 2 Mesh convergence (A-A)

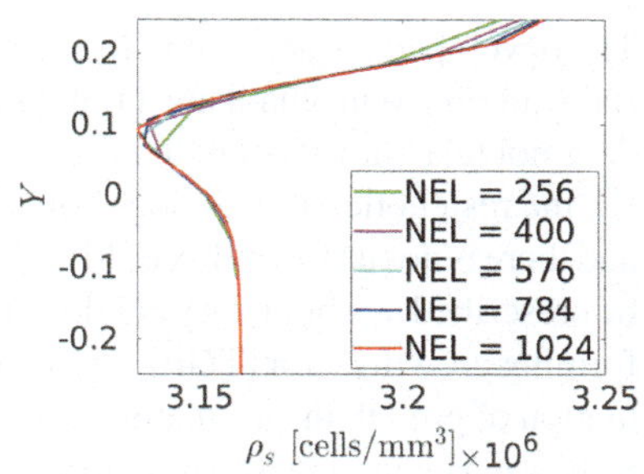

Fig. 3 Time-step convergence (A-A)

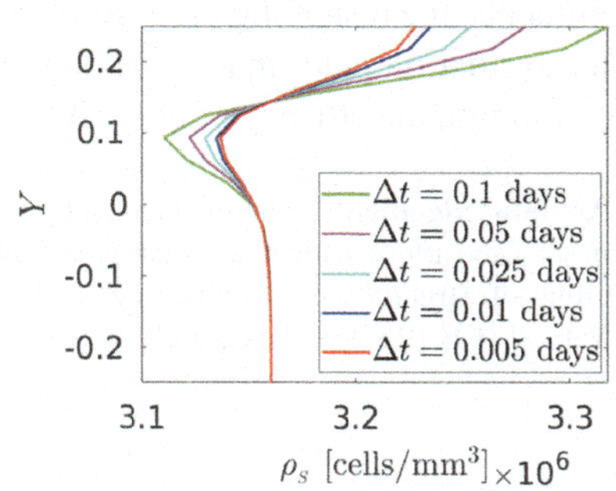

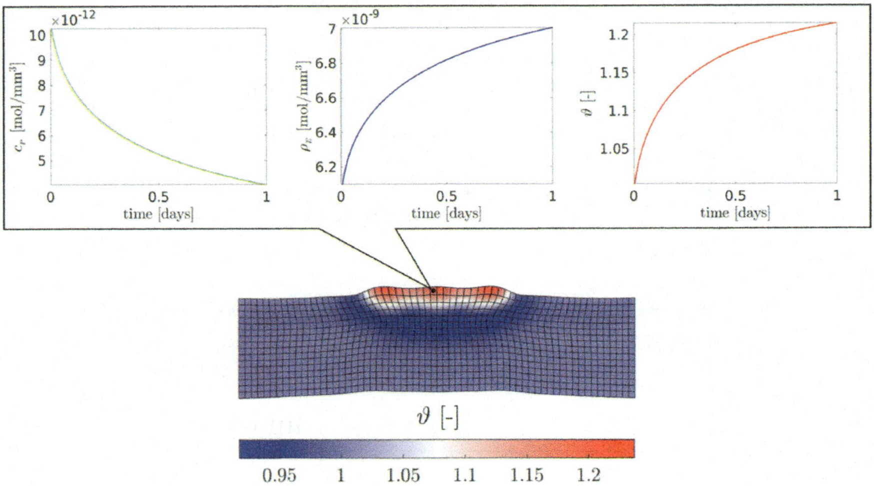

Fig. 4 Deformed arterial wall and evolution of quantities of interest

5 Conclusion and Outlook

The developed model is able to successfully capture the growth induced in the arterial wall due to mechanisms brought about by PDGF. It is to be noted that the model does not take into account the loss and healing processes of the endothelium. In reality, the restenotic process does not cease until the endothelium is completely healed and there is no further influx of PDGF. Wall shear stress dependent influx of PDGF can be prescribed as a boundary condition to achieve the aforementioned effect. The transforming growth factor (TGF)-β is another significant constituent responsible for modulation of growth in vasculatures. It considerably affects the proliferative response of SMCs and hence needs to be considered in the framework developed in this work. Furthermore, it remains to be seen how sensitive the model is to the set of model parameters and which of those can be determined experimentally.

Future work shall therefore entail consideration of additional wall constituents that influence the restenotic process, physiological experimentation, parameter sensitivity evaluation, and finally arriving at a high-fidelity in-silico model which aides in examining the efficacy of drugs intended to be used on drug-eluting stents.

Acknowledgements The financial support of DFG for the projects ‘Drug-eluting coronary stents in stenosed arteries: medical investigation and computational modelling’ (RE 1057/44-1) and ‘Modelling of Structure and Fluid-Structure Interaction during Tissue Maturation in Biohybrid Heart Valves’ (RE 1057/45-1) is gratefully acknowledged.

Appendix

Element matrices in the linearised decoupled system of transport equations are listed here. $\boldsymbol{N}$ refers to the shape function vector associated with linear finite elements.

$$\boldsymbol{M}^e = \int_{\Omega_e} \boldsymbol{N}\,\boldsymbol{N}^T\,dv$$

$$\boldsymbol{L}^e = \int_{\Omega_e} D_P \nabla\boldsymbol{N}\,\nabla\boldsymbol{N}^T\,dv$$

$$\boldsymbol{P}^e = \int_{\Omega_e} \alpha\rho_S^n \boldsymbol{N}\,\boldsymbol{N}^T\,dv$$

$$\boldsymbol{T}^e = \int_{\Omega_e} \left(\beta\,\frac{\rho_S^n}{\rho_{E,th}} + \gamma\,c_P^n\right)\boldsymbol{N}\,\boldsymbol{N}^T\,dv$$

$$\boldsymbol{K}^e = \int_{\Omega_e} \left[\chi\,c_P^n\left(1 - \frac{\rho_E^n}{\rho_{E,th}}\right)\nabla\boldsymbol{N}\,\nabla\rho_1^n\boldsymbol{N}^T\right]dv$$

$$\boldsymbol{Q}^e = \int_{\Omega_e} \kappa\,c_P^n\left(1 - \frac{\rho_E^n}{\rho_{E,th}}\right)\boldsymbol{N}\,\boldsymbol{N}^T\,dv$$

$$\boldsymbol{R}^e = \int_{\Omega_e} \beta\,\rho_S^n\,\boldsymbol{N}\,dv$$

References

1. Budu-Grajdeanu, P., Schugart, R. C., Friedman, A., Valentine, C., Agarwal, A. K., Rovin, B. H. (2008). A mathematical model of venous neointimal hyperplasia formation. *Theoretical Biology and Medical Modelling, 5*, 2.
2. Cyron, C. J., Aydin, R. C., Humphrey, J. D. (2016). A homogenized constrained mixture (and mechanical analog) model for growth and remodeling of soft tissue. *Biomechanics and Modeling in Mechanobiology, 15*, 1389–1403.
3. Escuer, J., Martínez, M. A., McGinty, S., Peña, E. (2019). Mathematical modelling of the restenosis process after stent implantation. *Journal of the Royal Society Interface, 16*, 20190313.
4. Gasser, T. C., Ogden, R. W., Holzapfel, G. A. (2006). Hyperelastic modelling of arterial layers with distributed collagen fibre orientations. *Journal of the Royal Society Interface, 3*, 15–35.
5. Himpel, G., Kuhl, E., Menzel, A., Steinmann, P. (2005). Computational modelling of isotropic multiplicative growth. *Computer Modeling in Engineering and Sciences, 8*(2), 119–134.
6. Holzapfel, G. A., Gasser, T. C., Ogden, R. W. (2000). A new constitutive framework for arterial wall mechanics and a comparative study of material models. *Journal of Elasticity, 61*, 1–48.
7. Humphrey, J. D., Rajagopal, K. R. (2002). A constrained mixture model for growth and remodeling of soft tissues. *Mathematical Models and Methods in Applied Sciences, 12*(3), 407–430.
8. Keller, E., Segel, L. A. (1971). Model for chemotaxis. *Journal of Theoretical Biology, 30*, 225–234.
9. Koskinas, K. C., Chatzizisis, Y. S., Antoniadis, A. P., Giannoglou, G. D. (2012). Role of endothelial shear stress in stent restenosis and thrombosis pathophysiologic. *Journal of the American College of Cardiology, 59*(15), 1337–1349.
10. Koyama, N., Hart, C. E., Clowes, A. W. (1994). Different functions of the platelet-derived growth factor-α and -β receptors for the migration and proliferation of cultured baboon smooth muscle cells. *Circulation Research, 75*, 682–691.
11. Kuzmin, D., Turek, S. (2002). Flux correction tools for finite elements. *Journal of Computational Physics, 175*, 525–558.
12. Li, S., Lei, L., Hu, Y., Zhang, Y., Zhao, S., Zhang, J. (2019). A fully coupled framework for in silico investigation of in-stent restenosis. *Computer Methods in Biomechanics and Biomedical Engineering, 22*(2), 217–228.
13. Nolan, D. R., Gower, A. L., Destrade, M., Ogden, R. W., McGarry, J. P. (2014). *Journal of the Mechanical Behavior of Biomedical Materials, 39*, 48–60.

14. Rodriguez, E. K., Hoger, A., McCulloch, A. D. (1994). Stress-dependent finite growth in soft elastic tissues. *Journal of Biomechanics, 21*(4), 455–467.
15. Strehl, R., Sokolov, A., Kuzmin, D., Horstmann, D., Turek, S. (2013). A positivity-preserving finite element method for chemotaxis problems in 3D. *Journal of Computational and Applied Mathematics, 239*, 290–303.
16. Thon, M. P., Hemmler, A., Glinzer, A., Mayr, M., Wildgruber, M., Zernecke-Madsen, A., Gee, M. W. (2018). A multiphysics approach for modeling early atherosclerosis. *Biomechanics and Modeling in Mechanobiology, 17*, 617–644.
17. Zahedmanesh, H., van Oosterwyck, H., Lally, C. (2012). A multi-scale mechanobiological model of in-stent restenosis: Deciphering the role of matrix metalloproteinase and extracellular matrix changes. *Computer Methods in Biomechanics and Biomedical Engineering, 17*(8), 813–828.

It's Too Stiff: On a Novel Mixed Finite Element Formulation for Nearly-Inextensible Materials

Michele Marino

Professor Wriggers has been my Mentor during my postdoctoral training from 2015 until 2019. I had the privilege to collaborate with Him on several topics, like multiphysical computational modelling for biomedical applications, the Virtual Element Method for homogenization, and mixed formulations for anisotropic materials. Discussions with Professor Wriggers have always been inspiring. His leadership and example have helped me grow into my potential and I would not be where and who I am today without Him. I am looking forward to many more meetings, especially those gathering scientific discussions with a glass of nice red wine, possibly on Hamburg's pier.

Abstract Engineering applications often comprise materials with an anisotropic response, eventually including nearly-inextensible fibers and/or small volumetric deformations. These behaviours can be regarded as internal constraints in material response. Numerical simulations comprising "constrained" materials show an over-stiff structural response, referred to as element locking. Implementations based on mixed variational methods can heal locking but several solutions adopted in the state-of-the-art are still non-optimal for anisotropic materials. This issue is here analysed by discussing a decomposition of anisotropic deformation modes based on kinematic and energy criteria. Low-order mixed finite element formulations for dealing with near-inextensibility and/or near-incompressibility constraints are also introduced, and analysed on novel numerical applications of engineering interest.

M. Marino (✉)
Department of Civil Engineering and Computer Science, University of Rome Tor Vergata, Rome, Italy
e-mail: m.marino@ing.uniroma2.it

F. Aldakheel et al. (eds.), *Current Trends and Open Problems in Computational Mechanics*, https://doi.org/10.1007/978-3-030-87312-7_31

1 Introduction

The robustness and accuracy of numerical simulations can be severely hampered by internal constraints in material response, e.g., when materials are characterized by small volumetric deformations (nearly-incompressible) or when material microstructure is made up by highly-stiff fibers (nearly-inextensible). Traditional finite element formulations exhibit poor convergence rates in the basic variables and element responses result overstiff. This phenomenon is generally referred to as locking [1, 2]. Different strategies were pursued in computational mechanics over the last decades in order to circumvent locking. Few examples are in [1, 2] for incompressibility and in [3–5] for inextensibility. Among available approaches, this work addresses mixed finite element formulations [6].

Constraints in material response introduce restrictions for the admissible kinematics in equilibrium problems. Ideal constraints impose that some components of the deformation field take a given value, while non-ideal behaviours are generally introduced by relaxing the constrained deformation metric introduced in the ideal case. Equilibrium and kinematic constraints can be enforced through a variational principle, read from stationary conditions of a mixed functional. The latter depends on displacements and additional variables which represent a reactive stress (associated with the internal constraint), and in the non-ideal case, its thermodynamically-conjugate (kinematic) variable. In the ideal case, the reactive stress depends only on the constrained kinematics and it can be defined *ab initio*, independently from the actual constitutive response. Hence, it is required to be workless in all deformations satisfying the constraint, that is energetic orthogonal to the non-constrained kinematics. For instance, in the presence of incompressibility, the classical volumetric-deviatoric split on the deformation is generally introduced and a hydrostatic reactive stress is postulated since orthogonal in energy to the deviatoric deformation. On the other hand, in the non-ideal case, the reactive component transforms into a stress term which cannot be postulated but depends on material constitutive response. Hence, if the reactive stress is energy orthogonal to the non-constrained kinematics depends on the material symmetries at hand. For instance, the classical volumetric-deviatoric split for nearly-incompressible materials is orthogonal in energy if and only if the material is isotropic. In the presence of anisotropies, the stress associated with a volumetric deformation does not indeed correspond to a hydrostatic pressure and energy orthogonality would be not respected by the classical volumetric-deviatoric split. Addressing near-inextensible fibers, the standard choice of removing fiber strain from the total one also does not lead to an energetic orthogonal formulation.

To overcome these issues, Marino and Wriggers have proposed in [7] a novel energetically-decoupled kinematic description of transversely isotropic linear elastic materials, which allows to consistently isolate the effects of near-inextensibility and/or near-incompressibility. A novel Hu-Washizu-like variational formulation based on this theory and its implementation in mixed finite element models have been also proposed. The present paper addresses novel numerical examples to test the convergence properties of the proposed formulation in problems of engineering interest, also referring to three-dimensional (3D) problems.

2 Theory

The main theoretical arguments introduced in [7] are here presented for the sake of completeness. The theory is developed in the framework of linear elasticity, with $\boldsymbol{\varepsilon}$ and $\boldsymbol{\sigma}$ being the small-strain and the Cauchy stress second-order tensors. For the sake of notation, let second-order tensors be denoted by bold capital letters (e.g., $\mathbf{A}$), while fourth-order tensors by capital double-struck letters (e.g., $\mathbb{A}$). Hence, let $\mathbf{A}:\mathbf{B}$ be the inner product between $\mathbf{A}$ and $\mathbf{B}$, and $\mathbb{A}:\mathbf{B}$ the right mapping of $\mathbb{A}$ with $\mathbf{B}$.

2.1 Nearly-Inextensible Transversely Isotropic Materials

The stiffness tensor $\mathbb{C}_{\mathrm{TI}}$ of a transversely isotropic material can be represented on the basis of the Walpole's formalism[1] as [8]:

$$\mathbb{C}_{\mathrm{TI}} = c_1\mathbb{B}_1 + 2\kappa_{23}\mathbb{B}_2 + \sqrt{2}c_2(\mathbb{B}_3 + \mathbb{B}_4) + 2G_{23}\mathbb{B}_5 + 2G_{12}\mathbb{B}_6, \tag{1}$$

where c_1 and c_2 represent axial (i.e., along the material preferred direction $\mathbf{a}$) and in-plane (i.e., in the isotropy plane $\mathcal{P}(\mathbf{a})$) stiffnesses, κ_{23} the in-plane (plane-strain) bulk modulus, G_{23} and G_{12} the in-plane and out-of-plane shear moduli.

This section addresses materials characterized by near-inextensibility along the preferred direction $\mathbf{a}$, that is $c_1 \rightarrow +\infty$. Near-inextensibility is formulated by enforcing that the normal component ε_{a} of $\boldsymbol{\varepsilon}$ along $\mathbf{a}$ respects:

$$\phi_{ine}(\boldsymbol{\varepsilon}) = \boldsymbol{\varepsilon}:\mathbf{A} = \varepsilon_{\mathrm{a}} \rightarrow 0 \qquad \text{for} \quad c_1 \rightarrow +\infty. \tag{2}$$

Near-inextensibility is treated by introducing the transversely isotropic projection operator $\mathbb{P}_{\mathrm{TI}}$, the projected component of the strain tensor $\boldsymbol{\varepsilon}^p_{\mathrm{TI}}$, and the complementary one $\boldsymbol{\varepsilon}^c_{\mathrm{TI}}$ as:

$$\boldsymbol{\varepsilon}^p_{\mathrm{TI}} = \mathbb{P}_{\mathrm{TI}}:\boldsymbol{\varepsilon}\,, \qquad \boldsymbol{\varepsilon}^c_{\mathrm{TI}} = (\mathbb{I} - \mathbb{P}_{\mathrm{TI}}):\boldsymbol{\varepsilon} \qquad \text{with} \quad \mathbb{P}_{\mathrm{TI}} = \mathbb{B}_1 + \mathbb{B}_2. \tag{3}$$

It is immediate to verify (see [7]) that it results $\boldsymbol{\varepsilon}^p_{\mathrm{TI}} + \boldsymbol{\varepsilon}^c_{\mathrm{TI}} = \boldsymbol{\varepsilon}$ and the split (3) is:

1. kinematically consistent with near-inextensibility, that is both $\boldsymbol{\varepsilon}^p_{\mathrm{TI}}$ and $\boldsymbol{\varepsilon}$ allow to enforce (2) since leading to the same value of ε_{a}, i.e.

$$\boldsymbol{\varepsilon}^p_{\mathrm{TI}}:\mathbf{A} = (\mathbb{P}_{\mathrm{TI}}:\boldsymbol{\varepsilon}):\mathbf{A} = \varepsilon_{\mathrm{a}} \Rightarrow \phi_{ine}(\boldsymbol{\varepsilon}) \rightarrow 0 \iff \phi_{ine}(\boldsymbol{\varepsilon}^p_{\mathrm{TI}}) \rightarrow 0; \tag{4}$$

[1] Introducing $\mathbf{A} = \mathbf{a}\otimes\mathbf{a}$, $\mathbf{B} = \mathbf{I} - \mathbf{A}$, and tensor products as in [7], the Walpole's basis reads:

$$\mathbb{B}_1 = \mathbf{A}\otimes\mathbf{A}\,, \quad \mathbb{B}_2 = \frac{1}{2}\mathbf{B}\otimes\mathbf{B}\,, \quad \mathbb{B}_3 = \frac{\sqrt{2}}{2}\mathbf{A}\otimes\mathbf{B}\,, \quad \mathbb{B}_4 = \frac{\sqrt{2}}{2}\mathbf{B}\otimes\mathbf{A},$$
$$\mathbb{B}_5 = \frac{1}{2}\left(\mathbf{B}\overline{\otimes}\mathbf{B} + \mathbf{B}\underline{\otimes}\mathbf{B} - \mathbf{B}\otimes\mathbf{B}\right), \quad \mathbb{B}_6 = \frac{1}{2}\left(\mathbf{A}\overline{\otimes}\mathbf{B} + \mathbf{A}\underline{\otimes}\mathbf{B} + \mathbf{B}\overline{\otimes}\mathbf{A} + \mathbf{B}\underline{\otimes}\mathbf{A}\right).$$

.

2. energetically consistent for transversely isotropic materials, that is $\boldsymbol{\varepsilon}^p_{\rm TI}$ is orthogonal to the stress associated with the complementary deformation mode $\boldsymbol{\sigma}^c_{\rm TI} = \mathbb{C}_{\rm TI} : \boldsymbol{\varepsilon}^c_{\rm TI}$ (and viceversa), i.e.

$$\boldsymbol{\sigma}^c_{\rm TI} : \boldsymbol{\varepsilon}^p_{\rm TI} = [\mathbb{C}_{\rm TI} : (\mathbb{I} - \mathbb{P}_{\rm TI}) : \boldsymbol{\varepsilon}] : (\mathbb{P}_{\rm TI} : \boldsymbol{\varepsilon}) = \mathbb{C}^c_{\rm TI} : \boldsymbol{\varepsilon} : (\mathbb{P}_{\rm TI} : \boldsymbol{\varepsilon}) = 0, \tag{5a}$$

$$\boldsymbol{\sigma}^p_{\rm TI} : \boldsymbol{\varepsilon}^c_{\rm TI} = [\mathbb{C}_{\rm TI} : \mathbb{P}_{\rm TI} : \boldsymbol{\varepsilon}] : (\mathbb{I} - \mathbb{P}_{\rm TI}) : \boldsymbol{\varepsilon} = \mathbb{C}^p_{\rm TI} : \boldsymbol{\varepsilon} : (\mathbb{I} - \mathbb{P}_{\rm TI}) : \boldsymbol{\varepsilon} = 0, \tag{5b}$$

with:

$$\mathbb{C}^p_{\rm TI} = \mathbb{C}_{\rm TI} : \mathbb{P}_{\rm TI} = c_1\, \mathbb{B}_1 + 2\kappa_{23}\, \mathbb{B}_2 + \sqrt{2} c_2 (\mathbb{B}_3 + \mathbb{B}_4), \tag{6a}$$

$$\mathbb{C}^c_{\rm TI} = \mathbb{C}_{\rm TI} : (\mathbb{I} - \mathbb{P}_{\rm TI}) = 2G_{23}\, \mathbb{B}_5 + 2G_{12}\, \mathbb{B}_6. \tag{6b}$$

The energy orthogonality condition in Eqs. (5), together with Eqs. (6), yields a fully decoupled split of the strain-energy of transversely isotropic materials into a projected $\Psi^p_{\rm TI}(\boldsymbol{\varepsilon}^p_{\rm TI})$ and complementary part $\Psi^c_{\rm TI}(\boldsymbol{\varepsilon}^c_{\rm TI})$:

$$\Psi_{\rm TI}(\boldsymbol{\varepsilon}) = \frac{1}{2}\boldsymbol{\varepsilon} : \mathbb{C}_{\rm TI} : \boldsymbol{\varepsilon} = \overbrace{\frac{1}{2}\boldsymbol{\varepsilon}^p_{\rm TI} : \mathbb{C}^p_{\rm TI} : \boldsymbol{\varepsilon}^p_{\rm TI}}^{\Psi^p_{\rm TI}(\boldsymbol{\varepsilon}^p_{\rm TI})} + \overbrace{\frac{1}{2}\boldsymbol{\varepsilon}^c_{\rm TI} : \mathbb{C}^c_{\rm TI} : \boldsymbol{\varepsilon}^c_{\rm TI}}^{\Psi^c_{\rm TI}(\boldsymbol{\varepsilon}^c_{\rm TI})}. \tag{7}$$

2.2 *Variational Formulation for Nearly-Inextensible Materials*

Near-inextensibility is introduced by means of a perturbed Lagrangian formulation based on the Hu-Washizu principle. From Eq. (7), the following mixed variational functional $\overline{\Psi}_{\rm TI}$ is introduced:

$$\overline{\Psi}_{\rm TI}(\boldsymbol{\varepsilon}, \boldsymbol{\Sigma}_{\rm TI}, \boldsymbol{\Theta}_{\rm TI}) = \Psi^p_{\rm TI}(\boldsymbol{\Theta}_{\rm TI}) + \Psi^c_{\rm TI}(\boldsymbol{\varepsilon}^c_{\rm TI}) + \boldsymbol{\Sigma}_{\rm TI} : (\boldsymbol{\varepsilon}^p_{\rm TI} - \boldsymbol{\Theta}_{\rm TI}), \tag{8}$$

where $\boldsymbol{\Sigma}_{\rm TI}$ is the stress variable thermodynamic-conjugate to $\boldsymbol{\varepsilon}^p_{\rm TI}$, and $\boldsymbol{\Theta}_{\rm TI}$ the kinematic variable controlling $\boldsymbol{\varepsilon}^p_{\rm TI}$. Stationary conditions of $\overline{\Psi}_{\rm TI}$ yield:

Balance of linear momentum: $$\delta_{\boldsymbol{\varepsilon}} \overline{\Psi}_{\rm TI} = \delta\boldsymbol{\varepsilon} : \left(\boldsymbol{\Sigma}_{\rm TI} : \mathbb{P}_{\rm TI} + \mathbb{C}^c_{\rm TI} : \boldsymbol{\varepsilon}^c_{\rm TI}\right) = 0\,, \tag{9a}$$

Compatibility: $$\delta_{\boldsymbol{\Sigma}_{\rm TI}} \overline{\Psi}_{\rm TI} = \delta\boldsymbol{\Sigma}_{\rm TI} : \left(\boldsymbol{\varepsilon}^p_{\rm TI} - \boldsymbol{\Theta}_{\rm TI}\right) = 0, \tag{9b}$$

Constitutive behaviour: $$\delta_{\boldsymbol{\Theta}_{\rm TI}} \overline{\Psi}_{\rm TI} = \delta\boldsymbol{\Theta}_{\rm TI} : \left(\mathbb{C}^p_{\rm TI} : \boldsymbol{\Theta}_{\rm TI} - \boldsymbol{\Sigma}_{\rm TI}\right) = 0. \tag{9c}$$

Accounting for Eqs. (6a) and (9c), it would result $\|\boldsymbol{\Sigma}_{\rm TI}\| \to +\infty$ for $c_1 \to +\infty$ when $\phi(\boldsymbol{\Theta}_{\rm TI}) = \theta_{\rm a} \nrightarrow 0$. An infinite stress would violate the equilibrium condition in Eq. (9a). Hence, it shall result $\phi(\boldsymbol{\Theta}_{\rm TI}) = \theta_{\rm a} \to 0$, such that $\boldsymbol{\Sigma}_{\rm TI}$ takes finite values. Since $\boldsymbol{\varepsilon}^p_{\rm TI} \approx \boldsymbol{\Theta}_{\rm TI}$ (in a weak sense) from Eq. (9b), it also results $\phi(\boldsymbol{\varepsilon}^p_{\rm TI}) = \varepsilon_{\rm a} \to 0$. These relationships introduce the penalized-kinematics in the proposed functional.

Remark 1 Kinematic and constitutive considerations lead to: $\mathbf{\Theta}_{\mathrm{TI}} = \theta_{\mathrm{a}}\mathbf{A} + \theta_{\mathrm{b}}\mathbf{B}$ and $\mathbf{\Sigma}_{\mathrm{TI}} = \sigma_{\mathrm{a}}\mathbf{A} + \sigma_{\mathrm{b}}\mathbf{B}$. Hence, only 4 additional scalar variables (θ_{a}, θ_{b}, σ_{a} and σ_{b}) are introduced with respect to a pure-displacement formulation.

Remark 2 The mixed functional $\overline{\Psi}_{\mathrm{TI}}$ is energetically decoupled in the sense that the complementary material strain energy term Ψ^c_{TI} does not depend on the penalized deformation mode $\boldsymbol{\varepsilon}^p_{\mathrm{TI}}$, i.e. $\Psi_c = \Psi_c(\boldsymbol{\varepsilon}^c_{\mathrm{TI}})$. This condition follows the properties of the kinematic split based on the transversely isotropic operator $\mathbb{P}_{\mathrm{TI}}$ in Eq. (3).

Since the mixed functional in Eq. (8) is energetically decoupled, the overstiff material constant c_1 do not explicitly enter in the balance of linear momentum (see Eq. (9a)). This occurrence avoids locking phenomena in numerical simulations based on finite element models derived with Eq. (8). As a matter of fact, if constant $c_1 \to +\infty$ explicitly enters in Eq. (9a), locking might occur since equilibrium can be satisfied with (practically) null deformation. This would lead to an overstiff/locked element response [1, 7].

2.3 On Nearly-Incompressible Transversely Isotropic Materials

This section addresses nearly-incompressible transversely isotropic materials, for which the material constraint acts on the volumetric deformation ε_{vol}, that is:

$$\phi_{inc}(\boldsymbol{\varepsilon}) = \mathrm{Tr}(\boldsymbol{\varepsilon}) = \varepsilon_{vol} \to 0 \qquad \text{for } \kappa \to +\infty. \tag{10}$$

Here, κ represents the bulk modulus of the material, which can be written as function of material constants in Eq. (1) as [7]:

$$\kappa = \frac{c_1\kappa_{23} - c_2^2}{c_1 - 2c_2 + \kappa_{23}}. \tag{11}$$

The split of the strain tensor in Eq. (3) is kinematically consistent with the near-incompressibility constraint since, as proven in [7], it results:

$$\mathrm{Tr}(\boldsymbol{\varepsilon}^p_{\mathrm{TI}}) = \varepsilon_{vol} \Rightarrow \phi_{inc}(\boldsymbol{\varepsilon}) \to 0 \quad \Longleftrightarrow \quad \phi_{inc}(\boldsymbol{\varepsilon}^p_{\mathrm{TI}}) \to 0. \tag{12}$$

Since the split (3) remains energetically-consistent as shown in Eqs. (5), the mixed functional (8) is energetically decoupled also for nearly-incompressible transversely isotropic materials. The same considerations made in Sect. 2.2 about locking hold true also in this case (see [7]). In fact, the overstiff material constant (κ for near-incompressibility) does not explicitly enter into the balance of linear momentum, since $\mathbb{C}^c_{\text{TI}}$ does not depend on constants involved in Eq. (11).

Clearly, these advantages are maintained also when near-incompressibility and near-inextensibility occur at the same time.

3 Mixed Finite Element Models and Numerical Results

Two numerical applications are introduced. Both applications consider a nearly-inextensible material, the second one considering also the coupling with near-incompressibility.[2] All quantities (e.g., lengths, forces and stiffnesses) are expressed in coherent SI-units.

3.1 Near-Inextensibility: A Bias-Extension Test

A bias-extension test of a rectangular $L \times H$ plate made up by $\pm$ 45 cross-plied unidirectional composite layers is firstly considered. The plate is clamped at $y = 0$ and an along-y displacement $\bar{v}$ is applied at $y = H$. Plane-stress is thus attained.

The bias-extension test has been originally designed for analysing the behavior of woven fabrics [9], which deform as pin-jointed-nets (PJNs). As shown in Fig. 1, the idealised PJN post-test geometry is characterized by 3 zones with different constant in-plane inter-fibers shear strains. The central region (A) is characterized by the maximum shear γ_s, approximated as:

$$\gamma_s \approx \gamma_{\text{PJN}} = \arcsin\left(\frac{\bar{v}}{\sqrt{2}(H-L)} + \frac{\sqrt{2}}{2}\right) - \frac{\pi}{2}. \tag{13}$$

Moreover, side regions (B) are characterized by intermediate shear $\gamma_s/2$, while underformed regions (C) are obtained at the boundary. This pattern determines a characteristic final shape of the specimen with sharp boundaries [9]. Although differences arise in local deformation mechanisms, experimental data show that the PJN model satisfactorily predicts the deformation response of ± 45 cross-plied composites under bias extension up to shear angles of $20° - 30°$ [10]. Thus, it is chosen $L = 50$, $H = 100$ and $\bar{v} = 7.923$, such that $\gamma_{\text{PJN}} = 20°$.

[2] Material constants in Eq. (1) for $\mathbb{C}_{\text{TI}}$ are obtained by employing a micromechanical approach which considers two isotropic constituents, i.e. fibers and matrix (see [7]). Fibers are significantly stiffer than the surrounding matrix (i.e., Young's moduli $E_f = 10^6 E_m$ with $E_m = 0.01$), leading to near-inextensibility. Fiber volume fraction is chosen equal to $V_f = 10\%$, fiber Poisson's ratio $\nu_f = 0.3$, while matrix Poisson's ratio ν_m is defined differently in the two addressed applications.

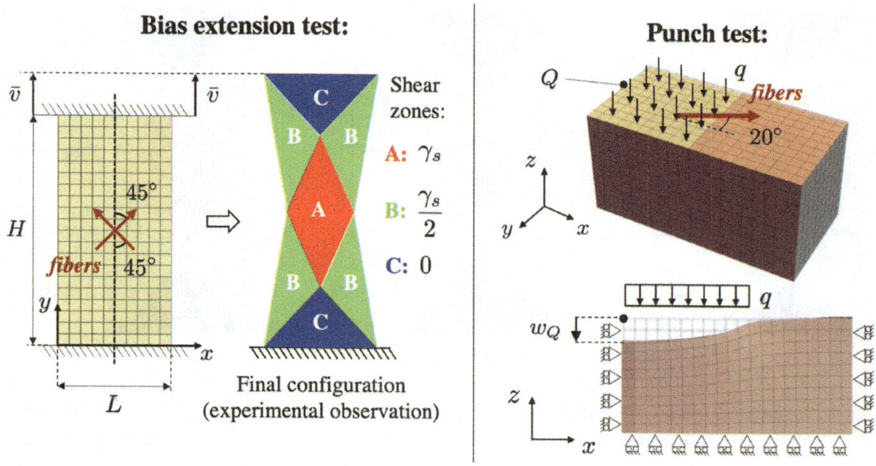

Fig. 1 Numerical applications: the bias-extension test (left) and the punch test (right). The orthogonal basis (**i**, **j**, **k**) is introduced and parametrized in x, y and z

The internal strain energy density $\Psi_c(\boldsymbol{\varepsilon})$ of the cross-plied composite with unidirectional fiber-reinforced layers is modelled as the superposition of two transversely isotropic materials $\Psi_{\mathrm{TI}}^{(\alpha)}$ with fibers inclined by $\alpha = \pm 45°$, namely

$$\Psi_c(\boldsymbol{\varepsilon}) = \frac{1}{2}\Psi_{\mathrm{TI}}^{(+45)}(\boldsymbol{\varepsilon}) + \frac{1}{2}\Psi_{\mathrm{TI}}^{(-45)}(\boldsymbol{\varepsilon}). \tag{14}$$

Here, $\Psi_{\mathrm{TI}}^{(\alpha)}$ is given by Eq. (7) where $\mathbb{C}_{\mathrm{TI}}$ is obtained from Eq. (1) with material preferred direction inclined by α. For this application, matrix Poisson's ratio is chosen $\nu_m = 0.3$. Therefore, only near-inextensibility is considered for this example.

The domain is discretized with quadrilateral (square) elements, with mesh density defined by the number of elements n_{el} along x. Three element formulations are considered: $\mathrm{Q1}_c$ and $\mathrm{Q2}_c$, where the strain-energy (14) is implemented in a pure-displacement formulation with bi-linear and bi-quadratic Lagrangian shape functions; $\overline{\mathrm{Q}}1\mathrm{A0}_c$, where each term $\Psi_{\mathrm{TI}}^{(\alpha)}$ in (14) is expressed as in Eq. (8) in a mixed formulation. For the latter, a bi-linear Lagrangian interpolation of the displacement field is considered, together with two sets of element-wise constant mixed variables $\boldsymbol{\Sigma}_{\mathrm{TI}}^{\alpha}$ and $\boldsymbol{\Theta}_{\mathrm{TI}}^{\alpha}$.

Figure 2 shows the distribution (obtained from finite element simulations) of the inter-fibers shear strains γ_s. Pure-displacement formulations $\mathrm{Q1}_c$ and $\mathrm{Q2}_c$ fail to recover the characteristic constant shear zones associated with a PJN mechanism. On the other hand, this is well-captured by the proposed mixed element $\overline{\mathrm{Q}}1\mathrm{A0}_c$, for which $\gamma_s \approx \gamma_{\mathrm{PJN}} = 20°$ in zone A, $\gamma_s \approx \gamma_{\mathrm{PJN}}/2 = 10°$ in zones B and $\gamma_s \approx 0$ in zones C.

The accurateness of the solution obtained with the mixed formulation is also proven by the convergence plot in Fig. 3a. The computed reaction force is fairly

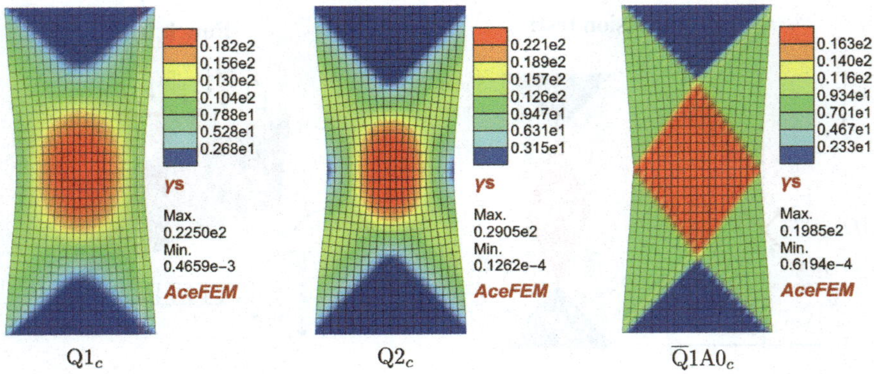

Fig. 2 Bias-extension test: distribution of the shear strain $\gamma_s = 2\|\mathbf{a}^{(-45)} \cdot \boldsymbol{\varepsilon}\, \mathbf{a}^{(+45)}\|$, with $\mathbf{a}^{(\pm 45)} = (1/\sqrt{2}, \pm 1/\sqrt{2})$, for the pure-displacement elements Q1$_c$ and Q2$_c$, and the mixed element $\overline{\text{Q}}$1A0$_c$

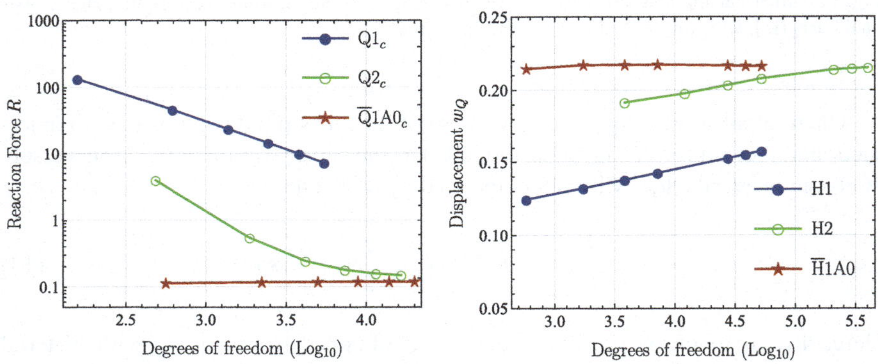

(a) Bias-extension test (b) Punch test

Fig. 3 Convergence analysis: **a** total reaction force R obtained in the bias-extension test for $n_{el} \in (5, 30)$ with Q1$_c$, Q2$_c$ and $\overline{\text{Q}}$1A0$_c$ elements; **b** displacement w_Q along $\mathbf{k}$ at the probe point $Q = (0, H/2, V)$ obtained in the punch test for $n_{el} \in (4, 20)$ with H$_1$, H$_1$ and $\bar{\text{H}}$1A0 elements

independent from the mesh size and equal to the converged value for $\overline{\text{Q}}$1A0$_c$, while highly sensitive to element size for Q1$_c$, but also for Q2$_c$.

3.2 Near-Inextensibility and Near-Incompressibility: A Punch Test

The second application addresses a punch test of a composite material made up by a soft nearly-incompressible matrix (with $\nu_m = 0.4995$) reinforced by stiff fibers. Simulations are conducted on a 3D domain $L \times H \times V$, with dimensions $L = 2$,

$H = 1$ and $V = 1$. As shown in Fig. 1, fibers lie in the $(\mathbf{i}, \mathbf{j})$-plane and are inclined by $\alpha = 20°$ with respect to $\mathbf{i}$. The punch test is simulated by considering a constant load q, acting along $\mathbf{k}$, on half of the top surface at $z = V$ for $x \leq L/2$.

The domain is discretized with hexahedral (cubic) elements, with mesh density defined by the number of elements n_{el} along z. Three element types are employed: H1 and H2, based on pure-displacement formulations with tri-linear and tri-quadratic Lagrangian shape functions; $\overline{\text{H}}$1A0, based on the mixed formulation in Eq. (8) with tri-linear Lagrangian shape functions for displacement and element-wise constant fields for stress and strain variables. Static condensation is performed for $\overline{\text{H}}$1A0.

Figure 3b clearly proves the outperformance of the mixed formulation with respect to pure-displacement ones in avoiding locking. The converged value of displacement w_Q is never reached with H1 for the range of investigated mesh densities, while H2 requires almost one million of nodal degrees of freedom (i.e., mesh density $n_{el} = 16$). On the other hand, $\overline{\text{H}}$1A0 is equally accurate with less than one thousand of nodal degrees of freedom (i.e., mesh density $n_{el} = 4$).

4 Conclusions

An energetically-decoupled mixed functional for dealing with near-inextensibility and/or near-incompressibility in linear elastic transversely isotropic materials has been recently presented by Marino and Wriggers in [7]. In the present work, the convergence properties of low-order finite element implementations based on the proposed formulation have been shown on two applications of engineering interest: a bias-extension test of composite laminates and a punch test of a fiber-reinforced rubber specimen. For the addressed applications, the proposed mixed formulation outperforms linear and quadratic pure-displacement formulations, being characterized by the absence of locking phenomena and optimal convergence properties.

It is worth highlighting that structures subjected to the bias-extension or punch tests are in many situations undergoing finite strains. The aim was not to reach conclusive results for applications but to discuss the significance of the proposed formulation in engineering-relevant problems. Although not final in terms of applications, simulations performed under linear elastic assumptions anyway give relevant indications. In fact, non-linear formulations would anyway be solved by means of a series of linear elastic steps via iterative algorithms like Newton-Raphson. The outperformance of the proposed formulation is thereby meaningful also with regards to possible extensions in large deformations, which is object of on-going work.

References

1. Zienkiewicz, O.C., Taylor, R.L., & Zhu, J.Z. (2013). *The finite element method: Its basis and fundamentals*. 7th ed. Butterworth-Heinemann, Oxford.
2. Wriggers, P. (2008). *Nonlinear finite element methods*. Berlin, Heidelberg: Springer.

3. Wriggers, P., Schröder, J., & Auricchio, F. (2016). Finite element formulations for large strain anisotropic material with inextensible fibers. *Advanced Modeling and Simulation in Engineering Sciences, 3*(1), 25.
4. Rasolofoson, F., Grieshaber, B. J., & Reddy, B. D. (2019). Finite element approximations for near-incompressible and near-inextensible transversely isotropic bodies. *International Journal for Numerical Methods in Engineering, 117*(6), 693–712.
5. Zdunek, A. (2020). On purely mechanical simple kinematic internal constraints. *Journal of Elasticity, 139*(1), 123–152.
6. Auricchio, F., Beirão da Veiga, L., Brezzi, F., & Lovadina, C. (2017) Mixed finite element methods. In *Encyclopedia of computational mechanics* (2nd ed., pp. 1–53). Wiley.
7. Marino, M., & Wriggers, P. (2020). Nearly-constrained transversely isotropic linear elasticity: Energetically consistent anisotropic deformation modes for mixed finite element formulations. *International Journal of Solids and Structures, 202,* 166–183.
8. Walpole, L. J. (1981). Elastic behavior of composite materials: Theoretical foundations. *Advances in Applied Mechanics, 21,* 169–242.
9. Hamila, N., & Boisse, P. (2013). Locking in simulation of composite reinforcement deformations. Analysis and treatment. *Composites Part A: Applied Science and Manufacturing, 53,* 109–117.
10. Potter, K. (2002). Bias extension measurements on cross-plied unidirectional prepreg. *Composites Part A: Applied Science and Manufacturing, 33*(1), 63–73.

A Comparison of Matrix-Free Isogeometric Galerkin and Collocation Methods for Karhunen–Loève Expansion

Michal L. Mika, René R. Hiemstra, Dominik Schillinger, and Thomas J. R. Hughes

> *Michal, René and I are part of the younger generation of researchers in Hannover. As a junior faculty member, my vision for the future is borne by the legacy of excellent research, teaching and mentorship that Peter Wriggers and the many colleagues he attracted to Hannover have maintained over the past 25 years. Michal is one of the many outstanding graduates of the educational programs in computational methods initiated by Peter. It was also Peter who established the contact between us at UT Austin, where the foundations for this work were laid. (Dominik Schillinger).*

Abstract Numerical computation of the Karhunen–Loève expansion is computationally challenging in terms of both memory requirements and computing time. We compare two state-of-the-art methods that claim to efficiently solve for the K–L expansion: (1) the matrix-free isogeometric Galerkin method using interpolation based quadrature proposed by us in [1] and (2) our new matrix-free implementation of the isogeometric collocation method proposed in [2]. Two three-dimensional benchmark problems indicate that the Galerkin method performs significantly better for smooth covariance kernels, while the collocation method performs slightly better for rough covariance kernels.

M. L. Mika (✉) · R. R. Hiemstra · D. Schillinger
Institute of Mechanics and Computational Mechanics, Leibniz University Hannover, Hannover, Germany
e-mail: mika@ibnm.uni-hannover.de

R. R. Hiemstra
e-mail: rene.hiemstra@ibnm.uni-hannover.de

D. Schillinger
e-mail: schillinger@ibnm.uni-hannover.de

T. J. R. Hughes
Oden Institute for Computational Engineering and Science, The University of Texas at Austin, Austin, USA
e-mail: hughes@oden.utexas.edu

F. Aldakheel et al. (eds.), *Current Trends and Open Problems in Computational Mechanics*, https://doi.org/10.1007/978-3-030-87312-7_32

1 Introduction

The Karhunen–Loève (K–L) expansion decomposes a random field into an infinite linear combination of L^2 orthogonal functions with decreasing energy content. Truncated representations have applications in stochastic finite element analysis (SFEM) [3–6], proper orthogonal decomposition (POD) [7, 8] and in image processing where the technique is known as principal component analysis (PCA) [9]. All these techniques are closely related and widely used in practice [10].

Numerical approximation of the K–L expansion by means of the Galerkin or collocation method leads to a generalized eigenvalue problem: *Find* $(\mathsf{v}_k^h, \lambda_k^h) \in \mathbb{R}^N \times \mathbb{R}^+$ *such that*

$$\mathsf{A}\mathsf{v}^h = \lambda_k^h \mathsf{Z}\mathsf{v}^h \qquad \text{for} \qquad k = 1, 2, \ldots, M. \tag{1}$$

This matrix eigenvalue problem is computationally challenging for the following reasons: (1) the matrix A is dense and thus memory intensive to store explicitly; (2) every iteration of an iterative eigenvalue solver requires a backsolve of a factorization of Z; and (3) the assembly of A is computationally expensive.[1]

In this work, we investigate and compare two state-of-the-art methods that were recently proposed to efficiently solve for the K–L expansion. The first method is the matrix-free isogeometric Galerkin method proposed by us in [1], which uses an advanced quadrature technique to gain high performance that is scalable with polynomial order. The second method is our new matrix-free implementation of the isogeometric collocation method proposed in [2]. As a collocation method it requires far fewer quadrature points than a standard Galerkin method such that the assembly of the collocation equations is simple and efficient.

This paper is structured as follows. In Sect. 2, we briefly review the basic aspects of the K–L expansion in the context of random field representations. In Sect. 3, we concisely present the two matrix-free solution methods and assess their algorithmic complexity. Three-dimensional numerical benchmark problems with comparisons in terms of accuracy and solution time are provided in Sect. 4. We summarize our conclusions in Sect. 5 and discuss future work.

2 Karhunen–Loève Expansion of Random Fields

Consider a complete probability space $(\Theta, \Sigma, \mathbb{P})$ where Θ denotes a sample set of random events and $\mathbb{P}$ is a probability measure $\mathbb{P} : \Sigma \to [0, 1]$. Let $\alpha(\cdot, \theta) : \Theta \mapsto L^2(\mathcal{D})$ denote a random field on a bounded domain $\mathcal{D} \in \mathbb{R}^d$ with mean $\mu(x) \in L^2(\mathcal{D})$ and covariance function $\Gamma(x, x') \in L^2(\mathcal{D} \times \mathcal{D})$. Without loss of generality we assume zero-mean random fields. The K–L expansion of the random field $\alpha(\cdot, \theta)$ requires the solution of an integral eigenvalue problem. Consider the self-adjoint

[1] Formation and assembly costs for a standard Galerkin method scale $O(N_e^2(p+1)^{3d})$, where N_e is the number of finite elements, p is the polynomial degree and d is the spatial dimension.

positive semi-definite linear operator $T : L^2(\mathcal{D}) \mapsto L^2(\mathcal{D})$,

$$(T\phi)(x) := \int_{\mathcal{D}} \Gamma(x, x')\phi(x')\,\mathrm{d}x'. \tag{2}$$

The eigenfunctions $\{\phi_i\}_{i\in\mathbb{N}}$ of T are defined by the homogeneous Fredholm integral eigenvalue problem of the second kind,

$$T\phi_i = \lambda_i\phi_i, \quad \phi_i \in L^2(\mathcal{D}) \text{ for } i \in \mathbb{N}. \tag{3}$$

The eigenfunctions ϕ_i are orthonormal in $L^2(\mathcal{D})$ and the corresponding eigenvalues form a non-increasing sequence $\lambda_1 \geq \lambda_2 \geq \cdots \geq 0$. The K–L expansion of the zero-mean random field $\alpha(\cdot, \theta)$ is given as

$$\alpha(x, \theta) = \sum_{i=1}^{\infty} \sqrt{\lambda_i}\phi_i(x)\xi_i(\theta), \qquad \xi_i(\theta) := \frac{1}{\sqrt{\lambda_i}} \int_{\mathcal{D}} \alpha(x, \theta)\phi_i(x)\,\mathrm{d}x. \tag{4}$$

Truncating the series in (4) after M terms leads to an approximation of α denoted by α_M. For practical computations in the context of stochastic finite element methods [3–6], the truncation order M is typically chosen between 20 and 30 terms [5, 11]. Each term in the expansion introduces one stochastic dimension, which is an example for the *curse of dimensionality*.

3 Numerical Methods

In this section we briefly review the matrix-free Galerkin method proposed in [1] and introduce our matrix-free implementation of the isogeometric collocation method proposed in [2].

In both approaches the generalized algebraic eigenvalue problem is first reformulated as a standard algebraic eigenvalue in terms of an invertible mapping C, which is a standard linear algebra technique [12]: *Find* $(\mathsf{v}_k^h, \lambda_k^h) \in \mathbb{R}^N \times \mathbb{R}^+$ *s.t.*

$$\begin{cases} \mathsf{A}'\mathsf{v}_k' = \lambda_k^h \mathsf{v}_k' \\ \mathsf{v}_k^h = \mathsf{C}\mathsf{v}_k' \end{cases} \quad \text{for} \quad k = 1, 2, \ldots, M. \tag{5}$$

Matrix-free isogeometric Galerkin method

A variational treatment of (3) reads: *Find* $(\phi, \lambda) \in L^2(\mathcal{D}) \times \mathbb{R}^+$ *s.t.* $\forall \psi \in L^2(\mathcal{D})$

$$\int_{\mathcal{D}} \left(\int_{\mathcal{D}} \Gamma(x, x')\phi(x')\,\mathrm{d}x' - \lambda\phi(x) \right) \psi(x)\,\mathrm{d}x = 0. \tag{6}$$

From (6), the Galerkin method is obtained by replacing $\phi, \psi \in L^2(\mathcal{D})$ by finite dimensional representations $\phi^h, \psi^h \in \mathcal{S}^h \subset L^2(\mathcal{D})$. Being posed in the variational setting, Galerkin methods inherit several advantageous properties such as exact L^2 orthogonality of the numerical eigenvectors and monotonic convergence of the numerical eigenvalues [3, 13]. Furthermore, powerful tools exist in the variational setting to study the stability and convergence of the method.[2]

With a trial space $\mathcal{S}^h := \text{span}\,\{N_i(x)\}_{i=1,\ldots,N}$ the Galerkin method leads to the eigenvalue problem defined in (1) with the system matrices

$$A_{ij} := \int_{\mathcal{D}} N_i(x) \left(\int_{\mathcal{D}} \Gamma(x, x') N_j(x')\,\mathrm{d}x' \right) \mathrm{d}x \tag{7a}$$

$$Z_{ij} := \int_{\mathcal{D}} N_i(x) N_j(x)\,\mathrm{d}x. \tag{7b}$$

Alternatively, the eigenvalue problem can be solved in the standard form introduced in (5) where $\mathsf{A}' := \mathsf{L}^{-1}\mathsf{A}\mathsf{L}^{-\top}$ and $\mathsf{C} := \mathsf{L}^{-\top}$. The matrix L is defined by the lower triangular matrix in the Cholesky decomposition of $\mathsf{Z} = \mathsf{L}\mathsf{L}^{\top}$.

Typically, the space $\mathcal{S}^h$ is spanned by piecewise C^0-continuous polynomial functions on quadrilateral, hexagonal or simplicial elements [3]. Recently, non-uniform rational B-splines (NURBS) have been applied in the context of an isogeometric Galerkin method [14]. These methods commonly evaluate the integrals in (7) using standard numerical quadrature rules. A Gauss–Legendre numerical quadrature rule leads, however, to an algorithmic complexity of $\mathcal{O}(N_e^2 \cdot (p+1)^{3d})$ [1], which becomes excessively expensive with the number of elements N_e, polynomial degree p and spatial dimension d. Furthermore, as mentioned in the introduction, the matrix A is dense and requires $\mathcal{O}(8 \cdot N^2)$ bytes of storage in double precision arithmetic, where N is the number of degrees of freedom in the trial space.

To overcome these limitations, the matrix-free Galerkin method proposed in [1] avoids storing the main system matrix A and achieves computational efficiency by utilizing a non-standard trial space in combination with a specialized quadrature technique, called *interpolation based quadrature*. This approach requires a minimum number of quadrature points and enables application of global sum factorization techniques [15]. In the following we sketch the main ideas of the method and refer to [1] for further details.

Let $\{B_i(\hat{x})\}_{i=1,\ldots,N}$ and $\{\tilde{B}_j(\hat{x})\}_{j=1,\ldots,\tilde{N}}$ denote two sets of tensor product B-splines of, for simplicity, uniform polynomial degree p. The first set is used in the definition of the trial space, whereas the second set is used in a projection of the kernel $\Gamma(x, x')$ and is a part of the *interpolation based quadrature*. Let $F : \hat{\mathcal{D}} \to \mathcal{D}$ be the geometric mapping from the reference domain to the physical domain. The trial space is defined as

$$\mathcal{S}^h := \text{span} \left\{ B_i(\hat{x}) / \sqrt{\det \mathrm{D}F(\hat{x})} \right\}_{i=1,\ldots,N}. \tag{8}$$

[2] In general the stability and convergence analysis are challenging in the context of collocation methods.

The advantage of this particular choice of the trial space is that the mass matrix in (7b) has a *Kronecker* structure and can be factored as $\mathsf{Z} = \mathsf{Z}_d \otimes \cdots \otimes \mathsf{Z}_2 \otimes \mathsf{Z}_1$, where $\{\mathsf{Z}_k\}_{k=1,2,\ldots,d}$ are univariate mass matrices. By leveraging this factorization the matrix-vector products of Kronecker matrices can be evaluated in nearly linear time complexity. This also holds for the matrix L in the Cholesky factorization of Z, which is factored as $\mathsf{L} = \mathsf{L}_d \otimes \cdots \otimes \mathsf{L}_2 \otimes \mathsf{L}_1$ from which the respective inverse follows as $\mathsf{L}^{-1} = \mathsf{L}_d^{-1} \otimes \cdots \otimes \mathsf{L}_2^{-1} \otimes \mathsf{L}_1^{-1}$.

The interpolation based quadrature in combination with the choice of the trial space in (8) leads to a factorization of the matrix A as $\mathsf{A} = \mathsf{M}^\top \tilde{\mathsf{B}}^{-1} \mathsf{J} \mathsf{G} \mathsf{J} \tilde{\mathsf{B}}^{-\top} \mathsf{M}$. Here $\mathsf{G} := \Gamma(x_i, x_j) \in \mathbb{R}^{\tilde{N} \times \tilde{N}}$ is the covariance kernel evaluated at the Greville abscissae, $\mathsf{J} \in \mathbb{R}^{\tilde{N} \times \tilde{N}}$ is the square root of a diagonal matrix of determinants of the Jacobian of the mapping at these points and the matrices $\tilde{\mathsf{B}} = \tilde{\mathsf{B}}_d \otimes \cdots \otimes \tilde{\mathsf{B}}_2 \otimes \tilde{\mathsf{B}}_1 \in \mathbb{R}^{\tilde{N} \times \tilde{N}}$ and $\mathsf{M} = \mathsf{M}_d \otimes \cdots \mathsf{M}_2 \otimes \mathsf{M}_1 \in \mathbb{R}^{\tilde{N} \times N}$ are Kronecker product matrices. In fact $\tilde{\mathsf{B}}_k$ and M_k, $k = 1, 2, \ldots, d$, are univariate collocation and mass matrices, respectively, which are introduced by the interpolation based quadrature. The computation of the eigenvalues and eigenvectors requires evaluation of matrix-vector products $\mathsf{v}' \mapsto \mathsf{A}'\mathsf{v}'$. This leads to a nine step algorithm presented in [1]. The matrix-vector products with the Kronecker structured matrices $\mathsf{L}^{-\top}$, M, $\mathsf{B}^{-\top}$ and the diagonal matrix J as well as all the respective transpose operations are performed in linear or nearly linear time complexity. The matrix-vector products with the matrix G are performed in quadratic time complexity. Hence, our matrix-free algorithm scales quadratically with the dimension of the interpolation space $\tilde{N}$. We note that in this algorithm, the matrix rows of G are computed on the fly, which saves memory by not explicitly storing the dense matrix G. Memory requirements for the remaining matrices are negligible, since they are either diagonal or Kronecker product matrices. For additional details about the matrix-free method, interpolation based quadrature and Kronecker products, we refer to [1].

Matrix-free isogeometric collocation method

In contrast to a Galerkin method, a collocation method does not treat the integral equation (3) in a variational manner. Instead, we require the discretized residual

$$r^h(x) := \int_{\mathcal{D}} \Gamma(x, x')\phi^h(x')\,\mathrm{d}x' - \lambda^h \phi^h(x) \tag{9}$$

to vanish at distinct points $x \in \mathcal{D}$. In [2], the geometry and trial spaces are discretized in terms of NURBS basis functions $\mathcal{S}^h := \{R_i(x)\}_{i=1,\ldots,N}$ in the sense of the isoparametric approach of isogeometric analysis.

In this study, we choose to collocate (9) at the Greville abscissae $\{x_i\}_{i=1,\ldots,N}$. The method is expressed concisely in matrix form (1) where the corresponding system matrices are given by

$$A_{ij} := \int_{\mathcal{D}} \Gamma(x_i, x') R_j(x') \,\mathrm{d}x' \qquad \text{and} \qquad Z_{ij} := R_j(x_i). \tag{10}$$

In primal form (5), this means that $\mathsf{A}' = \mathsf{Z}^{-1}\mathsf{A}$ and C is the identity matrix. The matrices A and Z are square and, in general, not symmetric. In contrast to variational methods, where the system matrices are symmetric and positive (semi-)definite by construction, collocation methods do not ensure a real-valued eigensolution for any element size $h > 0$. For an in-depth exposition of the collocation method, we refer the reader to [13], and to [16, 17] for details on the isogeometric formulation.

The matrix-free version of the collocation method is derived analogously to the matrix-free Galerkin method described above. Due to the properties of the system matrix Z, instead of the Cholesky decomposition employed in the Galerkin method, we use the *pivoted* LU decomposition, $\mathsf{PZQ} = \mathsf{LU}$, to arrive at the standard matrix form. We observed that without pivoting the matrix-free collocation method suffers from numerical instabilities at polynomial orders $p > 3$. We use the pivoted LU decomposition of Z to apply the inverse of Z to the matrix A and thus obtain A'. The standard algebraic eigenvalue problem is then given by

$$\mathsf{A}'\mathsf{v}' = \lambda \mathsf{v}' \qquad \text{where} \qquad \mathsf{A}' := \mathsf{Q}\mathsf{U}^{-1}\mathsf{L}^{-1}\mathsf{P}\mathsf{A} \tag{11}$$

Following [1], we choose a row-wise evaluation of the coefficient vector in the standard matrix-vector product $\mathsf{v}' \mapsto \mathsf{A}'\mathsf{v}'$. The optimal evaluation order and further details for each step are given in Algorithm 1.

Algorithm 1 Matrix-free evaluation of the matrix-vector product $\mathsf{v}' \mapsto \mathsf{A}'\mathsf{v}'$ emerging from collocation

Input: $v_j \in \mathbb{R}^N$, $R_{jk} \in \mathbb{R}^{N\times(N_e \cdot N_q)}$, $P_{ij}, Q_{ij}, U_{ij}, L_{ij} \in \mathbb{R}^{N\times N}$, $J_k \in \mathbb{R}^{N_e \cdot N_q}$, $W_k \in \mathbb{R}^{N_e \cdot N_q}$
Output: $v'_i \in \mathbb{R}^N$

1: $y_k \leftarrow R_{jk} v_j$ ▷ Interpolation at quadrature points
2: $y'_k \leftarrow y_k \odot J_k \odot W_k$ ▷ Scaling at quadrature points
3: $z_l \leftarrow G_{lk} y'_k$ ▷ Kernel evaluation one row at a time
4: $v'_i \leftarrow Q_{it} U^{-1}_{tr} L^{-1}_{rs} P_{sl} z_l$ ▷ Backsolve using LU of Z

3.1 Algorithmic Complexity

Matrix-free Galerkin method

Under the assumption of $\tilde{N} \propto N$, the formation and assembly costs are negligible compared to the matrix-vector products that scale independently of p as $\mathcal{O}(\tilde{N}^2)$ [1]. The total cost of the method scales as $\mathcal{O}(N_{\text{iter}} \cdot \tilde{N}^2)$, where N_{iter} is the number of iterations of the eigenvalue solver.

Matrix-free collocation method

First, we are interested in the algorithmic complexity of an element-wise assembly procedure for the system matrices that arise from the collocation method. We assume that (1) $\hat{\mathcal{D}}$ has N_e elements; (2) the products on every d-dimensional element $\square^d$ in $\hat{\mathcal{D}}$ are integrated with a quadrature rule $Q(f) := \sum_{k=1}^{N_q} w_k f(x_k)$ with $1 \leq N_q \leq (p+1)^d$ quadrature points; and (3) the number of collocation points N_c is equal to the number of degrees of freedom N. The leading term in the total cost of formation and assembly arises from the cost of forming the element matrices $\mathsf{A}^{\square}$,

$$A_{ij}^{\square} = \int_{\square^d} \Gamma(\hat{x}_i, \hat{x}') R_j(\hat{x}')\, d\hat{x}' \approx \sum_{k=1}^{N_q} w_k \Gamma(\hat{x}_i, \hat{x}_k') B_j(\hat{x}_k') = C_{ik} D_{kj}$$

with $C_{ik} = w_k \Gamma(\hat{x}_i, \hat{x}_k')$ and $D_{kj} = R_j(\hat{x}_k'), i = 1, \ldots, N, j = 1, \ldots, (p+1)^d$. The formation cost of C and D is negligible. The matrix-matrix product cost is of $O(N_c N_q (p+1)^d)$ and the cost for summation over all N_e is of $O(N_e N_c N_q (p+1)^d)$. Now, assuming a Gauss–Legendre quadrature rule with $N_q := (p+1)^d$ quadrature points and the proportionality relationship $N_e \propto N$, a collocation method with $N_c = N$ has a leading cost of $O(N^2 (p+1)^{2d})$.

The algorithmic complexity in the matrix-free formulation is driven by the most expensive steps in Algorithm 1. In a single iteration of the eigenvalue solver, steps 1 and 3 have a complexity $O(N \cdot N_e \cdot N_q)$. The element-wise multiplication in step 2 scales linearly with the number of quadrature points, $O(N_e \cdot N_q)$. The last step scales as $O(N^2)$. Evidently, steps 1 and 3 depend on the number of quadrature points. Since $N_e \cdot N_q \geq N$, they determine the overall cost of the method. Assuming a Gauss-Legendre quadrature rule with $N_q := (p+1)^d$ quadrature points in each element and $N_e \propto N$, the leading cost of a single iteration of the eigenvalue solver is $O(N^2 (p+1)^d)$. Hence, the total cost of the matrix-free isogeometric collocation method scales as $O(N_{\text{iter}} \cdot N^2 (p+1)^d)$, where N_{iter} is the number of iterations of the eigenvalue solver.

Comparison

Compared to the matrix-free Galerkin method with interpolation based quadrature, the collocation method scales unfavourably with the polynomial degree. Furthermore, due to the lack of Kronecker structure, it is necessary to compute the pivoted LU decomposition of the full matrix Z. The computational cost of this factorization increases with N as well as p, which is due to an increasing bandwidth of the matrix B.

4 Numerical Examples

In this section, we compare the accuracy and efficiency of the matrix-free isogeometric Galerkin and collocation methods. In [1], it was shown that the proposed Galerkin method performed especially well in the case of a smooth covariance kernel. For rough kernels, such as the C^0 exponential kernel, the interpolation based quadrature performed suboptimally. In our study, we benchmark both methods for two kernels of different smoothness and appropriate refinement strategies of the spaces involved: (1) the exponential kernel together with h-refinement and (2) the Gaussian kernel and k-refinement. In both variants, the solution space is equal for the Galerkin and collocation methods. The interpolation space used in the Galerkin method is defined on the same mesh as the solution space, but, as discussed in Remark 4.1 in [1], its continuity is one class lower than that of the solution space. All computations are performed sequentially on a laptop machine with an Intel(R) Core(TM) i7-9750H CPU @ 2.60GHz as well as 2×16 GB of DDR4 2666MHz RAM. Our reference solution is the standard isogeometric Galerkin solution computed on the finest possible mesh with a runtime of roughly 17 h, tabulated in [1].

Example 1 – Exponential covariance kernel

In Example 1, we compare the performance with respect to h-refinement assuming an exponential kernel on the half-cylindrical domain shown in Fig. 1. The polynomial order in each parametric direction is $p = 2$. We choose a tensor product Gauss–Legendre quadrature rule with $(p + 1)^3$ points per element of the domain in the collocation method. In accordance with Remark 4.1 made in [1] the continuity of the interpolation space of the Galerkin method at the element interfaces is reduced to C^0. Furthermore, at element interfaces where the geometry is C^0, the interpolation space of the Galerkin method is set to C^{-1}. Our comparative investigation is based on five different resolution cases with respect to the characteristic size h of the solution

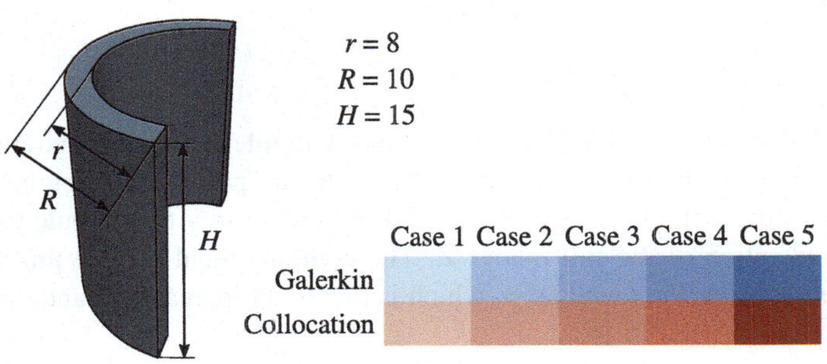

Fig. 1 Benchmark geometry of a half-cylinder. The correlation length $bR = \frac{1}{2}R$ is used throughout all cases. The provided color-coding is used to differentiate between five different cases and two different methods

Table 1 Mesh, solution space and interpolation space details in Example 1 and Example 2

	Example 1—Exponential kernel					*Example 2—Gaussian kernel*				
	Case 1	Case 2	Case 3	Case 4	Case 5	Case 1	Case 2	Case 3	Case 4	Case 5
h	2.857	1.719	1.556	1.423	1.142	2.857	2.857	2.857	2.857	2.857
p	2	2	2	2	2	2	3	4	5	6
N	1050	2108	2800	3772	5625	1050	1628	2340	3198	4214
$\tilde{N}$	1980	8990	12210	16770	28294	1080	1672	2400	3276	4312

h mesh size in the solution and interpolation mesh
p polynomial order of the solution and interpolation mesh
N number of degrees of freedom (dof) in the solution space
$\tilde{N}$ number of dof in the interpolation space (IBQ-Galerkin only)

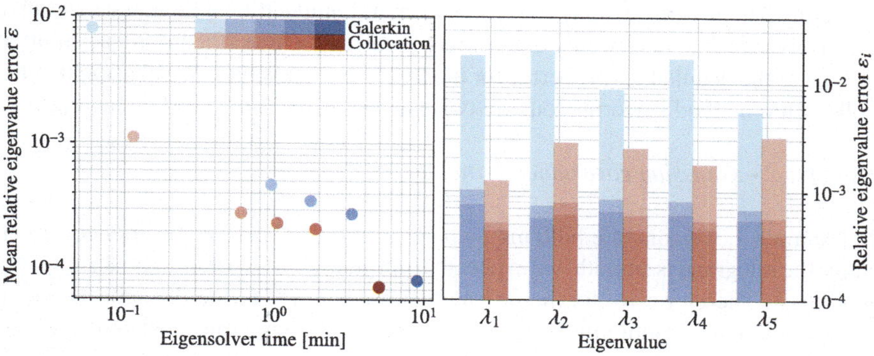

Fig. 2 Mean relative eigenvalue error computed with the first 20 eigenvalues versus the eigensolver time and the error of the first 5 eigenvalues plotted for Cases 1–3 (Example 1, exponential kernel)

and interpolation mesh. Our specific choices of mesh size and number of degrees of freedom in the interpolation and solution spaces are summarized in Table 1.

For Case 1, we visualize the first, second and fourth eigenfunctions computed by both methods, plotted in Fig. 3 on the half-cylinder domain. Already for the coarsest resolution, both methods produce results that are practically indistinguishable from each other when plotted along an arbitrary cut line.

For a quantitative comparison, let us introduce a relative eigenvalue error ε_i and a mean-relative error $\overline{\varepsilon}$ with respect to the reference solution as

$$\varepsilon_i := \varepsilon(\lambda_i^{\text{ref}}, \lambda_i^h) := \frac{|\lambda_i^{\text{ref}} - \lambda_i^h|}{\lambda_i^{\text{ref}}} \quad \text{and} \quad \overline{\varepsilon} := \frac{1}{M}\sum_{i=1}^{M} \varepsilon_i. \tag{12}$$

To enable a concise illustration with respect to the five cases defined in Table 1, we define the color coding shown in Fig. 1. Blue indicates results obtained with the Galerkin method, red indicates results obtained with the collocation method. The change in shading from light to full color indicates the increasing mesh resolution from Case 1 to Case 5.

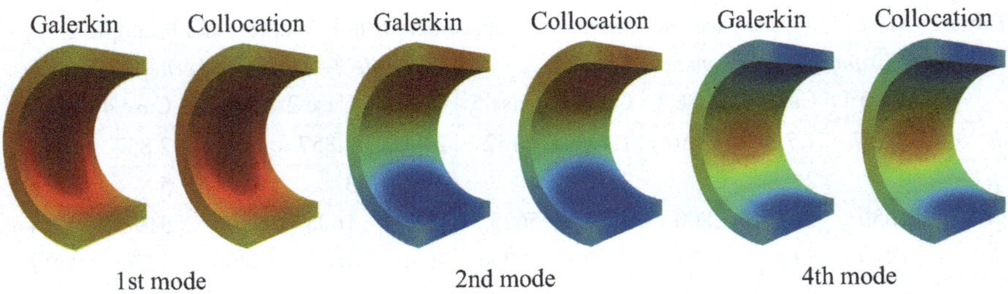

Fig. 3 First, second and fourth eigenfunctions (Example 1, Case 1).

Figure 2 depicts mean relative accuracy versus computational time of the iterative eigensolver for the first twenty eigenvalues measured against the reference solution as well as the detailed relative error for the first five eigenvalues. We observe that the collocation method performs roughly twice as fast at the same level of accuracy.

Example 2 – Gaussian covariance kernel

In Example 2, we compare both methods for a smooth Gaussian covariance kernel. Since the integrand is smooth, we expect that optimally smooth approximation spaces work best. Therefore, we fix the polynomial order p and refine the approximation spaces with C^{p-1} continuity between elements until a target mesh size of 2.857 is reached (k-refinement). The resulting five different cases are summarized in Table 1.

Comparing Case 1 in Example 1 with Case 1 in Example 2, we find that the number of degrees of freedom in the interpolation space is smaller. This is due to the increased continuity at element interfaces of the interpolation space of the Galerkin method. This trend is also characteristic for k-refinement and is observable in the remaining Cases 2–5.

We resort again to the color coding of Fig. 1 to concisely differentiate between the five different resolutions and the two methods.

Figure 4 plots the mean relative accuracy of the first twenty eigenvalues versus the eigensolver timings. It is evident that for the smooth Gaussian kernel, the Galerkin method outperforms the collocation method by more than one order of magnitude at the same level of accuracy. Furthermore, in line with the complexity analysis presented in Sect. 3.1, we observe that the performance gap increases with increasing polynomial order. Following the scheme of Fig. 2, we provide a more detailed account of the approximation accuracy of the first five eigenvalues in Fig. 4.

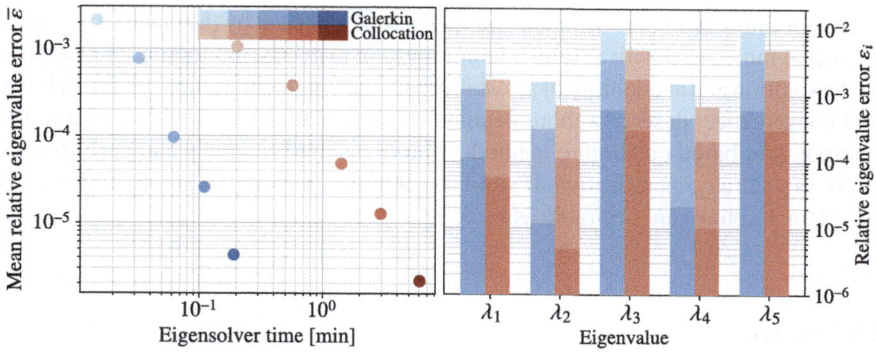

Fig. 4 Mean relative eigenvalue error computed with the first 20 eigenvalues versus the eigensolver time and the error of the first 5 eigenvalues plotted for Cases 1–3 (Example 2, smooth Gaussian kernel)

5 Conclusions

In this paper, we compared accuracy versus the computational time of two state-of-the-art isogeometric discretization methods for the numerical approximation of the truncated Karhunen–Loève expansion. The first method is the matrix-free isogeometric Galerkin method proposed by us in [1]. It achieves its computational efficiency by combining a non-standard trial space with a specialized quadrature technique called *interpolation based quadrature*. This method requires a minimum of quadrature points and relies heavily on global sum factorization. The second method is our new matrix-free version of the isogeometric collocation method proposed in [2]. This method achieves its computational performance by virtue of a low number of point evaluations.

On the one hand, our comparative study showed that for a C^0-continuous exponential kernel, the matrix-free collocation method was about twice as fast at the same level of accuracy as the Galerkin method. On the other hand, our comparative study showed that for a smooth Gaussian kernel, the matrix-free Galerkin method was roughly one order of magnitude faster than the collocation method at the same level of accuracy. Furthermore, the computational advantage of the Galerkin method over the collocation method increases with increasing polynomial degree. These results are not surprising, since it was already shown in [1] that interpolation based quadrature scales virtually independently of the polynomial degree. In our study, we also illustrated via complexity analysis that the matrix-free collocation method scales unfavorably with polynomial order. The suboptimal accuracy of the interpolation based quadrature for rough kernels is also known and was already discussed in [1]. Besides the aspect of computational performance, we also showed that both methods are highly memory efficient by virtue of their matrix-free formulation.

As for future work, the advantageous properties inherited by the Galerkin method, such as symmetric, positive (semi-)definite system matrices, monotonic convergence of the solution and availability of an established mathematical framework for stability and convergence, deserve a more detailed theoretical discussion with regard to the interpolation based quadrature method. A more detailed accuracy and performance study as well as generalization of similar techniques to complex geometric models based on T-splines are desirable as well.

References

1. Mika, M.L., Hughes, T.J.R., Schillinger, D., Wriggers, P., & Hiemstra, R.R. (2021). A matrix-free isogeometric Galerkin method for Karhunen-Loève approximation of random fields using tensor product splines, tensor contraction and interpolation based quadrature. *Computer Methods in Applied Mechanics and Engineering*, 379. June.
2. Jahanbin, R., & Rahman, S. (2019). An isogeometric collocation method for efficient random field discretization. *International Journal for Numerical Methods in Engineering, 117*(3), 344–369. January.
3. Ghanem, R. G., & Spanos, P. D. (1991). *Stochastic finite elements: A spectral approach*. New York, NY: Springer.
4. Keese, A. (2003). A review of recent developments in the numerical solution of stochastic partial differential equations (Stochastic Finite Elements). *Braunschweig, Institut für Wissenschaftliches Rechnen*.
5. Stefanou, G. (2009). The stochastic finite element method: Past, present and future. *Computer Methods in Applied Mechanics and Engineering, 198*, 1031–1051.
6. Sudret, B., & Kuyreghian, A. (2000). *Stochastic finite element methods and reliability: A state-of-the-art report*. Berkeley: Department of Civil and Environmental Engineering, University of California.
7. Lu, K., Jin, Y., Chen, Y., Yang, Y., Hou, L., Zhang, Z., Li, Z., & Fu, C. (2019). Review for order reduction based on proper orthogonal decomposition and outlooks of applications in mechanical systems. *Mechanical Systems and Signal Processing, 123*, 264–297. May.
8. Rathinam, M., & Petzold, L. R. (2003). A new look at proper orthogonal decomposition. *SIAM Journal on Numerical Analysis, 41*(5), 1893–1925. January.
9. Jolliffe, I. T., & Cadima, J. (2016). Principal component analysis: A review and recent developments. *Philosophical Transactions of the Royal Society A, 374*(2065). April.
10. Liang, Y. C., Lee, H. P., Lim, S. P., Lin, W. Z., Lee, K. H., & Wu, C. G. (2002). Proper orthogonal decomposition and its applications-Part I: Theory. *Journal of Sound and Vibration, 252*(3), 527–544. May.
11. Eiermann, M., Ernst, O. G., & Ullmann, E. (2007). Computational aspects of the stochastic finite element method. *Computing and Visualization in Science, 10*(1), 3–15. February.
12. Saad, Y. (2011). *Numerical methods for large eigenvalue problems*. In Number 66 in Classics in applied mathematics. Society for Industrial and Applied Mathematics, Philadelphia, rev. ed.
13. Atkinson, K. E. (1997). *The numerical solution of integral equations of the second kind* 1st ed. Cambridge University Press.
14. Rahman, S. (2018). A Galerkin isogeometric method for Karhunen-Loève approximation of random fields. *Computer Methods in Applied Mechanics and Engineering, 338*, 533–561.
15. Bressan, A., & Takacs, S. (2019). Sum factorization techniques in isogeometric analysis. *Computer Methods in Applied Mechanics and Engineering, 352*, 437–460. August.
16. Auricchio, F., Beirão Da Veiga, L., Hughes, T. J. R., Reali, A., Sangalli, G. (2010). Isogeometric collocation methods. *Mathematical Models and Methods in Applied Sciences, 20*(11), 2075–2107.

17. Schillinger, D., Evans, J. A., Reali, A., Scott, M. A., & Hughes, T. J. R. (2013). Isogeometric collocation: Cost comparison with Galerkin methods and extension to adaptive hierarchical NURBS discretizations. *Computer Methods in Applied Mechanics and Engineering, 267*, 170–232.

Forerunning and Bridging in Dry and Saturated Fracturing Solids

Enrico Milanese, Tao Ni, Carlo Peruzzo, Mirco Zaccariotto, Ugo Galvanetto, Gennady S. Mishuris, and Bernhard A. Schrefler

Our friendship started in the late 80ties. In 89 my former student, now Professor, Giorgio Zavarise visited Peter in Hanover and later in Darmstadt. In 97 our families met in front of Old Faithful in Yellowstone Park and my daughter Lorna visited Peter's home in Hanover. We also had a common doctoral candidate Hong Wu Zhang from Dalian, now Professor. In 2005, thanks to the efforts of Peter and the late Professor Erwin Stein, I was awarded an honorary Doctorate in Engineering at the Leibniz University of Hanover.

—Bernhard Schrefler

E. Milanese
Civil Engineering Institute, Materials Science and Engineering Institute, EPFL (Ècole Polytechnique Fèdèrale de Lausanne), Lausanne, Switzerland
e-mail: enrico.milanese@epfl.ch

T. Ni · M. Zaccariotto · U. Galvanetto
Industrial Engineering Department, University of Padova, via Venezia 1, 35131 Padova, Italy
e-mail: tao.ni@unipd.it

M. Zaccariotto
e-mail: mirco.zaccariotto@unipd.it

U. Galvanetto
e-mail: ugo.galvanetto@unipd.it

C. Peruzzo
Geo Energy Laboratory, GEL, EPFL (Ècole Polytechnique Fèdèrale de Lausanne), Lausanne, Switzerland
e-mail: carlo.peruzzo@epfl.ch

G. S. Mishuris
Department of Mathematics, IMPACS, Aberystwyth University, Ceredigion SY23 3BZ, Wales, UK
e-mail: ggm@aber.ac.uk

B. A. Schrefler (✉)
Department of Civil, Environmental and Architectural Engineering, University of Padova, via Marzolo 9, 35131 Padova, Italy
e-mail: bernhard.schrefler@dicea.unipd.it

F. Aldakheel et al. (eds.), *Current Trends and Open Problems in Computational Mechanics*, https://doi.org/10.1007/978-3-030-87312-7_33

Abstract We investigate crack forerunning and bridging with numerical models on three different structures. Bridging is observed in a heterogeneous saturated lattice model in quasi-static conditions under fluid injection. A single episode of forerunning in a dry double beam in dynamics confirms that forerunning increases the overall crack tip advancement speed with respect to a steady-state behaviour and finally we show on a two-dimensional specimen that in presence of a fluid forerunning increases the overall advancement speed even more.

1 Introduction

There are several instabilities observed in fracture propagation [1]. Those, related to our interest, are branching, bridging and forerunning. The first two monotonically change the geometry while the latter is a repetition of a fully continuous open crack and a topologically different situation with major and minor crack/s in front merging in time together. For forerunning to happen it is generally assumed that dynamics is necessarily involved, because dynamics itself may lead to a non-monotonic field distribution in space (and time) and thus, even in homogeneous brittle materials, may promote the mini-crack nucleation ahead of the front. However, forerunning can also appear in quasi-static situations, as for instance in fatigue, where weakening of the material in front of the crack may produce a mini-crack that on the next cycle will link with the major crack. The usual justification in quasi-statics for the existence of the mini-crack is damage. This is so because, in any quasi-static problems, stress strain fields rather monotonically decrease when moving away from the crack tip. We now discuss some results from the literature, disregarding the molecular dynamics approach to fracture where mostly branching was observed.

Hints of forerunning fracture behaviour have been documented in dynamic crack failure experiments on dried Darley Dale Sandstone specimens under triaxial compression with a range of confining pressures up to 200 MPa [2]. Measurements of acoustic wave velocities and acoustic statistics have permitted to identify a region prior to failure involving closure of pre-existing cracks, followed by nucleation and growth of new micro cracks predominantly parallel to the principal compression axis, their interaction and coalescence to produce macroscopic failure. P wave velocities up to 4300 m/s have been measured.

2 Forerunning in Dry Solids

Tvergaard and Needleman [3] numerically performed a fully transient dynamic crack growth analysis in a plane strain edge cracked specimen, subject to impulsive loading at one end, and observe that "sometimes a secondary crack develops ahead of the main crack and subsequently links up with it. This leads to a situation where the crack appears to be stationary for a relatively long time and then grows rather abruptly".

These authors also observed branching and no specific ad hoc criterion was used in their analysis. A crack advancement speed up to 1833 m/s has been found.

In the papers [4–9], fracture growth is treated as a transition by means of analytical and numerical approaches. Note that analytical solutions usually can only answer the question on whether instabilities (not steady-state movement) may occur and what are the conditions that promote such scenarios, and give an idea of what type of instability may occur. Clearly, complex problems with changing geometrical topology in time are extremely challenging for any analytical analysis and often numerical approaches can be used.

In [4–8], related to lattice fracture dynamics with lattices created from springs and beams, it was shown that for non-oscillating or oscillating loads the speed of the transition point increases approaching its critical value, associated with the incident wave group velocity, as the action of the force increases, or the speed of the transition point coincides with the phase speed of the incident wave up to some limiting amplitude of the sinusoidal loading. In this case, as long as an energy release is required for the transition, the incident wave group speed must exceed the phase speed since the energy flux velocity is equal to the group velocity. This is not always the case because in a discrete waveguide, as shown in [4], there is an infinite set of phase speeds corresponding to a given frequency, and a sufficiently low phase speed can always be found. In the steady-state transition, the lattice bonds on the crack path break one after another at regular time intervals.

However, under a sufficiently large wave amplitude, the transition wave mode is not steady-state anymore and a new regular movement of the transition point after a small increase of the amplitude appears. Namely, a two-bond clustering occurs with two alternating values of the local crack speed. Increasing the amplitude further, three, four or more cluster regimes may occur. In any of those clusters, the transition point still moves monotonically but not with the same speed inside the cluster. In case of the spring lattice, only one pure steady-state regime is possible while the clusters increase in size with amplitude growth [4, 5]. Finally, after reaching some critical speed, the monotonic movement of the transition point cannot be any longer supported and forerunning occurs. This is when one or a few points lying ahead of the tip on some distance from that point is/are destroyed and, only after this, other points in the interval between the previous transition point and the newly destroyed one will be crushed. Moreover, such movement repeats in time.

In case of the beam lattices [6–8], the movement of the transition point becomes more complex. Now, there are a few steady-state regimes, where between them, several cluster regimes are possible and there exists no rule that for larger amplitudes the size of the cluster should be larger, as it was for the springs. Nieves et al. [6] predicted analytically forerunning fracture modes, defining also the critical load, and observed them in simulations of the failure of discrete beam structures subjected to sinusoidal loads. The results were used to verify simplified analytical models of collapsing civil engineering structures exposed to vibration [7]. An analytical and numerical analysis of a wider class of failure regimes for similar flexural periodic systems has been presented in [8] where forerunning modes were also observed.

Finally, Slepyan et al. [9] investigated a continuous model, namely: the separation of a beam from an elastic foundation under a sinusoidal incident wave. In fact, it is also shown that any load would lead to the same result if applied on a proper distance from the transition point making the specification of the kind of the load less important. If the transition can occur instantaneously, again two ordered regimes of the transition wave are found: a steady-state one with the transition wave speed independent of the incident wave amplitude, and the forerunning. The steady-state regime exists only in a bounded parametrical domain of the incident wave parameters while outside this domain the steady-state mode is replaced by a set of local separation segments periodically emerging at a distance ahead of the main transition point. As above, the group speed is greater than the phase speed allowing the incident wave to deliver the energy required for the separation. In the example, it appears that the main crack continues to grow as does the separation segment, but the left part of this latter is fixed, so that the main crack catches up with the separation segment. These authors investigate also the case of supersonic incident wave, where it is found that no steady-state solution exists in such a case at all.

Interestingly, in both cases (lattice and continuous), while various propagation regimes occurs and the instantaneous (local) speeds are different, and thus are not a continuous function of the load parameter/s, the average speed is always a continuous function with jumps of the average acceleration at a set of discrete points.

Forerunning can also be obtained in quasi-static situations when a disordered material is assumed. In fact, Araujo et al. [11] investigate fatigue crack growth for the aim to determine the Paris exponent governing subcritical crack-growth at the macroscopic scale. Damage accumulation is assumed to follow a power law of the local stress amplitude. A single thin elliptic crack is initially produced in an infinite two-dimensional sample of a linear elastic material. The sample is subject to cyclic loading, with an external stress transverse to the major axis of the crack. It is assumed that the crack grows only along its major axis, hence crack propagation becomes essentially a one-dimensional problem. The relaxation time of the material is much shorter than the period of the loading cycle, so that crack propagation can be investigated within a quasi-static approximation, according to which the system always reaches its equilibrium state between two successive crack-growth events. When all fatigue thresholds are equal, i.e. in the uniform limit, the monotonic behaviour of the stress amplitude function ensures the existence of a single crack along the whole rupture process: the elements break sequentially, starting from the initial crack tips, and the crack advances symmetrically. When introducing disorder in the fatigue thresholds it is assumed that the element at position x along the crack line has a fatigue threshold chosen randomly from a uniform probability distribution with uncorrelated fatigue thresholds at different elements. In the presence of disorder, elements far from the crack tips may reach their fatigue thresholds, giving rise to secondary cracks as mentioned above. The authors focus on the growth of the initial crack or main crack which may involve secondary cracks when these coalesce with the main crack while the process developing.

Interestingly, in case of heterogenous lattice, it was shown that a bridge crack can be developed, and it may propagate with a speed faster than the wave speed in the lattice [4, 10].

None of the above cases consider the presence of a fluid. This will be addressed in two of the following examples.

3 Forerunning in Solids with Fluid Presence

3.1 2D Lattice Model Under Fluid Injection

We first consider crack bridging in a fully saturated lattice model in quasi-static conditions. Such a case appeared when investigating avalanching in a 2D central force lattice model for saturated disordered media at mesoscopic level carried out in [12]. For the solid phase, the stress thresholds of each element are picked randomly from a uniform distribution in the interval (0, 1) MPa. The elements break only under traction. When the stress in a truss of the lattice exceeds the local threshold, the elastic modulus of the truss is reduced adopting a continuous damage law. At each step, the thresholds of the damaged elements are updated using a uniform distribution in the same interval as above. The system is then solved again with unchanged assigned loads and these steps are repeated for as many times as needed to reach an equilibrium state in which no stress threshold is exceeded in any bond. Only then, the load is increased and the steps necessary to check if any and how many trusses are damaged at this stage are repeated. The process continues until the final failure of the lattice is reached when a full crack has developed.

The fluid phase is introduced in the model by coupling the equilibrium equation for the lattice with the mass balance equations for the solid and fluid phases, following Biot's theory [13] which introduces its own timescale. The presence of the fluid affects then the mechanics of the lattice. With respect to the dry scenario described above, there is now a nested time stepping scheme where at any time step two scenarios can take place: when at a time station the external load (injection in our case) is increased, Biot's coupled problem is solved and (i) equilibrium is reached without any rearrangement (avalanches); or (ii) the equilibrium is not reached without rearrangements and more sub-steps with rearrangements are needed to reach equilibrium where Biot's coupled problem is solved in every substep. Only then the external load is increased. In such a way, the timescale of the external loading is always larger than the timescale within the fracturing process and the longer the rearrangements, the larger the difference between the two timescales.

In this way, the fracture may propagate and the fluid has time to rearrange itself inside the sample before the lattice is loaded again. For the purpose of investigating the avalanche behaviour of the specimen i.e. the number of failing elements per step, being the material inherently heterogeneous, many simulations are run with the same boundary conditions but randomly varying the stress thresholds. With this model,

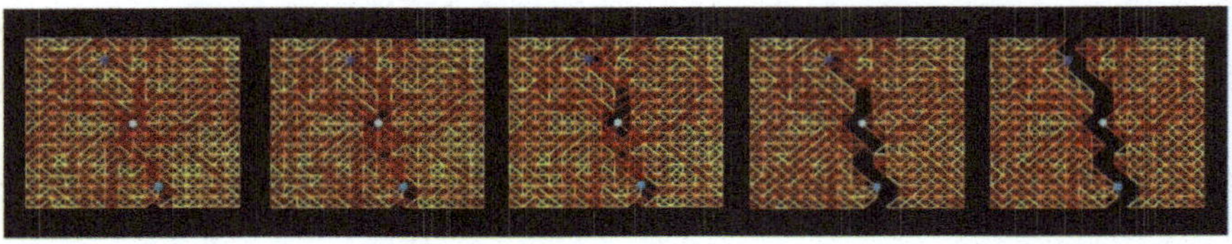

Fig. 1 Snapshots of crack bridging in a 2D lattice model under fluid injection at the centre. Yellow bars are undamaged, red ones damaged which are eliminated once a threshold for damage is reached

we have investigated the cases of assigned biaxial boundary tractions, pressures and fluid injection [12]. In case of fluid injection at the centre of the model, among the roughly 70 runs needed for the statistical analysis of about 55000 avalanches we have encountered at least one case where crack bridging takes place. This case is shown in Fig. 1 where the yellow bars indicate undamaged bars, while the red ones have some degree of damage. It can be clearly seen that secondary cracks develop at the lower and the upper part which coalesce with the main crack from the central part.

3.2 *Forerunning in a Dry Double Beam*

A single episode of forerunning on a double beam of finite size and in dynamics is chosen to confirm that forerunning increases the speed of the crack tip advancement in dry solids.

This problem has been investigated in [14]. Two beams are pulled apart from each other by a distributed load acting on their entire length. The load is asymptotically growing with time (t) with an arctan(t) - like behaviour. The link between the two beams is provided by a linearly weakening cohesive zone, characterized by a stiffness k and a maximum displacement $V_{\max}$ upon which, the connection between the two beams is lost. A linear penalty method with stiffness β is adopted to limit the overlapping of the two beams in the case of fracture closure. The following set of parameters led to a single episode of forerunning: $L = 10$ (beam length), $L_o = 0.8$ (length of the initial notch), $EJ = 0.15$ (stiffness), $\rho A = 500$ (mass per length unit), $V_{\max} = 0.008$, $k = 18$ and $\beta = 12000$. The corresponding crack length evolution is shown by the blue curve in Fig. 2. If we consider a slightly smaller stiffness, such that $EJ = 0.14$, no forerunning is observed. The fracture advancement in this case is shown by the purple curve in the same figure. At $t = 35s$, in both systems, a weakened patch of the cohesive zone is located ahead of the main crack tip. In the case of $EJ = 0.15$ (blue curve), this patch nucleates a new traction-free crack that coalesces with the main one. In the case of $EJ = 0.14$ no nucleation is observed.

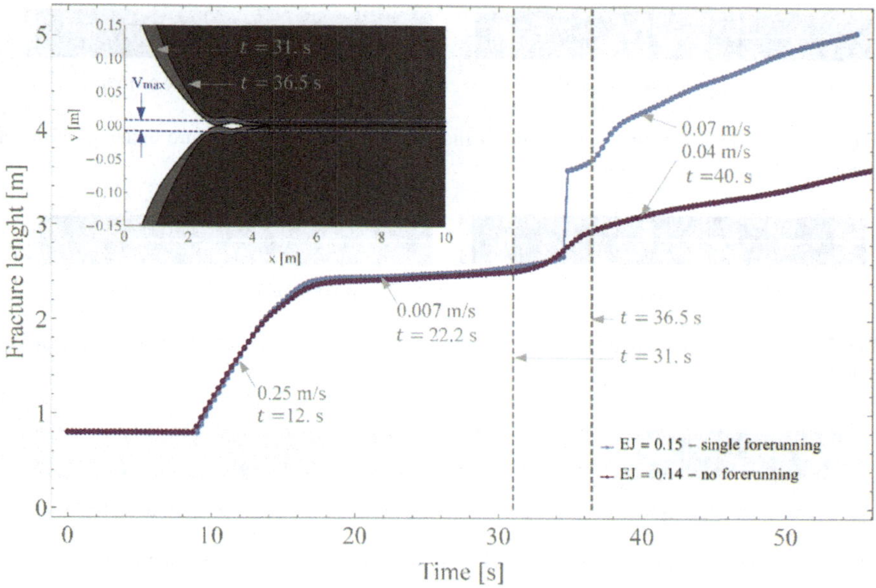

Fig. 2 Crack length versus time with and without forerunning. In the insert V_{max} indicates on the deformed beams the opening of the fracture beyond which cohesive forces are no longer active

3.3 Forerunning in a Dry and a Saturated Porous Medium Under Mechanical Load and Under Fluid Injection

Forerunning in a two-dimensional domain is investigated under mechanical loading and fluid injection, using a hybrid Finite Element/Peridynamic formulation [15]. The Peridynamic approach is used for the solid phase while the flow field is simulated with a FEM model. A staggered procedure is employed for the solution of the hybrid system. The geometry and constraints of the rectangular structure are shown in Fig. 3. The mechanical and fluid parameters used in the calculation are, Young's modulus: $E = 10GPa$, Poisson's ratio: $\nu = 0.2$, critical energy release rate: $G_c = 1J/m^2$, porosity: $n_r = 0.002$, mass density: $\rho_r = \rho_f = 1000kg/m^3$, Biot constant: $\alpha = 1$, bulk modulus and viscosity coefficient of the fluid: $K_w = 2.2GPa$ and $\mu_w = 10^{-3}Pa \cdot s$, permeability coefficient of the reservoir domain: $k_r = 10^{-12}m^2$.

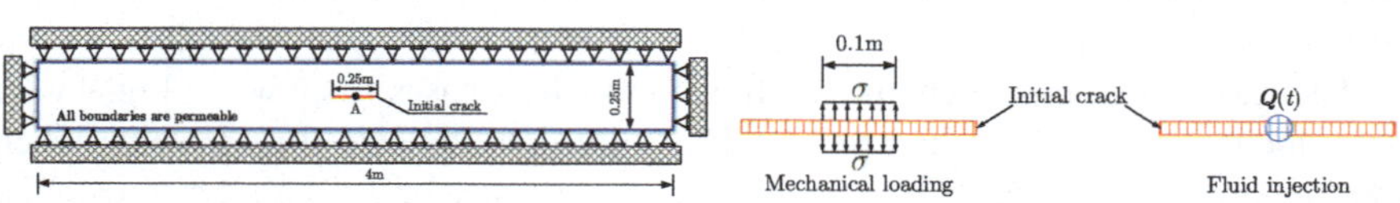

(a) Geometry and constraints. (b) Loading conditions.

Fig. 3 Geometry and boundary conditions of the rectangular structure with a central initial crack

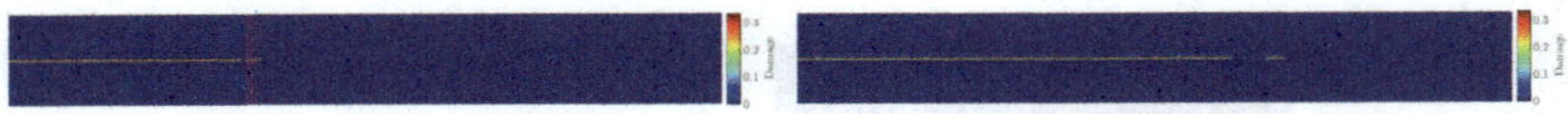

(a) Damage levels at 4.43×10⁻⁴s. (b) Damage levels at 9.9×10⁻⁴s.

Fig. 4 Forerunning fracture in the front of the crack tip in the dry structure under mechanical loading

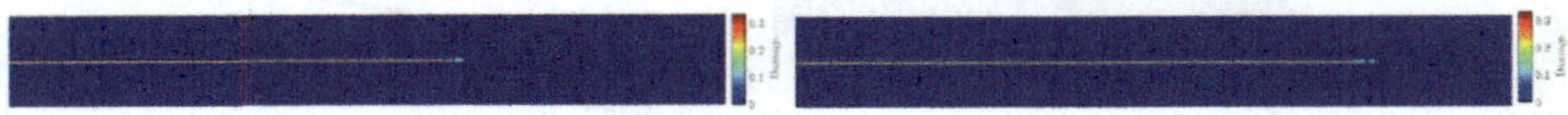

(a) Damage levels at 7.5×10⁻⁴s. (b) Damage levels at 9.38×10⁻⁴s.

Fig. 5 Forerunning fracture in the front of the crack tip in the saturated structure under mechanical loading

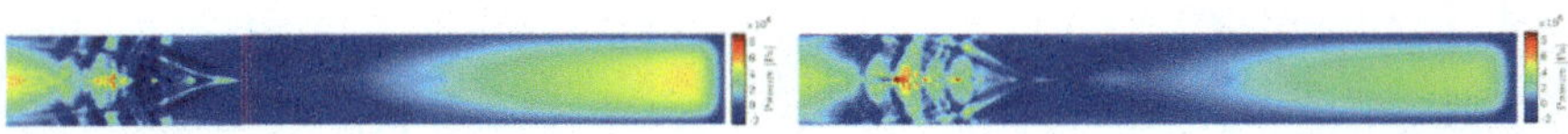

(a) Distribution of pore pressure at 2.65×10⁻⁴s. (b) Distribution of pore pressure at 2.85×10⁻⁴s.

Fig. 6 Pore pressure waves in the saturated cross section under mechanical loading

In the first case (case 1), a dry porous material under mechanical loading condition is analysed. As shown in Fig. 3b, a constant pressure $\sigma = 15MPa$ is applied on the surface of the initial crack near the central position to force crack opening. Under the action of this force, the crack propagates from the initial crack tip. Forerunning events are observed during the simulation with two of them shown in Fig. 4a and b.

The second case (case 2) deals with a saturated porous material with the same mechanical loading condition as above. Under the action of applied external force, the pore pressure in the whole domain changes and the crack propagation occurs under the combined action of external force and pore pressure. Again forerunning events occur in this case and two of them are shown in Fig. 5a and b. Note that the time instants differ from above because of the interaction with the fluid phase. The distributions of pore pressure at $2.65 \times 10^{-4}s$ and $2.85 \times 10^{-4}s$ are drawn in Fig. 6a and b. Wave propagation can be noticed.

The last case (case 3) is carried out with fluid injection at the centre of the initial crack with a constant volume rate of $Q = 1m^3/s$. Figure 7a and b depict the crack patterns at two different time instants where small cracks can also be seen in front of the main crack. The distributions of pore pressure at $2.3 \times 10^{-4}s$ and $2.55 \times 10^{-4}s$ are plotted in Fig. 8a and b. Compared to the mechanical loading case the wave structure is much simpler.

The variation of crack length with time in the three cases is shown in Fig. 9 with tip velocity values at selected points. The solid dots indicate time instants when forerunning occurs, either before or after the coalescence of the main crack with the small crack in front. It has been shown in [4, 5] and in the above example of the double beam that forerunning increases the overall fracturing speed in dry bodies. By comparing the diagrams for the dry structure under mechanical loading and the

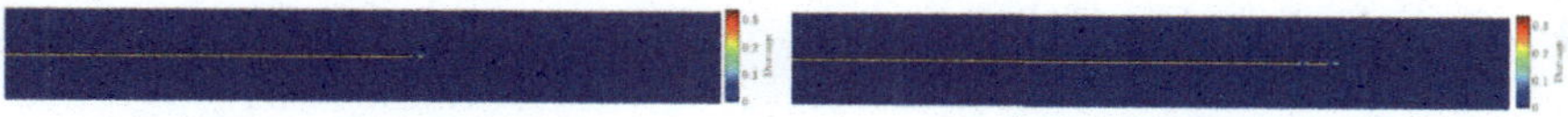

(a) Damage levels at 5.5×10^{-4}s. (b) Damage levels at 7.2×10^{-4}s.

Fig. 7 Forerunning fracture in the front of the crack tip in the saturated structure under fluid injection

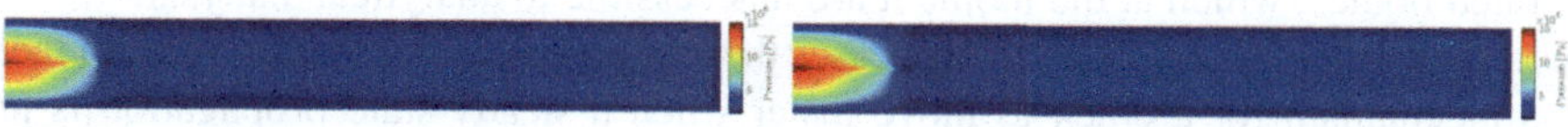

(a) Distribution of pore pressure at 2.35×10^{-4}s. (b) Distribution of pore pressure at 2.55×10^{-4}s.

Fig. 8 Pore pressure waves in the saturated cross section under fluid injection

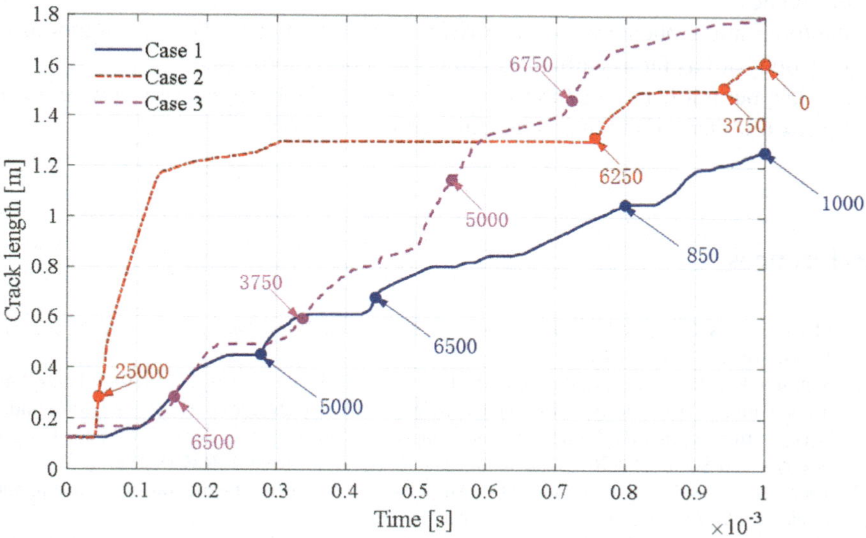

Fig. 9 Variation of crack length with time; case 1 is a dry structure under mechanical loading, case 2 a saturated structure under mechanical loading and case 3 a saturated structure with fluid injection. Velocity values in m/s

saturated structure under the same mechanical loading of Fig. 9 it appears that the presence of the fluid phase further increases the overall fracture speed. This can also be seen by comparing the respective positions of the crack tip in Figs. 4 and 5. This further increase is due to the interaction between waves in the solid and in the fluid domain.

4 Conclusions

Examples of a dry double beam, a saturated lattice and a saturated two-dimensional continuous specimen confirm that:

- numerical models allow for solution of more demanding problems such as saturated bodies, which at the moment are inaccessible to analytical analysis;
- in dry bodies, the forerunning increases the overall fracturing speed and is, in fact, a mechanism for a crack to move faster when a steady-state propagation is no longer supported by the body/structure due to a high level of external forces;
- in presence of the forerunning, interaction with the waves in the fluid phase increases the average speed even further comparing to the movement in the same dry bodies;
- the forerunning phenomenon deserves further scrutiny because of its importance in geophysics as far as earthquake events are concerned;
- and last but not least, the forerunning is an undeniable source of stepwise crack tip advancement of the main crack in continuum models.

References

1. Marder, M., & Cross, S. (1995). Origin of crack tip instabilities. *Journal of the Mechanics and Physics of Solids, 43*, 1–48.
2. Sammonds, P. R., Ayling, M. R., Meredith, P. G., Murrell, S. A. F., Jones, C. (1989). A laboratory investigation of acoustic emission and elastic wave velocity changes during rock failure under triaxial stresses. In ISRM International Symposium, 30 August-2 September, Pau, France, International Society for Rock Mechanics and Rock Engineering, ISRM-IS-1989-030.
3. Tvergaard, V., & Needleman, A. (1993). An analysis of the brittle-ductile transition in dynamic crack growth. *International Journal of Fracture, 59*, 53–67.
4. Mishuris, G. S., Movchan, A. B., & Slepyan, L. I. (2009). Localised knife waves in a structured interface. *Journal of the Mechanics and Physics of Solids, 57*, 1958–1979.
5. Slepyan, L. I., Mishuris, G. S., & Movchan, A. B. (2010). Crack in a lattice waveguide. *International Journal of Fracture, 162*, 91–106.
6. Nieves, M. J., Mishuris, G. S., & Slepyan, L. I. (2016). Analysis of dynamic damage propagation in discrete beam structures. *The International Journal of Solids and Structures, 97–98*, 699–713.
7. Nieves, M. J., Mishuris, G. S., & Slepyan, L. I. (2016). Transient wave in a transformable periodic flexural structure. *The International Journal of Solids and Structures, 112*, 185–208.
8. Garau, M., Nieves, M. J., & Jones, I. S. (2019). Alternating strain regimes for failure propagation in flexural systems. *The Quarterly Journal of Mechanics and Applied Mathematics, 72*, 305–339.
9. Slepyan, L., Ayzenberg-Stepanenko, M., & Mishuris, G. (2015). Forerunning mode transition in a continuous waveguide. *Journal of the Mechanics and Physics of Solids, 78*, 32–45.
10. Mishuris, G. S., Movchan, A. B., & Slepyan, L. I. (2008). Dynamics of a bridged crack in a discrete lattice. *The Quarterly Journal of Mechanics and Applied Mathematics, 61*, 151–160.
11. Araujo, M. S., Vieira, A. P., Andrade, J. S., & Herrmann, H. J. (2016). Mesoscopic approach to subcritical fatigue crack growth. *Physical Review E, 94*(4), 043003.

12. Milanese, E., Yılmaz, O., Molinari, J.-F., & Schrefler, B. (2016). Avalanches in dry and saturated disordered media at fracture. *Physical Review E, 93*, 043002.
13. Biot, M. A. (1956). Theory of propagation of elastic waves in a fluid-saturated porous solid, part I-low-frequency range. *The Journal of the Acoustical Society of America, 28*, 168–178.
14. Peruzzo, C., Simoni, L., & Schrefler, B. A. (2019). On stepwise advancement of fractures and pressure oscillations in saturated porous media. *Engineering Fracture Mechanics, 215*, 246–250.
15. Ni, T., Pesavento, F., Zaccariotto, M., Galvanetto, U., Schrefler , B. A. Numerical simulation of forerunning fracture in saturated porous solids with hybrid FEM/Peridynamic model (Computers and Geotechnics, published online https://doi.org/10.1016/j.compgeo.2021.104024).

An Optimized Material Removal Process

Jean-François Molinari and Son Pham-Ba

I am deeply honored to contribute a chapter to celebrate Peter Wriggers' 70th birthday. As my research is for a large part at the interface between contact mechanics and computational methods, Peter's work is evidently a great source of inspiration. Dedicated versions of his books sit on my book shelves. And even my first paper, published in 2001, cites and uses Peter's work. Beyond his immense and inspiring scientific contributions, I truly enjoy our discussions and long for the post Covid days in which we will sit again in a beer garden, during a summer school or a thesis defense. Peter is a good listener, he is compassionate, and cares as much about people than science.

Abstract We conduct boundary element simulations of a contact problem consisting of an elastic medium subject to tangential load. Using a particle swarm optimization algorithm, we find the optimal shape and location of the micro-contacts to maximize for a given load the stored elastic energy contributing to the removal of a spherical particle contained in between the micro-contacts. We propose an ice scream scoop as an application of this optimization process.

1 Introduction

Wear, the process of material removal when two solids are in sliding contact, comes in various forms, adhesive and abrasive wear being the most prominent [1]. The formation of debris particles is often thought as a probabilistic event. It is known that natural or man made surfaces are rough over a range of length scales [2–4]. It implies that the contact between nominally flat surfaces is in reality a contact between two rough surfaces when viewed microscopically, such that the real contact area is much smaller than the apparent contact area [5–8]. Protruding asperities from

J.-F. Molinari (✉) · S. Pham-Ba
Institute of Civil Engineering, Institute of Materials Science and Engineering, École polytechnique fédérale de Lausanne (EPFL), CH 1015 Lausanne, Switzerland
e-mail: jean-francois.molinari@epfl.ch

F. Aldakheel et al. (eds.), *Current Trends and Open Problems in Computational Mechanics*, https://doi.org/10.1007/978-3-030-87312-7_34

both rough surfaces make junctions and result in what are called *micro-contacts*. In the probabilistic view of wear, only a fraction of those micro-contacts form debris particles.

Recent advances have permitted a leap forward on establishing a deterministic criterion for wear particle formation, at least in the context of adhesive wear. This new understanding emerged thanks to recent numerical studies performed at the small near-atomic scale [9–11]. The formation of wear particles at an unlubricated tribological interface due to adhesive wear was first theorized to be driven by a competition between deformation energy and fracture energy in 1958 by Rabinovicz [12]. This Griffith (fracture mechanics) approach to wear particle formation was recently extended to account for plastic flow, and validated with molecular dynamics (MD) simulations [9]. The theory predicts the existence of a critical length scale d^*, dictating a transition between a ductile and a brittle behavior for a given material at a contact junction. Consequently, d^* also corresponds to the minimal wear particle size which can be formed under adhesive wear when two asperities located on two opposed sliding surfaces collide into each other. d^* was found to be mainly dependent on the material properties, with second order effects related to the geometry of the contacting asperities. While these works focused on adhesive wear, abrasive wear mechanisms can also be understood through the lens of fracture mechanics [13].

Later, these numerical simulations were extended to account for interactions between nearby micro-contacts, each micro-contact being susceptible to result in the formation of a wear particle under the application of shear load [14] (see Fig. 1). Micro-contact junctions that are far from each other result in the formation of separated wear particles (Fig. 1a). However, the simulations revealed that micro-contacts

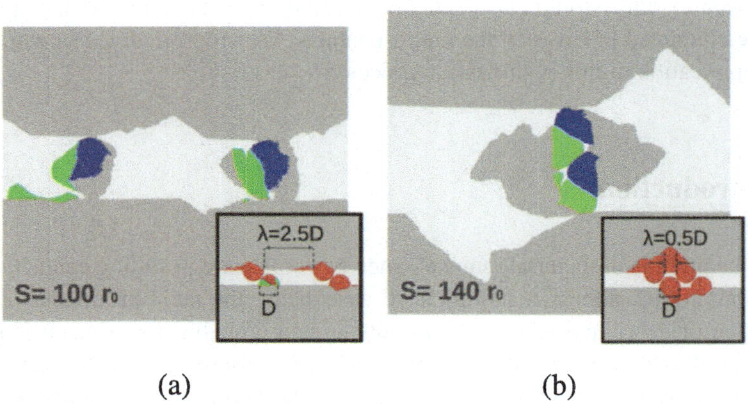

Fig. 1 2D MD simulations of asperity-level wear mechanisms. The insets show the initial setups and positions of contact junctions. Atoms in red color end up in the formed fragments seen in the main figures when the system is sheared. This figure is reproduced from Aghababaei et al. [14]. **a** Contact junctions are initially far apart and each form an individual wear particle. **b** Contact junctions are close and, through elastic interactions, result in the formation of a single, much larger, wear particle

that are close to each other, i.e. separated by a distance of the order (or less) than the junction size, result in the formation of a combined larger particle, due to crack shielding mechanisms (Fig. 1b). This simple observation provides a mechanistic argument for the transition from mild to severe wear observed at high loads, e.g. when the contact surface is populated by larger and denser micro-contacts thereby promoting elastic interactions between those. More recent theoretical considerations, supported by discrete MD simulations and simulations conducted in a continuum setting using the boundary element method, confirmed and extended these findings to multiple interacting junctions in a 2D setting [15]. Also noteworthy is the confirmation of the importance of crack shielding mechanisms for nearby contact junctions thanks to 2D finite-element simulations in which a phase-field formulation of fracture permitted a robust mesh-independent resolution of crack paths [16]. The extent of such interactions remains to be thoroughly studied in 3D.

With the general understanding that elastic interactions between contact patches can increase the possibility of forming wear particles of a larger volume, this paper explores the uncharted territory of elastic interactions in a 3D setting. We aim to exploit those interactions by searching for an adhesive contact shape that maximises the volume of a detached chunk of material. This shape would comprise of multiple adhesive regions, or in general, regions able to transmit a tangential load to the material to be carved. These tangential loads can be transmitted by a hard rigid tool indenting a soft elastic surface, thereby entering the realm of abrasive wear. Section 2 describes the wear criterion, which compares the adhesive energy required to create new surfaces to the stored elastic energy, evaluated using the boundary element method. Section 3 details the particle swarm optimization algorithm to probe contact patches shapes. Finally results are shown in Sect. 4. We propose as an application a novel ice cream scoop design.

2 Micro-Mechanics of Wear

2.1 Wear Criteria

We consider two surfaces that are sliding on each other, with the adhesive junctions formed between them being loaded tangentially.[1] The two bodies resist the sliding force, deforming, and accumulating elastic energy E_{el}. Considering one of the sliding bodies as a semi-infinite body, the elastic energy stored inside this body is

$$E_{\text{el}} = \frac{1}{2} \int_{\Gamma} \boldsymbol{u} \cdot \boldsymbol{p} \, d\Gamma, \tag{1}$$

[1] One can also consider a hard abrasive tool gripping an elastic body.

where Γ is the nominally flat surface upon which load is applied, $\boldsymbol{p}$ is the traction field applied on this surface and $\boldsymbol{u}$ is the displacement field caused by the traction field.

At the scale of the contact junctions, wear is reduced to the formation of debris particles under the junctions. The detachment of a wear particle from a body requires the creation of new surfaces, thus requiring surface energy, or adhesive energy E_{ad}, which is proportional to the total surface area created times a surface energy γ.

When a wear particle is detached, the tangential load it was carrying can no longer be transmitted between the two sliding surfaces, resulting in a drop ΔE_{el} in the amount of stored elastic energy. Similarly to Griffith's criterion for crack propagation, a criterion can be established for the possibility to fully detach a wear particle of a given shape: the drop in elastic energy obtained when detaching the particle must be equal or greater than the amount of adhesive energy required:

$$\Delta E_{\text{el}} \geqslant E_{\text{ad}}. \tag{2}$$

The other necessary condition for the detachment of a wear particle is to have a location where crack nucleation can occur. We assume that a crack can be initiated at a point if

$$\sigma_{\text{I}} \geqslant \sigma_{\text{m}}, \tag{3}$$

where σ_{I} is the first principal stress, or the maximum tensile stress if positive, and σ_{m} is the tensile strength of the material.

Consequently, we are left with two criteria for wear particle formation: a *crack initiation criterion* (3), and an *energetic feasibility criterion* (2).

2.2 Elastic Energy Computation

Let us consider a semi-infinite body Ω whose free surface Γ is in the (x, y) plane at $z = 0$. Some contact junctions are distributed on Γ and are described by a 'contact' field $c(x, y)$ equal to 1 where a junction is present and 0 otherwise. When a sliding force is applied on the body in the x direction, we assume that the adhesive junctions will carry a uniform tangential stress q also in the x direction. No normal load is applied. The x component of the surface traction field $\boldsymbol{p}$ on Γ is therefore

$$p_x(x, y) = c(x, y)q, \tag{4}$$

with the other components in the y and z direction equal to 0. The setup is shown in Fig. 2.

The surface displacements can be obtained from the surface tractions. A unit point load applied at the origin of Ω in the x direction results in a displacement field whose x component is [17]

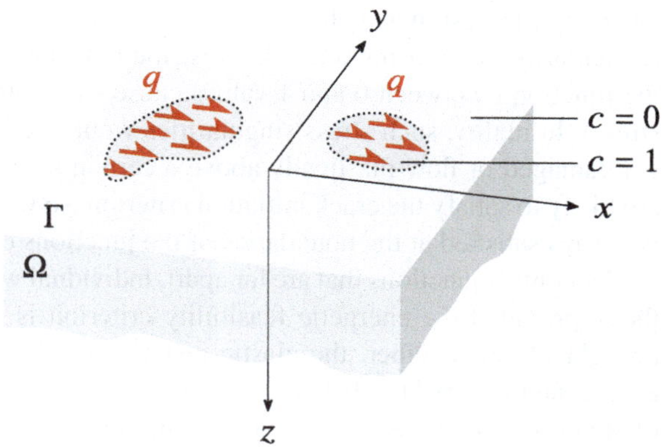

Fig. 2 Sheared adhesive junctions on a semi-infinite body. The function $c(x, y)$ describes the geometry of the junctions on the surface Γ

$$u^{\text{ker}}_{x \to x} = \frac{1}{4\pi G}\left[2(1-\nu)\frac{1}{r} + 2\nu\frac{x^2}{r^3}\right], \tag{5}$$

where G is the shear modulus of the material, ν the Poisson's ratio, and r is the distance from the origin: $r^2 = x^2 + y^2 + z^2$. There are also non-zero components of the displacement field in the y and z direction, but they are not relevant in this case, as shown below.

The surface displacements in the x direction due to the full traction field (4) is

$$u_x(x, y) = \iint u^{\text{ker}}_{x \to x}(x - \xi, y - \eta)\, p_x(\xi, \eta)\, d\xi\, d\eta \tag{6}$$

$$= [u^{\text{ker}}_{x \to x} * p_x](x, y), \tag{7}$$

which is a convolution (denoted by the $*$ symbol). The expression for elastic energy (1) becomes

$$E_{\text{el}} = \frac{1}{2}\int_{\Gamma} u_x p_x \, d\Gamma, \tag{8}$$

where u_x and p_x are obtained from (7) and (4). Since $p_y = 0$ and $p_z = 0$, the components of displacement in those directions do not intervene in (8).

When dealing with this setup computationally, Γ can be discretized into a finite grid, and the integral of (8) can be turned into a finite sum. Care must be taken when considering a system of finite size, since the displacement kernel (5) decreases when moving away from the origin but does not vanish before reaching infinity. Therefore, the contact region c must not have non-zero values near the boundaries of the discretized finite Γ to lower the impact of the finite size domain on the elastic energy computation. The discretization of the surface and the use of a kernel (5) to compute the displacements from the surface tractions are part of the boundary element method (BEM) [18].

The computation of the elastic energy allows to check if the energetic feasibility criterion is satisfied, when the adhesive energy is already known (it is easily calculated from the estimated shape of the wear particle to be potentially formed). One particularity of choosing a traction distribution such as (4) is that discontinuities in the function c between 0 and 1 values cause stress singularities (regions of infinite stress). In reality, such stress singularities would be regularized, because materials get damaged or flow plastically above a certain stress. Nevertheless, those regions are likely to satisfy the crack initiation criterion, so we will assume that this criterion is always satisfied at the boundaries of the junctions defined by the function c.

For contact junctions that are far apart, individual wear particles can form beneath them, provided the energetic feasibility criterion is satisfied. When junctions are brought closer together, the elastic energy stored in the system increases due to elastic interactions [15, 16]. This elastic energy increase can result in the formation of larger wear particles, encompassing multiple nearby junctions, as shown in Fig. 1 in the 2D case.

3 Material Removal

3.1 Problem Statement

We now wish to find the most efficient way to remove a piece of material from a body with a flat surface. We assume that the piece of removed material must have a roughly hemispherical shape of known diameter, so that E_{ad} is also known and fixed. To minimize the effort put into the detachment of material, one must maximize E_{el} by changing the shape of the contact junctions c while trying to decrease the imposed tangential load, where the total tangential load is

$$F_x = \int_{\Gamma} cq \, d\Gamma, \tag{9}$$

which is deduced from (4).

The optimization problem is the following: find the function c, which maximizes E_{el} for a given F_x (q is modified according to c to keep F_x constant).

To parameterize the function c, which is a binary representation of the shape of the sheared junctions, we use n_{m} *metaballs* [19], which are n-dimensional[2] circular objects usually used in computer graphics because of their organic appearance, as they smoothly merge with nearby metaballs. They allow us to create a complex shape c using simple circular objects with smooth connections between them. Each metaball has a parameterized center (x_i, y_i), with $1 \leqslant i \leqslant n_{\mathrm{m}}$, and a fixed width w. A metaball adds a $w^2/(4r_i^2)$ term to c, where r_i is the distance to its center: $r_i^2 = (x - x_i)^2 + (y - y_i)^2$. The whole function c is the sum of all metaballs contributions, binarized to only keep regions where it is greater than 1. Mathematically:

[2] In the present case, they are two-dimensional.

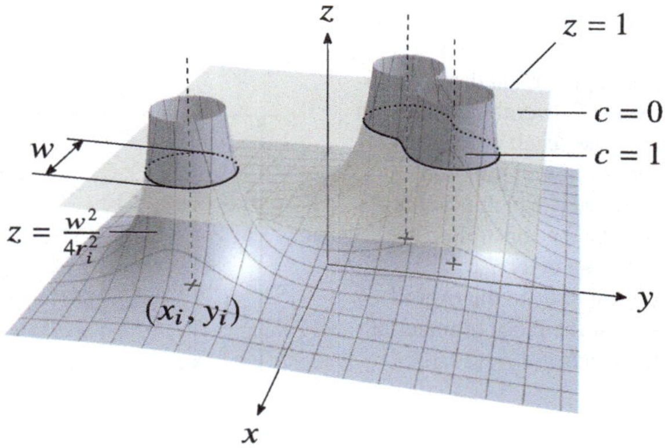

Fig. 3 Creation of metaballs from functions added together and clamped. Each metaball adds a contribution to the plotted surface. The contact function c is equal to 1 whenever the summed surface is above $z = 1$. On the left, an isolated metaball makes a circular shape in c. On the right, nearby metaballs are smoothly merged to construct the contact function c

$$c(x, y) = \begin{cases} 1 & \text{if } \sum_{i=1}^{n} \frac{w^2}{4((x - x_i)^2 + (y - y_i)^2)} \geqslant 1, \\ 0 & \text{otherwise.} \end{cases} \tag{10}$$

This expression can be verified to work properly when a single metaball is present: it results in c being non-zero in the region where $r_1 \leqslant w/2$, which is a circular region of width w centered on (x_1, y_1) and is the intended behavior. Figure 3 illustrates how metaballs merge to create the function c.

Geometrical constrains have to be put on the function c. It must have some edges coincident to the edge of the particle to be detached in order to satisfy the crack initiation criterion. Also, its overall size can be constrained to fit design limitations or to reduce the size of the search space, which has to be done carefully in order to maintain the performance of optimal solutions. To fit the geometrical constraints imposed on c, the metaballs are placed such that their centers are located in a ring of inner diameter d_{in} and outer diameter d_{out}. The inner diameter corresponds to the size of the piece of material to detach, and the outer diameter limits the size of the search space of the contact zone, without loss of generality. Indeed, choosing a too large outer diameter would not yield a better optimized design because contact junctions that are far from each other do not interact elastically with each other. Another advantage of using a search space delimited by radii is that it brings natural symmetries. An n_s fold symmetry can be imposed on the positions of the metaballs, and each sector can itself be symmetric. An orientation parameter φ can be added to control the angular position of the axes of symmetry with respect to the direction of shear. Those sector parameters are represented in Fig. 4.

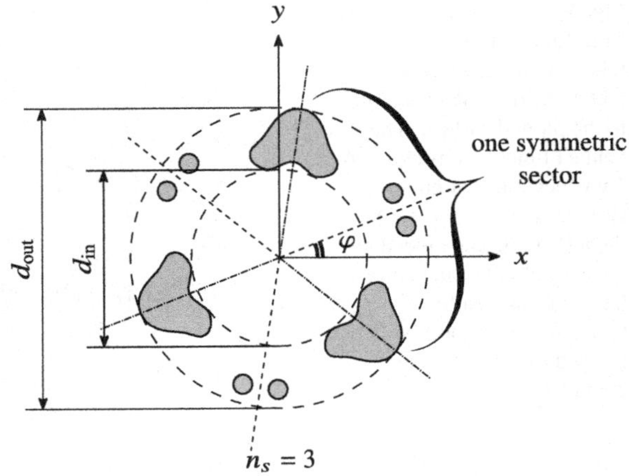

Fig. 4 Representation of the sectors of the contact function c. Each dark area is made of several metaballs. In this case, each of the three sectors has an imposed symmetry, which is not a mandatory constraint. The inner diameter d_{in} corresponds to the overall size of the piece of material to detach

3.2 *Optimization*

The elastic energy has to be maximized by finding the optimal metaball parameters: the position of their centers, x_i and y_i for $1 \leqslant i \leqslant n_m$, and the global orientation φ. If all symmetry conditions are used (forcing n_s identical and symmetric sectors), the total number of parameters is[3]

$$n_m/n_s + 1. \tag{11}$$

The number of metaballs n_m has to be taken large enough to have a fine control over the shape of c. To deal with the large number of parameters and the potential non-convexity of the problem, the choice of an evolutionary algorithm was opted for. Here, the *particle swarm optimization* (PSO) [20, 21] is used.

In PSO, a *swarm* (a population) of *particles* is considered. Each particle is a candidate solution (a shape of c) with a *position* and a *velocity*. The position[4] is the current set of parameters of the particle, and the velocity is the rate of change of each parameter between two iterations of the PSO. The position is bounded by the limits imposed on each parameter.

The swarm is initialized with n_p particles having random initial positions within the bounds. Each solution is randomized such that the centers of its metaballs are uniformly distributed in the (x, y) space (inside the ring of diameters d_{in} and d_{out}). The positions were parameterized in polar coordinates (r, θ) to facilitate the enforcement of bounds. The initial velocities are also randomized, but such that an iteration of the PSO does not create an off-bound position (more details about the update process below).

[3] There are n_m parameterized metaballs with two coordinates, so $2n_m$ parameters. There are n_s identical sectors, so this number is divided by n_s. Each sector is symmetric so this number is again divided by 2. The overall orientation adds 1.

[4] Not to be confused with the positions of the metaballs in c.

At each iteration t, the particles compute their objective function (the value of stored elastic energy) at their current position (in the space of all parameters) $p_j^t = [\varphi, r_1, \theta_1, r_2, \theta_2, \ldots]$ and update their memory of best visited position p_j^{b} if necessary. The particles have inertia, so their velocity v_j^t is conserved up to a factor $k_v \leqslant 1$. The particles are also attracted toward their own best visited position p_j^{b} and toward the overall best position visited by the swarm p^{B}, which have an influence on the particles velocity thanks to the hyperparameters $k_{\mathrm{b}} \leqslant 1$ and $k_{\mathrm{B}} \leqslant 1$. To summarize, the velocity of each particle is updated as follow:

$$v_j^t = k_v v_j^{t-1} + k_{\mathrm{b}}(p_j^{\mathrm{b}} - p_j^t) + k_{\mathrm{B}}(p^{\mathrm{B}} - p_j^t) \tag{12}$$

and their position is updated as

$$p_j^t = p_j^{t-1} + v_j^t \, \Delta t \tag{13}$$

with $\Delta t = 1$. The positions p_j^t are kept inside their bounds after each update by clamping their metaballs polar coordinates r_i inside a ring and θ_i inside a sector (see Fig. 4). k_v, k_{b} and k_{B} are hyperparameters of the optimization method and have to be fixed by the user of the method.

4 Results

4.1 Optimal Shape

Since the problem is a matter of maximizing the elastic energy while keeping other dimensions (such as tangential load and maximum overall size) constant, we can work with adimensionalized unitless quantities.

We use a discretized space of size 1×1 and with a resolution of 256×256 for the computation of E_{el} from c. The Young's modulus of the material is set to $E = 1$ and its Poisson's ratio to $\nu = 0.3$. The tangential load is set to $F_x = 0.3$ (its choice has no incidence on the results).

The whole function c is made of a total of $n_{\mathrm{m}} = 60$ metaballs of width $w = 0.019$ and has $n_{\mathrm{s}} = 3$ identical and symmetrical sectors. With those symmetry conditions on the metaballs, there are 10 independent metaballs per half-sector and therefore 21 parameters per solution according to (11). The centers of the metaballs are constrained in a ring of diameters $d_{\mathrm{in}} = 0.3$ and $d_{\mathrm{out}} = 0.5$. Figure 5b to f shows examples of initial positions (solutions).

For the PSO, $n_{\mathrm{p}} = 200$ particles are used, with the hyperparameters $k_v = 0.97$, $k_{\mathrm{b}} = 0.2$ and $k_{\mathrm{B}} = 0.2$. The optimization is run for 50 iterations (see Fig. 6) using a custom code, resulting in the solution shown in Fig. 5g. The optimization routine was performed five times with different randomized initial particles to ensure that it was not stuck in local optima.

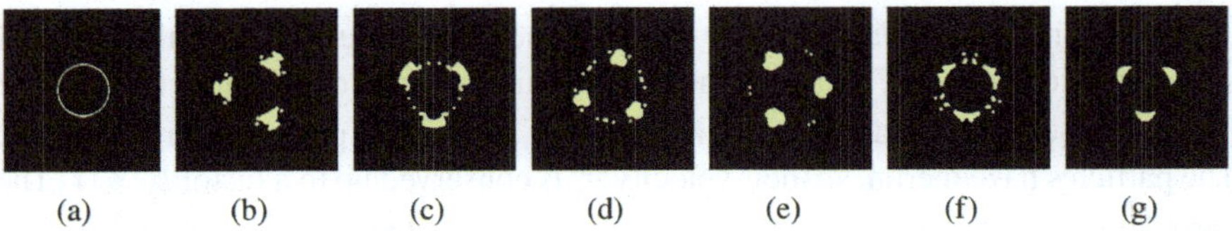

Fig. 5 Various shapes. **a** Trivial reference shape. **b–f** Examples of random initial solutions with $n_{\mathrm{m}} = 60$ metaballs of width $w = 0.019$ and $n_{\mathrm{s}} = 3$ identical and symmetrical sectors. **g** Optimized shape

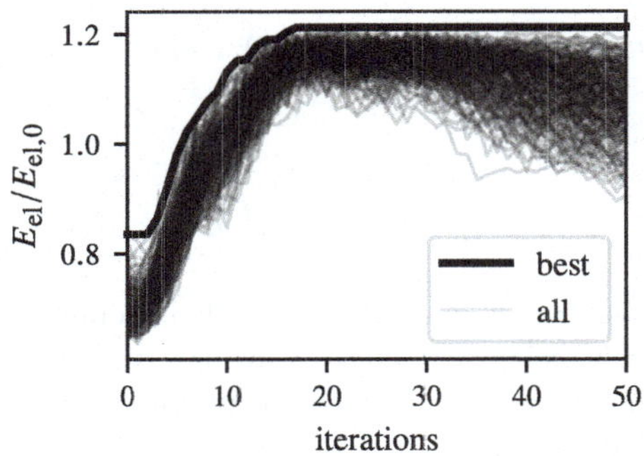

Fig. 6 Evaluations of the objective function for each particle during the optimization. The vertical axis shows the objective function relative to the evaluation of the shape Fig. 5a. The thickest line is the overall best at a given iteration, going from the random shape Fig. 5b to the optimized shape Fig. 5g

In the following text, we express the objective function of a particle as its elastic energy E_{el} divided by the elastic energy $E_{\mathrm{el,0}}$ of a trivial shape fulfilling the crack initiation criterion, i.e. a thin ring of diameter d_{in} (see Fig. 5a). In the considered discretized space, we have $E_{\mathrm{el,0}} = 12700$.

On average, the objective function $E_{\mathrm{el}}/E_{\mathrm{el,0}}$ of random initial solutions evaluates to 0.71. The optimized shape (Fig. 5g) has an objective function of 1.21, meaning it is 20% more efficient energetically than the thin ring. Therefore, we have an increased performance compared to trivial shapes. The overall orientation φ has a negligible influence on the results.

In order to check for convergence, a finer description of c is used, with $n_{\mathrm{m}} = 120$ and $w = 0.013$ to keep the overall surface area of c constant. The optimization is run for 100 steps, and results in a best objective function at 1.24 and visually indistinguishable shapes compared to $n_{\mathrm{m}} = 60$.

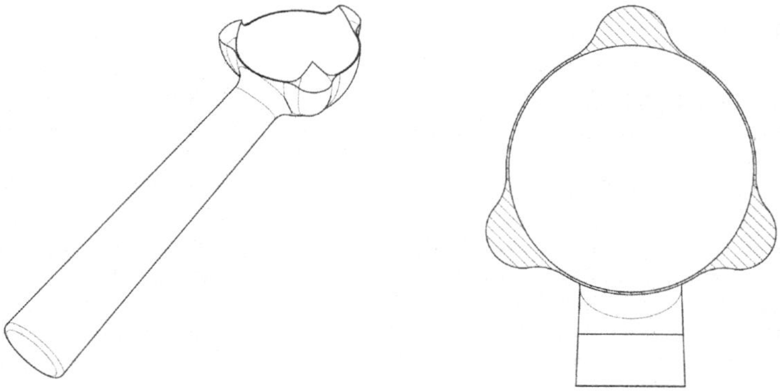

(a) Isometric view (b) Cross section of the claws

Fig. 7 **a** Design of optimized ice scoop. **b** The cross section of the claws matches the optimized sheared shape

4.2 Example of Application: Ice Cream Scoop

One practical way to use the newly found optimal contact shape is to design a tool for ice cream scooping. The size of the ball to create is fixed, and the amount of force needed to detach it can be minimized by utilizing the optimized shape. When the tool is used, it must pull on the surface of the ice cream while having the same contact shape as the one shown in Fig. 5g, instead of the usual one shown in Fig. 5a. To this end, claws can be added on a basic hemispherical scoop design, that will penetrate into the surface of the ice cream and create the desired pulling pattern. Figure 7a shows a simple design implementing this idea of penetrating claws, and Fig. 7b shows a cross-sectional view of the pattern formed by the penetration of the claws into the surface of the ice cream, matching the optimized shape found above. This design and the optimization method are patented [22].

5 Conclusion

We have explored the design space of contact junctions location and shape to maximize the energetic efficiency of material removal, thanks to elastic interactions. This study extends to a three-dimensional setting previous efforts that were limited to two dimensions, thereby providing a much richer design space. The numerical approach combines the efficient boundary element method to solve the contact problem to a particle swarm optimization algorithm to search the optimal location and shape of contact junctions. We have found a three claws design that increases the energetic efficiency of 20%. We propose an ice scream scoop application for which we have filed a patent.

References

1. Rabinowicz, E. (1995). *Friction and wear of materials* (2nd ed.). New York: Wiley.
2. Mandelbrot, B. B., Passoja, D. E., & Paullay, A. J. (1984). Fractal character of fracture surfaces of metals. *Nature, 308*(5961), 721.
3. Majumdar, A., & Tien, C. L. (1990). Fractal characterization and simulation of rough surfaces. *Wear, 136*(2), 313–327.
4. Thom, C. A., Brodsky, E. E., Carpick, R. W., Pharr, G. M., Oliver, W. C., & Goldsby, D. L. (2017). Nanoscale roughness of natural fault surfaces controlled by scale-dependent yield strength. *Geophysical Research Letters, 44*(18), 9299–9307.
5. Greenwood, J. A., & Williamson, J. B. P. (1966). Contact of nominally flat surfaces. *Proceedings of the Royal Society of London. Series A, Mathematical and Physical Sciences,295*(1442), 300–319.
6. Bush, A. W., Gibson, R. D., & Thomas, T. R. (1975). The elastic contact of a rough surface. *Wear, 35*(1), 87–111.
7. Persson, B. N. J. (2001). Elastoplastic contact between randomly rough surfaces. *Physical Review Letters,87*(11).
8. Hyun, S., Pei, L., Molinari, J.-F., & Robbins, M. O. (2004). Finite-element analysis of contact between elastic self-affine surfaces. *Physical Review E,70*(2).
9. Aghababaei, R., Warner, D. H., & Molinari, J. F. (2016). Critical length scale controls adhesive wear mechanisms. *Nature Communications,7*.
10. Aghababaei, R., Warner, D. H., & Molinari, J.-F. (2017). On the debris-level origins of adhesive wear. *Proceedings of the National Academy of Sciences, 114*(30), 7935–7940.
11. Brink, T., & Molinari, J. F. (2019). Adhesive wear mechanisms in the presence of weak interfaces: Insights from an amorphous model system. *Physical Review Materials,3*(5).
12. Rabinowicz, E. (1958). The effect of size on the looseness of wear fragments. *Wear, 2*(1), 4–8.
13. Harish, A. B., & Wriggers, P. (2019). Modeling of two-body abrasive wear of filled elastomers as a contact-induced fracture process. *Tribology International, 138,* 16–31.
14. Aghababaei, R., Brink, T., & Molinari, J. F. (2018). Asperity-level origins of transition from mild to severe wear. *Physical Review Letters,120*(18).
15. Pham-Ba, S., Brink, T., & Molinari, J.-F. (2020). Adhesive wear and interaction of tangentially loaded micro-contacts. *International Journal of Solids and Structures, 188–189,* 261–268.
16. Collet, S., Molinari, J. -F., & Brach, S. (2020). Variational phase-field continuum model uncovers adhesive wear mechanisms in asperity junctions. *Journal of the Mechanics and Physics of Solids, 145,* 104130.
17. Johnson, K. L. (1985). *Contact mechanics*. Cambridge: Cambridge University Press.
18. Bonnet, M. (1999). *Boundary integral equation methods for solids and fluids*. Wiley.
19. Blinn, J. F. (1982). A generalization of algebraic surface drawing. *ACM Transactions on Graphics, 1*(3), 235–256.
20. Kennedy, J., & Eberhart, R. (1995). Particle swarm optimization. In *Proceedings of ICNN'95—International Conference on Neural Networks* (Vol. 4, pp. 1942–1948).
21. Zhang, Y., Wang, S., & Ji, G. (2015). A comprehensive survey on particle swarm optimization algorithm and its applications. *Mathematical Problems in Engineering, 2015,* e931256.
22. Molinari, J. F., & Pham-Ba, S. (2020). Shovelling tool and method for designing the same.

How to Push Computational Bio-Mechanics to Clinical Application?

Udo Nackenhorst and Maximilian Bittens

It is dated back more than 20 years ago, when I joined Peter's group at the University of Hannover as an associate. Looking back, I don't want to miss any day of that time. Peter guided me to new perspectives of scientific work and we collaborated in many academic subjects, like the organization of scientific workshops and implementation of structural doctoral education programs, for example. In 2008 he handed over the chair of the internationally renowned Institute of Mechanics and Computational Mechanics, letting me step into very big footprints after Erwin Stein and Peter. Thank you Peter for the last 20 years of fruitful collaboration and friendship. In am looking ahead for the next decades, Udo.

Abstract Computational bio-mechanics is well established and sophisticated models have been developed to compute patient's individual predictions for medical treatments. Drawback of these models till today is the computational complexity. In addition, the simulations have to be done by experts with deep knowledge about the modeling and computing approaches behind. For that reasons, besides some scientific collaborations between physicians and engineers, that approaches to this day found only minor impact in clinical practice. A solution to that problem is signed by a new paradigm in computational mechanics, i.e. model order reduction. This contribution describes the recent progress made into that direction for the treatment of bones with implants.

U. Nackenhorst (✉) · M. Bittens
Leibniz University Hannover, Institute of Mechanics and Computational Mechanics, Hannover, Germany
e-mail: nackenhorst@ibnm.uni-hannover.de

M. Bittens
e-mail: maximilian.bittens@ibnm.uni-hannover.de

F. Aldakheel et al. (eds.), *Current Trends and Open Problems in Computational Mechanics*, https://doi.org/10.1007/978-3-030-87312-7_35

1 Introduction

Computational techniques based on (bio-chemo-) physical principles for medical intervention are under investigation since a couple of decades. Specific disciplines have been established documented by a series of scientific journals, international conferences and related monographs. Models and computational methods have been grown to maturity, thus nowadays very powerful techniques are available for the prediction of patient individual behavior. With a particular focus on bio-mechanics of bones, first simplistic computational modeling approaches came up in the late 20th century, e.g. [1–3]. With increasing computer power and more robust numerical methods the modeling strategy has become mature during the last 20 years. Nowadays, robust computational techniques and modeling strategies based on image data are available, which enable for patient individual predictions, e.g. [4]. Main focus of these computations has been on the implants long term stability, which is assessed in terms of bone-remodeling. Inspired from medical questions, these models have been refined recently for the investigation of primary implant stability by simulating the osseointegration of non-cemented stem-prosthesis, [5]. In parallel to these mainly phenomenological modeling, by which simply the bone remodeling is described as a function on the mechanical stress- or strain distribution, a research direction on the enrichment of these models with more bio-chemo-physical knowledge has been developed. Here, the cellular behavior and the inter-cellular communication is considered [6, 7]. Additional computational complexity on incorporating these processes has been discussed in literature, e.g. [8]. Recently, it has been realized, that the computational complexity of these state of the art models diverges to the need in clinical practice, [9, 10]. The computational complexity of existing modeling and numerical methods, even for the simple phenomenological approaches, is much too high for patient individual planning on the best treatment options as well as for follow up monitoring. A promising pathway on that direction is paved by a new branch in computational engineering, i.e. Model Order Reduction (MOR) techniques, also phrased as surrogate modeling, e.g. [9].

A recent state of the art review on MOR techniques is provided in [11]. In principle, one distinguishes between intrusive and non-intrusive methods. Because of the complexity of the simulation chain, refer Sect. 2, here only non-intrusive methods will be considered which call the solver as a black-box system. A remaining challenge is on efficient sampling techniques, especially in higher dimensional parametric spaces. In this work we decided for an adaptive sparse grid collocation method [12], refer to Sect. 3. The surrogate modeling approach is applied to a six-dimensional parametrization of hip-joint prosthesis implantation into the femoral bone, see Sect. 4.

2 The High-Fidelity Modeling Approach

The HF-modeling approach consists of several sequential steps as shown in Fig. 1. In a first step (1) the geometrical model of the bone has to be build. For patient individual modeling one will go for image data obtained from high-resolution CT or sophisticated X-ray technologies. The shape and the internal bone-mass density (bmd) distribution will be extracted from these data. The geometry is modeled by CAD surface reconstruction and after meshing, the distribution of bmd is mapped to the elements. A next step (3) is on the computation of the boundary conditions denoted as statically equivalent joint and muscle forces from an inverse optimization procedure. This is performed for the untreated bone first, which is indicated by the upper loop in Fig. 1. A next step (2) is on modeling the implant treated bone via Boolean operations of the CAD models. The treated bone and the prosthesis are meshed again and the obtained bmd-distribution is remapped to the new bone elements. The statically equivalent loads obtained before are applied to the treated bone model. The interface between the prosthesis and the bone is discretized with bio-active interface elements in order to mimic the osseo-integration process. Next the osseointegration of the non-cemented prosthesis is simulated (4) and eventually the long-term bone-remodeling (5) is computed. Finally the computational results are post-processed for visualization of the results with regard to the quantities of medical interest (QoI's), e.g. X-ray emulations.

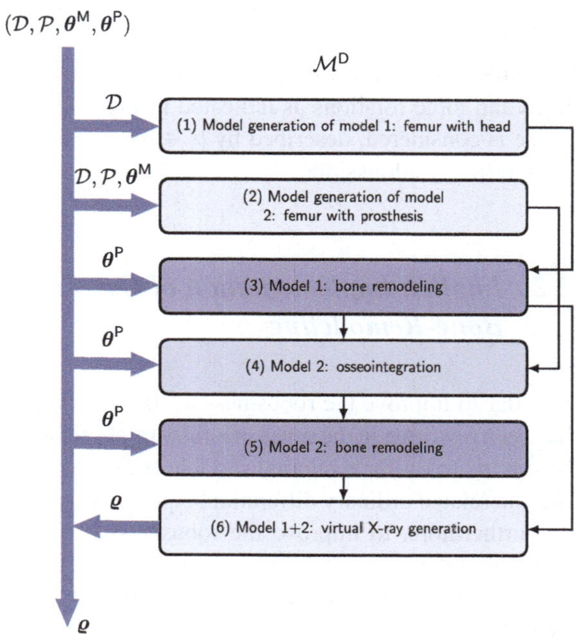

Fig. 1 The work-flow of the parametric high-fidelity model. (1) and (3) describe the untreated bone, while (2), (4) and (5) tackle the implanted bone

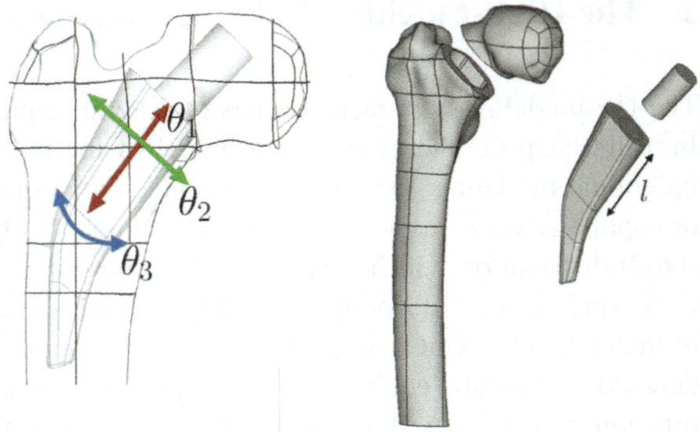

Fig. 2 Parametrization of the relative positioning of the prosthesis in the femur. In total the translation in three orthogonal directions relative to the bones mean axis and three rotations around these axis have been considered in this work, spanning a six-dimensional parameter space

This work-flow has been described in detail in [4]. However, at that time many of these steps have been performed manually. For an efficient computation of the surrogate model a parametrized approach that runs fully automatic without any further user input is needed, and the models should run robust for all points in the multi-dimensional parametric space, which will be described in the next subsections.

2.1 Model Parametrization

In this first approach we restrict ourselves to a moderate parameter space, i.e. to the relative placement of the prosthesis in the bone. Six parameters have been chosen to define the position of the implant relative to the bones axis, i.e. three translations and three rotations as indicated in Fig. 2. Thus, a six-dimensional parametric space is considered, described by $\theta_i \in [\theta_{i,\min}, \theta_{i,\max}]$, where the bounds have to be geometrically admissible.

2.2 Implicit Implementation of Osseointegration and Bone-Remodeling

In order to improve the robustness of the simulations the bone remodeling scheme has been rewritten in the sense of a fully implicit constitutive model, i.e. the change of bmd is treated as local internal variable at the integration points of the finite elements, and the related ordinary differential equations are integrated via implicit Euler rules.

Furthermore, to improve the robustness of the approach and to avoid explicit smoothing steps to prevent intensively discussed checker-board pattern, the stimulus-equation for the evolution of local bmd has been modified after a suggestion of [13]. Starting from the constitutive relationship between Young's-modulus E and bmd ϱ,

$$E = E_0 \left(\frac{\varrho}{\varrho_0} \right)^n , \tag{1}$$

where E_0 and ϱ_0 are reference values, e.g. for cortical bone. The exponent n has been discussed controversially in literature, however, within the thermodynamically consistent constitutive framework it should be $n = 2$ [14], and it has been shown based on experimental investigations that with this value of the exponent the best fit is obtained [10, 15].

Following [13], the stimulus function is slightly changed by a weighting factor

$$\dot{\varrho} = k \left(\left(\frac{\varrho}{\varrho_0} \right)^{-m} \Psi - \Psi_{\mathrm{ref}} \right) , \tag{2}$$

where k is a so far undetermined time constant, Ψ the mass-specific strain energy function of the linear elastic and isothermal bone tissue, depending on the local bone mass density after Eq. (1) and Ψ_{ref} describes a physiological target value.

It can be shown, that a value for $m > n$ stabilizes the scheme with regard to checker-boarding. However, without any additional smoothing step, i.e. using a C^0–smooth stimulus ψ, checker-boarding is still observable for linear tetrahedral finite elements. Thus we decided for tetrahedral elements with quadratic shape functions. This requires some additional effort for the post-processing, i.e. X-ray imaging from the computed 3D results.

3 The Surrogate-Model

For the construction of the surrogate model an Adaptive Sparse-Grid Collocation Method (ASGCM) has been chosen. Instead of using global Lagrangian polynomials we decided for piecewise linear interpolation of the parametric space following [12].

The overall scheme for the construction of the surrogate is depicted in Fig. 3. In the first step (1) the initial experimental design, denote by χ_{init} is defined. For each $\chi_{\mathrm{i}} \in \chi_{\mathrm{init}}$ a deterministic computation for the QoI is performed (2). By that, the initial surrogate model is constructed by 2 to 3 hierarchical levels of the ASGCM, step (3). By evaluation the hierarchical surpluses indicator $c_l^m > \epsilon_{\mathrm{ref}}$ a local refinement is evaluated and the experimental design is enriched in steps (5) and (6).

4 Numerical Example

As an example the surrogate model has been computed for the bone-remodeling of a treated femur bone in the six-dimensional parametric space,

$$\theta^{\mathrm{M}} = [\theta_1, \theta_2, \theta_3, \theta_4, \theta_5, \theta_6] , \tag{3}$$

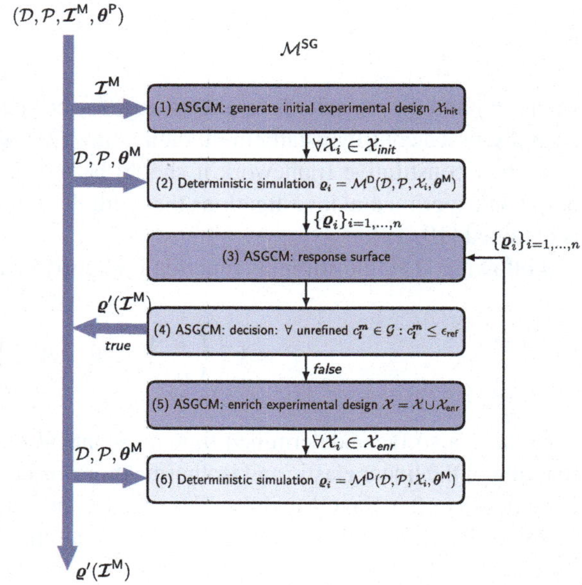

Fig. 3 Scheme of the surrogate model. Based on an initial design of experiments a first sparse-grid surrogate is constructed, (1) and (2). In further steps a quality criterion is evaluted and the surrogate model is improved by local refinement of the adaptive sparse grid, steps (3) to (6)

where

$$\theta_1, \theta_2, \theta_3 \in [-0.25\text{mm}, 0.25\text{mm}] \tag{4}$$

and

$$\theta_4, \theta_5, \theta_6 \in \left[-1^\circ, 1^\circ\right]. \tag{5}$$

The 64 extremal points of the hypercube have been spawn on level 7 of the six-dimensional grid, summing up to 15.121 collocation points across 7 hierarchical level of the sparse grid. The computation time took about 50 h on the cluster system of the Leibniz University IT Services (LUIS), where on average 600 worker had been employed. For a better approximation quality of the surrogate model, one or two more refinement levels might be beneficial, resulting to a sparse grid with 127.105 collocation point. For that, the computing time has been estimated to about 3 weeks. For more details on the applicability of sparse grids in higher dimensions, the reader is referred to [16].

Exemplary, for two extreme parameter sets the results are depicted in Fig. 4. Shown are X-ray emulations of the computed bmd in the long-term bio-mechanical equilibrium state after the prosthesis treatment, where due to the small changes in the parameters itself no mentionable differences are seen.

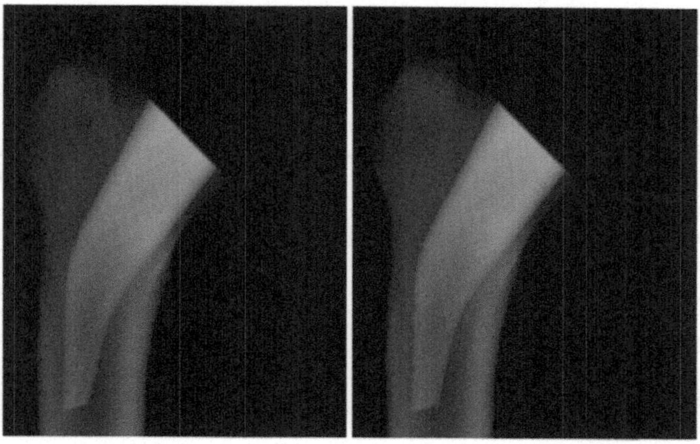

Fig. 4 Computed bone mass density at the long term bio-mechanical equilibrium state at two extreme points of the parametric space

5 Conclusion

Doubtless, model order reduction techniques pave the way for the next generation of computational mechanics. In this project we tried to build a reduced order modeling strategy for surgical treatment of hip-joint diseases via total hip-joint replacement. Final goal is to provide assistance tools based on well established simulation methods in computational bio-mechanics for physicians. Due to the complexity of the modeling chain we decided for a surrogate modeling technique based on sparse grid collocation methods. We experienced that (a) traditional software concepts, despite have been proven to work well for patient individual deterministic computations have to be redefined and (b) further improvements in sophisticated techniques for higher dimensional parametric spaces are needed. Nonetheless, a first step in that direction has been done and carefully investigated. Open topics, e.g. with regard to parametrized geometric modeling and more efficient strategies for surrogate modeling in higher dimensions have been addressed.

Acknowledgements This project started about five years ago with the government funding of Maximilian's research assistant position. We quickly realized that existing software developments have not been suitable for fully automatic computation of these models complexity, thus a plenty of effort has been spent on the development of a fully integrated simulation chain. Fortunately, we have been granted recently to run the new International Research Training Group (GRK 2657/1–2021) from the German Research Foundation (DFG), where the topic "model order reduction in high dimensions" is in the focus, for which we appreciate our acknowledgements. Furthermore, the first author would like to thank Peter for his constructive and patient guidance to prepare the proposals and defense sessions on that application.

References

1. Carter, D., Orr, T., & Fyhrie, D. (1989). Relationships between loading history and femoral cancellous bone architecture. *Journal of Biomechanics, 22*(3), 231–244.
2. Beaupre, G., Orr, T., & Carter, D. (1990). An approach for time-dependent bone modeling and remodeling-application: A preliminary remodeling simulation. *Journal of Orthopaedic Research, 8*(5), 662–670.
3. Weinans, H., Huiskes, R., & Grootenboer, H. (1992). The behavior of adaptive bone-remodeling simulation models. *Journal of Biomechanics, 25*(12), 1425–1441.
4. Lutz, A., & Nackenhorst, U. (2010). Numerical investigations on the biomechanical compatibility of hip-joint endoprostheses. *Archive of Applied Mechanics, 80*(5), 503–512.
5. Lutz, A., & Nackenhorst, U. (2012). Numerical investigations on the osseointegration of uncemented endoprostheses based on bio-active interface theory. *Computational Mechanics, 50*(3), 367–381.
6. Webster, D., & Müller, R. (2010). In silico models of bone remodeling fram macro to nano—from organ to cell. *WIREs Systems Biology and Medicine*.
7. Hambli, R. (2014). Connecting mechanics and bone cell activities in the bone remodeling process: An integrated finite element modeling. *Frontiers in Bioengineering and Biotechnology, 2*(6), 1–12.
8. Sapotnick, A., & Nackenhorst, U. (2015). A mechanically stimulated fracture healing model using a finite element framework. In *Biomedical Technology—Lecture Notes in Applied and Computational Mechanics* (pp. 41–53). Berlin: Springer.
9. Mohaghegh, K., Perez, M. A., & Garcí-Aznar, J. M. (2014). Accelerating numerical simulations of strain-adaptive bone remodelling predictions. *Computer Methods in Applied Mechanics and Engineering, 273*, 255–272.
10. Nackenhorst, U. (2018). Modeling of bone adaption processes. In H. Altenbach, & A. Öchsner (Eds.), *Encyclopedia of Continuum Mechanics*. Berlin: Springer.
11. Chinesta, F., Huerta, A., Rozza, G., & Willcox, K. (2018). Model reduction methods. In E. Stein, R. de Borst, & T. J. R. Hughes (Eds.), *Ecyclopedia of Computational Mechanics*. Wiley.
12. Ma, X., & Zabras, N. (2009). An adaptive hierarchical sparse grid collocation algorithm for the solution of stochastic partial differential equations. *Journal of Computational Physics, 228*(88), 3084–3113.
13. Harrigan, T., & Hamilton, J. (1992). Optimality conditions for finite element simulation of adaptive bone remodeling. *International Journal of Solids and Structures, 29*(23), 2897–2906.
14. Krstin, N., Nackenhorst, U., & Lammering, R. (2000). Zur konstitutiven Beschreibung des anisotropen beanspruchungsadaptiven Knochenumbaus. *Technische Mechanik*.
15. Lutz, A. (2011). *Ein integrales Modellierungskonzept zur numerischen simulation der osseointegration und der Langzeitstabilität von Endoprothesen*. Ph.D. thesis, Leibniz Universität Hannover.
16. Pflüger, D. M. (2010). *Spatially adaptive sparse grids for high-dimensional problems*. Ph.D. thesis, Technische Universität München.

VARTOP: A New Variational Approach to Structural and Thermal Topology Optimization Problems

Javier Oliver and Daniel Yago

X. Oliver's dedicatory: I met Prof. Peter Wriggers for the first time in November 1989 at Stanford University. I was there in a stay with, the prematurely lost, Juan Simo and Peter, a good friend of Juan, was visiting him. Since then I have always admired Peter's enormous scientific contributions in Computational Mechanics, his friendliness with everyone around him, his calmed attitude at all times and, last but not least,... his ability to make me laugh when sitting together at the conferences dinners. Happy birthday and long life Peter!!

Abstract This work summarizes the topology optimization technique (VARTOP) developed by the authors in previous papers. The proposed topology optimization techniques, using a *non-smoothed characteristic function* as design variable, is formulated on the basis of an *incremental pseudo-time-advancing scheme* ruled by the volume constraint, in addition of a consistent derivation of a cost function sensitivity. The *relaxed topological derivative* (RTD), considering the *ersatz material approach*, can be easily defined for any type of problem. In this setting, the optimization problem is analytically solved in a variational framework, leading to nonlinear, closed-form algebraic solutions for the design variable, which are then solved, at every time-step, via a fixed-point method termed *pseudo-energy cutting algorithm*. At each of these iterations, the volume constraint is exactly fulfilled via a bisection method, in the so-called *cutting* and *bisection* method.

J. Oliver (✉)
Centre Internacional de Mètodes Numèrics en Enginyeria (CIMNE), Campus Nord UPC, Barcelona, Spain
e-mail: oliver@cimne.upc.edu

E.T.S d'Enginyers de Camins, Canals i Ports de Barcelona (ETSECCPB) - Technical University of Catalonia (UPC/Barcelona Tech) - Campus Nord UPC, Departament of Civil and Environmental Engineering, Barcelona, Spain

D. Yago
Escola Superior d'Enginyeries Industrial, Aeroespacial i Audiovisual de Terrassa (ESEIAAT) - Technical University of Catalonia (UPC/Barcelona Tech) - Campus Terrassa UPC, Department of Physics - Division of Aerospace Engineering, Terrassa, Spain
e-mail: daniel.yago@upc.edu

F. Aldakheel et al. (eds.), *Current Trends and Open Problems in Computational Mechanics*, https://doi.org/10.1007/978-3-030-87312-7_36

1 Introduction

In the last three decades, topology optimization has become an active research field to seek new, optimal unintuitive designs in a wide range of designs problems, which are used more frequently by professionals for industrial applications. During this time a number of topology optimization approaches have been proposed by researchers, being two of the most important the SIMP and Level-set methods, among others.

Unlike these well-established methods, the *Variational Topology Optimization approach* tackles the original topology optimization problem via the consideration of the *characteristic function* χ, relaxed with the *ersatz material approach*, as the design variable and the implicit representation of the domain through a 0-level-set function, thus obtaining *smooth white-and-black* designs. In addition to this feature, the sensitivity of the cost function is obtained using the *relaxed topological derivative*, an efficient and simple approximation to the *exact topological derivative*. The RTD can be applied independently of the optimization problem.

The last two main characteristics of this new technique are the application of a *Laplacian regularization* to the sensitivity so that the minimum material filament size can be determined and the definition of a *fixed-point closed form algebraic system* to update the topology and fulfill a *volume constraint*, in contrast to other updating schemes based on Hamilton-Jacobi or Reaction-Diffusion equations. In the proposed scheme, the volume constraint, used as a *pseudo-time*, is iteratively increased thus obtaining intermediate converged optimal topologies (local minimums) in an *incremental time-advancing scheme*.

2 Problem Formulation

The following section defines the mathematical formulation of the VARTOP approach, focusing on the topology optimization problem and the derivation of the cost function sensitivity using the so-called *relaxed topological derivative* (RTD).

2.1 VARTOP: Variational Topology Optimization

Let us consider a fixed design domain, $\Omega \subset \mathbb{R}^n$, with $n = \{2, 3\}$, composed by two smooth subdomains Ω^+ and Ω^-, denoted as solid and void material domains. These two domains are defined via the *nonsmooth characteristic function*, $\chi(\mathbf{x}) : \Omega \to \{0, 1\}$, as

$$\begin{cases} \Omega^+ := \{\mathbf{x} \in \Omega \;/\; \chi(\mathbf{x}) = 1\} \\ \Omega^- := \{\mathbf{x} \in \Omega \;/\; \chi(\mathbf{x}) = 0\} \end{cases} . \tag{1}$$

The *characteristic function*, $\chi(\mathbf{x})$, can also be defined in terms of the *smooth discrimination function* (i.e., $\psi(\mathbf{x}) : \Omega \rightarrow \mathbb{R}$, $\psi \in H^1(\Omega)$) as $\chi_\psi(\mathbf{x}) = \mathcal{H}(\psi(\mathbf{x}))$, where $\mathcal{H}(\cdot)$ stands for the Heaviside function evaluated at $(\cdot)$.[1] Instead of using the *classical Heaviside function* $\{1, 0\}$, the image of this function is now relaxed to $\chi_\psi(\mathbf{x}) : \Omega \rightarrow \{\beta, 1\}$ employing the *ersatz material approach*, where the void material is replaced with a soft material, which presents a *relaxation factor* β in the material property relevant to the addressed problem, e.g., in Young modulus E or thermal conductivity κ.

The topology optimization seeks to minimize a cost function $\mathcal{J}(\mathbf{p}_\chi, \chi)$ subjected to the volume constraint $C(\chi)$ and governed by the corresponding state equation. Based on this definition, the mathematical formulation of the problem is given by

$$\left[\begin{array}{lr} \displaystyle\min_{\chi \in \mathscr{U}_{ad}} \;\; \mathcal{J}(\mathbf{p}_\chi, \chi) \equiv \int_\Omega j(\mathbf{p}_\chi, \chi, \mathbf{x})\, d\Omega & (a) \\ \text{subject to:} & \\ \qquad C(\chi, t) \equiv \displaystyle\int_\Omega c(\mathbf{x}, \chi)\, d\Omega = t - \frac{|\Omega^-|(\chi)}{|\Omega|} = 0 & (b) \\ \text{governed by:} & \\ \qquad \text{State equation} & (c) \end{array}\right. , \tag{2}$$

where $\mathscr{U}_{ad}$ stands for the set of admissible solutions for χ, and $\mathbf{p}_\chi$ and t are the state variable field (i.e., $\mathbf{u}_\chi$ or θ_χ depending on the state equation) and the *pseudo-time parameter* (used in the *pseudo-time-advancing strategy*), respectively.

According to Oliver et al. [1], the *Relaxed Topological Derivative* (RTD) used to computed the sensitivity of a given functional is defined as

$$\frac{\delta \mathcal{J}(\chi)}{\delta \chi}(\hat{\mathbf{x}}) = \left[\frac{\partial j(\mathbf{p}_\chi, \chi, \mathbf{x})}{\partial \chi}\right]_{\mathbf{x}=\hat{\mathbf{x}}} \Delta\chi(\hat{\mathbf{x}}), \text{ with } \Delta\chi(\hat{\mathbf{x}}) = \begin{cases} -(1-\beta) < 0 & \textit{for } \hat{\mathbf{x}} \in \Omega^+ \\ (1-\beta) > 0 & \textit{for } \hat{\mathbf{x}} \in \Omega^- \end{cases}, \tag{3}$$

which will depend on each specific cost function $\mathcal{J}(\mathbf{p}_\chi, \chi)$ and state equation. For the volume constraint (2-b), the RTD of the functional becomes

$$\frac{\delta C(\chi, t)}{\delta \chi}(\hat{\mathbf{x}}) = \left[\frac{\partial c(\mathbf{x}, \chi)}{\partial \chi}\right]_{\mathbf{x}=\hat{\mathbf{x}}} \Delta\chi(\hat{\mathbf{x}}) = \frac{1}{|\Omega|}\operatorname{sgn}(\Delta\chi(\hat{\mathbf{x}})), \tag{4}$$

where $\operatorname{sgn}(\cdot)$ denotes the sign function of $(\cdot)$.

Considering the *Lagrangian function* defined as $\mathcal{L}(\mathbf{p}_\chi, \chi) = \mathcal{J}(\mathbf{p}_\chi, \chi) + \lambda C(\chi, t)$, the optimality condition of the *original topology optimization problem* reads

[1] The subscript ψ in χ_ψ remarks the fact that the *characteristic function* is defined in terms of ψ.

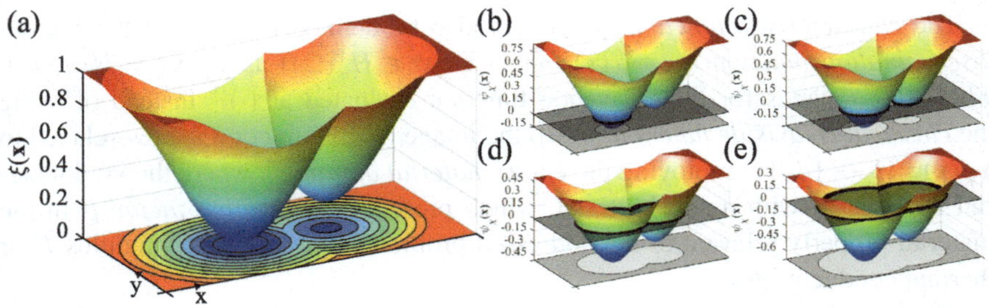

Fig. 1 Implicit topology representation: **a** *pseudo-energy* distribution ξ and **b–e** *discrimination function*, ψ, and the corresponding *characteristic function*, χ, for a set of Lagrange multiplier values: $\lambda = \{0.15, 0.25, 0.5, 0.7\}$, thus satisfying $|\Omega^-(\lambda_{(b)})| < |\Omega^-(\lambda_{(c)})| < |\Omega^-(\lambda_{(d)})| < |\Omega^-(\lambda_{(e)})|$

$$\frac{\delta \mathcal{L}(\chi, \lambda)}{\delta \chi}(\hat{\mathbf{x}}) = \left(\frac{\partial j(\hat{\mathbf{x}}, \chi)}{\partial \chi} \Delta\chi(\hat{\mathbf{x}}) + \lambda \operatorname{sgn}(\Delta\chi(\hat{\mathbf{x}})) \right) = \tag{5}$$

$$= \psi(\hat{\mathbf{x}}, \chi) = (\xi(\hat{\mathbf{x}}, \chi) - \lambda) \;\; \forall \hat{\mathbf{x}} \in \Omega \,, \tag{6}$$

where $\psi(\hat{\mathbf{x}}, \chi)$ corresponds to the *discrimination function* and $\xi(\hat{\mathbf{x}}, \chi)$ is termed the *pseudo-energy* and must be computed for each optimization problem. In contrast to other approaches, the *pseudo-energy* ξ is first shifted and normalized,[2] before updating the topology, and later regularized via a *Laplacian regularization*, thus providing black-and-white optimal layouts with crisp, smooth edges. The *smooth pseudo-energy function*, ξ_τ, corresponds to the regularized *pseudo-energy*, where τ stands for the dimensionless regularization parameter.

As aforementioned, the topology layout, defined via $\chi(\mathbf{x})$, is updated via a fixed point method termed *Cutting&Bisection* algorithm, where the *Lagrange multiplier* λ fulfilling the volume constraint (2-b) is computed. The *bisection algorithm* is represented in Fig. 1 where the *pseudo-energy* ξ is cut by 4 different values of λ, this resulting in different volume constraint values. Then, the *closed-form solution* of problem (2) is obtained as

$$\begin{cases} \psi_\tau(\hat{\mathbf{x}}) := \hat{\xi}_\tau(\hat{\mathbf{x}}, \chi) - \lambda \\ \chi(\hat{\mathbf{x}}) = \mathcal{H}(\psi_\tau(\hat{\mathbf{x}})) \qquad in\; \Omega \,, \\ C(\chi(\lambda), t) = 0 \end{cases} \tag{7}$$

which provides the optimal topology for each time-step t (percentage of void material), once the criteria in cost function and Lagrange multiplier have been met.[3]

[2] The values Δ_{shift} and Δ_{norm} are defined at the first iterations and kept constant throughout the optimization.

[3] The reader is referred to [1, 2] for further details of the topology optimization approach.

2.2 State Problem

The governing variational problem for linear elasticity and thermal problems, neglecting volumetric terms, can be written as

$$\left[\begin{array}{l} \text{Find the state field } \mathbf{p}_\chi \in \mathcal{U}(\Omega) \text{ such that} \\ a(\mathbf{w}, \mathbf{p}_\chi) = l(\mathbf{w}) \quad \forall \mathbf{w} \in \mathcal{V}(\Omega) \qquad (8) \\ \text{where} \\ a(\mathbf{w}, \mathbf{p}_\chi) = \int_\Omega \nabla^S \mathbf{w}(\mathbf{x}) : \mathbb{D}_\chi(\mathbf{x}) : \nabla^S \mathbf{p}_\chi(\mathbf{x})\, d\Omega\,, \qquad (9) \\ l(\mathbf{w}) = \int_{\partial_s\Omega} \mathbf{w}(\mathbf{x}) \cdot \bar{\boldsymbol{s}}(\mathbf{x})\, d\Gamma\,, \qquad (10) \end{array}\right.$$

where, for the elastic problem, the variables $\mathbb{D}_\chi$, $\mathbf{p}_\chi$ and $\bar{\boldsymbol{s}}$ are replaced with $\mathbb{C}_\chi$, $\mathbf{u}_\chi$ and $\bar{\boldsymbol{\sigma}}$, respectively. As for the thermal problem, these variables are $\boldsymbol{\kappa}_\chi$, θ_χ and $-\bar{\boldsymbol{q}}$ (see Sect. 3).

The relevant material property $\mathbb{D}_\chi$ can be expressed in terms of the nominal property $\mathbb{D}$ and the *characteristic function* χ as $\mathbb{D}_\chi(\mathbf{x}) = \chi^m(\mathbf{x})\mathbb{D}(\mathbf{x})$, where $m > 1$ stands for the exponential factor, and the *characteristic function* is relaxed with β being $\alpha^{1/m}$. Bear in mind that the elastic material is governed by Hooke's law, i.e., $\boldsymbol{\sigma}(\mathbf{x}, \chi) = \mathbb{C}(\mathbf{x}, \chi) : \boldsymbol{\varepsilon}(\mathbf{x}) = \mathbb{C}(\mathbf{x}, \chi) : \nabla^S \mathbf{u}(\mathbf{x})$, while the conductive material is ruled by Fourier's law, given by $\boldsymbol{q}(\mathbf{x}, \chi) = -\boldsymbol{\kappa}(\mathbf{x}, \chi) \cdot \nabla\theta_\chi(\mathbf{x})$.

The classical finite element method is used to discretize the topology optimization problem (2) as well the state equation (8). Specific details of the discretization can be found in [1, 3] and [2] for the structural and thermal optimization problems.

3 Applications

In the following subsections three different applications of the proposed topology optimization technique are presented, the first two applications ruled by the structural problem and the last one by the thermal problem.

3.1 Minimum Mean Compliance

The main goal of the *minimum mean compliance problem* is to find the optimal topology layout that maximizes the stiffness of the entire structure under specific boundary conditions. Therefore, the cost function (2-a) is defined as the work of the external forces, thus leading the *pseudo-energy density* to the *actual strain energy density*. For further details on the derivation of the sensitivity please refer to [1].

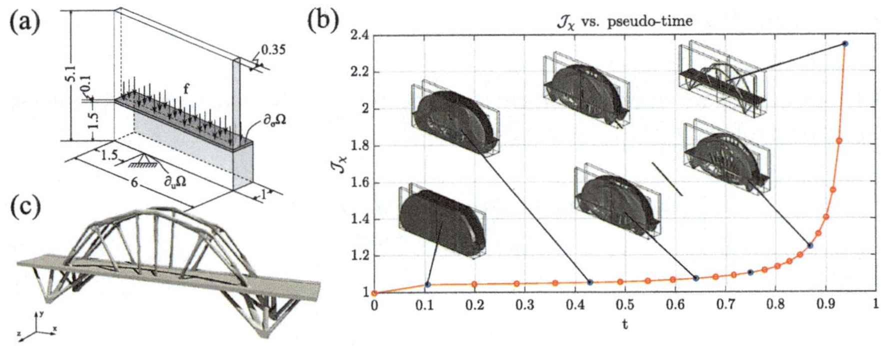

Fig. 2 Bridge design: **a** setup of the analysis domain, **b** cost function and topology evolution and **c** Optimal design for $t = 0.94$

Let us now address the topology optimization of a bridge structure, depicted in Fig. 2a, with a hole located above the road to allow the pass of vehicles. The domain is discretized with a structured mesh of $240 \times 204 \times 40$ eight-node hexahedral (Q_1) finite elements, considering a Young modulus $E = 210GPa$, a Poisson ratio $\nu = 0.3$ and a contrast factor $\alpha = 10^{-6}$ ($m = 5$) for the solid material. The regularization parameter τ is set to 0.316 while the time interval of interest [0, 0.96] is discretized in 22 steps following an exponential law. The topology evolution and the optimal design for $t = 0.96$ are displayed in Fig. 2b, c, respectively. As it can be observed, the cost function increases as the pseudo-time t increases, thus leading to designs easily recognizable by bridge engineers.

3.2 *Compliant Mechanism*

The second numerical application corresponds to the design of compliant mechanisms, in which the optimization objective is the maximization of the displacement at the *output port* due to an action at the *input port*. The cost function (2-a) can be defined in terms of the external work done by a *constant dummy force* applied only on the *output port*, thus defining the *mutual potential energy density* as the *pseudo-energy* sensitivity. The exact details on the derivations can be found in [1, 3].

Let us now use the topology optimization approach for the design of a 3D compliant gripper, where the compressive displacement at the jaws (vertical displacement at the top-right side) is maximized when an horizontal force is applied at the input-port (top-left side) (see Fig. 3). The design domain, discretized with $160 \times 80 \times 40$ regular hexahedra, is made of the same material as in Sect. 3.1, except that the contrast factor is now equal to $\alpha = 10^{-2}$ ($m = 3$). The optimization is performed using a regular-

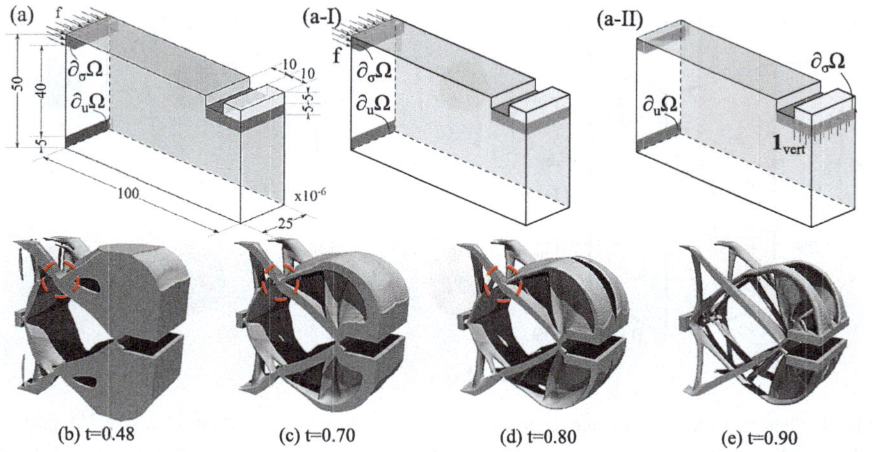

Fig. 3 Gripper design: **a** setup of the analysis domain, (a-I) *original state system*, (a-II) *auxiliary state system* with the *constant dummy force* applied on the *output port* and **b**–**e** Optimal design for different time-steps. Localized hinges are highlighted with red circles

ization parameter $\tau = 0.5$ and an exponential updating scheme. Some intermediate optimal designs are illustrated in Fig. 3b–e, where localized hinges are highlighted.

3.3 *Thermal Cloaking in Terms of Heat Flux*

The last application addresses the topology optimization of a *cloaking device* capable of thermally cloak a surrounded object within an homogeneous material, with different thermal properties, from being detected by an external thermal detecting device, measuring alterations in the heat flux field from a known homogeneous one. Therefore, the cost function is defined as the deviation between the known homogeneous field $\overline{\mathbf{q}}$ and the actual heat flux field $\mathbf{q}_\chi$ via a L_2 norm. For more details, the reader is referred to [2].

For the numerical solution of the topology optimization problem, the square prismatic domain Ω displayed in Fig. 4a, is discretized with a structured mesh of $100 \times 200 \times 100$ hexahedral elements. The temperature on the left and right surfaces of the domain are prescribed to $\overline{\theta}_h = 321.85$ K and $\overline{\theta}_c = 283.15$ K, respectively, thus resulting in a constant heat flux $\overline{\mathbf{q}} = [245.1,\ 0,\ 0]\,\mathrm{W/m^2}$ while the other surfaces are assumed adiabatic. The domain is partitioned in three distinct regions: the conductivity of regions 1 (an ellipsoid colored in green) and 3 being $\kappa = 0.57\,\mathrm{W/mK}$ (where the deviation is minimized), while $\kappa = 403\,\mathrm{W/mK}$ with a contrast factor $\alpha = 5.46 \cdot 10^{-4}$ ($m = 5$) for the cloaking device (an sphere colored in orange). Figures 4f–i depict the evolution of the isotherms throughout the optimization at the middle x-y plane. As the *pseudo-time* evolves, the isotherms resembles the constant

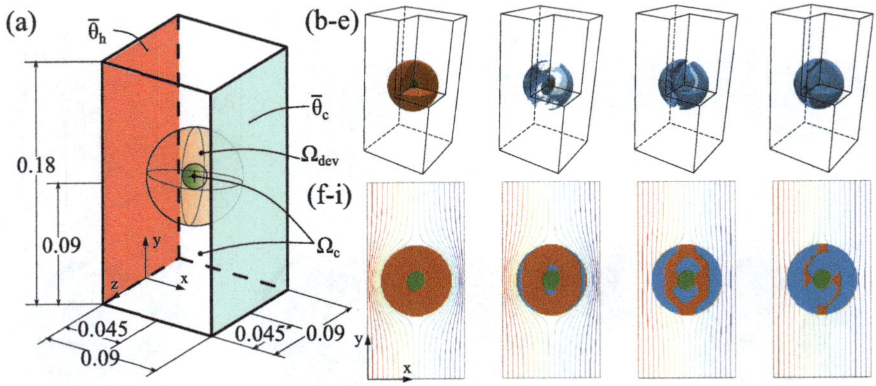

Fig. 4 Heat flux cloaking device: **a** analysis domain with boundary conditions, **b–e** 3D view of intermediate topologies (steps = {0, 1, 5, 8}) and **f–i** evolution of the isotherms and layout of the cloaking device at the middle x-y plane for the same representative steps

heat flux field targeted in the optimization thus cloaking the object surrounded by the device.

4 Conclusions

As demonstrated in previous work, the VARTOP approach exhibits great potential in topology optimization of structural and thermal problems, as illustrated here with several 3D numerical examples. As a result, complex smooth designs can be obtained with low computational cost when compared to equivalent Level-set methods.

The benefits of the method are based on the usage of a *nonsmooth characteristic function* as the design variable thus obtaining *black-and-white* designs, the derivation of the sensitivity via the *Relaxed Topological derivative* (RTD) using the *ersatz material approach*, and the definition of a *closed-form optimality criteria* for the topology optimization problem using a robust *cutting* and *bisection* algorithm in a *pseudo-time advancing scheme*. In addition, the manufacturability and smoothness of the design is controlled via a *Laplacian regularization*.

Acknowledgements This research has received funding from the European Research Council (ERC) under the European Union's Horizon 2020 research and innovation programme (Proof of Concept Grant agreement n 874481) through the project "Computational design and prototyping of acoustic metamaterials for target ambient noise reduction" (METACOUSTIC). Grant CEX2018-000797-S to CIMNE from MCIN/AEI /10.13039/501100011033 is also acknowledged

References

1. Oliver, J., Yago, D., Cante, J., & Lloberas-Valls, O. (Oct 2019). Variational approach to relaxed topological optimization: Closed form solutions for structural problems in a sequential pseudo-time framework. *Computer Methods in Applied Mechanics and Engineering, 355*, 779–819.
2. Yago, D., Cante, J., Lloberas-Valls, O., & Oliver, J. (2020). Topology optimization of thermal problems in a nonsmooth variational setting: Cosed-form optimality criteria. *Computational Mechanics, 66*(2), 259–286.
3. Yago, D., Cante, J., Lloberas-Valls, O., & Oliver, J. (2020). Topology optimization using the unsmooth variational topology optimization (UNVARTOP) method: an educational implementation in MATLAB. *Structural and Multidisciplinary Optimization*.

From the Pioneering Contributions by Wriggers to Recent Advances in Computational Tribology

Marco Paggi, Jacopo Bonari, and José Reinoso

This work is dedicated to Prof. Wriggers on his 70th birthday, for his outstanding and pioneering contributions to computational contact mechanics that inspired new progresses in computational methods for tribology.

Abstract This review article provides an outlook over the pioneering contributions by Peter Wriggers in the field of computational contact mechanics that have inspired the development of new computational methods for tribology. The first development concerns with methods to efficiently account for any real roughness or texture in the finite element discretization of the contact surface, that is one of the major issues limiting the capabilities of the finite element method when applied to tribology. The second development regards a concurrent coupling of BEM and FEM, to upscale the analysis to realistic component sizes, but preserving the microstructural topology information for contact. Finally, the last topic regards a novel fluid-structure interaction scheme proposed within the smoothed particle hydrodynamics code SPHERA v.9.0.0 to simulate hydrodynamic lubrication with complex rough boundaries.

M. Paggi (✉) · J. Bonari
IMT School for Advanced Studies Lucca, Lucca, Italy
e-mail: marco.paggi@imtlucca.it

J. Bonari
e-mail: jacopo.bonari@imtlucca.it

J. Reinoso
Elasticity and Strength of Materials Group, School of Engineering, Universidad de Sevilla, Camino de los Descubrimientos s/n, 41092 Seville, Spain
e-mail: jreinoso@us.es

F. Aldakheel et al. (eds.), *Current Trends and Open Problems in Computational Mechanics*, https://doi.org/10.1007/978-3-030-87312-7_37

1 Introduction

Since his PhD dissertation in 1981 [1], Peter Wriggers made pioneering contributions on contact mechanics, choosing the finite element method (FEM) as the main framework to cast his new theories and algorithms. Due to the versatility of FEM in handling finite-size geometries, finite elasticity, and complex material constitutive relations, he has pioneeringly foreseen the potential of FEM to advance over analytical methods for contact mechanics. He made significant progresses on contact search algorithms [2, 3], mathematical methods to impose the satisfaction of unilateral contact constraints [4, 5], thermoelastic contact problems [6, 7], discretization methods [8–11], and rigorous implicit nonlinear solution schemes [12, 13], that are all detailed in his highly cited publications and well summarized in his textbook [14]. He also devoted major efforts in developing numerical schemes for the simulation of particle flow and contact in fluids based on the discrete element method [15]. Although applications were mostly concerned with particulate flow for civil engineering problems or biomechanics, the methodology has an innovation potential also in tribology.

Nowadays, computational contact mechanics is a mature research area and further impact is expected as far as tribological applications are concerned. Tribology is the science and engineering of interacting components in relative motion. Friction, wear, lubrication, heat and electrical conduction are just some of the key problems relevant for this field which is progressively expanding its scope, becoming increasingly interdisciplinary [16]. Materials scientists, engineers and physicists actively cooperate towards the design and the realization of novel tribofilms, coatings, surface textures, new lubricants. The applications are also rapidly expanding from the historical mechanical engineering field to biomechanics and other high-tech domains.

In this context, trial-and-error experimental design can be extremely expensive and time consuming. Therefore, computational methods, one of the leading enabling technologies for Industry 4.0, may offer the advantage of reducing time from design to production, exploring and scouting new solutions, providing the basis for virtual testing of customized tribological components.

The solution of contact problems is complicated by the inherent nonlinearity stemming from the fact that the extent of the contact area is a-priori unknown. It is clear that practical applications can be efficiently solved provided that an accurate description of the component geometry (the macro-scale) and loading conditions are well modelled. Moreover, at the same time, a detailed resolution of any surface feature relevant for the surface interactions is requested.

Therefore, computational contact mechanics is nowadays facing new challenges, which are related to the need of solving a multi-scale problem in real structural applications. The involved size-scales range from the component size down to the smallest length scale representative of roughness [17].

In this article, three recent developments are outlined along this line of research. The first one relates to a novel finite element discretization method for the interface, which allows for an exact representation of any rough surface topology with a significant reduced complexity in contact detection and contact geometry modeling. This method presents several advantages for the simulation of high-tech applications requiring a detailed description of tribological interactions and, at the same time, modelling of bulk dissipative phenomena (viscoelasticity, plasticity, fracture, etc.). The second exploits scale separation arguments to establish a computational technique to address contact problems at the macroscale using FEM, but with a detailed analysis of contact at the microscale using the boundary element method (BEM). Finally, the third advancement regards the application of the smoothed particle hydrodynamics (SPH) meshless method to the simulation of hydrodynamic lubrication regimes with complex rough boundaries.

2 High Precision Contact Solution Using Interface Finite Elements with Embedded Roughness

The prediction of the contact response is often strongly affected by the interface topology. Therefore, assuming the interface as a topologically flat domain and introducing average tribological parameters (heat resistance, friction coefficient, wear coefficient, etc.) is often inadequate for the accurate simulation of the physical problem. However, the exact representation of surface textures and/or roughness is posing several challenges for contact search algorithms when classical surface discretization schemes are employed. Regularization of the surface geometry, for instance with Bezier or NURBS interpolation, is not a viable strategy, since relevant fine scale details might be erroneously filtered out. For this reason, the boundary element method, although primarily applicable to simplified half-plane geometries and linear elastic materials, has seen a discrete popularity for the solution of contact problems with roughness, since it can accommodate any realistic rough topology in input [18, 19].

In order to overcome the major limitations of conventional FEM discretization schemes, as well as the limited capabilities of BEM for the analysis of nonlinear material constitutive laws and multi-field problems, a novel interface finite element with eMbedded Profile for Joint Roughnes (MPJR interface finite element) has been proposed in [20], see also [21] for a comparison with BEM. The MPJR interface finite elements can be used to discretize the interface with a uniform spacing dictated by the profilometer resolution, to achieve a one-to-one correspondence between finite element nodes and profilometer sampling points. However, instead of following the actual rough surface elevations, the interface is treated as nominally flat. The deviation from planarity, on the other hand, is considered within the element as an exact correction of the normal gap function. In case of an analytical expression of surface texture, for instance modelled as a Weierstrass-Mandelbrot (WM) function, the global coordinate of the interface finite element nodes is used to determine the surface elevation through the WM function and, in its turn, the correction to impose to the normal displacement computed from the standard kinematics of the interface finite element [22]. Alternatively, if the surface topology is provided as a data file containing nodal-wise AFM or profilometric elevations, then a searching algorithm is run once at the beginning of the simulation to assign the elevation to the interface node based on its global coordinate. For the next steps, such elevation data are efficiently stored in a history variable of the element, to avoid repeated reading from external files.

The methodology has been proved to be efficient for solving frictionless contact problems with nominally flat surfaces with microscale roughness in [20]. The solution of a frictionless normal contact problem for a rigid indenter with a profile modelled according to the WM function with $n = 1$, 2 or 3 length scales of roughness is shown in Fig. 1. The upper plot shows the microscopically rough profile elevations, z/g_0, where g_0 is a scaling parameter. Periodic boundary conditions are applied at $x/\lambda = 0$ and 1, which lead to symmetric profiles with respect to an axis passing through $x/\lambda = 0.5$, where λ is the longest wavelength of the WM function. The corresponding predicted normal contact tractions, p/E, are shown in the lower plot, for the same imposed far-field displacement in the normal direction. The surface refinement, by adding additional wavelength components to the WM function, leads

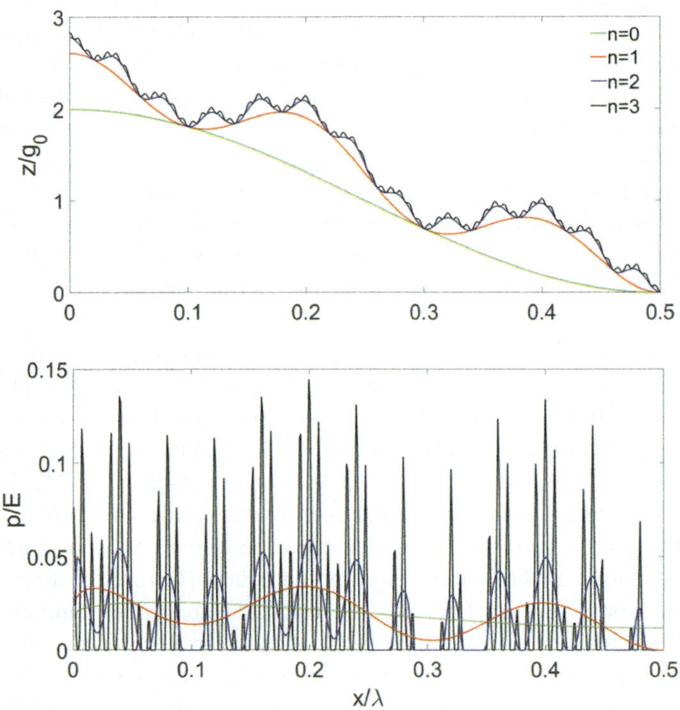

Fig. 1 Geometry of the embedded rough profiles (above) derived from WM functions with different resolution n, and predicted normal contact tractions (below), for the same applied far-field normal displacement

to an increased intensity of contact tractions which do localize near the fine scale asperities. Overall, the progressive rarefaction of the real contact area by increasing the resolution parameter n is observed, as for fractal-like boundaries.

Moreover, due to the possibility of introducing any nonlinear interface constitutive relation, contact mechanics simulations with roughness and also interface adhesion were successfully performed in [20]. A further extension of the approach to deal with frictional contact problems and in particular partial slip regimes under the action of infinitesimal sliding displacements has been proposed in [31], where a regularized Coulomb frictional law has been included to model the contact response in the shearing direction. Finally, very recently, the methodology has been extended to deal with finite relative sliding displacements, to simulate also full sliding regimes and dissipation mechanisms in viscoelastic contacts [32].

3 Upscaling of Contact Problems with a Concurrent BEM-FEM Coupling

In order to address tribological problems involving mechanical components of complex shape, it can be extremely useful to exploit the capabilities of FEM in discretizing 2D or 3D bodies. However, in performing this upscaling procedure, the physics of contact is usually simplified by introducing mathematical models to describe contact interactions that are no longer dictated from the emergent response of micromechanical surface interactions. In this context, for instance, it has been proved in [6] that the standard penalty approach with constant penalty parameter could be enhanced by introducing a nonlinear interface constitutive relation provided by micromechanical contact theories. As a further step forward, in order to overcome the assumptions of micromechanical contact theories on the statistical distribution of asperities and on their elastic interactions [28], Bonari et al. [29] proposed a concurrent integration of BEM into FEM. The formulation, based on the interface finite element kinematics proposed in [22] for fracture mechanics, implicitely assumes scale separations, where FEM discretization is used to model the component geometry at the macroscale. In each integration point of the interface finite element, the normal contact response is no longer provided in closed form, but it is now the result of BEM contact predictions for a statistically representative contact surface, see Fig. 2. Such a contact surface can now be given in input directly from confocal or AFM measurements. The micro- and macro-scales are linked by the common relative displacements, computed at the FEM level and passed to the BEM routine. The BEM code, which in the present implementation assumes elastic deformations, provides the exchanged normal contact force and the numerically computed tangent stiffness matrix as a feedback quantity to FEM.

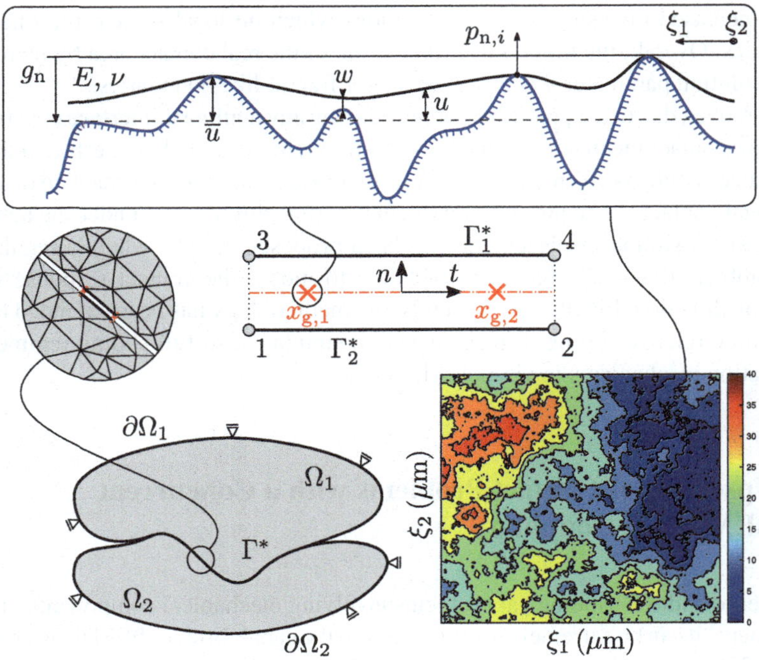

Fig. 2 FEM-BEM concurrent coupling scheme

4 Hydrodynamic Lubrication with Complex Rough Boundaries

The topic of hydrodynamic lubrication with complex rough boundaries is also a very important research area for technological applications. In this field, most of the state-of-the-art models are 2D codes based on Reynolds' equation for fluid films [23, 24], while there is the general need for a computational framework which can comprehensively consider the full 3D form of the Navier–Stokes equations for incompressible fluids with uniform viscosity, being able to be validated on both local and global quantities (e.g. pressure and velocity profiles, load-bearing capacity), and could simulate any 3D topology.

To this aim, the computational fluid dynamics (CFD) code SPHERA v.9.0.0 based on smoothed particle hydrodynamics (SPH), which was originally designed for the simulation of water-related natural hazards [25] (floods and earth landslides, transport of solid bodies in fluid flows), has been extended in [26] to deal with fluid-solid body interactions under no-slip conditions and laminar regimes for the simulation of hydrodynamic lubrication. The formulation has been validated in [26] for two classical problems related to a uniform slider bearing (i.e., for a constant lubricant film depth) and a linear slider bearing (i.e., for a film depth with a linear profile variation along the main flow direction) by comparing CFD predictions with ana-

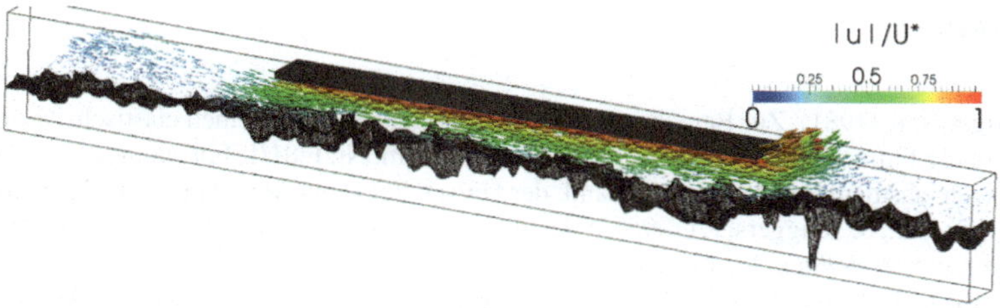

Fig. 3 3D view of the scalar (absolute value) and vector fields of the non-dimensional velocity for a linear slider bearing separated by a lubricant from a complex rough surface

lytical solutions. The methodology presents the advantage of simulating the inertia and the 3D effects close to the slider edges, and it opens new research directions overcoming the limitations of the codes based on the Reynolds' equation for fluid films. Moreover, through a pre-processor which allows considering realistic rough surface geometries in input, it can effectively predict the effect of roughness on local and global parameters relevant for hydrodinamic lubrication [27], see e.g. Fig. 3.

5 Conclusions

This review article has provided a synopsis of recent advancements in the field of computational tribology, with special attention to the accurate modelling and representation of complex rough interfaces and their efficient discretization schemes to enhance contact mechanics simulations. Future developments are expected for the simulation of coupled problems thanks to the finite element method. Reliability of coatings will also require the implementation of wear models and, for the simulation of fretting fatigue, coupling the contact mechanics formulations with phase field fracture models for the continuum. Finally, as far as hydrodynamic lubrication is concerned, fluid-structure interaction with deformable boundaries and the effect of particles embedded into the fluid flow, to model wear debris, represent natural research directions for advanced predictive tribological simulations.

Acknowledgements The research results have been achieved thanks to the support from MIUR-DAAD to the 2017 Joint Mobility Project "Multi-scale modeling of friction for large scale engineering problems", and from MIUR to the Research Project of National Interest (PRIN) 2017 "XFAST-SIMS: Extra fast and accurate simulation of complex structural systems" (D68D19001260001). The authors would like to acknowledge the collaborations with Prof. A. Popp (Bundeswehr Universität München) on FEM-BEM coupling and with Dr. A. Amicarelli (RSE S.p.A.) on SPH.

References

1. Wriggers, P. (1981). Zur Berechnung von Kontakt- und Stoßproblemen elastischer Körper mit Hilfe der finite-element-methode, Ph.D. Dissertationm, Nr. F 81/1, Forschungs- und Seminarberichte aus dem Bereich der Mechanik der Universität Hannover, Hannover, Germany.
2. Zavarise, G., & Wriggers, P. (1998). A segment-to-segment contact strategy. *Mathematical and Computer Modelling, 28,* 497–515.
3. Zavarise, G., & Wriggers, P. (2000). Contact with friction between beams in 3-D space. *International Journal for Numerical Methods in Engineering, 49,* 977–1006.
4. Simo, J. C., Wriggers, P., & Taylor, R. L. (1985). A perturbed Lagrangian formulation for the finite element solution of contact problems. *Computer Methods in Applied Mechanics and Engineering, 50,* 163–180.
5. Wriggers, P. (1995). Finite element algorithms for contact problems. *Archives of Computational Methods in Engineering, 2,* 1–49.
6. Zavarise, G., Wriggers, P., Stein, E., & Schrefler, B. A. (1992). Real contact mechanisms and finite element formulation-A coupled thermomechanical approach. *International Journal for Numerical Methods in Engineering, 35,* 767–785.
7. Wriggers, P., & Miehe, C. (1994). Contact constraints within coupled thermomechanical analysis-A finite element model. *Computer Methods in Applied Mechanics and Engineering, 113,* 301–319.
8. Wriggers, P., & Reinelt, J. (2009). Multi-scale approach for frictional contact of elastomers on rough rigid surfaces. *Computer Methods in Applied Mechanics and Engineering, 198,* 1996–2008.
9. Temizer, I., Wriggers, P., & Hughes, T. J. R. (2011). Contact treatment in isogeometric analysis with NURBS. *Computer Methods in Applied Mechanics and Engineering, 200,* 1100–1112.
10. Wriggers, P., Rust, W. T., & Reddy, B. D. (2016). A virtual element method for contact. *Computational Mechanics, 58,* 1039–1050.
11. Fischer, K. A., & Wriggers, P. (2006). Mortar based frictional contact formulation for higher order interpolations using the moving friction cone. *Computer Methods in Applied Mechanics and Engineering, 195,* 5020–5036.
12. Wriggers, P., & Simo, J. C. (1985). Note on tangent stiffness for fully nonlinear contact problems. *Communications in Numerical Methods in Engineering, 1,* 199–203.
13. Wriggers, P., Van Vu, T., & Stein, E. (1990). Finite element formulation of large deformation impact-contact problems with friction. *Computers and Structures, 37,* 319–331.
14. Wriggers, P. (2006). *Computational Contact Mechanics*, Wiley, Chichester, 2002 (2nd ed., p. 2006). Berlin, Heidelberg: Springer.
15. Avci, B., & Wriggers, P. (2012). A DEM-FEM coupling approach for the direct numerical simulation of 3D particulate flows. *The Journal of Applied Mechanics, 79,* 010901.
16. Vakis, A. I., et al. (2018). Modeling and simulation in tribology across scales: An overview. *Tribology International, 125,* 169–199.
17. Borri, C., & Paggi, M. (2015). Topological characterization of antireflective and hydrophobic rough surfaces: Are random process theory and fractal modeling applicable? *Journal of Physics D, 48,* 1–12.
18. Paggi, M., & Barber, J. R. (2011). Contact conductance of rough surfaces composed of modified RMD patches. *International Journal of Heat and Mass Transfer, 54,* 4664–4672.
19. Bemporad, A., & Paggi, M. (2015). Optimization algorithms for the solution of the frictionless normal contact between rough surfaces. *The International Journal of Solids and Structures, 69–70,* 94–105.
20. Paggi, M., & Reinoso, J. (2020). A variational approach with embedded roughness for adhesive contact problems. *Mechanics of Advanced Materials and Structures, 20,* 1731–1747.
21. Paggi, M., Bemporad, A., & Reinoso, J. (2020). Computational methods for contact problems with roughness. In M. Paggi, Hills, D. (Eds.), *Modeling and Simulation of Tribological Problems in Technology* (pp. 131–178). Cham: Springer International Publishing.

22. Paggi, M., & Wriggers, P. (2016). Node-to-segment and node-to-surface interface finite elements for fracture mechanics. *Computer Methods in Applied Mechanics and Engineering, 300,* 540–560.
23. Almqvist, A., & Dasht, J. (2006). The homogenization process of the Reynolds equation describing compressible liquid flow. *Tribology International, 39,* 994–1002.
24. Almqvist, A. (2011). Homogenization of the Reynolds equation governing hydrodynamic flow in a rotating device. *The Journal of Tribology, 133*, 021705.
25. Amicarelli, A., et al. (2020) SPHERA v.9.0.0: A computational fluid dynamics research code, based on the smoothed particle hydrodynamics mesh-less method. Computer Physics Communications 250, 107157
26. Paggi, M., Amicarelli, A., & Lenarda, P. (2020). SPH modelling of hydrodynamic lubrication: Laminar fluid flow-structure interaction with no-slip conditions for slider bearings. *Computational Particle Mechanics, 54,* 4664–4672.
27. Paggi, M., Amicarelli, A., & Lenarda, P. (2020). SPH modelling of hydrodynamic lubrication along rough surfaces. *Lubricants 2019, 7*(12), 103.
28. Zavarise, G., Borri-Brunetto, M., & Paggi, M. (2004). On the reliability of microscopical contact models. *Wear, 257*, 229–245.
29. Bonari, J., et al. (2019). A multi-scale FEM-BEM formulation for contact mechanics between rough surfaces. *Computational Mechanics, 65,* 731–749.
30. Carneiro Couto, A. M., Pinto Carvalho, R., & Andrade Pires, F. M. (2020). Representative contact element size determination for micromechanical contact analysis of self-affine topographies. *The International Journal of Solids and Structures, 206,* 262–281.
31. Bonari, J., Paggi, M., & Reinoso, J. (2020). A framework for the analysis of fully coupled normal and tangential contact problems with complex interfaces, submitted.
32. Bonari, J., & Paggi, M. (2020). Viscoelastic effects during tangential contact analyzed by a novel finite element approach with embedded interface profiles, submitted.

Assessment of the Structural Response of Steel Reinforced and Steel-Fibre Reinforced Concrete Structures with 3D Detailed Modeling: Limitations and Remedies

Manolis Papadrakakis and George Markou

This paper is dedicated to my long-term colleague and friend Peter Wriggers on his 70th birthday. Peter caught my attention 30 years ago at the first EURODYN conference in Bochum, where he delivered a lecture shortly after his appointment at a young age, as Full Professor at the Institute of Mechanics at TH Darmstadt. In the following years, his research work and his lectures at international conferences have been instrumental and a point of reference for many researchers worldwide. Peter has made profound and ground-breaking contributions to the field of computational mechanics, by bringing together concepts of mechanics and mathematics for addressing intractable engineering problems. I had the pleasure to collaborate with him in a number of joint research projects, as well as in the Boards of ECCOMAS and IACM where I was enjoying his easygoing character and his constructive interventions to our discussions. I wish him to continue pursuing his research interests with the same vigor and vitality for many years to come.

Abstract The assessment of the nonlinear response of existing structures, or of new structural designs, to extreme loading conditions is of significant importance in achieving a safe and sustainable built environment. The accurate prediction of structural response of steel and steel-fibre reinforced concrete structures, that are expected to undergo ultimate limit state, monotonic and cyclic loading, is a challenging task, which is hindered by numerous modeling and numerical obstacles. As a consequence to this, the assessment of retrofitted reinforced concrete structures, is still conducted with the use of simplistic numerical models and semi-empirical design formulae. Although, quite a few 3D detailed modeling approaches have been proposed over the last two decades, they have a number of limitations that prevent

M. Papadrakakis (✉)
Institute of Structural Analysis & Antiseismic Research, National Technical University of Athens, Zografou Campus, Athens, Greece
e-mail: mpapadra@central.ntua.gr

G. Markou
Department of Civil Engineering, University of Pretoria, Pretoria, South Africa
e-mail: george.markou@up.ac.za

F. Aldakheel et al. (eds.), *Current Trends and Open Problems in Computational Mechanics*, https://doi.org/10.1007/978-3-030-87312-7_38

them from their implementation, without simplifications, to the design of steel and fiber/steel reinforced concrete structures. In this work, the capabilities as well as the limitations of the recently published work on the subject will be critically presented, while different open problems will be discussed and remedies will be proposed to overcome these limitations. Furthermore, the development of a simulation tool with the ability to accurately and efficiently predict the structural performance of new designs and assess the bearing capacity of existing or retrofitted structures under extreme loading conditions, will also be discussed.

1 Introduction

The revolutionary developments that have occurred in the field of modeling and simulation-based engineering over the last decades have provided computational tools with the ability to assess the design of new structures and the bearing capacity of existing or retrofitted structures under extreme loading conditions. In particular, for the case of steel reinforced concrete (SRC) and steel fibre reinforced concrete (SFRC) structures, a number of modeling approaches has been proposed [1–13] in an effort to develop reliable and robust numerical tools to perform these tasks.

An attempt to assess the range of applications of the proposed simulation methods was performed in [14, 15] for SRC structures and in [16] for SFRC structures. These studies revealed that even for monotonic loading conditions the assessment of the bearing capacity of this type of structures requires further investigation. For the case of cyclic static or dynamic loading conditions, the ability of existing simulation approaches in capturing the experimental results was found to be less reliable and more case sensitive.

During the last three decades, 3D detailed finite element modeling approaches have been proposed [1, 2, 5, 17] for the analysis of structures with any geometry and type of loading. It is well known that beam-column and plane finite elements have significant simulation limitations when used to discretize 3D structures, as being unable to capture joint deformations and normal-shear stress coupling in three dimensions during a nonlinear analysis. Therefore, these simplistic models usually provide acceptable results when the overall structural behavior is bending dominated, where shear, joint deformation and the development of torsion at an overall structural level are assumed to be insignificant.

Before discussing the modeling limitations for SRC and SFRC structures, it is worth mentioning that multiscale modeling of composite materials seems to be a promising approach for modeling SRC and SFRC structures [18, 19]. Nevertheless, the current literature shows that applying this method to full-scale structures even for simple monotonic loading conditions is not practically possible due to the excessive computational effort required even for today's available computational resources. For these reasons, this manuscript will concentrate on the open problems in predicting the structural response of SRC and SRFC structures through nonlinear 3D detailed finite element modeling combined with the smeared crack approach [1], where the concrete domain is discretized with hexahedral elements and the constitutive material laws are described with 3D material models. This modeling approach also accounts for microcracking and macrocracking in concrete, whereas the discretization of the standard steel reinforcement is performed through the use of embedded rebar finite elements (rods or beams) and appropriate cyclic stress-strain relationships to account for the pinching effects.

2 Modeling Remarks of Steel Reinforced Concrete (SRC) Structures

One of the challenges in modeling SRC structures is the nonlinearities that are present at an early stage due to pre-existing cracks and the microcracking that develops during the loading procedure. Predicting the local and global structural response of SRC structures becomes more cumbersome when macrocracks occur given that discontinuities appear in the physical problem. When modeling SRC structures through the use of solid finite element models [2, 5, 6, 14, 20–22] the simulation of the physical gaps triggers additional numerical problems that lead to the need of simplification assumptions or assumptions that are not in line with actual mechanical behavior of SRC structures.

For instance, the material models proposed in [20, 21] accept a remaining tensile and shear strength (softening branch) after a macrocrack has developed at a Gauss point. This is not always true given that concrete cracks are not always found within a confined region of the SRC structural member. Figure 1 shows the case of a cantilever RC beam that is loaded at the tip, while developing a crack at the area of support. It is easy to observe that, the pure tensile crack that develops does not have any remaining strength in tension nor in compression. Given that the methods proposed in [20, 21] require finer meshes (10–50 mm) to achieve numerical convergence and reliable results, this means that there is at least one layer of finite elements which is used to discretize the concrete cover of any RC structural member. Therefore, in real physical problems these elements would never develop softening branches at the material level. Furthermore, the numerical response of these material models creates numerical instabilities requiring a larger number of internal iterations during the nonlinear solution. This is a simulation limitation that still needs to be addressed when using this type of material models.

On the other hand, the use of material laws that assume a complete loss of strength (brittle behavior) at the material level when a macrocrack occurs [14, 22] is also inconsistent with the behavior of cracked concrete, when it is well confined due to the presence of stirrups. The concrete material model presented in these studies [15, 23] can represent the remaining strength at the crack plane due to friction and gravel interlocking with the use of a remaining shear strength factor β. This factor is assumed to be constant throughout the numerical simulation and is affecting the performance of the model. However, the development of a formula to perform the calculation of β within each finite element is still an open problem.

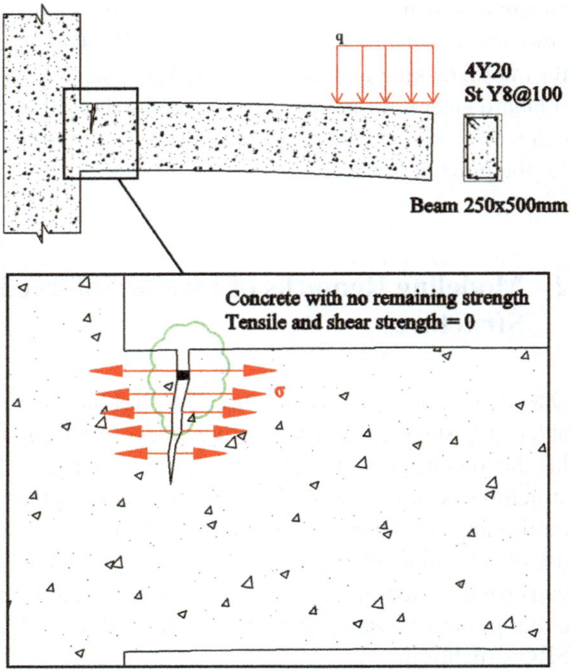

Fig. 1 Cracked SRC cantilever beam. Concrete area without any remaining strength

To illustrate the numerical performance of models using this type of material laws, a comparison between the experimental and numerical results of an SRC joint retrofitted with carbon fibre reinforced polymer (CFRP) sheets and designed to develop extreme reinforcement slippage and cracking within the joint, is shown in Fig. 2 [1]. The developed model was able to capture extreme nonlinearities and extreme pinching, where the displacement control cyclic analysis was found to be able to converge at each displacement increment with relatively low numbers of linearized internal iterations. The same performance was observed in the case of bare SRC joints, where it was concluded that the proposed damage factors for both steel and concrete material models were able to capture these extreme nonlinearities. These proposed damage factors are directly connected to the number of opening and closing of cracks during the cyclic loading of the model [1].

Based on the obtained results shown in Fig. 2, as well as on a number of simulation tests performed in [1], it is reasonable to conclude that the proposed 3D detailed modeling, with material models that assume a brittle type of behavior, appears to be more stable, accurate and robust than the models using softening branches for the simulation of the cyclic post-crack response of concrete.

It has to be mentioned that this concrete material model does not satisfy the 2nd law of thermodynamics, since it was derived through regression on a large dataset of experimental results. However, it was found to be able to provide with acceptable numerical results when used to predict the structural response of RC structures [2, 5, 17, 22] under monotonic and cyclic loading conditions. In this case,

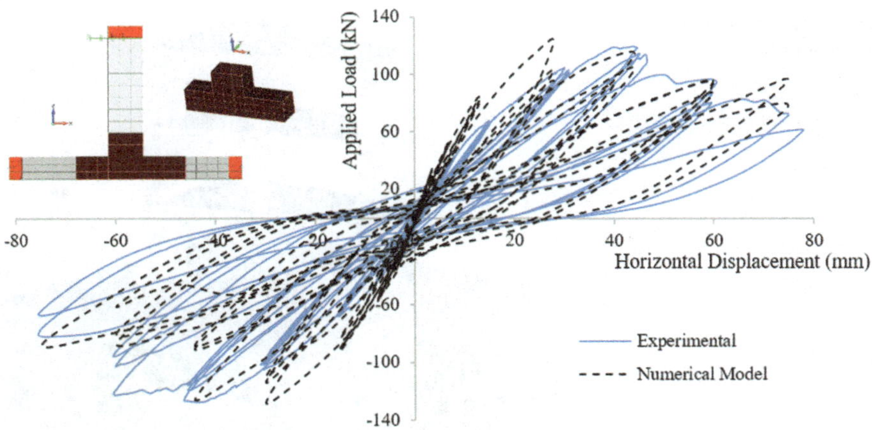

Fig. 2 Cracked SRC cantilever beam. Concrete area without any remaining strength

the physical laws were substituted by a large experimental dataset that incorporates the actual physical laws, instead of rigorous concepts or artificial physical laws. This approach has similarities with the implementation of artificial intelligence (AI) and machine learning (ML) algorithms [22–24, 26], which pursue the correspondence between large datasets of input and expected outcome. It is envisaged that the use of ML-based algorithms will prevail in the near future in predicting the structural response of SRC structures under extreme loading conditions [27].

Given that the overall objective is to derive a reliable and robust simulation methodology for predicting the ultimate nonlinear static and dynamic response of RC structures, the scientific community should be further involved in:

1. Developing reliable concrete material models able to capture phenomena such us opening and closing of cracks, material deterioration and slippage.
2. Producing reliable experimental data representing the actual mechanical behavior of the material, free from the influence of specimen-testing equipment interaction.
3. Obtaining robust experimental data to the extent that "identical" specimens attain ultimate capacities that do not significantly differ (i.e. less than 15%).

In this context, a 3D model shown in Fig. 3, was investigated in the SMART 2013 project [15], where an asymmetric SRC structure was tested under pure torsion. This specimen was found to have significant interactions with the relatively flexible seismic table that was used to investigate the nonlinear dynamic structural behavior of this scaled structure. For the purpose of this study the SRC model was discretized with 20-noded hexahedral isoparametric finite elements and was tested on a relatively flexible seismic table that was also included in the augmented structural model, as shown in Fig. 3. The modal analysis was performed with the Reconan FEA software [28] and its corresponding modal analysis algorithm [2], and the numerically obtained results were compared to the respective experimental data.

Table 1 shows the comparison between the numerical and experimental frequencies [15], where it can be observed that the numerical model manages to capture

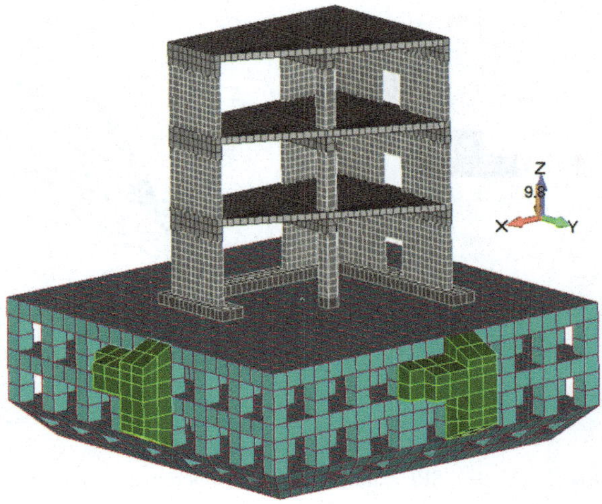

Fig. 3 Hexahedral finite element mesh

Table 1 Comparison between the numerically and experimentally obtained frequencies

Mode	Numerical f_z (Hz)	Experimental [15] f_z (Hz)	$(f_{exp} - f_{num})$ / f_{exp}
1	6.35	6.28	−1.12%
2	7.66	7.86	2.50%
3	15.43	16.5	6.51%
		Average	2.63%
		Abs. Av.	3.38%

the experimental data within an acceptable accuracy. The overall computed average error was 2.63%, where the respective absolute average error was 3.38%. A graphical representation of the computed and experimentally obtained modes are shown in Fig. 4.

This SRC structure with shear walls shown in Fig. 3 is depicted to underline the ability of a detailed solid element model to discretize any type of SRC structure, as opposed to discretization with a beam-column finite element mesh. Even in cases where shell finite elements are used to model the behavior of SRC shear walls, the limitations in capturing the 3D stress state that develops at the material level hinders their implementation. Even though shell finite element simulation has the ability to discretize 2D slender domains in a relatively accurate manner, it remains unable to provide accurate simulation of joints and connections between different structural members.

Conclusively, the use of 3D solid elements with appropriate material laws is currently one of the most promising modeling approaches, achieving a reliable and robust simulation of the structural response of real-scale SRC structures under extreme loading conditions. Any limitation of its range of application due to excessive computations resulted frame, the refined modeling can be overcome with the implementation

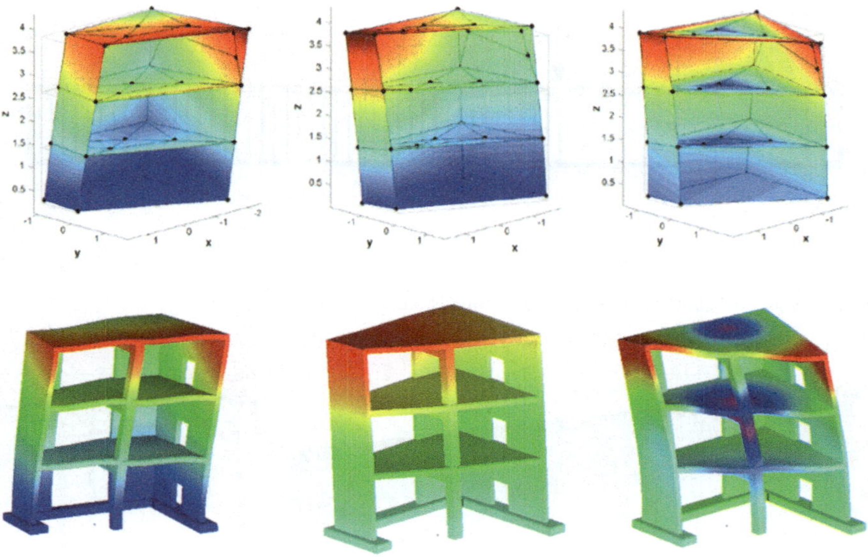

Fig. 4 Comparison between the experimentally (up) and numerically (down) obtained modal shapes. From left to right: 1st, 2nd and 3rd modal shapes

of high-performance computing as long as the physical problem is accurately simulated.

Methods such as X-FEM and multiscale modeling still require significant improvement and further investigation until they reach to a point of analyzing specimens under cyclic loading and then implemented for the numerical investigation of large-scale SRC structures similar to those investigated in [29–31]. It is noteworthy to state that the same modeling approach [1] was applied successfully for the simulation of the overall dynamic response of SRC structures including the soil-structure interaction effect [31].

3 Limitations in Modeling of Steel Fibre Reinforced Concrete (SFRC) Structures

A recent blind prediction competition [16] was performed to provide an inside on the modelling of fibre reinforced concrete structures and in particular on the ability of current numerical models in predicting the structural response of SRFC T-beams that are designed to fail in shear. Figure 5 shows a SFRC T-beam specimen and its respective reinforcement, as it was reported in [16], whereas Fig. 6 shows the 3D detailed finite element that the authors developed during this competition. This finite element model adopts the smeared crack approach [22] for simulating macro-cracks, where the reinforcement was modeled as embedded rebar finite elements.

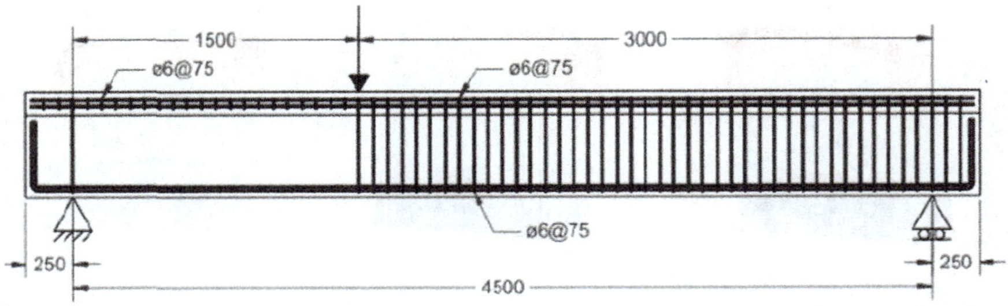

Fig. 5 SFRC T-beam specimen [16]. All dimensions in mm

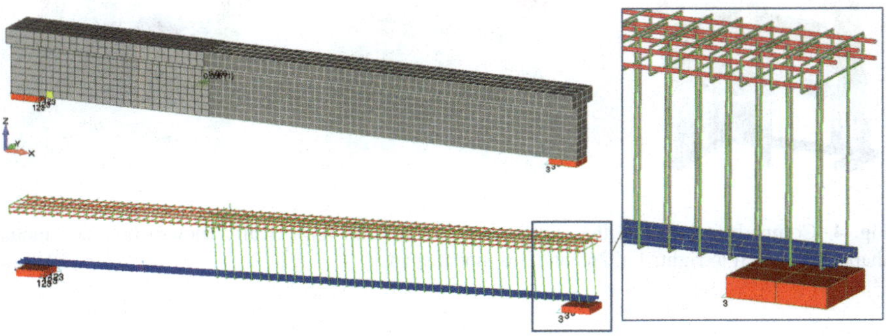

Fig. 6 SFRC T-beam finite element mesh. 8-noded hexahedral elements and embedded rebars (Ø6 stirrups in green, Ø10 compressive bars in magenta, Ø25 tensile bars in blue)

Even though Reconan FEA was not yet integrated with a 3D SFRC material model, it was decided to use the concrete material model implemented in [22] and submit a prediction. A total of 36 teams participated in this competition from both academia and industry.

In Fig. 7, the predictions of both force and strain vs deflection curves of the 36 teams were plotted against the derived experimental data. It can be observed that most of the teams failed to predict both the strength and the strain at maximum strength of the beam. This is a clear indication of the need for improving current numerical models in predicting the ultimate capacity of SFRC structures under monotonic and cyclic loading conditions, albeit the latter is currently not receiving much attention.

It is also worth noting that, the prediction submitted by the authors received the 9th overall place out of the 36 teams, even though the material used to predict the mechanical response of the SFRC domain was not developed for SFRC materials, but for standard and high strength concrete. In the case of the maximum strength prediction the authors received the 3rd overall place (with a 7.97% error; 309.23 kN, which was in favor of safety), while in the case of the maximum computed strain at the ultimate load was the best out of all the submitted predictions. Therefore, if the objective was to compute the maximum expected force and the respective strain, the authors' prediction was the best, with an average error of around 10%.

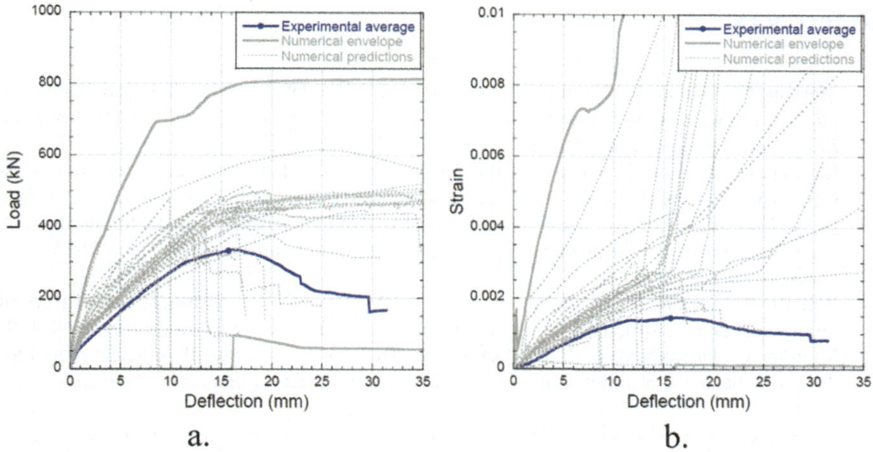

Fig. 7 Numerical verses experimental data. **a** Force verses deflection and **b** Strain verses deflection [16]

The predicted loading submitted by the authors for the first crack opening was 48 kN, where the experimentally observed first cracks occurred for a total of 50 kN. It has to be noted that, the use of 8-noded hexahedral finite elements combined with the concrete material model implemented by the authors led to a stiffer overall response underestimating the maximum deflection.

This blind competition [16] indicated that the 3D concrete material model [22] had to be integrated with a relevant tensile behavior, able in capturing the SFRC material's post-crack response and thereafter to be integrated within the 20-noded hexahedral element. This implementation is a work that is currently under development and will be discussed in a future communication.

4 Conclusions

From the brief discussion in this paper it may be concluded that, developing an accurate, objective and robust numerical model for assessing the structural behavior of SRC structures under extreme loading conditions is a challenging task that still has many open problems to be addressed. It is evident that there is some debate in relation to which simulation approach is the best in predicting the ultimate capacity of SRC structural members, while the ability to capture the cyclic behavior of SRC structural members is still a research topic that requires further investigation, especially when extreme nonlinearities and pinching effects are present [1].

When dealing with SFRC structural members, it was demonstrated that not only the 3D models need to be further improved, but most of the current material models do not have the ability to capture the shear capacity of SFRC material. Therefore, the development of a model able to predict the ultimate cyclic behavior of SFRC structural members is still in need by the international scientific community.

References

1. Markou, G., Garcia, R., Mourlas, C., Guadagnini, M., Pilakoutas, K., & Papadrakakis, M. (2021). *A New Damage Factor for Seismic Assessment of Deficient Bare and FRP-Retrofitted RC Structures. Engineering Structures 248*, 113152.
2. Mourlas, C., Markou, G., & Papadrakakis, M. (2019). A New damage factor for seismic assessment of deficient bare and FRP-retrofitted RC structures. *Engineering Structures, 178*, 258–285.
3. Li, X., Gao, W., & Liu, W. (2019). A mesh objective continuum damage model for quasi-brittle crack modelling and finite element implementation. *International Journal of Dams Mechanics, 28*(9), 1299–1322.
4. Belletti, B., Scolari, M., & Vecchi, F. (2017). PARC_CL 2.0 crack model for NLFEA of reinforced concrete structures under cyclic loadings. *Composite Structures, 191*(9), 165–179
5. Mourlas, C., Papadrakakis, M., & Markou, G. (2017). A computationally efficient model for the cyclic behavior of reinforced concrete structural members. *Engineering Structures, 141*, 97–125.
6. Kotsovos, M. D. (2015). *Finite-element modelling of structural concrete: Short-term static and dynamic loading conditions*, 1st edn. CRC Press
7. Yuchuan, L., Shaoqian, X., & Xuechao, G. (2011). An energy-based damage model for concrete structures under cyclic loading. *Procedia Engineering, 14*, 460–469.
8. Richard, B., Ragueneau, F., Cremona, C., & Adelaide, L. (2011). Isotropic continuum damage mechanics for concrete under cyclic loading: Stiffness recovery, inelastic strains and frictional sliding. *Engineering Fracture Mechanics, 77*, 1203–1223.
9. Červenka V., Jendele L., & Červenka J. (2008) ATENA program documentation. Part 1: Theory. In: Červenka Consulting, Prague, Czech Republic.
10. Rots, J. G., Belletti, B., & Invernizzi, S. (2008). Robust modeling of RC structures with an "event-by-event" strategy. *Engineering Fracture Mechanics, 75*, 590–614.
11. Kwak, H.-G., & Kim, D.-Y. (2006). Cracking behavior of RC panels subject to biaxial tensile stresses. *Composites and Structures, 84*, 305–317.
12. Jason, L., Huerta, A., Pijaudier-Cabot, G., & Ghavamian, S. (2006). An elastic plastic damage formulation for concrete: Application to elementary tests and comparison with an isotropic damage model. *Computer Methods in Applied Mechanics and Engineering, 195*, 7077–7092.
13. Shiohara, H., & Kusuhara, F. (2006). Benchmark Test for Validation of Mathematical Models for Nonlinear and Cyclic Behaviour of R/C Beam-Column Joints. Department of Architecture, School of Engineering, University of Tokyo.
14. Collins, M. P., Bentz, E. C., Quach, P. T., & Proestos, G. T. (2015). The challenge of predicting the shear strength of very thick slabs. *ACI Concrete International, 37*, 29–36.
15. Richard, B., Cherubini, S., Voldoire, F., Charbonnel, P. E., Chaudat, T., Abouri, S., & Bonfils, N. (2016). SMART 2013: Experimental and numerical assessment of the dynamic behavior by shaking table tests of an asymmetrical reinforced concrete structure subjected to high intensity ground motions. *Engineering Structures, 109*, 99–116.
16. Barros J. A. O., & Merino B. S. (2020). Blind competition on numerical stimulation of FRC beams Results. https://www.fib-international.org/blog/412-blind-competition-on-numerical-stimulation-of-frc-beams-results.html. Retrieved 29 Oct 2020
17. Markou, G., Mourlas, C., & Papadrakakis, M. (2019). A hybrid finite element model (HYMOD) for the non-linear 3D cyclic simulation of RC structure. *International Journal for Computational Methods, 16*, 1–40.
18. Wu, L., Liu, P., Zhang, Z., Zhu, D., & Wang, H. (2018). Multiscale modeling for high-performance concrete: A review. *International Journal for Multiscale Computational Engineering, 16*, 1–17.
19. Wenjuan, S., Ya, W., Dong, W., & Linbing, W.: review of multiscale characterization techniques and multiscale modeling methods for cement concrete: From atomistic to continuum. *Multi-Scale Modeling and Characterization of Infrastructure Materials RILEM, 8*, 325–341.
20. Evan, C. B., & Stephen, J. F. (2019). On shear in members without stirrups and the application of energy-based methods in light of 30 years of test observations. *Str. Conc., 20*, 1481–1489.

21. Dönmez, A. A., Carloni, C., Cusatis, G., & Zdeněk, P. B. (2020). Size effect on shear strength of reinforced concrete: Is CSCT or MCFT a viable alternative to energy-based design code? *Journal of Engineering Mechanics, 146*, 1–13.
22. Markou, G., & Papadrakakis, M. (2013). Accurate and computationally efficient 3D finite element modeling of RC structures. *Composites & Concrete, 12*, 443–498.
23. Iruansi, O., Guadagnini, M., Pilakoutas, K., & Neocleous, K. (2012). Predicting the shear resistance of rc beams without shear reinforcement using a Bayesian neural network. *International Journal of Reliability and Safety, 6*, 82–109.
24. Dutta, S., Samui, P., & Kim, D. (2018). Comparison of machine learning techniques to predict compressive strength of concrete. *Composites and Concrete, 21*, 463–470.
25. Kaya, M. (2018). Developing a new mutation operator to solve the RC deep beam. *Composites and Concrete, 22*, 493–500.
26. Gravett, Z.D., Mourlas, C., Taljaard V.L., Bakas P.N., Markou, G. and Papadrakakis, M. (2021), New fundamental period formulae for soil-reinforced concrete structures interaction using machine learning algorithms and ANNs, soil dynamics and earthquake engineering, *144*, 106656.
27. Markou, G., & Bakas, N. P. (2019) Artificial Intelligence (A.I.) Reinforced Concrete. Algorithms predict concrete strength without training on experimental data. Innovate, 3903187.
28. Reconan FEA v2.0 In: User's Manual, (2020) Available via DIALOG. https://www.researchgate.net/publication/342361609_ReConAn_v200_Finite_Element_Analysis_Software_User' s_Manual. Retrieved 20 Oct 2020
29. Markou, G., Mourlas, C., Bark, H., & Papadrakakis, M. (2018). Simplified HYMOD non-linear simulations of a full-scale multistory retrofitted RC structure that undergoes multiple cyclic excitations - an infill RC wall retrofitting study. *Engineering Structures, 176*, 892–916.
30. Markou, G.,& Genco, F. (2019). Seismic assessment of small modular reactors: Nuscale case study for the 8.8 mw earthquake in chile. *Nuclear Engineering and Design, 342*, 176–204.
31. Mourlas, C., Khabele, N., Bark, H. A., Karamitros, D., Taddei, F., Markou, G. et al. (2019). The effect of soil-structure interaction on the nonlinear dynamic response of reinforced concrete structures. Int. Journal of Structural Stability and Dynamics. https://doi.org/10.1142/S0219455420410138.

Dynamic Fracture of Brittle Shells in a Space-Time Adaptive Isogeometric Phase Field Framework

Karsten Paul, Thomas J. R. Hughes, Chad M. Landis, and Roger A. Sauer

Peter Wriggers has contributed immensely to the field of computational mechanics, and he has been a source of knowledge and inspiration to countless researchers. It was he who first brought me in touch with advanced fracture simulations, it was his textbook that first brought me in touch with shell finite elements, and it was he who first brought me in close touch with Tom Hughes. It is with gratitude that I dedicate this work to Peter Wriggers on the occasion of his 70th birthday. (Roger A. Sauer).

Abstract Phase field models for fracture prediction gained popularity as the formulation does not require the specification of ad-hoc criteria and no discontinuities are inserted in the body. This work focuses on dynamic crack evolution of brittle shell structures considering large deformations. The energy contributions from in-

K. Paul (✉)
Aachen Institute for Advanced Study in Computational Engineering Science (AICES), RWTH Aachen University, Aachen, Germany
e-mail: paul@aices.rwth-aachen.de

T. J. R. Hughes
Oden Institute for Computational Engineering and Sciences, The University of Texas at Austin, Austin, TX, USA
e-mail: hughes@oden.utexas.edu

C. M. Landis
Department of Aerospace Engineering and Engineering Mechanics, The University of Texas at Austin, Austin, TX, USA
e-mail: landis@utexas.edu

R. A. Sauer
Aachen Institute for Advanced Study in Computational Engineering Science (AICES), RWTH Aachen University, Aachen, Germany
e-mail: sauer@aices.rwth-aachen.de

Faculty of Civil and Environmental Engineering, Gdańsk University of Technology, Gdańsk, Poland

Department of Mechanical Engineering, Indian Institute of Technology Kanpur, Kanpur, India

F. Aldakheel et al. (eds.), *Current Trends and Open Problems in Computational Mechanics*, https://doi.org/10.1007/978-3-030-87312-7_39

plane and out-of-plane deformations are separately split into tensile and compressive components and the resulting coupled system is discretized within the isogeometric analysis framework. The resulting system is solved fully monolithically and adaptive local refinement is used in space and time.

1 Introduction

Thin-walled structures are characterized by low weight and high strength, making them interesting for many engineering designs. Especially for high slenderness ratios, these shells can be modeled based on the assumptions of Kirchhoff-Love theory. In these, no rotational degrees-of-freedom (dofs) are used, but only displacement dofs are considered. The resulting equation of motion thus, includes fourth-order derivatives. Isogeometric analysis [10] is used to obtain the required C^1-continuity in the corresponding weak form. In this work, the shell formulation of [7] is used.

The prediction of fracture and failure is of crucial importance for the design of engineering structures. Phase field models for the prediction of fracture gained popularity as they do not require ad-hoc criteria and do not insert discontinuities in the body, e.g. in the displacement field. Phase field models for brittle fracture are based on the theory of Griffith [9] and its variational reformulation as an energy minimization problem [8]. An additive energy split is required to model the anisotropic behavior of crack evolution, i.e. such that there occurs no cracking in compression. The membrane and bending energies are split separately and the split based on surface stretches [1] is employed for the split of the membrane part. Based on the idea of [12], the bending energy is decomposed based on thickness integration, but here, the split is also based on surface stretches [14]. The resulting formulation allows for large deformations and avoids the expensive computations of spectral decompositions. A higher-order fracture energy [2] is employed. The small length scale parameter of the phase field is resolved by using local spatial refinement based on LR NURBS [6, 17] and the mesh is adaptively refined during the computation [14]. The resulting discretized coupled system is implicitly integrated in time using an adaptive time-stepping scheme and a monolithic solution approach. Patch constraints are used to allow for multi-patch discretizations [13].

2 Fracture of Deforming Surfaces

The mapping $\boldsymbol{x} = \boldsymbol{x}(\xi^\alpha, t)$ is used to describe a curved surface $\mathcal{S}$. Here, $\xi^\alpha, \alpha = 1, 2$ are the convected coordinates and t denotes time. Based on this mapping, a co- and contra-variant basis can be associated with each surface point, i.e. $\{\boldsymbol{a}_\alpha, \boldsymbol{n}\}$ and $\{\boldsymbol{a}^\alpha, \boldsymbol{n}\}$, with covariant tangent vectors $\boldsymbol{a}_\alpha = \partial\boldsymbol{x}/\partial\xi^\alpha$, surface normal $\boldsymbol{n} = (\boldsymbol{a}_1 \times \boldsymbol{a}_2)/||\boldsymbol{a}_1 \times \boldsymbol{a}_2||$, and contra-variant tangent vectors $\boldsymbol{a}^\alpha = a^{\alpha\beta}\boldsymbol{a}_\beta$.

The co- and contra-variant surface metrics are given by $a_{\alpha\beta} = \boldsymbol{a}_\alpha \cdot \boldsymbol{a}_\beta$ and $[a^{\alpha\beta}] = [a_{\alpha\beta}]^{-1}$. Using the second parametric derivative $\boldsymbol{a}_{\alpha,\beta} = \boldsymbol{x}_{,\alpha\beta} = \partial \boldsymbol{a}_\alpha / \partial \xi^\beta$, the surface curvature is described via $b_{\alpha\beta} = \boldsymbol{a}_{\alpha,\beta} \cdot \boldsymbol{n}$.

2.1 Thin Shell Theory

The equation of motion can be written as [7]

$$\boldsymbol{T}^\alpha_{;\alpha} + \boldsymbol{f} = \rho \dot{\boldsymbol{v}} \quad \text{on } \mathcal{S} \times (0, \bar{T}) , \tag{1}$$

with traction vector $\boldsymbol{T}^\alpha$, surface force $\boldsymbol{f}$ and final time $\bar{T}$. The stresses and moments follow from constitution. Here, hyperelastic material behavior is assumed and the elastic energy density is given by

$$\Psi_{\text{el}}(a_{\alpha\beta}, b_{\alpha\beta}) = \Psi_{\text{mem}}(a_{\alpha\beta}) + \Psi_{\text{bend}}(b_{\alpha\beta}) , \tag{2}$$

with the membrane energy density being composed of dilatational and deviatoric contributions, i.e. $\Psi_{\text{mem}} = \Psi_{\text{dil}} + \Psi_{\text{dev}}$, where [16]

$$\Psi_{\text{dil}} = \frac{K}{4}\Big(J^2 - 1 - 2\ \ln J\Big) , \quad \text{and} \quad \Psi_{\text{dev}} = \frac{G}{2}\Big(I_1/J - 2\Big) . \tag{3}$$

Here, $I_1 := A^{\alpha\beta} a_{\alpha\beta}$ is the first invariant of the surface Cauchy-Green strain tensor and $J := \sqrt{\det[A^{\alpha\beta}]\det[a_{\alpha\beta}]}$ is the surface stretch.[1] Using $b_0^{\alpha\beta} = A^{\alpha\gamma} b_{\gamma\delta} A^{\beta\delta}$, the bending energy in (2) is given by [5]

$$\Psi_{\text{bend}} = \frac{c}{2}\Big(b_{\alpha\beta} - B_{\alpha\beta}\Big)\Big(b_0^{\alpha\beta} - B^{\alpha\beta}\Big) . \tag{4}$$

2.2 Brittle Fracture

The phase field $\phi = \phi(\xi^\alpha, t) \in [0, 1]$ ranges from the undamaged state ($\phi = 1$) to the fully fractured state ($\phi = 0$). Phase evolution is described by a partial differential equation, which stems from the minimization of the Helmholtz free energy

$$\Pi_{\text{int}} = \int_{\mathcal{S}_0} \Psi\, \mathrm{d}A = \int_{\mathcal{S}_0} \Big[g(\phi)\, \mathcal{H} + \Psi_{\text{el}}^{-} + \Psi_{\text{frac}} \Big]\, \mathrm{d}A . \tag{5}$$

The higher-order fracture energy density in (5) is given by [2, 14]

---

[1] Quantities on the reference surface are either indicated by the subscript '0' or by a capital symbol.

$$\Psi_{\text{frac}} = \frac{\mathcal{G}_c}{4\ell_0}\Big[(\phi - 1)^2 + 2\ell_0^2\, \nabla_S\phi \cdot \nabla_S\phi + \ell_0^4\, (\Delta_S\phi)^2\Big] , \tag{6}$$

with fracture toughness $\mathcal{G}_c$ and length scale ℓ_0 of the phase field model. The surface gradient and Laplacian are indicated by ∇_S and Δ_S, respectively. The degradation of the bulk material is described by the cubic degradation function [3]

$$g(\phi) = (3 - s)\phi^2 - (2 - s)\phi^3 , \quad \text{with } s = 10^{-4} . \tag{7}$$

Irreversibility of the fracture process is ensured by keeping track of the *fracture driving energy* by means of the history field

$$\mathcal{H}(\boldsymbol{x}, t) := \max_{\tau \in [0,t]} \Psi_{\text{el}}^{+}(\boldsymbol{x}, \tau) . \tag{8}$$

The 'positive' and 'negative' energies in (5) and (8) are based on the additive energy split

$$\Psi_{\text{el}} = \Psi_{\text{el}}^{+} + \Psi_{\text{el}}^{-} , \quad \text{with } \Psi_{\text{el}}^{\pm} = \Psi_{\text{mem}}^{\pm} + \Psi_{\text{bend}}^{\pm} , \tag{9}$$

which splits the energy into a part that contributes to crack evolution ('+'), and a part that does not ('−'). The contributions are given by [1, 14]

$$\Psi_{\text{mem}}^{+} = \begin{cases} \Psi_{\text{dev}} + \Psi_{\text{dil}} , & J \geq 1 \\ \Psi_{\text{dev}} , & J < 1 \end{cases} , \qquad \Psi_{\text{mem}}^{-} = \begin{cases} 0 , & J \geq 1 \\ \Psi_{\text{dil}} , & J < 1 \end{cases} , \tag{10}$$

and

$$\Psi_{\text{bend}}^{\pm} = \int_{-\frac{T}{2}}^{\frac{T}{2}} \xi^2 \frac{12}{T^3} \frac{c}{2} \operatorname{tr}(\boldsymbol{K}^2)\, \chi^{\pm}(\tilde{J}(\xi))\, \mathrm{d}\xi , \quad \text{with } \chi^{+}(\tilde{J}(\xi)) = \begin{cases} 1 , & \tilde{J}(\xi) \geq 1 , \\ 0 , & \tilde{J}(\xi) < 1 , \end{cases} \tag{11}$$

and $\chi^{-}(\tilde{J}(\xi))$ analogously. In (11), $\boldsymbol{K} = (b_{\alpha\beta} - B_{\alpha\beta})\, \boldsymbol{A}^{\alpha} \otimes \boldsymbol{A}^{\beta}$ denotes the relative curvature tensor, T is the shell thickness, and $\tilde{J} = \sqrt{\det[\tilde{A}^{\alpha\beta}] \det[\tilde{a}_{\alpha\beta}]}$ is the surface stretch of a shell layer [7]. Using the Euler-Lagrange equation and standard arguments of variational calculus, the strong form of the fracture framework is found as

$$\begin{aligned} 2\ell_0/\mathcal{G}_c\, g'(\phi)\, \mathcal{H} + \phi - 1 - 2\,\ell_0^2\, \Delta_S\phi + \ell_0^4\, \Delta_S(\Delta_S\phi) &= 0 && \text{on } \mathcal{S} \times (0, \bar{T}) , \\ \Delta_S\phi &= 0 && \text{on } \partial\mathcal{S} \times (0, \bar{T}) , \\ \nabla_S(\ell_0^4\, \Delta_S\phi - 2\,\ell_0^2\, \phi) \cdot \boldsymbol{\nu} &= 0 && \text{on } \partial\mathcal{S} \times (0, \bar{T}) , \\ \phi &= \phi_0 && \text{on } \mathcal{S} \times 0 . \end{aligned} \tag{12}$$

2.3 Computational Aspects

Isogeometric analysis [10] is used to obtain the required C^1-continuity in the weak formulations of (1) and (12), also see [14]. LR NURBS [6, 17] are used to allow for local refinement, such that the small length scale of the model (ℓ_0) can be resolved properly. The FE mesh is adaptively refined based on the phase field: If a control value fulfills $\phi < 0.975$, all elements that lie in the support domain of the corresponding basis function will be flagged for refinement. The refinement is performed using the *structured mesh* strategy [11, 14].

The generalized-α method [4] is used for temporal discretization and the discretized coupled system is solved within a monolithic Newton-Raphson scheme. In experiments, it has been shown that the crack tip velocity stays below 60% of the Rayleigh wave speed c_R [15], such that a physical upper bound can be imposed on the time step, i.e. $\Delta t \leq \Delta t_{max} < \Delta x_{min}/(0.6\, c_R)$, where Δx_{min} denotes the minimum element length. The time step is adjusted based on the number of required Newton-Raphson iterations n_{NR} during the previous time step, i.e.

$$\Delta t_{n+1} = \begin{cases} 1.5\,\Delta t_n\,, & n_{NR} < 4\,, \\ 1.1\,\Delta t_n\,, & n_{NR} = 4\,, \\ 0.5\,\Delta t_n\,, & n_{NR} > 4\,, \\ 0.2\,\Delta t_n\,, & \text{local spatial refinement}\,. \end{cases} \tag{13}$$

3 Numerical Examples

The material parameters K, G and c in (3)–(4) are set like in [14]. Crack patterns are visualized by means of a red ($\phi = 0$) to blue ($\phi = 1$) colorscale. Further, the non-dimensionaliztion scheme by [14] is adopted such that all quantities are normalized by the reference time T_0, length L_0 and stiffness E_0.

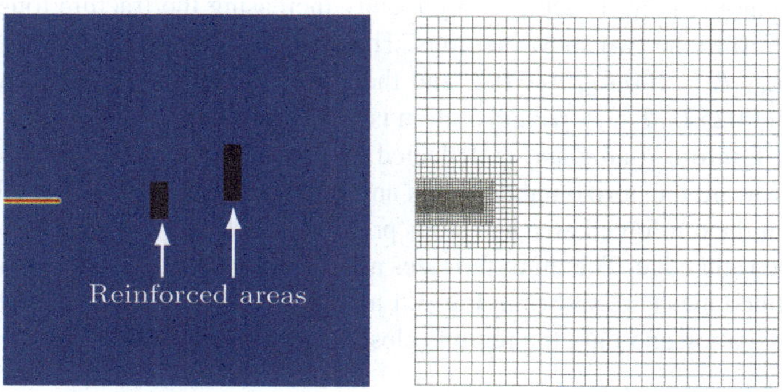

Fig. 1 Crack propagation around obstacles: Initial phase field and initial LR mesh. The dark colored rectangles mark the regions of increased fracture toughness. The initial mesh consists of 32 × 32 elements and three local refinement levels are used

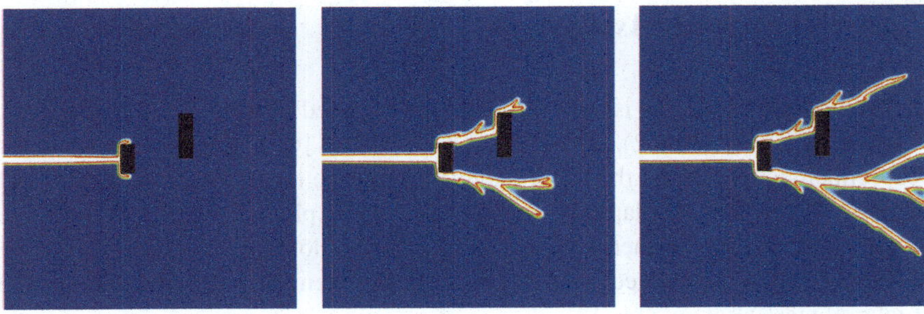

(a) $t \approx 2.98\, T_0$ (b) $t \approx 3.09\, T_0$ (c) $t \approx 3.20\, T_0$

Fig. 2 Crack propagation around obstacles: Crack pattern at different snapshots. Regions with $\phi < 0.01$ are not visualized

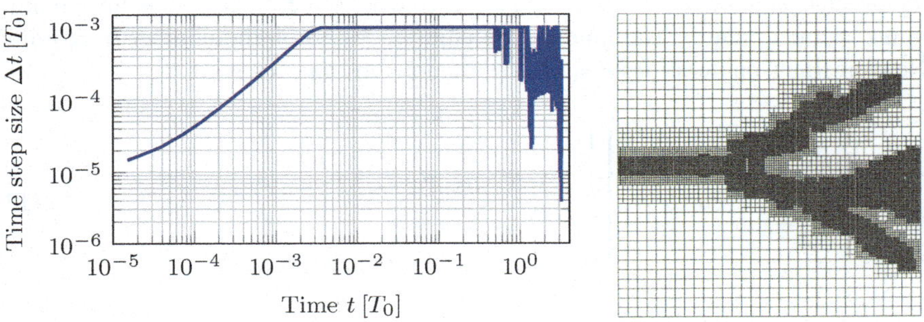

Fig. 3 Crack propagation around obstacles: Adaptivity in time (left) and space (right, final LR mesh)

3.1 Crack Propagation Around Obstacles

This example investigates dynamic crack branching, kinking and deflection in a square two-dimensional domain. The initial state is shown in Fig. 1 and a displacement is imposed on the top edge upwards, and on the bottom edge downwards. Branching and kinking is achieved by locally increasing the fracture toughness by a factor of 10 in the two shown regions. The parameters are $\ell_0 = 0.0025\, L_0$, $\mathcal{G}_c = 0.001\, E_0\, L_0$, $E = 100\, E_0$, $\nu = 0.3$, and the displacement increment is $\Delta\bar{u} = \bar{v}\Delta t$ with $\bar{v} = 0.0025\, L_0\, T_0^{-1}$. Crack evolution is illustrated in Fig. 2. The crack branches at the first reinforced area and is deflected in vertical direction. The two branches then kink toward the horizontal direction and start branching. The top crack is again deflected at the reinforced area and stops propagating shortly after the bottom crack reaches the right edge. The time step sizes according to (13) and the final adaptively locally refined mesh are shown in Fig. 3. Large time steps are used when there is no crack propagation and only the regions close to the cracks are refined.

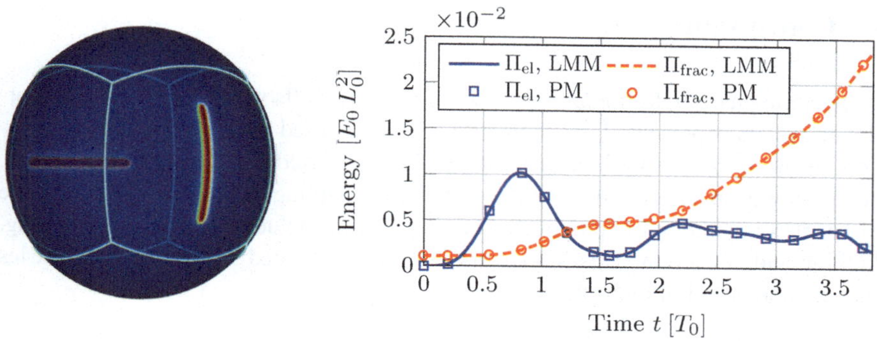

Fig. 4 Fracturing balloon: Initial state with patch interfaces (left) and energy-time-curves (right)

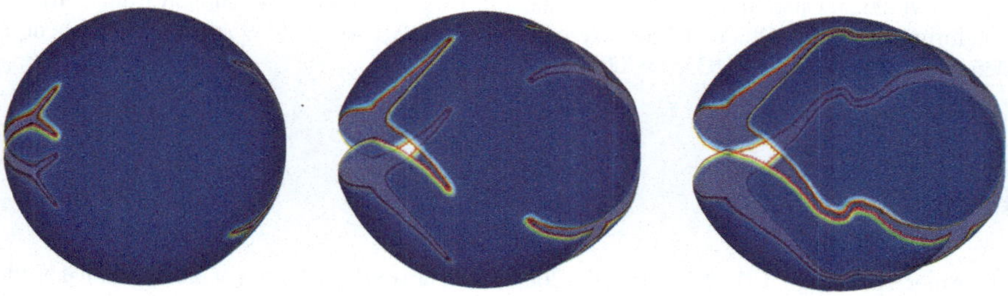

(a) $t \approx 1.6\,T_0$ (b) $t \approx 3.16\,T_0$ (c) $t \approx 3.81\,T_0$

Fig. 5 Fracturing balloon: Crack pattern at different snapshots (visualized transparently). Regions with $\phi < 0.005$ are not visualized

3.2 *Fracturing Balloon*

This section investigates crack evolution on a spherical geometry that is imposed to the internal pressure $p(\phi) = 0.1\,E_0\,L_0^{-1}\phi$. The geometry is composed of six patches, as shown in Fig. 4 on the left (patch interfaces are indicated by the cyan-colored lines). Quadratic NURBS are used within each patch, and C^1-continuity across patch interfaces is restored by imposing patch constraints with the Lagrange multiplier (LMM) (constant interpolation) or the penalty method (PM), which is further elaborated in [13]. The radius is L_0 and the parameters are $\ell_0 = 0.015\,L_0$, $\mathcal{G}_c = 0.0005\,E_0\,L_0$, $E = 10\,E_0$, $\nu = 0.3$, $T = 10^{-4}L_0$, and bending stiffness $c = 10^{-5}E_0\,L_0^2$. The elastic and fracture energies (Π_{el} and Π_{frac}, respectively) over time for the two enforcement techniques are shown in Fig. 4 on the right. Crack evolution is visualized in Fig. 5. The cracks start branching and merge at the end. The first drop of the elastic energy occurs after the onset of crack propagation, and the last occurs when the geometry is fully fractured. Excellent agreement between the two enforcement techniques is achieved.

4 Conclusion

A dynamic phase field fracture framework for thin shells within a convective coordinate system is presented. Isogeometric analysis is used to obtain the required C^1-continuity in the weak form and LR NURBS are used for the local refinement. Adaptivity in space and time and the monolithic coupling of both PDEs ensures the efficient computation of complicated crack patterns, including branching, merging, kinking and deflection. This framework is further extended to consider geometries that are composed of multiple patches.

Acknowledgements The authors acknowledge funding by the Deutsche Forschungsgemeinschaft (DFG, German Research Foundation)—33849990/GRK2379 (IRTG Modern Inverse Problems). Simulations were performed with computing resources granted by RWTH Aachen University under project rwth0625. Chad M. Landis and Thomas J.R. Hughes acknowledge support from ONR and Coreform through the Office of Naval Research grant N00014-17-1-2039 and the Department of Defense, Navy, Contract N6833518C0014.

References

1. Amor, H., Marigo, J.-J., & Maurini, C. (2009). Regularized formulation of the variational brittle fracture with unilateral contact: Numerical experiments. *Journal of the Mechanics and Physics of Solids, 57*(8), 1209–1229.
2. Borden, M. J., Hughes, T. J. R., Landis, C. M., & Verhoosel, C. V. (2014). A higher-order phase-field model for brittle fracture: Formulation and analysis within the isogeometric analysis framework. *Computer Methods in Applied Mechanics and Engineering, 273,* 100–118.
3. Borden, M. J., Hughes, T. J. R., Landis, C. M., Anvari, A., & Lee, I. J. (2016). A phase-field formulation for fracture in ductile materials: Finite deformation balance law derivation, plastic degradation, and stress triaxiality effects. *Computer Methods in Applied Mechanics and Engineering, 312,* 130–166.
4. Chung, J., & Hulbert, G. M. (1993). A time integration algorithm for structural dynamics with improved numerical dissipation: The generalized-alpha method. *Journal of Applied Mechanics, 60*(2), 371–375.
5. Ciarlet, P. G. (1993). *Mathematical elasticity: three dimensional elasticity*. North-Holland
6. Dokken, T., Lyche, T., & Pettersen, K. F. (2013). Polynomial splines over locally refined box-partitions. *Computer Aided Geometric Design, 30*(3), 331–356.
7. Duong, T. X., Roohbakhshan, F., & Sauer, R. A. (2017). A new rotation-free isogeometric thin shell formulation and a corresponding continuity constraint for patch boundaries. *Computer Methods in Applied Mechanics and Engineering, 316,* 43–83.
8. Francfort, G., & Marigo, J.-J. (1998). Revisiting brittle fracture as an energy minimization problem. *Journal of the Mechanics and Physics of Solids, 46*(8), 1319–1342.
9. Griffith, A. A. (1921). VI. The Phenomena of Rupture and Flow in Solids. *Philosophical Transactions of the Royal Society of London Series A, 221*, 163–198.
10. Hughes, T. J. R., Cottrell, J. A., & Bazilevs, Y. (2005). Isogeometric analysis: CAD, finite elements, NURBS, exact geometry and mesh refinement. *Computer Methods in Applied Mechanics and Engineering, 194*(39–41), 4135–4195.
11. Johannessen, K. A., Kvamsdal, T., & Dokken, T. (2014). Isogeometric analysis using LR B-splines. *Computer Methods in Applied Mechanics and Engineering, 269,* 471–514.

12. Kiendl, J., Ambati, M., De Lorenzis, L., Gomez, H., & Reali, A. (2016). Phase-field description of brittle fracture in plates and shells. *Computer Methods in Applied Mechanics and Engineering, 312,* 374–394.
13. Paul, K., Zimmermann, C., Duong, T. X., & Sauer, R. A. (2020). Isogeometric continuity constraints for multi-patch shells governed by fourth-order deformation and phase field models. *Computer Methods in Applied Mechanics and Engineering, 370,* 113219.
14. Paul, K., Zimmermann, C., Mandadapu, K. K., Hughes, T. J. R., Landis, C. M., & Sauer, R. A. (2020). An adaptive space-time phase field formulation for dynamic fracture of brittle shells based on LR NURBS. *Computational Mechanics, 65,* 1039–1062.
15. Ravi-Chandar, K., & Knauss, W. G. (1984) An experimental investigation into dynamic fracture: III. On steady-state crack propagation and crack branching. *International Journal of Fracture, 26*(2), 141–154.
16. Sauer, R. A., & Duong, T. X. (2017). On the theoretical foundations of thin solid and liquid shells. *Mathematics & Mechanics of Solids, 22*(3), 343–371.
17. Zimmermann, C., & Sauer, R. A. (2017). Adaptive local surface refinement based on LR NURBS and its application to contact. *Computational Mechanics, 60,* 1011–1031.

Higher Order Geometrically Exact Shear-Rigid Beam Finite Elements

Paulo M. Pimenta, Cátia da Costa e Silva, and Carlos Tiago

This chapter is dedicated to Peter Wriggers in recognition of his fantastic scientific achievements and as an appreciation of our long-lasting friendship. He has undoubtedly been one of the leading figures in the field of nonlinear computational mechanics, to which this contribution belongs. It has been a privilege and pleasure to know Peter and to share with him many moments in different places around the world (P.M. Pimenta).

Abstract This work presents a polynomial refinement for our geometrically exact shear-rigid rod model presented in [1]. The objective is to compare different polynomials interpolations that could have a practical use. Displacements and rotations can be unlimited large. Linear elastic constitutive equations are considered in the numerical examples. Energetically conjugated cross-sectional generalized stresses and strains are defined. A straight reference configuration is assumed for the rod, but initially curved rods can be easily accomplished. The rotation is incrementally described using the Rodrigues rotation vector as parameterization. Six different interpolation schemes for the finite elements are tested in our examples, which are suitable for connection with the C^1 TUBA shell elements developed in [2].

P. M. Pimenta (✉)
Polytechnic School at the University of São Paulo, São Paulo, Brazil
e-mail: ppimenta@usp.br

C. da Costa e Silva
Federal Institute of Science and Technology Education of São Paulo, São Paulo, Brazil
e-mail: catiacosta@ifsp.edu.br

C. Tiago
Instituto Superior Técnico, University of Lisbon, Lisbon, Portugal
e-mail: carlos.tiago@tecnico.ulisboa.pt

F. Aldakheel et al. (eds.), *Current Trends and Open Problems in Computational Mechanics*, https://doi.org/10.1007/978-3-030-87312-7_40

1 Introduction

This work presents simple finite elements for geometrically exact shear-rigid rod formulation. It also presents polynomial refinement for the finite elements. So that the transversal shear deformation is not accounted for, the class of admissible motions is obtained by imposing that the cross sections of the rod, that are initially orthogonal to the axis, remain rigid and orthogonal to it after deformation. This theory is geometrically exact because no approximation is employed after the basic kinematical assumption. Finite rotations is treated here by the Euler–Rodrigues formula in an updated Lagrangian way.

As framework the theory presented in [1] is used. This approach defines energetically conjugated generalized cross section stress and strains. The tangent bilinear form is always symmetric for hyper-elastic materials and conservative loadings, even far from an equilibrium state. New interpolation polynomials were developed with the objective to connect these elements to TUBA elements from [2].

The models are implemented using the finite element method with different polynomial interpolation schemes for the finite element approximation. Linear elastic constitutive equations for small strains are considered in the numerical examples of this paper.

Throughout the text, italic Greek or Latin lowercase letters $(a, b, \ldots, \alpha, \beta, \ldots)$ denote scalars, bold italic Greek or Latin lowercase letters $(\boldsymbol{a}, \boldsymbol{b}, \ldots, \boldsymbol{\alpha}, \boldsymbol{\beta}, \ldots)$ denote vectors and bold italic Greek or Latin capital letters $(\boldsymbol{A}, \boldsymbol{B}, \ldots)$ denote second-order tensors. Summation convention over repeated indices is adopted in the entire text, whereby Greek indices range from 1 to 2, while Latin indices range from 1 to 3. $\|\boldsymbol{v}\| = \sqrt{\boldsymbol{v} \cdot \boldsymbol{v}}$ is the norm of vector $\boldsymbol{v}$ where $\cdot$ denotes the scalar product. The operator $\otimes$ denotes the dyadic or tensor product of two vectors. For instance, $\boldsymbol{a} \otimes \boldsymbol{b}$ is a second-order tensor such that $(\boldsymbol{a} \otimes \boldsymbol{b})\boldsymbol{c} = (\boldsymbol{b} \cdot \boldsymbol{c})\boldsymbol{a}$. Note that $(\boldsymbol{a} \otimes \boldsymbol{b})^T = (\boldsymbol{b} \otimes \boldsymbol{a})$, where $(\cdot)^T$ denotes the transpose. The operator axial$(\cdot)$ is such that, if $\boldsymbol{v} = \text{axial}(\boldsymbol{V})$, with $\boldsymbol{V}$ skew-symmetric, then $\boldsymbol{V}\boldsymbol{x} = \boldsymbol{v} \times \boldsymbol{x}$, where $\times$ denotes the cross product of two vectors. $\text{Skew}(\boldsymbol{V}) = \frac{1}{2}(\boldsymbol{V} + \boldsymbol{V}^T)$ is the skew-symmetric part of $\boldsymbol{V}$.

2 The Geometrically-Exact Shear Rigid Rod Theory

Figure 1 shows the kinematics considered in this theory. $\{\boldsymbol{e}_1^r, \boldsymbol{e}_2^r, \boldsymbol{e}_3^r\}$ is the orthogonal basis at the reference configuration of the rod. The superscript $(\cdot)^r$ indicates the material or back-rotated counterpart of $(\cdot)$ and is not affected by superimposed rigid body motions. Back-rotated means that the quantity is calculated according to $(\cdot) = \boldsymbol{Q}(\cdot)^r \Leftrightarrow (\cdot)^r = \boldsymbol{Q}^T(\cdot)$ where $\boldsymbol{Q}$ is the cross section rotation tensor.

Rodrigues parameters are employed for the description of finite rotations. Considering $\boldsymbol{\alpha}$ as the vector of Rodrigues parameters associated to the rotation tensor, and $\boldsymbol{A} = \text{Skew}(\boldsymbol{\alpha})$, this is given by

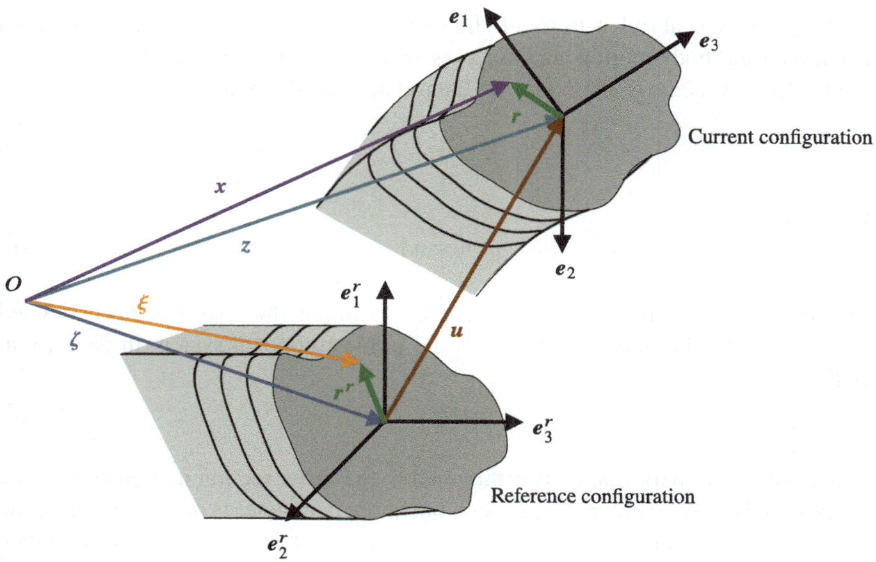

Fig. 1 Rod description and basic kinematical quantities

$$\hat{Q}(\alpha) = I + \frac{4}{4+\alpha^2}\left(A + \frac{1}{2}A^2\right), \tag{1}$$

where $\alpha^2 = \boldsymbol{\alpha} \cdot \boldsymbol{\alpha}$. The e_3^r vector is placed along the shear centers of the rod and is orthogonal the cross-section plane. The vectors e_α^r with $\alpha = 1, 2$ describe the basis on the cross-section plane. $\boldsymbol{\zeta} = \zeta e_3^r$ defines the position along the axis rod on the reference configuration. $\boldsymbol{r}^r = \xi_\alpha e_\alpha^r$ is the director that describes the relative position of the material points on the cross-section at the reference configuration.

Thus, the position of the rod material points in the reference configuration can be described by

$$\boldsymbol{\xi} = \boldsymbol{\zeta} + \boldsymbol{r}^r. \tag{2}$$

The coordinates $\zeta = \boldsymbol{\zeta} \cdot e_3^r$ and $\xi_\alpha = \boldsymbol{r}^r \cdot e_\alpha^r$ are introduced, as such that ξ_1, ξ_2 and ζ build a Cartesian coordinate system. $\zeta \in \Omega = (0, \ell)$, where ℓ is the rod length at reference configuration. The boundary of the domain Ω is denoted by Γ. So, Γ contains the two ends of the rod, that is $\Gamma = \{0, \ell\}$.

In the current configuration, as it can be seen in Fig. 1, the position of the material points is given by

$$\boldsymbol{x} = \boldsymbol{z} + \boldsymbol{r}, \tag{3}$$

where $\boldsymbol{z} = \hat{\boldsymbol{z}}(\zeta)$ describes the position of the rod axis at the current configuration and $\boldsymbol{r}$ is the current director given by

$$\boldsymbol{r} = \boldsymbol{Q}\boldsymbol{r}^r. \tag{4}$$

This is the typical Euler–Bernoulli kinematical assumption, i.e. the cross sections undergo a rigid body motion, and no cross-section change is assumed.

The displacements of the points on the rod axis are defined by

$$\boldsymbol{u} = \boldsymbol{z} - \boldsymbol{\zeta}. \tag{5}$$

Also note that

$$\boldsymbol{z}' = \boldsymbol{e}_3^r + \boldsymbol{u}' \quad \text{and} \quad \boldsymbol{z}'' = \boldsymbol{u}''. \tag{6}$$

whereby the following notation for derivative along the axis has been defined $(\cdot)' = d(\cdot)/d\zeta$. The basis $\{\boldsymbol{e}_1, \boldsymbol{e}_2, \boldsymbol{e}_3\}$ is the local orthogonal system in the current configuration, with

$$\boldsymbol{e}_2 = \left\|\boldsymbol{z}'\right\|^{-1} \boldsymbol{z}'. \tag{7}$$

The authors remark that $\boldsymbol{e}_3$ is orthogonal to the cross section and tangent to the rod axis. To be able to define $\{\boldsymbol{e}_1, \boldsymbol{e}_2, \boldsymbol{e}_3\}$, one more parameter is needed, because the cross section can rotate around the axis. Thus, the scalar $\varphi = \hat{\varphi}(\zeta)$ is defined, which is called herein torsion parameter. Thus, the cross-sectional rotation tensor can be expressed by

$$\boldsymbol{Q} = \hat{\boldsymbol{Q}}(\boldsymbol{e}_3, \varphi) = \hat{\boldsymbol{Q}}(\boldsymbol{u}', \varphi). \tag{8}$$

Note that $\boldsymbol{e}_i = \boldsymbol{Q}\boldsymbol{e}_i^r$, where $\boldsymbol{Q} = \boldsymbol{e}_i \otimes \boldsymbol{e}_i^r$.

The use of Rodrigues parameters is restricted to $-\pi < \theta < \pi$. To overcome this drawback, the rotation is described by an incremental approach, as in [1]. The limitation affects only a load increment in Statics or a time increment in Dynamics.

Let $(\cdot)_i$ and $(\cdot)_{i+1}$ denote a quantity $(\cdot)$ at instants t_i and t_{i+1}, respectively. And let $(\cdot)_\Delta$ be an incremental quantity. Thus, one gets for the rotation tensor the following relations

$$\boldsymbol{Q}_{i+1} = \boldsymbol{Q}_\Delta \boldsymbol{Q}_i, \tag{9}$$

where $\boldsymbol{Q}_{i+1} = \hat{\boldsymbol{Q}}\left(\boldsymbol{\alpha}_{i+1}\right)$, $\boldsymbol{Q}_\Delta = \hat{\boldsymbol{Q}}\left(\boldsymbol{\alpha}_\Delta\right)$, and $\boldsymbol{Q}_i = \hat{\boldsymbol{Q}}\left(\boldsymbol{\alpha}_i\right)$. We recall the following relevant result by Rodrigues

$$\boldsymbol{\alpha}_{i+1} = \frac{4}{4 - \boldsymbol{\alpha}_i \cdot \boldsymbol{\alpha}_\Delta} \left(\boldsymbol{\alpha}_i + \boldsymbol{\alpha}_\Delta - \frac{1}{2} \boldsymbol{\alpha}_i \times \boldsymbol{\alpha}_\Delta \right). \tag{10}$$

At instants t_i and t_{i+1} the triad of vectors $\{\boldsymbol{e}_1, \boldsymbol{e}_2, \boldsymbol{e}_3\}$ is denoted by $\{\boldsymbol{e}_1^i, \boldsymbol{e}_2^i, \boldsymbol{e}_3^i\}$ and $\{\boldsymbol{e}_1^{i+1}, \boldsymbol{e}_2^{i+1}, \boldsymbol{e}_3^{i+1}\}$, respectively. The incremental torsion parameter is denoted by φ_Δ. With $\boldsymbol{e}_3^{i+1} = \boldsymbol{Q}_\Delta \boldsymbol{e}_3^i$, one arrives at the important result

$$\boldsymbol{e}_3^{i+1} - \boldsymbol{e}_3^i = \boldsymbol{\alpha}_\Delta \times \boldsymbol{e}_3^m, \quad \text{where} \quad \boldsymbol{e}_3^m = \frac{1}{2}\left(\boldsymbol{e}_3^{i+1} + \boldsymbol{e}_3^i\right). \tag{11}$$

It is remarked that e_3^m is not a unitary vector, but $\left\|e_3^m\right\|^{-1} e_3^m$ is. From [1] we recover the following definition

$$\alpha_\Delta = \left\|e_3^m\right\|^{-2} \left(e_3^i \times e_3^{i+1}\right) + \varphi_\Delta \left\|e_3^m\right\|^{-1} e_3^m. \tag{12}$$

Note that $\varphi_\Delta = \left\|e_3^m\right\|^{-1} \alpha_\Delta \cdot e_3^m$.

The deformation gradient can then be expressed by

$$F = QF^r, \qquad \text{where} \qquad F^r = I + \gamma^r \otimes e_3^r \tag{13}$$

is the back-rotated deformation gradient, I is the identity tensor and

$$\gamma^r = \eta^r + \kappa^r \times r^r, \tag{14}$$

are back-rotated cross section strains of Biot type. In (14) the following generalized back-rotated strain has been introduced

$$\eta^r = Q^T \left(z' - e_3\right) \qquad \text{and} \qquad \kappa^r = \text{axial}\left(Q^T Q'\right). \tag{15}$$

We remark that $\eta^r \cdot e_\alpha^r = 0$ due to the Bernoulli–Euler assumption. Note that

$$\eta^r = \varepsilon e_3^r, \qquad \text{where} \qquad \varepsilon = \left\|z'\right\| - 1. \tag{16}$$

Equation (14) and (15) are the back-rotated counterparts of the following cross-sectional generalized strains,

$$\gamma = \eta + \kappa \times r, \qquad \eta = z' - e_3 \qquad \text{and} \qquad \kappa = \text{axial}\left(Q' Q^T\right). \tag{17}$$

κ is the curvature vector of the rod axis at the current configuration, which, with $\alpha_\Delta^2 = \alpha_\Delta \cdot \alpha_\Delta$, is given by

$$\kappa = \Xi_\Delta \alpha'_\Delta, \qquad \text{where} \qquad \Xi_\Delta = \frac{4}{4 + \alpha_\Delta^2}\left(I + \frac{1}{2} A_\Delta\right). \tag{18}$$

The back-rotated-curvature vector needs to be updated at instant t_{i+1} from the curvature vector at instant κ_i. This is done through the following expression

$$\kappa_{i+1}^r = \kappa_i^r + Q_i^T \Xi_\Delta^T \alpha'_\Delta. \tag{19}$$

Let the 1st Piola-Kirchhoff stress tensor be expressed by its columns as follows

$$P = \tau_i \otimes e_i^r = Q\left(\tau_i^r \otimes e_i^r\right) \tag{20}$$

One can now introduce the back-rotated 1st Piola–Kirchhoff stress tensor by

$$\boldsymbol{P}^r = \boldsymbol{Q}^T \boldsymbol{P} = \boldsymbol{\tau}_i^r \otimes \boldsymbol{e}_i^r, \qquad \text{where} \qquad \boldsymbol{\tau}_i^r = \boldsymbol{Q}^T \boldsymbol{\tau}_i, i = 1, 2, 3 \tag{21}$$

are the back-rotated nominal stress vectors.

The following cross-sectional resultants are obtained by integration of the stresses $\boldsymbol{\tau} = \boldsymbol{\tau}_3$ on the cross section

$$\boldsymbol{n} = \int_A \boldsymbol{\tau} dA \qquad \text{and} \qquad \boldsymbol{m} = \int_A \boldsymbol{r} \times \boldsymbol{\tau} dA. \tag{22}$$

with, $\boldsymbol{n}$ the true forces and $\boldsymbol{m}$ the true moments that are acting on a cross-section. The back-rotated counterparts of the cross-section forces and moments are, respectively

$$\boldsymbol{n}^r = \int_A \boldsymbol{\tau}^r dA \qquad \text{and} \qquad \boldsymbol{m}^r = \int_A \boldsymbol{r}^r \times \boldsymbol{\tau}^r dA. \tag{23}$$

The internal virtual work on a domain $\Omega \subset \mathbb{R}$ is given by

$$\delta W_{\text{int}}^{\Omega} = \int_{\Omega} \boldsymbol{\sigma}^r \cdot \delta \boldsymbol{\varepsilon}^r d\Omega, \tag{24}$$

where

$$\boldsymbol{\sigma}^r = \begin{bmatrix} \boldsymbol{n}^r \\ \boldsymbol{m}^r \end{bmatrix} \qquad \text{and} \qquad \delta \boldsymbol{\varepsilon}^r = \begin{bmatrix} \delta \boldsymbol{\eta}^r \\ \delta \boldsymbol{\kappa}^r \end{bmatrix}. \tag{25}$$

The external virtual work on a domain is given by

$$\delta W_{\text{ext}}^{\Omega} = \int_{\Omega} \left(\bar{\boldsymbol{n}}^{\Omega} \cdot \delta \boldsymbol{u} + \bar{\boldsymbol{m}}^{\Omega} \cdot \boldsymbol{\Xi}_{\Delta} \delta \boldsymbol{\alpha}_{\Delta} \right) d\Omega. \tag{26}$$

The local equilibrium equations of the rod are obtained by applying the Virtual Work Theorem valid for statics as follows

$$\delta W_{\text{int}}^{\Omega} - \delta W_{\text{ext}}^{\Omega} = \delta W_{\text{ext}}^{\Gamma}, \qquad \forall \delta \boldsymbol{u}, \delta \boldsymbol{\alpha}_{\Delta} \qquad \text{in } \Omega. \tag{27}$$

where $\delta W_{\text{ext}}^{\Gamma}$ is the external virtual work on the boundary.

3 Finite Element Implementation

The simulations presented in this paper were performed within the AceFEM finite element software. Both AceGen and AceFEM programs are developed and maintained by Jože Korelc (University of Ljubljana) [3].

The rod elements proposed in the present work were designed to seamless couple with both primary and reduced geometrically exact shell elements of the TUBA family [2]. These are designated by BE Pn-Pm, meaning Bernoulli–Euler element

Table 1 Family of rod finite elements and corresponding TUBA elements

$\boldsymbol{u}$ order	Primary element			Reduced element		
	BE	TUBA	Configuration	BE	TUBA	Configuration
5th	P5–P4	6	○ 1 – ☆ 3 – ○ 2	P5–P3	3	○ 1 – ○ 2
6th	P6–P5	13	○ 1 – ☆ 3 – • 4 – ☆ 5 – ○ 2	P6–P4	7	○ 1 – □ 3 – ○ 2
7th	P7–P6	15	○ 1 – ◇ 3 – ☆ 4 – ◇ 5 – ○ 2	P7–P5	12	○ 1 – ◇ 3 – ◇ 4 – ○ 2

Legend: ○ $\boldsymbol{u}, \boldsymbol{u}', \boldsymbol{u}'', \varphi_\Delta, \varphi'_\Delta$ • $\boldsymbol{u}$ ☆ φ_Δ □ $\boldsymbol{u}, \varphi_\Delta$ ◇ $\boldsymbol{u}', \varphi_\Delta$

with polynomial approximations of degree n and m for the displacement vector $\boldsymbol{u}$ and incremental torsion angle φ_Δ, respectively, see Table 1.

For the connection between elements only the two end-nodes of the finite element are considered. For this purpose 7 DOF's at each end-node are elected, specifically $\boldsymbol{u}$, $\boldsymbol{\alpha}_\Delta$ and ε, as in [1]. At nodes $I = 1, 2$, with the nodal values $\boldsymbol{\alpha}^i_I, \boldsymbol{\alpha}_{\Delta I}, \varepsilon^i_I$ and ε^{i+1}_I, one gets $\boldsymbol{e}^i_{iI}, \boldsymbol{e}^{i+1}_{iI}, \boldsymbol{e}^m_{3I}, \varphi_{\Delta I}, \boldsymbol{u}'^i_I$ and $\boldsymbol{u}'^{i+1}_I$. Displacements $\boldsymbol{u}$ and the torsion parameter φ_Δ are approximated by different polynomials shown in Table 1. Hence, an element has in total $2 + 3(n + 1) + (m + 1)$ DOF's. Along the rod, one can compute $\boldsymbol{u}^{i+1}, \boldsymbol{u}'^{i+1}, \boldsymbol{u}''^{i+1}, \boldsymbol{u}^i, \boldsymbol{u}'^i, \boldsymbol{u}''^i, \varphi_\Delta$ and φ'_Δ with the aid of Table 1 and their derivatives. Afterwards, $\boldsymbol{\alpha}_\Delta$ and $\boldsymbol{\alpha}'_\Delta$ are obtained. Finally, on the domain, with aid of (10) and (18), one can compute ε^{i+1}, $\boldsymbol{\eta}^r_{i+1}$ and $\boldsymbol{\kappa}^r_{i+1}$. At each integration point, $\boldsymbol{\alpha}_i$ and $\boldsymbol{\kappa}^r_i$ must be stored. In general, only $\boldsymbol{u}$ and $\boldsymbol{\alpha}_\Delta$ must be connected by neighboring elements. ε can be shared only in the case of smooth connection of 2 elements, with no change of cross-sectional properties and no nodal external forces. In the example a 4 Gauss–Lobatto integration point scheme was used, but other schemes can be used without altering the results.

4 Numerical Example—Lateral Buckling Analysis

The cantilever problem in Fig. 2 is considered in the following. This is a classical buckling problem. Determining the critical point appears to be a standard benchmark problem in this context and is thus considered in many publications. The geometrical and material parameters within this paper were adapted from [1]. To enter the post buckling equilibrium path, a perturbation load is applied. In all the presented forms of this problem, the perturbation load is active during the whole simulation but with a rather small magnitude of 10^{-4} times the primal load. The lateral buckling (displacement in z-direction) is shown in Fig. 3. The critical load can be determined at approximately 2.2 kN. Hence a good agreement with the literature can be observed.

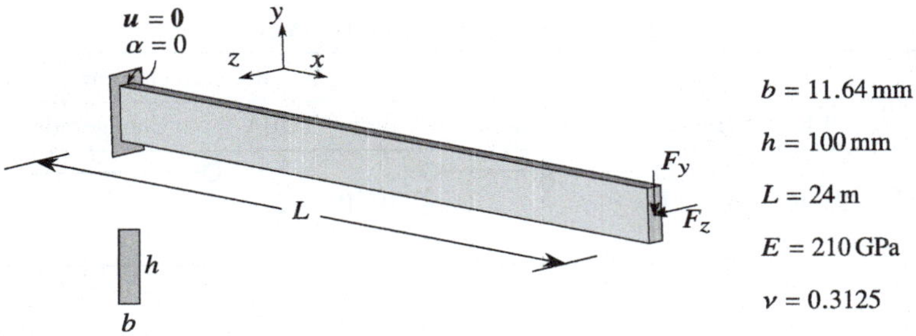

Fig. 2 3D cantilever problem

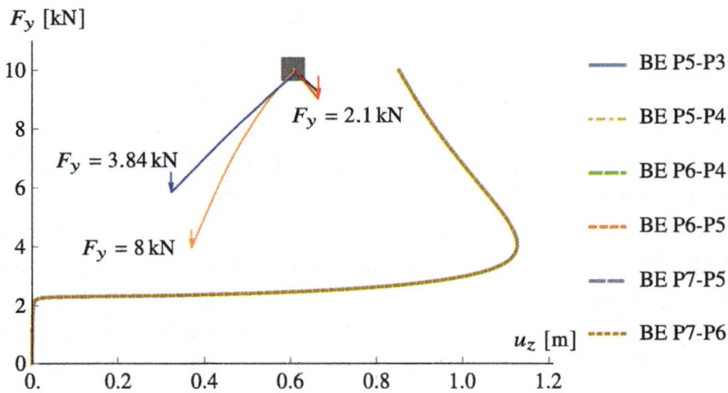

Fig. 3 Lateral Buckling of a cantilever: results for 16 elements discretization

5 Conclusions

The geometrically exact rod formulation presented in [1] was extended to different polynomial approximations that are similar to those used in TUBA elements of [2].

References

1. da Costa e Silva, C., Maassen, S. F., Pimenta, P. M., & Schröder, J. (2020). A simple finite element for the geometrically exact analysis of Bernoulli-Euler rods. *Computational Mechanics, 65*, 905–923. https://doi.org/10.1007/s00466-019-01800-5.
2. Ivannikov, V., Tiago, C., & Pimenta, P. M. (2015). Generalization of the C^1 TUBA plate finite elements to the geometrically exact Kirchhoff-Love shell model. *Computer Methods in Applied Mechanics and Engineering, 294*, 210–244. https://doi.org/10.1016/j.cma.2015.05.018.
3. Korelc, J., & Wriggers, P. (2016). *Automation of finite element methods*. Berlin: Springer.

Finite Element Formulations for Beam-to-Solid Interaction–from Embedded Fibers Towards Contact

Alexander Popp and Ivo Steinbrecher

Dedicated to Prof. Peter Wriggers on the occasion of his 70th birthday. Since the very first day of his Ph.D. research in computational contact mechanics, the first author has been inspired by the internationally acknowledged, innovative work of Prof. Peter Wriggers in this field (as in many other fields). Later, a successful collaboration has emerged in the organization of scientific events such as the CISM Advanced Course "Computational Contact and Interface Mechanics" in 2016 and the biennial "International Conference on Computational Contact Mechanics (ICCCM)" under the auspices of ECCOMAS.

Abstract Contact and related phenomena, such as friction, wear or elastohydrodynamic lubrication, remain as one of the most challenging problem classes in nonlinear solid and structural mechanics. In the context of their computational treatment with finite element methods (FEM) or isogeometric analysis (IGA), the inherent non-smoothness of contact conditions, the design of robust discretization approaches as well as the implementation of efficient solution schemes seem to provide a never ending source of hard nuts to crack. This is particularly true for the case of beam-to-solid interaction with its mixed-dimensional 1D-3D contact models. Therefore, this contribution gives an overview of current steps being taken, starting from state-of-the-art beam-to-beam (1D) and solid-to-solid (3D) contact algorithms, towards a truly general 1D-3D beam-to-solid contact formulation.

A. Popp (✉) · I. Steinbrecher
Institute for Mathematics and Computer-Based Simulation (IMCS), University of the Bundeswehr Munich, Werner-Heisenberg-Weg 39, 85577 Neubiberg, Germany
e-mail: alexander.popp@unibw.de

I. Steinbrecher
e-mail: ivo.steinbrecher@unibw.de

F. Aldakheel et al. (eds.), *Current Trends and Open Problems in Computational Mechanics*, https://doi.org/10.1007/978-3-030-87312-7_41

1 Introduction

Computational contact mechanics is a particularly challenging research field in nonlinear solid and structural mechanics that has been shaped over the course of almost four decades now by Professor Peter Wriggers and his many co-workers as well as collaborators [1, 2]. The interest of most researchers has in large parts been directed towards 3D solid-to-solid contact with the aim of accurately resolving large deformation contact problems of elastic and inelastic engineering components, manufacturing processes or biomechanical systems. Nevertheless, a smaller yet constant stream of innovative contributions on 1D beam-to-beam contact can also be observed, which focus more on the significance of contact interaction of slender rod-wise components such as ropes, cables, fiber webbings or again biological structures on different length scales.

Among the most important research topics for 3D solid-to-solid contact has always been the choice of a robust contact discretization scheme suitable for large deformation kinematics, with some of the famous variants being node-to-segment, Gauss-point-to-segment or segment-to-segment schemes to name only a few [3, 4]. Within the last two decades, so-called mortar methods originally proposed in the context of domain decomposition have emerged as widely accepted state-of-the-art contact discretization approach [5–10]. Equally important is the question of contact constraint enforcement within the underlying variational formulation, where some common options are penalty methods, Lagrange multiplier methods, the Augmented Lagrangian approach and, more recently, Nitsche's method [11, 12]. In recent years, NURBS-based isogeometric analysis has beome increasingly popular also in computational contact mechanics due to its higher-order continuity, which can be advantageous in the accurate resolution of curved contact boundaries [13–16]. Without any claim of completeness, other relevant research directions in the field of 3D solid-to-solid contact include mesh adaptivity [17], contact smoothing [18], third-medium contact [19] and virtual element methods (VEM) [20].

When considering contact interaction of 1D slender rod-like structures, the associated challenges are quite different, both from a mechanical and from a computational perspective. In particular, the underlying 1D Cosserat continuum theory of beams requires a thorough re-formulation of contact and friction kinematics, which becomes rather complex in the fully nonlinear realm with finite deformations and finite cross-section rotations. Early contributions can be found in [21–23]. In recent years, such 1D beam-to-beam contact formulation have again been investigated more closely, partly due to progress in the development of geometrically exact beam finite elements and partly due to the high relevance of fiber-based structures and materials for many modern engineering applications [24–26].

It is striking that the combined treatment of solid-based contact and beam-based contact in the sense of a 1D-3D beam-to-solid contact formulation has received much less attention over the years, which can of course be attributed to the fact that it combines the complexities associated with both building blocks [27–30]. The authors of this contribution have recently started research efforts to develop a fully nonlinear 1D-3D beam-to-solid contact formulation including friction. In the following, an overview of the first steps taken in this direction is given, i.e. tied contact formulations for 1D fibers being embedded into 3D solid bodies [31]. Moreover, several numerical examples illustrate the interplay of the resulting 1D-3D beam-to-solid volume coupling approach with well-established solid-based contact and beam-based contact models, thus showing the path towards truly general 1D-3D beam-to-solid contact interaction.

2 Governing Equations

We consider a quasi-static 3D finite deformation beam-to-solid volume coupling problem, cf. Fig. 1. The weak form is derived by the principle of virtual work with contributions from the solid, the beam, and the coupling terms.

The solid is modeled as a 3D continuum, represented by the open set $\Omega_0^S \subset \mathbb{R}^3$ in the reference configuration. The volume of the embedded beam is not explicitly subtracted from the solid volume, thus resulting in a modeling error due to overlapping material points. However, the high ratio of beam stiffness to solid stiffness typically alleviates the impact of this error for the envisaged applications. The deformed position $\underline{\boldsymbol{x}}^S = \underline{\boldsymbol{X}}^S + \underline{\boldsymbol{u}}^S$ is related to the reference position $\underline{\boldsymbol{X}}^S$ through the displacement field $\underline{\boldsymbol{u}}^S$. Virtual work contributions ∂W^S of the solid are given by

$$\partial W^S = \int_{\Omega_0^S} \boldsymbol{S} : \partial \boldsymbol{E} \mathrm{d}V - \partial W_{\mathrm{ext}}^S, \tag{1}$$

where δ denotes the variation of a quantity, $\boldsymbol{S}$ the second Piola-Kirchhoff stress tensor, $\boldsymbol{E}$ the energy-conjugate Green-Lagrange strain tensor and $\partial W_{\mathrm{ext}}^S$ the virtual work of the external forces. The Green-Lagrange strain tensor is given as $\boldsymbol{E} = \frac{1}{2}(\boldsymbol{F}^{\mathrm{T}}\boldsymbol{F} - \boldsymbol{I})$, with $\boldsymbol{F} = \partial \underline{\boldsymbol{x}}^S / \partial \underline{\boldsymbol{X}}^S$ being the material deformation gradient and $\boldsymbol{I} \in \mathbb{R}^{3\times 3}$ the identity tensor, respectively. For the compressible or nearly incompressible solid, we assume a hyperelastic strain energy function $\Psi(\boldsymbol{E})$, which is related to the second Piola-Kirchhoff stress tensor via $\boldsymbol{S} = \partial \Psi(\boldsymbol{E})/\partial \boldsymbol{E}$.

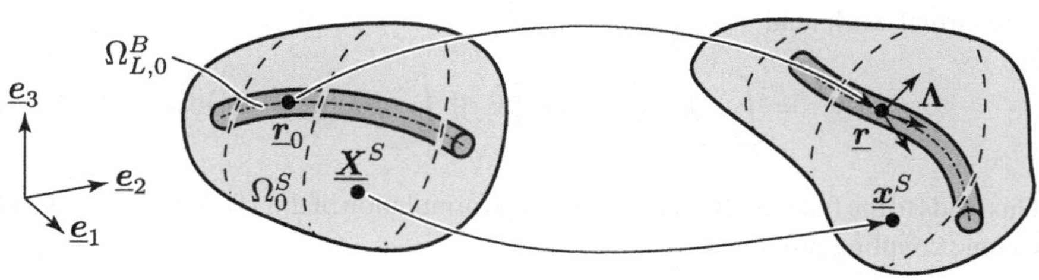

Fig. 1 Notation of the finite deformation beam-to-solid volume coupling problem

The beams used in this work are based on the fully nonlinear, geometrically exact beam theory, which in turn builds upon the kinematic assumption of plane, rigid cross-sections. The complete beam kinematics can be defined by a centerline curve $\underline{\boldsymbol{r}}(s) \in \mathbb{R}^3$, connecting the cross-section centroids, and a field of right-handed orthonormal triads $\boldsymbol{\Lambda}(s) \in SO^3$ defining the rotation of the cross-sections. Here, $s \in [0, L] =: \Omega^B_{L,0} \subset \mathbb{R}$ is the arc-length along the undeformed beam centerline and $\boldsymbol{\Lambda}(s)$ is a rotation tensor, which maps the global Cartesian basis vectors onto the local cross-section basis vectors. The beam centerline displacement $\underline{\boldsymbol{u}}^B_r$ relates the undeformed beam centerline position $\underline{\boldsymbol{r}}_0$ to the deformed position $\underline{\boldsymbol{r}} = \underline{\boldsymbol{r}}_0 + \underline{\boldsymbol{u}}^B_r$. The beam contribution to the global virtual work reads

$$\delta W^B = \delta \Pi^B_{\text{int}} - \delta W^B{}_{\text{ext}}, \tag{2}$$

where $\delta \Pi^B_{\text{int}}$ is the variation of the beam's internal elastic energy and $\delta W^B{}_{\text{ext}}$ is the virtual work of external forces and moments on the beam. The elastic beam energy Π^B_{int} depends on the employed beam theory, namely either Simo-Reissner, Kirchoff-Love or Euler-Bernoulli (torsion free) theories, cf. [32].

In the real physical problem, the beam surface (defined by the centerline fields $\underline{\boldsymbol{r}}$ and $\boldsymbol{\Lambda}$) is tied to an underlying, internal solid surface. Modeling this type of surface-to-surface interaction would result in a computationally expensive evaluation of the coupling terms, thus canceling out the advantages of employing a 1D beam theory. For the inherent assumption in the beam-to-solid volume coupling method, that the beam cross-section dimensions are small compared to the other dimensions of the problem, we approximate the coupling of the beam surface and the solid volume as a coupling between the beam centerline and the solid volume. This is a significant change in the mathematical description of the mechanical model, and for the implications of this choice on the applicability and spatial convergence of the proposed method the interested reader is referred to [31]. The kinematic coupling constraints formulated along the beam centerline $\Omega^B_{L,0}$ in the reference configuration read

$$\underline{\boldsymbol{u}}^B_r - \underline{\boldsymbol{u}}^S = \underline{\boldsymbol{0}} \quad \text{on} \quad \Omega^B_{L,0}. \tag{3}$$

The constraints are enforced weakly via a Lagrange multiplier field $\underline{\boldsymbol{\lambda}}(s) \in \mathbb{R}^3$ defined on the beam centerline. We point out that from a physical point of view $\underline{\boldsymbol{\lambda}}$ represents a distributed line load along the beam centerline. Coupling contributions ∂W^C to the total virtual work read

$$\partial W^C = \int_{\Omega^S_{L,0}} \underline{\boldsymbol{\lambda}} \left(\delta \underline{\boldsymbol{u}}^B_r - \delta \underline{\boldsymbol{u}}^S\right) \mathrm{d}s + \int_{\Omega^S_{L,0}} \delta \underline{\boldsymbol{\lambda}} \left(\underline{\boldsymbol{u}}^B_r - \underline{\boldsymbol{u}}^S\right) \mathrm{d}s. \tag{4}$$

This leads to the final saddle point-type weak formulation of the 1D-3D beam-to-solid volume coupling problem,

$$\partial W^S + \delta W^B + \partial W^C = 0. \tag{5}$$

3 Spatial Discretization

We build upon the finite element method here for spatial discretization. In the solid field either standard C^0-continuous finite elements or a C^1-continuous isogeometric approach based on second-order NURBS is used. The beam centerlines are discretized using C^1-continuous finite elements based on third-order Hermite polynomials. For details on the objective and path-independent interpolation of the rotational field along the beam centerline, see [32]. Employing a mortar-type coupling approach [33], the Lagrange multipliers are also approximated with a finite element interpolation. In this work, linear Lagrange polynomials are used to interpolate the Lagrange multiplier field, cf. [34], thus yielding discrete mortar coupling matrices $\boldsymbol{D}$ and $\boldsymbol{M}$ associated with the beam and solid side, respectively. In the spirit and nomenclature of classical mortar methods, the beam is treated as the slave side and the solid as the master side, respectively. The linearized system of equations to be solved in every Newton iteration exhibits saddle-point structure and reads

$$\begin{bmatrix} \boldsymbol{K}_{SS} & \boldsymbol{0} & -\boldsymbol{M}^{\mathrm{T}} \\ \boldsymbol{0} & \boldsymbol{K}_{BB} & \boldsymbol{D}^{\mathrm{T}} \\ -\boldsymbol{M} & \boldsymbol{D} & \boldsymbol{0} \end{bmatrix} \begin{bmatrix} \Delta\boldsymbol{d}^S \\ \Delta\boldsymbol{d}^B \\ \boldsymbol{\lambda} \end{bmatrix} = \begin{bmatrix} -\boldsymbol{r}^S \\ -\boldsymbol{r}^B \\ \boldsymbol{M}\boldsymbol{d}^S - \boldsymbol{D}\boldsymbol{d}^B \end{bmatrix}, \tag{6}$$

where $\boldsymbol{d}^S$, $\boldsymbol{d}^B$, $\Delta\boldsymbol{d}^S$ and $\Delta\boldsymbol{d}^B$ are the displacements and their increments of solid and beam, $\boldsymbol{r}^S$ and $\boldsymbol{r}^B$ denote the discrete residual vectors, $\boldsymbol{K}_{SS} = \partial\boldsymbol{r}^S/\partial\boldsymbol{d}^S$ and $\boldsymbol{K}_{BB} = \partial\boldsymbol{r}^B/\partial\boldsymbol{d}^B$ are the tangent stiffness matrices, and $\boldsymbol{\lambda}$ are the discrete Lagrange multiplier values. The system (6) is solved by introducing a weighted penalty regularization.

4 Numerical Examples

The following numerical examples are set up using the beam finite element pre-processor MeshPy [36] and are simulated with our in-house parallel multi-physics research code BACI [35].

First, we investigate the twisting of a coated rope consisting of 7×7 individual fibers. This is an extension of the rope example proposed in [26] and, unless stated otherwise, all parameters are taken from said reference.

The initial configuration of the 49 straight fibers is shown in Fig. 2a. Each fiber has the length $L = 2.5\,\mathrm{m}$ and is discretized using 10 torsion free beam elements. The All-Angle Beam Contact (ABC) formulation [26] is used to model the beam-to-beam contact which arises in this example. In the first stage of the simulation the fibers are loaded in axial direction and each of the seven sub-bundles of fibers is twisted around its center fiber by two full rotations. This twisting process is realized in a Dirichlet controlled manner. The solution is obtained within 40 quasi-static load steps and is displayed in Fig. 2b. In the next stage the sub-bundles themselves are rotated around the rope axis by an additional full rotation, cf. Fig. 2c. Again this

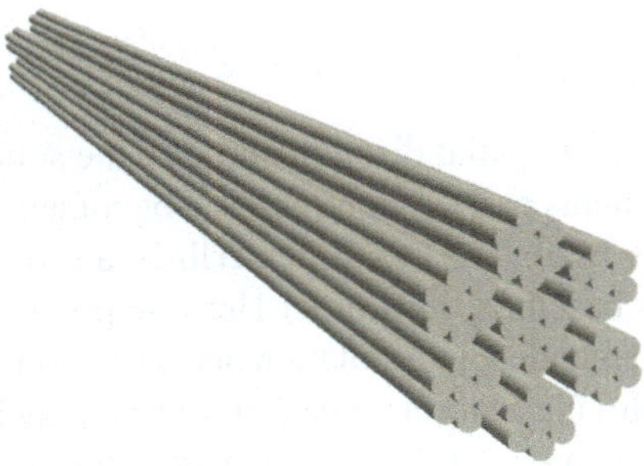

(a) Reference configuration.

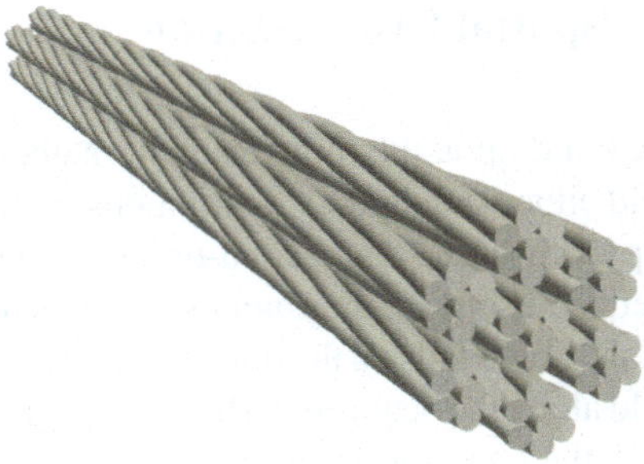

(b) Configuration at load step 40.

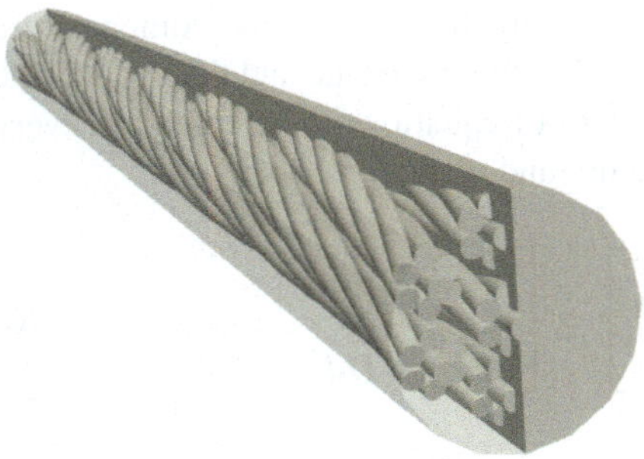

(c) Configuration at load step 60.

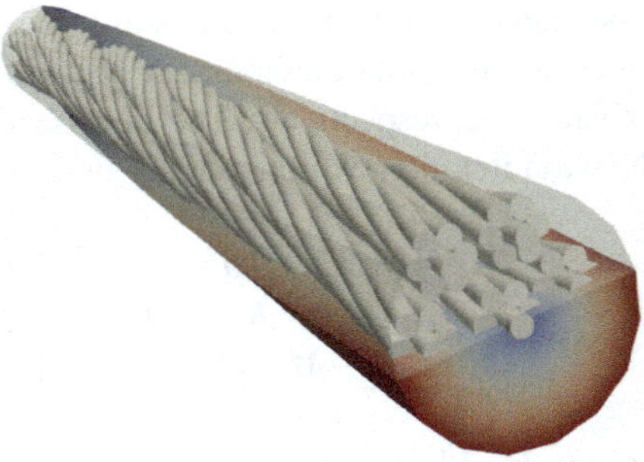

(d) Configuration at load step 80.

Fig. 2 Configuration of the coated rope example at different stages during the simulation

is realized with Dirichlet boundary conditions and the solution is obtained within 20 additional quasi-static load steps. Up to this point the model consists only of beam elements. In the next and final stage the twisted wires are covered with a solid coating ($E_{\text{solid}} = E_{\text{beam}}/200$). The coating and the (deformed) beams are coupled to each other via the beam-to-solid volume coupling method using first-order interpolation of the Lagrange multipliers. The Dirichlet conditions at the end of the fibers are replaced by Neumann boundaries matching the reaction forces at the Dirichlet conditions in the last step of the twisting process. Therefore, the bundle of wires is in equilibrium with itself and the external loads, i.e. there is no initial interaction between the solid coating and the wires. The Neumann load at the end of the fibers is now linearly reduced to zero within 20 quasi-static load steps. Now the solid coating interacts with the beam fibers in the sense of a pre-stressed composite material. This results in a back-twisting of the rope and coating of about a quarter rotation as displayed in Fig. 2d. The normalized (to a maximum value of 1) reaction torque at the fixed end of the wire is plotted in Fig. 3. One can observe that even tough the external loads are decreased linearly (in load steps 60–80) the resulting reaction torque exhibits a nonlinear behavior, thus highlighting the complexity of this problem. This example illustrates the combination of beam-to-beam contact with beam-to-solid volume coupling and also shows the maturity of the beam-to-solid

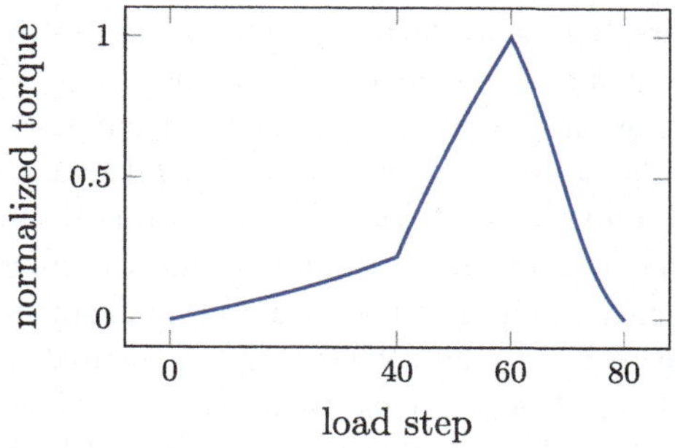

Fig. 3 Resulting (normalized) torque at the fixed end of the rope over the course of the simulation

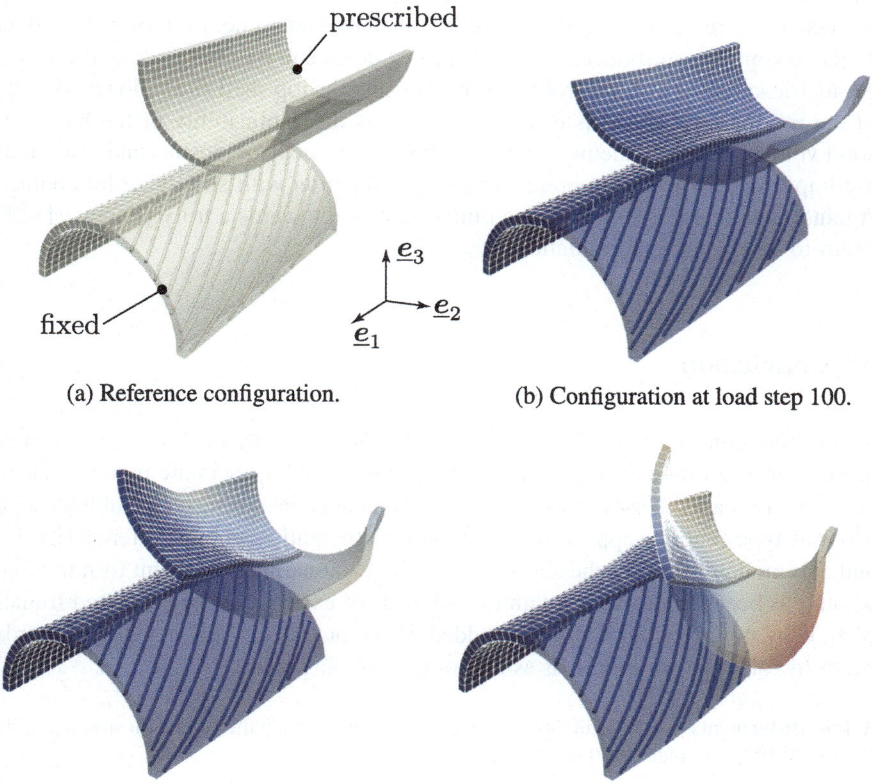

(a) Reference configuration.

(b) Configuration at load step 100.

(c) Configuration at load step 150.

(d) Configuration at load step 200.

Fig. 4 Configuration of the reinforced shell example during different stages of the simulation

volume coupling method for complex beam and solid geometries as well as real life engineering problems. Therefore, it can be seen as an important step towards a truly general 1D-3D beam-to-solid contact formulation.

In the second example, two shells with the shape of hollow half-cylinders are pressed against each other. The two shells have the same spatial dimensions as well as material parameters, but one of them is reinforced with stiff ($E_{\text{fiber}} = 200E_{\text{shell}}$) fibers, which are oriented at 45° relative to the cylinder axis. The reference configuration is shown in Fig. 4a. The reinforced shell is fixed at one end and the other shell is moved towards the fixed one with prescribed displacements in negative $\underline{e}_3$ direction. Solid-shell finite elements [37, 38] are used to discretize the shells, and both shells employ the same hyperelastic Saint Venant–Kirchhoff material. The reinforcements are explicitly discretized with beam finite elements and coupled to the shell elements via the beam-to-solid volume coupling method. The 3D solid-to-solid (surface-to-surface) contact between the shells is modeled with a state-of-the-art mortar approach using standard Lagrange multipliers [7, 9]. The problem is solved using 200 quasi-static load steps and Fig. 4b-d show the configuration of the two shells at different stages during the simulation. The reinforcement effects can clearly be seen as the reinforced shell deforms less than the other one. Moreover, the effect of the asymmetry introduced by the reinforcements can be observed, as the initial geometric symmetry of the problem is broken and the top shell slides down one side of the reinforced shell. This example showcases the applicability of the beam-to-solid volume coupling method to model fiber-reinforced composites and also gives a glimpse on the possible applications in combination with solid-to-solid contact. Again, it can therefore be seen as an important step towards a truly general 1D-3D beam-to-solid contact formulation.

5 Conclusion

Within this contribution we have presented state-of-the-art finite element formulations for beam-to-solid interaction. Specifically, slender fibers are modeled using efficient 1D beam theories and subsequently embedded inside 3D solid volumes with a mortar-type coupling approach. This allows to efficiently combine different (i.e. 1D and 3D) modeling approaches into a mixed-dimensional finite element formulation, which has been demonstrated with two illustrative examples. The presented framework is by no means limited to embedded fibers, but is currently extended towards beam-to-solid contact problems as part of our ongoing research.

Acknowledgements Sketches in this work have been created using the Adobe Illustrator plug-in LaTeX2AI (https://github.com/latex2ai/latex2ai).

References

1. Wriggers, P. (2006). *Computational contact mechanics*. Berlin, Heidelberg: Springer.
2. Popp, A., & Wriggers, P. (eds.) (2018). *contact modeling for solids and particles CISM international centre for mechanical sciences* (vol. 585). Springer International Publishing.
3. Simo, J. C., Wriggers, P., & Taylor, R. L. (1985). A perturbed Lagrangian formulation for the finite element solution of contact problems. *Computer Methods in Applied Mechanics and*

Engineering, 50, 163–180. August.
4. Wriggers, P., Vu Van, T., & Stein, E. (1990). Finite element formulation of large deformation impact-contact problems with friction. *Computers & Structures, 37*, 319–331.
5. Fischer, K. A., & Wriggers, P. (2005). Frictionless 2D contact formulations for finite deformations based on the mortar method. *Computational Mechanics, 36*, 226–244. August.
6. Popp, A., Gee, M. W., & Wall, W. A. (2009). A finite deformation mortar contact formulation using a primal-dual active set strategy. *International Journal for Numerical Methods in Engineering, 79*, 1354–1391.
7. Popp, A., Gitterle, M., Gee, M. W., & Wall, W. A. (2010). A dual mortar approach for 3D finite deformation contact with consistent linearization. *International Journal for Numerical Methods in Engineering, 83*, 1428–1465.
8. Popp, A., Wohlmuth, B. I., Gee, M. W., & Wall, W. A. (2012). Dual quadratic mortar finite element methods for 3d finite deformation contact. *SIAM Journal on Scientific Computing, 34*, B421–B446.
9. Popp, A., & Wall, W. A. (2014). Dual mortar methods for computational contact mechanics - overview and recent developments. *GAMM-Mitteilungen, 37*, 66–84.
10. Farah, P., Popp, A., & Wall, W. A. (2015). Segment-based vs. element-based integration for mortar methods in computational contact mechanics. *Computational Mechanics, 55*, 209–228.
11. Zavarise, G., & Wriggers, P. (1999). A superlinear convergent augmented Lagrangian procedure for contact problems. *Engineering Computations, 16*, 88–119.
12. Wriggers, P., & Zavarise, G. (2008). A formulation for frictionless contact problems using a weak form introduced by Nitsche. *Computational Mechanics, 41*, 407–420.
13. Temizer, I., Wriggers, P., & Hughes, T. J. R. (2011). Contact treatment in isogeometric analysis with NURBS. *Computer Methods in Applied Mechanics and Engineering, 200*, 1100–1112.
14. De Lorenzis, L., Wriggers, P., & Hughes, T. J. R. (2014). Isogeometric contact: a review. *GAMM-Mitteilungen, 37*, 85–123.
15. Seitz, A., Farah, P., Kremheller, J., Wohlmuth, B. I., Wall, W. A., & Popp, Alexander. (2016). Isogeometric dual mortar methods for computational contact mechanics. *Computer Methods in Applied Mechanics and Engineering, 301*, 259–280.
16. Wunderlich, L., Seitz, A., Deniz Alaydin, M., Wohlmuth, B., & Popp, A. (2019). Biorthogonal splines for optimal weak patch-coupling in isogeometric analysis with applications to finite deformation elasticity. *Computer Methods in Applied Mechanics and Engineering, 346*, 197–215.
17. Wriggers, P., & Scherf, O. (1995). An adaptive finite element algorithm for contact problems in plasticity. *Computational Mechanics, 17*, 88–97.
18. Wriggers, P., Krstulovic-Opara, L., & Korelc, J. (2001). Smooth C1-interpolations for two-dimensional frictional contact problems. *International Journal for Numerical Methods in Engineering, 51*, 1469–1495.
19. Wriggers, P., Schröder, J., & Schwarz, A. (2013). A finite element method for contact using a third medium. *Computational Mechanics, 52*, 837–847.
20. Wriggers, P., Rust, W. T., & Reddy, B. D. (2016). A virtual element method for contact. *Computational Mechanics, 58*, 1039–1050.
21. Wriggers, P., & Zavarise, G. (1997). On contact between three-dimensional beams undergoing large deflections. *Communications in Numerical Methods in Engineering, 13*, 429–438.
22. Zavarise, G., & Wriggers, P. (2000). Contact with friction between beams in 3-D space. *International Journal for Numerical Methods in Engineering, 49*, 977–1006.
23. Litewka, P., & Wriggers, P. (2002). Frictional contact between 3D beams. *Computational Mechanics, 28*, 26–39.
24. Gay Neto, A., Pimenta, P. M., & Wriggers, P. (2015). Self-contact modeling on beams experiencing loop formation. *Computational Mechanics, 55*, 193–208.
25. Meier, C., Popp, A., & Wall, W. A. (2016). A finite element approach for the line-to-line contact interaction of thin beams with arbitrary orientation. *Computer Methods in Applied Mechanics and Engineering, 308*, 377–413.

26. Meier, C., Wall, W. A., & Popp, A. (2017). A unified approach for beam-to-beam contact. *Computer Methods in Applied Mechanics and Engineering, 315*, 972–1010.
27. Ninic, J., Stascheit, J., & Meschke, G. (2014). Beam-solid contact formulation for finite element analysis of pile-soil interaction with arbitrary discretization. *International Journal for Numerical and Analytical Methods in Geomechanics, 38*, 1453–1476.
28. Konyukhov, A., & Schweizerhof, K. (2015). On some aspects for contact with rigid surfaces: Surface-to-rigid surface and curves-to-rigid surface algorithms. *Computer Methods in Applied Mechanics and Engineering, 283*, 74–105.
29. Steinbrecher, I., Humer, A., & Vu-Quoc, L. (2017). On the numerical modeling of sliding beams: A comparison of different approaches. *Journal of Sound and Vibration, 408*, 270–290.
30. Humer, A., Steinbrecher, I., & Vu-Quoc, L. (2020). General sliding-beam formulation: A non-material description for analysis of sliding structures and axially moving beams. *Journal of Sound and Vibration, 480*, 115341.
31. Steinbrecher, I., Mayr, M., Grill, M. J., Kremheller, J., Meier, C., & Popp, A. (2020). A mortar-type finite element approach for embedding 1D beams into 3D solid volumes. *Computational Mechanics, 66*, 1377–1398.
32. Meier, C., Popp, A., & Wall, W. A. (2019). Geometrically exact finite element formulations for slender beams: Kirchhoff-Love theory versus Simo-Reissner theory. *Archives of Computational Methods in Engineering, 26*, 163–243.
33. Wohlmuth, B. I. (2000). A mortar finite element method using dual spaces for the Lagrange multiplier. *SIAM Journal on Numerical Analysis, 38*, 989–1012.
34. Puso, M. A. (2004). A 3D mortar method for solid mechanics. *International Journal for Numerical Methods in Engineering, 59*, 315–336.
35. BaciWebsite (2021) BACI: A comprehensive multi-physics simulation framework. https://baci.pages.gitlab.lrz.de/website
36. Steinbrecher, I., & Popp, A. (2021) MeshPy—A general purpose 3D beam finite element input generator. https://compsim.gitlab.io/codes/meshpy
37. Bischoff, M., & Ramm, E. (1997). Shear deformable shell elements for large strains and rotations. *International Journal for Numerical Methods in Engineering, 40*, 4427–4449.
38. Vu-Quoc, L., & Tan, X. G. (2003). Optimal solid shells for non-linear analyses of multilayer composites. I. Statics. *Computer Methods in Applied Mechanics and Engineering, 192*, 975–1016.

Alternative Approaches to the Stabilization of Virtual Element Formulations for Hyperelasticity

B. Daya Reddy and Daniel van Huyssteen

For Peter Wriggers, on the occasion of his 70th birthday: in high esteem, and in appreciation of more than three decades of friendship, collaboration, generosity, and unwavering support of computational mechanics in South Africa.

Abstract This work constitutes a further contribution to developing convergent, locking-free virtual element formulations for problems of large-deformation nonlinear elasticity. Earlier work by the authors and others has introduced simple and effective forms of stabilization. This work investigates the approximation properties of the virtual element method for problems with domains having a high aspect ratio, and for alternative forms of stabilization: an existing approach, and two modifications. The results in all cases demonstrate the locking-free convergence of the approach, with the newer modifications leading to improvements in accuracy.

1 Introduction

The virtual element method (VEM), a relatively recent extension of the finite element method, has received considerable attention in recent years, with applications in solid mechanics that range from contact [1] and fracture [2, 3] to those involving nonlinear material behaviour with large deformations [4–7]. The attraction of the method lies in its flexibility in being able to accommodate polygonal or polyhedral elements that might deviate considerably from convex, as well as the ease in constructing the matrices necessary for solution. Furthermore, the method exhibits robust behaviour

B. D. Reddy (✉) · D. van Huyssteen
Centre for Research in Computational and Applied Mechanics, University of Cape Town, Rondebosch, South Africa
e-mail: daya.reddy@uct.ac.za

D. van Huyssteen
e-mail: Daniel.vanhuyssteen@uct.ac.za

F. Aldakheel et al. (eds.), *Current Trends and Open Problems in Computational Mechanics*, https://doi.org/10.1007/978-3-030-87312-7_42

and leads to convergent locking-free solutions in situations of near-incompressibility [8] and near-inextensibility [9–11] for transversely isotropic materials.

The typical lowest-order VEM formulation comprises a consistency term that depends on a projection of gradients of the unknown variable on to constants, together with a stabilization term that is required to address the rank-deficiency inherent in the consistency term. There exist various standard approaches to the construction of stabilizations for linear problems. In the case of nonlinear problems, recent works have shown the effectiveness of some simple approaches to stabilization [8, 10, 12], based on the use of a modified strain energy function. This approach has been extended in recent work [13], with a view to seeking greater robustness and improved convergence properties for problems of near-incompressibility. In the present work the approach adopted in [13] is used to study VEM approximations of a problem with high aspect ratio. This represents a severe test of the robustness of the VEM, especially for problems of near-incompressibility, and while the method yields results that are acceptable, two modifications to the stabilization term are explored as alternatives to arrive at results of greater accuracy. Results are presented with the use of the standard stabilization as well as the two modifications. Error plots demonstrate the locking-free behaviour as well as rates of convergence that are somewhat better than linear.

2 The Governing Equations

We summarize in this section the equations that govern equilibrium of an isotropic hyperelastic body [14], which occupies a polygonal domain $\Omega \subset \mathbb{R}^2$ in its undeformed configuration, with boundary Γ. The body undergoes the motion $\varphi(x, t) = X + u(X, t)$ where $\boldsymbol{X}$ and $\boldsymbol{x}$ are respectively the initial and current positions of a material point, and $\boldsymbol{u}$ is the displacement. The deformation gradient $\boldsymbol{F}$ is defined by $\boldsymbol{F} = \text{Grad}\, \boldsymbol{\varphi} = \boldsymbol{I} + \text{Grad}\, \boldsymbol{u}$ where $\boldsymbol{I}$ is the identity and Grad denotes the referential gradient.

The body is subjected to a body force $\boldsymbol{f}(\boldsymbol{X}, t)$ on its domain and boundary conditions

$$\boldsymbol{u} = \bar{\boldsymbol{u}} \quad \text{on}\ \Gamma_u\,, \qquad \boldsymbol{P}\boldsymbol{n} = \bar{\boldsymbol{t}} \quad \text{on}\ \Gamma_t\,, \tag{1}$$

where Γ_u and Γ_t are complementary subsets of the boundary Γ with unit outward normal $\boldsymbol{n}$; $\bar{\boldsymbol{u}}$ and $\bar{\boldsymbol{t}}$ are respectively a prescribed displacement and surface traction, and $\boldsymbol{P}$ is the first Piola-Kirchhoff stress.

The material response is specified through a strain energy function Ψ which depends on the right Cauchy-Green tensor $\boldsymbol{C} := \boldsymbol{F}^T\boldsymbol{F}$ through its invariants I_k $(k = 1, 2, 3)$, where $I_1 = \text{tr}\, C,\ \ I_2 = \frac{1}{2}[(\text{tr}\, \boldsymbol{C})^2 - \text{tr}\, \boldsymbol{C}^2]$ and $I_3 = \det \boldsymbol{C} = J^2$ and $J = \det \boldsymbol{F}$ is the Jacobian of the deformation gradient.

An equilibrium solution $\boldsymbol{u}$ of the hyperelasticity problem is then found as a minimizer of the total potential energy

$$U(\boldsymbol{v}, \text{Grad}\,\boldsymbol{v}) = \int_{\Omega} [\Psi(I_k(\text{Grad}\,\boldsymbol{v})) - \boldsymbol{f} \cdot \boldsymbol{v}]\, dx - \int_{\Gamma_t} \bar{\boldsymbol{t}} \cdot \boldsymbol{v}\, ds\,. \tag{2}$$

3 The Virtual Element Method

The domain Ω is partitioned into a mesh of non-overlapping polygons E with $\overline{\cup E} = \overline{\Omega}$. A generic element has a total of n_v vertices, with edges e_i connecting vertices V_i and V_{i+1} $(i = 1, \ldots, n_v)$ (mod n_v) (see Fig. 1a).

We construct a conforming approximation in a space $V^h \subset V$, comprising functions that are continuous on Ω, piecewise linear on edges e, and such that the Laplacian $\Delta \boldsymbol{v}_h := \text{Div}\,\text{Grad}\,\boldsymbol{v}$ vanishes on each element E:

$$V^h = \{\boldsymbol{v}_h \in V \mid \boldsymbol{v}_h \in [C(\Omega)]^2,\ \Delta \boldsymbol{v} = \boldsymbol{0} \text{ on } E,\ \boldsymbol{v}_h|_e \in P_1(e)\}\,. \tag{3}$$

Nodal values or degrees of freedom of the function $\boldsymbol{v}$ are defined at the element vertices, and it is convenient to write $\boldsymbol{v}_h|_{\partial E} = \mathsf{N}\,\mathsf{d}$ where d is the vector of degrees of freedom of element E and N is a matrix of local Lagrangian basis functions defined on the element boundary ∂E. The basis functions on elements are not defined explicitly, nor are they required in the computations.

We define the projection $\Pi_\nabla \boldsymbol{u}_h$ of the displacement gradient by its restriction $\Pi_\nabla \boldsymbol{u}_h|_E$ to element E, given by

$$\Pi_\nabla \boldsymbol{u}_h|_E = \frac{1}{|E|} \int_E \text{Grad}\,\boldsymbol{u}_h\, dx\,. \tag{4}$$

This is clearly constant on elements.

The VEM approximation to the minimization problem (2) is then obtained by introducing a suitably constructed energy $\widehat{\Psi}$ [8] so that

$$\Psi(\text{Grad}\,\boldsymbol{u}_h) \simeq \Psi_{\text{VEM}}(\text{Grad}\,\boldsymbol{u}_h) := \Psi(\Pi_\nabla \boldsymbol{u}_h) + \widehat{\Psi}(\text{Grad}\,\boldsymbol{u}_h) - \widehat{\Psi}(\Pi_\nabla \boldsymbol{u}_h)\,. \tag{5}$$

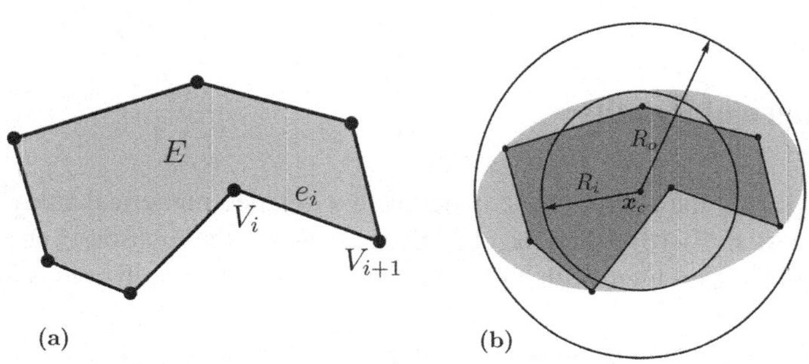

Fig. 1 **a** Element geometry; **b** quantities used in definition of aspect ratio

The first term on the right-hand side is a consistency term and the second and third serve as a stabilization term. The modified problem is now one of minimizing the functional

$$\widehat{U}(\boldsymbol{u}_h, \nabla \boldsymbol{u}_h) = \int_\Omega \Psi_{\mathrm{VEM}}(\mathrm{Grad}\, \boldsymbol{u}_h)\, dx - \int_\Omega \boldsymbol{f} \cdot \boldsymbol{u}_h\, dx - \int_{\Gamma_t} \bar{\boldsymbol{t}} \cdot \boldsymbol{u}_h\, ds\,. \tag{6}$$

Integration of the term involving $\widehat{\Psi}$ in (5) is carried out as proposed in [8], using one-point quadrature on triangular subdomains.

The strain energy function $\widehat{\Psi}$ has been further modified in [13]. Based on the neo-Hookean strain energy function, it is defined by

$$\widehat{\Psi} = \tfrac{\widehat{\mu}}{2}(I_1 - 3 - 2\ln J) + \tfrac{\widehat{\lambda}}{2}(J-1)^2\,, \tag{7}$$

where $\widehat{\lambda}$ and $\widehat{\mu}$ are modified Lamé parameters, defined by

$$\widehat{\lambda} = \frac{\Theta}{1+\Theta} T_5(\lambda), \qquad \widehat{\mu} = \frac{\Theta}{1+\Theta}(1+\alpha)^2 \mu\,, \qquad \Theta = \frac{2(1+\nu)}{\rho^2} \tag{8}$$

where $\rho = R_o/R_i$ is the element aspect ratio (see Fig. 1b), $T_5(\lambda)$ is the 5th-order Taylor series expansion of λ as a function of ν, about $\nu_0 = -0.25$, and $\alpha = T_5(\lambda)/E$ is a factor that modifies the stabilization energy for the case of near-incompressibility and for severe deformations. All quantities are evaluated and applied elementwise.

Two further modifications of the stabilization energy (7) will be considered in the numerical investigation, to account in alternative ways for high values of element aspect ratio. The first, denoted by $\widehat{\Psi}^*$, is defined by

$$\widehat{\Psi}^* = \widehat{\Psi}/\rho\,, \tag{9}$$

while the second, denoted by $\widehat{\Psi}^{**}$, is obtained by replacing Θ in $(8)_3$ by

$$\Theta^{**} = \frac{2(1+\nu)}{\rho^3}\,. \tag{10}$$

4 Numerical Results

In the work [13] the authors have undertaken a detailed numerical study based on the stabilization (7)–(8). Here, the objective is to assess the robustness and accuracy of the VEM for a problem with high aspect ratio, using the stabilizations introduced earlier.

This problem, previously studied in [8], concerns the behaviour of a thin cantilever of width 100 m and height 1 m that is subjected to a point load at one end and is fully constrained at the other. A neo-Hookean relation is assumed, with Young's modulus

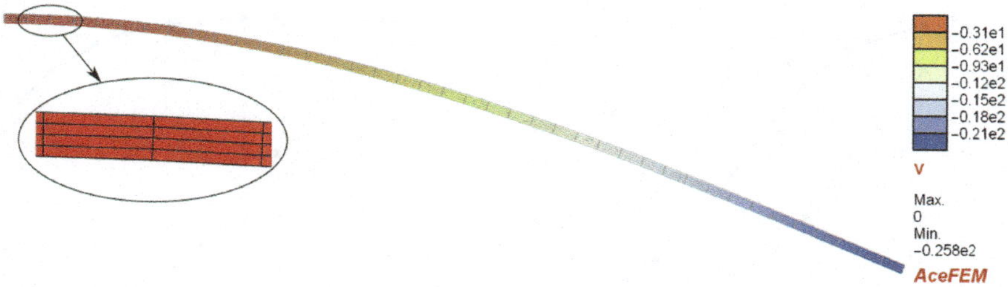

Fig. 2 Cantilever

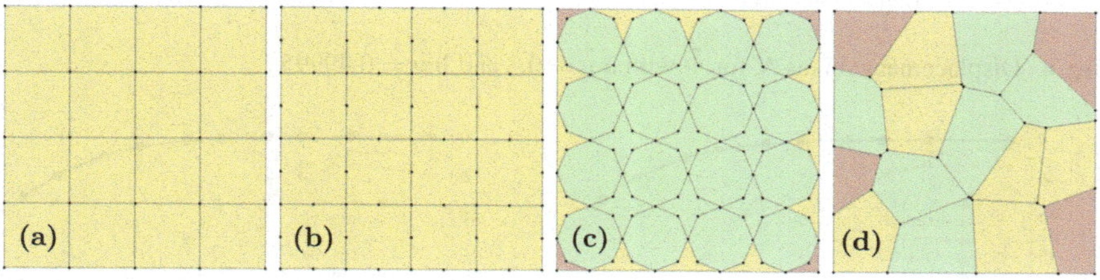

Fig. 3 Meshes considered

$E = 200$ Pa, and for various choices of Poisson's ratio. A point load $P = 0.002$ N is applied at the free end. Meshes of $10 \cdot 2^N \times 2^N$ elements are used, where N is the mesh refinement level. The nominal aspect ratio of the elements is thus $\rho = 10$. The deformed cantilever is shown in Fig. 2 for $\nu = 0.3$ with a mesh of four-noded rectangular elements for $N = 2$.

Figure 3 depicts the meshes employed: (a) structured four-noded rectangular virtual and finite elements (SQ1-VE and SQ1-FE), (b) structured eight-noded rectangular elements (SQ2S-VE), (c) a sun-and-star arrangement of elements (S&S-VE), and (d) elements constructed by Voronoi tessellation (VRN-VE). Results are also shown for the case of rectangular biquadratic finite elements (Q2-FE).

Figure 4 shows the vertical displacement of the free end of the cantilever versus mesh refinement using the standard stabilization approach $\widehat{\Psi}$ and for the cases of (a) compressibility, with $\nu = 0.3$, and (b) near-incompressibility, with $\nu = 0.49995$. In both cases the SQ1-VE and SQ2S-VE formulations exhibit significantly greater accuracy than the Q1-FE formulation. The S&S-VE and VRN-VE formulations, however, exhibit slower convergence and poorer accuracy than the other VEM formulations. In the case of near-incompressibility the Q1-FE formulation exhibits well-known volumetric locking while the VEM formulations are locking-free.

Figure 5 shows the displacement versus mesh refinement for the nearly incompressible case with modified approaches to stabilization for (a) $\widehat{\Psi}^*$ and (b) $\widehat{\Psi}^{**}$. In both cases the VEM formulations exhibit greater accuracy than that observed in Fig. 4b for every mesh refinement level. Most notably, the slow convergence of the S&S-VE and VRN-VE formulations observed in Fig. 4 is improved significantly by

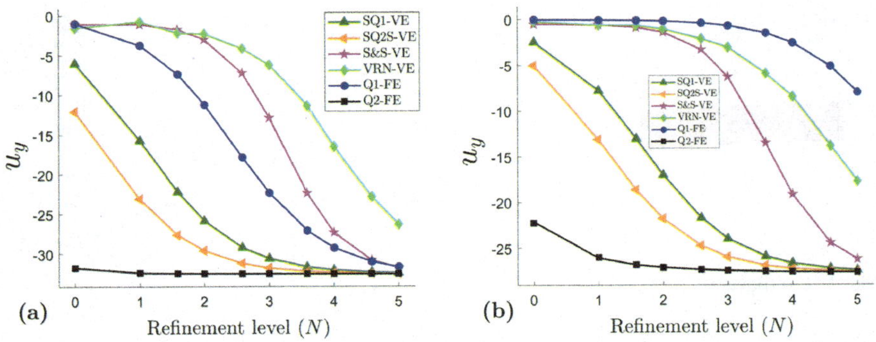

Fig. 4 Displacement versus N for $\widehat{\Psi}$ with **a** $\nu = 0.3$ and **b** $\nu = 0.49995$

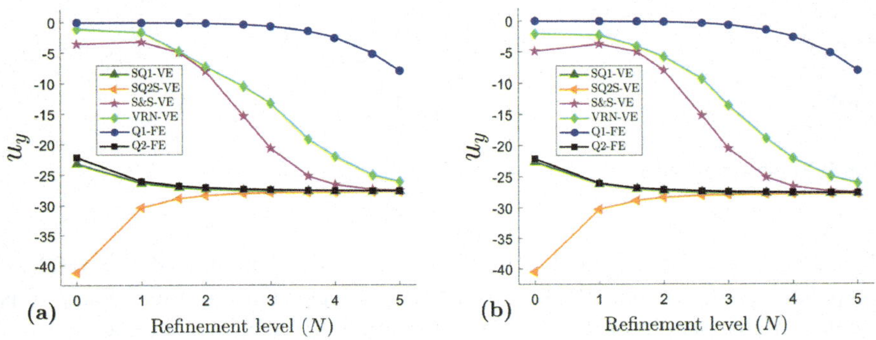

Fig. 5 Displacement versus N with $\nu = 0.49995$ for **a** $\widehat{\Psi}^*$ and **b** $\widehat{\Psi}^{**}$

the modifications to $\widehat{\Psi}$ with the approach $\widehat{\Psi}^{**}$ exhibiting a slightly greater improvement than that of $\widehat{\Psi}^*$.

Figure 6 shows plots of $\mathcal{H}^1$ error versus mean element diameter $\bar{h}$ for the standard stabilization approach $\widehat{\Psi}$ and for (a) various mesh types with $\nu = 0.3$ and (b) the SQ1-VE formulation in the cases of compressibility and near-incompressibility. The $\mathcal{H}^1$ error is computed as described in [13]. In Fig. 6a the SQ1-VE and SQ2S-VE formulations exhibit superlinear convergence behaviour. The convergence rates of the S&S-VE and VRN-VE formulations are initially slow but increase as element size decreases. For finer meshes the VRN-VE formulation exhibits a convergence rate of approximately one, while that of the S&S-VE formulation is superlinear. In Fig. 6b superlinear convergence is again exhibited by the SQ1-VE formulation.

Figure 7 shows plots of the $\mathcal{H}^1$ error vs mean element diameter $\bar{h}$ for the nearly incompressible case and for the two modified stabilization approaches. The convergence behaviour exhibited for the two is very similar. Superlinear convergence behaviour is again exhibited by the SQ1-VE formulation, while the convergence rate of the SQ2S-VE formulation is initially superlinear but approaches approxi-

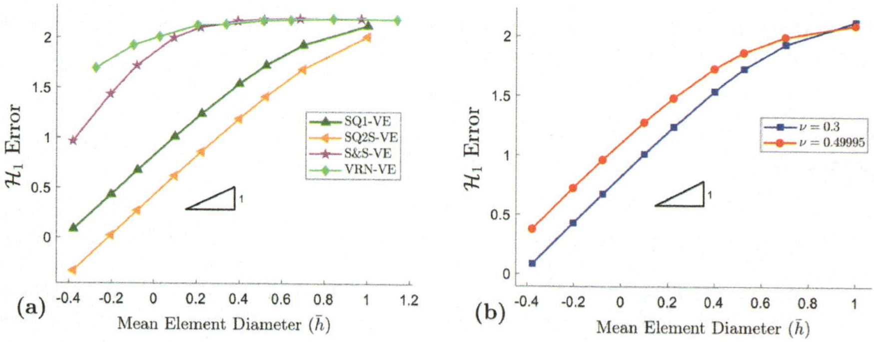

Fig. 6 $\mathcal{H}^1$ error versus mean element diameter for **a** a variety of mesh types with $\nu = 0.3$ and **b** a variety of Poisson's ratios for the SQ1-VE formulation

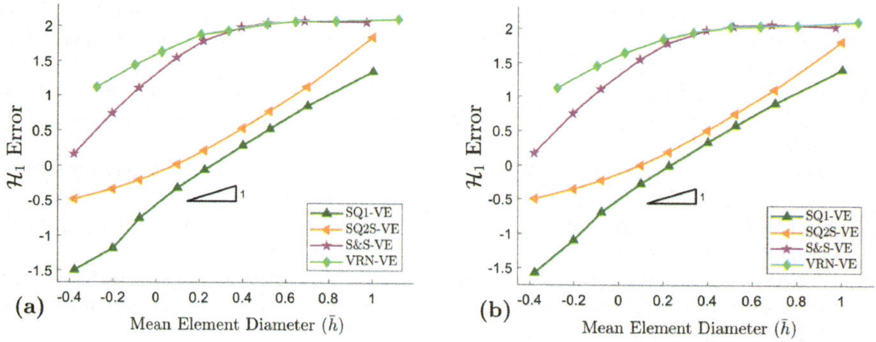

Fig. 7 $\mathcal{H}^1$ error versus mean element diameter with $\nu = 0.49995$ for **a** $\widehat{\Psi}^*$ and **b** $\widehat{\Psi}^{**}$

mately one as element size decreases. The convergence behaviour of the S&S-VE and VRN-VE formulations is qualitatively similar to that observed in Fig. 6; however, the magnitude is significantly decreased.

5 Concluding Remarks

The results in this work demonstrate the ability of the VEM to approximate with high accuracy problems involving large deformations and near-incompressibility, with standard shapes such as rectangular elements showing greater accuracy than more complex polygons. In addition to a stabilization approach that has previously been introduced and used, two modifications are shown to lead to improvements in accuracy of the approximate solution. Subsequent work [11] shows similar accuracy and robustness in relation to near-incompressibility and near-inextensibility in the case of problems involving transversely isotropic material behaviour.

Acknowledgements The authors acknowledge the support of the National Research Foundation through the South African Research Chair in Computational Mechanics.

References

1. Wriggers, P., Rust, W. T., & Reddy, B. D. (2016). A virtual element method for contact. *Computational Mechanics, 58*(6), 1039–1050.
2. Aldakheel, F., Hudobivnik, B., Hussein, A., & Wriggers, P. (2018). Phase-field modeling of brittle fracture using an efficient virtual element scheme. *Computer Methods in Applied Mechanics and Engineering, 341*, 443–466.
3. Aldakheel, F., Hudobivnik, B., & Wriggers, P. (2019). Virtual element formulation for phase-field modeling of ductile fracture. *International Journal for Multiscale Computational Engineering, 17*(2), 181–200.
4. Chi, H., Beirão da Veiga, L., & Paulino, G. (2017). Some basic formulations of the virtual element method (VEM) for finite deformations. *Computer Methods in Applied Mechanics and Engineering, 318*, 148–192.
5. Wriggers, P., & Hudobivnik, B. (2017). A low order virtual element formulation for finite elasto-plastic deformations. *Computer Methods in Applied Mechanics and Engineering, 327*, 459–477.
6. Hudobivnik, B., Aldakheel, F., & Wriggers, P. (2018). A low order 3D virtual element formulation for finite elasto–plastic deformations. *Computational Mechanics, 63*(2), 253–269.
7. Aldakheel, F., Hudobivnik, B., & Wriggers, P. (2019). Virtual elements for finite thermo-plasticity problems. *Computational Mechanics, 64*(5), 1347–1360.
8. Wriggers, P., Reddy, B. D., Rust, W., & Hudobivnik, B. (2017). Efficient virtual element formulations for compressible and incompressible finite deformations. *Computational Mechanics, 60*(2), 253–268.
9. Reddy, B. D., & van Huyssteen, D. (2019). A virtual element method for transversely isotropic elasticity. *Computational Mechanics, 64*(4), 971–988.
10. Wriggers, P., Hudobivnik, B., & Korelc, J. (2018). Efficient low order virtual elements for anisotropic materials at finite strains. In *Advances in computational plasticity* (pp. 417–434). Springer.
11. van Huyssteen, D., & Reddy, B. D. (2021). A virtual element method for transversely isotropic hyperelasticity. *Computer Methods in Applied Mechanics and Engineering, 386*, 114108.
12. De Bellis, M. L., Wriggers, P., & Hudobivnik, B. (2019). Serendipity virtual element formulation for nonlinear elasticity. *Computers & Structures, 223*, 106094.
13. van Huyssteen, D., & Reddy, B. D. (2020). A virtual element method for isotropic hyperelasticity. *Computer Methods in Applied Mechanics and Engineering, 367*, 113134.
14. Wriggers, P. (2008). *Nonlinear finite element methods*. Springer-Verlag GmbH.

Finite Element Formulations for Gradient Damage at Finite Strains

Johannes Riesselmann and Daniel Balzani

My collaboration with Peter Wriggers on finite element formulations for finite strain problems started in 2009 while working at Leibniz University Hannover as substitute professor for one year. At this rather early stage of my carreer, I appreciated a lot to learn from the experience of Peter in programming finite element code. Since this paper addresses a new finite element formulation for gradient-enhanced damage, it is dedicated to Peter Wriggers on the occasion of his 70th birthday in recognition of our past collaboration.

(Daniel Balzani)

Abstract Gradient enhanced damage formulations present a remedy for the loss of ellipticity, lack of robustness and mesh dependency of finite element simulations present in local damage formulations. Corresponding mixed gradient damage formulations may incorporate the gradient enhancement through introduction of a mixed solution variable corresponding to the damage variable and enforcing compatibility through a penalty or Lagrange multiplier term leading to either a numerical parameter or additional variables, respectively. In our approach proposed here, a mixed gradient damage formulation without additional penalty or Lagrange-multiplier term is obtained by solving the balance of linear momentum and the damage evolution equation simultaneously in a corresponding coupled finite element system of equations.

J. Riesselmann (✉) · D. Balzani
Ruhr University Bochum, Chair of Continuum Mechanics, Bochum, Germany
e-mail: johannes.riesselmann@rub.de

D. Balzani
e-mail: daniel.balzani@rub.de

F. Aldakheel et al. (eds.), *Current Trends and Open Problems in Computational Mechanics*, https://doi.org/10.1007/978-3-030-87312-7_43

1 Introduction

Damage models usually aim at the description of strain softening effects due to material failure such as e.g. micro-crack formation (see [1–3]). Since the partial differential equations resulting from the balance of linear momentum and local damage models suffer from a loss of ellipticity once a certain amount of damage has evolved, corresponding finite element simulations may suffer from mesh dependency and a lack of robustness with respect to convergence of the numerical solution procedure. Various regularization approaches such as a relaxed variational approach (see e.g. [4]) present a remedy through retaining convexity of the underlying incremental stress potential. Another approach is gradient damage, in which, through gradient enhancement of the strain energy density, well-posedness of the mathematical equations and thus increased numerical robustness and mesh-independent simulations may be obtained. For a staggered gradient damage approach see [5–7], in which the solution of the balance of linear momentum and the update of the internal damage variable incorporating the gradient enhancement are performed in an alternating manner. Another approach is to incorporate the gradient enhancement through introducing additional mixed solution variables corresponding to the damage variable (see e.g. [8] for small strains and [9] for large strains). In these approaches compatibility between the gradient-extended mixed variable and the internally updated damage variable is enforced by a penalty term which introduces the necessity of a numerical parameter, the penalty parameter. Unfortunately, the alternative, to accommodate compatibility via a Lagrange multiplier, requires additional variables resulting in an increased number of degrees of freedom in associated finite element formulations. To avoid these drawbacks, a mixed approach is proposed here, in which the elastic potential is formulated exclusively in terms of the displacements and the gradient-extended mixed damage variable, thereby avoiding the direct need for compatibility of the mixed variable in the gradient-extended part and the damage variable obtained from an approximation of the displacements. Thermodynamic consistency can be ensured through incorporation of the condition $\dot{\alpha} \geq 0$, where α is the damage variable.

2 Continuum Damage Mechanics

In this section, the continuum mechanical framework for the finite strain gradient damage approach is given. Let $\boldsymbol{\varphi} = \boldsymbol{X} + \boldsymbol{u}(\boldsymbol{X}, t)$ describe the finite strain deformation map $\boldsymbol{\varphi} : \mathcal{B} \rightarrow \mathcal{S}$, where $\boldsymbol{X}$ describes the position of a material point of the considered body in the reference configuration $\mathcal{B}$. The body in deformed configuration is denoted by $\mathcal{S}$ and the displacement is given by $\boldsymbol{u}$. The deformation gradient is given by $\boldsymbol{F} := \nabla\boldsymbol{\varphi} = \mathbf{1} + \nabla\boldsymbol{u}$. The gradient-damage-extended strain energy density function following the approach of [9] based on [8] reads

$$\Psi(\boldsymbol{F}, \alpha) := (1 - D(\alpha))\, \Psi_0(\boldsymbol{F}) + \frac{c_d}{2}(\nabla\alpha)^2. \tag{1}$$

Material softening due to microscopic damage is modelled through the damage function $D : \boldsymbol{R} \to [0, 1)$ with $D(\alpha) := 1 - f(\alpha)$ and $f(\alpha) := \exp(-\alpha)$, wherein α denotes the damage variable. Ψ_0 denotes the hyperelastic strain energy density function corresponding to the fictively undamaged state. Furthermore, differentiation with respect to the deformation gradient gives the first Piola Kirchhof stress tensor $\boldsymbol{P} := \partial_F \Psi = f\, \boldsymbol{P}_0$ with the fictively undamaged part $\boldsymbol{P}_0 = \partial_F \Psi_0$. The balance of linear momentum and the stress flux on the Neumann boundary Γ_N read

$$-\operatorname{Div} \boldsymbol{P} = \boldsymbol{f} \quad \text{in } \mathcal{B} \qquad \text{and} \qquad \boldsymbol{P} \cdot \boldsymbol{N} = \boldsymbol{t} \quad \text{on } \Gamma_N, \tag{2}$$

where $\boldsymbol{f}$ is the vector of body forces and $\boldsymbol{t}$ denotes the vector of surface tractions. Discontinuous damage models (see e.g.[3, 10]) which only consider an evolution of damage when reaching a damage surface incorporate the Karush-Kuhn-Tucker conditions

$$\dot{\alpha} \geq 0 \qquad \Phi \leq 0 \qquad \Phi\, \dot{\alpha} = 0 \tag{3}$$

$$\text{with} \quad \Phi := f\, \Psi_0 + c_d \Delta\alpha - f\, \alpha \leq 0 \tag{4}$$

$$\text{and} \quad \nabla\alpha \cdot \boldsymbol{N} = 0 \quad \text{on } \partial\mathcal{B} \quad \text{(cf. [9,8])}. \tag{5}$$

Based on the damage surface Φ, the evolution of the damage variable is described, such that $\Phi = 0$ characterizes the case where damage evolution takes place. Note that in the following mathematical derivations, we only consider this case for the sake of simplicity. If no damage evolution occurs, then the whole formulation reduces to a purely elastic problem with fixated values for α. Multiplying relations (2) with a test function $\delta\boldsymbol{u}$ and integration over $\mathcal{B}$, as well as multiplying (4) with $\delta\alpha$, integrating over $\mathcal{B}$, and inserting (5) yields[1]:

$$\int_{\mathcal{B}} \boldsymbol{P} \cdot \delta\boldsymbol{F} \mathrm{d}V - \int_{\mathcal{B}} \delta\boldsymbol{u} \cdot \boldsymbol{f} \mathrm{d}V - \int_{\Gamma_N} \delta\boldsymbol{u} \cdot \boldsymbol{t} \mathrm{d}A = 0 \tag{6}$$

$$\int_{\mathcal{B}} \big(- f\, \Psi_0\, \delta\alpha + c_d \nabla\alpha \cdot \nabla\delta\alpha + f\, \alpha\, \delta\alpha\big) \mathrm{d}V = 0. \tag{7}$$

By making use of the relation $\partial_\alpha f = -f$ (due to the specific choice of $f(\alpha)$) we can also obtain the weak forms (6) and (7) by variation of the potential Π defined as

[1] Analogously to deriving the weak form of the balance of linear momentum, the divergence theorem applied to the integrals of (4) and (5) yields $-\int_{\mathcal{B}} c_d\ \Delta\alpha\, \delta\alpha \mathrm{d}V + \int_{\partial\mathcal{B}} \nabla\alpha \cdot \boldsymbol{N}\, \delta\alpha \mathrm{d}A = \int_{\mathcal{B}} \nabla\alpha \cdot \nabla\delta\alpha \mathrm{d}V$.

$$\Pi := \Pi^{\text{int}}[\boldsymbol{F}, \alpha, \nabla\alpha] + \Pi^{\text{diss}}[\alpha] + \Pi^{\text{ext}}[\boldsymbol{u}] \quad \text{with} \tag{8}$$

$$\Pi^{\text{int}} := \int_{\mathcal{B}} \Psi(\boldsymbol{F}, \alpha, \nabla\alpha)\mathrm{d}V, \tag{9}$$

$$\Pi^{\text{diss}} := \int_{\mathcal{B}} -f\,(\alpha + 1)\mathrm{d}V \text{ and} \tag{10}$$

$$\Pi^{\text{ext}} := -\int_{\mathcal{B}} \boldsymbol{u} \cdot \boldsymbol{f}\mathrm{d}V - \int_{\Gamma_N} \boldsymbol{u} \cdot \boldsymbol{t}\mathrm{d}A. \tag{11}$$

Here, Π^{diss} can be constructed from the relation $\int_t f\,\alpha\,\dot{\alpha}\,\mathrm{d}t = -f\,(\alpha + 1)$, where the integrand is the last term of (7). The formulation of the potential (8)-(11) is the starting point of the finite element formulation in the following subsection.

3 Finite Element Formulation

In this section a straightforward finite element discretization based on the previously introduced relations is discussed and corresponding discrete matrix equations are derived. For this, a partition of the body $\mathcal{B}$ into tetrahedra $\mathcal{T} = \bigcup_e T_e$ is considered.

$$\mathcal{V}^h := \{\delta\boldsymbol{u}^h \in H^1_{\Gamma_D}(\mathcal{B}) \cap P_2(\mathcal{T}) : \delta\boldsymbol{u}^h|_{\Gamma_D} = 0\}, \tag{12}$$

$$\mathcal{Q}^h := \{\delta\alpha^h \in H^1(\mathcal{B}) \cap P_1(\mathcal{T}) \cap \mathbb{R}^+ : \textstyle\int \delta\alpha^h \mathrm{d}X = 0\}. \tag{13}$$

Here, $H^1_{\Gamma_D}(\mathcal{B})$ denotes the Sobolev space of functions, which are square-integrable over $\mathcal{B}$ and whose gradients are also square-integrable over $\mathcal{B}$. Moreover, $P_k(\mathcal{T})$ denotes the space of piecewise polynomial functions of degree $\leq k$. For the following finite element formulation, which will be denoted by P2$_{\boldsymbol{u}}$-P1$_{\bar{\alpha}}$ we introduce the matrix interpolation operators for the discrete solution variables $\boldsymbol{u}^h \in \mathcal{V}^h$ and $\bar{\alpha}^h \in \mathcal{Q}^h$:

$$\underline{\boldsymbol{u}}^h = \underline{\boldsymbol{N}}_u \underline{\boldsymbol{d}}_u, \qquad \delta\underline{\boldsymbol{u}}^h = \underline{\boldsymbol{N}}_u \delta\underline{\boldsymbol{d}}_u, \qquad \nabla\underline{\boldsymbol{u}}^h = \underline{\boldsymbol{B}}_u \underline{\boldsymbol{d}}_u, \qquad \nabla\delta\underline{\boldsymbol{u}}^h = \underline{\boldsymbol{B}}_u \delta\underline{\boldsymbol{d}}_u, \tag{14}$$

$$\underline{\bar{\alpha}}^h = \underline{\boldsymbol{N}}_{\bar{\alpha}} \underline{\boldsymbol{d}}_{\bar{\alpha}}, \qquad \delta\underline{\bar{\alpha}}^h = \underline{\boldsymbol{N}}_{\bar{\alpha}} \delta\underline{\boldsymbol{d}}_{\bar{\alpha}}, \qquad \nabla\underline{\bar{\alpha}}^h = \underline{\boldsymbol{B}}_{\bar{\alpha}} \underline{\boldsymbol{d}}_{\bar{\alpha}}, \qquad \nabla\delta\underline{\bar{\alpha}}^h = \underline{\boldsymbol{B}}_{\bar{\alpha}} \delta\underline{\boldsymbol{d}}_{\bar{\alpha}}. \tag{15}$$

Here, $\underline{\boldsymbol{N}}_u$, $\underline{\boldsymbol{N}}_{\bar{\alpha}}$, $\underline{\boldsymbol{B}}_u$ and $\underline{\boldsymbol{B}}_{\bar{\alpha}}$ denote interpolation matrices containing the corresponding Lagrange shape functions and their derivatives respectively. Furthermore, $\underline{\boldsymbol{d}}_u$ and $\underline{\boldsymbol{d}}_{\bar{\alpha}}$ denote the corresponding nodal degrees of freedom.[2] We formulate the potential (8) in terms of the discrete functions $\Pi^h := \Pi^h\,[\underline{\boldsymbol{u}}^h, \nabla\underline{\boldsymbol{u}}^h, \underline{\bar{\alpha}}^h, \nabla\underline{\bar{\alpha}}^h]$ and after variation, the following discrete terms are obtained

[2] In order to simplify the notation with respect to the assembly of the global system the introduced quantities are assumed to be in the dimension of the global system.

$$\delta_u \Pi^h = \sum_{T\in\mathcal{T}} \delta \underline{d}_u^T \Big(\underbrace{\int_T f(\underline{d}_{\bar\alpha}) \frac{\partial \Psi_0(\underline{d}_u)}{\partial \underline{d}_u} \mathrm{d}X}_{\underline{r}_u(\underline{d}_u,\underline{d}_{\bar\alpha})} - \underbrace{\int_T \underline{N}_u^T \underline{f} \mathrm{d}X - \int_{\delta T} \underline{N}_u^T \underline{t} \mathrm{d}A}_{\underline{r}_u^{\mathrm{ext}}} \Big), \quad (16)$$

$$\delta_{\bar\alpha} \Pi^h = \sum_{T\in\mathcal{T}} \delta \underline{d}_{\bar\alpha}^T \Big(\underbrace{\int_T \big(- f(\underline{d}_{\bar\alpha}) \underline{N}_{\bar\alpha}^T \Psi_0(\underline{d}_u) + c_d \underline{B}_{\bar\alpha}^T \underline{B}_{\bar\alpha} \underline{d}_{\bar\alpha} + f(\underline{d}_{\bar\alpha}) \underline{N}_{\bar\alpha}^T \big) \mathrm{d}X}_{\underline{r}_{\bar\alpha}(\underline{d}_u,\underline{d}_{\bar\alpha})} \Big), \quad (17)$$

Since both, Ψ_0 and f, are nonlinear in the corresponding solution fields, the residual terms $\underline{r}_u$ and $\underline{r}_{\bar\alpha}$ are both nonlinear in $\underline{d}_u$ and $\underline{d}_{\bar\alpha}$. Thus, (16) and (17) are linearized in the context of a Newton-Raphson-iteration scheme yielding the equation

$$\mathrm{Lin}[\delta_u \Pi^h + \delta_{\bar\alpha} \Pi^h] = \sum_{T\in\mathcal{T}} \begin{bmatrix} \delta \underline{d}_u \\ \delta \underline{d}_{\bar\alpha} \end{bmatrix}^T \left(\begin{bmatrix} \underline{k}_u & \underline{k}_{u\bar\alpha} \\ \underline{k}_{\bar\alpha u} & \underline{k}_{\bar\alpha} \end{bmatrix} \begin{bmatrix} \Delta \underline{d}_u \\ \Delta \underline{d}_{\bar\alpha} \end{bmatrix} + \begin{bmatrix} \underline{r}_u(\bar{\underline{d}}_u, \bar{\underline{d}}_{\bar\alpha}) + \underline{r}_u^{\mathrm{ext}} \\ \underline{r}_{\bar\alpha}(\bar{\underline{d}}_u, \bar{\underline{d}}_{\bar\alpha}) \end{bmatrix} \right) = 0, \quad (18)$$

which is solved in each iteration for arbitrary test variations of the solution field. Herein, $\bar{\underline{d}}_u$ and $\bar{\underline{d}}_{\bar\alpha}$ denote the nodal values from the previous Newton iteration. The tangent submatrices are given through differentiation

$$\underline{k}_u = \frac{\partial \underline{r}_u}{\partial \bar{\underline{d}}_u} = \int_T f(\bar{\underline{d}}_{\bar\alpha}) \frac{\partial^2 \Psi_0(\bar{\underline{d}}_u)}{\partial \bar{\underline{d}}_u^2} \mathrm{d}X \quad (19)$$

$$\underline{k}_{\bar\alpha} = \frac{\partial \underline{r}_{\bar\alpha}}{\partial \bar{\underline{d}}_{\bar\alpha}} = \int_T \big(c_d \underline{B}_{\bar\alpha}^T \underline{B}_{\bar\alpha} - f(\bar{\underline{d}}_{\bar\alpha}) \underline{N}_{\bar\alpha}^T \underline{N}_{\bar\alpha} \big) \mathrm{d}X \quad (20)$$

$$\underline{k}_{u\bar\alpha} = \frac{\partial \underline{r}_u}{\partial \bar{\underline{d}}_{\bar\alpha}} = - \int_T f(\bar{\underline{d}}_{\bar\alpha}) \frac{\partial \Psi_0(\bar{\underline{d}}_u)}{\partial \bar{\underline{d}}_u} \underline{N}_{\bar\alpha} \mathrm{d}X = \underline{k}_{u\bar\alpha}^T \quad (21)$$

Instead of additional internal update iterations as in [9] for thermodynamic consistency, an algorithmic incorporation of the condition $\dot{\alpha} \geq 0$ is used. As a starting point we consider the simple update procedure for each timestep t:

- Solve repeatedly (18) within the Newton-Raphson procedure.
- Check condition $\dot{\alpha} \geq 0$:
  For each node j:
  - If $(\underline{d}_{\bar\alpha})_j^{t+1} < (\underline{d}_{\bar\alpha})_j^t$, set $(\underline{d}_{\bar\alpha})_j^{t+1} = (\underline{d}_{\bar\alpha})_j^t$

Comparative formulation: As comparative model, the approach following [9] and [8] is considered. Here, the mixed variable $\bar{\alpha}$ is introduced in addition to the internal variable α and compatibility between the variables is enforced with a penalty term. Thus, the internal potential is modified to

$$\Pi^{\mathrm{int}}[\boldsymbol{F}, \alpha, \bar{\alpha}, \nabla\bar{\alpha}] := \int_{\mathcal{B}} (1 - D(\alpha)) \Psi_0(\boldsymbol{F}) + \frac{p}{2}(\bar{\alpha} - \alpha)^2 + \frac{c_d}{2}(\nabla\bar{\alpha})^2 \mathrm{d}X, \quad (22)$$

where p denotes a penalty parameter. The penalty parameter can also be interpreted as micromorphic coupling parameter as investigated by [11] (see also the recent contribution [12]). The corresponding finite element discretization considers $\boldsymbol{u}^h \in \mathcal{V}^h, \bar{\alpha}^h \in Q^h, \alpha^h \in \boldsymbol{R}^+$ and will in the following be denoted by P2$_{\boldsymbol{u}}$-P1$_{\bar{\alpha}}$-pen. The corresponding linearized discrete system of equations is denoted by

$$\mathrm{Lin}[\delta_{\boldsymbol{u}}\Pi^{h,\mathrm{pen}} + \delta_{\bar{\alpha}}\Pi^{h,\mathrm{pen}}] = 0. \tag{23}$$

Furthermore, the condition $\Phi = f\,\Psi_0 + p(\bar{\alpha} - \alpha) - f\,\alpha \leq 0$ is considered in an internal update process in order to remain thermodynamic consistency. The procedure reads, for each timestep t:

- Solve (23) repeatedly within the Newton-Raphson procedure.
- Check condition $\Phi \leq 0$:
  For each gauss point j:
  - If $(\Phi)_j^{t+1} > 0$, solve $(\Phi(\alpha))_j^{t+1} = 0$ with Newton-Raphson sub-iterations

4 Numerical Tests

In the following, the introduced finite element formulations are numerically tested. For the implementation, the automatic-differentiation-based software package AceGen/AceFEM has been used (cf. [13]). The incremental solution procedure incorporates an adaptive load step scheme. As elastic constitutive model we use the Neo-Hooke type hyperelastic energy function

$$\Psi_0 = \frac{\mu}{2}(I_1 - 3) + g(J) \tag{24}$$

with $g(J) = \lambda/4(J^2 - 1) - \lambda/2 \ln J - \mu \ln J$ (cf. [14]), $\boldsymbol{C} = \boldsymbol{F}^T\boldsymbol{F}$, $I_1 = \mathrm{tr}\,\boldsymbol{C}$ and $J = \det \boldsymbol{F}$. An overview of the parameter values chosen for the numerical tests are given in Fig. 1b. For the numerical tests the additional damage propagation parameter η of [9] is included in the damage function $D(\alpha) = 1 - \exp(-\eta\alpha)$ and corresponding derivatives. The considered benchmark test is the boundary value problem of a plate with a hole as shown in Fig. 1a. Due to its symmetry, a consideration of only the upper right quarter is sufficient with symmetry boundary conditions $u_X = 0$ at $X = 0$ and $u_Y = 0$ at $Y = 0$. Moreover, at the upper face the displacement is prescribed with $\hat{\boldsymbol{u}} = (0, \hat{u}, 0)^T$ and $\hat{u} = 25$ mm. The geometric dimensions are $L = 100$ mm, $H = 10$ mm and $R = 50$ mm (cf. Fig. 1a).

In Fig. 2a, force displacement curves of the P2$_{\boldsymbol{u}}$-P1$_{\bar{\alpha}}$ computation are displayed for various mesh refinement stages. In order to visualize the tendency to converge, the plot on the right hand side shows an enlargement of the section marked with a dashed box on the left. Contour plots of the distribution of the damage function D in Fig. 1c correspond to the load stages marked with bullets in Fig. 2a. For the computation

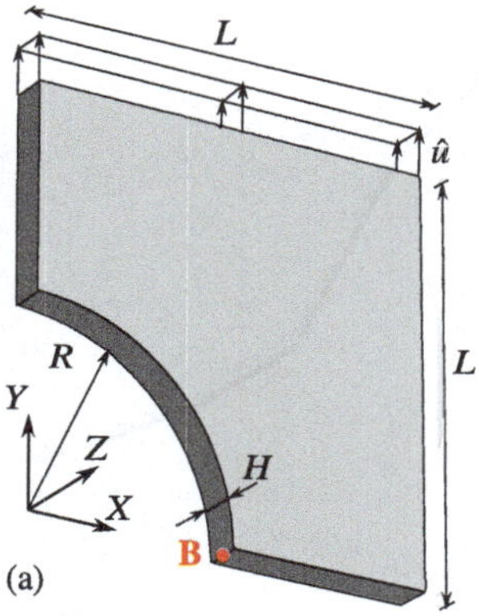

(a)

Material and Boundary Data

Symbol	Description	Value	Unit
E	E-modulus	1000	MPa
ν	poisson ratio	0.3	-
η	damage propagation parameter	0.1	MPa
c_d	nonlocal parameter	5	$\text{mm}^2/\text{MPa}^{-1}$
p	penalty parameter	2	MPa^{-1}
$\hat{u}$	prescribed y-displacement	25	mm

(b)

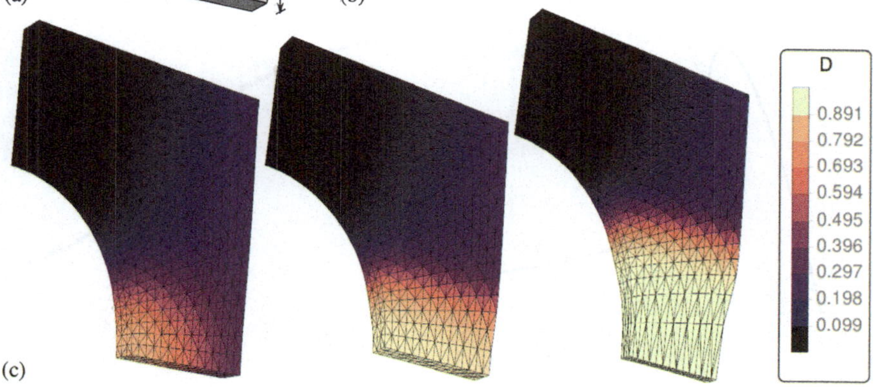

(c)

Fig. 1 **a** Plate with hole problem description. **b** Parameters used for the computation. **c** Contourplots of the distribution of the damage field D corresponding to the load stages marked with bullets in Fig. 2a

of the comparative P2$_u$-P1$_{\bar{\alpha}}$-pen element the force displacement curves are given in Fig. 2b. From comparing the plots in Figs. 2a, b , a similar convergence behavior with respect to the refinement step can be observed. In Fig. 2c, the convergence of the reaction force corresponding to the final load step with respect to the total computing time for each mesh refinement step is shown for the compared elements. Here, a speed-up factor of ≈ 2 of the computing time of the P2$_u$-P1$_{\bar{\alpha}}$ compared to the P2$_u$-P1$_{\bar{\alpha}}$ can be observed (a visualization is shown in the barplot of Fig. 2c. Evaluated is the computing time of the last refinement step and it is marked with colored bullets in the left plot of Fig. 2c. The increased efficiency of the P2$_u$-P1$_{\bar{\alpha}}$ may be due to the fact that while the size and condition number of the tangent matrices of the compared elements should be similar, no subiterations are performed in the P2$_u$-P1$_{\bar{\alpha}}$ computation. However, the proposed formulation does not require the choice of a numerical penalty parameter.

In order to investigate the ability of the P2$_u$-P1$_{\bar{\alpha}}$ element to model damage evolution in loading-unloading scenarios, in a second test, the prescribed displacement $\hat{\boldsymbol{u}} = (0, 5, 0)^T$ mm is applied with an increasing sine function as load step multiplier

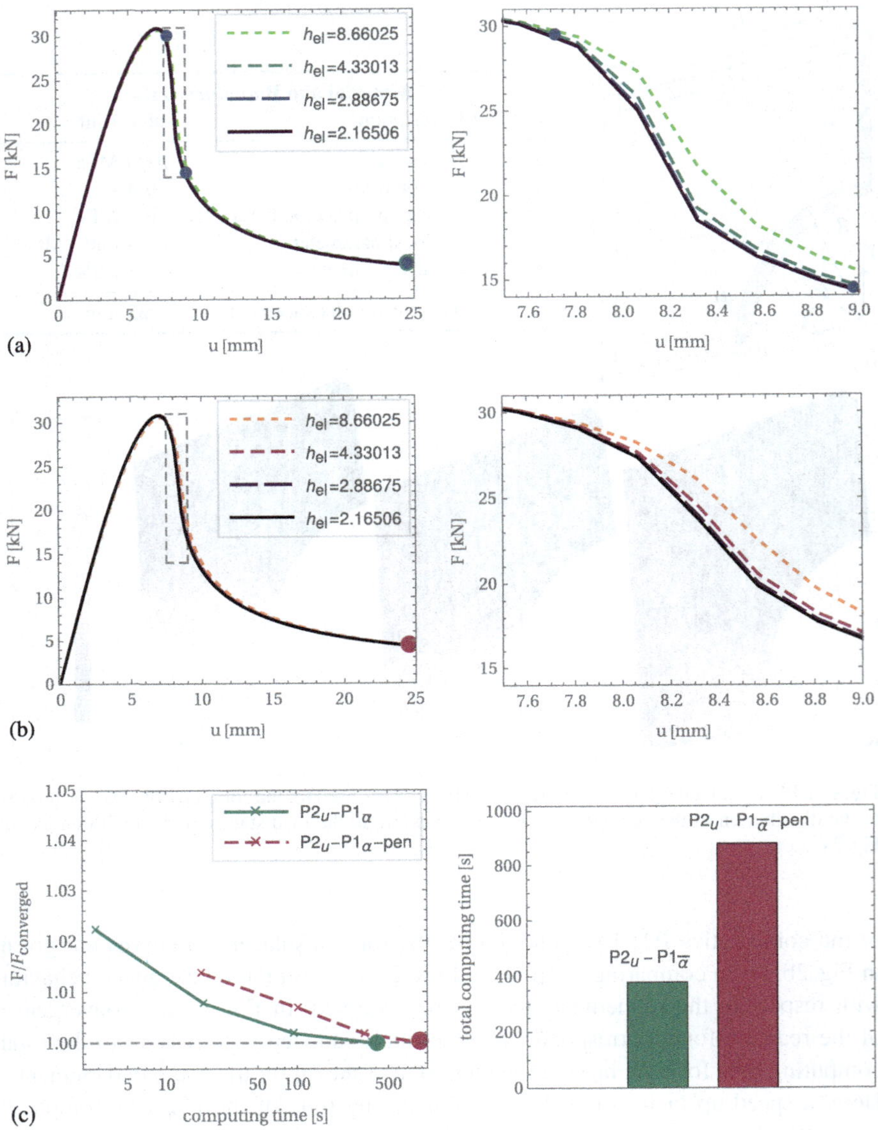

Fig. 2 Force displacement curves **a** corresponding to the P2_u-P$1_{\bar{\alpha}}$ and **b** to the P2_u-P$1_{\bar{\alpha}}$-pen formulation. Images with enlarged resolution are shown on the right hand side. **c** Convergence of the reaction force with respect to computing time

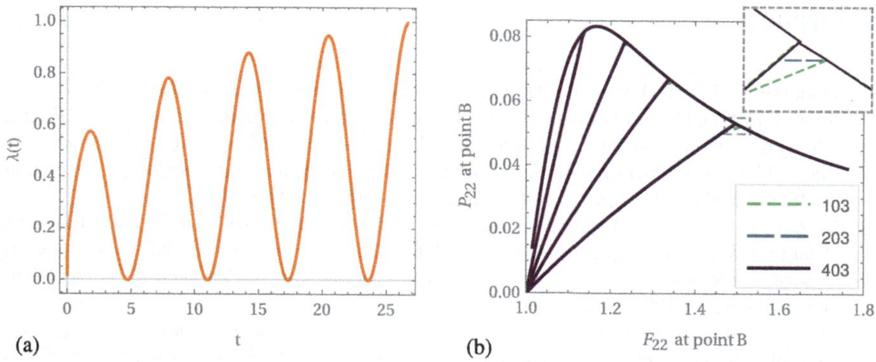

Fig. 3 **a** Cyclic function of the loadstep multiplier $\lambda(t)$ with $\hat{u}(t) = \lambda(t)\hat{u}_{\max}$. **b** Corresponding stress-deformation curves evaluated at point $B = (50, 5, 0)$ mm for varying number of loadsteps

$\lambda(t)$ with $\hat{u}(t) = \lambda(t)\hat{u}_{\max}$ shown in Fig. 3a. Figure 3b shows the corresponding stress deformation response of the 22-entries of $(\boldsymbol{P})_{22}$ and $(\boldsymbol{F})_{22}$ evaluated at the localization point $B = (50, 0, 5)$ mm (cf. Fig. 1a). Shown are the curves for a fixed mesh ($h_{el} = 8.66$ mm) and varying number of steps of the time step solution procedure. For the chosen number of solution steps, the curves are observed to coincide.

5 Conclusion

A new mixed finite element formulation has been presented, in which the governing equations are formulated in terms of the displacements and a mixed damage variable. The derived finite element system of equations solves the balance of linear momentum and the damage evolution equation analogously and thus it avoids additional sub-iterations and the penalty coupling present in the comparative model. Numerical tests have unveiled the ability to produce mesh-independent simulations and further, an increase in numerical efficiency compared to the competitive approach. The ability to model discontinuous damage evolution in the case of unloading scenarios has been shown. The investigation of more refined algorithmic treatments in order to further increase robustness and efficiency for unloading scenarios is left for future research.

Acknowledgements The authors highly appreciate financial funding from the German Science Foundation (Deutsche Forschungsgemeinschaft DFG) within the Priority Program 1748 "Reliable simulation techniques in solid mechanics. Development of non-standard discretization methods, mechanical and mathematical analysis" under the project "Robust and Efficient Finite Element Discretizations for Higher-Order Gradient Formulations" (Project number 392564687, project ID BA2823/15-1.

References

1. Lemaitre, J. (1984). How to use damage mechanics. *Nuclear engineering and design.*
2. Simo, J. (1987). On a fully three-dimensional finite-strain viscoelastic damage model: Formulation and computational aspects. *Computer Methods in Applied Mechanics and Engineering.*
3. Miehe, C. (1995). Discontinuous and continuous damage evolution in ogden-type large-strain elastic materials. *European Journal of Mechanics - A/Solids, 14*, 697–720.
4. Balzani, D., & Ortiz, M. (2012). Relaxed incremental variational formulation for damage at large strains with application to fiber-reinforced materials and materials with truss-like microstructures. *International Journal for Numerical Methods in Engineering, 92*, 551–570. https://doi.org/10.1002/nme.4351
5. Junker, P., Schwarz, S., Jantos, D. R., & Hackl, K. (2019). A fast and robust numerical tretatment of a gradient-enhanced model for brittle damage. *International Journal for Multiscale Computational Engineering, 17*(2), 151–180.
6. Vogel, A., & Junker, P. (2019). Adaptive and highly accurate numerical treatment for a gradient-enhanced brittle damage model. *International Journal for Numerical Methods in Engineering.*
7. Junker, P., Riesselmann, J., & Balzani, D. (2021). Efficient numerical treatment of a gradient-enhanced damage model for hyperelastic materials underunder large deformations. *(submitted)).*
8. Dimitrijevic, B. J., & Hackl, K. (2008). A method for gradient enhancement of continuum damage models. *Technology Mechanics, 28*(1).
9. Waffenschmidt, T., Polindara, C., Menzel, A., & Blanco, S. (2013). A gradient-enhanced large-deformation continuum damage model for fibre-reinforced materials. *Computer Methods in Applied Mechanics and Engineering, 268*, 801–842.
10. Liebe, T., Steinmann, P., & Benallal, A. (2001). Theoretical and computational aspects of a thermodynamically consistent framework for geometrically linear gradient damage. *Computer Methods in Applied Mechanics and Engineering.*
11. Forest, S. (2009). Micromorphic approach for gradient elasticity, viscoplasticity and damage. *Journal of Engineering Mechanics, 135*(3).
12. Brepols, T., Wulfinghoff, S., & Reese, S. (2020). A gradient-extended two-surface damage plasticity model for large deformations. *International Journal of Plasticity.*
13. Korelc , J., & Wriggers, P. (2016). *Automation of finite element methods.* Springer.
14. Wriggers, P. (2008). *Nonlinear finite element methods.* Springer.

New Approaches for Progressive Damage Analysis of Fiber Reinforced Composites and Fiber Metal Laminates

Raimund Rolfes, Christian Gerendt, and Martin Brod

Prof. Raimund Rolfes: "I have been knowing Peter Wriggers for more than 40 years. He was about to finish his Ph.D. at IBNM of LUH, when I joined the institute as a young student assistant. Later he changed to TU Darmstadt and I spent a long time at DLR. We met again after having returned to LUH and have been cooperating since then. I always admired Peter Wriggers' very quick comprehension as well as the width and depth of his tremendous scientific work in computational mechanics."

Abstract In this contribution, novel modeling approaches for progressive damage analyses of fiber reinforced polymer composites (FRP) and hybrid fiber metal laminates (FML) under static and cyclic loading conditions are presented. The finite element-based models are implemented as user-defined material subroutines (UMAT) in the commercial software ABAQUS/Implicit. First, the respective functionalities and the special features of the FRP and FML damage models are explained. Subsequently, the model predictions are compared to first-hand experimental results, demonstrating their predictive capabilities. In this context, special attention is paid to bolted-joint applications, representing a challenging use-case due to complex 3D-stress states. The comparisons of numerical and experimental results demonstrate the predictive capabilities of the proposed modeling approaches for designing complex load-bearing FRP and FML components.

R. Rolfes (✉) · C. Gerendt · M. Brod
Leibniz University Hannover, Institute of Structural Analysis, Appelstraße 9A, 30167 Hannover, Germany
e-mail: r.rolfes@isd.uni-hannover.de

C. Gerendt
e-mail: c.gerendt@isd.uni-hannover.de

M. Brod
e-mail: m.brod@isd.uni-hannover.de

F. Aldakheel et al. (eds.), *Current Trends and Open Problems in Computational Mechanics*, https://doi.org/10.1007/978-3-030-87312-7_44

1 Introduction

Components made of FRP offer numerous advantages compared to conventional materials like metals. Their low weight and high strength are particularly relevant for lightweight requirements in aerospace, automotive and wind energy applications. In addition, the anisotropic character of FRP offers sophisticated possibilities to match a structure's design to the expected load requirements.

Despite these benefits related to FRP composites, their usage in combination with bolted joints is comparatively challenging. By drilling the bolt holes, load-carrying fibers are cut, which causes an initial pre-damage of the FRP composite. To compensate this pre-damage and to account for the complex stress states near the bolt, quasi-isotropic laminates are usually applied, which often results in additional weight of the structure. Additionally, rather brittle materials, like FRP composites, are generally not capable to withstand stress concentrations near holes or cut-outs without damage, resulting in a low joint efficiency compared to more ductile materials [1]. In this context, so-called fiber metal laminates (FML) represent a promising concept to strengthen FRP composites for bolted joint applications. Here, thin metallic inlays are used to substitute single plies of the FRP laminate in the bolt's periphery. By doing so, the laminate's isotropy and plasticity is increased, resulting in a significant increase of the joint's load capacity [2].

In order to design FRP and FML components, reliable simulation tools are required in addition to preliminary experimental investigations. Therefore, the aim of this contribution is to present novel modeling approaches for the progressive damage and failure analysis of FRP and FML components. In this context, Sect. 2 gives a theoretical overview of the latest progressive damage models for FRP composites [3–5] and FML [6]. Section 3 demonstrates the predictive abilities of the modeling approaches, comparing numerical and experimental results. Finally, Sect. 4 contains some concluding remarks based on the work carried out.

2 Theoretical Aspects

2.1 A Progressive Damage Model for FRP Composites Under Cyclic Loading

The fatigue damage model presented in this subsection is designed for FRP composites (in the following denoted as FRP-FDM) and can be applied from coupon- to component-level. The model includes a nonlinear damage accumulation law and analyzes the damage evolution in each layer of the laminate (layer-based approach). The nonlinear damage accumulation law allows to account for load-sequence effects, typical for FRP composites, whereas the layer-based analysis allows to capture complex load-redistribution effects. Additionally, in case of static failure during the fatigue analysis, material properties are degraded according to the mode-discrete Puck fracture theory [7]. The FRP-FDM has been implemented as a user-defined material subroutine (UMAT) in the commercial finite element software ABAQUS/Implicit. The model is based on the assumption of a transversal-isotropic material behavior. In relation to the local material orientations j and the applied loads k, strength and stiffness degradation factors are introduced ($\eta_{R_j^k}$ and $\eta_{E_j^k}$, respectively), which describe the state of damage for each Gauss Point of the FE-model. The degraded stiffness and strength parameters ($E_{j,da}^k$ and $R_{j,da}^k$, respectively) can be expressed as follows:

$$E_{j,da}^k = \eta_{E_j^k} \cdot E_j^k, \text{ and } R_{j,da}^k = \eta_{R_j^k} \cdot R_j^k \,, \tag{1}$$

where $\eta_{R_j^k,E_j^k} = 0$ represents complete damage and $\eta_{R_j^k,E_j^k} = 1$ the pristine material state.

The main feature of the FRP-FDM is the application of an energy-based damage hypothesis [8], originally developed for the fatigue analysis of reinforced concrete. This hypothesis implies that the state of damage under static loading (superscript st) is comparable with the state of fatigue damage under cyclic loading (superscript fat),

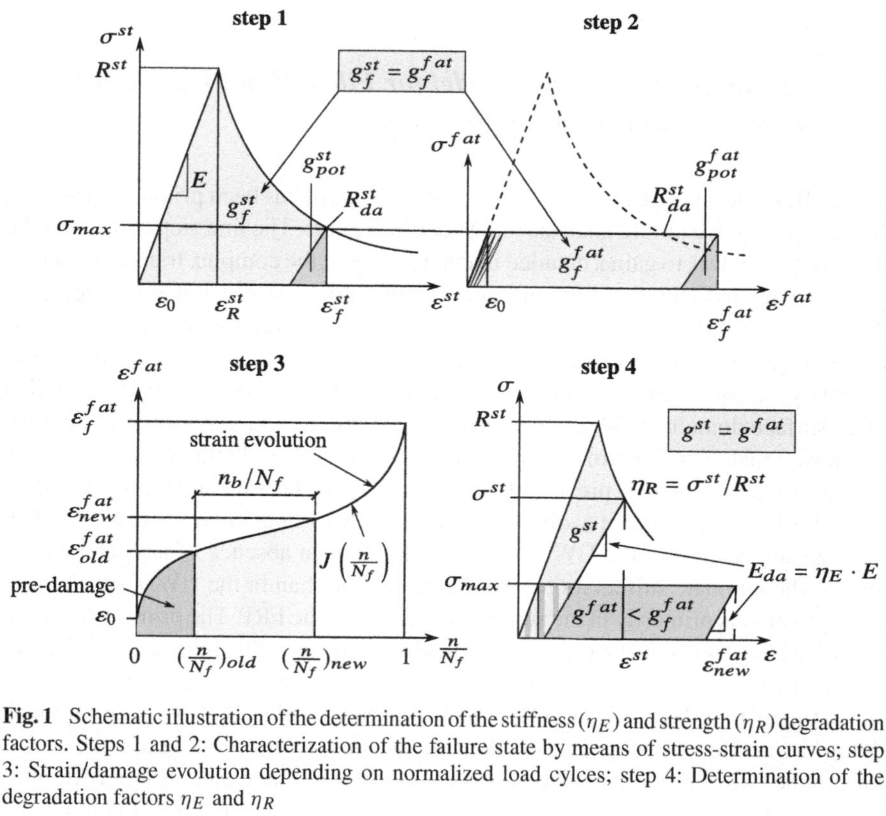

Fig. 1 Schematic illustration of the determination of the stiffness (η_E) and strength (η_R) degradation factors. Steps 1 and 2: Characterization of the failure state by means of stress-strain curves; step 3: Strain/damage evolution depending on normalized load cylces; step 4: Determination of the degradation factors η_E and η_R

if the amount of energy g_j^k dissipated by a material point for both loading scenarios is equal. When adopting this hypothesis to fiber composites, the damage state of a certain material orientation j depends only on the amount of energy dissipated in this orientation. Consequently, the damage state in relation to stiffness and strength properties (with identical material orientation j) is comparable ($g_j^{st,k} = g_j^{fat,k} \rightarrow E_{j,da}^{st,k} = E_{j,da}^{fat,k}$, and $R_{j,da}^{st,k} = R_{j,da}^{fat,k}$). The application of this energy hypothesis also allows the fatigue analysis to be performed load-block-wise, thus bypassing a time-consuming and computationally intensive cycle-by-cycle analysis. Figure 1 explains the main steps during the damage calculation procedure. In steps 1 an 2, the ultimate failure state is characterized (for the quasi-static and cyclic load case) considering the magnitude of stresses σ_{max} of the current load block n_b. Here, the basic idea of the above energy hypothesis ($g_f^{st} = g_f^{fat}$) is applied. In step 3, the fatigue damage increase, represented by the fatigue strain evolution, is calculated based on the number of load cycles n_b applied virtually for the current load block. Finally, in step 4, the updated degradation factors $\eta_{E/R}$ for strength and stiffness, respectively, are determined iteratively by ensuring the balance of energies ($g^{st} = g^{fat}$) for the set parameters σ_{max} and ε_{new}^{fat}. A detailed description of the FRP-FDM including all assumptions and mathematical expressions can be found in [3–5, 8–11].

2.2 A Progressive Damage Model for Fiber Metal Laminates in Static Bolted Joint Applications

The FML static damage model recently presented in [6] has been primarily designed for the progressive failure analysis of FML bolted joints. The first step to develop the damage model was to gain a detailed understanding of the complex transient damage processes related with the different joint failure modes. As illustrated by Fig. 2, the damage of an FML bolted joint in case of a typical bearing failure mode develops differently, depending on whether the laminate is located in the area laterally constraint by the bolt's head (so-called "under washer domain", UWD) or in the so-called "free surface domain" (FSD). First, the damage initiates with a crushing of the FRP composite matrix in front of the bolt in load direction. Beside that crushing, severe damage of the laminate is prevented at this stage by the lateral constraint induced by the bolt's head. As damage evolves further, the state of final failure is reached when damage spreads out of the UWD, entering the FSD. In absence of any lateral constraint, the laminate suffers more severe degradation than in the UWD, resulting in buckling effects primarily in the zero-degree layers of the FRP. The primary buckling of the FRP layers results in a secondary buckling of the metallic inlays, which feature only a thickness of 0.25 mm. The FML damage model accounts for all these effects by distinguishing between the UWD and the FSD, see Fig. 3. Note, that laminate deformation due to loading is also taken into account while defining the two zones. Within the two zones, different configurations of the damage models (for FRP plies and the metallic inlays) are used to account for the above phenomena. To predict the

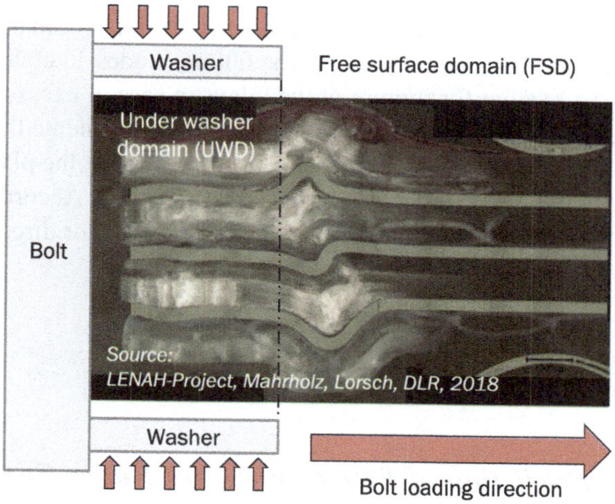

Fig. 2 Side-view of an FML bolted joint after bearing failure. The final damage of the joint is caused by a buckling of the zero-degree FRP plies, followed by a secondary buckling of the metallic inlays. The laterally constrained "UWD" features no severe damage beside crushing of the matrix material

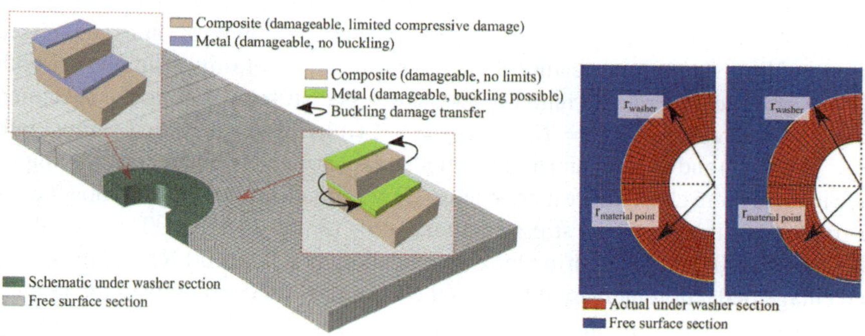

Fig. 3 Illustration of the joint sections distinguished by the FML damage model for bolted joint applications. Different damage algorithms are provided for the FML constituents (FRP and metallic inlays) depending on the material point's location with respect to the bolt's head or washer. The actual bolt is not illustrated for the sake of simplicity

state of damage for the FRP plies, a static damage model based on the failure theory of Puck [7] is applied [5, 6]. For material points located in the laterally constraint UWD, compressive FRP damage is limited, so the material point's stiffness can not drop below the stiffness of the pure matrix material [5]. By doing so, the previously explained effect of crushing can be taken into account effectively in a continuum mechanical manner. In case compressive fiber failure is detected by the FRP damage model in the area not constrained by the bolt's head or washer (FSD), the model reduces the material stiffness not only for the corresponding FRP material point, but also for neighboring metallic inlay material points. Here, the assumption is, that the inlays can not carry any load without surrounding intact (zero degree) FRP plies, see Fig. 2. However, in addition to buckling, the metallic inlays can also be damaged by excessive plastic deformation, typical for failure modes like shear-out or net-tension failure. To account for rupture of the inlays in case of excessive straining, a strain-based damage model according to [12] has been implemented. This inlay rupture model is based on a so-called fracture locus, representing the plastic equivalent strain at failure initiation as a function of the stress triaxility. Accordingly, an inlay Gauss Point can be damaged either indirectly (buckling) and/or directly (rupture).

3 Application Examples and Discussion

3.1 *Fatigue Prediction of FRP Composites Under Complex Stress States*

This subsection demonstrates the predictive capability of the FRP-FDM under cyclic loading with complex stress states that generally prevail for many technical applications. A modified three-point bending test serves as an application example (see Fig. 4a). Here, the special feature is that the specimen is firmly clamped on both ends in order to generate an inhomogeneous stress distribution along the specimen length l_s. A force ratio of $R = F_{min}/F_{max} = -1$ (fully reversed cyclic loading, with $F_{max} = 400$ N) and a maximum number of load cycles $n = 10^5$ under constant amplitude loading were chosen as test parameters. Figure 4b illustrates the inhomogeneous distribution of the transverse stress component σ_{22} for the different 90° layers of the [90/0/90/0/90] laminate during maximum loading ($F_{max} = 400$ N). In Fig. 4c, the percentage deflection increase during cyclic loading is used to compare the numerical prediction with experimental measurements. The deflection increase is mainly induced by inter-fiber cracking. As can be seen, the FRP-FDM predicts the damage behavior throughout the entire test period in close accordance with experimental measurements. More details on this example can be found in [4].

3.2 *Progressive Fatigue Damage Analysis of a GFRP T-Bolted Joint*

In this example, a recently published version of the FRP fatigue damage model (see [5] for further information) is validated for a bolted joint component made from GFRP multi-axial laminate, see Fig. 6. This so-called T-bolt joint is typically used in the wind energy industry for joining the hub and the rotor blades of the turbine.

A comparison between the experimentally measured global stiffness degradation and the corresponding numerical prediction is displayed in Fig. 5. The comparison in Fig. 5 shows, that the evolution of the global degradation D is predicted with sufficient

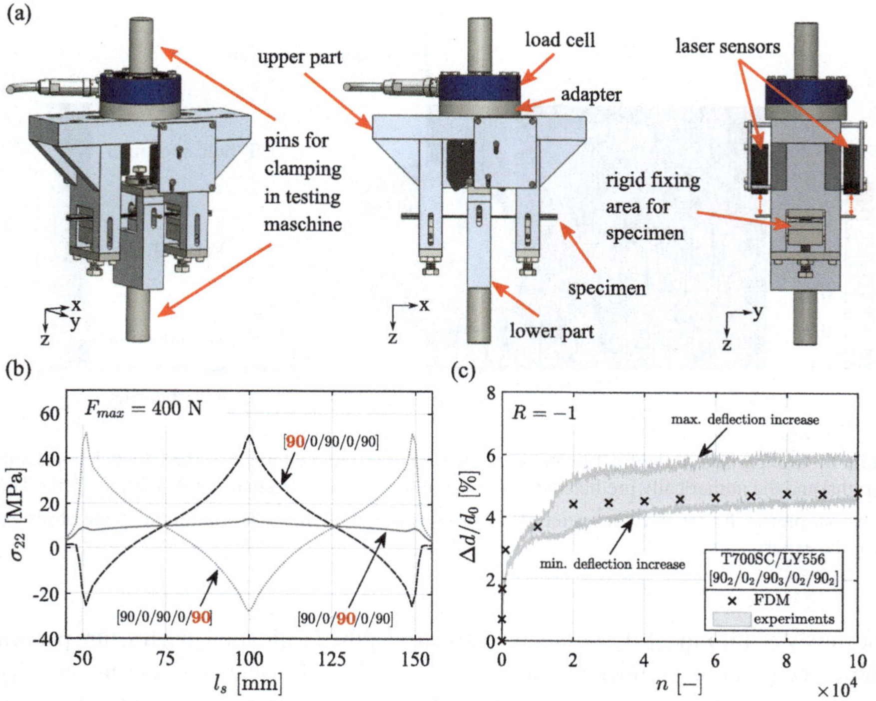

Fig. 4 **a** Schematic illustration of the bending test set-up; **b** Varying local stress component σ_{22} in different 90° plies, **c** Comparison of predicted and experimentally determined fatigue caused deflection evolution of the bending specimen $\Delta d/d_0$ (with $\Delta d = d_{da} - d_0$). For details see [4]

Fig. 5 Comparison between experimental measurement and numerical prediction of the global stiffness degradation D for two T-bolt bearing setups tested under tension-tension cyclic loading conditions

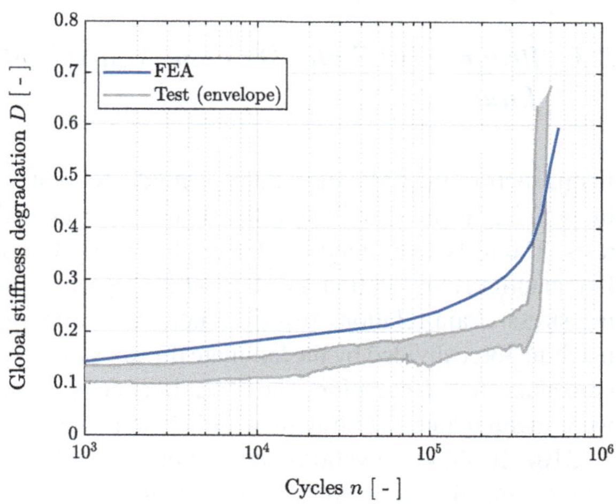

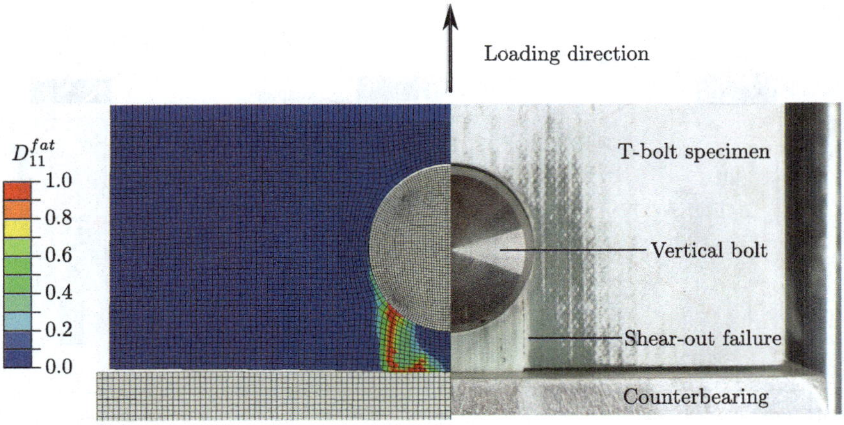

Fig. 6 Comparison of the fracture pattern observed after fatigue failure of the GFRP T-bolt setup (right) and the numerically predicted distribution of the damage parameter D_{11}^{fat} at failure (left). The contour plot of the damage parameter D_{11}^{fat} matches well with the shear-out failure mode observed for the experiment

accuracy by the model. To further increase the depth of validation, the fracture pattern observed for the experimental tests is compared to the distribution of the numerically predicted decisive fatigue damage parameter, see Fig. 6. The comparison shows, that the fatigue damage model accurately predicts the so-called shear-out failure mode of the T-bolt joint setup under cyclic loading.

3.3 Progressive Static Damage Analysis of an FML Bolted Joint

In this section, the FML static damage model for bolted joint applications, introduced in Sect. 2.2, is validated using first-hand experimental results of static bolted joint tests. The tests were conducted according to up-to-date standards (AITM 1-0009). For quantitative validation, experimentally measured and numerically predicted bearing stress-strain relations (σ_B-ε_B relations) are compared. Here, σ_B represents the machine force divided by the laminate thickness and the hole diameter. ε_B defines the percentage hole elongation. The FML specimens used for the experimental campaign were manufactured from regular GFRP and stainless steel sheet inlays (grade: EN 1.4310). To show the validity of the model for different degrees of metal hybridization, i.e. metal volume content (MVC), the MVC has been increased systematically for two base-laminates (zero-degree unidirectional and a cross-ply laminate), covering a total range from 0% MVC (pure GFRP reference laminate) up to 18% MVC (three inlays of 0.25 mm thickness each). As Fig. 7 illustrates, the model predictions are in sufficiently close accordance with the experimental measurements for

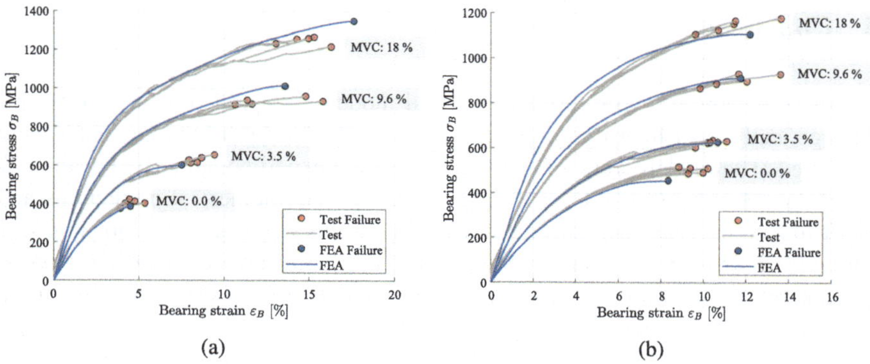

Fig. 7 Comparison of experimentally measured and numerically predicted bearing stress-strain curves for two different base laminates (unidirectional - left, cross-ply - right) with different levels of metal hybridization (0% $\leq$ MVC $\leq$ 18%). For further information refer to [6]

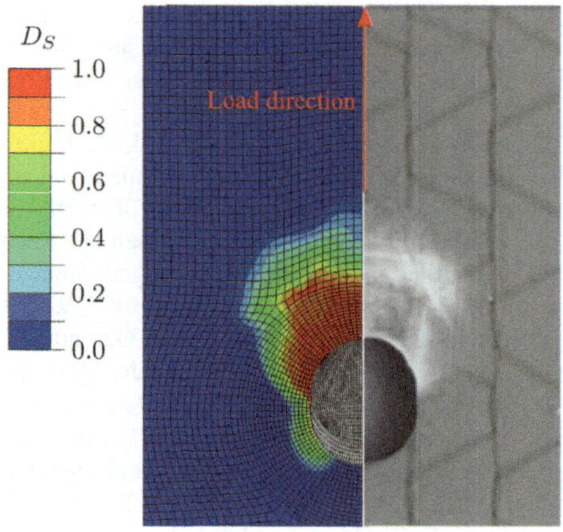

Fig. 8 Comparison of the damage pattern obtained from an FML bolt bearing test after ultimate failure. The failure mode "bearing failure" is predicted correctly by the model

both base laminates and the different levels of hybridization (MVC), respectively. To demonstrate the capabilities of the novel FML damage model in terms of failure mode prediction, the damage pattern, as photographed in one of the FML bearing tests, is compared with the numerically predicted distribution of the maximum FRP damage $0 \leq D_S \leq 1$. As Fig. 8 shows, the predicted damage distribution matches quite well with the experimental observation. Accordingly, the herein presented model [6] is not only capable of predicting the progressive damage process until final failure of the joint, but gives also further information which damage mode is to expect for a specific joint design.

4 Conclusion

In this contribution, two recently developed modeling approaches for progressive damage analysis of fiber reinforced plastics and fiber-metal hybrid laminates were presented. The models were validated by first-hand experimental tests. For the different use-cases considered, close agreements between numerical predictions and experimental measurements were achieved for various static and cyclic loading scenarios. The rich information provided by the numerical predictions offer promising capabilities to adapt the herein presented models in virtual test rigs for future computer-aided design of high-performance composites.

References

1. Hart-Smith, L. J. (1980). Mechanically-fastened joints for advanced composites—phenomenological considerations and simple analyses: The theory of Alfred Puck. In: E. M. Lenoe, D. W. Oplinger, J. J. Burke (Eds.), *Fibrous composites in structural design*. Boston, MA: Springer.
2. Fink, A., Camanho, P. P., Andrés, J. M., Pfeiffer, E., & Obst, A. (2010). Hybrid CFRP/titanium bolted joints: Performance assessment and application to a spacecraft payload adaptor. *Composites Science and Technology, 70*(2), 305–317.
3. Brod, M., Just, G., Dean, A., Jansen, E., Koch, I., Rolfes, R., & Gude, M. (2019). Numerical modelling and simulation of fatigue damage in carbon fibre reinforced plastics at different stress ratios. *Thin-Walled Structures, 139,* 219–231.
4. Brod, M., Dean, A., Scheffler, S., Gerendt, C., & Rolfes, R. (2020). Numerical modeling and experimental validation of fatigue damage in cross-ply CFRP composites under inhomogeneous stress states. *Composites Part B: Engineering, 200*, 108050.
5. Gerendt, C., Dean, A., Mahrholz, T., Englisch, N., Krause, S., & Rolfes, R. (2020). On the progressive fatigue failure of mechanical composite joints: Numerical simulation and experimental validation. *Composite Structures, 248*, 112488.
6. Gerendt, C., Dean, A., Mahrholz, T., & Rolfes, R. (2019). On the progressive failure simulation and experimental validation of fiber metal laminate bolted joints. *Composite Structures, 229*, 111368.
7. Knops, M. (2008). Analysis of failure in fiber polymer laminates: The theory of Alfred Puck. Springer Science and Business Media.
8. Pfanner, D. (2003). Zur Degradation von Stahlbetonbauteilen unter Ermüdungsbeanspruchung. Ph.D. Thesis, VDI-Verlag.
9. Krüger, H. (2012). Ein physikalisch basiertes Ermüdungsschädigungsmodell zur Degradationsberechnung von Faser-Kunststoff-Verbunden. Ph.D. Thesis, Leibniz-Universität Hannover.
10. Krüger, H., & Rolfes, R. (2015). A physically based fatigue damage model for fibre-reinforced plastics under plane loading. *International Journal of Fatigue, 70,* 241–251.
11. Madhusoodanan, H., Jansen, E., & Rolfes, R. (2018). A physically based fatigue damage model for simulating three-dimensional stress states in composites under very high cycle fatigue loading. In H.-J. Christ (Ed.), *Fatigue of materials at very high numbers of loading cycles* (pp. 533–559). Heidelberg: Springer.
12. Song, Q., Heidarpour, A., Zhao, X., & Han, L. (2018). Experimental and numerical investigation of ductile fracture of carbon steel structural components. *Journal of Constructional Steel Research, 145,* 425–437.

Differential Geometry of Surfaces with Application to Shell Structures

Shahab Sahraee and Meisam Soleimani

It is our honor and pleasure to contribute to this Festschrift dedicated to Prof. P. Wriggers on the occasion of his 70th birthday. We, both, were privileged to accomplish our PhD under his supervision during 2010–2018. His open-mindedness as a leader is a unique and significantly precious characteristic. It is of great importance since Freedom is an inalienable part of research when it comes to creativity. At the same time, we benefited from his wealth of knowledge and deep insight, which is something beyond the knowledge, in computational mechanics. In short, one could enjoy both spreading his/her wings and being under his wings at the same time! Prof. Wriggers is one of the pioneers in computational contact mechanics in which curvilinear coordinates can be exploited as an elegant and powerful tool if the contact is perceived as a geometrical interaction between deformable surfaces. Similarly, it can be applied to classical shell theories. This motivated us to write a short monograph on the application of differential geometry in analyzing geometrically exact shells.

Abstract This work deals with infinitesimal deformations of an isotropic elastic shell in a differential geometry framework. It provides an introduction to differential geometry of embedded surfaces in the Euclidean three-dimensional space and serves as a tangible and practical "recipe" for those, particularly students, who are interested in invoking the old art of working with curvilinear coordinates. It seems that the fundamental concepts in this field have remained obscure and vague for many students and they do not dare employ this elegant tool. With regard to this, a shell structure composed of only a single element is taken into consideration represented geometrically exact in a curvilinear setting. The stress/strain are computed at the

S. Sahraee (✉)
Faculty of Mechanical Engineering, University of Guilan, Rasht, Iran
e-mail: shahab_sahraee@guilan.ac.ir

M. Soleimani
Leibniz University Hannover, Institute of Continuum Mechanics, Garbsen, Germany
e-mail: soleimani@ikm.uni-hannover.de

F. Aldakheel et al. (eds.), *Current Trends and Open Problems in Computational Mechanics*, https://doi.org/10.1007/978-3-030-87312-7_45

Gauss points using the framework of differential geometry and the results are validated against ones predicted by ANSYS as a commercial software.

One should notice that, nowadays, this approach is not implemented in commercial software such as ANSYS due to its complications. Rather, such software prefers to use the so-called solid degenerated shell elements. Using such approach, one can circumvent dealing directly with objects such as Christoffel symbols which naturally emerge as soon as the derivative are taken in a curvilinear coordinate system.

1 Surfaces in Euclidean Three-Dimensional Space

Working with curvilinear coordinates in a curved space and flat space differs substantially. Recall that the forth-order Riemann-Christoffel curvature tensor (or equivalently the Gaussian curvature) acts as a decisive criterion for flatness of space. The Euclidean space represents a flat space with this fourth-order tensor being vanished all over the space. Indeed, the Euclidean space is one for which all well-known postulates of Euclidean geometry - expressing relationships between primitive quantities such as points and lines - perfectly applies. Straightness is its key characteristic and it is usually coordinated with the very popular Cartesian coordinate system. For such a coordinate system, the matrix form of metric coefficients becomes diagonal with all diagonal elements being +1. Accordingly, all of its Christoffel elements vanish. When the space is Euclidean and referred to a Cartesian coordinate frame, any general basis that may be resulted from embedding a curvilinear coordinate system can be expressed as a linear combination of the standard basis. Thus, the metric coefficients for such basis and subsequently the Christoffel elements can consistently be computed. For more studies, the reader is referred to [3, 6, 7, 10].

Of interest in this work is to consider two-dimensional surfaces embedded in the ambient three-dimensional space. Any surface is assumed to be a sufficiently smooth two-dimensional subspace of the three-dimensional space. Moreover, it is assumed that there exists tangent and normal spaces at each point of the surface. Finally, suppose that the standard properties of the real vector spaces remain valid in this context.

A two-dimensional surface in the ambient space is defined by the following vector function [5]

$$\mathbf{x} = \mathbf{x}^{s}\left(t^{1}, t^{2}\right), \tag{1}$$

where $\mathbf{x}$ denotes the position of an arbitrary point on the surface and the real numbers $t^{\alpha},\ \alpha = 1, 2$, are referred to as surface (or Gaussian) coordinates. It should be noted that all Greek indices in this work assume values 1 and 2. Moreover, the Einstein summation convention is implied when these indices are repeated. Suppose that the ambient space under consideration is coordinated by Cartesian coordinates. The sufficiently smooth vector function (1), which parametrically describes the surface, then represents

$$x_1 = x_1^{s}\left(t^1, t^2\right) \quad , \quad x_2 = x_2^{s}\left(t^1, t^2\right) \quad , \quad x_3 = x_3^{s}\left(t^1, t^2\right). \tag{2}$$

1.1 Surface Basis Vectors and Metric Coefficients

The surface covariant basis vectors (or tangent vectors) at an arbitrary point on the surface are defined by

$$\mathbf{a}_\alpha = \frac{\partial \mathbf{x}}{\partial t^\alpha}. \tag{3}$$

These two linearly independent vectors are basically tangential to the coordinate lines corresponding to the surface coordinates and, thus, define a plane known as tangential plane. These vectors can be completed to represent a basis in the ambient space by using the following surface normal vector

$$\widehat{\mathbf{n}} = \pm \frac{\mathbf{a}_1 \times \mathbf{a}_2}{|\mathbf{a}_1 \times \mathbf{a}_2|} \quad \text{satisfying} \quad \widehat{\mathbf{n}} \cdot \mathbf{a}_\alpha = 0 \;\; \text{and} \;\; \widehat{\mathbf{n}} \cdot \widehat{\mathbf{n}} = 1. \tag{4}$$

The appropriate sign in $(4)_1$ is chosen according to the well-known discussions on orientation of surfaces, see Fig. 1. One can now express any arbitrary tensorial variable in the ambient space relative to the covariant basis

$$\{\mathbf{a}_i\} := \{\mathbf{a}_1, \mathbf{a}_2, \mathbf{a}_3\} \quad \text{where} \quad \mathbf{a}_3 = \widehat{\mathbf{n}}. \tag{5}$$

Let

$$\left\{\mathbf{a}^i\right\} := \left\{\mathbf{a}^1, \mathbf{a}^2, \mathbf{a}^3\right\}, \tag{6}$$

be the unique dual basis of $\{\mathbf{a}_i\}$. This set of vectors provides another basis for the ambient space. And it is called contravariant basis. The covariant and contravariant bases are related by

$$\mathbf{a}_\alpha = a_{\alpha\beta}\mathbf{a}^\beta \;\; \text{and} \;\; \mathbf{a}^\alpha = a^{\alpha\beta}\mathbf{a}_\beta, \tag{7}$$

where $a_{\alpha\beta}$ and $a^{\alpha\beta}$ are called the surface covariant and contravariant metric coefficients, respectively. They are given by

$$a_{\alpha\beta} = \mathbf{a}_\alpha \cdot \mathbf{a}_\beta = a_{\beta\alpha} \quad \text{with the matrix form} \quad \left[a_{\alpha\beta}\right] = \begin{bmatrix} \mathbf{a}_1 \cdot \mathbf{a}_1 & \mathbf{a}_1 \cdot \mathbf{a}_2 \\ \mathbf{a}_1 \cdot \mathbf{a}_2 & \mathbf{a}_2 \cdot \mathbf{a}_2 \end{bmatrix}, \tag{8}$$

and

$$a^{\alpha\beta} = \mathbf{a}^\alpha \cdot \mathbf{a}^\beta = a^{\beta\alpha} \quad \text{with the matrix form} \quad \left[a^{\alpha\beta}\right] = \begin{bmatrix} \mathbf{a}^1 \cdot \mathbf{a}^1 & \mathbf{a}^1 \cdot \mathbf{a}^2 \\ \mathbf{a}^1 \cdot \mathbf{a}^2 & \mathbf{a}^2 \cdot \mathbf{a}^2 \end{bmatrix}. \tag{9}$$

The surface metric coefficients are related by

$$a_{\alpha\gamma}a^{\gamma\beta} = \delta_{\alpha}^{\beta} \quad \text{or} \quad \left[a_{\alpha\beta}\right] = \left[a^{\alpha\beta}\right]^{-1}, \tag{10}$$

where δ_{α}^{β} presents the mixed two-dimensional Kronecker delta. It is then easy to see that

$$\mathbf{a}^{\alpha} \cdot \mathbf{a}_{\beta} = \delta_{\beta}^{\alpha} = \delta_{\alpha}^{\beta} = \mathbf{a}_{\alpha} \cdot \mathbf{a}^{\beta}. \tag{11}$$

Moreover,

$$\mathbf{a}^{3} = \mathbf{a}_{3} = \widehat{\mathbf{n}} \quad \text{satisfying} \quad \widehat{\mathbf{n}} \cdot \mathbf{a}^{\alpha} = 0 \text{ and } \widehat{\mathbf{n}} \cdot \mathbf{a}_{\alpha} = 0. \tag{12}$$

The determinant of the surface covariant metric coefficients, $a = \det\left[a_{\alpha\beta}\right]$, may be obtained by

$$\sqrt{a} = |\mathbf{a}_{1} \times \mathbf{a}_{2}| \quad \text{and, therefore,} \quad \mathbf{a}_{1} \times \mathbf{a}_{2} = \sqrt{a}\,\widehat{\mathbf{n}}. \tag{13}$$

In a similar manner, the determinant of the surface contravariant metric coefficients, according to $a^{-1} = \det\left[a^{\alpha\beta}\right]$, can also be determined from

$$\frac{1}{\sqrt{a}} = \left|\mathbf{a}^{1} \times \mathbf{a}^{2}\right| \quad \text{and, therefore,} \quad \mathbf{a}^{1} \times \mathbf{a}^{2} = \frac{1}{\sqrt{a}}\widehat{\mathbf{n}}. \tag{14}$$

Let $\mathbf{w}$ be a vector with the following representations

$$\mathbf{w} = w^{\alpha}\mathbf{a}_{\alpha} = w_{\alpha}\mathbf{a}^{\alpha}. \tag{15}$$

Such a vector which lies in tangent plane is referred to as the surface vector. The contravariant component w^{α} is usually referred to as vector in the context of differential geometry and the covariant component w_{α} is called covector. They are related by

$$w^{\alpha} = a^{\alpha\beta}w_{\beta} \quad \text{or} \quad w_{\alpha} = a_{\alpha\beta}w^{\beta}. \tag{16}$$

In a similar manner, a surface tensor $\mathbf{A}$ intrinsically admits the following decompositions

$$\mathbf{A} = A^{\alpha\beta}\mathbf{a}_{\alpha} \otimes \mathbf{a}_{\beta} = A^{\alpha}_{\cdot\beta}\mathbf{a}_{\alpha} \otimes \mathbf{a}^{\beta} = A_{\alpha}^{\cdot\beta}\mathbf{a}^{\alpha} \otimes \mathbf{a}_{\beta} = A_{\alpha\beta}\mathbf{a}^{\alpha} \otimes \mathbf{a}^{\beta}. \tag{17}$$

Note that the 'dot' utilized in the mixed components $A^{\alpha}_{\cdot\beta}$ or $A_{\alpha}^{\cdot\beta}$ presents a placeholder (which means that the other index is the first index). By index juggling, one then has

$$A^{\alpha\beta} = a^{\alpha\gamma}A_{\gamma}^{\cdot\beta} = A^{\alpha}_{\cdot\gamma}a^{\gamma\beta} = a^{\alpha\gamma}A_{\gamma\delta}a^{\delta\beta}. \tag{18}$$

1.2 Surface Christoffel Symbols. Gauss and Weingarten Formulas

Consider the following family of vectors

$$\boldsymbol{\Gamma}_{\alpha\beta} := \frac{\partial \mathbf{a}_\alpha}{\partial t^\beta}, \tag{19}$$

possessing the symmetry in α and β owing to

$$\boldsymbol{\Gamma}_{\alpha\beta} = \frac{\partial \mathbf{a}_\alpha}{\partial t^\beta} = \frac{\partial^2 \mathbf{x}}{\partial t^\beta \partial t^\alpha} = \frac{\partial^2 \mathbf{x}}{\partial t^\alpha \partial t^\beta} = \frac{\partial \mathbf{a}_\beta}{\partial t^\alpha} = \boldsymbol{\Gamma}_{\beta\alpha}. \tag{20}$$

Then, the elements

$$\Gamma^\gamma_{\alpha\beta} = \frac{\partial \mathbf{a}_\alpha}{\partial t^\beta} \cdot \mathbf{a}^\gamma \quad \text{satisfying} \quad \Gamma^\gamma_{\alpha\beta} = \Gamma^\gamma_{\beta\alpha}, \tag{21}$$

are referred to as surface Christoffel symbols of the second kind (or simply surface Christoffel symbols). They represent an array of 6 independent quantities for curved surfaces.

The sensitivity of surface covariant basis vectors with respect to the Gaussian coordinates is referred to as the Gauss formulas:

$$\frac{\partial \mathbf{a}_\alpha}{\partial t^\beta} = \Gamma^\gamma_{\alpha\beta} \mathbf{a}_\gamma + b_{\alpha\beta} \widehat{\mathbf{n}}, \tag{22}$$

where $b_{\alpha\beta}$ is known as the surface covariant curvature tensor; given by,

$$b_{\alpha\beta} = \frac{\partial \mathbf{a}_\alpha}{\partial t^\beta} \cdot \widehat{\mathbf{n}} = \frac{\partial \mathbf{a}_\beta}{\partial t^\alpha} \cdot \widehat{\mathbf{n}} = b_{\beta\alpha}. \tag{23}$$

The surface covariant curvature tensor admits the following forms

$$b_{\alpha\beta} = \frac{\partial^2 \mathbf{x}}{\partial t^\alpha \partial t^\beta} \cdot \widehat{\mathbf{n}} = -\frac{\partial \mathbf{x}}{\partial t^\alpha} \cdot \frac{\partial \widehat{\mathbf{n}}}{\partial t^\beta} = -\frac{\partial \mathbf{x}}{\partial t^\beta} \cdot \frac{\partial \widehat{\mathbf{n}}}{\partial t^\alpha}. \tag{24}$$

The sensitivity of normal surface vector relative to the Gaussian coordinates is referred to as the Weingarten formulas:

$$\frac{\partial \widehat{\mathbf{n}}}{\partial t^\alpha} = -b_\alpha^{\cdot\beta} \mathbf{a}_\beta = -b_{\alpha\beta} \mathbf{a}^\beta \quad \text{where} \quad b_\alpha^{\cdot\beta} = b_{\alpha\gamma} a^{\gamma\beta} = a^{\beta\gamma} b_{\gamma\alpha} = b^\beta_{\cdot\alpha}, \tag{25}$$

is termed the surface mixed curvature tensor. The principal invariants of the matrix $[b_\alpha^{\cdot\beta}]$ play an important role in differential geometry of curved surfaces. The so-called mean curvature is defined to be the half of the trace of that matrix and what is called the Gaussian curvature is basically its determinant. They are given by

$$\mathrm{H} = \frac{1}{2}\mathrm{tr}\left[b_{\alpha}^{\cdot\beta}\right] = \frac{1}{2}\left(b_{1}^{\cdot 1} + b_{2}^{\cdot 2}\right) \quad , \quad \mathrm{K} = \det\left[b_{\alpha}^{\cdot\beta}\right] = b_{1}^{\cdot 1} b_{2}^{\cdot 2} - b_{1}^{\cdot 2} b_{2}^{\cdot 1}. \tag{26}$$

It can be shown that the Gaussian curvature, which is of fundamental importance in differential geometry of surfaces, can only be captured by measuring distances on the surfaces. In another words, this object has basically an intrinsic attribute.

The surface Christoffel symbols can also be written as

$$\Gamma^{\alpha}_{\beta\gamma} = -\frac{\partial \mathbf{a}^{\alpha}}{\partial t^{\gamma}} \cdot \mathbf{a}_{\beta} \ \text{ satisfying } \ \Gamma^{\alpha}_{\beta\gamma} = \Gamma^{\alpha}_{\gamma\beta} \,. \tag{27}$$

The partial derivative of the surface metric coefficients are

$$\frac{\partial a_{\alpha\beta}}{\partial t^{\gamma}} = \Gamma^{\rho}_{\gamma\alpha}\, a_{\rho\beta} + \Gamma^{\rho}_{\gamma\beta}\, a_{\rho\alpha} \quad , \quad \frac{\partial a^{\alpha\beta}}{\partial t^{\gamma}} = -\Gamma^{\alpha}_{\gamma\rho}\, a^{\rho\beta} - \Gamma^{\beta}_{\gamma\rho}\, a^{\rho\alpha}. \tag{28}$$

It is not then difficult to see that

$$\frac{\partial \mathbf{a}^{\alpha}}{\partial t^{\beta}} = -\Gamma^{\alpha}_{\beta\gamma}\, \mathbf{a}^{\gamma} + b^{\alpha}_{\cdot\beta}\, \widehat{\mathbf{n}} \ \text{ where } \ b^{\alpha}_{\cdot\beta} = a^{\alpha\theta} b_{\theta\beta} = b_{\beta}^{\cdot\alpha}. \tag{29}$$

1.3 Surface First Covariant Derivative

Let $\mathbf{w} = \mathbf{w}\left(t^1, t^2\right)$ be a given surface vector field. Then, by using the expressions (15), (22) and (29) along with the product rule, the rate of change in this vector field can be represented with respect to the (ambient) bases $\{\mathbf{a}_1, \mathbf{a}_2, \widehat{\mathbf{n}}\}$ and $\left\{\mathbf{a}^1, \mathbf{a}^2, \widehat{\mathbf{n}}\right\}$ as

$$\frac{\partial \mathbf{w}}{\partial t^{\beta}} = \frac{\partial\left(w^{\alpha}\mathbf{a}_{\alpha}\right)}{\partial t^{\beta}} = w^{\alpha}|_{\beta}\, \mathbf{a}_{\alpha} + w^{\alpha} b_{\alpha\beta}\, \widehat{\mathbf{n}} = \frac{\partial\left(w_{\alpha}\mathbf{a}^{\alpha}\right)}{\partial t^{\beta}} = w_{\alpha}|_{\beta}\, \mathbf{a}^{\alpha} + w_{\alpha} b^{\alpha}_{\cdot\beta}\, \widehat{\mathbf{n}} \,, \tag{30}$$

where $w^{\alpha}|_{\beta}$ and $w_{\alpha}|_{\beta}$ present the surface first-order covariant derivative of w^{α} and w_{α}, respectively. They are given by [4]

$$w^{\alpha}|_{\beta} = \frac{\partial w^{\alpha}}{\partial t^{\beta}} + \Gamma^{\alpha}_{\beta\gamma} w^{\gamma} \quad , \quad w_{\alpha}|_{\beta} = \frac{\partial w_{\alpha}}{\partial t^{\beta}} - \Gamma^{\gamma}_{\alpha\beta} w_{\gamma}. \tag{31}$$

Note that the surface covariant derivative is the only rate of change that the two-dimensional entities living on the surface can measure. And it should thus be viewed as an intrinsic operation.

2 Application of Differential Geometry in Shells

Differential geometry of two-dimensional smooth surfaces was briefly introduced in the previous section. The goal here is to utilize the formulations developed so far for the kinematics of a shell mathematical model. Shells are basically thin-walled structures whose thickness is small in comparison with their other two dimensions. This specific geometric feature of these three-dimensional creatures allows one to separate the thickness variable from the in-plane ones. As a result, one can describe the geometry as well as deformation of shells by using proper functions of two in-plane variables corresponding to their midsurface. In another words, the problem is simplified to finding some unknown two-variable functions. That is how differential geometry of surfaces finds its way in mathematical description of shell structures, see [8] and references therein.

The three-dimensional geometry of the shell model under consideration is described by [2]

$$\mathbf{x}^*\left(t^1,t^2,t^3\right)=\mathbf{x}\left(t^1,t^2\right)+t^3\mathbf{a}_3\left(t^1,t^2\right) \quad \text{with} \quad t^3\in\left]-\frac{h}{2},+\frac{h}{2}\right[, \tag{32}$$

where $\mathbf{x}\left(t^1,t^2\right)$ presents the position vector of an arbitrary point on the midsurface, $\mathbf{a}_3=\mathbf{a}_1\times\mathbf{a}_2/\left|\mathbf{a}_1\times\mathbf{a}_2\right|$ denotes the midsurface normal vector wherein $\mathbf{a}_\alpha=\partial\mathbf{x}/\partial t^\alpha$ and h is the shell thickness. See Fig. 1 for a geometrical interpretation. Using (3) and (25), the sensitivity of $\mathbf{x}^*$ with respect to t^i, $i=1,2,3$, will provide the new covariant basis vectors

$$\mathbf{g}_\alpha=\frac{\partial\left(\mathbf{x}+t^3\mathbf{a}_3\right)}{\partial t^\alpha}=\mathbf{a}_\alpha-t^3b_{\alpha}^{\cdot\beta}\mathbf{a}_\beta \quad \text{and} \quad \mathbf{g}_3=\frac{\partial\left(\mathbf{x}+t^3\mathbf{a}_3\right)}{\partial t^3}=\mathbf{a}_3. \tag{33}$$

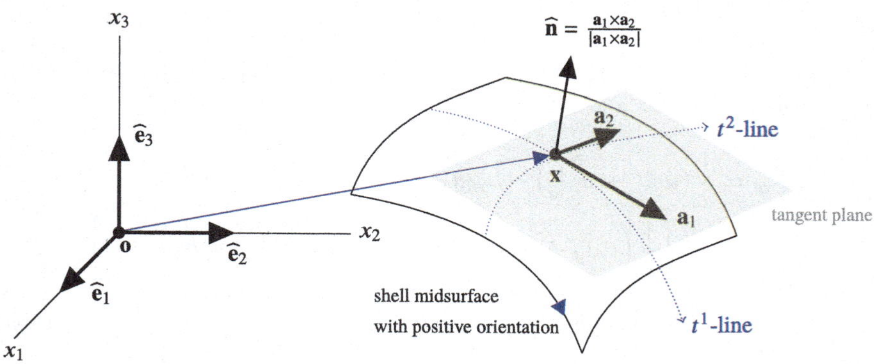

Fig. 1 Geometric description of the shell midsurface

This helps one to construct [9]

$$g_{ij} = \mathbf{g}_i \cdot \mathbf{g}_j \quad \text{and, subsequently,} \quad \left[g^{ij}\right] = \left[g_{ij}\right]^{-1}, \tag{34}$$

to arrive at the dual vectors

$$\mathbf{g}^i = g^{ij}\mathbf{g}_j. \tag{35}$$

Consistent with (32), the displacement field of a typical particle in the shell continuum is represented by

$$\mathbf{u}^*\left(t^1, t^2, t^3\right) = \mathbf{u}\left(t^1, t^2\right) + t^3\boldsymbol{\theta}\left(t^1, t^2\right), \tag{36}$$

where $\mathbf{u}$ denotes the translational displacement of a material point on the midsurface and $\boldsymbol{\theta}$ presents the infinitesimal rotation of a material line perpendicular to the midsurface. With respect to the dual basis $\{\mathbf{a}^i\}$, the ambient vector $\mathbf{u}$ and the surface vector $\boldsymbol{\theta}$ are decomposed according to

$$\mathbf{u} = u_\alpha \mathbf{a}^\alpha + u_3\mathbf{a}^3 \quad , \quad \boldsymbol{\theta} = \theta_\alpha \mathbf{a}^\alpha + 0\mathbf{a}^3. \tag{37}$$

Denoting by ^{T} the transpose operator and $\otimes$ the tensor product, the infinitesimal strain tensor for a linear shell reads

$$\epsilon = \frac{1}{2}\left[\left(\frac{\partial \mathbf{u}^*}{\partial \mathbf{x}^*}\right) + \left(\frac{\partial \mathbf{u}^*}{\partial \mathbf{x}^*}\right)^{\mathrm{T}}\right] \quad \text{where} \quad \frac{\partial \mathbf{u}^*}{\partial \mathbf{x}^*} = \frac{\partial \mathbf{u}^*}{\partial t^k} \otimes \mathbf{g}^k. \tag{38}$$

The covariant components of this symmetric second-order tensor, ϵ_{ij}, with respect to the contravariant basis $\{\mathbf{g}^i\}$ render

$$\epsilon_{ij} = \frac{1}{2}\mathbf{g}_i \cdot \left[\frac{\partial \mathbf{u}^*}{\partial t^k} \otimes \mathbf{g}^k + \mathbf{g}^k \otimes \frac{\partial \mathbf{u}^*}{\partial t^k}\right]\mathbf{g}_j = \frac{1}{2}\left[\mathbf{g}_i \cdot \frac{\partial \mathbf{u}^*}{\partial t^j} + \frac{\partial \mathbf{u}^*}{\partial t^i} \cdot \mathbf{g}_j\right]. \tag{39}$$

Using $(12)_1$, $(25)_1$, $(30)_4$, (33), (36) and (37), the covariant components (39) may be represented by

$$\begin{aligned}
\epsilon_{\alpha\beta} = & \frac{1}{2}\left(u_\alpha|_\beta + u_\beta|_\alpha\right) - b_{\alpha\beta}u_3 \\
& + t^3\left[\frac{1}{2}\left(\theta_\alpha|_\beta + \theta_\beta|_\alpha - b_\alpha^{\cdot\gamma}\, u_\gamma|_\beta - b_\beta^{\cdot\gamma}\, u_\gamma|_\alpha\right) + b_\alpha^{\cdot\gamma} b_{\gamma\beta}u_3\right] \\
& - \left(t^3\right)^2\left[\frac{1}{2}\left(b_\alpha^{\cdot\gamma}\, \theta_\gamma|_\beta + b_\beta^{\cdot\gamma}\, \theta_\gamma|_\alpha\right)\right],
\end{aligned} \tag{40a}$$

$$\epsilon_{\alpha 3} = \frac{1}{2}\left(\theta_\alpha + \frac{\partial u_3}{\partial t^\alpha} + b_\alpha^{\cdot\gamma} u_\gamma\right), \tag{40b}$$

$$\epsilon_{33} = 0\,. \tag{40c}$$

Accordingly, by using the plane stress assumption, the contravariant components of the stress tensor for an isotropic linear elastic shell take the following form

$$\sigma^{\alpha\beta} = \frac{E}{2\,(1+\nu)}\left[g^{\alpha\gamma}g^{\beta\delta} + g^{\alpha\delta}g^{\beta\gamma} + \frac{2\nu}{1-\nu}g^{\alpha\beta}g^{\gamma\delta}\right]\epsilon_{\gamma\delta}, \tag{41a}$$

$$\sigma^{\alpha 3} = \frac{E}{1+\nu}\left[g^{\alpha\beta}\right]\epsilon_{\beta 3}, \tag{41b}$$

$$\sigma^{33} = 0\,, \tag{41c}$$

where E denotes Young's modulus and ν is Poisson's ratio.

3 Results and Discussions

To investigate the performance of the formulation developed in the previous sections, an isotropic linear elastic shell model is numerically implemented by means of finite element method (FEM). It is important to point out that the developed formulation doesn't rely on standard isotropic concept of FEM. Indeed, the geometry of shell is exactly described by parametrization of the shell midsurface. And this leaves the displacement field as the only variable to be interpolated. In this work, it is interpolated by using the bilinear ansatz functions of the popular quadrilateral element, see [1, 11].

The only example here regards a hyperboloid shell of one sheet; defined by,

$$x_1 = R_1 \cosh t^1 \cos t^2 \quad , \quad x_2 = R_2 \cosh t^1 \sin t^2 \quad , \quad x_3 = R_3 \sinh t^1 \;, \tag{42}$$

where R_1, R_2, R_3 are nonzero real numbers, $-\infty < t^1 < \infty$ and $0 \le t^2 < 2\pi$. Figure 2 displays the overall shape of the model under consideration as well as the geometrically exact representation of a single element (in red) that is selected for investigation. Moreover, the global deformation of the shell as well as the stress magnitude in the chosen element are depicted therein. In this example, $E = 2.0 \times 10^5$Mpa and $\nu = 0.333$. And the numerical value of the parameters are reported in Table 1. Note that the parametrization (42) serves as the keystone to exactly represent the basis vectors, metric coefficients, curvature tensor and Christoffel symbols in a completely analytical manner. The procedure to calculate the strain and stress tensors at every point within the specific finite element is then straightforward. This has been carried out in an in-house MATLAB code.

One can see that the difference in the maximum stress amounts to 10% for this size of mesh. To understand the impact of element size on the error, one can take a look at Fig. 3. It reveals that for ANSYS SHELL181 the results are improved if the surface curvature and metrics are properly resolved upon refining the mesh, while our geometrically exact shell is almost insensitive to the mesh size due to the usage

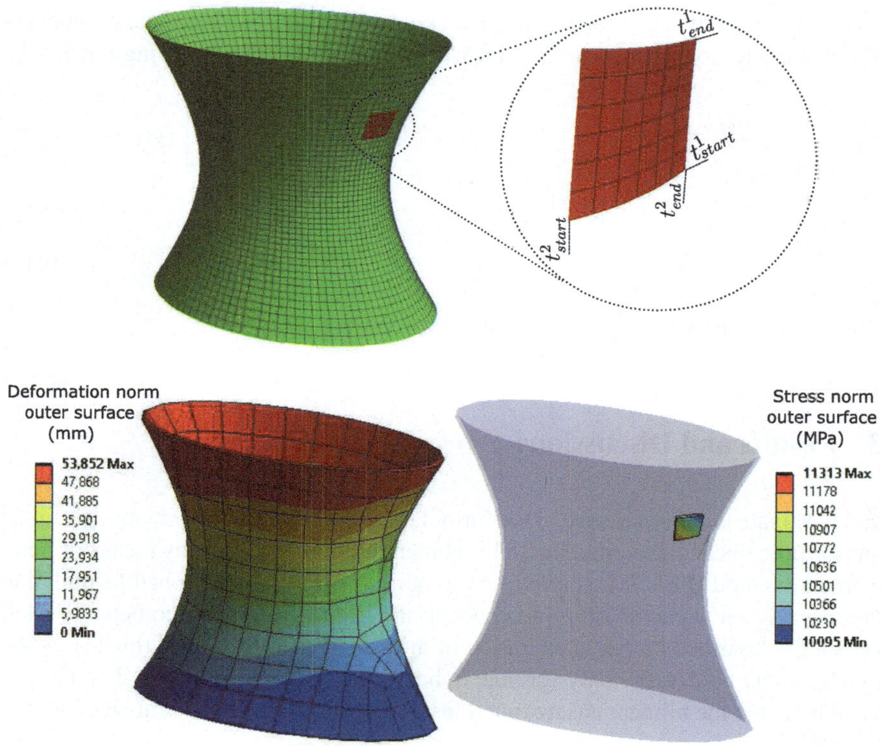

Fig. 2 (Top) Geometrically exact shell parametrization and, (Bottom) ANSYS results for the shell using standard element SHELL181

Table 1 The parameters for the geometrically exact shell element selected from the hyperboloid shell body

Parameter	Value	Unit
t^1_{start}	0.544171	—
t^1_{end}	0.791615	—
t^2_{start}	0.000000	—
t^1_{end}	0.288944	—
R_1	200	mm
R_2	300	mm
R_3	350	mm

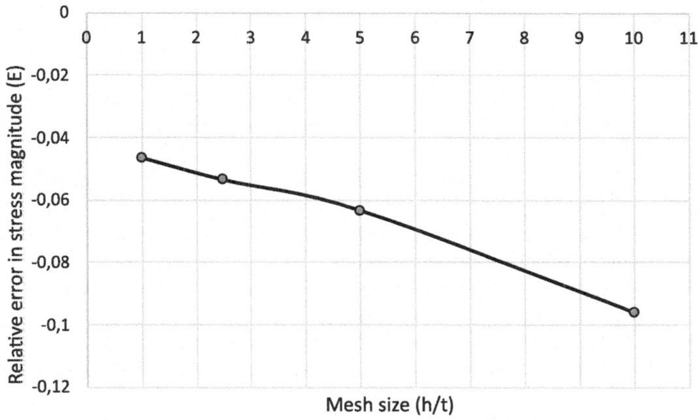

Fig. 3 Relative error (E) in maximum stress versus mesh refinement index

of natural curvilinear coordinate. The refinement index (horizontal axis) is the ratio between the element size (h) and element thickness (t). The relative error for the selected element is defined as $\mathrm{E} = \left(\sigma_{max}^{\mathrm{ANSYS}} - \sigma_{\max}^{*}\right) / \sigma_{\max}^{*}$.

4 Conclusions

In this work, a shell structure was treated as a two-dimensional manifold equipped with a curvilinear coordinate system. The geometrically exact parametrization of the shell was utilized instead of commonly used parent coordinate of the standard FEM shape functions. Clearly speaking, the concept of isoparametric mapping was dismissed unlike the standard approach in FE analysis. The covariant derivative concept was utilized to compute the strain and consequently the stress tensor fields. Covariant derivative of vector fields on a surface (typically, the unknown displacement vector field of the middle surface of a shell) was plugged into the constitutive equation of the shell. The emergence of Christoffel symbols, first and second fundamental forms is natural in this framework and hence they have to be evaluated numerically. As an indicator, the maximum stress is compared with those of predicted by ANSYS using 4-noded shell element named SHELL181. This element is naturally flat with straight wedges due to the total number of nodes. The extent of agreement between the two was investigated in case of more refined mesh. The ANSYS solution shows a stress and energy 10% smaller than our geometrically exact shell model, revealing that the metrics of a curved surface are not captured properly using a coarse mesh. Once the mesh is refined the error shrinks. One should keep in mind that the constitutive behavior of the shell is assumed to be linear. A possible direction for the extension of this work is to reformulate the geometrically exact shell mathematical model for structures undergoing finite strains.

References

1. Bathe, K. J. (1996). *Finite element procedures*. New York: Prentice Hall.
2. Chapelle, D., & Bathe, K. J. (2003). *The finite element analysis of shells-fundamentals*. Berlin: Springer.
3. Do Carmo, M. P. (2016). *Differential geometry of curves and surfaces* (2nd ed.). Dover Publications.
4. Grinfeld, P. (2013). *Introduction to tensor analysis and the calculus of moving surfaces*. New York, NY: Springer Science + Business Media.
5. Itskov, M. (2019). *Tensor algebra and tensor analysis for engineers—With applications to continuum mechanics* (5th ed.). Berlin: Springer.
6. Kreyszig, E. (1991). *Differential geometry* (2nd ed.). Dover Publications.
7. Lovelock, D., & Rund, H. (1989). *Tensors, differential forms, and variational principles*. Dover Publications.
8. Naghdi, P. M. (1973). The theory of shells and plates. In C. Truesdell (Ed.), *Linear theories of elasticity and thermoelasticity.* Heidelberg: Springer, Berlin.
9. Nguyen-Schäfer, H., & Schmidt, J. P. (2017). *Tensor analysis and elementary differential geometry for physicists and engineers* (2nd ed.). Berlin, Heidelberg: Springer.
10. Steinmann, P. (2015). *Geometrical foundations of continuum mechanics—An application to first- and second-order elasticity and elasto-plasticity*. Berlin, Heidelberg: Springer.
11. Wriggers, P. (2008). *Nonlinear finite element methods*. Berlin: Springer.

Phase Field Modeling of Fatigue Fracture

Christoph Schreiber, Ralf Müller, and Fadi Aldakheel

This paper is dedicated to Professor Peter Wriggers in acknowledgement of his seminal contributions to Computational Mechanics and his instructive works in Engineering Mechanics. I got in contact with Peter Wriggers during his first semester in Darmstadt; he in his first semester as a full professor, me in the first semester as a student of Mechanical Engineering. I have always enjoyed Peter's structured and efficient way, especially during writing Engineering Mechanics books together with him. For his 70th birthday I wish Peter all the best and many successful, healthy and happy years to come. (Ralf Müller)

Abstract In this contribution we introduce a phase field model for fatigue crack growth. The model is based on a diffuse formulation of quasi static brittle fracture. In order to account for the fatigue phenomenon an additional energy contribution is incorporated. This additional component represents the amount of accumulated energy associated to irreversibilities of cyclic loading and unloading. The evolution of a fracture phase field is governed by an appropriate Ginzburg-Landau type equation. To enable efficient computation the integration scheme is transferred into the cycle domain. Finally, by showing results of different fatigue crack growth scenarios the model behaviour in terms of crack growth rate, mean stress effect and also growth direction is illustrated.

C. Schreiber · R. Müller (✉)
University of Kaiserslautern, Institute of Applied Mechanics, Kaiserslautern, Germany
e-mail: ram@rhrk.uni-kl.de

C. Schreiber
e-mail: schreibc@rhrk.uni-kl.de

F. Aldakheel
Leibniz University Hannover, Institute of Continuum Mechanics, Garbsen, Germany
e-mail: aldakheel@ikm.uni-hannover.de

F. Aldakheel et al. (eds.), *Current Trends and Open Problems in Computational Mechanics*, https://doi.org/10.1007/978-3-030-87312-7_46

1 Introduction

The fact that the phenomenon of cyclic mechanical fatigue is one of the most crucial fields of structural design of technical components is uncontroversial. A majority of structures undergo cyclic loads and react to this type of loading with fatigue failure. This effect may not emerge in some cases where the state of fatigue stays noncritical during the complete time of service. However, it must always be considered to avoid devastating failures of machines or structures like airplanes, trains, bridges, turbines, pressure vessels and others, which happened in the past. As it is outlined in a number of textbooks as e.g. [13] the damage of a material increases during cyclic loading, where the load maximum may be far below static design limits. Within practical applications mainly empirical laws are considered as e.g. the Miner rule [10] for fatigue damage accumulation or Paris' law [12] for quantification of fatigue crack growth rates. The issue in terms of numerical modeling is on one hand the lag of a general mechanistic basis of the phenomenon and on the other hand the high number of load cycles that must be considered. Also the basic approximation of cracks is related to issues concerning element deletion or adaptive remeshing. In contrast to conventional sharp interface approaches the phase field modeling paradigm provides several benefits (see e.g. [1, 8]). A crack is modeled via an order parameter indicating broken or intact material. This order parameter is coupled to the mechanical boundary value problem. The method was frequently applied and extended within the last decade. However the fatigue phenomenon is considered only recently in first contributions [2, 4, 16]. The resistance against cracking is decreased within these studies in order to enable crack growth under static uncritical loads. A different approach was established in [14], where an additional fatigue related energy contribution is considered as driving force of fatigue cracks. In what follows we introduce our phase field model for fatigue fracture and provide several results from numerical simulation, which will be assessed.

2 Model Formulation

The temporal evolution of the phase field variable s can be obtained by the time dependent Ginzburg-Landau equation:

$$\dot{s} = -M \frac{\delta F}{\delta s}, \tag{1}$$

where $F[s, \nabla s]$ is a free energy potential. In this equation M is a viscous regularization coefficient to control the relaxation towards stationary states. The potential F is defined by an energy density ψ with.

$$F[s, \nabla s] = \int_{\Omega} \psi(s, \nabla s) \mathrm{d}V. \tag{2}$$

The functional derivative is defined similar to the Euler-Lagrange equations of calculus of variations by

$$\frac{\delta F}{\delta s} = \frac{\partial \psi}{\partial s} - \nabla \cdot \left(\frac{\partial \psi}{\partial \nabla s} \right). \tag{3}$$

Accordingly (1) yields

$$\frac{1}{M} \dot{s} = \nabla \cdot \left(\frac{\partial \psi}{\partial \nabla s} \right) - \frac{\partial \psi}{\partial s}. \tag{4}$$

The state $\delta F/\delta s = 0$ is a necessary condition for an extreme distribution of s. Therefore, this characterizes stationary states and (4) can be viewed as relaxation to reach stationary states, where the transition rate is governed by the viscosity parameter $M > 0$.

The phase field model for brittle fracture from Kuhn and Müller [8] utilizes the energy potential

$$F[\boldsymbol{\varepsilon}, s, \nabla s] = \int_{\Omega} \Bigg[\underbrace{(g(s) + \eta) \frac{1}{2} \boldsymbol{\varepsilon} : [\mathbb{C}\boldsymbol{\varepsilon}]}_{\psi_e} + \underbrace{\mathcal{G}_{\mathrm{c}} \left(\frac{(1-s)^2}{4\epsilon} + \epsilon |\nabla s|^2 \right)}_{\psi_c} \Bigg] \mathrm{d}V. \tag{5}$$

The phase field variable s interpolates between broken ($s = 0$) and intact ($s = 1$) material. The energy functional (5) was proposed in [3] as a regularization of the variational formulation of brittle fracture (see [6]). In (5) ψ_c is the specific crack energy and the specific elastic energy is modeled by ψ_e. In ψ_e the elastic energy density for small deformations with the linearized second order strain tensor $\boldsymbol{\varepsilon} = \frac{1}{2}(\nabla \boldsymbol{u} + \nabla^{\mathrm{T}} \boldsymbol{u})$ and the fourth order stiffness tensor $\mathbb{C}$ is used. In order to account for the degradation of the material's stiffness in areas with $s < 1$ the strain energy is multiplied with $g(s) + \eta$. The scalar function $g(s)$ is referred to as a degradation function and models the material response to a variation of the phase field s. The parameter η with $0 < \eta \ll 1$ is introduced for numerical convenience and ensures a residual stiffness $\eta\mathbb{C}$. The second term in (5) may be interpreted as regularization of Griffith's crack surface energy, and $\mathcal{G}_{\mathrm{c}}$ can be considered as a crack resistance.

The potential (5) covers linear elastic fracture events rather well. The peculiarity for fatigue fracture is that the static fracture limit can be significantly higher than the threshold for the onset of fatigue crack growth. But certainly failure occurs under cyclic loading. In other words, it will become necessary to decrease the phase field s, than to allow for more strain energy. Due to the significantly lower maximum load value in the cyclic situation, an evolution of the phase field solely by (5) is not appropriate. However, a cyclic load with repetition of loading and unloading indeed consumes a significant amount of work by irreversible processes associated with fatigue (see e.g. [11]). Obviously, the classical model [8] does not take into account the presence of these irreversibilities. Hence, a consequent conclusion is to include an additional contribution into the free energy. The new enhanced regularized formulation of the energy is proposed as

$$
\begin{aligned}
F^{\mathrm{f}} &= F(\boldsymbol{\varepsilon}, s, \nabla s) + F^{\mathrm{ac}}(D_{\mathrm{f}}, s) \\
&= \int_{\Omega} \Big[(g(s) + \eta) W + \mathcal{G}_c \Gamma \Big] \,\mathrm{d}V + \int_{\Omega} \underbrace{h(s) P(D_{\mathrm{f}})}_{\psi_{\mathrm{f}}} \,\mathrm{d}V
\end{aligned}
\tag{6}
$$

in order to account for the discussed fatigue related work contribution. This functional of the total internal energy consists of the basic formulation of brittle fracture and an additional energy contribution characterized by the specific energy term ψ_{f}. This contribution is governed by the current state of fatigue damage D_{f} and provides the necessary sum of additional driving forces. The function $h(s)$ is a degradation function with an analog purpose as $g(s)$. Assuming an increase of the energy density $P(D_{\mathrm{f}})$ this degradation function enables a decline of the total energy by decreasing the phase field s. This mechanism generally enables crack growth driven by fatigue. A piece wise defined power function is proposed to govern the energy contribution ψ_{f} by

$$
P(D_{\mathrm{f}}) = q \Big\langle D_{\mathrm{f}} - D_{\mathrm{fcrit}} \Big\rangle^{b} \tag{7}
$$

with the Macaulay brackets, which are defined as $\langle \cdot \rangle^n = 0$ for $(\cdot) \leq 0$ and $\langle \cdot \rangle^n = (\cdot)^n$ for $(\cdot) > 0$. This function is designed such that the fatigue driving force increases rapidly once the critical value for damage D_{fcrit} is exceeded. A rapid increase is crucial as otherwise cracks are prone to grow in lateral direction, i.e. “fat”. To evaluate the fatigue damage D_{f} Miner’s rule is employed in a first approach. However, more complex models can be incorporated if desired. The critical cycle number N_i for a certain stress amplitude can be obtained from appropriate S-N curves. Accordingly, the estimate for the fatigue damage D_{f} in (7) becomes

$$
dD_{\mathrm{f}} = D_0 + \left(n^D\right)^{-1} \left(\frac{\tilde{\sigma}}{A_D}\right)^{k} \mathrm{d}N. \tag{8}
$$

The actual driving force entering to evaluate the $S - N$ curves in (8) is the stress $\tilde{\sigma}$, which is a scalar amplitude stress measure depending on the stress state. The exact choice may depend on several properties like for instance the material or the type of loading. Other important quantities are the fatigue limit A_D, the knee point cycle number n^D and the slope of the $S - N$ curve k. Within the cyclic resolved simulation scheme, the previous state of fatigue damage D_0 is treated as a history variable updated for every simulation step. The energetically associated stresses provide a good insight into the physical interpretation of this enhanced phase field formulation:

$$\boldsymbol{\sigma} = \frac{\partial\left(\psi_{\mathrm{e}} + \psi_{\mathrm{c}} + \psi_{\mathrm{f}}\right)}{\partial \boldsymbol{\varepsilon}} = \underbrace{(g(s) + \eta)\mathbb{C}\boldsymbol{\varepsilon}}_{\boldsymbol{\sigma}_{\mathrm{e}}} + \underbrace{h(s)\frac{\partial P}{\partial D_{\mathrm{f}}}\frac{\partial D_{\mathrm{f}}}{\partial \tilde{\sigma}}\frac{\partial \tilde{\sigma}}{\partial \boldsymbol{\varepsilon}}}_{\boldsymbol{\sigma}_{\mathrm{f}}}, \tag{9}$$

with the fatigue contribution to the stresses:

$$\boldsymbol{\sigma}_{\mathrm{f}} = h(s) q b \Big\langle D_{\mathrm{f}} - D_{\mathrm{fcrit}} \Big\rangle^{b-1} \frac{k \tilde{\sigma}^{k-1}}{n^{D} A_{D}} \frac{\partial \tilde{\sigma}}{\partial \boldsymbol{\varepsilon}}, \tag{10}$$

where the derivative $\frac{\partial \tilde{\sigma}}{\partial \boldsymbol{\varepsilon}}$ depends on the choice for the respective driving stress. The first term in Eq. (9) is the usual degraded tensor of the static stresses. The second term can be interpreted as additional stress component accounting for accumulated micro stresses as consequence of cyclic slip irreversibilities caused by repetition of loading and unloading.

As fatigue damage is evaluated only with regard to the cycle number the evolution Eq. (4) is transferred into the cycle domain

$$\frac{\mathrm{d}s}{\mathrm{d}N} = -\hat{M}\left\{g(s)'\frac{1}{2}\boldsymbol{\varepsilon} : [\mathbb{C}\boldsymbol{\varepsilon}] - \mathcal{G}_c\left(2\epsilon\nabla \cdot \nabla s - \frac{s-1}{2\epsilon}\right) + h(s)'P(D_{\mathrm{f}})\right\}, \tag{11}$$

with a mobility constant $\hat{M}$ related to the cycle number.

3 Numerical Examples

The phase field model was implemented as a quadrilateral element routine within the finite element software FEAP. The examples explained in the following were set up to illustrate some features of the proposed model.

3.1 R-ratio and Sequence Effects

To characterize and measure fracture mechanical properties the CT-specimen is often used. The geometric setup as well as the loading and the discretization is sketched in Fig. 1a. The loading introduces a pure mode I loading. Thus a straight crack occurs, as is depicted in Fig. 1b by a contour plot of the fracture field s. The simulations were conducted with a force control. As also indicated in Fig. 2a different load ratios (R-ratios)

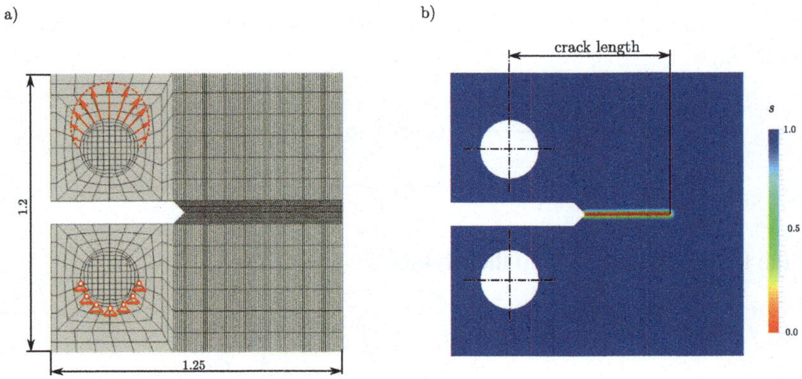

Fig. 1 Set up and result of phase field fatigue crack growth simulation: **a** discretization **b** contour plot of the phase field parameter

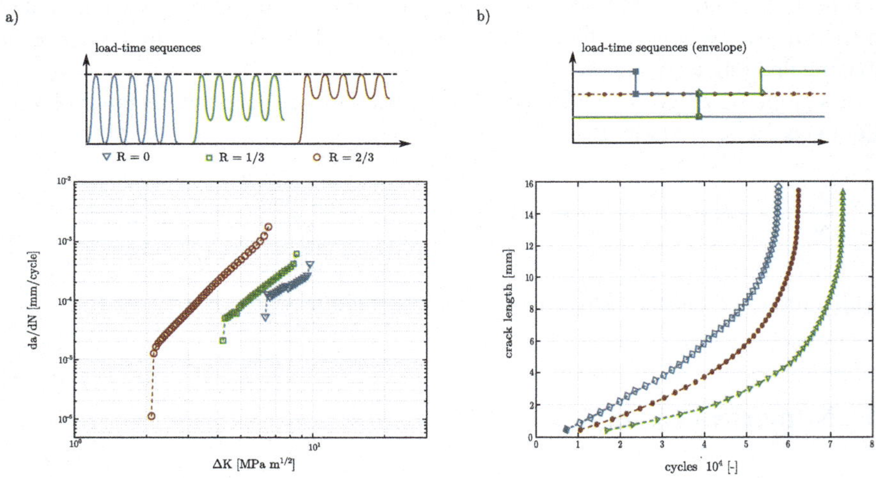

Fig. 2 Evaluation of crack propagation: **a** different R-ratios and **b** different load sequences

$$R = \frac{F_{\min}}{F_{\max}} \tag{12}$$

were simulated. Simulations with different parameter sets were run. Within these simulations the normal stress in vertical direction was used as driving stress $\tilde{\sigma}$. For the different R-ratios the minimal force is increased, see Fig. 2a (top), resulting in an increased crack propagation speed. This is reported in the $\mathrm{d}a/\mathrm{d}N$ curves. The almost linear relation in the logarithmic plots indicates, that the fracture phase field model predicts a Paris' law of the type

$$\frac{\mathrm{d}a}{\mathrm{d}N} = C(\Delta K)^n , \tag{13}$$

where ΔK is the difference in the stress intensity factor (K-factor) for the applied forces F_{max} and F_{min}. The model thus captures an influence of the R-ratio on the parameters of Paris' law, which is a frequently observed phenomena (see e.g. [13]). For a detailed analyse of this property of the phase field model see [15]. In Fig. 2b the influence of load sequences is analysed. In a cyclic setup with $R = 0$ the maximal load is distributed differently throughout the load cycles, see sketch in Fig. 2b (top). It is interesting to note, that from Fig. 2b the sequence with high loads first and lower loads later has the fastest crack propagation, as compared to a constant load or a sequence of first low loads and then high loads. Thus the fracture phase field model using Miner's rule, which has no sequence sensibility, shows a sequence effect by the solution of a complete time dependent boundary value problem.

3.2 Mode II Loading

In order to analyse not only the features for a straight crack propagation, which occurs under mode I loading conditions, a mode II situation is analyzed. Therefore, in the situation depicted in Fig. 3a the driving stress $\tilde{\sigma}$ was taken to be the first (largest) principal stress. This is in agreement with the maximum tangential stress criterion proposed in [5]. It can be seen from Fig. 3b and c that the crack now propagates under an angle of around 70° from the horizontal direction. This feature is in agreement with analytical findings from linear elastic fracture mechanics (see e.g. [7]) and experimental observations [9]. Thus using the correct driving stress the phase field model for fatigue fracture not only predicts the crack propagation in terms of growth rate but also in terms of direction.

4 Conclusion

The proposed model is an extension of the fracture phase field model for quasi static brittle fracture to fatigue scenarios. In contrast to other approaches [2, 4, 16] in which the crack resistance $\mathcal{G}_c$ is reduced as function of the load history, we propose a modification of the crack driving force. This assumption allows the parameter $\mathcal{G}_c$ to be interpreted as the crack resistance related to the experimentally determined fracture toughness. It is mentioned that the static fracture toughness is measured on sharp cracks which are introduced by a cyclic loading phase.

To drive the crack propagation in the phase field setup Miner's rule is used. Although being rather simple this results in complex model response. The model predicts different well known phenomena from fatigue fracture mechanics, among them R-ratio dependent Paris' law parameters, load sequence effects as well as

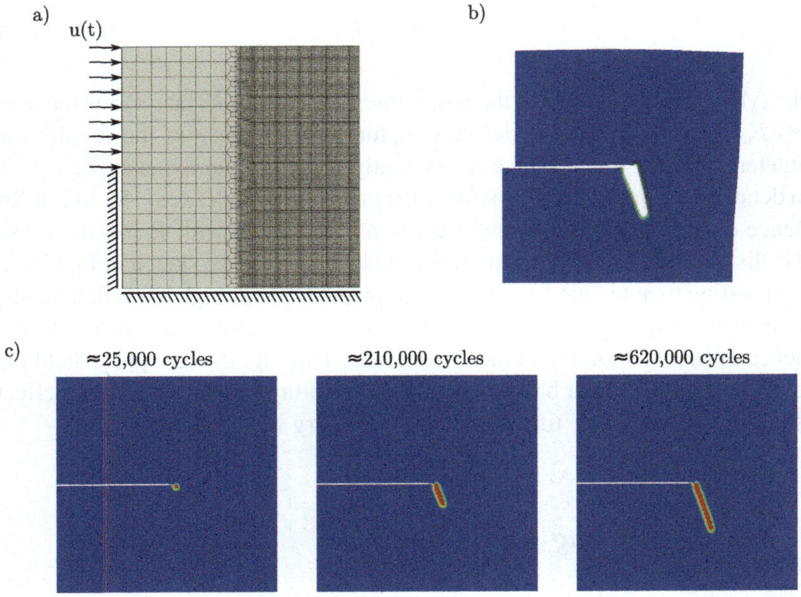

Fig. 3 Numerical example of a mode II load case: **a** set up, **b** deformation of cracked specimen after 620,000 cycles and **c** contour plot of phase field variable for different number of load cycles

prediction of the crack path. It is emphasized that in contrast to classical fatigue fracture simulations Paris' law is not needed to simulated crack propagation, but an outcome of the model. Thus Paris' law serves here more as a model validation due to its wide range of acceptance in fatigue fracture mechanics.

Acknowledgements Christoph Schreiber and Ralf Müller gratefully acknowledge the funding for this research by the German Science Foundation (DFG) within IRTG 2057-2524083 and SPP 1748-255846293. Fadi Aldakheel gratefully acknowledges the support by (DFG) within SPP 2020-WR 19/58-2.

References

1. Aldakheel, F., Wrigger, P., & Miehe, C. (2018). A modified Gurson-type plasticity model at finite strains: Formulation, numerical analysis and phase-field coupling. *Computational Mechanics, 62,* 815–833.
2. Alessi, R., Vidoli, S., & DeLorenzis, L. (2017). A phenomenological approach to fatigue with a variational phase-field model: The one-dimensional case. *Engineering Fracture Mechanics, 190*, 53–73.
3. Bourdin, B., Francfort, G. A., & Marigio, J.-J. (2000). Numerical experiments in revisited brittle fracture. *Journal of the Mechanics and Physics of Solids, 48*, 797–826.
4. Carrara, P., Marreddy, A., Alessi, R., & DeLorinzis, L. (2018). A novel framework to model the fatigue behavior of brittle materials based on a variational phase-field approach. arXiv: Materials Science

5. Erdogan, F., & Sih, G. C. (1963). On the crack extension in plates un-der plane loading and-transverse shear. *Journal of Basic Engineering, 85*, 519–525.
6. Francfort, G. A., & Marigo, J. J. (1998). Revisiting brittle fracture as an energy minimization problem. *Journal of the Mechanics and Physics of Solids, 8*, 1319–1342.
7. Gross, D., & Seelig, T. (2011). *Bruchmechanik-Mit einer Einfürung in die Mikromechanik*. Berlin: Springer.
8. Kuhn, C., & Müller, R. (2010). A continuum phase field model for fracture. *Engineering Fracture Mechanics, 77*, 3625–3634.
9. Mahajan, R. V., & Ravi Chandar, K. (1989). An experimental investigation of mixed-mode fracture. *International Journal of Fracture, 41*, 235–252.
10. Miner, M. A. (1945). Cumulative damage in fatigue. *Journal of Applied Mechanics, 12*, 159–164.
11. Mughrabi, H. (2013). Microstructural fatigue mechanisms: Cyclic slip irreversibility, crack initiation, non-linear elastic damage analysis. *International Journal of Fatigues, 57*, 2–8.
12. Paris, P. C., & Erdogan, F. (1963). A critical analysis of crack propagation laws. *Journal of Basic Engineering, 85*, 528–539.
13. Schijve, J. (2009). *Fatigue of structures and materials*. Berlin: Springer.
14. Schreiber, C., Kuhn, C., Müller, R., & Zohdi, T. (2020). A phase field modeling approach of cyclic fatigue crack growth. *International Journal of Fracture, 225*, 89–100.
15. Schreiber, C., Kuhn, C., & Müller, R. (2020). Phase field simulation of fatigue crack propagation under complex load situations. *Archive of Applied Mechanics*. https://doi.org/10.1007/s00419-020-01821-0
16. Seiler, M., Hantschke, P., Brosius, A., Kaästner, M. (2018). A numerically efficient phase-field model for fatigue fracture—1D analysis. PAMM 18

Challenges for the Least-Squares Finite Element Method in Solid Mechanics

Jörg Schröder and Maximilian Igelbüscher

Dear Peter,
on the occasion of your 70th birthday we wish you all the best, health, happiness and high efficiency for all your future plans. We would like to express our sincere thanks for your continuous efforts to gain recognition for our scientific discipline. In particular, we would like to thank you for the trusting and successful cooperation on the scientific front of Finite Element Method in the past years. Our friendly and scientific ties were and are a constant incentive for us.
Jörg & Max Igelbüscher

Abstract In this contribution we discuss the mixed least-squares finite element method (LSFEM) for applications in solid mechanics. The LSFEM is characterized by the minimization of the sum of the squared $\mathcal{L}^2(\mathcal{B})$ norms of the residuals of a first order system of differential equations. For an analysis of the LSFEM, we discuss two applications in solid mechanics, with a focus on the challenges of the method. Here, we investigate the moderate performance for low order elements and the sensitive point of choosing the weighting factors and their balancing. Furthermore, we discuss the crucial point of the recalculation of support reactions. In this context we introduce two formulations, one classical approach and one with the balance of angular momentum as an additional constraint instead of simple introducing a condition for the symmetry of the stress tensor in order to fulfill the balance of angular momentum. It should be noted, that the symmetry of the stress tensor is not a priori fulfilled due to the application of Raviart-Thomas functions. A further aspect is the consistent approximation of stresses at material interfaces when utilizing Lagrange ansatz functions within the LSFEM. Therefore, we present a hybrid mixed formulation on the basis of a LS functional.

J. Schröder (✉) · M. Igelbüscher
University of Duisburg-Essen, Institute of Mechanics, Essen, Germany
e-mail: j.schroeder@uni-due.de

M. Igelbüscher
e-mail: maximilian.igelbuescher@uni-due.de

F. Aldakheel et al. (eds.), *Current Trends and Open Problems in Computational Mechanics*, https://doi.org/10.1007/978-3-030-87312-7_47

1 Challenges in the LSFEM

The LSFEM has been considered in many applications over the last decades and especially the advantages of the method have been exploited. These advantages are well known and given by the provision of an a posteriori error estimator, the LSFEM is unconstraint by the LBB condition and exhibits symmetric and positive (semi-) definite matrices, cf. [1, 2]. Beside this, the challenges of the LSFEM method are, among others, the poor performance for low order elements and especially the choice of the weighting parameters of the single functionals and their balancing, see e.g. [3–5]. For a stress-displacement least-squares formulation, conforming finite element approximations are given by $\boldsymbol{\sigma} \in \mathcal{H}(\text{div}, \mathcal{B})$ and $\boldsymbol{u} \in \mathcal{H}^1(\mathcal{B})$.

The application of higher-order Raviart-Thomas functions is not straight-forward, especially in 3D applications, since they are constructed based on outer and inner moments. As an alternative, a straight-forward utilization of Lagrange type approximations for the stresses can be applied, but this leads to unphysical stress distributions at material interfaces. This observation is due to the properties of the function space $\mathcal{H}^1(\mathcal{B})$, which enforce continuity of the stresses in the normal and tangential directions. However, from a physical point of view, stresses are only normal continuous. This can be overcome by a hybrid mixed formulation, in which all components of the stress tensor are approximated with Lagrange type ansatz functions on element level and the continuity conditions are enforced by Lagrange multipliers on the inter-element boundaries in a weak sense. An overview on mixed and hybrid finite element formulations is given e.g. in [6–9].

Here, two aspects within the LSFEM are discussed for linear elasticity. The first one presents the comparison of a standard LS formulation and an extended LS formulation. The extended formulation is constructed with explicit consideration of the balance of angular momentum. For these two formulations, we analyze the recalculation of support reactions (forces and moments) for the LSFEM, as a crucial point in engineering applications, and check the global equilibrium conditions for forces and moments. Furthermore, a scale independent formulation is obtained by the chosen weighting parameter setups. The complete discussion on this topic and the results are closely related to the publication [5].

As a second point, a hybrid mixed formulation is constructed. This formulation is based on an extended least-squares functional with discontinuous stress approximations and a variational formulation for the enforcement of the traction continuity using Lagrange multipliers. This approach provides physically consistent approximations of stress values at material interfaces using continuous piecewise polynomial functions of Lagrange type. Furthermore, an application of higher order Lagrange ansatz functions for the stress approximation can be preformed in a straight-forward manner in contrast to the more sophisticated Raviart-Thomas functions.

2 Approximation of Reaction Forces Within the LSFEM

The body of interest $\mathcal{B}$, parameterized in $\boldsymbol{x} \in \mathbb{R}^d$, is introduced, with the boundary $\partial\mathcal{B}$ subdivided into Neumann $\partial\mathcal{B}_N$ and Dirichlet $\partial\mathcal{B}_D$ boundaries with $\partial\mathcal{B} = \partial\mathcal{B}_D \bigcup \partial\mathcal{B}_N$ and $\partial\mathcal{B}_D \bigcap \partial\mathcal{B}_N = \emptyset$. A least-squares formulation is characterized by the minimization of the sum of the squared $\mathcal{L}^2(\mathcal{B})$ norms of the residuals $\mathcal{R}_i$ of a first order system of differential equations, see e.g. [1, 2].

For investigating the approximation quality and reliability of support reactions two LS formulations are introduced. The first formulation $\mathcal{F}(\boldsymbol{\sigma}, \boldsymbol{u})$ is defined in terms of balance of momentum and the constitutive equation by

$$\mathcal{F}(\boldsymbol{\sigma}, \boldsymbol{u}) = \frac{1}{2}\left(\left\|\omega_1(\operatorname{div}\boldsymbol{\sigma} + \boldsymbol{f})\right\|^2_{\mathcal{L}^2(\mathcal{B})} + \left\|\omega_2(\boldsymbol{\sigma} - \mathbb{C} : \nabla^s\boldsymbol{u})\right\|^2_{\mathcal{L}^2(\mathcal{B})}\right), \tag{1}$$

with the fourth-order elasticity tensor $\mathbb{C}$, the body force $\boldsymbol{f}$, the symmetric displacement gradient $\nabla^s\boldsymbol{u}$ and the weighting parameters ω_i, which are discussed later on, cf. [5]. The stress symmetry condition $\boldsymbol{\sigma} = \boldsymbol{\sigma}^T$ is not fulfilled a priori due to the application of $\mathcal{RT}_m$ functions and only enforced in a weak sense, see e.g. [10, 11]. The stress symmetry condition is in general controlled through the constitutive equation, cf. [4].

For the second formulation $\mathcal{F}^\star(\boldsymbol{\sigma}, \boldsymbol{u})$, the system of equations in $\mathcal{F}$ is enhanced by the balance of angular momentum, which is only represented implicitly in (1). An additional control of the balance of angular momentum, as shown e.g. in [12, 13], leads to an improved formulation, especially for low order elements. Therefore, the balance of angular momentum is introduced as an additional residual equation, which states that the material time derivative of the moment of momentum, w.r.t. a fixed reference point $\boldsymbol{x}_0$, is equal to the resultant moment of all forces acting on $\mathcal{B}$. The balance of angular momentum reads

$$\int_{\mathcal{B}} \left((\boldsymbol{x} - \boldsymbol{x}_0) \times (\operatorname{div}\boldsymbol{\sigma} + \boldsymbol{f}) + \operatorname{axl}[\boldsymbol{\sigma} - \boldsymbol{\sigma}^T]\right) \mathrm{d}V = \boldsymbol{0}, \quad \operatorname{axl}[\boldsymbol{\sigma} - \boldsymbol{\sigma}^T] := \begin{pmatrix} \sigma_{32} - \sigma_{23} \\ \sigma_{13} - \sigma_{31} \\ \sigma_{21} - \sigma_{12} \end{pmatrix}. \tag{2}$$

From the continuum mechanical point of view the balance of angular momentum is ensured by the symmetry of the Cauchy stresses, if in addition the balance of momentum is exactly fulfilled, that is, $\operatorname{div}\boldsymbol{\sigma} + \boldsymbol{f} = \boldsymbol{0}$. The second LS functional reads

$$\begin{aligned} \mathcal{F}^\star(\boldsymbol{\sigma}, \boldsymbol{u}) = \frac{1}{2}\Bigg(&\left\|\omega_1^\star(\operatorname{div}\boldsymbol{\sigma} + \boldsymbol{f})\right\|^2_{\mathcal{L}^2(\mathcal{B})} + \left\|\omega_2^\star(\boldsymbol{\sigma} - \mathbb{C} : \nabla^s\boldsymbol{u})\right\|^2_{\mathcal{L}^2(\mathcal{B})} \\ &+ \left\|\omega_3^\star((\boldsymbol{x} - \boldsymbol{x}_0) \times (\operatorname{div}\boldsymbol{\sigma} + \boldsymbol{f}) + \operatorname{axl}[\boldsymbol{\sigma} - \boldsymbol{\sigma}^T]\right\|^2_{\mathcal{L}^2(\mathcal{B})}\Bigg). \end{aligned} \tag{3}$$

The weighting parameters, ω_i, $\omega_i^\star$ are chosen w.r.t. the Lamé constant μ and a characteristic length $\tilde{l}$ to obtain a scale independent formulation, i.e. a dimensionless formulation, see [14]. Therefore, we consider $\omega_i = \alpha_i/\mu, \omega_i^\star = \alpha_i^\star/\mu$ for $i = 2, 3$ and $\omega_1^\star = \tilde{l}\alpha_1^\star/\mu$. Therein, α_i and $\alpha_i^\star$ denote the new dimensionless (i.e. unit 1) weighting factors. For $\tilde{l}$ a reasonable choice is demanded, which is here set as the square root of the area of the boundary value problem, cf. [5].

For the solution of the linear problem, the underlying functional is minimized w.r.t. appropriate finite-dimensional spaces, i.e. find $(\boldsymbol{\sigma}, \boldsymbol{u}) \in \mathcal{S}^m \times \mathcal{V}^k$ s.t. $\delta_{\sigma,u}\mathcal{F} = \mathbf{0}\,\forall(\delta\boldsymbol{\sigma}, \delta\boldsymbol{u}) \in \mathcal{S} \times \mathcal{V}$ with $\mathcal{S} := [\mathcal{H}(\text{div}, \mathcal{B})]^{d\times d}$ and $\mathcal{V} := [\mathcal{H}^1(\mathcal{B})]^d$. For a conforming discretization we choose the function spaces

$$\begin{aligned} \mathcal{S}_h^m &:= \{\boldsymbol{\sigma}_h \in [\mathcal{H}(\text{div}, \mathcal{B})]^{d\times d} : \boldsymbol{\sigma}_h|_K \in [\mathcal{RT}_m]^{d\times d}\, \forall K \in \mathcal{T}\}, \\ \mathcal{V}_h^k &:= \{\boldsymbol{u}_h \in [\mathcal{H}^1(\mathcal{B})]^d : \boldsymbol{u}_h|_K \in [\mathcal{P}_k]^d\, \forall K \in \mathcal{T}\}, \end{aligned} \tag{4}$$

with the dimension d, $\mathcal{T}$ denotes the triangulation of $\mathcal{B}$ with finite elements and K is a finite element. Here, conforming approximation functions for $\mathcal{S}_h^m$ and $\mathcal{V}_k^m$ are functions whose normal components are continuous across element interfaces of order m, given by vector-valued Raviart-Thomas functions ($\mathcal{RT}_m$), see e.g. [15] and $\mathcal{V}_h^k$ is the space of continuous piecewise polynomial functions of order $k \geq 1$, chosen as Lagrange type functions ($\mathcal{P}_k$).

Numerical Analysis of Reaction Forces Within the LSFEM

The recalculation of support reactions, as a critical value in engineering applications e.g. in the design and the structural stability of components, is analyzed for a clamped cantilever beam, see Fig. 1 with $E = 70\,\text{kN/mm}^2$, $\mu = 26.12\,\text{kN/mm}^2$ and loaded by a boundary traction $\boldsymbol{\sigma} \cdot \boldsymbol{n} = (0, 0.1)^T \text{kN/mm}^2$, cf. [5].

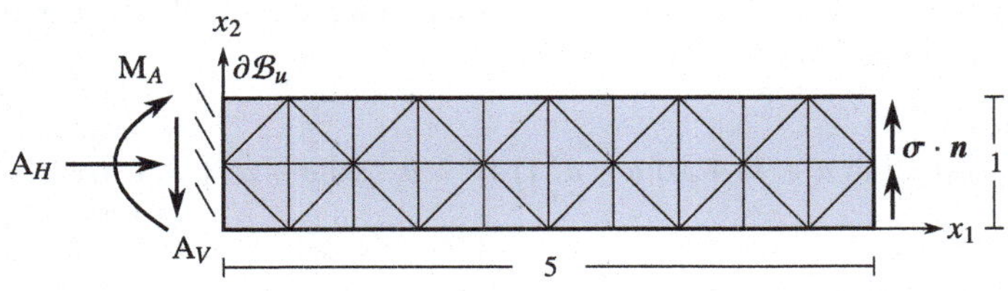

Fig. 1 Clamped cantilever: setup and exemplary coarse mesh density

For the evaluation of support reactions $\{\mathrm{A}_H, \mathrm{A}_V, \mathrm{M}_A\}$ both LS formulations (1) and (3) are compared to a linear displacement element ($\mathcal{P}_1$). A simple calculation yields the analytical solutions

$$\mathrm{A}_H = 0\text{ kN}, \quad \mathrm{A}_V = 0.1\text{ kN} \quad \text{and} \quad \mathrm{M}_A = 0.5\text{ kNmm}. \tag{5}$$

The determination for A_H, A_V and M_A for the LSFEM is explained in [5].

The investigated weighting parameter setups for $\mathcal{F}$ in (1) and $\mathcal{F}^\star$ in (3) are given for fixed $\alpha_1 = 1$ as well as $\alpha_{1,3}^\star = 1$ and a decreasing α_2 and $\alpha_2^\star$ from 1 to 0.01. An unsatisfying convergence of the support reactions and moments is obtained by the LSFEM (1), with exception of the convergence of A_H, which is exactly fulfilled. However, the LS formulation (3) leads to an improved convergence of A_V and M_A and further give a satisfying solution for A_H with $\alpha_2^\star \leq 0.1$. For the results in Fig. 2 no improvement for α_2 and $\alpha_2^\star$ below 0.1 is observed, whereas for lower order elements an improvement for $\alpha_2^\star < 0.1$ is shown, see Fig. 3. The influence of weights can be reduced by use of higher order polynomial functions as visualized in Figs. 2 and 3 for the choice of $\mathcal{RT}_m\mathcal{P}_k$ element types with $m = 0, 1, 2$ and $k = 1, 2, 3$. However,

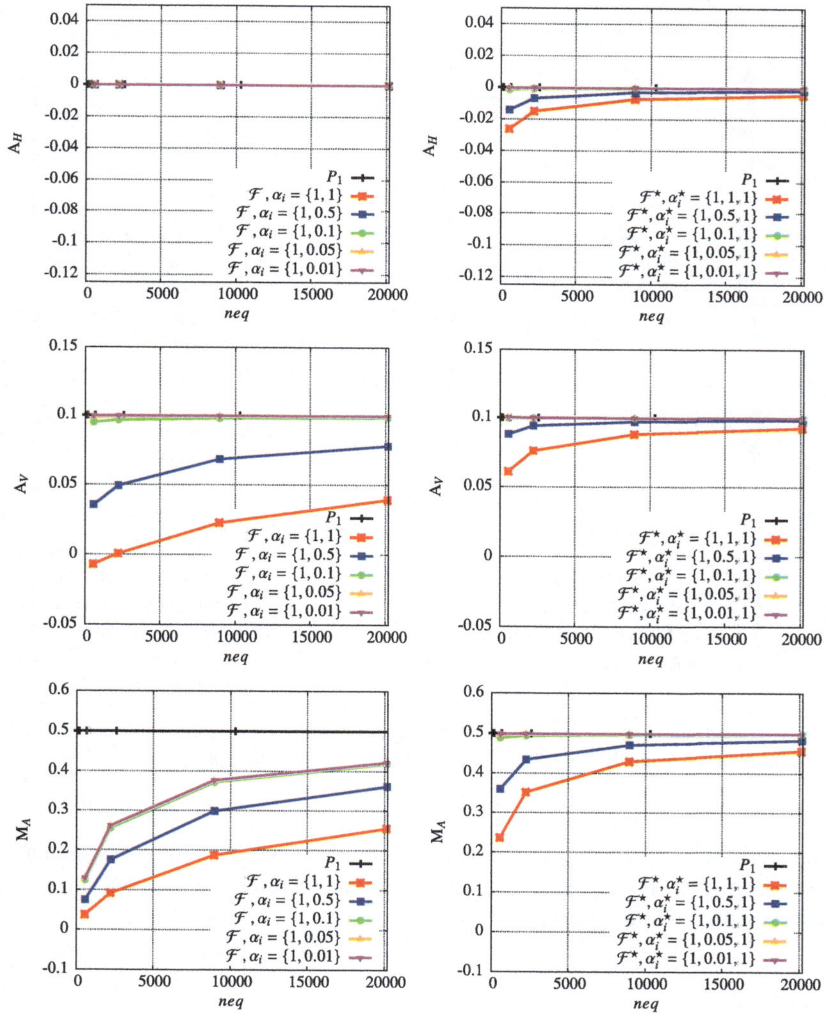

Fig. 2 Clamped cantilever: results for the fulfillment of equilibrium of forces and moments for the LS formulation $\mathcal{F}$ (left) and $\mathcal{F}^\star$ (right) utilizing a $\mathcal{RT}_1\mathcal{P}_2$ element type, taken from [5]

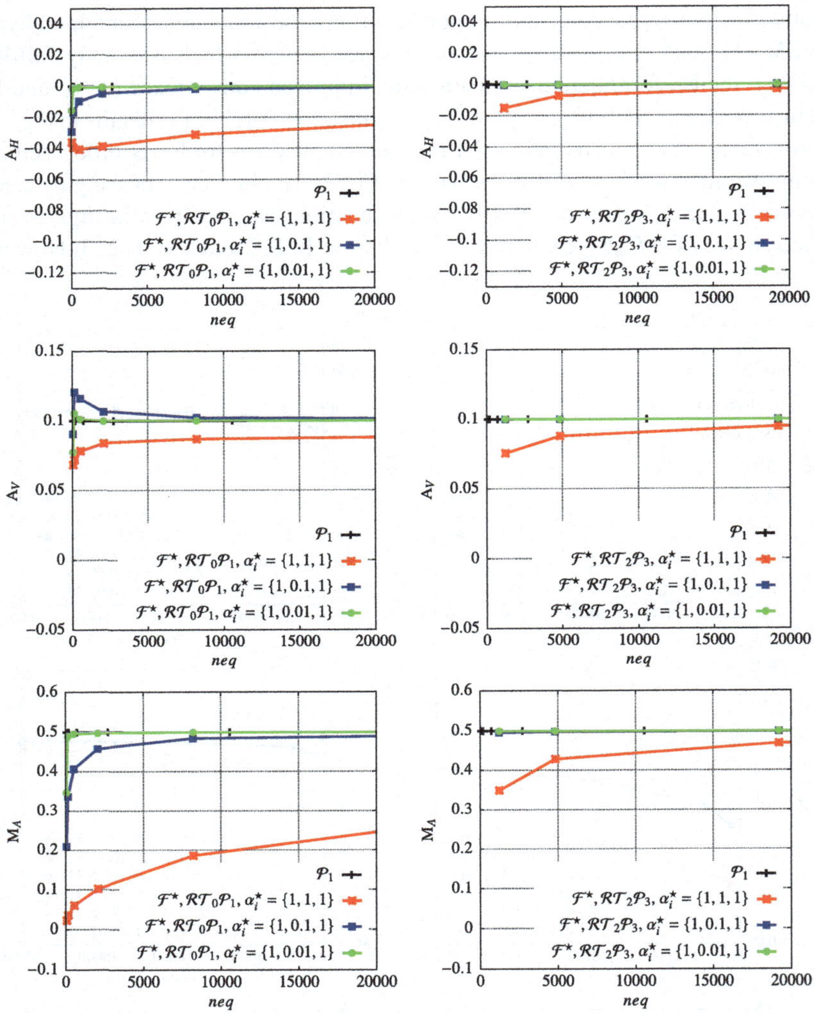

Fig. 3 Clamped cantilever: results for the fulfillment of equilibrium of forces and moments for the LS formulation $\mathcal{F}^\star$ using $\mathcal{RT}_0\mathcal{P}_1$ (left) and $\mathcal{RT}_2\mathcal{P}_3$ (right) element types

even when choosing a $\mathcal{RT}_2\mathcal{P}_3$ element, a sufficiently accurate solution is shown only for weightings with $\alpha_2^\star \leq 0.1$. The linear displacement element yields directly the correct results, see Eq. (5) and Fig. 2.

3 Hybrid Mixed FEM Based on a Least-Squares Approach

An alternative approach for a continuous stress approximation is given by $\boldsymbol{\sigma} \in \mathcal{H}^1(\mathcal{B})$ using Lagrange type functions $\mathcal{P}_m$, consisting of complete polynomials of degree $\leq m$. The approach is straightforward applicable to a LS functional, e.g. (3), here given by

$$\mathcal{F}_c = \sum_{K \in \mathcal{T}} \frac{1}{2}\Big(\| \operatorname{div} \boldsymbol{\sigma} + \boldsymbol{f} \|^2_{\mathcal{L}^2(K)} + \| \mathbb{C}^{-1} : \boldsymbol{\sigma} - \boldsymbol{\varepsilon} \|^2_{\mathcal{L}^2(K)} + \| \mathbb{C}^{-1} : (\boldsymbol{\sigma} - \boldsymbol{\sigma}^T) \|^2_{\mathcal{L}^2(K)} \Big), \quad (6)$$

where $\mathcal{T}$ denote the triangulation of the body $\mathcal{B}$ with finite elements K. This continuous formulation $\mathcal{F}_c$ (6), is approximated using continuous piecewise polynomial functions of Lagrange type for both fields, i.e.,

$$\boldsymbol{\sigma}_h \in [\mathcal{H}^1(\mathcal{T})]^{d \times d} \quad \text{and} \quad \boldsymbol{u}_h \in [\mathcal{H}^1(\mathcal{T})]^d, \quad (7)$$

with

$$\sigma_{ij} = \sum_{I=1}^{nel_\sigma} \mathbb{N}^I \cdot \beta^I_{ij} \quad \in \mathcal{P}_m, \quad (8)$$

where nel_σ denote the stress nodes per element, $\mathbb{N}^I$ the shape functions and β^I_{ij} the nodal stress degrees of freedom. This element type is denoted by $\mathcal{P}_m\mathcal{P}_k$. Based on this approximation, the stress continuity is given by $[\![\boldsymbol{\sigma}]\!] = \boldsymbol{0}$, i.e. the jump of the stress components at inter-element boundaries is zero, which is inconsistent, since only traction continuity ($[\![\boldsymbol{\sigma} \cdot \boldsymbol{n}]\!] = [\![\boldsymbol{t}]\!] = \boldsymbol{0}$) have to be fulfilled by the approximation.

In order to achieve this, utilizing a stress approximation with Lagrange type functions, a hybrid mixed finite element formulation is introduced based on a least-squares formulation with discontinuous stress approximation $\boldsymbol{\sigma} \in \mathcal{L}^2(\mathcal{B})$. For conforming discretization of $\boldsymbol{\sigma}$ and $\boldsymbol{u}$ different continuity conditions have to be fulfilled by the solution spaces. These conditions can be enforced in a weak sense by application of Lagrange multipliers. A discontinuous stress approximation can be used in combination with a Lagrange multiplier enforcing the normal continuity of the stress field. For further steps, $\mathcal{E}$ denotes the set of all sides of the triangulation $\mathcal{T}$. Furthermore, $\partial\mathcal{T} = \partial\mathcal{T}_N \cup \partial\mathcal{T}_D$ is the triangulation of the outer boundary with the Neumann $\partial\mathcal{T}_N$ and Dirichlet $\partial\mathcal{T}_D$ specifications and the interior boundary is $\partial\mathcal{T}_i = \mathcal{E} \backslash \partial\mathcal{T}$, where $\partial\mathcal{T}_i \cap \partial\mathcal{T}_N = \emptyset$ as well as $\partial\mathcal{T}_i \cap \partial\mathcal{T}_D = \emptyset$ holds.

For the hybrid mixed formulation a LS formulation, e.g. (6) is extended by the relaxed continuity conditions multiplied with a Lagrange multiplier. Here, the balance of momentum, constitutive equation and stress symmetry condition is solved on each local element without enforcement of traction continuity. Since we relax the traction continuity, the condition

$$[\![\boldsymbol{t}]\!] = (\boldsymbol{\sigma} \cdot \boldsymbol{n})^+ + (\boldsymbol{\sigma} \cdot \boldsymbol{n})^- = \boldsymbol{0} \quad \text{on } \mathcal{E} \quad (9)$$

is not fulfilled a priori; the characters (+) and (–) denote the two sides of an arbitrary inter-element boundary. We enforce this traction condition in a weak sense by the Lagrange multiplier $\boldsymbol{\lambda}$. This is applied to the jump of the traction vector $[\![\boldsymbol{\sigma} \cdot \boldsymbol{n}]\!]$ on the skeleton $\partial\mathcal{T}_i$ and for the boundary tractions on the outer Neumann boundary $\partial\mathcal{T}_N$:

$$\boldsymbol{\lambda} \cdot [\![\boldsymbol{\sigma} \cdot \boldsymbol{n}]\!] = \boldsymbol{0} \quad \text{on } \partial\mathcal{T}_i \quad \text{and} \quad \boldsymbol{\lambda} \cdot (\boldsymbol{\sigma} \cdot \boldsymbol{n} - \bar{\boldsymbol{t}}) = \boldsymbol{0} \quad \text{on } \partial\mathcal{T}_N. \tag{10}$$

For a detailed discussion of the enforcement of these continuity conditions see e.g. [16]. The resulting hybrid mixed formulation $\mathcal{F}^h$ based on a mixed LS approach is given by

$$\mathcal{F}^h(\boldsymbol{\sigma}, \boldsymbol{u}, \boldsymbol{\lambda}) = \mathcal{F}_c(\boldsymbol{\sigma}, \boldsymbol{u}) + \mathcal{F}_t(\boldsymbol{\sigma}, \boldsymbol{\lambda}), \tag{11}$$

where the continuous LS functional is denoted by $\mathcal{F}_c$ (6) and $\mathcal{F}_t$ is the functional part enforcing the traction continuity (9) on the inter-element boundaries as well as for the boundary conditions on the outer boundary $\partial\mathcal{T}_N$. Here, $\mathcal{F}_t$ is given by

$$\mathcal{F}_t = \sum_{E \in \partial\mathcal{T}_i} \int_E [\![\boldsymbol{\sigma} \cdot \boldsymbol{n}]\!] \cdot \boldsymbol{\lambda} \, \mathrm{d}A + \sum_{E \in \partial\mathcal{T}_N} \int_E (\boldsymbol{\sigma} \cdot \boldsymbol{n} - \bar{\boldsymbol{t}}) \cdot \boldsymbol{\lambda} \, \mathrm{d}A. \tag{12}$$

Furthermore, the resulting finite element type is denoted by $\mathrm{d}\mathcal{P}_m\mathcal{P}_k\mathcal{P}_n$ with the stress approximation

$$\sigma_{ij} = \sum_{I=1}^{nel_\sigma} \mathbb{N}^I \cdot \beta_{ij}^I \quad \in \mathrm{d}\mathcal{P}_m. \tag{13}$$

Thus, the discontinuity is realized by not assembling the stress nodes for the $\mathrm{d}\mathcal{P}_m\mathcal{P}_k\mathcal{P}_n$ element. The problem is solved, considering the first variation $\delta\mathcal{F}^h$, by finding $(\boldsymbol{\sigma}, \boldsymbol{u}, \boldsymbol{\lambda}) \in \bar{\mathcal{S}} \times \bar{\mathcal{V}} \times \bar{\mathcal{X}}$ s.t.

$$\delta_{\sigma,u,\lambda}\mathcal{F}^h = \boldsymbol{0} \quad \forall \, (\delta\boldsymbol{\sigma}, \delta\boldsymbol{u}, \delta\boldsymbol{\lambda}) \in \bar{\mathcal{S}} \times \bar{\mathcal{V}} \times \bar{\mathcal{X}} \tag{14}$$

with

$$\bar{\mathcal{S}} := [\mathcal{L}^2(\mathcal{T})]^{d \times d}, \quad \bar{\mathcal{V}} := [\mathcal{H}^1(\mathcal{T})]^d \quad \text{and} \quad \bar{\mathcal{X}} := [\mathcal{H}^{-1/2}(\mathcal{E})]^d \tag{15}$$

For the solution spaces a conforming choice is considered demanding continuity of $\boldsymbol{u}$ and $\boldsymbol{\lambda}$ and allowing jumps of $\boldsymbol{\sigma}$ at inter-element boundaries, specified by

$$\begin{aligned}
\bar{\mathcal{S}}_h^m &:= \{\boldsymbol{\sigma}_h \in [\mathcal{L}^2(\mathcal{T})]^{3\times 3} : \boldsymbol{\sigma}_h|_K \in [\mathrm{d}\mathcal{P}_m]^{3\times 3} \, \forall \, K \in \mathcal{T}\}, \\
\bar{\mathcal{V}}_h^k &:= \{\boldsymbol{u}_h \in [\mathcal{H}^1(\mathcal{T})]^3 : \boldsymbol{u}_h|_K \in [\mathcal{P}_k]^3 \, \forall \, K \in \mathcal{T}\}, \\
\bar{\mathcal{X}}_h^n &:= \{\boldsymbol{\lambda}_h \in [\mathcal{H}^{-1/2}(\mathcal{E})]^3 : \boldsymbol{\lambda}_h|_E \in [\mathcal{P}_n]^3 \, \forall \, E \in \mathcal{E}\}.
\end{aligned} \tag{16}$$

For $\boldsymbol{u}$ and $\boldsymbol{\lambda}$ continuous Lagrange type functions are applied and the stresses are approximated discontinuous. Here, m, k, n denote the polynomial order of the functions spaces.

Numerical Analysis of Hybrid Mixed FEM Based on a LS Approach
The hybrid mixed finite element formulation $\mathcal{F}^h$ is investigated for a square plate ($x_1 \in [-1, 1], x_2 \in [-1, 1]$) under displacement controlled boundary conditions with four material setups ($\nu = 0.35$, $E_{1,2,3,4} = \{100, 200, 300, 400\}\text{kN/mm}^2$), see Fig. 4. The plate is subjected to a uniform elongation, where the displacements in normal direction of the outer edges are set to 0.1 mm and the shear stresses on the edges are set to 0. Furthermore, $\boldsymbol{\lambda}$ in normal direction on the outer edges is 0. The influence of the different materials can be clearly seen in the deformed configuration with a scaling factor of 10 in Fig. 4, on a finite element mesh with 20 elements per side. The hybrid mixed formulation is compared to the continuous LS formulation $\mathcal{F}_c$ in (6) with an approximation of $\boldsymbol{\sigma} \in \mathcal{P}_m$ and $\boldsymbol{u} \in \mathcal{P}_k$ denoted as a $\mathcal{P}_m\mathcal{P}_k$ element. Figure 5 shows the stress distribution of σ_{22} on the undeformed configuration and the σ_{22} stresses over a section of the plate (at line A-B). The physically correct σ_{22} stresses have to be continuous in x_2-direction and discontinuous across the vertical material interface, cf. [17]. For the $\mathcal{P}_m\mathcal{P}_k$ element, the stress component σ_{22} at the material interface is incorrect, since the continuity requirements in $\mathcal{H}^1(\mathcal{B})$ yield not only normal continuous solutions, see Fig. 5(a). However, the hybrid mixed formulation $\text{d}\mathcal{P}_k\mathcal{P}_m\mathcal{P}_n$ represents the jump of stresses over the plate at line A-B correctly, see Fig. 5(b). As mentioned before, these results can be also achieved by an approximation of $\boldsymbol{\sigma} \in \mathcal{RT}_m$, which directly yield normal continuous stresses.

The finite element implementations and computations have been done using the *AceGen* and *AceFEM* packages (version 6.503), see [18–20] and *Mathematica* (version 10.1), see [21]. For the visualization *Paraview* (version 4.3.1), see [22], has been used.

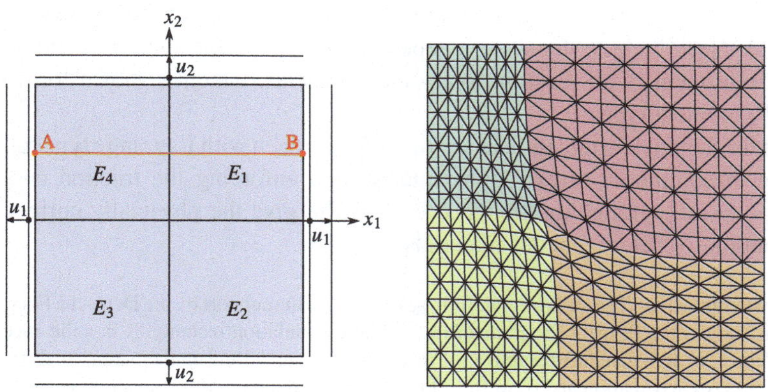

Fig. 4 BVP of the square plate, definition of intersection line A-B (left) and deformed configuration with a scaling factor of 10 (right)

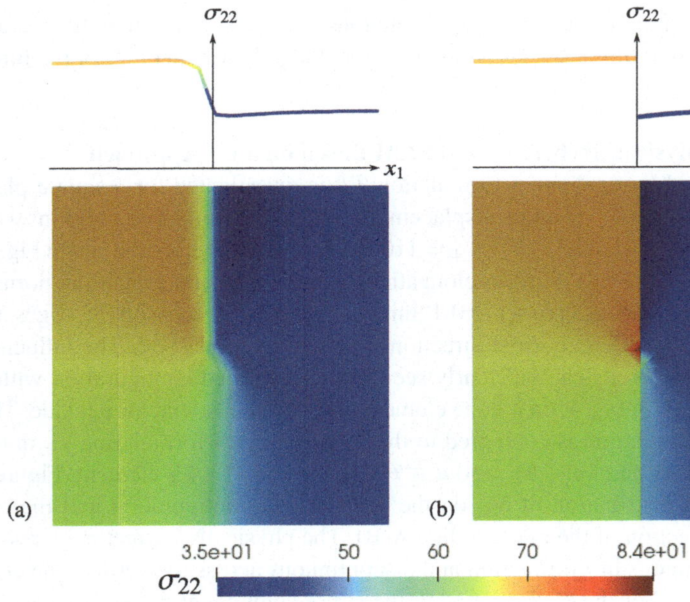

Fig. 5 σ_{22} stress distribution over the square plate (bottom) and along the intersection line A-B (top) for the FE discretization **(a)** $\mathcal{P}_2\mathcal{P}_3$ and **(b)** $d\mathcal{P}_2\mathcal{P}_3\mathcal{P}_1$ with 20 elements per side

4 Conclusion

The presented investigations of a classical and an extended LS formulation, with an explicit consideration of the balance of angular momentum, show that the recalculation of reaction forces and moments, as a crucial point in engineering applications, requires special consideration. However, the considered scale independent approach is sensitive to the choice of weighting factors, which is one of the challenges of the LSFEM. Nevertheless, the scale independent extended formulation yields a satisfying fulfillment of all reaction forces and moments, which is shown for different $\mathcal{RT}_m\mathcal{P}_k$ element types.

A hybrid mixed formulation based on a LS approach with Lagrange type functions for the discontinuous stress approximation and enforcing the traction continuity condition via Lagrange multiplier $\boldsymbol{\lambda}$ is shown to give the physically correct stress distribution for heterogeneous materials.

Acknowledgements The authors gratefully acknowledge the support by the Deutsche Forschungsgemeinschaft in the Priority Program 1748 "Reliable simulation techniques in solid mechanics. Development of non-standard discretization methods, mechanical and mathematical analysis" under the project "Approximation and reconstruction of stresses in the deformed configuration for hyperelastic material models" (Project number 392587488, project ID SCHR 570/34-1).

References

1. Jiang, B.-N. (1998). *The least-squares finite element method*. Berlin: Springer.
2. Bochev, P. B., & Gunzburger, M. D. (2009). Least-squares finite element method. Berllin: Springer.
3. Pontaza, J. P. (2003). Least-squares variational principles and the finite element method: Theory, form, and model for solid and fluid mechanics. Ph.D. thesis, Texas A&M University.
4. Cai, Z., & Starke, G. (2004). Least-squares methods for linear elasticity. *SIAM Journal on Numerical Analysis, 42*, 826–842.
5. Igelbüscher, M., Schröder, J., & Schwarz, A. (2020). A mixed least-squares finite element formulation with explicit consideration of the balance of moment of momentum, a numerical study. *GAMM-Mitteilungen, 43*, e202000009. https://doi.org/10.1002/gamm.202000009
6. Atluri, S. N., Gallagher, R. H., & Zienkiewicz, O. (1983). *Hybrid and mixed finite element methods*. Chichester: Wiley.
7. Carey, G. F., & Oden, J. T. (1983). Finite elements: A second course. Prentice Hall, Inc.
8. Roberts, J. E., & Thomas, J. -M. (1991). Mixed and hybrid methods. In P. G. Ciarlet, & J. L. Lions (Eds.), *Handbook of numerical analysis* (Vol. 2). Elsevier Science.
9. Brezzi, F., & Fortin, M. (1991). Mixed and hybrid finite element methods. Berlin: Springer.
10. Boffi, D., Brezzi, F., & Fortin, M. (2009). Reduced symmetry elements in linear elasticity. *Communications on Pure and Applied Analysis, 8*, 95–121.
11. Cockburn, B., Gopalakrishnan, J., & Guzmán, J. (2010). A new elasticity element made for enforcing weak stress symmetry. *Mathematics of Computation, 79*, 1331–1349.
12. Schwarz, A., Steeger, K., & Schröder, J. (2014). Weighted overconstrained least-squares mixed finite elements for static and dynamic problems in quasi-incompressible elasticity. *Computational Mechanics, 54*(1), 603–612.
13. Schwarz, A., Steeger, K., Igelbüscher, M., & Schröder, J. (2018). Different approaches for mixed lsfems in hyperelasticity: Application of logarithmic deformation measures. *International Journal of Numerical Methods in Engineering, 115*, 1138–1153.
14. Bell, B. C., & Surana, K. S. (1994). Time coupled p-version least-squares element formulation for unsteady fluid dynamics problems. *International Journal for Numerical Methods in Engineering, 37*, 3545–3569.
15. Raviart, P. A., & Thomas, J. M. (1977). A mixed finite element method for 2-nd order elliptic problems. *Mathematical Aspects of Finite Element Mehtods. Lecture Notes in Mathematics* (Vol. 606, pp. 292–315) New York: Springer.
16. Xue, W.-M., Karlovitz, L. A., & Atluri, S. N. (1985). On the existence and stability conditions for mixed-hybrid finite element solutions based on reissner's variational principle. *International Journal of Solids and Structures, 21*, 97–116.
17. Steeger, K. (2017). Least-squares mixed finite elements for geometrically nonlinear solid mechanics. Ph.D. thesis, Univesity of Duisburg-Essen.
18. Korelc, J. (1997). Automatic generation of finite-element code by simultaneous optimization of expressions. *Theoretical Computer Science, 187*(1), 231–248.
19. Korelc, J. (2002). Multi-language and multi-environment generation of nonlinear finite element codes. *Engineering with Computers, 18*, 312–327.
20. Korelc, J., & Wriggers, P. (2016). *Automation of finite element methods*. Springer International Publishing Switzerland.
21. Inc. Wolfram Research. (2015). *Mathematica*. Wolfram Research, Inc., version 10.1 ed. Champaign, Illinois.
22. Ahrens, J., Geveci, B. & Law, C. (2005). *ParaView: An end-user tool for large data visualization, visualization handbook*. Elsevier, version 10.1 ed. Champaign, Illinois.

On Two-Scale Modelling of Softening Material Responses

Jurica Sorić, Tomislav Lesičar, Zdenko Tonković, and Filip Putar

Peter is not only the best scientist, he is also the best of friends. Wishing him good health and further success in the years ahead, sincerely Jurica.

Abstract The paper deals with the two-scale approaches for modelling of quasi-brittle and ductile softening responses of heterogeneous materials. The damage is induced at the microstructural level and after the homogenization procedure it is mapped via the constitutive stiffness at the macrolevel material point. In the case of quasi-brittle softening the nonlocal continuum theory employed in the C^1 continuous finite element formulation is applied at both micro- and macrolevel, and a second-order computational homogenization has been performed. The ductile damage is modelled by means of the first- and the second-order computational homogenization. At the first order homogenization the macrostructure is discretized by the regular displacement finite element formulation, while the mixed finite elements employing elastoplastic algorithm are used at microlevel, where the three microstructural volume elements are considered. On the contrary, only one microstructural volume element discretized by the mixed finite elements is used at the second-order homogenization, while the discretization at macrolevel has been performed by the C^1 continuous finite element formulation. All approaches proposed are verified in the standard benchmark examples.

J. Sorić (✉) · T. Lesičar · Z. Tonković
Faculty of Mechanical Engineering and Naval Architecture, University of Zagreb, Ivana Lučića 5, 10000 Zagreb, Croatia
e-mail: jurica.soric@fsb.hr

T. Lesičar
e-mail: tomislav.lesicar@fsb.hr

Z. Tonković
e-mail: zdenko.tonkovic@fsb.hr

F. Putar
AVL-AST d.o.o., Strojarska cesta 22, 10000 Zagreb, Croatia

F. Aldakheel et al. (eds.), *Current Trends and Open Problems in Computational Mechanics*, https://doi.org/10.1007/978-3-030-87312-7_48

Keywords Heterogeneous material · Two-scale approach · Computational homogenization · Quasi-brittle damage · Ductile damage

1 Introduction

Softening phenomena characterized by decrease in material stiffness or so-called damage are common in engineering materials and can decrease structural load-carrying capacity, and lead to loss of mechanical integrity. A lot of engineering materials can be treated as heterogeneous, particularly if they are observed at microscale. Therefore, in order to assess structural integrity and to predict structural lifetime, an analysis evolving microstructure is necessary. Derivation of an efficient multiscale approach is still an important challenge in the computational mechanics community.

A more accurate modelling of the macrostructural softening evolution could be made if the assessment of damage is firstly done at the microstructural level, where multiscale approaches enable consistent bridging of the material behaviour from micro to macroscale. Therein, homogenization techniques are applied in which certain properties are averaged over a representative volume element (RVE), assumed to be statistically representative for the macroscopic material point [1–3]. The computational homogenization scheme is shown to be most accurate and versatile. The two boundary value problems (BVPs) have to be mostly solved simultaneously during the calculation, one for the macroscale and another for the underlying microstructure, where the transfer of state variables represents a crucial and most challenging part, particularly when softening responses are considered. The transition scheme of a two-scale approach is displayed in Fig. 1. As shown, the strain measures computed at the macrolevel material point are transformed to the boundary displacement of microstructural volume element (MVE). After MVE homogenization procedure, the constitutive relation and other homogenized microstructural state variables are upscaled to the macrolevel.

Over the past decades, the studies on multiscale numerical methods for modelling of quasi-brittle failure processes have been mostly performed. If the conventional

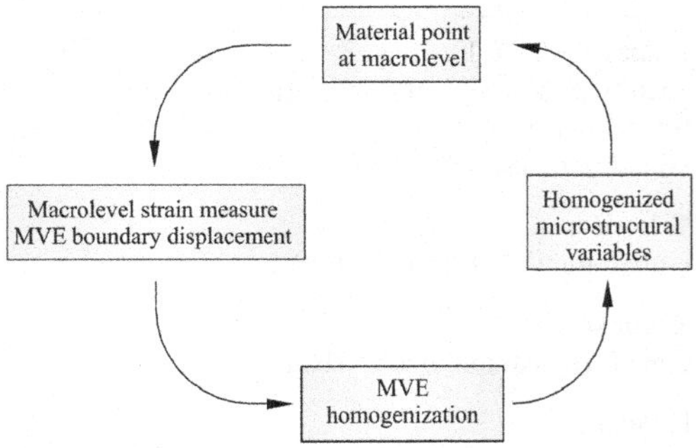

Fig. 1 Two-scale computational scheme

computational homogenization is used for the simulation of heterogeneous material softening responses, some issues arise such as ill-posedness of boundary value problems and lack of separation of scales. To alleviate these undesired phenomena, some remedies are proposed in [4, 5]. Furthermore, the relatively efficient failure zone averaging schemes can be found in [6, 7]. Considering constitutive behaviour, brittle damage and linear elasticity simulations have been more fruitful compared to the modelling of ductile damage employing elastoplastic material responses.

A multiscale method dealing with ductile damage responses of polycrystalline materials has been proposed in [8], where the integral nonlocal terms are introduced in order to preserve objectivity of the results. In [9], the ductile damage in heterogeneous materials is described by means of the mean field homogenization with an addition of isotropization procedure, but this method is suited only for mild damage. The "Failure Oriented Multiscale Formulation" for the consistent upscaling of ductile softening behaviour has been presented in [10], with an emphasis on the RVE boundary conditions during the localization. A multiscale approach which employs strong coupling between scales has been presented in [11], based on the principle of the operator split on a two-phase material.

An efficient multiscale method for modelling of damage responses at microlevel still remains an open question. In the present contribution the damage responses of both quasi-brittle and ductile materials are considered using two-scale computational procedures. To model quasi-brittle damage, a second-order computational homogenization approach [12] is applied. Therein, the C^1 continuous triangular finite element formulation based on the nonlocal continuum theory is used for the discretization at both micro- and macroscales. The damage enhanced constitutive relations [13] are employed at the microlevel, where an appropriate RVE, representing a sample of heterogeneous material, is considered. The ductile damage is modelled using both the first- and the second-order computational homogenization schemes. The first-order homogenization employs three MVEs with the failure averaging procedure and the macrolevel discretization is performed by means of the regular displacement finite element formulation. The MVE is discretized by the newly developed mixed finite element, where the gradient-enhanced elastoplasticity is employed. The second-order homogenization deals with the C^1 finite element discretization at macrolevel and the standard averaging procedure over the MVE. All developed algorithms are implemented into the finite element software ABAQUS via user subroutines. All proposed computational models are verified by means of several benchmark examples.

2 Modelling of Quasi-brittle Damage

As mentioned above, the C^1 continuous plane strain triangular finite element is used for the discretization at both the microlevel and the macrolevel. It consists of three nodes, each having 12 degrees of freedom, which are two displacement components and their first- and second-order derivatives. The element formulation is based on the nonlocal continuum theory under assumption of small strain. The constitutive relations at the macrolevel and the finite element derivation are presented in [12]. The state variables computed at the macrostructural material point are the strain and the strain gradient which are transformed to the MVE boundary displacement. The constitutive relations at the microscale employ the damage variable as presented in the following incremental expressions

$$\Delta\boldsymbol{\sigma} = \left(1 - D^{i-1}\right)\mathbf{C}\Delta\boldsymbol{\varepsilon} - \mathbf{C}\boldsymbol{\varepsilon}^{i-1}\Delta D, \tag{1}$$

$$\Delta\boldsymbol{\mu}_{x_1} = l^2\left(1 - D^{i-1}\right)\mathbf{C}\Delta\boldsymbol{\varepsilon}_{x_1} - l^2\mathbf{C}\boldsymbol{\varepsilon}_{x_1}^{i-1}\Delta D, \tag{2}$$

$$\Delta\boldsymbol{\mu}_{x_2} = l^2\left(1 - D^{i-1}\right)\mathbf{C}\Delta\boldsymbol{\varepsilon}_{x_2} - l^2\mathbf{C}\boldsymbol{\varepsilon}_{x_2}^{i-1}\Delta D. \tag{3}$$

Here, $\boldsymbol{\sigma}$ and $\boldsymbol{\varepsilon}$ are the Cauchy stress and the strain tensors, respectively. The values $\boldsymbol{\varepsilon}_{x_1}$ and $\boldsymbol{\varepsilon}_{x_2}$ stand for the strain gradients with respect to the Cartesian coordinates x_1 and x_2, and $\boldsymbol{\mu}_{x_1}$ and $\boldsymbol{\mu}_{x_2}$ are their work conjugates. l represents the microstructural parameter, and $\mathbf{C}$ is the elasticity matrix which describes the stiffness behaviour of the MVE bulk material. D is the damage variable expressed by the exponential softening law [13]. The exponent $(i-1)$ refers to the last converged equilibrium state.

Implementation of the constitutive relations (1)–(3) into the incremental principle of virtual work, as described in [14], leads to the finite element equation containing the particular element stiffness matrices by means of which the homogenization procedure has been performed, as displayed in [15]. The constitutive relations have been computed in terms of the damage variable at microlevel, and then they are mapped at the integration points of the macroscopic C^1 finite element discretization. More details on the scale transition can be found in [14]. Therein, an example of heterogeneous plate subjected to tensile load is computed. In order to trigger the localization, the Young's modulus is slightly reduced in the middle of the plate. The heterogeneity is described using an academic MVE. The damage distribution

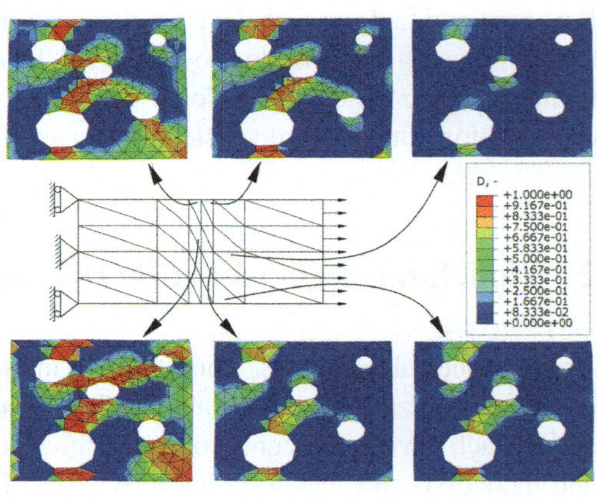

Fig. 2 Distribution of damage variable D over several characteristic MVEs [14]

over several characteristic MVEs forming the macrostructural localization zone is presented in Fig. 2. It can be observed that the most intense damage bands are formed in the middle of the plate, where the strongest localization is exhibited, as expected.

3 Modelling of Ductile Damage

The modelling of ductile damage has been performed by using both the first- and second-order homogenizations. Therein the discretizations at macrolevel employ the different finite element formulations, and the same gradient elastoplastic formulation has been applied for the modelling of microstructural softening responses.

3.1 *Three MVE First-Order Homogenization*

According to the first-order computational homogenization, the discretization at the macrolevel is performed by using a standard 4-node quadrilateral finite element formulation. Here the stiffness matrix $\mathbf{K}$ has the standard form $\mathbf{K} = \int_V \mathbf{B}^T \mathbf{C}_\mathrm{M} \mathbf{B} \mathrm{d}V$, where the constitutive matrix $\mathbf{C}_\mathrm{M}$ is expressed in terms of the damage variable D computed at the microlevel, as

$$\mathbf{C}_\mathrm{M} = \left(1 - D^{i-1}\right) \mathbf{C}_\mathrm{B} - \left(\frac{\mathrm{d}D}{\mathrm{d}\boldsymbol{\varepsilon}}\right)^{i-1} \mathbf{C}_\mathrm{B} \boldsymbol{\varepsilon}^{i-1}. \tag{4}$$

Herein $\mathbf{C}_\mathrm{B}$ denotes the constitutive matrix of the bulk material of a heterogeneous structure. Both $\mathbf{C}_\mathrm{B}$ and D are computed at the microlevel using the first-order computational homogenization which has been performed over the two microstructural samples representing the two BVPs, as presented in the scheme in Fig. 3. One (the bulk microstructural volume element - BMVE) is without damage, where only elastoplastic response has been computed, and another (the damaged microstructural volume element - DMVE) employs the nonlocal ductile damage model to compute softening evolution. In order to numerically compute the derivatives of damage variable with respect to three macrostrain components quoted in (4), the additional third MVE associated with three microstructural BVPs has to be solved. It means, the five BVPs computations should be performed at every macrolevel material point. More can be found in [16]. In Fig. 3., $\Delta\mathbf{u}$ and $\Delta\boldsymbol{\varepsilon}_\mathrm{M}$ are the incremental displacement and strain imposed at the MVE boundaries, respectively, $\boldsymbol{\sigma}_\mathrm{B}$ denotes the stress tensor obtained for the bulk material, while ξ is a scalar perturbation parameter needed for the derivatives computation.

To compute the damage variable, the microstructural homogenization based on the failure averaging has been carried out, where the mixed finite element formulation with nonlocal equivalent plastic strain interpolation has been employed. Therein,

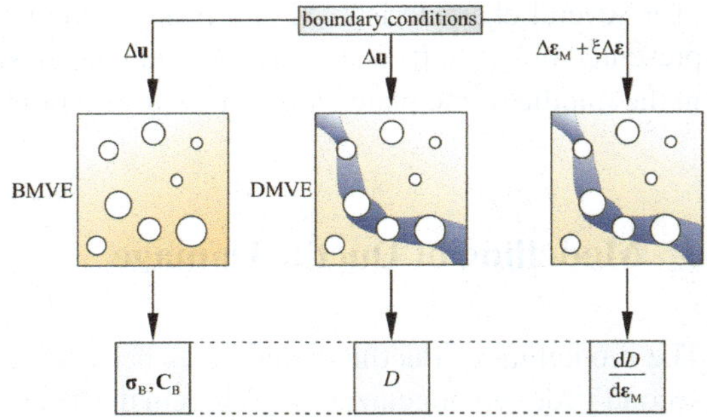

Fig. 3 Scheme of three MVE homogenization

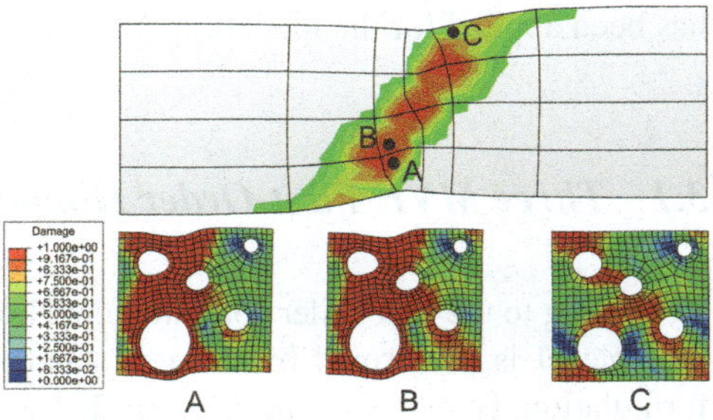

Fig. 4 Distribution of damage over macromodel and MVEs at points A, B and C

besides the standard equilibrium equation, an additional partial differential equation of the Helmholtz type is solved

$$\bar{\varepsilon}_p - l^2 \nabla^2 \bar{\varepsilon}_p = \varepsilon_p, \tag{5}$$

where ε_p is the local equivalent plastic strain, $\bar{\varepsilon}_p$ denotes the nonlocal equivalent plastic strain measure which governs the damage response, and l^2 represents the regularizing microstructural parameter. The gradient elastoplastic formulation based on the following von Mises yield function has been employed

$$F\left(\boldsymbol{\sigma}, \varepsilon_p, D\right) = \sigma_e\left(\boldsymbol{\sigma}\right) - (1 - D)\,\sigma_y\left(\varepsilon_p\right), \tag{6}$$

where σ_e represents the equivalent von Mises stress, and $\sigma_y(\varepsilon_p)$ expressed the linear isotropic hardening. Herein the standard elastoplastic algorithm as well as the standard homogenization procedure using Hill-Mandel energy equivalence have been applied. An exponential damage low has been used [17].

The softening response has again been demonstrated in a simple example of heterogeneous plate subjected to tensile load using the same MVE as in the case of quasi-brittle damage presented above. The damage distribution over the macromodel as well as over the MVEs positioned at some characteristic points of the macrostructure are shown in Fig. 4. It is shown that the softening evolution has been captured correctly, and the physically realistic structural responses have been modelled.

3.2 *Second-Order Homogenization*

Here a second-order homogenization procedure using the standard averaging approach is employed for the computation of ductile damage in heterogeneous materials. Instead of the formulation presented in the previous section, the nonlocal theory embedded into the triangular C^1 finite element formulation, described in [12], is used for discretization at the macrolevel. The constitutive matrices are upscaled from the microlevel, where they are computed by the second-order homogenization, which is in a general form presented also in [12]. Instead of three MVEs, here only one MVE at the microscale is used. As in the previous section, the same implicit gradient-enhanced elastoplasticity employing von Mises yield function is applied for the consideration of softening behaviour. The constitutive matrix of the bulk material of a heterogeneous microstructure is not computed and homogenized separately, and it is not needed the computation of derivatives of damage variable with respect to the macrostrain components. The MVE discretization has been performed by using the mixed quadrilateral formulation mentioned in the previous section, and the macrostructural regularization has been achieved by using the nonlocal strain gradient theory at macrolevel. In [14], it is shown that the mesh independent damage evolution at the macrolevel may be obtained, when nonlocal continuum theory is applied. It is to note that the second-order homogenization procedure is much simpler because the averaging only over one MVE has to be performed. However, the disadvantage of here proposed approach is the computation of the state variables at greater number of material points at macrolevel, as a consequence of using high order triangular finite element satisfying C^1 continuity. Instead of 4 integration points associated to the classical quadrilateral finite elements, 13 integration points are used in the C^1 finite element formulation. The performance of the computational procedure described is demonstrated in the computation of a plate subjected to compression. The spreading of the softening at the microscale as well as the distribution of the homogenized damage at the macrolevel are displayed in Fig. 5. As evident,

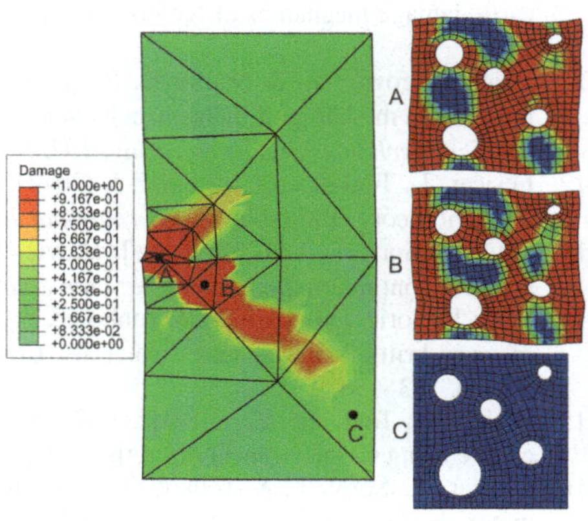

Fig. 5 Damage distribution at micro- and macrolevel

the weakness is initiated only in a small portion in the middle of the model, and thereafter the localization propagates towards the opposite boundary, as expected.

Despite the fact that the above presented formulations deliver the results which are physically correct and realistic, the computations are complex and time demanding. Therefore, the computationally more efficient formulations are desirable.

References

1. Kouznetsova, V. G., Geers, M. G. D., & Brekelmans, W. A. M. (2004). Multi-scale second-order computational homogenization of multi-phase materials: a nested finite element solution strategy. *Computer Methods in Applied Mechanics and Engineering, 193,* 5525–5550.
2. Lehmann, E., Loehnert, S., & Wriggers, P. (2012). Computational homogenization of polycrystalline elastoplastic microstructures at finite deformation. *Technische Mechanik, 32,* 369–379.
3. Wu, T., Temizer, I., & Wriggers, P. (2013). Computational thermal homogenization of concrete. *Cement and Concrete Composites, 35,* 59–70.
4. Bosco, E., Kouznetsova, V. G., Coenen, E. W. C., Geers, M. G. D., & Salvadori, A. (2014). A multiscale framework for localizing microstructures towards the onset of macroscopic discontinuity. *Computational Mechanics, 54,* 299–319.
5. Verhoosel, C. V., Remmers, J. J. C., de Gutiérrez, M. A., & Borst, R. (2010). Computational homogenization for adhesive and cohesive failure in quasi-brittle solids. *International Journal for Numerical Methods in Engineering, 83,* 1155–1179.
6. Sluys, L. J., Nguyen, V. P., & Stroeven, M. (2011). Multiscale continuos and discountinuous modeling of heterogeneous materials: A review on recent developments. *Journal of Multiscale Modelling, 03,* 229–270.
7. Nguyen, V. P., Lloberas-Valls, O., Stroeven, M., & Sluys, L. J. (2010). On the existence of representative volumes for softening quasi-brittle materials - A failure zone averaging scheme. *Computer Methods in Applied Mechanics and Engineering, 199,* 3028–3038.
8. Benedetti, I., & Aliabadi, M. H. (2015). Multiscale modeling of polycrystalline materials: A boundary element approach to material degradation and fracture. *Computer Methods in Applied Mechanics and Engineering, 289,* 429–453.
9. Tchalla, A., Azoti, W. L., Koutsawa, Y., Makradi, A., Belouettar, S., & Zahrouni, H. (2015). Incremental mean-fields micromechanics scheme for non-linear response of ductile damaged composite materials. *Composites Part B: Engineering, 69,* 169–180.
10. Fernandino, D. O., Cisilino, A. P., Toro, S., & Sanchez, P. J. (2017). Multi-scale analysis of the early damage mechanics of ferritized ductile iron. *International Journal of Fracture, 207*(1), 1–26.
11. Ibrahimbegović, A., & Markovič, D. (2003). Strong coupling methods in multi-phase and multi-scale modeling of inelastic behavior of heterogeneous structures. *Computer Methods in Applied Mechanics and Engineering, 192,* 3089–3107.
12. Lesičar, T., Tonković, Z., & Sorić, J. (2017). Two-scale computational approach using strain gradient theory at microlevel. *International Journal of Mechanical Sciences, 126,* 67–78.
13. Putar, F., Sorić, J., Lesičar, T., & Tonković, Z. (2017). Damage modelling employing strain gradient continuum theory. *International Journal of Solids and Structures, 120,* 171–185.
14. Putar, F., Sorić, J., Lesičar, T., & Tonković, Z. (2019). A multiscale method for damage analysis of quasi-brittle heterogeneous materials. *Computer Modeling in Engineering and Sciences, 200*(1), 123–156.
15. Lesičar, T., Tonković, Z., & Sorić, J. (2014). A second-order two-scale homogenization procedure using C1 macrolevel discretization. *Computational Mechanics, 54*(2), 425–441.
16. Lesičar, T., Sorić, J., & Tonković, Z. (2019). Ductile damage modelling of heterogeneous materials using a two-scale computational approach. *Computer Methods in Applied Mechanics and Engineering, 355,* 113–134.

17. Engelen, R. A. B., Geers, M. G. D., & Baaijens, F. P. T. (2003). Nonlocal implicit gradient-enhanced elasto-plasticity for the modelling of softening behaviour. *International Journal of Plasticity, 19*(4), 403–433.

A Novel Approach to Phasefield-Fracture for Inelastic Materials and Finite Deformations

Johannes Storm, Bo Yin, and Michael Kaliske

Peter Wriggers and I have a similar origin and a long lasting connection. Both of us were born in Hamburg. Main parts of our academic education were obtained in Hannover (Diploma, Doctorate, Habilitation) associated with late Prof. Erwin Stein. Over the decades, I have always enjoyed Peter's stimulating support and steady driving force. Various coinciding scientific interests and activities with respect to computational mechanics like research on constitutive modelling, fracture mechanics, contact mechanics, in order just to name some areas, let to frequent dialogues. Currently, we collaborate within the national Priority Programme "Cyclic deterioration of high-performance concrete in an experimental-virtual lab". I remember many nice and enjoyable social events in which we participated together associated with scientific activities all over the community. Moreover, strong interaction took place due to our common service to national scientific societies as Gesellschaft für Angewandte Mathematik und Mechanik (GAMM) and German Association for Computational Mechanics (GACM). Peter led both organizations in a dedicated and calm manner. I appreciate very much his company (Michael Kaliske).

Abstract An increasing number of areas in the field of fracture mechanics is addressed by the phasefield method for fracture. The fast development and improvement of the method in the last decade led to successful applications in brittle fracture for elastic and inelastic materials, for fracture at impact loads, fatigue, cohesive fracture and many more. However, several fundamental issues, like low convergence rate and consistent irreversibility formulation, require further investigations in order to obtain a powerful tool for general application to fracture mechanics. The manuscript at hand summarises recent developments in phasefield-fracture towards a realistic deformation kinematics at cracks, which yields the framework of *Representative Crack Elements*.

J. Storm · B. Yin · M. Kaliske (✉)
Technische Universität Dresden, Institute for Structural Analysis, Dresden, Germany
e-mail: Michael.Kaliske@tu-dresden.de

F. Aldakheel et al. (eds.), *Current Trends and Open Problems in Computational Mechanics*, https://doi.org/10.1007/978-3-030-87312-7_49

1 Introduction

In history of engineering, cracks in solids remain continuously a field of investigation on the state and functionality of a material. Cracks occur in nowadays masonry and concrete structures as well as in the Great Wall of China, Egyptian pyramids and burial chambers thousands of years old. Starting with the construction phase, the existence, nucleation and propagation of cracks needs to be considered in many simple and complex structures, like air-planes, trains and bridges. Engineers have developed fail save concepts, component monitoring strategies and even smart self-healing composite solutions. Not only for the restoration of a 300 years old violin or historical paintings, but also for the use of ceramic turbine blades at 1500 °C, facade anchoring at a 800 m heigh skyscraper as well as for opening tabs in plastic packagings, realistic crack models define our technical abilities and progress (Fig. 1).

The list of models, methods and approaches applied to cracks and material degradation is long and growing. Famous and successful concepts from the early days of computational mechanics are for example

- fracture mechanics based on the pioneering works of Griffith and Irwin on linear elastic fracture mechanics,
- damage mechanics which origins go back to Kachanov and Krajcinovic, and
- cohesive zone models.

Limitations and challenges led to ongoing developments and new methods for investigations of cracks and with cracks. The progress in computational fracture mechanics

Fig. 1 Cracks in a historical pianoforte (above) [1] and in a painting of the 16th century (right) [2] due to periodic swelling processes in the wood caused by cyclic moisture changes

is the result of a persistent search for a generic description of cracks and fracture processes. Researchers and engineers are interested in a generic, mathematical framework which overcomes the artificial and empirical assumptions made in previous approaches towards fracture, e.g. for nucleation, propagation and branching. Material forces, eigen-erosion, Extended Finite Element Method, Discrete Particle Method etc. attracted great interest in the global community of computational mechanics for the numerical calculation of solids with cracks.

The publication at hand addresses the phasefield method for fracture. Based on the variational description of fields with discontinuities in [3], and the regularised formulations developed in [4, 5], Francfort and Marigo [6] have presented a variational model for brittle fracture. Many people identify large potential in phasefield-fracture which results in a rapid development of the method. Recently, approaches towards phasefield-fracture for ductile, viscous and fatigue material behaviour, for finite deformations and for cohesive fracture have been developed. However, the findings in [7, 8] have demonstrated that the prediction of those models for the crack kinematics, i.e. for force transfer through the crack and for the re-contact of crack surfaces, is not precise even for elastic materials and small deformations. The novel framework of *Representative Crack Elements* (RCE) allows to derive phasefield models for fracture with physically realistic predictions for the crack kinematics. The introduction of representative crack elements in [9] is applied to linear, anisotropic elasticity and thermo-elasticity. The concept is extended to linear visco-elasticity in [10], where path-dependent deformations by means of internal variables are considered. A generalised, variational formulation for the RCE framework for non-linear material behaviour and finite deformations is presented in [11]. Further developments will follow in the near future.

2 Concept of Representative Crack Elements

The variational formulation for phasefield-fracture is based on the first law of thermodynamics and can be expressed in terms of the total virtual power $\delta\mathcal{P}^{\text{tot}}$. The total virtual power balances internal and external virtual power. The virtual power of external forces is equal to the dual product of the variation of the state variables $\boldsymbol{u}$ and p with their thermodynamically conjugate forces $\mathbf{f}^u$ and $\mathbf{f}^p$ in the domain $\mathcal{B}$

$$\delta\mathcal{P}^{\text{tot}} = \int\limits_{\mathcal{B}} \left[\delta\dot{\boldsymbol{u}} + \mathbf{f}^u + \partial\dot{p}\,\mathbf{f}^p\right]\mathrm{d}V. \tag{1}$$

The phasefield variable p continuously represents the presence of a crack or not at the material point. Conservative forces for the external virtual power are fully defined by boundary conditions. The principle of local actions is used to formulate the internal virtual power in terms of the displacement gradient $\boldsymbol{H}$, the phasefield p

and the phasefield gradient ∇p and their thermodynamically conjugate stresses $\boldsymbol{\Sigma}^{\boldsymbol{H}}$, $\boldsymbol{\Sigma}^{p}$ and $\boldsymbol{\Sigma}^{\nabla p}$

$$\partial \mathcal{P}^{\text{int}} = \int_{\mathcal{B}} \left[\delta \dot{\boldsymbol{H}} : \boldsymbol{\Sigma}^{\boldsymbol{H}} + \partial \dot{p}\, \boldsymbol{\Sigma}^{p} + \partial \nabla \dot{p} \cdot \boldsymbol{\Sigma}^{\nabla p} \right] \mathrm{d}V. \tag{2}$$

The unknown stresses are determined by a constitutive law in terms of the state variables and additional internal variables α, which represent the deformation history of the material. Frequently, a material is postulated by means of a HELMHOLTZ free energy potential $\psi(\boldsymbol{u}, p, \alpha)$. Then, the laws of state yield the stresses $\boldsymbol{\Sigma}^{\boldsymbol{H}}$, $\boldsymbol{\Sigma}^{p}$, $\boldsymbol{\Sigma}^{\nabla p}$ and conjugate forces β to the internal variables

$$\begin{aligned} \boldsymbol{\Sigma}^{\boldsymbol{H}} &= \frac{\partial \psi(\boldsymbol{u}, p, \alpha)}{\partial \boldsymbol{H}}, & \boldsymbol{\Sigma}^{p} &= \frac{\partial \psi(\boldsymbol{u}, p, \alpha)}{\partial p}, \\ \boldsymbol{\Sigma}^{\nabla p} &= \frac{\partial \psi(\boldsymbol{u}, p, \alpha)}{\partial \nabla p}, & \beta &= \frac{\partial \psi(\boldsymbol{u}, p, \alpha)}{\partial \alpha}. \end{aligned} \tag{3}$$

The constitutive structure of phasefield models for fracture reads

$$\psi(\boldsymbol{u}, p, \alpha) = \psi^{c}(\boldsymbol{u}, \alpha) + g(p) \left[\psi^{0}(\boldsymbol{u}, \alpha) - \psi^{c}(\boldsymbol{u}, \alpha) \right] + \psi^{\Gamma}(p, \nabla p). \tag{4}$$

ψ^{Γ} is the crack surface energy given as regularised formulation in terms of p and ∇p. The mechanical degradation of the material is described as interpolation between intact material, given by ψ^{0}, and fully broken material, given by ψ^{c}, by means of a degradation function $g(p)$. Intact material can be provided by classical constitutive descriptions. In the concept of representative crack elements, the fully degraded material state is characterised by a crack model as ordinary boundary value problem, compare [9] for a detailed description. The coupling of the representative crack model and the constitutive formulation ψ^{c} is obtained by means of computational homogenisation adopting the formulation in [12]. For the following examples, the kinematic coupling yields the relations

$$\overline{\boldsymbol{H}}(\overline{\mathbf{x}}) = \boldsymbol{H} - [\![\overline{\mathbf{h}}]\!] \otimes \mathbf{N}_1, \qquad \boldsymbol{\Sigma}^{\boldsymbol{H}} = \frac{1}{\overline{V}} \int_{\overline{\mathcal{B}}} \overline{\boldsymbol{\Sigma}} \mathrm{d}V \tag{5}$$

for the displacement gradient in the RCE and the stress homogenisation, where $[\![\overline{\mathbf{h}}]\!]$ is the normalised displacement discontinuity at the crack, $\mathbf{N}_1$ is the normal vector at of crack surface and $\overline{V}$ is the volume of the representative crack model.

3 Applications

Modelling the deformation kinematics at a crack by means of representative crack elements gives further access to processes which take place inside the crack. In many cases, friction between crack surfaces in contact is of particular interest, especially for failure at compressive loads. The friction behaviour between surfaces at compression is divided into sticking (static friction) and sliding (dynamic friction). Adopting the friction concept of COULOMB, the friction force acts as tangential reaction force which prevents the surfaces from a relative translation during static friction. The maximal friction force is proportional to the normal force at the surface. When the maximal friction force is exceeded, the surfaces can slide on each other while the friction force remains constant.

The precise prediction of the transition from sticking to sliding is demonstrated in [13, 14] using an elastic sliding block on a rigid foundation applied to compression and tangential load, cf. Fig. 2a. In both publications, frictional contact formulations are presented using discrete interfaces between block and foundation. COULOMB friction is considered in the contact zone except close to the block corners, which are considered friction free. Phasefield-fracture using DIRICHLET boundary conditions is applied to model the contact zone in Fig. 2b. The elastic material parameters are $E = 1$ MPa and $\nu = 0.3$. All three friction states,

- friction free ends ($x \in [0, 200]$ mm, $x \in [3800, 4000]$ mm),
- sliding zone ($x \in [\sim 2800, 3800]$ mm) and
- frictional sticking,

can be identified in Fig. 2c, where the computed normal and shear stresses along the contact area are compared. The results of the phasefield model based on the RCE framework agree well with those of the discrete contact models.

The realistic prediction of the deformation kinematics is not only important for the deformation at fully evolved cracks, but defines also the crack driving force in the variational phasefield formulation. Next to unphysical crack deformations, also incorrect predictions for crack nucleation and crack paths can be the result. In the

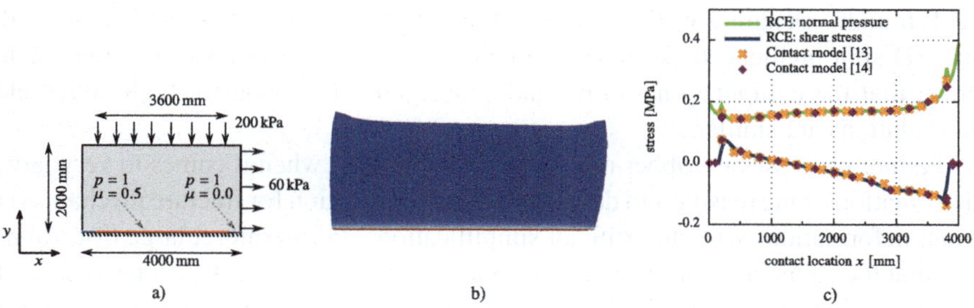

Fig. 2 **a** Sketch of the sliding block, **b** deformed block with phasefield distribution and **c** normal and shear stresses by phasefield model [11] compared to those of discrete contact models from [13, 14]

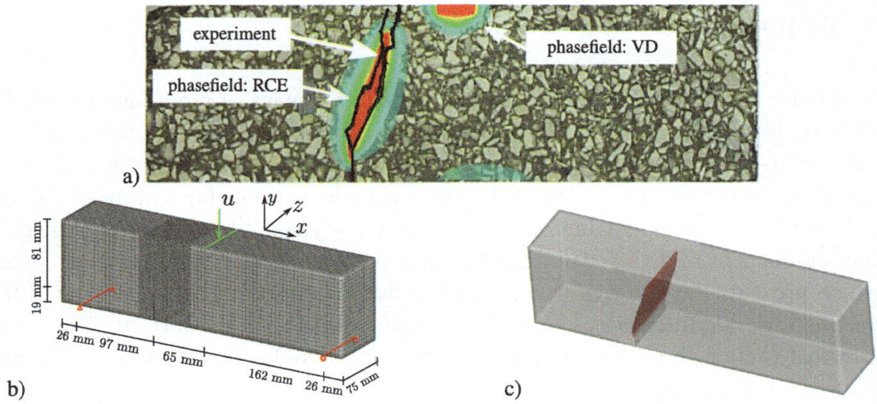

Fig. 3 Three-point bending test with notch, **a** experimental crack paths of [15] compared to those of the phasefield simulations based on RCE and VD-split [10], **b** sketch of the boundary value problem and **c** three-dimensional visualisation of the phasefield crack

following example, the novel phasefield formulation and a phasefield model based on the volumetric-deviatoric strain split are applied to a three-point bending test of asphalt concrete with a notch, cf. Fig. 3b. The visco-elastic material behaviour is considered by means of the generalised MAXWELL model. The numerical time integration of the stress history yields the time discrete formulation

$$\boldsymbol{\sigma}_{t^{n+1}} = \left[1 + \sum_{I=1}^{m} \chi^I \exp\left(-\frac{\Delta t}{2\,\tau^I}\right)\right] \boldsymbol{\sigma}^{eq}_{t^{n+1}}\left(\boldsymbol{\varepsilon}_{t^{n+1}}\right) + \boldsymbol{h}_{t^n}\left(\boldsymbol{\sigma}^{eq}_{t^n}, \boldsymbol{\sigma}^{ne}_{t^n}\right) \tag{6}$$

with stress of the elastic branch $\boldsymbol{\sigma}^{eq}$ and the stresses of the viscous branches $\boldsymbol{\sigma}^{ne,I}$. The material parameters of the m MAXWELL elements are τ^I and χ^I. The time increment from time t^n to t^{n+1} is Δt. The deformation history is represented by the internal variable $\boldsymbol{h}_{t^n}$.

The visualisation of the experimental crack paths from [15] in Fig. 3a is supplemented by the two different phasefield results. While the crack path predicted by the RCE framework agrees well with those of the experiment, the phasefield model with the VD-split predicts crack nucleation below the applied load instead of the notch. Note, that the computational effort and convergence behaviour of both phasefield formulations are similar.

Failure processes in rubber material are challenging when it comes to very large deformations. One reason is to develop a stable formulation for fracture mechanics at finite deformations without artificial simplifications. Furthermore, large distortions occur at the elements near the crack tip which can prevent successful convergence of the simulation. In order to circumvent unacceptable element distortions, an adaptive discretisation scheme can address this issue in future investigations. The variational RCE framework for phasefield-fracture can be directly developed into a kinematics

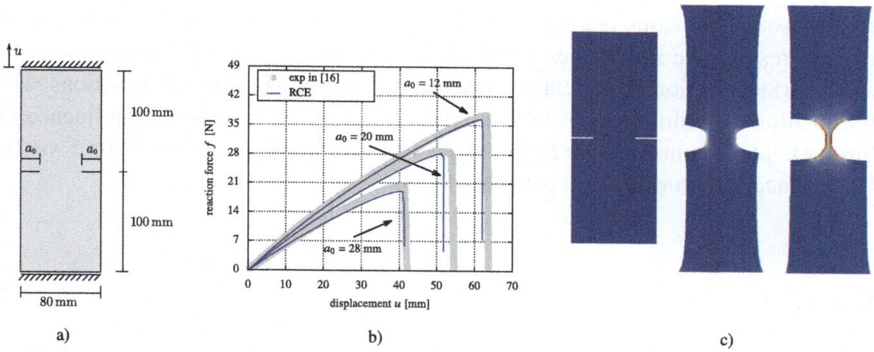

Fig. 4 Tensile test of a rubber block with two notches, **a** sketch of the boundary value problem, **b** force-displacement characteristics from the experiments in [16] and from the phasefield simulation, **c** visualisation of the deformation state before the test, at the beginning of crack propagation and close to final rupture. The blanking technique where $p \geq 0.95$ is applied to the deformation plots

for finite deformations. However, the computational effort for the material model increases in a finite deformation framework due to the formulation of the RCE model in the reference configuration and the necessary transformations towards the current configuration.

Crack propagation at large tensile deformations is studied at the rubber block with two notches in Fig. 4a. A nearly incompressible NEO-HOOKEAN model

$$\psi = \kappa\,(J - \ln(J) - 1) + \frac{\mu}{2}\,(\mathrm{tr}\,(\boldsymbol{C}_{\mathrm{iso}}) - 3)\,. \tag{7}$$

is considered for the bulk material, where J is the determinant of the deformation gradient and $\boldsymbol{C}_{\mathrm{iso}}$ is the isochoric part of the right CAUCHY-GREEN tensor. The elastic parameters κ and μ are bulk and shear modulus. Deformations of the rubber block are given at the beginning or the crack propagation and close to final rupture in Fig. 4c. The force-displacement characteristics of the phasefield simulation with the three notch lengths $a_0 = \{12, 20, 28\}$ mm agree well with the experimental findings in [16].

4 Conclusions

The fast development, recent improvements and many successful applications of the phasefield model for fracture have demonstrated large potential to obtain a very generalised, powerful method for fracture mechanics. By means of *Representative Crack Elements*, a variational framework is introduced which allows to formulate phasefield models to realistically predict the deformation at the crack. The framework gives further access to model processes which take place inside the crack. In comparison to previous phasefield models, to discrete crack models and to experimental results, the advantages of the RCE approach is shown by applications on crack surface friction, visco-elastic material behaviour and large deformations. The formulation of multi-physical RCE models will allow to consider the influence of further fields like moisture on the crack nucleation and propagation in future works, for instance in historical objects made of wood.

References

1. Kaliske, M., Konopka, D., & Stöcklein, J. (2019). Neue numerische simulation für alte Holzkonstruktionen. In *23. Dresdner Baustatik-Seminar, Realität – Modellierung – Tragwerksplanung*, pp. 67–88.
2. Gebhardt, C., Konopka, D., Börner, A., Mäder, M., & Kaliske, M. (2018). Hygro-mechanical numerical investigations of a wooden panel painting from "Katharinenaltar" by Lucas Cranach the Elder. *Journal of Cultural Heritage, 29,* 1–9.
3. Mumford, D., & Shah, J. (1989). Optimal approximations by piecewise smooth functions and associated variational problems. *Communications on Pure and Applied Mathematics, 42,* 577–685.
4. de Giorgi, E., Carriero, M., & Leaci, A. (1989). Existence theorem for a minimum problem with free discontinuity set. *Archive for Rational Mechanics and Analysis, 108,* 195–218.
5. Ambrosio, L., & Maria Tortorelli, V. (1990). Approximation of functional depending on jumps by elliptic functional via t-convergence. *Communications on Pure and Applied Mathematics, 43*, 999–1036.
6. Francfort, G. A., & Marigo, J. J. (1998). Revisiting brittle fracture as an energy minimisation problem. *Journal of the Mechanics and Physics of Solids, 46,* 1319–1342.
7. Strobl, M., & Seelig, T. (2016). On constitutive assumptions in phase field approaches to brittle fracture. *Procedia Structural Integrity, 2,* 3705–3712. 21st European Conference on Fracture, ECF21, 20-24 June 2016, Catania, Italy.
8. Steinke, C., & Kaliske, M. (2018). A phase-field crack model based on directional stress decomposition. *Computational Mechanics, 63,* 1019–1046.
9. Storm, J., Supriatna, D., & Kaliske, M. (2020). The concept of representative crack elements for phase-field fracture: Anisotropic elasticity and thermo-elasticity. *International Journal for Numerical Methods in Engineering, 121,* 779–805.
10. Yin, B., Storm, J., & Kaliske, M. (2021). Viscoelastic phase-field fracture using the framework of representative crack elements. *International Journal of Fracture*. https://doi.org/10.1007/s10704-021-00522-1
11. Storm, J., Yin, B., & Kaliske, M. (submitted). The concept of representative crack elements (rce) for phase-field fracture - non-linear materials and finite deformations.
12. Blanco, P. J., Sánchez, P. J., Souza Neto, E. A., & Feijóo, R. A. (2016). Variational foundations and generalized unified theory of rve-based multiscale models. *Archives of Computational Methods in Engineering*, *23,* 1–63.
13. Simo, J. C., & Laursen, T. A. (1992). An augmented lagrangian treatment of contact problems involving friction. *Computers and Structures, 42,* 97–116.
14. Oden, J. T., & Pires, E. B. (1984). Algorithms and numerical results for finite element approximations of contact problems with non-classical friction laws. *Computers and Structures, 19,* 137–147.
15. Song, S. H., Paulino, G. H., & Buttlar, W. G. (2006) . A bilinear cohesive zone model tailored for fracture of asphalt concrete considering viscoelastic bulk material. *Engineering Fracture Mechanics*, *73,* 2829–2848.

16. Hocine, N. A., Abdelaziz, M. N., & Imad, A. (2002). Fracture problems of rubbers: J-integral estimation based upon η factors and an investigation on the strain energy density distribution as a local criterion. *International Journal of Fracture, 117,* 1–23.

Space–Time Flow Computation with Contact Between the Moving Solid Surfaces

Kenji Takizawa, Takuya Terahara and Tayfun E. Tezduyar

I met Professor Peter Wriggers in 2011, at the conference dedicated to his 60th birthday. It was an honor to give a presentation at the birthday conference of a giant in computational mechanics research. I have been proud of serving as an Assistant Editor of Computational Mechanics from 2012 to 2019 and as an Editorial Board Member after that. It was always a pleasure having research discussions with Professor Wriggers at conferences. It is an honor to contribute a chapter to a book dedicated to his 70th birthday. Kenji Takizawa

Abstract In computation of flow problems with moving boundaries and interfaces, including fluid–structure interaction, moving the fluid mechanics mesh to follow the fluid–solid interface enables mesh-resolution control near the interface. Therefore moving-mesh methods, such as the Space–Time Variational Multiscale (ST-VMS) method, enable high-resolution boundary-layer representation near fluid–solid interfaces and thus higher accuracy in such critical flow regions. In flow problems with contact between solid surfaces, until recently, one had to either give up on representing the actual contact and leave a small gap or give up on using a moving-mesh method and thus give up on having high-fidelity flow solution near the solid surfaces. The ST Topology Change (ST-TC) method changed all that. Now we can both represent the actual contact and have high-fidelity flow solution near the solid surfaces. With the ST-VMS, which serves as the core method, and the ST-TC and two other special methods, the ST Slip Interface method and ST Isogeometric Analysis, we have created a powerful computational framework. The new framework is enabling high-fidelity computational flow analysis of some of the most complex problems, such as the ventricle-valve-aorta sequence. This chapter is a description and demonstration of that framework.

K. Takizawa (✉) · T. Terahara · T. E. Tezduyar
Waseda University, Tokyo, Japan
e-mail: Kenji.Takizawa@tafsm.org

T. E. Tezduyar
Rice University, Houston, TX, USA
e-mail: tezduyar@tafsm.org

F. Aldakheel et al. (eds.), *Current Trends and Open Problems in Computational Mechanics*, https://doi.org/10.1007/978-3-030-87312-7_50

1 Introduction

In computation of flow problems with moving boundaries and interfaces (MBI), including fluid–structure interaction (FSI), high-resolution boundary-layer representation near fluid–solid interfaces requires mesh-resolution control near the interface. Moving-mesh methods, such as the Space–Time Variational Multiscale (ST-VMS) method [1], meet that requirement. In an FSI or MBI problem with contact between solid surfaces, until recently, one had to either give up on representing the actual contact and leave a small gap or give up on using a moving-mesh method and thus give up on having high-fidelity flow solution near the solid surfaces. The ST Topology Change (ST-TC) method [2] changed all that. Now we can both represent the actual contact and have high-fidelity flow solution near the solid surfaces.

With the ST-TC and two other special ST methods around the core method ST-VMS, we have created a powerful computational framework, the "ST-SI-TC-IGA" [3]. The two other methods are the ST Slip Interface (ST-SI) method [4] and the ST Isogeometric Analysis (ST-IGA) [1, 5, 6]. The ST-SI-TC-IGA is enabling high-fidelity flow analysis in some of the most complex problems. We briefly describe it and demonstrate it in ventricle-valve-aorta flow analysis. We include the left ventricle (LV) in the model to make the flow into the valve anatomically more realistic, which, in turn, makes the flow into the aorta more realistic.

2 Methods

The **ST-VMS** is the VMS version of the Deforming-Spatial-Domain/Stabilized ST (DSD/SST) method [7]. The DSD/SST is a moving-mesh method, introduced for computation of flows with MBI, including FSI. Because the stabilization components of the original DSD/SST are the Streamline-Upwind/Petrov-Galerkin (SUPG) [8] and Pressure-Stabilizing/Petrov-Galerkin (PSPG) [7] methods, it is called "**ST-SUPS**." The VMS components of the ST-VMS are from the residual-based VMS (RBVMS) method [9]. The ST-SUPS and ST-VMS have been applied to many classes of FSI, MBI and fluid mechanics problems (see [10] and the chapter contributed by the third author). For more on the ST-VMS and ST-SUPS, see [11].

The **ST-SI** was introduced to retain the desirable moving-mesh features of the ST-VMS and ST-SUPS in computations involving spinning solid surfaces, such as a turbine rotor. The mesh covering the spinning surface spins with it, retaining the high-resolution representation of the boundary layers, while the mesh on the other side of the SI remains unaffected. This is accomplished by adding to the ST-VMS formulation interface terms similar to those in the version of the ALE-VMS [12] for computations with sliding interfaces [13]. The interface terms accurately connect the two sides of the solution. An ST-SI version where the SI is between fluid and solid domains was also presented in [4]. The SI in that case is a "fluid–solid SI" rather than a standard "fluid–fluid SI" and enables weak enforcement of the Dirichlet boundary conditions for the fluid. The ST-SI has been applied to many MBI and fluid mechanics problems (see [10] and the chapter contributed by the third author). For more on the ST-SI, see [4].

The **ST-TC** made moving-mesh computations possible even when there is an actual contact between solid surfaces or other TC, not just a near contact. The ST-TC, by collapsing elements as needed, without changing the connectivity of the "parent" mesh, can handle an actual TC while maintaining high-resolution boundary layer representation near solid surfaces. The ST-TC has been applied to several MBI and fluid mechanics problems (see [10] and the chapter contributed by the third author). For more on the ST-TC, see [2].

The **ST-SI-TC** [14] is the integration of the ST-SI and ST-TC. A fluid–fluid SI requires elements on both sides of the SI. When part of an SI needs to coincide with a solid surface, which happens for example when the solid surfaces on two sides of an SI come into contact or when an SI reaches a solid surface, the elements between the coinciding SI part and the solid surface need to collapse with the ST-TC mechanism. The collapse switches the SI from fluid–fluid SI to fluid–solid SI. With that, an SI can be a mixture of fluid–fluid and fluid–solid SIs. With the ST-SI-TC, the elements collapse and are reborn independent of the nodes representing a solid surface. The ST-SI-TC enables high-resolution flow representation even when parts of the SI are coinciding with a solid surface. It also enables dealing with contact location change and contact sliding. The ST-SI-TC has been applied to several MBI and fluid mechanics problems (see [10] and the chapter contributed by the third author). For more on the ST-SI-TC, see [14].

The **ST-IGA** is the integration of the ST framework and IGA discretization. It was motivated by the success with using IGA basis functions in space [15]. In the ST-IGA, the IGA basis functions are used also in temporal representation. Computations with the ST-VMS and ST-IGA were first reported in [1] in a 2D context, with IGA basis functions in space and time. The ST-IGA with IGA basis functions in time enables a more accurate representation of the motion of the solid surfaces and a mesh motion consistent with that. It also enables more efficient temporal representation of the motion and deformation of the volume meshes, and more efficient remeshing. These motivated the development of the ST/NURBS Mesh Update Method (STNMUM) [5]. The STNMUM has a wide scope that includes spinning solid surfaces. With the spinning motion represented by quadratic NURBS in time, and with sufficient number of temporal patches for a full rotation, the circular paths are represented exactly. A "secondary mapping" [1] enables also specifying a constant angular velocity for invariant speeds along the circular paths. The ST framework and NURBS in time also enable, with the "**ST-C**" method, extracting a continuous representation from the computed data and, in large-scale computations, efficient data compression [16, 17]. The ST-IGA with IGA basis functions in space enables more accurate representation of the geometry and increased accuracy in the flow solution. It accomplishes that with fewer control points, and consequently with larger effective element sizes. That in turn enables using larger time-step sizes while keeping the Courant number at a desirable level for good accuracy. The ST-IGA has been applied to many MBI and fluid mechanics problems (see [10] and the chapter contributed by the third author). For more on the ST-IGA, see [5, 6].

The **ST-SI-IGA** [6] is the integration of the ST-SI and ST-IGA. As a fluid–fluid SI, it enables, in the context of IGA discretization, retaining the desirable moving-mesh features of the ST-VMS and ST-SUPS in computations involving spinning solid surfaces. As a fluid–solid SI, it enables, in the context of IGA discretization, weak enforcement of the Dirichlet boundary conditions for the fluid. An SI provides mesh generation flexibility in a general context by accurately connecting the two sides of the solution computed over nonmatching meshes. This flexibility is especially valuable in complex-geometry flow computations with IGA discretization, removing the matching requirement between the NURBS patches without loss of accuracy. The ST-SI-IGA has been applied to many MBI and fluid mechanics problems (see [10] and the chapter contributed by the third author). For more on the ST-SI-IGA, see [6].

The **ST-SI-TC-IGA** [3] is the integration of the ST-SI, ST-TC and ST-IGA. It provides, in the context of IGA discretization, the desirable features of the ST-SI-TC. The ST-SI-TC-IGA was applied to heart-valve flow analysis with a moving-mesh method, IGA discretization and full valve closure [3], IGA-discretization moving-mesh tire aerodynamic analysis with actual tire geometry, road contact, tire deformation, road roughness and fluid film [18, 19], and ventricle-valve-aorta flow analysis with a moving-mesh method, IGA discretization and full valve closure [20]. We describe the ventricle-valve-aorta flow analysis in the next section. For more on the ST-SI-TC-IGA, see [3].

The ST-SI-TC-IGA is supplemented with two other special methods. (a) A structural mechanics computation method generates the **LV motion** from the CT scans of the LV and anatomically realistic values for the LV volume ratio. The structural mechanics computations, performed in different ways for the diastole and systole, generate a "table" of LV volumes and shapes. From that and the volume ratio given, the cardiac-cycle representation of the LV motion is obtained by using cubic B-splines in time and the ST-C. (b) The **Constrained-Flow-Profile (CFP) Traction** provides flow stability at the inflow boundary. This is done by placing adjacent to the inflow boundary a special-purpose element with 27 basis functions. The ST-SI connects the flow solutions over that element and the rest of the mesh. The special-purpose element, with only one unspecified control-point velocity at the inflow, results in a constrained flow profile, which is quadratic. The solution obtained for the unspecified velocity, together with the quadratic profile, represents the flow rate generated by the traction conditions specified at the inflow and outflow boundaries.

3 Ventricle-Valve-Aorta Flow Analysis

This section is from [20]. The flow computation model consists of the LV, aortic valve with sinuses, and the aorta. We do not include the mitral valve in the model. The boundary between the LV and the left atrium becomes our inflow boundary. We use the CFP Traction there when the mitral valve is open, and zero-velocity when it is closed. The aorta main outlet is a regular outflow boundary with traction condition, and the three smaller outlets are prescribed-velocity outflow boundaries.

3.1 Geometry and Motion

The entire model is shown in Fig. 1. The quadratic NURBS meshes for the three parts are generated separately and the SIs connect the three solution parts. The LV shrinks and expands with a cardiac cycle of $T = 0.9$ s, the valve opens and closes, and the aorta remains stationary. The starting fluid mechanics volume mesh is generated using the method in [21]. Then the mesh is moved, with the mesh-Jacobian-based stiffening [22], to conform to the LV shapes obtained in the structural mechanics computations. The valve-mesh motion is obtained by transformation from the mesh motion used in the valve computation reported in [23]. We note that the mesh motion in [23] is based on the ST-SI-TC-IGA, which deals with the TC created by the contact between the leaflets while maintaining high-resolution representation near them. The aorta geometry is based on a different set of CT scans than the LV scans and is represented by cubic T-splines. From that, the fluid mechanics volume mesh is generated using the method in [21]. The aorta part remains rigid.

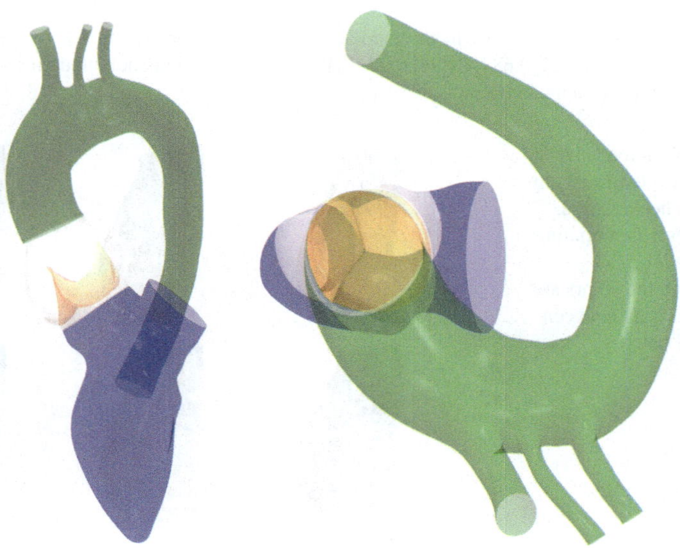

Fig. 1 LV-valve-aorta model. LV (*blue*), leaflets (*orange*), and aorta (*green*). The *left* picture shows the front view, and the *right* picture shows the view along the valve axis

3.2 Mesh, Boundary Conditions, Blood Properties, and Computational Conditions

To the mesh composed of the three parts, we add, for the CFP Traction, the SI and the special-purpose element. We set $\rho = 1{,}050\,\mathrm{kg/m^3}$ and $\mu = 4.2\times10^{-3}\,\mathrm{Pa \cdot s}$. Figures 2 and 3 show the mesh at different instants in the cardiac cycle. We use the ST-SUPS, with the stabilization parameters given by Eqs. (4)–(9) in [18]. The time-step size is 2.81×10^{-3} s. The number of nonlinear iterations per time step is 3, and the number of GMRES iterations per nonlinear iteration is 300.

Fig. 2 Mesh at different instants in the cardiac cycle. The frames (from *left* to *right*) are for $t/T = 0.003, 0.250, 0.625, 0.684, 0.894, 0.981$. The lines are the element boundaries. The sinuses are transparent to make the valve motion visible

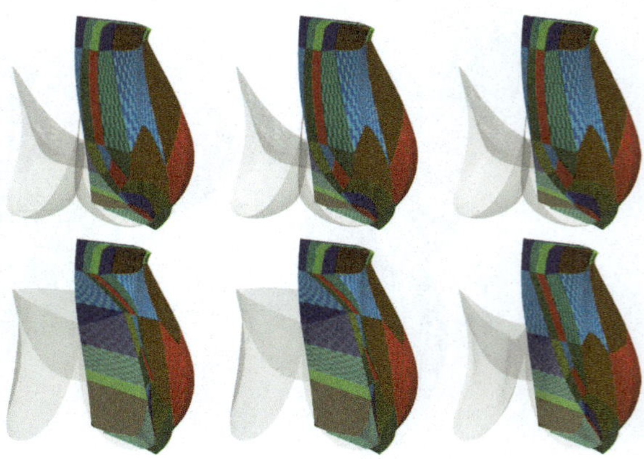

Fig. 3 Mesh in the valve at the cardiac-cycle instants in Fig. 2. The checkerboard pattern is for differentiating between the NURBS elements, and the colors are for differentiating between the NURBS patches

Fig. 4 Isosurfaces corresponding to a positive value of the second invariant of the velocity gradient tensor, colored by the velocity magnitude (m/s). The frames are for the same instants as in Fig. 2

3.3 Results and Concluding Remarks

Figure 4 shows the flow patterns. We capture the spiral flow in the valve and aorta and have a reasonable flow field even when the leaflets come into contact.

References

1. Takizawa, K., & Tezduyar, T. E. (2011). Multiscale space-time fluid-structure interaction techniques. *Computational Mechanics, 48*, 247–267.

2. Takizawa, K., Tezduyar, T. E., Buscher, A., & Asada, S. (2014). Space-time interface-tracking with topology change (ST-TC). *Computational Mechanics, 54*, 955–971.
3. Takizawa, K., Tezduyar, T. E., Terahara, T., & Sasaki, T. (2017). Heart valve flow computation with the integrated Space-Time VMS, Slip Interface, Topology Change and Isogeometric Discretization methods. *Computers and Fluids, 158,* 176–188.
4. Takizawa, K., Tezduyar, T. E., Mochizuki, H., Hattori, H., Mei, S., Pan, L., & Montel, K. (2015). Space-time VMS method for flow computations with slip interfaces (ST-SI). *Mathematical Models and Methods in Applied Sciences, 25*, 2377–2406.
5. Takizawa, K., Henicke, B., Puntel, A., Spielman, T., & Tezduyar, T. E. (2012). Space-time computational techniques for the aerodynamics of flapping wings. *Journal of Applied Mechanics, 79*, 010903.
6. Takizawa, K., Tezduyar, T. E., Otoguro, Y., Terahara, T., Kuraishi, T., & Hattori, H. (2017). Turbocharger flow computations with the Space-Time Isogeometric Analysis (ST-IGA). *Computers and Fluids, 142,* 15–20.
7. Tezduyar, T. E. (1992). Stabilized finite element formulations for incompressible flow computations. *Advances in Applied Mechanics, 28*, 1–44.
8. Brooks, A. N., & Hughes, T. J. R. (1982). Streamline upwind/Petrov-Galerkin formulations for convection dominated flows with particular emphasis on the incompressible Navier-Stokes equations. *Computer Methods in Applied Mechanics and Engineering, 32*, 199–259.
9. Hughes, T. J. R. (1995). Multiscale phenomena: Green's functions, the Dirichlet-to-Neumann formulation, subgrid scale models, bubbles, and the origins of stabilized methods. *Computer Methods in Applied Mechanics and Engineering, 127*, 387–401.
10. Tezduyar, T. E., & Takizawa, K. (2019). Space-time computations in practical engineering applications: A summary of the 25-year history. *Computational Mechanics, 63*, 747–753.
11. Bazilevs, Y., Takizawa, K., Tezduyar, T. E. (2013). *Computational Fluid–Structure Interaction: Methods and Applications*. Wiley.
12. Bazilevs, Y., Calo, V. M., Hughes, T. J. R., & Zhang, Y. (2008). Isogeometric fluid-structure interaction: theory, algorithms, and computations. *Computational Mechanics, 43*, 3–37.
13. Bazilevs, Y., & Hughes, T. J. R. (2008). NURBS-based isogeometric analysis for the computation of flows about rotating components. *Computational Mechanics, 43*, 143–150.
14. Takizawa, K., Tezduyar, T. E., Asada, S., & Kuraishi, T. (2016). Space-time method for flow computations with slip interfaces and topology changes (ST-SI-TC). *Computers and Fluids, 141,* 124–134.
15. Hughes, T. J. R., Cottrell, J. A., & Bazilevs, Y. (2005). Isogeometric analysis: CAD, finite elements, NURBS, exact geometry, and mesh refinement. *Computer Methods in Applied Mechanics and Engineering, 194*, 4135–4195.
16. Takizawa, K., & Tezduyar, T. E. (2014). Space-time computation techniques with continuous representation in time (ST-C). *Computational Mechanics, 53*, 91–99.
17. Takizawa, K., Tezduyar, T. E., & Kuraishi, T. (2015). Multiscale ST methods for thermo-fluid analysis of a ground vehicle and its tires. *Mathematical Models and Methods in Applied Sciences, 25*, 2227–2255.
18. Kuraishi, T., Takizawa, K., & Tezduyar, T. E. (2019). Tire aerodynamics with actual tire geometry, road contact and tire deformation. *Computational Mechanics, 63*, 1165–1185.
19. Kuraishi, T., Takizawa, K., & Tezduyar, T. E. (2019). Space-time computational analysis of tire aerodynamics with actual geometry, road contact, tire deformation, road roughness and fluid film. *Computational Mechanics, 64*, 1699–1718.
20. Terahara, T., Takizawa, K., Tezduyar, T. E., Tsushima, A., & Shiozaki, K. (2020). Ventricle-valve-aorta flow analysis with the Space-Time Isogeometric Discretization and Topology Change. *Computational Mechanics, 65*, 1343–1363.
21. Otoguro, Y., Takizawa, K., & Tezduyar, T. E. (2017). Space-time VMS computational flow analysis with isogeometric discretization and a general-purpose NURBS mesh generation method. *Computers and Fluids, 158,* 189–200.
22. Tezduyar, T., Aliabadi, S., Behr, M., Johnson, A., & Mittal, S. (1993). Parallel finite-element computation of 3D flows. *Computer, 26*(10), 27–36.

23. Takizawa, K., Tezduyar, T. E., Uchikawa, H., Terahara, T., Sasaki, T., Shiozaki, K.., Yoshida, A., Komiya, K., & Inoue, G. (2018). Aorta flow analysis and heart valve flow and structure analysis. In Tezduyar, T. E. ed., *Frontiers in Computational Fluid–Structure Interaction and Flow Simulation: Research from Lead Investigators under Forty – 2018*, Modeling and Simulation in Science, Engineering and Technology, pp. 29–89. Springer.

Higher-Order Finite Element Methods for Kohn-Sham Density Functional Theory

İlker Temizer

Recalling that Peter Wriggers is from Hamburg, I would like to quote what John Lennon said about this city – after adapting it to my academic life that began as a Ph.D. student: "I was born in Berkeley, but I grew up in Hannover." I am grateful to Prof. Wriggers for many opportunities that he provided for scientific growth and I can think of no better way to thank him than to choose a topic that actually started as a hobby of sorts while I was still at his institute. I extend my best wishes to him on the occasion of his 70th birthday. –İ.T.

Abstract Higher-order finite element methods are applied to electronic structure calculation in the context of the finite element method. For this purpose, the Kohn-Sham formalism of density functional theory is cast in a setting that is amenable to a finite element discretization. Both all-electron and pseudopotential formulations are presented, the latter incorporating both local and nonlocal contributions. Some of the outstanding challenges in applying this numerical framework to such ab initio methods are discussed. Finally, the approach is demonstrated with higher-order finite element basis sets that are associated with classical Lagrange discretizations as well as more with more recent isogeometric ones based on NURBS and B-splines.

1 Introduction

Over the past decade, finite element methods have been applied successfully and competitively in the context of electronic structure calculation [1]. In particular within the context of the Kohn-Sham density functional theory, efficient ab initio methods are desirable in order to access nonphenomenological predictions of material and interface behavior at small length scales [2]. From an engineering perspective, one aim is to subsequently upscale this information by connecting different length scales

İ. Temizer (✉)
Department of Mechanical Engineering, Bilkent University, Ankara, Turkey
e-mail: temizer@bilkent.edu.tr

F. Aldakheel et al. (eds.), *Current Trends and Open Problems in Computational Mechanics*, https://doi.org/10.1007/978-3-030-87312-7_51

using numerical methods which operate efficiently at each scale in order to reach at multiscale descriptions of deformation and failure for material and interface design. The triggering role of ab initio methods in this hierarchy of transitions is pivotal because it is able to initiate this process with a truly nonphenomenological seed. Therefore, there is an ongoing need for numerical methods which can attain desired levels of chemical accuracy faster and more efficiently, particularly in view of the very high computational cost of the problems involved [3–5]. From this perspective, the advantages of the finite element method are numerous [6]. First, in comparison to plane-wave approaches, it is a real-space method that does not require transforms that typically scale up rapidly in cost with increasing numerical resolution. Moreover, its structure naturally accommodates both periodic material systems such as crystals in addition to nonperiodic ones such as isolated molecules, unlike plane waves which assume periodicity from the outset. Second, in comparison to a real-space approach such as the finite-difference method, it preserves the variational structure that underlies density functional theory, similar to plane-wave discretizations. When combined with systematic improvability due to the completeness of the basis sets, this property ensures monotonic convergence in total energy–a property that is critical when one wishes to assess the solution quality. Such systematic improvability is missing in widely employed real-space methods based on Gaussian basis sets which retain the variational structure but cannot ensure that convergence achieved through the addition of multiple basis members is indeed due to having achieved the minimum total energy and not an artifact of incompleteness.

Density functional theory is an approach to solving the Schrödinger equation and the Kohn-Sham formalism is a theoretical framework that renders this approach feasible in a numerical setting [7]. The finite element method has recently been applied as a particular numerical setting in various contexts which range from time-independent periodic cases to time-dependent nonperiodic ones [6, 8–14]. Overall, the level of efficiency achieved is now competitive with, and in several cases already beyond, time-honored practices based on plane-waves for periodic problems and Gaussian basis sets for nonperiodic ones. The sparse structure of the matrices which emanate from finite element basis sets as well as the suitability of the framework to parallelization are contributing factors to this competitiveness. Moreover, higher-order basis sets are crucial if such problems are to be solved in reasonable times due to the insufficient convergence rate of linear ones, and the finite element method is naturally amenable to the incorporation of such discretizations. In the majority of these studies, and in all recent applications, basis sets are derived from Lagrange-type elements. As an alternative discretization scheme, following various earlier studies [15–17] but by building upon and benefiting from efficient eigensolvers to address comparatively larger material systems [18], the use of NURBS and B-splines was explored in [19], motivated by the isogeometric analysis approach [20]. The goal of this contribution is to summarize the finite element formulation of this ab initio problem and to present examples which indicate routes to even higher efficiency if isogeometric basis sets are employed as higher-order finite elements.

2 Density Functional Theory

The focus of the present study is on an isolated material system $\mathcal{M}$ consisting of N electrons and M nuclei, each with charge Z_A ($A \in 1, \ldots, M$). In a pseudopotential setting, the set of electrons is associated with the valence structure only and the nuclei are replaced by ions that are assigned a net charge that is augmented by those of the core electrons. The configuration of $\mathcal{M}$ is determined through the nucleus locations $\boldsymbol{R}_A$ that define $\boldsymbol{r}_A = \boldsymbol{r} - \boldsymbol{R}_A$ where the spatial position vector in unbounded space is indicated as $\boldsymbol{r}$ and integration over this space will be denoted by $\langle \cdot \rangle$. The external local potential generated by the nuclei can then be expressed as $v_{\text{ext}} = \sum_A v_A(r_A)$ where, introducing $r_A = |\boldsymbol{r}_A|$, $v_A = -Z_A m_A(r_A)$ are spherically symmetric potentials. For the all-electron setting $m_A(x) = 1/x$ delivers the classical Coulomb expression of the potential whereas in the local pseudopotential setting $m_A(x)$ acts as an ion-dependent mollifier which is chosen such that it delivers the regularized form of $1/x$, i.e. $m(0)$ is well-defined and $m(x)$ matches or rapidly approaches $1/x$ beyond a prescribed distance. In the case of nonlocal pseudopotentials, the electron-ion interaction is nonlocal such that the external potential as well as the corresponding energy contribution will entail additional terms that will be shortly commented upon.

Assuming a closed-shell structure without consideration for spin effects and degeneracy, $N/2$ real orthonormal spatial orbitals $\psi_i(\boldsymbol{r})$ are introduced which describe a noninteracting reference system of N electrons in the Kohn-Sham formalism of density functional theory. The electron density can then be expressed as $\rho(\boldsymbol{r}) = 2\sum_i \psi_i^2$ and, in atomic units, the energy as a functional of the density takes the form

$$E = T_s + E_H + E_{en} + E_{nn} + E_{xc} \ . \tag{1}$$

Here, T_s is the exact kinetic energy of the reference system, E_H is the Hartree energy that corresponds to the classical electrostatic interactions among the electrons, E_{en} is associated with the electron-nucleus interactions while E_{nn} is associated with nucleus-nucleus ones and, finally, E_{xc} is the exchange-correlation energy which accounts for errors in the kinetic energy and electron-electron interaction expressions:

$$T_s = 2\sum_i \langle\, \psi_i \, (-\tfrac{1}{2}\nabla^2) \, \psi_i \rangle \quad , \quad E_H = \tfrac{1}{2}\langle \rho \, v_H \rangle \quad , \quad E_{en} = \langle \rho \, v_{\text{ext}} \rangle \quad , \tag{2}$$

$$E_{nn} = \frac{1}{2}\sum_A \sum_{B \neq A} \frac{Z_A Z_B}{|\boldsymbol{R}_A - \boldsymbol{R}_B|} \quad , \quad E_{xc} = \langle \rho \, \varepsilon_{xc} \rangle \ . \tag{3}$$

Within these expressions, ε_{xc} is the exchange-correlation energy per electron and v_H is the Hartree potential that is defined through the Poisson equation

$$-\tfrac{1}{4\pi}\nabla^2 v_H = \rho \ . \tag{4}$$

Each contribution to the total energy E depends on the set of orbitals ψ_i, either directly or indirectly through ρ. The ground state electronic structure is determined through the minimization of E over ψ_i, leading to the Kohn-Sham equation

$$(-\tfrac{1}{2}\nabla^2 + v_{\text{eff}})\,\psi_i = \epsilon_i\,\psi_i \tag{5}$$

where $v_{\text{eff}} = v_H + v_{\text{ext}} + v_{xc}$, ϵ_i are the orbital energies and the exchange-correlation potential v_{xc} is defined through the variation $\delta E_{xc} = \langle v_{xc}\,\delta\rho\rangle$. Although the definition of v_{xc} is general, the present study will be restricted to the local density approximation where ε_{xc} will be a function of ρ only and therefore $v_{xc} = \varepsilon_{xc} + \rho\,\partial\varepsilon_{xc}/\partial\rho$. Overall, the summarized theory can also easily be extended to account for open-shell problems and degeneracies through the incorporation of fractional occupancy, although this will not be explicitly indicated in this presentation.

In comparison to the all-electron setting, the pseudopotential setting delivers a significantly more efficient numerical approach because it eliminates the explicit presence of the core electrons, thereby reducing the stringent requirements on the discretization as well as the number of orbitals which need to be calculated [21]. Although local pseudopotentials may suffice for some material systems [22], most cases require nonlocal pseudopotentials for accuracy and transferability [23]. For this purpose, the total energy (1) is augmented by a contribution

$$E_{NL} = \sum_i \langle \psi_i(\boldsymbol{r})\, v_{NL}(\boldsymbol{r},\boldsymbol{r}')\, \psi_i(\boldsymbol{r}')\rangle \quad , \quad v_{NL} = \sum_A \Lambda_A(\boldsymbol{r}_A,\boldsymbol{r}'_A) \tag{6}$$

where nonlocality explicitly manifests itself. Similarly, the effective potential v_{eff} in the Kohn-Sham equation (5) is augmented by the nonlocal part v_{NL} of the pseudopotential that is composed of a sum of atom-dependent nucleus-centered nonlocal contributions Λ_A. For an efficient numerical implementation, the preferred form of these contributions is based on a separable form. Presently, this implementation will not be explicitly discussed.

It is noted that it is possible to evaluate the exact Hartree potential v_H corresponding to a given electron density ρ, albeit at a high numerical cost. This cost is circumvented by solving the Poisson equation (4) instead. Now, however, v_H will be obtained to within the numerical error that is associated with the chosen discretization. Consequently, the overall problem for rendering the energy stationary towards the ground state electronic structure takes on a saddle-point form rather than one of minimization [24]. The practical consequence of this is that nonvariational results may be obtained if the finite element discretization is not chosen carefully, thereby rendering difficulties in the assessment of the solution quality. The finite element formulation to be discussed next assumes a single discretization that is chosen judiciously for the solution of both the Poisson and Kohn-Sham equations.

3 Finite Element Discretization

Because $v_{\rm eff}$ within (5) is nonlinear in the set of orbitals ψ_i, (4) and (5) must be solved iteratively to self-consistency in order to determine the ground state electronic structure, typically together with a mixing scheme in order to tune convergence. Presently, this solution approach will be built on the finite element method. Introducing test functions $\{\eta, \varphi\}$ and carrying out standard manipulations, the weak forms of the Poisson and Kohn-Sham equations may be expressed as

$$\tfrac{1}{4\pi}\langle \nabla\eta \cdot \nabla v_H\rangle = \langle \eta\, \rho\rangle \quad , \quad \tfrac{1}{2}\langle \nabla\varphi \cdot \nabla\psi_i\rangle + \langle \varphi\, v_{\rm eff}\, \psi_i\rangle = \epsilon_i \langle \varphi\, \psi_i\rangle \tag{7}$$

Invoking a single discretization for both problems that is described by a set of shape functions N_I, these expressions take the form

$$[P]\{v_H\} = \{d\} \quad , \quad [H]\{\psi_i\} = \epsilon_i [M]\{\psi_i\} \; . \tag{8}$$

Here, defining $L_{IJ} = \langle \nabla N_I \cdot \nabla N_J\rangle$,

$$P_{IJ} = \tfrac{1}{4\pi} L_{IJ} \quad , \quad d_I = \langle N_I\, \rho\rangle \quad , \quad H_{IJ} = \tfrac{1}{2} L_{IJ} + \langle N_I\, v_{\rm eff}\, N_J\rangle \tag{9}$$

while, based on the discretizations $v_H = \sum_I v_H^I N_I$ and $\psi_i = \sum_I \psi_i^I N_I$, $\{v_H\}$ and $\{\psi_i\}$ collect the degrees of freedom v_H^I and ψ_i^I, respectively. Overall, in view of the spherical symmetry of the potentials involved in all-electron and pseudopotential settings for a single atom, virtually the same formulation applies to the radial case as well, which is advantageous in comparing different finite element discretizations as well as in generating discrete orbital solutions for possible use as enrichment functions in the three-dimensional case [12, 13].

The solution of the generalized eigenvalue problem $(8)_2$ can scale in a prohibitively unfavorable fashion depending on the solution algorithm. Presently, a subspace iteration method will be employed based on the Chebyshev filtering of the spectrum through a modification of the original algorithm that was proposed for a standard eigenvalue problem [18]. Clearly, the ability to modify the finite element discretization or to process the mass matrix in order to arrive at a standard version is desirable to attain faster solutions. For isogeometric discretizations, such as those that are based on NURBS or B-splines, row-sum lumping retains the positive-definiteness of the mass matrix. However, without additional modification of the original solution algorithm for a standard version, this leads to significantly reduced accuracies. On the other hand, lumping is not suitable for classical higher-order Lagrange elements but the combination of reduced-order Gauss-Lobatto quadrature schemes in combination with spectral finite elements delivers a diagonal mass matrix which allows invoking the original algorithm [6]. Because the aim is to carry out the electronic structure calculation at a level that matches the desired chemical accuracy, these aspects additionally influence how different discretizations compete in terms of efficiency, in particular in view of the fact that isogeometric discretizations typically deliver significantly higher accuracy per degree of freedom. A detailed comparison of various competing factors remains an open issue.

4 Numerical Investigations

As a radial case, the Magnesium atom will be considered. In the all-electron setting, Slater exchange is employed together with Vosko-Wilk-Nusair parametrization of correlation. The resolution is controlled through a parameter e_o which would scale with the number of elements in a linear discretization, not shown due to low accuracy. The error in the total energy E (in hartrees) is assessed with respect to the reference result $E_o = -199.139\,406\,315$ that is converged to 10^{-9} accuracy. In the pseudopotential setting, Perdew-Zunger parameterization of correlation is employed together with the evanescent core local pseudopotential [22]. The reference result that is converged to 10^{-11} accuracy is $E_o = -0.847\,598\,456\,26$. The target chemical accuracy is 10^{-3} in the all-electron setting and 2×10^{-4} in the pseudopotential one.

The results in Fig. 1 indicate the significantly less stringent discretization requirements in the pseudopotential case where the target accuracy is already achieved at very coarse discretizations. Here, classical Lagrange discretizations are denoted by L whereas N denotes an isogeometric discretization based on NURBS or B-splines, the following number being associated with the order chosen in each case. The asymptotic convergence rates are measured from the last three points in the data set for a given discretization and noted next to the corresponding line as half of the negative slope. The optimal value of the convergence rate measured in this fashion is equal to the discretization order and it is clear that this value has been achieved in all cases. However, the factor in the finite element error estimate expression also significantly contributes to the efficiency with which target accuracy is achieved. Indeed, it is observed that in both cases $N3$ delivers a faster route to convergence than both $L4$ and $L5$ choices and this trend is preserved until significantly low errors are observed.

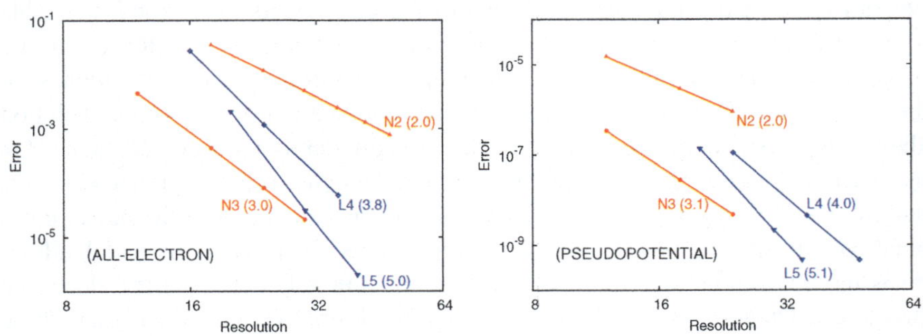

Fig. 1 All-electron and local pseudopotential results for the Magnesium atom in the radial case

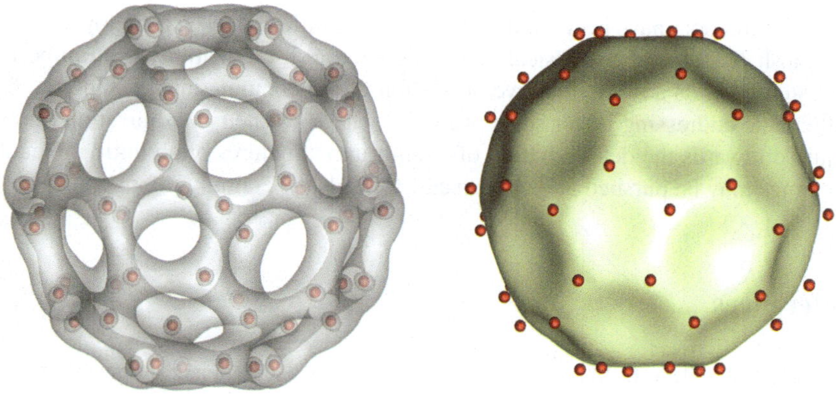

Fig. 2 Isosurfaces for ρ (left) and v_H (right) for the buckyball where red spheres indicate the ions

Because the number of elements does not decrease in an isogeometric discretization as the order increases in a standard implementation that is based on the classical finite element structure, it is advantageous to limit the order. In this respect, cubic NURBS and B-splines appear to offer competitive accuracy in return for the computational burden in the context of electronic structure calculation [19].

As a large-scale three-dimensional example with nonlocal pseudopotentials, the buckyball (C_{60} molecule) is considered. The setup of this analysis follows the details in [19] and requires the calculation of 120 orbitals. Therein, however, it was not possible to approach the target accuracy. Presently, an error of 1.07×10^{-4} Ha per atom has been achieved with an $N3$ discretization that contains 790,773 degrees of freedom, which satisfies the stringent pseudopotential calculation accuracy requirement. The corresponding solution is visualized in Fig. 2. It is noted that the error for an $L5$ discretization with 1,357,579 degrees of freedom on the same mesh structure remained above 4×10^{-4} Ha per atom. Recalling that the Kohn-Sham formalism leads to a generalized nonlinear eigenvalue problem, the efficient extraction of a large number of eigenpairs towards the solution of such problems on discretizations with $O(10^6)$ degrees of freedom continues to pose a computational challenge.

5 Conclusion

Kohn-Sham density functional theory provides a rich foundation for constructing nonphenomenological multiscale approaches through which the understanding of material and interface behavior can be advanced. The finite element method offers an effective numerical framework in addressing the corresponding computational complexity, specifically in the context of higher-order discretizations. Numerous challenges must still be addressed in order to further explore and harness its potential advantages in electronic structure calculation. Presently, the accuracy gains per degree of freedom have been limited due to nearly uniform meshes employed. Adaptive mesh refinement in classical and isogeometric discretizations is an important next step towards a more effective utilization of the computational resources. This will enable addressing larger scale problems where geometry optimization may be required towards the determination of equilibrium structures or ab initio molecular dynamics may be pursued at the next scale towards the continuum.

References

1. Pask, J. E., Klein, B. M., Sterne, P. A., & Fong, C. Y. (2001). Finite-element methods in electronic structure theory. *Computer Physics Communications, 135*, 1–34.
2. Martin, R. M. (2004). *Electronic Structure: Basis Theory and Practical Methods*. Cambridge.
3. Beck, T. L. (2000). Real-space mesh techniques in density-functional theory. *Reviews of Modern Physics, 72*, 1041–1080.
4. Helgaker, T., Jørgensen, P., & Olsen, J. (2000). *Molecular Electronic-Structure Theory*. Wiley.
5. Lin, L., Lu, J., & Ying, L. (2019). Numerical methods for Kohn-Sham density functional theory. *Acta Numerica, 28*, 405–539.
6. Motamarri, P., Nowak, M. R., Leiter, K., Knap, J., & Gavini, V. (2013). Higher-order adaptive finite-element methods for Kohn-Sham density functional theory. *Journal of Computational Physics, 253*, 308–343.
7. Parr, R. G., & Yang, W. (1989). *Density-Functional Theory of Atoms and Molecules*. Oxford.
8. Fattebert, J.-L., Hornung, R. D., & Wissing, A. M. (2007). Finite element approach for density functional theory calculations on locally-refined meshes. *Journal of Computational Physics, 223*, 759–773.
9. Zhang, D., Shen, L., Zhou, A., & Gong, X.-G. (2008). Finite element method for solving Kohn-Sham equations based on self-adaptive tetrahedral mesh. *Physics Letters A, 372*, 5071–5076.
10. Lehtovaara, L., Havu, V., & Puska, M. (2009). All-electron density functional theory and time-dependent density functional theory with higher-order elements. *The Journal of Chemical Physics, 131*, 054103.
11. Fang, J., Gao, X., & Zhou, A. (2012). A Kohn-Sham equation solver based on hexahedral finite elements. *Journal of Computational Physics, 231*, 3166–3180.
12. Kanungo, B., & Gavini, V. (2017). Large-scale all-electron density functional theory calculations using an enriched finite-element basis. *Physical Review B, 95*, 035113.
13. Pask, J. E., & Sukumar, N. (2017). Partition of unity finite element method for quantum mechanical materials calculations. *Extreme Mechanics Letters, 11*, 8–17.
14. Hu, G., Xie, H., & Xu, F. (2018). A multilevel correction adaptive finite element method for Kohn-Sham equation. *Journal of Computational Physics, 355*, 436–449.
15. Bachau, H., Cormier, E., Decleva, P., Hansen, J. E., & Mart ın, F. (2001). Applications of B-splines in atomic and molecular physics. *Reports on Progress in Physics, 64*, 1815–1942.
16. Masud, A., & Kannan, R. (2012). B-splines and NURBS based finite element methods for Kohn-Sham equations. *Computer Methods in Applied Mechanics and Engineering, 241–244*, 112–127.
17. Masud, A., Al-Naseem, A. A., Kannan, R., & Gajendran, H. (2018). B-splines and NURBS based finite element methods for strained electronic structure calculations. *Journal of Applied Mechanics, 85*, 091009.
18. Zhou, Y., Saad, Y., Tiago, M. L., & Chelikowsky, J. R. (2006). Self-consistent-field calculations using Chebyshev-filtered subspace iteration. *Journal of Computational Physics, 219*, 172–184.
19. Temizer, İ, Motamarri, P., & Gavini, V. (2020). NURBS-based non-periodic finite element framework for Kohn-Sham density functional theory. *Journal of Computational Physics, 410*, 109364.

20. Hughes, T. J. R., Cottrell, J. A., & Bazilevs, Y. (2005). Isogeometric analysis: CAD, finite elements, NURBS, exact geometry and mesh refinement. *Computer Methods in Applied Mechanics and Engineering, 194,* 4135–4195.
21. Pickett, W. E. (1989). Pseudopotential methods in condensed matter applications. *Computer Physics Reports, 9,* 115–198.
22. Fiolhais, C., Perdew, J. P., Armster, S. Q., Maclaren, J. M., & Brajczewska, M. (1994). Dominant density parameters and local pseudopotentials for simple metals. *Physical Review B, 51,* 14 001.
23. Goedecker, S., Teter, M., & Hutter, J. (1996). Relativistic separable dual-space Gaussian pseudopotentials from H to Rn. *Physical Review B, 54,* 1703–1710.
24. Beigi, S. I., & Arias, T. A. (2000). New algebraic formulation of density functional calculation. *Computer Physics Communications, 128,* 1–45.

Space–Time Computational FSI and Flow Analysis: 2004 and Beyond

Tayfun E. Tezduyar, Kenji Takizawa, and Takashi Kuraishi

I met Peter Wriggers in 2004, when he asked me to serve as an editor of Computational Mechanics. It has been a privilege to support his editorial leadership of the journal. When my five-year term on the ASME Applied Mechanics Division Executive Committee ended in 2011, he replaced me on the committee. I was proud of that, and we served together on six award committees for five years, including the year he chaired the division. It has been an honor to be friends with him. Tayfun E. Tezduyar

Abstract The space–time (ST) computational fluid–structure interaction (FSI) and flow analysis started in 1990, with the inception of the Deforming-Spatial-Domain/Stabilized ST (DSD/SST) method. In 1990–2003, the DSD/SST enabled computational FSI and flow analysis in many complex engineering problems, including parachute FSI and fluid–particle interaction with 1000 spheres. In 2004, the DSD/SST enabled some of the earliest cardiovascular FSI analyses, and, in combination with the "quasi-direct coupling," enabled more robust FSI analysis for very light structures, such as large parachutes. New core and special ST methods introduced in 2006 and 2007 enabled computational FSI analysis of the Orion spacecraft parachutes, with hundreds and gaps and slits that the flow goes through. In 2004, the first author also met the second author, which eventually led to a unique research collaboration and a new generation of ST methods. It also led to some of the most complex computational FSI and flow analyses, ranging from clusters of spacecraft parachutes with contact between the parachutes to heart valves to flow around tires with road contact and tire deformation. This chapter is the story of the ST computational FSI and flow analysis in 2004 and beyond.

T. E. Tezduyar (✉) · T. Kuraishi
Rice University, Houston, USA
e-mail: tezduyar@tafsm.org

T. E. Tezduyar · K. Takizawa
Waseda University, Tokyo, Japan
e-mail: Kenji.Takizawa@tafsm.org

F. Aldakheel et al. (eds.), *Current Trends and Open Problems in Computational Mechanics*, https://doi.org/10.1007/978-3-030-87312-7_52

1 Introduction

The space–time (ST) computational fluid–structure interaction (FSI) and flow analysis goes back to 1990, when the Deforming-Spatial-Domain/Stabilized ST (DSD/SST) method [1] was introduced for computation of flows with moving boundaries and interfaces (MBI), including FSI. The DSD/SST is a moving-mesh method. Moving the mesh to follow a fluid–solid interface enables mesh-resolution control near the interface and high-resolution boundary-layer representation. The stabilization components of this original DSD/SST are the Streamline-Upwind/Petrov-Galerkin (SUPG) [2] and Pressure-Stabilizing/Petrov-Galerkin (PSPG) [1] stabilizations, hence it is now called "ST-SUPS." The compressible-flow ST-SUPG method [3] is essentially the same as the compressible-flow DSD/SST method. It is a straightforward mixture of the DSD/SST concept and the compressible-flow SUPG [4].

Good moving-mesh methods require good mesh moving methods. In the mesh moving method introduced in 1992 [5], the motion of the internal nodes is determined by solving the equations of linear elasticity. The stiffness is inversely proportional to the Jacobian of the transformation from the element domain to the physical domain, hence the name "Mesh-Jacobian-based stiffening (MJBS)."

In 1990–2003, the ST-SUPS and the compressible-flow ST-SUPG enabled computational FSI and flow analysis in many complex problems. In fluid–particle interaction, the ST-SUPS was the first in 3D computations [6], first in reaching 100 spheres in 3D [7], first in reaching 1,000 spheres in 3D [8], and first in reaching 125 spheres in 3D spatially periodic domains [9]. The ST-SUPS was the first in 3D parachute FSI [10] and first in 3D FSI of a parachute crossing an aircraft wake [11]. The compressible-flow ST-SUPG computation of two high-speed trains in a tunnel [12] was the first 3D compressible-flow MBI computation with the SUPG.

2 2004

In 2004, the first author met Peter Wriggers. Other good things happened that year. The ST-SUPS was among the first in patient-specific full FSI analysis of arteries [13]. The "quasi-direct (QD) coupling" [14] was introduced in conjunction with the ST-SUPS, replacing the "block-iterative (BI) coupling" in ST FSI computations. The QD coupling is applicable to cases with nonmatching fluid and structure meshes at the interface and becomes equivalent to monolithic methods when the interface meshes are matching. When the fluid mass in the FSI dynamics is much larger than the structure mass, such as in large parachutes, the QD coupling is more robust than BI coupling, which was already far more robust than loose coupling.

The first author also meeting the second author in 2004 led to a unique research collaboration starting in 2007, and a new generation of ST FSI and flow analysis methods starting in 2008. It also led to some of the most complex computational FSI and flow analyses, ranging from spacecraft parachute clusters to heart valves to flow around tires with road contact and tire deformation.

3 2005–2010

In 2005, the QD coupling enabled ST FSI analysis of a new, complex parachute design [15]. New core and special ST methods introduced in 2006 and 2007 enabled FSI analysis of computationally challenging problems like a cloth piece falling over a rigid rod [16], flow through and around a flapping windsock [16], and sails [17]. Thanks to the homogenized modeling of geometric porosity (HMGP) [17], they also enabled the FSI analysis of the Orion spacecraft parachutes [17], with hundreds and gaps and slits that the flow goes through because of the ringsail construction. With that, in 2007, the ST-SUPS was the first in spacecraft parachute FSI.

In 2010, the ST-SUPS was the first in spacecraft parachute cluster FSI [18], for both the 2 and 3-parachute clusters, with contact between the parachutes, and the first in supporting NASA design studies [19] with computational FSI methods. The ST Variational Multiscale (ST-VMS) method [20], VMS version of the DSD/SST, was also introduced in 2010. Its VMS components are from the residual-based VMS (RBVMS) method [21]. Motivated by the success with using isogeometric analysis (IGA) basis functions in space [22], the idea of using IGA basis functions also in time in the ST framework was introduced in 2010 [20]. This is part of the integration of the ST framework and IGA discretization, broadly called "ST-IGA." Computations with the ST-VMS and ST-IGA were first reported in [20] in a 2D context, with IGA basis functions in space and time. The ST-IGA with IGA basis functions in time enables a more accurate representation of solid-surface motions and a mesh motion consistent with that. It also enables more efficient temporal representation of the mesh motion. The spinning motion of a solid surface is represented by quadratic NURBS in time, and with sufficient number of enough temporal patches for a full rotation, the circular paths are represented exactly. A "secondary mapping" [20] enables also specifying a constant angular velocity for invariant speeds along the circular paths.

4 2011–2020

In 2011, the second author gave a presentation at Peter Wriggers's 60th birthday conference and met him for the first time. Other good things happened that year. The ST/NURBS Mesh Update Method (STNMUM) [23] was introduced in 2011, where the mesh motion is represented with NURBS basis functions in time. With that, the ST-VMS was the first in flapping-wing aerodynamics with the wing motion coming from an actual locust [23], recorded in a wind tunnel with two high-speed cameras.

In 2012, the ST-VMS was the first in flow analysis of stent treatment of cerebral aneurysms [24] with resolved flow across the stent. The ST-SUPS was the first in FSI analysis of the Orion spacecraft 2-parachute cluster opening from Stage 2 to 3 and "MP" parachutes with modified geometric porosity [25]. The ST-SUPS was the first in FSI analysis of the Orion spacecraft drogue parachutes [26], with over 1,200 gaps that the flow goes through because of the ribbon construction, with detailed analysis for the 2 and 3-parachute clusters.

Several good things happened in 2013. The ST-VMS with STNMUM was applied to wind-turbine rotor and tower aerodynamics [27], with exact temporal representation of the rotation. The ST-SUPS was the first in FSI analysis of the clusters of the Orion spacecraft MP parachutes, for both the 2 and 3-parachute clusters [28]. The introduction of the ST Topology Change (ST-TC) method [29] enabled moving-mesh computation of flow problems with TC, such as contact between moving solid surfaces. The ST-TC made moving-mesh computations possible even when there is an actual contact between solid surfaces or other TC, not just a near contact. That was the first for a moving-mesh method, and that made it possible to have both an actual TC and high-resolution boundary layer representation near solid surfaces.

In 2014, the ST-VMS with ST-TC was the first in heart-valve flow analysis [30] with a moving-mesh method and full valve closure. The ST-SUPS was the first in FSI analysis of the opening of the Orion spacecraft MP and drogue parachutes from Stage 1 to 2 to 3 and opening of the 2-parachute MP cluster from Stage 2 to 3 [31].

Several good things happened also in 2015. The thermo-fluid ST-VMS [32] was introduced for computational analysis with the coupled incompressible-flow and thermal-transport equations. The ST-VMS was one of the first, perhaps the first, thermo-fluid VMS method. The computational thermo-fluid analysis presented in [32] for a ground vehicle and its tires was the first of its kind. The ST Slip Interface (ST-SI) method [33] was introduced to retain the desirable moving-mesh features of the ST-VMS in computations involving spinning solid surfaces. The ST-SI is essentially the ST version of the ALE-VMS for sliding interfaces [34]. An ST-SI version where the SI is between a thin porous structure and the fluid on its two sides was also introduced in [33]. The thermo-fluid analysis presented in [35] for a disk brake was also the first of its kind, thanks to the thermo-fluid ST-VMS and the thermo-fluid ST-SI introduced in [35]. The ST-VMS, ST-SI and the ST-IGA with IGA basis functions in space were integrated in [36], making the ST-SI-IGA the first of its kind. That enabled the turbocharger turbine flow analysis [36] with the ST-VMS and IGA discretization, the first of its kind.

In 2016, the ST-VMS, ST-SI and ST-TC were integrated in [37], making the ST-SI-TC the first of its kind, and enabling, for the first time, moving-mesh tire aerodynamic analysis with road contact and tire deformation [37]. The ST-SI-IGA with the porosity-version ST-SI enabled ram-air parachute internal/external flow analysis [38] with the ST-VMS and IGA discretization, the first of its kind. The ST-VMS, ST-SI, ST-TC and the ST-IGA with IGA basis functions in space were integrated in [39], making the ST-SI-TC-IGA the first of its kind. The ST-SI-TC-IGA was the first in heart-valve flow analysis [39] with a moving-mesh method, IGA discretization and full valve closure. In 2017, the compressible-flow ST-SUPG, porosity-version ST-SI, and a new HMGP enabled computational analysis of the drogue parachute at Mach 0.3 [40], the first compressible-flow computation for a spacecraft parachute with ribbon construction.

Several good things happened in 2018. The integration of the ST-VMS, ST-SI, STNMUM and ST-IGA enabled turbocharger turbine and exhaust manifold flow

Fig. 1 Tire aerodynamics with actual tire geometry and road roughness [41]. Road surface with dents and bumps (*left*). The *gray* plane represents the tire surface. Road surface with dents only (*center*). In our mesh, the road representation is relatively coarse, not suitable for roughness modeling. Therefore, we flip the situation and the tire surface represents the dents. Modified tire surface representing the road dents (*right*)

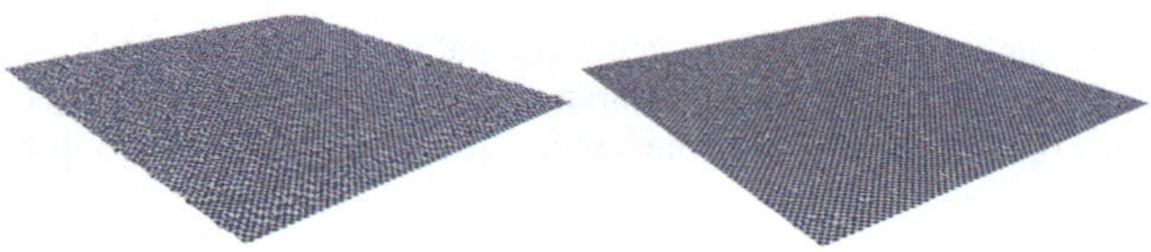

Fig. 2 Tire aerodynamics with actual tire geometry and road roughness [41]. The road roughness is added in a random fashion. Road surface (*left*) and road surface with dents only (*right*). The dents are projected to the tire surface. For visualization purposes the roughness is amplified in the height direction

analysis [42] with the ST-VMS and IGA discretization, the first of its kind. It also enabled flow-driven string dynamics in a pump [43] with the ST-VMS and IGA discretization, also the first of its kind. The integration of the compressible-flow ST-SUPG, porosity-version ST-SI, ST-IGA, and the new HMGP enabled computational analysis of the drogue parachute at Mach 0.3 [44], the first IGA-discretization compressible-flow computation for a ribbon-construction spacecraft parachute. The ST-SI-TC-IGA enabled, for the first time, IGA-discretization moving-mesh tire aerodynamic analysis with actual tire geometry, road contact and tire deformation [45]. It was shown, for the first time, in [46] that the ST-SI-IGA has a built-in Reynolds-equation limit. The method enables solution of lubrication fluid dynamics problems with a computational cost comparable to that of the Reynolds-equation model, but without the limitations of the Reynolds-equation model.

In 2019, the ST-SI-TC-IGA, with built-in Reynolds-equation limit, enabled, for the first time, IGA-discretization moving-mesh tire aerodynamic analysis with actual tire geometry, road contact, tire deformation, road roughness and fluid film [41] (see Figs. 1, 2 and 3). The ST-SI-TC-IGA was the first in ventricle-valve-aorta flow analysis [47] with a moving-mesh method, IGA discretization and full valve closure.

In 2020, a low-distortion mesh moving method based on fiber-reinforced hyperelasticity and optimized zero-stress state was introduced in [48]. With the fibers placed in multiple directions, the element is stiffened in those directions to reduce distortion during the mesh deformation. The integration of the ST-VMS, ST-SI, STNMUM and ST-IGA with IGA basis functions in space enabled computational flow analysis of a tsunami-shelter vertical-axis wind turbine [49].

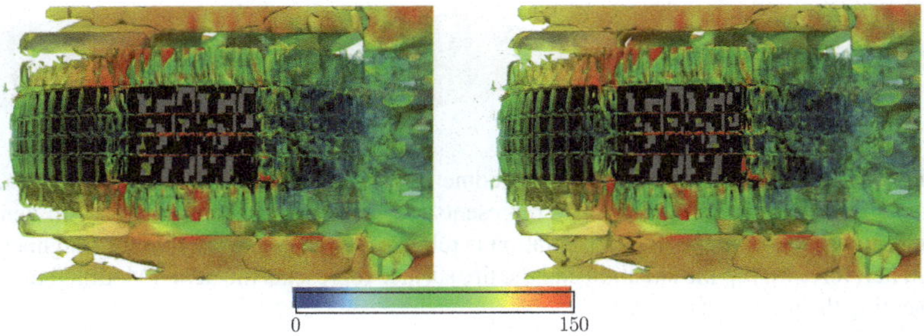

Fig. 3 Tire aerodynamics with actual tire geometry and road roughness [41]. Isosurfaces corresponding to a positive value of the second invariant of the velocity gradient tensor, colored by the velocity magnitude (km/h), viewed from the bottom. The *gray* zones are the full-contact areas

References

1. Tezduyar, T. E. (1992). Stabilized finite element formulations for incompressible flow computations. *Advances in Applied Mechanics, 28*, 1–44.
2. Brooks, A. N., & Hughes, T. J. R. (1982). Streamline upwind/Petrov-Galerkin formulations for convection dominated flows with particular emphasis on the incompressible Navier-Stokes equations. *Computer Methods in Applied Mechanics and Engineering, 32*, 199–259.
3. Tezduyar, T. E., Aliabadi, S. K., Behr, M., & Mittal, S. (1994). Massively parallel finite element simulation of compressible and incompressible flows. *Computer Methods in Applied Mechanics and Engineering, 119*, 157–177.
4. Hughes, T. J. R., & Tezduyar, T. E. (1984). Finite element methods for first-order hyperbolic systems with particular emphasis on the compressible Euler equations. *Computer Methods in Applied Mechanics and Engineering, 45*, 217–284.
5. Tezduyar, T., Aliabadi, S., Behr, M., Johnson, A., & Mittal, S. (1993). Parallel finite-element computation of 3D flows. *Computer, 26*(10), 27–36.
6. Johnson, A. A., & Tezduyar, T. E. (1996). Simulation of multiple spheres falling in a liquid-filled tube. *Computer Methods in Applied Mechanics and Engineering, 134*, 351–373.
7. Johnson, A. A., & Tezduyar, T. E. (1997). 3D simulation of fluid-particle interactions with the number of particles reaching 100. *Computer Methods in Applied Mechanics and Engineering, 145*, 301–321.
8. Johnson, A. A., & Tezduyar, T. E. (1999). Advanced mesh generation and update methods for 3D flow simulations. *Computational Mechanics, 23*, 130–143.
9. Johnson, A., & Tezduyar, T. (2001). Methods for 3D computation of fluid-object interactions in spatially-periodic flows. *Computer Methods in Applied Mechanics and Engineering, 190*, 3201–3221.
10. Kalro, V., & Tezduyar, T. (1998). A parallel finite element methodology for 3D computation of fluid–structure interactions in airdrop systems. In *Proceedings of the 4th Japan-US symposium on finite element methods in large-scale computational fluid dynamics*, Tokyo, Japan.
11. Tezduyar, T., & Osawa, Y. (2001). Fluid-structure interactions of a parachute crossing the far wake of an aircraft. *Computer Methods in Applied Mechanics and Engineering, 191*, 717–726.
12. Tezduyar, T., Aliabadi, S., Behr, M., Johnson, A., Kalro, V., & Litke, M. (1996). Flow simulation and high performance computing. *Computational Mechanics, 18*, 397–412.
13. Torii, R., Oshima, M., Kobayashi, T., Takagi, K., & Tezduyar, T. E. (2004). Influence of wall elasticity on image-based blood flow simulations. *Transactions of the Japan Society of Mechanical Engineers Series A, 70*, 1224–1231. in Japanese.

14. Tezduyar, T. E., Sathe, S., Keedy, R., & Stein, K. (2006). Space-time finite element techniques for computation of fluid-structure interactions. *Computer Methods in Applied Mechanics and Engineering, 195*, 2002–2027.
15. Sathe, S., Benney, R., Charles, R., Doucette, E., Miletti, J., Senga, M., Stein, K., & Tezduyar, T. E. (2007). Fluid-structure interaction modeling of complex parachute designs with the space-time finite element techniques. *Computers & Fluids, 36*, 127–135.
16. Tezduyar, T. E., & Sathe, S. (2007). Modeling of fluid-structure interactions with the space-time finite elements: Solution techniques. *International Journal for Numerical Methods in Fluids, 54*, 855–900.
17. Tezduyar, T. E., Sathe, S., Pausewang, J., Schwaab, M., Christopher, J., & Crabtree, J. (2008). Interface projection techniques for fluid-structure interaction modeling with moving-mesh methods. *Computational Mechanics, 43*, 39–49.
18. Takizawa, K., Wright, S., Moorman, C., & Tezduyar, T. E. (2011). Fluid-structure interaction modeling of parachute clusters. *International Journal for Numerical Methods in Fluids, 65*, 286–307.
19. Takizawa, K., Moorman, C., Wright, S., Spielman, T., & Tezduyar, T. E. (2011). Fluid-structure interaction modeling and performance analysis of the Orion spacecraft parachutes. *International Journal for Numerical Methods in Fluids, 65*, 271–285.
20. Takizawa, K., & Tezduyar, T. E. (2011). Multiscale space-time fluid-structure interaction techniques. *Computational Mechanics, 48*, 247–267.
21. Hughes, T. J. R. (1995). Multiscale phenomena: Green's functions, the Dirichlet-to-Neumann formulation, subgrid scale models, bubbles, and the origins of stabilized methods. *Computer Methods in Applied Mechanics and Engineering, 127*, 387–401.
22. Hughes, T. J. R., Cottrell, J. A., & Bazilevs, Y. (2005). Isogeometric analysis: CAD, finite elements, NURBS, exact geometry, and mesh refinement. *Computer Methods in Applied Mechanics and Engineering, 194*, 4135–4195.
23. Takizawa, K., Henicke, B., Puntel, A., Spielman, T., & Tezduyar, T. E. (2012). Space-time computational techniques for the aerodynamics of flapping wings. *Journal of Applied Mechanics, 79*, 010903.
24. Takizawa, K., Schjodt, K., Puntel, A., Kostov, N., & Tezduyar, T. E. (2012). Patient-specific computer modeling of blood flow in cerebral arteries with aneurysm and stent. *Computational Mechanics, 50*, 675–686.
25. Takizawa, K., Fritze, M., Montes, D., Spielman, T., & Tezduyar, T. E. (2012). Fluid-structure interaction modeling of ringsail parachutes with disreefing and modified geometric porosity. *Computational Mechanics, 50*, 835–854.
26. Takizawa, K., Montes, D., Fritze, M., McIntyre, S., Boben, J., & Tezduyar, T. E. (2013). Methods for FSI modeling of spacecraft parachute dynamics and cover separation. *Mathematical Models and Methods in Applied Sciences, 23*, 307–338.
27. Takizawa, K., Tezduyar, T. E., McIntyre, S., Kostov, N., Kolesar, R., & Habluetzel, C. (2014). Space-time VMS computation of wind-turbine rotor and tower aerodynamics. *Computational Mechanics, 53*, 1–15.
28. Takizawa, K., Tezduyar, T. E., Boben, J., Kostov, N., Boswell, C., & Buscher, A. (2013). Fluid-structure interaction modeling of clusters of spacecraft parachutes with modified geometric porosity. *Computational Mechanics, 52*, 1351–1364.
29. Takizawa, K., Tezduyar, T. E., Buscher, A., & Asada, S. (2014). Space-time interface-tracking with topology change (ST-TC). *Computational Mechanics, 54*, 955–971.
30. Takizawa, K., Tezduyar, T. E., Buscher, A., & Asada, S. (2014). Space-time fluid mechanics computation of heart valve models. *Computational Mechanics, 54*, 973–986.
31. Takizawa, K., Tezduyar, T. E., Boswell, C., Kolesar, R., & Montel, K. (2014). FSI modeling of the reefed stages and disreefing of the Orion spacecraft parachutes. *Computational Mechanics, 54*, 1203–1220.
32. Takizawa, K., Tezduyar, T. E., & Kuraishi, T. (2015). Multiscale ST methods for thermo-fluid analysis of a ground vehicle and its tires. *Mathematical Models and Methods in Applied Sciences, 25*, 2227–2255.

33. Takizawa, K., Tezduyar, T. E., Mochizuki, H., Hattori, H., Mei, S., Pan, L., & Montel, K. (2015). Space-time VMS method for flow computations with slip interfaces (ST-SI). *Mathematical Models and Methods in Applied Sciences, 25*, 2377–2406.
34. Bazilevs, Y., & Hughes, T. J. R. (2008). NURBS-based isogeometric analysis for the computation of flows about rotating components. *Computational Mechanics, 43*, 143–150.
35. Takizawa, K., Tezduyar, T. E., Kuraishi, T., Tabata, S., & Takagi, H. (2016). Computational thermo-fluid analysis of a disk brake. *Computational Mechanics, 57*, 965–977.
36. Takizawa, K., Tezduyar, T. E., Otoguro, Y., Terahara, T., Kuraishi, T., & Hattori, H. (2017). Turbocharger flow computations with the Space-Time Isogeometric Analysis (ST-IGA). *Computers & Fluids, 142*, 15–20.
37. Takizawa, K., Tezduyar, T. E., Asada, S., & Kuraishi, T. (2016). Space-time method for flow computations with slip interfaces and topology changes (ST-SI-TC). *Computers & Fluids, 141*, 124–134.
38. Takizawa, K., Tezduyar, T. E., & Terahara, T. (2016). Ram-air parachute structural and fluid mechanics computations with the space-time isogeometric analysis (ST-IGA). *Computers & Fluids, 141*, 191–200.
39. Takizawa, K., Tezduyar, T. E., Terahara, T., & Sasaki, T. (2017). Heart valve flow computation with the integrated Space-Time VMS, Slip Interface, Topology Change and Isogeometric Discretization methods. *Computers & Fluids, 158*, 176–188.
40. Takizawa, K., Tezduyar, T. E., & Kanai, T. (2017). Porosity models and computational methods for compressible-flow aerodynamics of parachutes with geometric porosity. *Mathematical Models and Methods in Applied Sciences, 27*, 771–806.
41. Kuraishi, T., Takizawa, K., & Tezduyar, T. E. (2019). Space-time computational analysis of tire aerodynamics with actual geometry, road contact, tire deformation, road roughness and fluid film. *Computational Mechanics, 64*, 1699–1718.
42. Otoguro, Y., Takizawa, K., Tezduyar, T. E., Nagaoka, K., & Mei, S. (2019). Turbocharger turbine and exhaust manifold flow computation with the Space-Time Variational Multiscale Method and Isogeometric Analysis. *Computers & Fluids, 179*, 764–776.
43. Kanai, T., Takizawa, K., Tezduyar, T. E., Komiya, K., Kaneko, M., Hirota, K., Nohmi, M., Tsuneda, T., Kawai, M., & Isono, M. (2019). Methods for computation of flow-driven string dynamics in a pump and residence time. *Mathematical Models and Methods in Applied Sciences, 29*, 839–870.
44. Kanai, T., Takizawa, K., Tezduyar, T. E., Tanaka, T., & Hartmann, A. (2019). Compressible-flow geometric-porosity modeling and spacecraft parachute computation with isogeometric discretization. *Computational Mechanics, 63*, 301–321.
45. Kuraishi, T., Takizawa, K., & Tezduyar, T. E. (2019). Tire aerodynamics with actual tire geometry, road contact and tire deformation. *Computational Mechanics, 63*, 1165–1185.
46. Kuraishi, T., Takizawa, K., & Tezduyar, T. E. (2019). Space-Time Isogeometric flow analysis with built-in Reynolds-equation limit. *Mathematical Models and Methods in Applied Sciences, 29*, 871–904.
47. Terahara, T., Takizawa, K., Tezduyar, T. E., Tsushima, A., & Shiozaki, K. (2020). Ventricle-valve-aorta flow analysis with the Space-Time Isogeometric Discretization and Topology Change. *Computational Mechanics, 65*, 1343–1363.
48. Takizawa, K., Tezduyar, T. E., & Avsar, R. (2020). A low-distortion mesh moving method based on fiber-reinforced hyperelasticity and optimized zero-stress state. *Computational Mechanics, 65*, 1567–1591.
49. Otoguro, Y., Mochizuki, H., Takizawa, K., & Tezduyar, T. E. (2020). Space-time variational multiscale isogeometric analysis of a tsunami-shelter vertical-axis wind turbine. *Computational Mechanics, 66*, 1443–1460.

Synthesis of Computational Mesoscale Modeling of Cementitious Materials and Coda Wave Based Damage Identification

Jithender J. Timothy, Giao Vu, Leslie Saydak, Erik H. Saenger, and Günther Meschke

Prof. Peter Wriggers' manifold pioneering contributions to computational mechanics have been an inspiration already since my early years as a Ph.D. student at TU Vienna. In particular, he demonstrated the potential of computational mechanics to better understand the deformation behavior and degradation processes of heterogeneous materials at the mesoscale, a topic, which is also adopted in the present paper dedicated to Prof. Peter Wriggers on the occasion of his 70th birthday.

Abstract High costs for repair of concrete and reinforced structures can be prevented if damage at an early stage of degradation is detected and precautionary measures are applied. The multiple-scattered late arriving signals (so-called coda waves) contain rich information that can be used to detect weak changes in complex heterogeneous materials such as concrete. When subjected to external loads, microcracks in concrete initiate around aggregates and evolve until crack-coalescence and complete failure. In this contribution, a virtual testing environment using a synthesis of the mesoscale simulation of damage in concrete, wave propagation and deep learning is presented.

J. J. Timothy · G. Vu · G. Meschke (✉)
Ruhr University Bochum, Institute for Structural Mechanics, Bochum, Germany
e-mail: guenther.meschke@rub.de

J. J. Timothy
e-mail: timothy.jithenderjaswant@rub.de

G. Vu
e-mail: Thi.VU-h6d@ruhr-uni-bochum.de

L. Saydak
Bochum University of Applied Sciences, Bochum, Germany
e-mail: leslie-anne.saydak@hs-bochum.de

E. H. Saenger
Bochum University of Applied Sciences, Fraunhofer IEG, Institute for Geology, Mineralogy and Geophysics, Bochum, Germany
e-mail: Erik.Saenger@hs-bochum.de

F. Aldakheel et al. (eds.), *Current Trends and Open Problems in Computational Mechanics*, https://doi.org/10.1007/978-3-030-87312-7_53

1 Introduction

If concrete damage at an early stage of degradation is detected and precautionary maintenance measures are applied, costs for repair and maintenance can be considerably reduced. Moreover, irreparable consequences, such as loss of life, due to sudden catastrophic failure, can be prevented. Damage processes in concrete under compression initiate as microcracking around aggregates. For an accurate description of damage of concrete, a mesoscale model in which the aggregates and the cement paste are spatially resolved is essential [4–6]. Damage in concrete subjected to external loads initiates with the growth of microcracks that are much smaller than the aggregate size, whose detection is not possible using conventional health monitoring techniques. However, diffuse ultrasonic waves, i.e., multiple-scattered late arriving signals (so-called coda waves [1]), are potentially able to detect weak changes due to multiple sampling of the material. Coda Wave Interferometry (CWI) can been used to study the influence of mechanical loadings on concrete structures in terms of the stretch factor α (equal to relative velocity change) that is a sensitive indicator of microstructural changes occurring at low loading levels, i.e., when damage in the concrete is characterized by diffuse microcracking and/or the decorrelation coefficient (DC) that is a strong indicator for macrocracking [2]. While there is consensus that diffuse ultrasonic waves are very sensitive to changes in the material, it is, however still a significant challenge to reliably correlate and classify the measured coda signal with the corresponding level of damage (from diffuse microcracking to localized macrocracks). Thus, the problem reduces to that of a multiclass classification problem that can be solved by a variety of machine learning algorithms such as Random Forests, Naive Bayes, Support Vector Machines, and Neural networks [3]. In this contribution, we use a combination of the Discrete Element Method, Finite Difference wave propagation, and Convolutional Neural Network based supervised learning to identify damage in concrete.

2 Simulation of Concrete Damage at the Mesoscale

2.1 The Concrete Mesostructure Generator

A concrete mesostructure generator (CMG) has been developed, using a Semi-Random (spatial) Sequential Algorithm to generate synthetic realistic concrete specimens in voxel format. The main algorithm involves two parts. Firstly, a concrete aggregate-like inclusion is generated, followed by an assembly onto the main specimen geometry. In order to create a realistic aggregate inclusion geometry, a cube with the maximum required size of the aggregate is reduced to a polyhedron by making cuts tangential to an inscribed imaginary ellipsoid (that approximates the aggregate shape). In order to create concave surfaces on the inclusion, a Gaussian surface equation as a function of depth and width of concavity is used. The inclusion

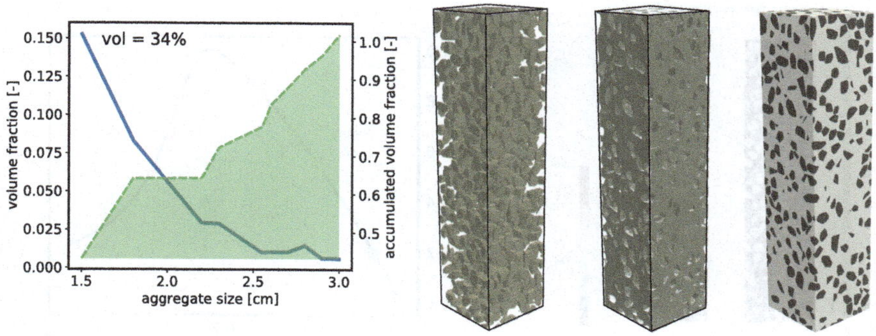

Fig. 1 The concrete mesostructure generator. Left: The aggregate size distribution. Right: The aggregate phase, the mortar matrix and the concrete mesostructure

is then assembled with a random orientation angle in a 3-D array, corresponding to the geometry of the required concrete specimen, using a semi-random-sequential absorption algorithm, where a new position for the aggregate is chosen randomly only after scanning a complete surface. This process is repeated until the required packing density is achieved. With this optimized semi-random-sequential algorithm, one can achieve up to 60% packing density that is 8 times faster than the standard random sequential absorption algorithm. Figure 1 shows a synthetic concrete specimen generated from a given size distribution.

2.2 DEM Simulation of Uniaxial Compression

The behavior of the concrete mesostructure subjected to uniaxial compression is simulated using the Discrete Element Method (DEM). According to the DEM framework, the mortar matrix phase and the aggregates are described as an assembly of particles that can collide, interact and exert cohesive frictional forces on each other. The dynamics of these particles, when subjected to external loadings, is governed by Newton's second law. The inter-particle forces are computed using the relative displacements with respect to the reference configuration and the corresponding contact physics. The resulting interaction forces are used in conjunction with the applied external forces as input for the equations of motion in the Leapfrog time integration step to solve for the new position of all particles. The DEM model used in this paper is based on the model calibrated in [7]. Figure 2 shows the results of the DEM simulation of the concrete mesostructure subjected to uniaxial compression showing three distinctive regimes: (i) the elastic deformation regime, characterized by the absence of damage; (ii) the microcracking regime before the peak nominal stress in the specimen is reached, characterized by diffusely distributed precursor damage; (iii) the crack localization regime beyond the peak nominal stress characterized by the growth of multiple cracks.

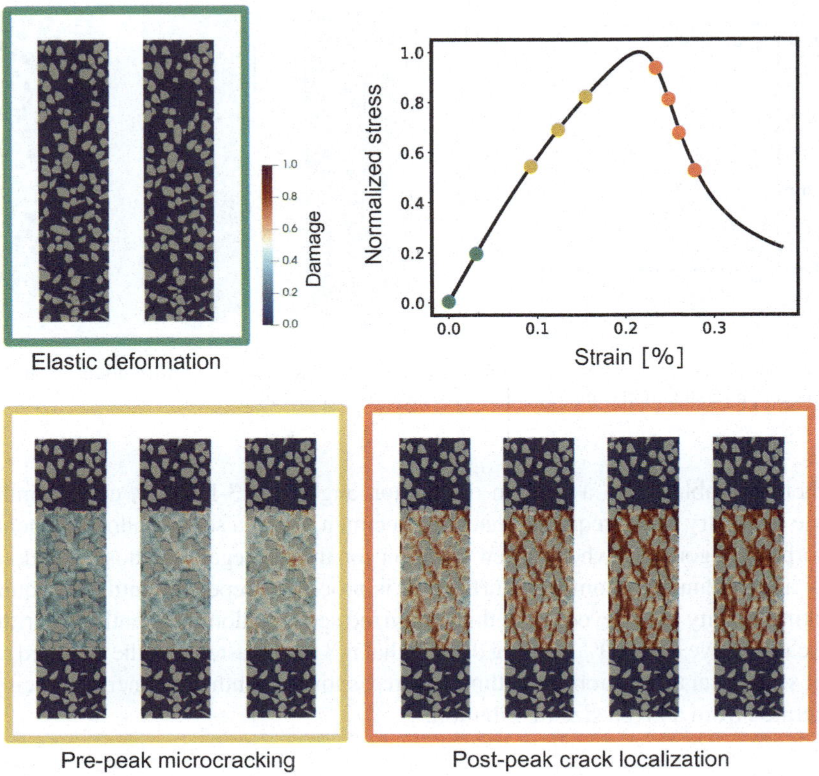

Fig. 2 DEM simulation of concrete subjected to uniaxial compression showing three distinctive regimes: the elastic deformation regime, the pre-peak microcracking regime and the post-peak crack localization regime. The stress and strain levels corresponding to the three regimes are shown as colored points in the normalized stress-strain curve

3 Finite Difference Simulation of Wave Propagation

A rotated staggered-grid (RSG) finite-difference scheme [8] is used to propagate the seismic wavefield in the forward simulations. The RSG uses rotated finite-difference operators, leading to a distribution of modeling parameters in an elementary cell where all components of one physical property are located only at one single position. This can be advantageous for modeling wave propagation in anisotropic media or complex media, including high-contrast discontinuities, because no averaging of elastic moduli is needed [11]. Concrete is a strongly heterogeneous and densely packed composite material. Due to the high density of scattering constituents and inclusions, ultrasonic wave propagation in this material consists of a complex mixture of multiple scattering, mode conversion and diffusive energy transport. With previous studies in 2D and 3D [9, 10] it was shown that the RSG-technique is well-suited for such applications. A finite-difference operator of second order is used in time

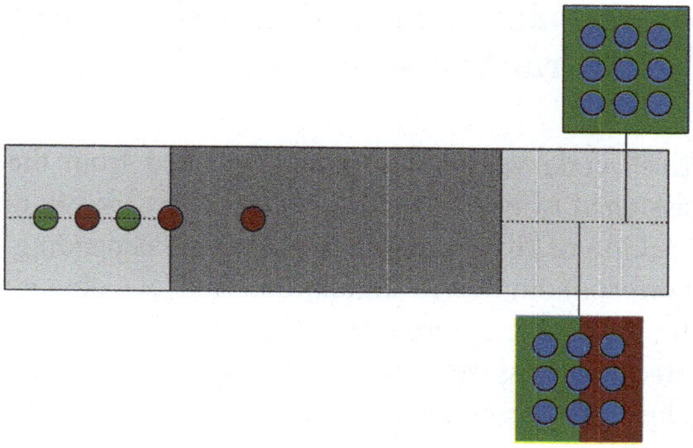

Fig. 3 Sensors: The blue circles correspond to receivers and the red and green circles correspond to the sender positions. The dark grey region is the location at which the mesostructures obtained from the DEM simulation are placed

as well as in space. The surfaces of the concrete model are implemented as a free surface using two layers of vacuum. These vacuum layers mimic the impedance of the concrete specimen-atmosphere boundary. A typical simulation with a model size of $40 \times 10 \times 10\,\text{cm}^3$ consists of 32 million grid points (spatial increment 5×10^{-4} m) and a simulation with 80,000 time steps takes about 12 h on three nodes on a mid-size cluster computer. For all simulations, a wavelet with a central frequency of $f_c =$ 60 kHz is used. Sources are implemented as body force sources. The time increment is set to 5×10^{-8} s to ensure stability. For the mortar matrix we use a P-wave velocity of $v_p = 3950\,\text{m/s}$, a S-wave velocity of 2250 m/s and a density of 2050 kg/m^3. For the aggregates we use a P-wave velocity of $v_p = 6230\,\text{m/s}$, a S-wave velocity of 3330 m/s and a density of 2950 kg/m^3. The sender-receiver configuration is shown in Fig. 3. Two sets of sender-receiver configurations are used. The first set consists of the senders shown in dark red. The corresponding receivers are shown in blue with the dark red background. The second set of senders are shown in green with the corresponding receivers shown in blue with a green background. The 9 different concrete mesostructures obtained from the DEM simulation (see Fig. 2) are placed in the dark region in Fig. 3. The receivers farthest from the dark grey region receive signals from the green senders while the receivers closest to the dark grey region receive signals from both the green and the red senders. Each receiver measures three displacement components in discrete amplitudes of time step size 5×10^{-8} s. Thus, from the first set of sender-receiver pairs we obtain $3 \times 9 \times 9 \times 3 = 729$ time series data. From the second set of sender-receiver pairs we obtain $2 \times 9 \times 18 \times 3 = 972$ time series data. In total, 1701 time series data is simulated. The time series is standardized with respect to each sender-receiver and a time window of up to 2 ms is used for the classification procedure. This time window includes the late-arriving coda waves. Each standardized and time-windowed coda wave consists of 40,000 discrete amplitude values (time step size 5×10^{-8} s).

4 Damage Classification Using a Deep Convolutional Neural Network

Each coda wave (data point) obtained from the wave propagation simulation is assigned to one of the three classes (i) Elastic deformation, (ii) Pre-peak microcracking or (iii) Post-peak localization depending on the mesostructure used in the corresponding wave simulation. Thus, we obtain in total 378 simulated coda waves for the elastic deformation class, 567 simulated coda waves for the pre-peak microcracking class and 768 simulated coda waves for the post-peak crack localization class. As the number of coda waves (data points) is not even across the classes, the data is first randomized and 189 coda waves from the pre-preak microcracking class and 390 coda waves from the post-peak localization class are discarded. The final size of the dataset is 1134 labeled coda waves. Using this dataset, the goal is now to construct a classifier which, given a new (unseen) coda wave (data point), should correctly predict the class (elastic deformation, pre-peak microcracking or post-peak localization) to which the (new) unseen coda wave (data point) belongs. To this end, we use a deep convolutional neural network to classify the material state. First, the data is randomized and 75% of the data set is used for training and the rest is used for validation. Each discrete coda wave data is reshaped into a 2D array of dimension 200×200. The 2D wave image is fed to a stack of three 2D convolutional layers with 8, 16 and 16 filters followed by three fully connected dense layers. Batch normalization layers, a flatten layer and a dropout layer are used to regularize and reduce overfit to the training data respectively. ReLU is used as the activation function across all the layers except the last layer where a softmax activation function is used in order to normalize the output of the network to a probability distribution over the predicted classes. Categorical cross-entropy is used as the loss function and weight update is performed using stochastic gradient descent with a learning rate of 0.001. Figure 4 shows the network architecture. The accuracy, loss and confusion matrix of the neural network classifier are shown in Fig. 5.

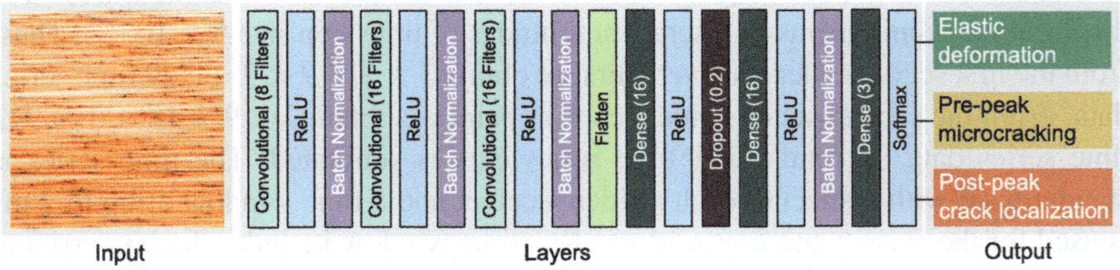

Fig. 4 The deep learning architecture used for damage identification: The input is a time-series image and the output is the material state

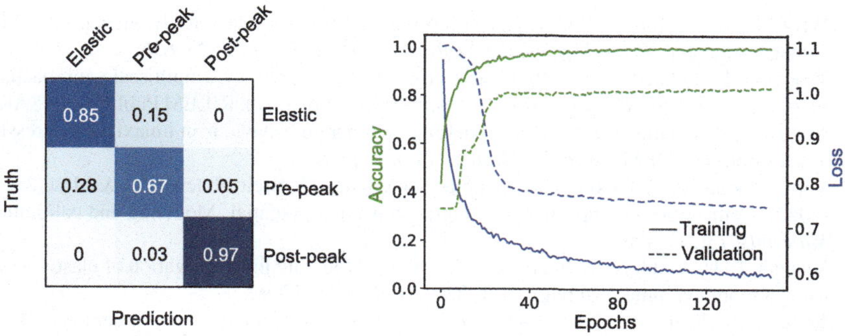

Fig. 5 Left: Confusion matrix of the classifier. Right: Training and validation accuracy and loss of the classifier up to 150 epochs

5 Conclusion

In this contribution, a computational methodology for the identification of damage in concrete using coda waves is presented. First, realistic concrete specimens are generated using a given aggregate size distribution. These synthetic specimens are subjected to uniaxial compression for simulating damage ranging from pre-peak microcracking to post-peak crack localization. The damaged mesostructures obtained from different loading levels are subjected to wave propagation analysis mimicking the signals emitted by the actuators of the Coda-monitoring system. Three distinct states of the mesostructure associated with elastic deformation, pre-peak microcracking and post-peak crack localization phenomena are identified for classification. The coda signals corresponding to the three classes and the corresponding material state is used to train a deep convolutional neural network classifier. The overall accuracy of the classifier is 83%.

Acknowledgements This work has been supported by the German Research Foundation (DFG) in the framework of Subproject RUB1 and BU of the Research Unit FOR 2825 (project number 398216472). This support is gratefully acknowledged. We also thank our student assistants Vijaya Holla and Divya Shyam Singh for programming the mesostructure generator and the neural network in Python, respectively.

References

1. Herraiz, M., & Espinosa, A. F. (1987). Coda waves: A review. *Pure and applied geophysics, 125*(4), 499–577.
2. Zhang, Y., Planes, T., Larose, E., Obermann, A., Rospars, C., & Moreau, G. (2016). Diffuse ultrasound monitoring of stress and damage development on a 15-ton concrete beam. *The Journal of the Acoustical Society of America, 139*(4), 1691–1701.
3. Bishop, C. M. (2006). Pattern recognition and machine learning. Berlin: Springer.

4. Wriggers, P., & Moftah, S. O. (2006). Mesoscale models for concrete: Homogenisation and damage behaviour. *Finite Elements in Analysis and Design, 42*(7), 623–636.
5. Wriggers, P., Löhnert, S. (2010). Characterization of concrete by a multiscale approach. In *International RILEM Conference on Material Science* (pp. 3–12). RILEM Publications SARL.
6. Nitka, M., & Tejchman, J. (2015). Modelling of concrete behaviour in uniaxial compression and tension with DEM. *Granular Matter, 17*(1), 145–164.
7. Vu, G., Iskhakov, T., Timothy, J. J., Schulte-Schrepping, C., Breitenbücher, R., & Meschke, G. (2020). Cementitious composites with high compaction potential: Modeling and calibration. *Materials, 13*(21), 4989.
8. Saenger, E. H., Gold, N., & Shapiro, S. A. (2000). Modeling the propagation of elastic waves using a modified finite-difference grid. *Wave Motion, 31*, 77–92.
9. Kocur, G. K., Saenger, E. H., & Vogel, T. (2010). Elastic wave propagation in a segmented X-ray computed tomography model of a concrete specimen. *Construction and Building Materials*, 2393–2400.
10. Saenger, E. H. (2011). Time reverse charachterization of sources in heterogeneous media. *NDT & E International, 44*, 751–759.
11. Saenger, E. H., & Bohlen, T. (2004). Finite-difference modeling of viscoelastic and anisotropic wave propagation using the rotated staggered grid. *Geophysics, 69*, 583–591.

On a Nonlinear Elastic Composite Shell Model with a Refined 3D Stress Analysis

Werner Wagner and Friedrich Gruttmann

We met Peter for the first time at the Institut für Baumechanik und Numerische Mechanik, where we were colleagues in the group of Professor Erwin Stein. This was the beginning of a long interesting and fruitful time in Hannover. Although we worked later in different places like Karlsruhe and Darmstadt this collaboration lead to a close personal friendship besides our discussions, joint papers and other activities on computational mechanics over the years.

Abstract In this contribution layered shells subjected to static loading are considered. The theory is based on a variational formulation, where the associated Euler-Lagrange equations include besides the usual shell equations formulated in stress resultants, the equilibrium of higher order stress resultants, which follow from the thickness integration of the local equilibrium equations. Within representative volume elements the displacement field is enriched with warping displacements and relative thickness displacements which allow for thickness strains. Elimination of a set of parameters by static condensation leads to a material matrix for the stress resultants. It is used in displacement based elements or in mixed hybrid elements with the usual 5 or 6 nodal degrees of freedom. The three-dimensional stress state is obtained with a back substitution of the eliminated parameters.

W. Wagner (✉)
Institut für Baustatik, Karlsruher Institut für Technologie, Kaiserstr. 12, 76131 Karlsruhe, Germany
e-mail: werner.wagner@kit.edu

F. Gruttmann
Fachgebiet Festkörpermechanik, Technische Universität Darmstadt, Franziska-Braun-Str. 7, 64287 Darmstadt, Germany
e-mail: gruttmann@mechanik.tu-darmstadt.de

F. Aldakheel et al. (eds.), *Current Trends and Open Problems in Computational Mechanics*, https://doi.org/10.1007/978-3-030-87312-7_54

1 Introduction

Shell elements which account for the layer sequence of a laminated structure are able to predict the deformation behaviour of the reference surface in an accurate way, when accurate shear correction factors are used. Also the assumption of a linear shape of the in–plane strains through the thickness is sufficient good enough, if the shell is not too thick. In contrast to that only constant transverse shear strains through the thickness are obtained within the Reissner-Mindlin theory. As a consequence only the average of the transverse shear stresses is accurate. Within the Kirchhoff theory the transverse shear strains are set to zero. By assumption the thickness normal stresses are neglected in a standard shell theory. This is necessary to avoid unphysical stresses due to inextensibility assumptions in thickness direction.

Several authors exploit the equilibrium equations within post–processing procedures to obtain the interlaminar stresses, e.g. [4, 18] for the transverse shear stresses and e.g. [17] for the thickness normal stresses. The essential restriction of the approach is the fact that these stresses are not embedded in the variational formulation and an immediate extension to geometrical and physical nonlinearity is not possible.

In [15] the displacement field through the thickness of layered plates is enhanced by so-called zig-zag functions. Another approach for multilayer anisotropic plates is based on the dimension reduction method in combination with the Hellinger-Reissner principle [1].

Higher order plate and shell formulations and layerwise approaches represent a wide class of advanced models, e.g. [3, 14]. For geometrical nonlinear formulations we refer to e.g. [7, 16]. As these theories are associated with global layerwise degrees of freedom, the general handling is complicated e.g. for shell intersection problems.

The use of so-called solid shell elements represents a computationally expensive approach, e.g. [11, 12]. For a sufficient accurate evaluation of the interlaminar stresses each layer must be discretized with several elements ($\approx 4 - 10$) in thickness direction. Especially for large scale industrial problems with a multiplicity of load steps and several iterations in each load step this is not a feasible approach.

The purpose of this paper is to present a shell model which is able to deliver the three–dimensional stress state of geometrical nonlinear elastic composite shells which have an in-plane homogeneous structure. The overall goal of the associated element formulation is the retention of the standard 5 or 6° of freedom at the nodes of the finite element mesh. It is build on our previous publications [5, 6, 9, 10].

2 Governing Equations

The displacements of the Reissner-Mindlin kinematics are enriched by warping displacements and thickness changes. This yields additional terms for the material deformation gradient and the Green–Lagrange strain tensor. In so-called representative volume elements (RVEs), which are located at the integration points of the shell reference surface, the introduced quantities are approximated with appropriate shape functions. Elimination of a set of parameters by static condensation leads to a material matrix for the stress resultants. The stresses are obtained with back substitution of the eliminated parameters.

2.1 Kinematics

Let $\mathcal{B}_0$ be the three–dimensional Euclidean space occupied by the shell with thickness h in the undeformed configuration. The position vector $\mathbf{X}$ of any point $P \in \mathcal{B}_0$ is parametrized with convected coordinates ξ^i

$$\mathbf{X}(\xi^1, \xi^2, \xi^3) = \mathbf{X}_0(\xi^1, \xi^2) + \xi^3\,\mathbf{N}(\xi^1, \xi^2) \qquad |\mathbf{N}| = 1, \tag{1}$$

where $\mathbf{X}_0$ and $\mathbf{N}$ denote the position vector and normal vector of the reference surface Ω, respectively. The thickness coordinate ξ^3 is defined in the range $h^- \leq \xi^3 \leq h^+$, where h^- and h^+ denote the coordinates of the outer surfaces. The coordinate on the boundary $\Gamma = \Gamma_u \cup \Gamma_\sigma$ is s. In the following the usual summation convention is used, where Latin indices range from 1 to 3 and Greek indices range from 1 to 2. Commas denote partial differentiation with respect to the coordinates ξ^i.

For the position vector of the deformed shell the kinematic assumption

$$\mathbf{x}(\xi^1, \xi^2, \xi^3) = \mathbf{x}_0(\xi^1, \xi^2) + \xi^3\,\mathbf{d}(\xi^1, \xi^2) + \tilde{\mathbf{u}}(\xi^3) \qquad |\mathbf{d}| = 1 \tag{2}$$

is chosen. Here, $\mathbf{x}_0$ describes the position vector of the current reference surface Ω_t. The director vector $\mathbf{d}(\boldsymbol{\varphi})$ of the current configuration is not perpendicular to the deformed reference surface, thus transverse shear deformations are accounted for. We introduce the vector field $\mathbf{v}(\xi^1, \xi^2) = [\mathbf{u}, \boldsymbol{\varphi}]^T$, where $\mathbf{u} = \mathbf{x}_0 - \mathbf{X}_0$ and $\boldsymbol{\varphi}$ contains rotational parameters. In Eq. (2) the assumptions of the Reissner-Mindlin theory are extended by the displacement fluctuation field

$$\tilde{\mathbf{u}}(\xi^3) = \boldsymbol{\Phi}(\xi^3)\,\boldsymbol{\alpha}. \tag{3}$$

In this context the shell is subdivided in thickness direction in N numerical layers. For laminated shells N usually corresponds to the number of physical layers. The vector $\boldsymbol{\alpha}$ contains displacements at nodes in thickness direction and is area by area constant. The matrix $\boldsymbol{\Phi}$ contains layer–wise cubic hierarchic functions

$$\begin{aligned}
\boldsymbol{\Phi}(\xi^3) &= \left[\, \phi_1\,\mathbf{1}_3\;\phi_2\,\mathbf{1}_3\;\phi_3\,\mathbf{1}_3\;\phi_4\,\mathbf{1}_3 \,\right] \mathbf{a}^i \\
\phi_1 &= \frac{1}{2}\,(1-\zeta) \qquad \phi_2 = 1-\zeta^2 \\
\phi_3 &= \frac{8}{3}\,\zeta\,(1-\zeta^2)\;\; \phi_4 = \frac{1}{2}\,(1+\zeta).
\end{aligned} \tag{4}$$

Here, $\mathbf{a}^i$ is an assembly matrix, which relates the 12 degrees of freedom of layer i to the M components of $\boldsymbol{\alpha}$. For N layers this leads to $M = 9 \cdot N + 3$. Furthermore, ζ is a normalized coordinate of layer i defined in the range $-1 \leq \zeta \leq 1$ and $\mathbf{1}_n$ denotes a unit matrix of order n.

With the parametrization of the shell (1) and kinematic assumption (2) one can express the material deformation gradient as

$$\mathbf{F} = \mathbf{g}_i \otimes \mathbf{G}^i \quad \mathbf{g}_i = \mathbf{x},_i \quad \mathbf{G}_i = \mathbf{X},_i \quad \mathbf{G}^i = G^{ij}\, \mathbf{G}_j \tag{5}$$

with the contravariant metric coefficients G^{ij} and

$$\begin{aligned} \mathbf{X},_\alpha &= \mathbf{X}_0,_\alpha + \xi^3\, \mathbf{N},_\alpha \qquad & \mathbf{x},_\alpha &= \mathbf{x}_0,_\alpha + \xi^3\, \mathbf{d},_\alpha \\ \mathbf{X},_3 &= \mathbf{N} & \mathbf{x},_3 &= \mathbf{d} + \tilde{\mathbf{u}},_3 \,. \end{aligned} \tag{6}$$

The vector $\tilde{\mathbf{u}},_3$ is obtained with $\tilde{\mathbf{u}},_3 = \boldsymbol{\Phi},_3\, \boldsymbol{\alpha}$. Inserting the deformation gradient into the Green–Lagrange strain tensor $\mathbf{E} = \frac{1}{2}\,(\mathbf{F}^T\mathbf{F} - \mathbf{1})$ yields

$$\mathbf{E} = E_{ij}\, \mathbf{G}^i \otimes \mathbf{G}^j \quad E_{ij} = \frac{1}{2}(\mathbf{x},_i \cdot \mathbf{x},_j - \mathbf{X},_i \cdot \mathbf{X},_j\,). \tag{7}$$

With $\mathbf{d} \cdot \mathbf{d} = \mathbf{N} \cdot \mathbf{N} = 1$ and $\mathbf{N},_\alpha \cdot \mathbf{N} = \mathbf{d},_\alpha \cdot \mathbf{d} = 0$ as well as Eq. (6) the components read

$$\begin{aligned} E_{\alpha\beta} &= \varepsilon_{\alpha\beta} + \xi^3\, \kappa_{\alpha\beta} + (\xi^3)^2\, \rho_{\alpha\beta} \\ 2E_{\alpha 3} &= \gamma_\alpha + \mathbf{g}_\alpha \cdot \tilde{\mathbf{u}},_3 \\ E_{33} &= \tilde{\mathbf{g}}_3 \cdot \tilde{\mathbf{u}},_3 \end{aligned} \tag{8}$$

where $\tilde{\mathbf{g}}_3 = \mathbf{d} + \frac{1}{2}\, \boldsymbol{\Phi},_3\, \boldsymbol{\alpha}$. The membrane strains $\varepsilon_{\alpha\beta}$, curvatures $\kappa_{\alpha\beta}$ and transverse shear strains γ_α correspond to the well-known strains of the Reissner-Mindlin shell theory with inextensible director kinematic. The higher order curvatures $\rho_{\alpha\beta}$ are neglected. This is admissible for sufficiently thin structures with $L/h \gg 1$, e.g. [2]. Here, L is a characteristic length of a plate or lowest curvature radius of a shell. Hence, the in–plane strains $\{E_{11}, E_{22}, 2E_{12}\}$ are linear functions of ξ^3. Our numerical investigations show that for $L/h \geq 10$ this is a good approximation. Using Voigt notation the Green-Lagrange strains of a point in shell space with coordinate ξ^3 are obtained with

$$\mathbf{E} = \mathbf{A}_1\, \boldsymbol{\varepsilon} + \tilde{\mathbf{A}}_2\, \boldsymbol{\alpha} \tag{9}$$

where

$$\boldsymbol{\varepsilon}(\mathbf{v}) = [\varepsilon_{11}, \varepsilon_{22}, 2\varepsilon_{12}, \kappa_{11}, \kappa_{22}, 2\kappa_{12}, \gamma_1, \gamma_2]^T$$

$$\mathbf{E} = \begin{bmatrix} E_{11} \\ E_{22} \\ E_{33} \\ 2E_{12} \\ 2E_{13} \\ 2E_{23} \end{bmatrix} \quad \mathbf{A}_1 = \begin{bmatrix} 1 & 0 & 0 & \xi^3 & 0 & 0 & 0 & 0 \\ 0 & 1 & 0 & 0 & \xi^3 & 0 & 0 & 0 \\ 0 & 0 & 0 & 0 & 0 & 0 & 0 & 0 \\ 0 & 0 & 1 & 0 & 0 & \xi^3 & 0 & 0 \\ 0 & 0 & 0 & 0 & 0 & 0 & 1 & 0 \\ 0 & 0 & 0 & 0 & 0 & 0 & 0 & 1 \end{bmatrix} \quad \tilde{\mathbf{A}}_2 = \begin{bmatrix} \mathbf{0} \\ \mathbf{0} \\ \tilde{\mathbf{g}}_3^T \\ \mathbf{0} \\ \mathbf{g}_1^T \\ \mathbf{g}_2^T \end{bmatrix} \boldsymbol{\Phi}_{,3}\,. \tag{10}$$

The variation of the Green–Lagrange strain tensor yields

$$\begin{aligned} \delta\mathbf{E} &= \delta E_{ij}\, \mathbf{G}^i \otimes \mathbf{G}^j \\ \delta E_{ij} &= \frac{1}{2}(\delta\mathbf{x}_{,i} \cdot \mathbf{x}_{,j} + \delta\mathbf{x}_{,j} \cdot \mathbf{x}_{,i}) \\ \delta\mathbf{x}_{,\alpha} &= \delta\mathbf{x}_{0,\alpha} + \xi^3\, \delta\mathbf{d}_{,\alpha} \\ \delta\mathbf{x}_{,3} &= \delta\mathbf{d} + \boldsymbol{\Phi}_{,3}\, \delta\boldsymbol{\alpha} \end{aligned} \tag{11}$$

and in Voigt notation

$$\begin{aligned} \delta\mathbf{E} &= \mathbf{A}_1\, \delta\boldsymbol{\varepsilon} + \mathbf{A}_2\, \delta\boldsymbol{\alpha} \\ \delta\mathbf{E} &= \begin{bmatrix} \delta E_{11} \\ \delta E_{22} \\ \delta E_{33} \\ 2\delta E_{12} \\ 2\delta E_{13} \\ 2\delta E_{23} \end{bmatrix} \qquad \mathbf{A}_2 = \begin{bmatrix} \mathbf{0} \\ \mathbf{0} \\ \mathbf{g}_3^T \\ \mathbf{0} \\ \mathbf{g}_1^T \\ \mathbf{g}_2^T \end{bmatrix} \boldsymbol{\Phi}_{,3} \\ \delta\boldsymbol{\varepsilon} &= [\delta\varepsilon_{11}, \delta\varepsilon_{22}, 2\delta\varepsilon_{12}, \delta\kappa_{11}, \delta\kappa_{22}, 2\delta\kappa_{12}, \delta\gamma_1, \delta\gamma_2]^T . \end{aligned} \tag{12}$$

The associated finite element equations are solved applying a Newton-Raphson iteration. For this purpose the second variation of the Green–Lagrange strain tensor has to be derived. One obtains

$$\begin{aligned} \Delta\delta\mathbf{E} &= \Delta\delta\mathbf{E}_1 + \Delta\delta\mathbf{E}_2 \\ \Delta\delta\mathbf{E}_1 &= \mathbf{A}_1\, \Delta\delta\boldsymbol{\varepsilon} \qquad \Delta\delta\mathbf{E}_2 = [0, 0, \Delta\delta E_{33}, 0, 0, 0]^T \\ \Delta\delta E_{33} &= \delta\boldsymbol{\alpha}^T (\boldsymbol{\Phi}^T_{,3}\, \boldsymbol{\Phi}_{,3})\, \Delta\boldsymbol{\alpha}. \end{aligned} \tag{13}$$

The components of the shell strains $\boldsymbol{\varepsilon}$, virtual shell strains $\delta\boldsymbol{\varepsilon}$ and associated linearizations $\Delta\delta\boldsymbol{\varepsilon}$ are specified e.g. in Ref. [20].

2.2 *Weak Form of the Boundary Value Problem*

For the constitutive law in shell space we assume elastic orthotropic material behaviour. The strain energy density may be a quadratic form in terms of $\mathbf{E}$ and the elasticity matrix for orthotropy $\mathbf{C} = \mathbf{C}^T$, thus $\Psi(\mathbf{E}) = \frac{1}{2}\, \mathbf{E}^T\, \mathbf{C}\, \mathbf{E}$. Hence, one obtains

$$\partial_{\mathbf{E}} \Psi := \mathbf{S} = \mathbf{C}\,\mathbf{E}\,. \tag{14}$$

The relation of the stress resultants to the Second Piola-Kirchhoff stresses $\mathbf{S}$ is defined by thickness integration of the internal virtual work per unit reference area considering $\delta\mathbf{E} = \mathbf{A}_1\,\delta\boldsymbol{\varepsilon} + \mathbf{A}_2\,\delta\boldsymbol{\alpha}$. This yields

$$\int\limits_{h^-}^{h^+} \delta\mathbf{E}^T \mathbf{S}\,\bar{\mu}\mathrm{d}\xi^3 = \delta\boldsymbol{\varepsilon}^T \partial_{\boldsymbol{\varepsilon}} W + \delta\boldsymbol{\alpha}^T \partial_{\boldsymbol{\alpha}} W\,, \tag{15}$$

where $\bar{\mu}$ denotes the determinant of the shifter tensor and

$$\partial_{\boldsymbol{\varepsilon}} W := \int\limits_{h^-}^{h^+} \mathbf{A}_1^T\,\mathbf{S}\,\bar{\mu}\mathrm{d}\xi^3 \quad \partial_{\boldsymbol{\alpha}} W := \int\limits_{h^-}^{h^+} \mathbf{A}_2^T\,\mathbf{S}\,\bar{\mu}\mathrm{d}\xi^3\,. \tag{16}$$

The components of $\partial_{\boldsymbol{\varepsilon}} W = [n^{11}, n^{22}, n^{12}, m^{11}, m^{22}, m^{12}, q^1, q^2]^T$ are membrane forces, bending moments and shear forces, whereas $\partial_{\boldsymbol{\alpha}} W$ is a vector of higher order stress resultants.

In the following the weak form of the boundary value problem and the associated linearization are specified. For a compact representation we introduce the vector $\boldsymbol{\theta} = [\mathbf{v}, \boldsymbol{\vartheta}]^T$ and admissible variations $\delta\boldsymbol{\theta} = [\delta\mathbf{v}, \delta\boldsymbol{\vartheta}]^T$ along with

$$\boldsymbol{\vartheta} := \begin{bmatrix} \boldsymbol{\varepsilon} \\ \boldsymbol{\alpha} \\ \boldsymbol{\lambda} \end{bmatrix} \qquad \delta\boldsymbol{\vartheta} := \begin{bmatrix} \delta\boldsymbol{\varepsilon} \\ \delta\boldsymbol{\alpha} \\ \delta\boldsymbol{\lambda} \end{bmatrix}. \tag{17}$$

Furthermore,

$$\boldsymbol{\psi} := \begin{bmatrix} \partial_{\boldsymbol{\varepsilon}} W \\ \partial_{\boldsymbol{\alpha}} W + \mathbf{D}_{23}\,\boldsymbol{\lambda} - \bar{\mathbf{q}} \\ \mathbf{D}_{32}\,\boldsymbol{\alpha} \end{bmatrix} \tag{18}$$

summarizes the vector of stress resultants with two further vectors. The latter, when set to zero, describe the equilibrium of higher order stress resultants and a constraint which enforces the correct shape of $\tilde{\mathbf{u}}$ through the thickness. The derivation of both equations is performed in detail in Ref. [10]. In this context the components of the vectors $\boldsymbol{\lambda}$, $\bar{\mathbf{q}}$ and of the matrix $\mathbf{D}_{23} = \mathbf{D}_{32}^T$ are specified.

The shell is loaded statically by surface loads $\mathbf{p}^+$ and $\mathbf{p}^-$ acting at the outer surfaces with coordinates $\xi^3 = h^+$ and $\xi^3 = h^-$ as well as by boundary forces $\bar{\mathbf{t}}$ on the boundary Γ_σ. Hence, the weak form of the boundary value problem can be written as

$$
\begin{aligned}
g(\boldsymbol{\theta}, \delta\boldsymbol{\theta}) &= \int_{\Omega} \delta\boldsymbol{\vartheta}^T \boldsymbol{\psi} \mathrm{d}A + g_{ext} = 0 \\
g_{ext} &= -\int_{\Omega} (\bar{\mathbf{p}} \cdot \delta\mathbf{u} + \bar{\mathbf{m}} \cdot \delta\mathbf{d}) \mathrm{d}A - \int_{\Gamma_\sigma} \bar{\mathbf{t}} \cdot \delta\mathbf{u} \mathrm{d}s.
\end{aligned}
\tag{19}
$$

with $\bar{\mathbf{p}} = \mathbf{p}^+ + \mathbf{p}^-$ and couple loads $\bar{\mathbf{m}} = h^+ \mathbf{p}^+ + h^- \mathbf{p}^-$.

With application of integration by parts to the integral of $\delta\boldsymbol{\varepsilon}^T \partial_{\boldsymbol{\varepsilon}} W$ in Eq. (19) and use of standard arguments of variational calculus one obtains as Euler–Lagrange equations the equilibrium of stress resultants, the equilibrium of higher order stress resultants and the constraint in Ω along with the static boundary conditions $\mathbf{t} = \bar{\mathbf{t}}$ on Γ_σ.

The associated nonlinear finite element equations are iteratively solved using Newton's method. For this purpose variational equation (19) is linearized. Considering Eqs. (13), (16) and (18) we define

$$
\begin{aligned}
\frac{\partial \boldsymbol{\psi}}{\partial \boldsymbol{\vartheta}} &:= \mathbf{D} = \begin{bmatrix} \mathbf{D}_{11} & \mathbf{D}_{12} & \mathbf{0} \\ \mathbf{D}_{21} & \mathbf{D}_{22} & \mathbf{D}_{23} \\ \mathbf{0} & \mathbf{D}_{32} & \mathbf{0} \end{bmatrix} \\
\mathbf{D}_{\alpha\beta} &= \int_{h^-}^{h^+} \mathbf{A}_\alpha^T \, \mathbf{C} \, \mathbf{A}_\beta \, \bar{\mu} \, \mathrm{d}\xi^3 \quad \alpha\beta \in \{11, 12, 21\} \\
\mathbf{D}_{22} &= \int_{h^-}^{h^+} (\mathbf{A}_2^T \, \mathbf{C} \, \mathbf{A}_2 + S^{33} \, \boldsymbol{\Phi}^T_{,3} \, \boldsymbol{\Phi}_{,3}) \, \bar{\mu} \, \mathrm{d}\xi^3.
\end{aligned}
\tag{20}
$$

The integration of the submatrices $\mathbf{D}_{ij}$ is performed by summation over N layers and three point Gauss integration in each layer. With displacement independent loads $\bar{\mathbf{p}}$, $\bar{\mathbf{m}}$, $\bar{\mathbf{t}}$ and consideration of (20) one obtains the linearization of (19)

$$
\begin{aligned}
\mathrm{L}[g(\boldsymbol{\theta}, \delta\boldsymbol{\theta}), \Delta\boldsymbol{\theta}] &= \int_{\Omega} \delta\boldsymbol{\vartheta}^T (\mathbf{D} \, \Delta\boldsymbol{\vartheta} + \boldsymbol{\psi}) \mathrm{d}A + \Delta\delta W^* + g_{ext} \\
\Delta\delta W^* &= \int_{\Omega} (\bar{\mathbf{m}} \cdot \Delta\delta\mathbf{d} + \Delta\delta\boldsymbol{\varepsilon}^T \partial_{\boldsymbol{\varepsilon}} W) \mathrm{d}A.
\end{aligned}
\tag{21}
$$

The second variation $\Delta\delta\mathbf{d}$ of the current director vector has been derived in Ref. [20]. To maintain quadratic convergence in the Newton-Raphson scheme the vectors $[\mathbf{d}, \mathbf{g}_1, \mathbf{g}_2]$ of the last converged load increment are used in $\tilde{\mathbf{A}}_2$ and $\mathbf{A}_2$.

2.3 Representative Volume Element

The evaluation of $\boldsymbol{\psi}$ and $\mathbf{D}$ is carried out in RVEs which are located at the integration points of the shell reference surface Ω, see Fig. 1. An orthonormal coordinate system is introduced in the center of the square Ω_i with edge length ℓ. Concerning the predefinition $\ell = 100\, h$ we refer to the investigations in Ref. [10].

The approximation for $\boldsymbol{\vartheta} = [\boldsymbol{\varepsilon}, \boldsymbol{\alpha}, \boldsymbol{\lambda}]^T$ in Ω_i is chosen as follows

$$\boldsymbol{\vartheta} = \mathbf{N}_\vartheta\, \hat{\boldsymbol{\vartheta}}. \tag{22}$$

The shape function matrix $\mathbf{N}_\vartheta$ and the number of parameters in $\hat{\boldsymbol{\vartheta}}$ is specified in detail in [10]. We insert the ansatz (22) as well as the corresponding equations $\delta\boldsymbol{\vartheta} = \mathbf{N}_\vartheta\, \delta\hat{\boldsymbol{\vartheta}}$ and $\Delta\boldsymbol{\vartheta} = \mathbf{N}_\vartheta\, \Delta\hat{\boldsymbol{\vartheta}}$ into the linearized variational equation (21), which now reads with $A_i = \ell^2$

$$\begin{aligned} \mathrm{L}[g(\boldsymbol{\theta}, \delta\boldsymbol{\theta}), \Delta\boldsymbol{\theta}] &= \int_\Omega \frac{1}{A_i} \int_{\Omega_i} \delta\boldsymbol{\vartheta}^T (\mathbf{D}\,\Delta\boldsymbol{\vartheta} + \boldsymbol{\psi})\mathrm{d}A_i \mathrm{d}A + \Delta\delta W^* + g_{ext} \\ &= \int_\Omega \delta\hat{\boldsymbol{\vartheta}}^T (\mathbf{H}\,\Delta\hat{\boldsymbol{\vartheta}} + \mathbf{F}_L)\mathrm{d}A + \Delta\delta W^* + g_{ext} = 0 \end{aligned} \tag{23}$$

where

$$\mathbf{H} = \frac{1}{A_i} \int_{\Omega_i} \mathbf{N}_\vartheta^T\, \mathbf{D}\, \mathbf{N}_\vartheta\, \mathrm{d}A_i \qquad \mathbf{F}_L = \frac{1}{A_i} \int_{\Omega_i} \mathbf{N}_\vartheta^T\, \boldsymbol{\psi}\, \mathrm{d}A_i. \tag{24}$$

The integrals in (24) are computed numerically applying a 3×3 Gauss integration scheme.

The parameter vector $\Delta\hat{\boldsymbol{\vartheta}}$ is subdivided in $\Delta\hat{\boldsymbol{\vartheta}} = [\Delta\hat{\boldsymbol{\varepsilon}}, \Delta\hat{\boldsymbol{\beta}}]^T$. The subvector $\Delta\hat{\boldsymbol{\beta}}$ contains independent parameters in the RVEs of the reference surface. Furthermore, $\Delta\delta W^*$ and g_{ext} do not depend on $\Delta\hat{\boldsymbol{\beta}}$. For this reason elimination of $\Delta\hat{\boldsymbol{\beta}}$ from the following set of equations is possible.

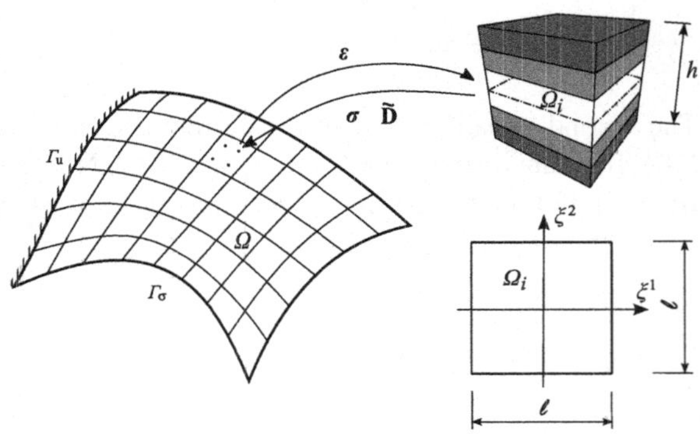

Fig. 1 RVE at an integration point of a shell and reference surface of the RVE

$$\begin{bmatrix} \mathbf{H}_{11} & \mathbf{H}_{12} \\ \mathbf{H}_{21} & \mathbf{H}_{22} \end{bmatrix} \begin{bmatrix} \Delta\hat{\boldsymbol{\varepsilon}} \\ \Delta\hat{\boldsymbol{\beta}} \end{bmatrix} + \begin{bmatrix} \mathbf{F}_1 \\ \mathbf{F}_2 \end{bmatrix} = \begin{bmatrix} \mathbf{0} \\ \mathbf{0} \end{bmatrix}. \tag{25}$$

Here, $\mathbf{H}_{\alpha\beta}$ and $\mathbf{F}_\alpha$ are submatrices of $\mathbf{H}$ and $\mathbf{F}_L$, respectively. Static condensation yields

$$\begin{aligned} \Delta\hat{\boldsymbol{\beta}} &= -\mathbf{H}_{22}^{-1}\,(\mathbf{F}_2 + \mathbf{H}_{21}\,\Delta\hat{\boldsymbol{\varepsilon}}) \\ \hat{\boldsymbol{\sigma}} := \tilde{\mathbf{F}}_1 &= \mathbf{F}_1 - \mathbf{H}_{12}\,\mathbf{H}_{22}^{-1}\,\mathbf{F}_2 \\ \tilde{\mathbf{D}} := \tilde{\mathbf{H}}_{11} &= \mathbf{H}_{11} - \mathbf{H}_{12}\,\mathbf{H}_{22}^{-1}\,\mathbf{H}_{21}. \end{aligned} \tag{26}$$

The components of the vector $\Delta\hat{\boldsymbol{\varepsilon}} \in \mathcal{R}^8$ can be identified with the incremental shell strains at the particular Gauss point. Thus it holds $\Delta\hat{\boldsymbol{\varepsilon}} = \Delta\boldsymbol{\varepsilon}$ as well as $\delta\hat{\boldsymbol{\varepsilon}} = \delta\boldsymbol{\varepsilon}$ and Eq. (23) leads with static condensation (26) to

$$\mathrm{L}[g(\mathbf{v}, \delta\mathbf{v}), \Delta\mathbf{v}] = \int_\Omega \delta\boldsymbol{\varepsilon}^T(\tilde{\mathbf{D}}\,\Delta\boldsymbol{\varepsilon} + \tilde{\boldsymbol{\sigma}})\mathrm{d}A + \Delta\delta W^* + g_{ext} = 0. \tag{27}$$

Equation (27) represents the linearized principle of virtual work as variational basis for geometrical nonlinear shell elements based on the displacement method. The computation of the stresses is performed with evaluation of the three–dimensional material law (14). This requires a back substitution of $\Delta\hat{\boldsymbol{\beta}}$ via $(26)_1$. The update of $\hat{\boldsymbol{\vartheta}}$ within the Newton-Raphson iteration and evaluation of Eq. (22) yields $\boldsymbol{\vartheta} = [\boldsymbol{\varepsilon}, \boldsymbol{\alpha}, \boldsymbol{\lambda}]^T$ and thus allows the computation of the layer strains $\mathbf{E}$ according to (9) using the subvectors $\boldsymbol{\varepsilon}$ and $\boldsymbol{\alpha}$. In this context we refer to Ref. [10] where an effective decoupling of the global and local model for the computation of the displacements and stresses is described.

The basis for mixed hybrid shell elements is given with an extension of equation (19) introducing $\boldsymbol{\theta} := [\mathbf{v}, \boldsymbol{\sigma}, \boldsymbol{\vartheta}]^T$ and the modified vector

$$\boldsymbol{\psi} := \begin{bmatrix} \partial_{\varepsilon} W - \boldsymbol{\sigma} \\ \partial_\alpha W + \mathbf{D}_{23}\,\boldsymbol{\lambda} - \bar{\mathbf{q}} \\ \mathbf{D}_{32}\,\boldsymbol{\alpha} \end{bmatrix}. \tag{28}$$

Here, $\boldsymbol{\sigma}$ denotes the vector of independent stress resultants. Furthermore, $\boldsymbol{\varepsilon}_g(\mathbf{v})$ and $\boldsymbol{\varepsilon}$ are geometrical and physical shell strains, respectively. Hence, the extension of variational equation (19) reads

$$g(\boldsymbol{\theta}, \delta\boldsymbol{\theta}) = \int_\Omega \delta\boldsymbol{\vartheta}^T\boldsymbol{\psi}\mathrm{d}A + \int_\Omega [\delta\boldsymbol{\sigma}^T(\boldsymbol{\varepsilon}_g - \boldsymbol{\varepsilon}) + \delta\boldsymbol{\varepsilon}_g^T\boldsymbol{\sigma}]\mathrm{d}A + g_{ext} = 0. \tag{29}$$

One obtains the geometric field equations and the material law for the stress resultants as further Euler-Lagrange equations. The associated mixed hybrid element formulation follows with elimination of $\boldsymbol{\sigma}$ and $\boldsymbol{\vartheta}$ from the set of equations. The basic version of the finite element formulation is specified in [20] and in an extended version for laminated shells in [8]. In both papers a remarkable robustness of the mixed hybrid element in nonlinear applications is shown. In comparison with displacement based elements and enhanced strain elements much bigger load steps with less iterations are possible. Present element formulation has been implemented in an extended version of the general finite element program FEAP [19].

3 Example

A cylindrical shell according to Fig. 2 is considered as an example. The figure shows a cross–section of the shell and a finite element mesh of the structure. Radius and length of the cylinder are $R = 1000$ mm, $L = 2000$ mm and the shell thickness is $h = 10$ mm. The shell is free at $x_2 = x_3 = 0$ and clamped at $x_2 = L$. A concentrated load F acts at $(x_1, x_2, x_3) = (0, 0, R)$. The skin of the structure consists of a $[0°/+45°/-45°/90°]_s$ lay-up, where 0° refers to the tangential direction and 90° to the length direction of the cylinder. The stiffeners with measurements $d = 50$ mm and $h = 10$ mm are arranged in radial direction and are homogeneous. Here, the fibre direction coincides with the length direction. The stiffener in the ridge line has a thickness $2\,h$. The material data for CFRP–layers are introduced as

$$\begin{aligned} E_1 &= 125000 \text{ N/mm}^2 \; G_{12} = 4800 \text{ N/mm}^2 \;\; \nu_{12} = 0.34 \\ E_2 &= 7400 \quad \text{ N/mm}^2 \; G_{23} = 2700 \text{ N/mm}^2 . \end{aligned} \tag{30}$$

The mesh density is denoted by $l \times m \times n$, where $l = 96$ is the number of elements in circumferential direction, $m = 32$ the number in length direction and $n = 4$ the number of elements for a stiffener in radial direction. The laminate of the skin is modeled with 8 layers, whereas 3 layers are used for the homogeneous stiffeners. In Fig. 2 a mesh of $l \times m \times n = 48 \times 16 \times 2$ elements is depicted.

Comparative results are computed with a 3D discretization of the skin using solid shell element [11]. In thickness direction of the skin each layer is discretized with

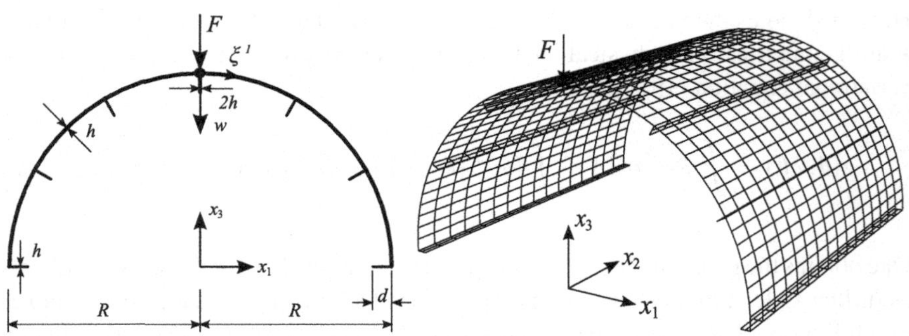

Fig. 2 Cylindrical shell and finite element mesh

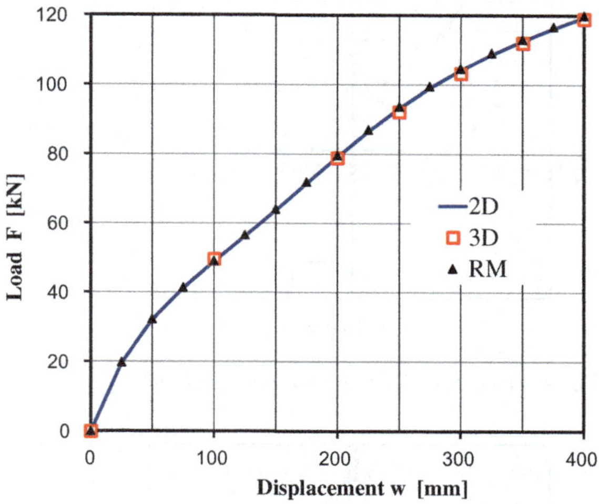

Fig. 3 Load deflection curve of the cylindrical shell

2 elements. For the discretization of the stiffeners present shell element is used. To obtain sufficient accurate results for the thickness normal stresses an in-plane refined mesh with $l \times m \times n = 192 \times 64 \times 4$ is used. Furthermore, we add results computed with shell element [13]. It is based on standard Reissner-Mindlin kinematic assumptions enhanced swith thickness strains which have a constant and linear shape through the thickness. Thus, an interface for three-dimensional constitutive laws can be used. The element is designed for homogeneous shell structures. With application to layered shells it leads to restraints especially for the thickness normal stresses. The geometrical nonlinear computations are performed with displacement control and a step size $\Delta w = 20$ mm. In Fig. 3 load F is plotted versus the prescribed displacement w. The step size Δw can be enlarged for all models in the first increments. For two deformed configuration $w = 200$ mm and $w = 400$ mm the stresses S_{11}, S_{22}, S_{12}, S_{13} and S_{33} at a point P of the reference surface with coordinates $\xi^1_p = (13/96 \cdot \pi/2) \cdot R$, $\xi^2_p = x^p_2 = 3/64 \cdot L$ are displayed in Figs. 4, 5, 6, 7 and 8 with respect to $z = \xi^3$. 2D denote results using present element, 3D is outcome from solid shell element [11] and RM is introduced for the analysis using shell element [13]. On the left-hand side results are depicted for $w = 200$ mm whereas results for $w = 400$ mm are presented on the right-hand side of the Figs. In all diagrams there is good agreement with the reference solution. Figure 9 shows the final deformed configuration using present element formulation.

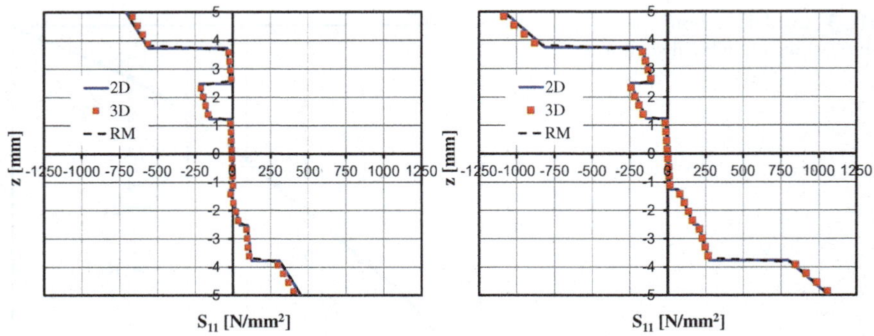

Fig. 4 Stress $S_{11}(\xi_p^1, \xi_p^2, z)$ in a cylindrical shell

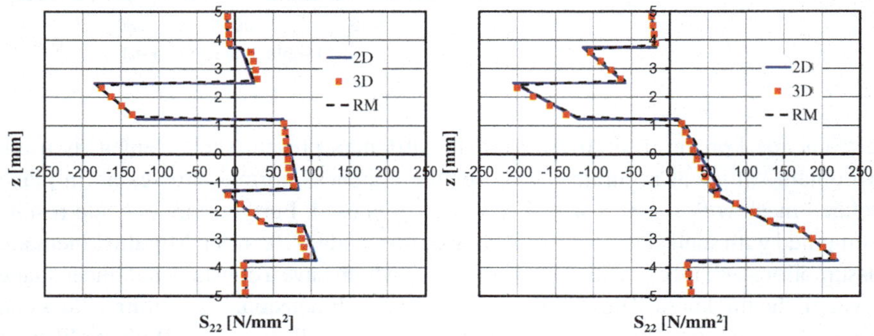

Fig. 5 Stress $S_{22}(\xi_p^1, \xi_p^2, z)$ in a cylindrical shell

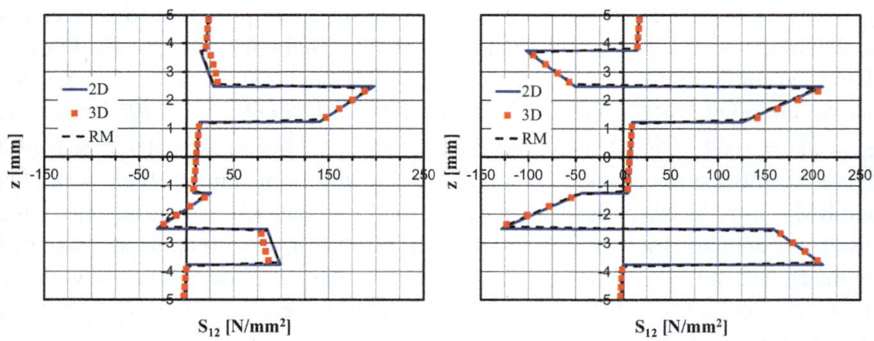

Fig. 6 Stress $S_{12}(\xi_p^1, \xi_p^2, z)$ in a cylindrical shell

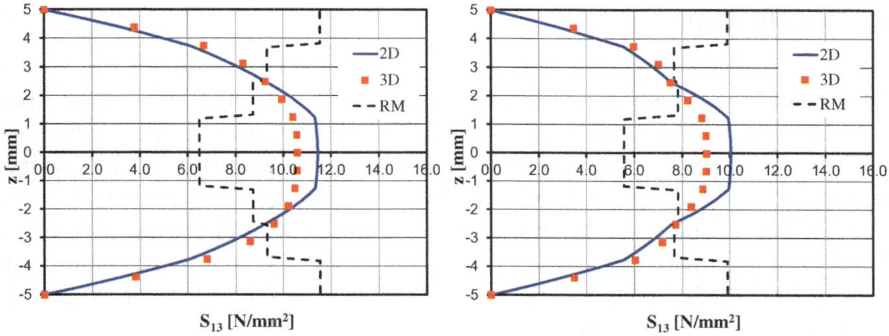

Fig. 7 Stress $S_{13}(\xi_p^1, \xi_p^2, z)$ in a cylindrical shell

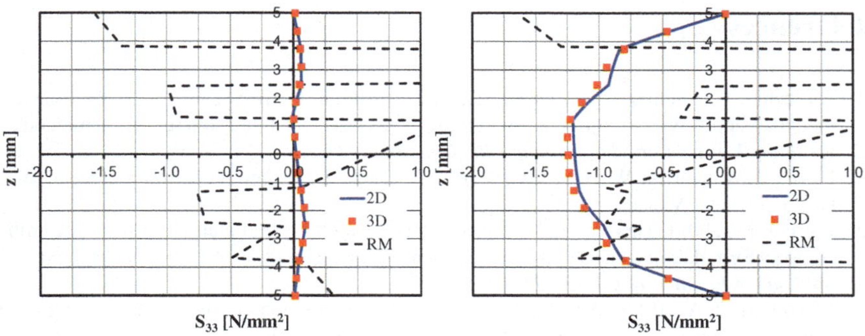

Fig. 8 Stress $S_{33}(\xi_p^1, \xi_p^2, z)$ in a cylindrical shell

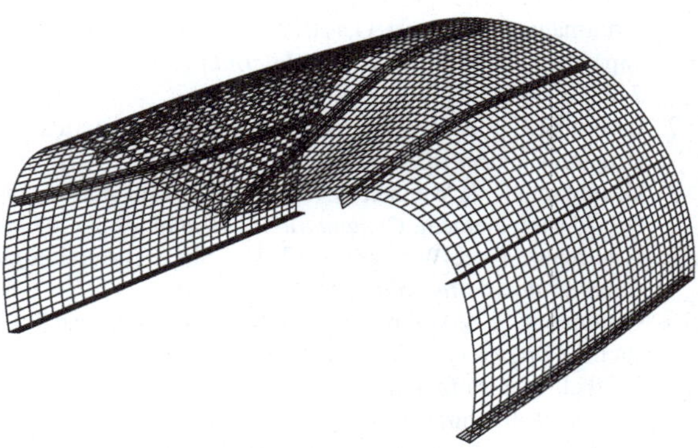

Fig. 9 Final deformed configuration of a cylindrical shell

4 Conclusions

An advanced shell model for elastic layered shells with in-plane homogeneous structure is presented. The stresses are evaluated via the three–dimensional orthotropic material law in RVEs, which are located at the integration points of the reference surface. Good agreement of displacements and stresses in comparison with costly 3D reference solutions can be shown. The essential feature of present approach is the fact that the usual 5 or 6 degrees of freedom are present at the nodes of the finite element mesh. This allows the application of standard boundary conditions and the possibility to discretize shells with intersections.

References

1. Auricchio, F., Balduzzi, G., Khoshgoftar, M. J., Rahimi, G., & Sacco, E. (2014). Enhanced modeling approach for multilayer anisotropic plates based on dimension reduction method and Hellinger-Reissner principle. *Composite Structures, 118,* 622–633.
2. Başar, Y., & Krätzig, W. B. (1985). *Mechanik der Flächentragwerke*. Braunschweig, Wiesbaden: Springer Vieweg.
3. Carrera, E. (2003). Theories and finite elements for multilayered plates and shells: A unified compact formulation with numerical assessment and benchmarking. *Archives of Computational Methods in Engineering, 10*(3), 215–296.
4. Chaudhuri, R. A. (1986). An equilibrium method for prediction of transverse shear stresses in a thick laminated plate. *Computers & Structures, 23*(2), 139–146.
5. Gruttmann, F., Knust, G., & Wagner, W. (2017). Theory and numerics of layered shells with variationally embedded interlaminar stresses. *Computer Methods in Applied Mechanics and Engineering, 326,* 713–738.
6. Gruttmann, F., & Knust, G. (2019). A shell element for the prediction of residual load-carrying capacities due to delamination. *International Journal for Numerical Methods in Engineering, 118,* 132–158.
7. Gruttmann, F., & Wagner, W. (1994). On the numerical analysis of local effects in composite structures. *Composite Structures, 29,* 1–12.
8. Gruttmann, F., & Wagner, W. (2006). Structural analysis of composite laminates using a mixed hybrid shell element. *Computational Mechanics, 37,* 479–497.
9. Gruttmann, F., & Wagner, W. (2017). Shear correction factors for layered plates and shells. *Computational Mechanics, 59*(1), 129–146.
10. Gruttmann, F., & Wagner, W. (2020). On an improved 3D stress analysis for elastic composite shells. *Computers & Structures, 231*, 106172.
11. Klinkel, S., Gruttmann, F., & Wagner, W. (1999). A continuum based 3D-shell element for laminated structures. *Computers & Structures, 71,* 43–62.
12. Klinkel, S., Gruttmann, F., & Wagner, W. (2006). A robust non-linear solid shell element based on a mixed variational formulation. *Computer Methods in Applied Mechanics and Engineering, 195,* 179–201.
13. Klinkel, S., Gruttmann, F., & Wagner, W. (2008). A mixed shell formulation accounting for thickness strains and finite strain 3D material models. *International Journal for Numerical Methods in Engineering, 74,* 945–970.
14. Kulikov, G. M., & Plotnikova, S. V. (2017). Strong sampling surfaces formulation for layered shells. *International Journal of Solids and Structures, 121*(15), 75–85.
15. Murakami, H. (1986). Laminated composite plate theory with improved in-plane response. *The Journal of Applied Mechanics, 53,* 661–666.

16. Rammerstorfer, F. G., Dorninger, K., & Starlinger, A. (1992). Composite and sandwich shells. In F. G. Rammerstorfer (Ed.), *Nonlinear analysis of shells by finite elements*. Wien: Springer.
17. Rolfes, R., Rohwer, K., & Ballerstaedt, M. (1998). Efficient linear transversal normal stress analysis of layered composite plates. *Computers & Structures, 68,* 643–652.
18. Schürg, M., Wagner, W., & Gruttmann, F. (2009). An enhanced FSDT model for the calculation of interlaminar shear stresses in composite plate structures. *Computational Mechanics, 44*(6), 765–776.
19. Taylor, R. L. (2020). FEAP. http://www.ce.berkeley.edu/projects/feap/
20. Wagner, W., & Gruttmann, F. (2005). A robust nonlinear mixed hybrid quadrilateral shell element. *International Journal for Numerical Methods in Engineering, 64,* 635–666.

Computational Homogenization Using Convolutional Neural Networks

Henning Wessels, Christoph Böhm, Fadi Aldakheel, Markus Hüpgen, Michael Haist, Ludger Lohaus, and Peter Wriggers

Dear Peter, we wish you a good health, further success in the years ahead and happy birthday.

Abstract The classic tasks of computational engineers are to investigate and optimize structures in terms of their mechanical behavior. This iterative process usually requires a large number of calculations of different macroscopic structures of the same material. The computational time in this design-loop directly affects the time to market. Depending on the model complexity, describing the interaction between micro- and macro-scale can be computationally expensive and even prohibitive for engineering practice. This holds especially true if the physics on the micro-scale is complex involving inelastic behavior, fracture and/or phase change. In this paper,

H. Wessels (✉)
Institute of Computational Modeling in Civil Engineering, Technical University of Braunschweig, Braunschweig, Germany
e-mail: h.wessels@tu-braunschweig.de

C. Böhm · F. Aldakheel · P. Wriggers
Institute of Continuum Mechanics, Leibniz University Hannover, Garbsen, Germany
e-mail: boehm@ikm.uni-hannover.de

F. Aldakheel
e-mail: aldakheel@ikm.uni-hannover.de

P. Wriggers
e-mail: wriggers@ikm.uni-hannover.de

M. Hüpgen · M. Haist · L. Lohaus
Institute of Building Materials Science, Leibniz University Hannover, Hannover, Germany
e-mail: m.huepgen@baustoff.uni-hannover.de

M. Haist
e-mail: haist@baustoff.uni-hannover.de

L. Lohaus
e-mail: lohaus@baustoff.uni-hannover.de

F. Aldakheel et al. (eds.), *Current Trends and Open Problems in Computational Mechanics*, https://doi.org/10.1007/978-3-030-87312-7_55

recent trends in Scientific Machine Learning (SciML), which may advance computational homogenization in the sense of the digital twin paradigm, are reviewed. We believe that SciML techniques for computational homogenization will make micro-macro simulations become applicable at low extra cost in engineering practice. This work is partially funded by the DFG Priority Program SPP 2020 *Experimental-Virtual-Lab* and the DFG Collaborative Research Center SFB 1153 *Tailored Forming*.

1 Introduction

Scientific Machine Learning (SciML) aims to build physical models from data. Once trained, a data-driven model can make predictions in near real-time. One possible SciML approach is to directly incorporate physical laws in the training stage of a data-driven model, that is for example a neural network. This approach is known as Physics-Informed Neural Networks [1], which have been proven successful for forward and inverse problems ranging from fluid [2] to solid mechanics [3]. Another SciML approach is to build models from physical data obtained from measurements or simulation. In this second class fall recent approaches for computational homogenization with Convolutional Neural Networks (CNN) [4–7].

CNNs are well-established in image and video recognition, analysis and classification, natural language processing and time series forecasting. The concept of computational CNN homogenization can be summarized as follows: In an offline stage, first a dataset has to be build. It consists of two or three-dimensional images of the micro-structure (e.g. from high resolution X-ray tomography (μCT) scans) and a macroscopic constitutive tensor. The latter is a result of computational homogeniza-

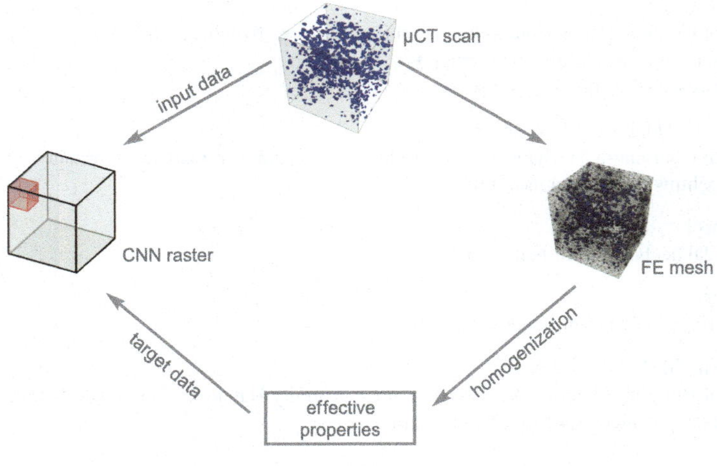

Fig. 1 Workflow for generating the required image-property dataset. The μCT outputs a voxel grid which is (i) meshed for FE simulation and (ii) filtered to yield a reduced raster as CNN input data. The CNN target data is obtained via computational homogenization using FE simulations

tion of the imaged micro-structure. The concept for creating the necessary image-property dataset for CNN training is sketched in Fig. 1. After training, the CNN can predict the constitutive tensor of a micro-structural image in near real time.

In [4] a CNN has been trained that predicts a single entry of the anisotropic elasticity tensor for artificially generated, heterogeneous 3D micro-structures. The CNN predictions were compared to sophisticated physics models and showed promising results. Also in [5] a scalar property has been learned from artificially generated 2D micro-structural images. Therein, the stress at a strain very close to the yield point has been chosen as a target property. The necessary training data was generated using crystal plasticity simulations. While the aforementioned CNNs only predict scalar properties, it has also been demonstrated that an orthotropic constitutive tensor can be learned [6]. Thanks to the fast evaluation after training, CNN homogenization has been shown to be appealing for uncertainty quantification using Monte Carlo sampling. In [7] a hybrid neural network architecture is presented which directly predicts stress of oligocrystals instead of material constants. For this purpose, a Recurrent Neural Network is designed to input a CNN encoded image along with the strain history. The latter is needed to account for the plastic behavior (see also [8] for an alternative approach to data-driven plasticity). However, in [7] only the loading path has been considered while results for the unloading path were not presented.

From the aforementioned contributions it becomes apparent that the optimal CNN architecture is problem dependent and topic of ongoing research. Especially the size of the dataset required for training still presents a bottleneck. Nevertheless, CNN homogenization seems to be a powerful tool to accelerate the structure-property interaction. It can also be integrated into a digital twin framework. Digital twins aim to collect data during the entire product life cycle of a physical asset. The data is collected from various sources and interpreted, thereby facilitating communication. An illustration of the concept can be found in Fig. 2.

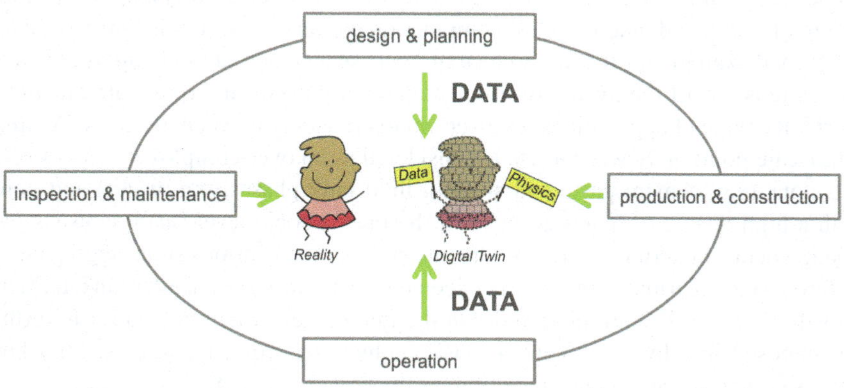

Fig. 2 The digital twin is a digital representation of a physical asset in all its aspects. It builds on data originating from experiment and numerical simulation throughout all stages of the product life cycle, from conception to operation

In the context of digital twins, CNN homogenization can be regarded as a data preparation method, where data from the conception phase is prepared for further use in operation, inspection and maintenance. As an illustrative example, consider a wind plant whose foundation must be designed to withstand harsh wind and weather conditions. In order to certify a novel concrete mixture, probes will be produced whose micro-structure can be imaged. These scans can be analyzed in a virtual lab using micro-mechanical simulations. The link of the micro-structural properties to the macro-structural behavior is obtained through computational homogenization. In the digital twin paradigm, training a CNN to map micro-structural images to computed homogenized macro-structural properties is not only a method which can accelerate macro-structural simulations in the design phase. It is also a way to save the computed data for later use. If during inspection a micro-structural image can be obtained from the actual foundation, this image could be directly fed into the CNN in order to monitor actual properties under operating conditions.

The remainder of this contribution is organized as follows: Sect. 2 gives an overview of the experimental, numerical and data-driven methods that will be employed in this project. Section 3 illustrates the workflow that we envision and prospective benchmark problems. We conclude with a summary in Sect. 4.

2 Methods

The performance of macroscopic structures is often determined by its underlying micro-structure, typically invisible to the human eye. Nowadays, modern imaging techniques such as μCT scanners or scanning electron microscopes offer the possibility to visualize complex micro-structures in 3D and high definition. This allows to investigate the mechanical performance of micro-structures using well-known numerical techniques like the Finite Element Method. To efficiently display the interaction of micro and macro scales, computational homogenization is indispensable.

A good example for the importance of multi-scale material modeling is concrete. Concrete is one of the most cost-efficient and durable construction materials in the world. Its range of applications has been enlarged widely in recent decades. Within a multiscale point of view, concrete is considered as an over-complex system of solid skeletons (e.g. cement paste and stones), fluid bulk phases (e.g. water) and pores with a high degree of heterogeneity. At the macroscopic level (m) it consists of a homogeneous material, whereas on the mesoscale (mm) various stone aggregates of different sizes distributed in a matrix of cement paste and surrounded by an interfacial transition zone (ITZ) are observed. On the microlevel (μm), this matrix is further decomposed into hydrated cement paste, unhydrated clinker particles, fluid bulk phases (water/air) and their interaction, as illustrated in Fig. 3.

The outline of this section is as follows: First, our experimental setup for μCT imaging of concrete samples is presented. Next, our previous work on detailed micro-mechanical modeling of experimentally obtained micro-structures is summa-

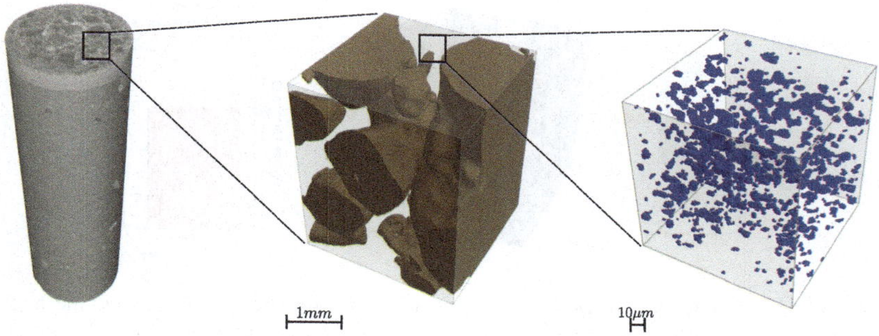

Fig. 3 Concrete strcuture through the scales. Left: cylindrical test specimen at the macro-level; Middle: representative volume element (RVE) at the meso-scale with stones in brown color and matrix in gray; Right: RVE at the micro-scale as a result of the zooming in the hydrated/unreacted cement paste (gray color) with pores in blue voxels. Figure taken from [9]

rized. We also briefly review the computational homogenization approach we seek to employ. The section ends with a short introduction to artificial neural networks.

2.1 μCT Imaging

In the last decade, increasing spatial and temporal resolutions allowed more detailed images of samples, as shown in the rising number of publications regarding CT-measurements. These measurements allow analyzing the 3D-structure of different samples in-situ and non-destructively. For concrete, CT-scans are used e.g. to segment different structures of concrete at micro-/mesoscale [10] and to evaluate the damage evolution of loaded structures [11]. Based on the same principles as an axial tomography, CT comprises an X-ray source, a sample holder and a detector. The sample gets penetrated by a beam of X-rays, projecting a 2D image onto a detector. After a predetermined rotation of the sample, a new image gets projected. For reconstructing a final 3D image, a computer reconstructs the full dataset out of each image for each rotation. Based on the iterative reconstruction algorithm, nowadays the complexity of algorithms ranges from different hybrid to fully iterative algorithms, offering improved reconstruction time and quality. Additionally, multiple steps to reduce scanning artefacts due to beam hardening and high density differences enable to display the micro-/mesostructure of samples in more detail and higher contrast, see [12].

As shown in Fig. 4, CT-scans of concrete samples are used to segment the micro-/mesostructure of concrete samples and to separate pores, matrix and agglomerates. Furthermore, a mesh can be applied to. This allows an in-situ evaluation of changes of the cement structure at a chosen scale.

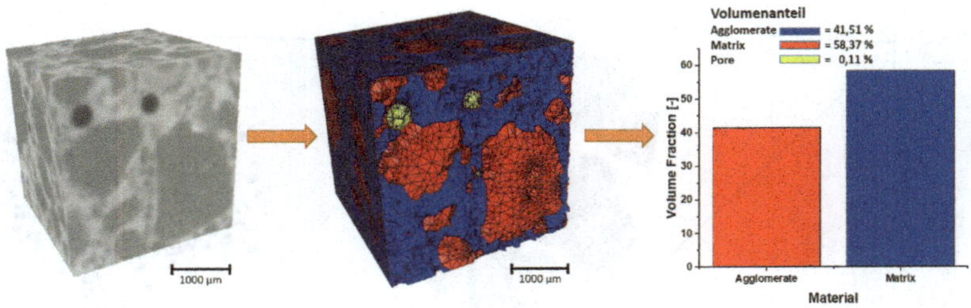

Fig. 4 Left: CT-scanned and reconstructed cement volume; middle: segmented and meshed concrete volume; right: 3D volume fractions of agglomerates, matrix and pores (pores not shown due to small volume fraction)

2.2 *Micro-Mechanical Modeling of Concrete*

At the micro-scale, the mechanical behavior of concrete is strongly affected by micro-cracking. A great number of macro-meso-micro-mechanically motivated approaches exist in the literature to model concrete failure behavior, see for example [9, 13–20] and the citations therein. The simulation of fracture processes at the micrometer length-scale can be achieved by utilizing the continuum phase-field method (PFM), which is based on the regularization of sharp crack discontinuities. This avoids the use of complex discretization methods for crack discontinuities and can account for multi-branched cracks within the solid skeleton [21–24].

Figure 5 demonstrates a contour plot of the fracture phase-field d at the final deformation state. The cracks start to initiate from the pores (c.f. Fig. 5) when a threshold energy is reached. Thereafter, they propagate in random directions inside the cement matrix and join other cracks till final rupture. Hereby a very fine mesh was required for the phase-field simulation to produce a sharp crack surface. Hence,

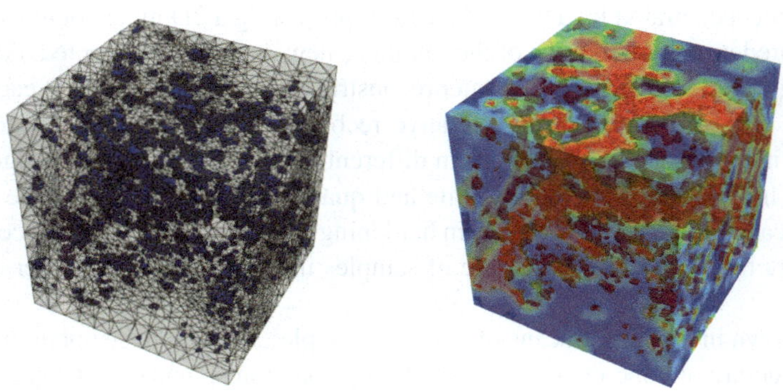

Fig. 5 Three-dimensional microstructure of HPC. Left: finite element discretization (blue colour depicts the capillary pores); Right: contour plot of the crack phase-field d at the final failure (blue and red contour colours correspond to the unbroken state with $d = 0$ and fully broken state with $d = 1$ of the material, respectively)

a simulation comes along with huge computational costs. Therefore, in order to investigate the micro-macro interaction in an engineering design loop, acceleration strategies such as computational homogenization (Sect. 2.3) in combination with artificial neural networks (Sect. 2.4) are required.

2.3 Computational Homogenization

Simulations, related to micromechanical modeling and being embedded within a lower scale, i.e. μm, are unable to be performed for a full macroscopic structure, e.g. m. Thus, the concept of the representative volume element (RVE) is widely used as an assumption to exhibit all statistical relevant information of the microstructure attached to a particular material point of the macrostructure [25]. The concept of computational homogenization demonstrates an appropriate tool to bridge the scales from micro- to macroscopic environments and to relate the micromechanical response to macroscopic loading conditions. Hereby, a macroscopic gradient of deformation induces microscopic stress such that $\boldsymbol{F}_M \rightarrow \boldsymbol{P}_m$, where $\boldsymbol{P}_m$ denotes the microscopic $1st$ Piola-Kirchhoff stress. The link between the macro- and micro- quantities is accomplished by the volume average, denoted by $\bar{(\cdot)} = \langle \cdot \rangle_V = \frac{1}{V} \int_V (\cdot) \mathrm{d}V$. Two-scale schemes, reported in the literature, are then i.e. the FE^2-scheme [26] or the global-local approach [27, 28]. Recent developments also extend the concept of computational homogenization towards a virtual element framework, e.g. [29].

The computational homogenization in this work is utilized to construct the target-data for the learning process of the convolutional neural network, connecting each RVE with its homogenized material matrix $\bar{\mathbb{A}}$ (e.g. Fig. 1). The cubiod shape of the RVE allows a straightforward application of periodic boundary-conditions [30], based on a prescribed displacement of the corner nodes $\boldsymbol{u}_{bc} = (\boldsymbol{F}_M - \mathbf{1})\, \boldsymbol{X}_C$, where $\mathbf{1}$ is the $2nd$ order unity tensor and $\boldsymbol{X}_C$ being the coordinates of the corner-nodes with respect to the initial configuration of the RVE. The computation of the homogenized material matrix $\bar{\mathbb{A}}$ is accomplished by the evaluation of the sensitivity [31]:

$$\bar{\mathbb{A}} = \frac{\partial \bar{\boldsymbol{P}}_M}{\partial \boldsymbol{F}_M} = \frac{1}{V} \int_V \frac{\partial \boldsymbol{P}_m}{\partial \boldsymbol{F}_M} \mathrm{d}V. \qquad (1)$$

The numerical computation of this sensitivity is achieved by utilization of automatic differentiation tools, using the software AceGen/AceFEM [32].

Recent approaches for computational homogenization with Convolutional Neural Networks (CNN) have been discussed in the introduction of this work. The next section briefly introduces the concept of artificial neural networks.

2.4 Artificial Neural Networks

Besides well-known applications in classification (e.g. image recognition), Artificial Neural Networks (ANNs) can be used for regression tasks: an ANN maps an input quantity x to a continuous output y. Input and output are connected through so-called neurons. Each neuron j is associated with a bias b_j and each connection from neuron i to neuron j with a weight w_{ij}. While this corresponds to a linear ansatz, non-linearity is introduced through an activation function σ. An ANN consisting only of a single input and a single output neuron obeys the equation:

$$y = \sigma\left(w \cdot x + b\right) \tag{2}$$

In practice, multiple neurons are arranged layerwise between input and output. The weights w_{ij} and b_j are the trainable parameters that must be optimized in order to obtain a best fit w.r.t to a given dataset (x_i, y_i). The best fit is defined by means of a loss function $\mathcal{L}$, e.g. the mean squared error norm:

$$\mathcal{L} = \frac{1}{n}\sum_{i}^{n}\left(y_{ANN}\left(x_i, w_{ij}, b_j\right) - y_{data}\left(x_i\right)\right)^2 \tag{3}$$

The loss function is minimized by adjusting the weights and the biases, usually using backpropagation and its variants.

For a homogenization network which inputs an image, the number of input neurons is equivalent to the number of the image's pixels or voxels, respectively. In a fully-connected network (each neuron of a layer is connected to each neuron in the next layer), this would result in high number of trainable parameters. Beside the prohibitive computational cost associated with high-resolution images, it has been shown that such networks are prone to overfitting. Convolutional Neural Networks (CNN) remedy these problems by introducing additional convolution and pooling layers. The latter reduce the dimensionality of the input while making use of the image's spatial strcture. For further details on CNN and a graphical illustration, the reader is referred to [4–7, 33].

3 Integrating CNN Homogenization into Virtual Design Loops

A possible integration of CNN homogenization into an engineering design loop is sketched in Fig. 6. First, simple probes of a novel concrete mixture are manufactured under representative processing conditions. These probes are analyzed in a μCT scanner (Sect. 2.1). As illustrated in Fig. 1, the micro-structural images are then used to build our dataset for CNN training. After training, complex concrete components of the same mixture will be produced. Micro-structural images will be taken at selected

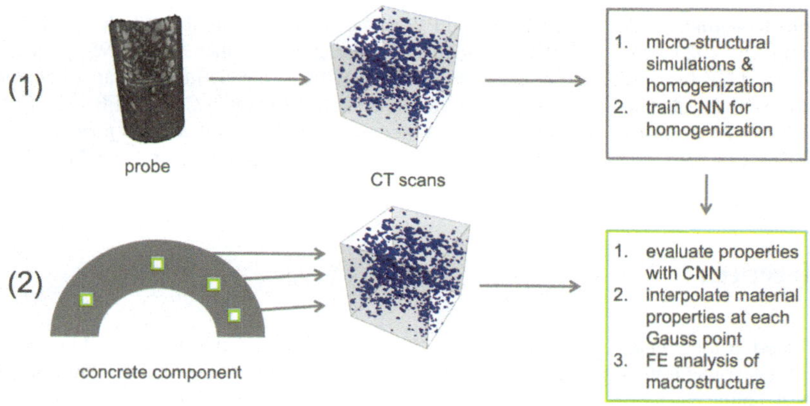

Fig. 6 Application of CNN homogenization in an engineering design loop. (1) A CNN is trained based on experiments and simulations of simple concrete probes. (2) After training, the CNN can be used to evaluate properties of any other macro-structural geometry, provided micro-structural images are available

positions of the component. The CNN can be used to predict the local properties on the fly. A suitable interpolation method must be defined in order to describe the material behavior in the entire component.

Note that innovative 3D printing processes increase the flexibility in geometric design tremendously compared to standard concrete formwork. This freedom in design significantly enlarges the design space for optimal structures and therefore the computational cost of micro-macro simulations. CNN homogenization contributes to a computationally efficient and accurate micro-macro analysis for such highly complex problems of practical interest.

4 Summary and Outlook

Computational homogenization with Convolutional Neural Networks (CNN) offers the potential to accelerate micro-macro simulations of materials with complex micro-structures. Beyond that, if measurements become available during operation or service and maintenance, a trained CNN can be used to predict properties in real time. By linking information from different stages of the product life cycle, CNN homogenization thereby is at the core of the digital twin paradigm.

In this contribution, we focused on the experimental and virtual characterization of concrete. A future field of application may be Additive Manufacturing (AM). In metal AM, the process-structure-property (PSP) interaction plays a crucial role in final part quality. Efficient computational methods like CNN homogenization may enable us in the future to decode the PSP interaction and to exploit it for the realization of AM fabricated functionally graded materials.

Acknowledgements FA, MH, MH, LL and PW acknowledge funding through the DFG Priority Program SPP 2020 *Experimental-Virtual-Lab* under the grants number 373757395; project WR 19/58-2 and GZ-LO 751/22-2 (668132). CB, FA and PW thank the German Research Foundation (DFG) for financial support to this work in the Collaborative Research Center SFB 1153 *Process chain for the production of hybrid high-performance components through tailored forming* with the subproject C04 *Modelling and simulation of the joining zone*, project number 252662854.

References

1. Raissi, M., Perdikaris, P., & Karniadakis, G. E. (2019). Physics-informed neural networks: A deep learning framework for solving forward and inverse problems involving nonlinear partial differential equations. *Journal of Computational Physics, 378*, 686–707.
2. Wessels, H., Weißenfels, C., & Wriggers, P. (2020). The neural particle method—An updated lagrangian physics informed neural network for computational fluid dynamics. *Computer Methods in Applied Mechanics and Engineering, 368*, 113–127.
3. Haghighat, E., Raissi, M., Moure, A., Gomez, H., & Juanes, R. (2020). A deep learning framework for solution and discovery in solid mechanics: Linear elasticity. arXiv:2003.02751.
4. Yang, Z., Yabansu, Y. C., Al-Bahrani, R., Liao, W.-K., Choudhary, A. N., Kalidindi, S. R., & Agrawal, A. (2018). Deep learning approaches for mining structure-property linkages in high contrast composites from simulation datasets. *Computational Materials Science, 151*, 278–287.
5. Beniwal, A., Dadhich, R., & Alankar, A. (2019). Deep learning based predictive modeling for structure-property linkages. *Materialia, 8*, 100435.
6. Rao, C., & Liu, Y. (2020). Three-dimensional convolutional neural network (3D-CNN) for heterogeneous material homogenization. *Computational Materials Science, 184*, 109850.
7. Frankel, A. L., Jones, R. E., Alleman, C., & Templeton, J. A. (2019). Predicting the mechanical response of oligocrystals with deep learning. *Computational Materials Science, 169*, 109099.
8. Huang, D., Fuhg, J. N., Weißenfels, C., & Wriggers, P. (2020). A machine learning based plasticity model using proper orthogonal decomposition. *Computer Methods in Applied Mechanics and Engineering, 365*, 113008.
9. Wriggers, P., Aldakheel, F., Lohaus, L., & Heist, M. (2020). Water-induced damage mechanisms of cyclically loaded high-performance concretes. *Bauingenieur, 95*(4), 126–132.
10. Obara, Y., Tanikura, I., Jung, J., Shintani, R., & Watanabe, S. (2016). Evaluation of micro-damage of concrete specimens under cyclic uniaxial loading by X-ray CT method. *Journal of Advanced Concrete Technology, 14*(8), 433–443.
11. Carrara, P., Kruse, R., Bentz, D. P., Lunardelli, M., Leusmann, T., Varady, P. A., & De Lorenzis, L. (2018). Improved mesoscale segmentation of concrete from 3D X-ray images using contrast enhancers. *Cement and Concrete Composites, 93*, 30–42.
12. Du Plessis, A., & Boshoff, W. P. (2019). A review of X-ray computed tomography of concrete and asphalt construction materials. *Construction and Building Materials, 199*, 637–651.
13. Wriggers, P., & Moftah, S. O. (2006). Mesoscale models for concrete: Homogenisation and damage behaviour. *Finite Elements in Analysis and Design, 42*(7), 623–636.
14. Hain, M., & Wriggers, P. (2008). Numerical homogenization of hardened cement paste. *Computational Mechanics, 42*(2), 197–212.
15. Hain, M., & Wriggers, P. (2008). Computational homogenization of micro-structural damage due to frost in hardened cement paste. *Finite Elements in Analysis and Design, 44*(5), 233 – 244. The Nineteenth Annual Robert J. Melosh Competition.
16. Lohaus, L., Oneschkow, N., & Wefer, M. (2012). Design model for the fatigue behaviour of normal-strength, high-strength and ultra-high-strength concrete. *Structural Concrete, 13*(3), 182–192.

17. Aldakheel, F., Tomann, C., Lohaus, L., & Wriggers, P. (2019). Water-induced failure mechanics for concrete. *Proceedings in Applied Mathematics and Mechanics, 19*(1), e201900140.
18. Tomann, C., Lohaus, L., Aldakheel, F., & Wriggers, P. (2019). Influence of water-induced damage mechanisms on the fatigue deterioration of high-strength concrete. *Proceedings of 6th International fib Congress: Concrete—Innovations in Materials, Design and Structures*.
19. Yang, S., Aldakheel, F., Caggiano, A., Wriggers, P., & Koenders, E. (2020). A review on cementitious self-healing and the potential of phase-field methods for modeling crack-closing and fracture recovery. *Materials, 13*(22), 5265.
20. Aldakheel, F. (2020). A microscale model for concrete failure in poro-elasto-plastic media. *Theoretical and Applied Fracture Mechanics, 107*, 102517.
21. Aldakheel, F., Mauthe, S., & Miehe, C. (2014). Towards phase field modeling of ductile fracture in gradient-extended elastic-plastic solids. *PAMM, 14*(1), 411–412.
22. Aldakheel, F., Hudobivnik, B., & Wriggers, P. (2019). Virtual element formulation for phase-field modeling of ductile fracture. *International Journal for Multiscale Computational Engineering, 17*(2).
23. Kienle, D., Aldakheel, F., & Keip, M.-A. (2019). A finite-strain phase-field approach to ductile failure of frictional materials. *International Journal of Solids and Structures, 172*, 147–162.
24. Dittmann, M., Aldakheel, F., Schulte, J., Schmidt, F., Krüger, M., Wriggers, P., & Hesch, C. (2020). Phase-field modeling of porous-ductile fracture in non-linear thermo-elasto-plastic solids. *Computer Methods in Applied Mechanics and Engineering, 361*, 112730.
25. Zohdi, T. I., & Wriggers, P. (2008). *An introduction to computational micromechanics*. Springer Science & Business Media.
26. Schröder, J. (2014). A numerical two-scale homogenization scheme: The FE2-method. In *Plasticity and Beyond* (pp. 1–64). Berlin: Springer.
27. Aldakheel, F., Noii, N., Wick, T., & Wriggers, P. (2020). A global–local approach for hydraulic phase-field fracture in poroelastic media. *Computers & Mathematics with Applications*. https://doi.org/10.1016/j.camwa.2020.07.013.
28. Noii, N., Aldakheel, F., Wick, T., & Wriggers, P. (2020). An adaptive global-local approach for phase-field modeling of anisotropic brittle fracture. *Computer Methods in Applied Mechanics and Engineering, 361*, 112744.
29. Böhm, C., Hudobivnik, B., Marino, M., & Wriggers, P. (2020). Electro-magneto-mechanically response of polycrystalline materials: Computational homogenization via the virtual element method. arXiv:2008.01516.
30. Terada, K., Hori, M., Kyoya, T., & Kikuchi, N. (2000). Simulation of the multi-scale convergence in computational homogenization approaches. *International Journal of Solids and Structures, 37*(16), 2285–2311.
31. Šolinc, U., & Korelc, J. (2015). A simple way to improved formulation of FE^2 analysis. *Computational Mechanics, 56*(5), 905–915.
32. Korelc, J., & Wriggers, P. (2016). *Automation of Finite Element Methods*. Berlin: Springer.
33. O'Shea, K., & Nash, R. (2015). An introduction to convolutional neural networks. arXiv:1511.08458.

Residual Stress Formation on the Powder Scale of Metal Powder Bed Fusion Processes

Henning Wessels and Christian Weißenfels

I will never forget my first interview as an employee. On a sheet of paper, I sketched out a few ideas that were written down loosely and rather jumbled up. At first glance, Prof. Wriggers understood the individual approaches and gave me a few more tips along the way. His quick grasp and solution-oriented thinking, coupled with a lot of empathy, great trust and respect characterized all the years we spent together. I will always have special memories of this wonderful time. Happy birthday, Peter Wriggers! Christian Weißenfels.
When I told you about my dream to research abroad, you immediately offered your support. And only a few months later I found myself on the way to California. For me, too, the years at IKM were an unforgettable time. Thank you for the creative freedom and support you have given me for my personal development. Happy birthday, Peter Wriggers! Henning Wessels.

Abstract Powder scale simulations are of utmost importance for scientists eager to decode the process-structure-property interaction inherent to powder bed fusion processes. In the literature, powder scale models are often reduced to a CFD problem where only the melt pool dynamics are considered but the solid phase physics are neglected. On the contrary, in this contribution a framework able to incorporate all relevant fluid and solid mechanical aspects of melting and solidification is derived from first principles. Since an implicit solution scheme is employed, relatively large time steps can be employed. Numerical examples illustrate that residual stress formation is concentrated at the solidification front which agrees with intuitive expectation.

H. Wessels (✉)
Institute of Computational Modeling in Civil Engineering, Technical University of Braunschweig, Braunschweig, Germany
e-mail: h.wessels@tu-braunschweig.de

C. Weißenfels
Institute of Materials Resource Management, University of Augsburg, Augsburg, Germany
e-mail: christian.weissenfels@mrm.uni-augsburg.de

F. Aldakheel et al. (eds.), *Current Trends and Open Problems in Computational Mechanics*, https://doi.org/10.1007/978-3-030-87312-7_56

1 Phase Change at Finite Deformations

Residual stress formation in thermo-mechanical problems is mainly affected by the inhibited thermal expansion. Its amount is affected by the temperature dependency of the heat expansion coefficient. Various thermo-mechanical material models for phase change problems exist in the literature. These models were primarily designed for welding [1, 2] or casting [3–5]. The deformations in both welding and casting are usually small and therefore an additive split of strains was sufficient in the aforementioned publications. However, the small strain assumption is not applicable in the fluid phase on the powder scale of powder bed fusion (PBF). In order to ensure a consistent description for the transition from fluid to solid, a finite deformation phase change formulation for PBF powder scale simulations is developed in [6]. This paper reviews the extension of the latter approach to an elasto-visco-plastic formulation for the solid phase [7].

The governing equations are discretized using the stabilized Optimal Transportation Meshfree Method (OTM) [8]. For plasticity, in each global step a local Newton Raphson algorithm is required in order to update the internal variables at the material point level.

1.1 Kinematics of Finite Deformation Phase Change Problems

The phase change approach presented in [6] is based on the following observations: Melting of metal results in an untying of internal bonds. Consequently, internal stresses are released. Once the metal becomes liquid, irreversible viscous deformation occurs. The mechanical volumetric deformation in the liquid phase due to external forces can be considered negligible compared to the thermal expansion. During consolidation from liquid to solid, internal links allowing the material to store isochoric elastic energy are created.

First, the kinematics of the solid phase are described. In the lower temperature regime, the plastic deformations can be regarded as rate independent. At higher temperatures, the yield limit gradually decreases and the plastic behavior becomes more rate dependent. As suggested in [1], this behavior can be captured with an elasto-plastic model up to a homologous temperature $\theta / T_{melt} = 0.5$ and from there on with an elasto-visco-plastic model until the melting temperature is reached. Following the concept of multiplicative decomposition, the deformation gradient $\mathbf{F}$ is split into a reversible part $\mathbf{F}_{rev}$ and an irreversible plastic part $\mathbf{F}_p$. The reversible part consists of an elastic and a thermal contribution as stated e.g. in [9]:

$$\mathbf{F} = \mathbf{F}_{rev}\,\mathbf{F}_p = \mathbf{F}_e\,\mathbf{F}_\theta\,\mathbf{F}_p \tag{1}$$

The thermal deformation is only related to thermal expansion and hence purely volumetric. Assuming plastic incompressibility, the plastic deformation gradient is only isochoric, i.e. $J_p = 1$. Hence, the Jacobi determinant J is split into a thermal part J_θ and an elastic part J_e:

$$J = J_e \, J_\theta \qquad J_\theta = e^{3\alpha_\theta (\theta - \theta_0)} \tag{2}$$

The thermal part of the Jacobi determinant J_θ incorporates the thermal expansion coefficient α_θ, the temperature θ and the reference temperature θ_0 [10]. In order to account for phase change related straining, the thermal expansion coefficient α_θ is defined as a function of temperature [1]. Making use of (1) and (2), the split of the deformation gradient into a volumetric component $\mathbf{F}_{vol}$ and an isochoric part $\mathbf{F}_{iso}$ is defined for the solid phase by:

$$\mathbf{F} = \mathbf{F}_{vol} \, \mathbf{F}_{iso} \qquad \mathbf{F}_{vol} = (J_\theta \, J_e)^{\frac{1}{3}} \, \mathbf{1} \qquad \mathbf{F}_{iso} \, (\phi = 1) = J_\theta^{-\frac{1}{3}} J_e^{-\frac{1}{3}} \, \mathbf{F}_e \, \mathbf{F}_p \tag{3}$$

Here, the phase indicator ϕ has been introduced and $\mathbf{1}$ denotes the unity tensor. Powder and consolidated material are assigned $\phi = 1$ and the liquid phase $\phi = 0$. The mechanical volumetric deformation in the powder as well as in the liquid phase are deemed negligible. In the powder phase, no external mechanical loads exist while the fluid phase is modeled as mechanically weakly incompressible. Hence, volumetric deformation is mainly caused by thermal expansion which must be preserved in the presence of a phase change. Consequently, this yields to the assumption that only isochoric deformation is phase dependent [6]. In the liquid phase, the isochoric deformation is purely viscous and therefore irreversible:

$$\mathbf{F}_{iso} \, (\phi = 0) = \mathbf{F}_v \tag{4}$$

At this stage, the following phase change treatment is applied: once metal changes phase, the isochoric deformation gradient is set equal to the unity tensor. This corresponds to equating the deformation gradient with its volumetric part [6]:

$$\mathbf{F}_{iso} = \mathbf{1} \qquad \mathbf{F} = \mathbf{F}_{vol} = J^{\frac{1}{3}} \mathbf{1} \tag{5}$$

In other words, the isochoric deformation gradient has a fading memory due to the phase change. A graphical illustration of the concept can be found in Fig. 1. Note that with the suggested approach, isochoric free energy is destroyed during the phase change from powder to liquid. However, the deformation in the powder phase is mostly related to volumetric thermal expansion. Hence, in the presence of latent heat the loss of isochoric free energy is deemed negligible. The latent heat characterizes the energy required to dissolve the lattice structure [11].

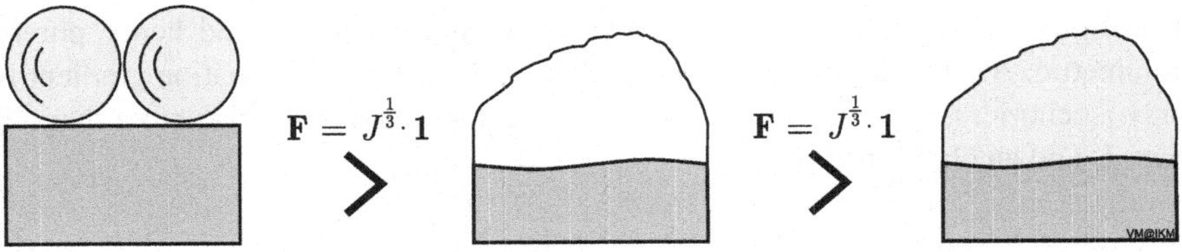

Fig. 1 When a phase change occurs, the isochoric deformation gradient $\mathbf{F}_{iso}$ is set equal to one. Only the volumetric part $\mathbf{F}_{vol}$ remains. Left: Metal powder. Centered: Molten metal. Right: Consolidated metal

1.2 A Free Energy Functional for the Phase Change Problem

Following the concept of [10], the free energy potential ψ is decomposed into a volumetric, an elastic isochoric, a plastic hardening and a thermal part ψ_{vol}, ψ_{iso}, ψ_{plas} and T, respectively. The thermo-elastic material has been described in [12] and an extension to associative plasticity can be found in [13]. The formulation of an uncoupled free energy potential requires an uncoupling of the strain measures. Reversible strains are described with the left Cauchy Green tensor $\mathbf{b}_{rev}$ defined in the current configuration [13]:

$$\mathbf{b}_{rev} = \mathbf{F}_{rev}\mathbf{F}_{rev}^T = \mathbf{F}\,\mathbf{C}_p^{-1}\,\mathbf{F}^T \qquad \mathbf{C}_p = \mathbf{F}_p^T\,\mathbf{F}_p \tag{6}$$

The isochoric reversible left Cauchy Green tensor $\bar{\mathbf{b}}_e$ is purely elastic and defined as:

$$\bar{\mathbf{b}}_e = J^{-\frac{2}{3}}\mathbf{b}_{rev} \tag{7}$$

The free energy potential ψ can now be expressed as a function of the uncoupled strain measures J_e and $\bar{\mathbf{b}}_e$, the temperature θ and the plastic hardening variable α:

$$\rho_0\,\psi = \rho_0\,\psi_{vol,e}\,(J_e) + \rho_0\,\psi_{iso,e}\left(\bar{\mathbf{b}}_e\right) + T\,(\theta) + \rho_0\psi_{plas}\,(\alpha) \tag{8}$$

Here, ρ_0 is the initial or reference density and α the plastic hardening variable. In the elastic region, the material is assumed to behave like a Neo-Hookean solid. The thermal contribution $T\,(\theta)$ is identified from the heat capacity and includes the latent heat in the presence of a phase change. The plastic part describes the influence of linear isotropic hardening. With these assumptions, the free energy potential (8) becomes:

$$\rho_0\psi\left(\mathbf{b}_{rev},\theta,\alpha\right)=\underbrace{\frac{1}{2}K(\ln J_e)^2}_{\rho_0\psi_{vol,e}}+\underbrace{\frac{1}{2}\phi\,\mu\left(I_{\bar{\mathbf{b}}_e}-3\right)}_{\rho_0\psi_{iso,e}}+\underbrace{\frac{1}{2}\,H_\alpha\alpha^2}_{\rho_0\psi_{plas}}$$
$$\underbrace{-\frac{1}{2}\left(3\alpha\right)^2K\left(\theta-\theta_0\right)^2-\rho_0\left(c+c_{L_\theta}\right)\left(\theta\ln\frac{\theta}{\theta_0}-\left(\theta-\theta_0\right)\right)}_{T} \tag{9}$$

The isochoric viscous deformation in the liquid phase ($\phi = 0$) dissipates energy and therefore does not yield any contribution to the free energy potential.

In the above expression, the first invariant $I_{\bar{\mathbf{b}}_e} = \text{tr}\,\bar{\mathbf{b}}_e$ of the isochoric elastic left Cauchy Green tensor and the second Lamé constant $\mu(\theta)$ have been introduced. Using a temperature dependent compression modulus $K(\theta)$, the volumetric free energy function is valid throughout all phases. For liquid metal, K is chosen as high as necessary to ensure mechanical incompressibility in a weak sense. Note that the temperature dependency of material parameters has been neglected in the derivation of the above expression, but is included in the simulations.

The second law of thermodynamics requires the dissipation $D_{int} \geq 0$. Using the material time derivative of the free energy potential (9) and the property of the double scalar product to filter antisymmetric terms leads to the Gibbs relation, see e.g. [13]:

$$D_{int}=\left(\boldsymbol{\sigma}-2\rho\frac{\partial\psi}{\partial\mathbf{b}_{rev}}\mathbf{b}_{rev}\right):\mathbf{d}-\rho\left(s+\frac{\partial\psi}{\partial\theta}\right)\dot{\theta}-\rho\frac{\partial\psi}{\partial\mathbf{b}_{rev}}:\mathscr{L}_v\mathbf{b}_{rev}-\rho\frac{\partial\psi}{\partial\alpha}\dot{\alpha}\geq 0 \tag{10}$$

The constitutive equations for the stress and the entropy are identified by equating the terms in the brackets to zero. In the presence of laser irradiation, plastic and viscous heating only play a minor role in the overall energy balance and are hence neglected.

1.2.1 Elasto-Plasticity

When the corresponding stress value reaches the yield limit, permanent deformations occur. Linear isotropic hardening causes the yield limit to increase proportionally to the accumulated plastic strain. At higher temperatures, thermal softening can be observed. A comprehensive overview of thermo-elasto-plasticity is given in [13] and a condensed review on elasto-plasticity can be found in [14].

The evolution of plastic flow needs to be modeled with an additional constitutive equation. Using standard von-Mises plasticity with isotropic strain hardening and thermal softening, the yield function $f(\boldsymbol{\tau},\theta,\alpha)$ is expressed in terms of the Kirchhoff stress $\boldsymbol{\tau} = J\boldsymbol{\sigma}$, the temperature θ and an internal hardening variable α:

$$f\left(\boldsymbol{\tau},\theta,\alpha\right)=\sqrt{\frac{3}{2}}||\,\text{dev}\,\boldsymbol{\tau}||-\left[\sigma_{Y0}\left(\theta\right)-\hat{q}\left(\theta\right)\right]\leq 0 \tag{11}$$

Here, $\sigma_{Y0}(\theta)$ is the temperature dependent yield limit and $\hat{q}(\theta)$ denotes the hardening stress. The latter is identified from the last term in (10) as the first derivative of the free energy potential with respect to α. Since f is written in terms of Kirchhoff stress $\boldsymbol{\tau}$, the last term in (10) is also multiplied with the Jacobian J:

$$\hat{q} = -J\rho \frac{\partial \psi_{plas}(\alpha)}{\partial \alpha} = -H_\alpha(\theta)\,\alpha \tag{12}$$

Both $\sigma_{Y0}(\theta)$ and $\hat{q}(\theta)$ are assumed to depend linearly on temperature. The slope is prescribed by the thermal softening modulus $H_\theta = (T_m - \theta_0)^{-1}$ which is chosen such that the yield limit vanishes at the melting point:

$$\begin{aligned} \sigma_{Y0}(\theta) &= \sigma_{Y0}(\theta_0)\,[1 - H_\theta(\theta - \theta_0)] \\ \hat{q}(\theta) &= H_\alpha(\theta)\,\alpha = H_\alpha(\theta_0)\,[1 - H_\theta(\theta - \theta_0)]\,\alpha \end{aligned} \tag{13}$$

Evolution equations for $\mathbf{d}_p$ and $\dot{\alpha}$ are derived from the material time derivative of $\mathbf{C}_p^{-1}$ using the yield criterion (11). The problem is complemented by the Kuhn-Tucker conditions. Details can be found in [7] or standard text books like [15].

1.2.2 Elasto-Visco-Plasticity

Beyond the temperature of $0.5T_{melt}$ the melt shows a rate dependent plastic response behavior. In this work, the Bingham approach is used to model the physical phenomena in this temperature regime. The incremental plastic strain can then be directly computed from the relation:

$$\langle \dot{\gamma} \rangle = \frac{1}{\eta_p} f(\boldsymbol{\tau}, \alpha, \theta) \tag{14}$$

The Macaulay bracket $\langle \dot{\gamma} \rangle$ ensures that the expression on the right hand side only holds if the yield function is greater or equal to zero. If $f(\boldsymbol{\tau}, \alpha, \theta) < 0$, the strain is assumed to be purely elastic and no plastic flow occurs.

Note that the viscous flow rule may be further adapted within the temperature range from $\theta \in [0.5,\ 0.8]\,T_{melt}$ and $\theta \in [0.8,\ 1.0]\,T_{melt}$ to model specific material behavior [1]. An overview of visco-plastic flow rules can be found in [15]. To illustrate the capability of the numerical framework to display the most important effects of melting and consolidation, the Bingham model is deemed sufficient. The overall approach is summarized in Fig. 2.

From the energy functional (9), heat capacity and entropy are defined by:

$$c + c_{L_\theta} = -\theta \frac{\partial^2 \psi}{\partial \theta^2}, \qquad s = -\frac{\partial \psi}{\partial \theta}$$

Isotropic Fourier type heat conduction

$$\tau = -k \,\text{grad}\, \theta$$

Kinematics: Multiplicative split of $\mathbf{F}$ and fading memory of $\mathbf{F}_{iso}$

$$\mathbf{F}_{vol} = J^{\frac{1}{3}} \mathbf{1} \qquad \text{solid: } \mathbf{F}_{iso} = J_e^{-\frac{1}{3}} \mathbf{F}_e \mathbf{F}_p \qquad \text{liquid: } \mathbf{F}_{iso} = \mathbf{F}_v \qquad \text{If change of phase: } \mathbf{F}_{iso} = \mathbf{1}$$

Constitutive law as a function of temperature

- If $(\theta > T_{liq})$: Viscous Kirchhoff stress in liquid phase

  $$\tau = K \,\ln J_e + 2\,\eta \,\text{dev}\, \mathbf{d}$$

- Else if $(\theta < T_{liq})$:
  1. Elastic Kirchhoff stress in solid phase

     $$\tau = K \,\ln J_e + \mu \,\text{dev}\, \bar{\mathbf{b}}_e$$

  2. Yield function:

     $$f\,(\tau, \theta, \alpha) = \sqrt{\frac{3}{2}} ||\text{dev}\,\tau|| - [\sigma_{Y0}\,(\theta) - \hat{q}\,(\theta)] \leq 0$$

  3. Plastic flow rule:

     $$\mathbf{d}_p = \dot{\gamma} \frac{\partial f}{\partial \tau}$$

  4. If $\theta \in [0.5 T_m, T_m]$: Evolution of plastic strain for visco-plasticity

     $$\langle \dot{\gamma} \rangle = \frac{1}{\eta_p} f\,(\tau, \alpha, \theta)$$

     If $(\theta < 0.5 T_m)$: Loading/ unloading conditions for elasto-plasticity

     $$f \leq 0, \qquad \dot{\gamma} \geq 0, \qquad \dot{\gamma} f\,(\tau, \alpha, \theta) = 0$$

  5. Evolution of accumulated plastic strain

     $$\dot{\alpha} = \dot{\gamma}$$

Fig. 2 Constitutive equations varying with temperature. Heat capacity, elastic stress, entropy and plastic hardening stress are derived from the free energy functional (9)

2 Consolidation Analysis

In order to investigate the formation of residual stresses, it is necessary to simulate until the ambient temperature is reached in the entire domain. To ensure fast cooling, Dirichlet boundary conditions are imposed on the temperatures at the bottom of the solid substrate. The ambient temperature is set to 823.15 K, which corresponds also to the boundary temperature. This setting mimics a preheated build chamber,

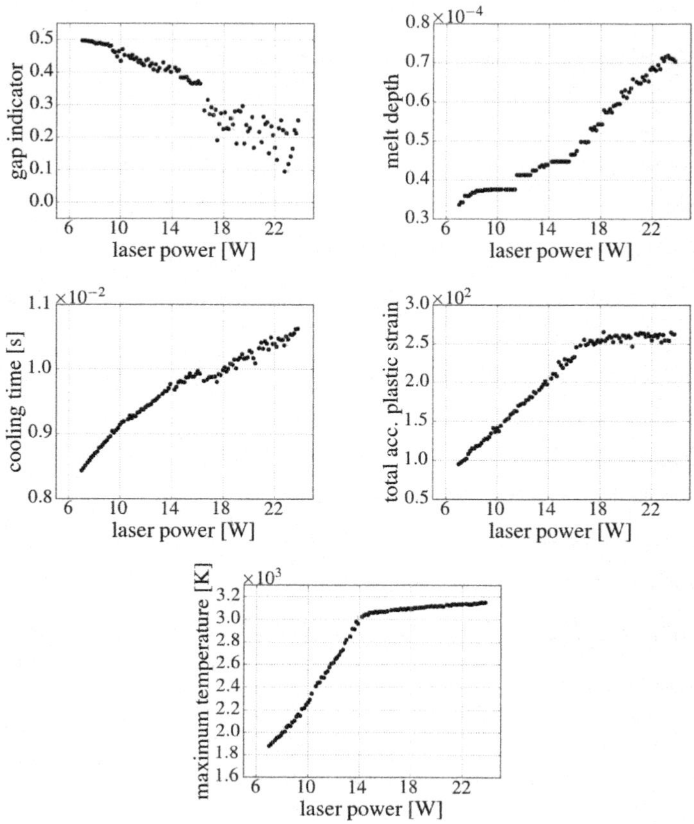

Fig. 3 Gap function, melt depth, cooling time, total accumulated plastic strain and maximum temperature as functions of laser power

which reduces residual stress formation caused by sharp thermal gradients in PBF processes. The solid substrate beneath the two metal powder particles (diameter 40 μm) possesses a depth of 120 μm. The spatial discretization consists of 2508 nodes and 9165 material points. The heat source is of Gusarov type. Details on the parameters can be found in [7]. An alternative ray tracing scheme is described in [16].

The simulations are conducted until all nodal temperatures have dropped below 10 K over ambient temperature. Besides the cooling time, the total accumulated plastic strain, the gap indicator, the maximum temperature and the melt depth are evaluated. Details on the gap indicator can be found in [6]. The melt depth is measured from the top of the two particles, i.e. the solid substrate is molten for a melt depth larger than 40 μm.

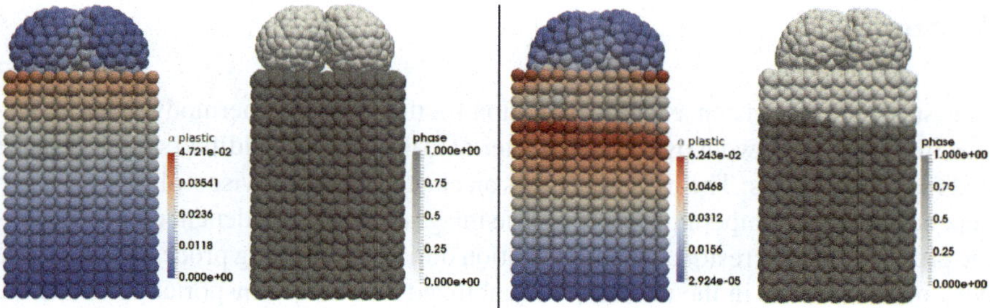

(a) $P = 9.85$ W (b) $P = 21.85$ W

Fig. 4 Accumulated plastic strain and phase fractions of solid and resolidified regions at the cooled stated for two different laser powers. Plastic deformation concentrates unbeneath the resolidified zone

Figure 4 illustrates that the accumulated plastic strain concentrates in the proximity of the re-solidified zone. This is in line with experimental observations, see e.g. [17, 18]. The prevented expansion of the melt creates a compressive stress on the surrounding solid material. The yield stress decreases with increasing temperatures and plastic deformations occur more likely. During cooling, the prevented shrinkage generates tensile stresses. Other peaks in accumulated plastic strain are observed at the topmost corners of the solid bloc. These originate from the Dirichlet type displacement boundaries on the faces. The boundaries inhibit thermal expansion, which favors residual stress formation and plastic deformation.

The measures of interest are plotted as functions of laser power in Fig. 3. In the power interval from 16 to 18 W, the cooling time remains nearly constant. The reason is a significant drop of the gap indicator from 0.35 to about 0.2 within this interval. The improved fusion bond allows the heat to escape faster.

Above 16 W, the melt depth increases linearly with laser power. Interestingly, from this point on the total accumulated plastic strain becomes independent of the melt depth and remains constant. This may be explained from the maximum temperature, which enters the absorption interval of latent heat at a laser power of 14 W. From there on, the maximum temperature increases only slightly. While the maximum temperature occurs at the top of the domain, the temperature on the ground plate is fixed to ambient conditions. When top and bottom temperatures are constant, the thermal gradients which are responsible for the residual stress formation vary only due to geometrical changes. A measure for the geometrical change is the gap indicator, which is constant for laser powers greater than 16 W. Since above 16 W both maximum temperature and gap indicator are constant, the same applies to the total accumulated plastic strain.

3 Summary

This study presents a consistent formulation for the complete thermodynamic transition from particle powder over melted material to the final re-solidified state in powder bed fusion processes. The material shows an elastic, plastic or visco-plastic behavior depending on the temperature regime. The integration of these dependencies enables the prediction of the residual stress formation during the printing process. Since large deformations occur in the process, the stabilized Optimal Transportation Meshfree method is used to discretize the resulting equations. The results of a representative study in which two powder particles fuse with the substrate show that the predicted residual stresses are in good qualitative agreement with experimental observations.

References

1. Goldak, J. A., & Akhlaghi, M. (2005). *Computational welding mechanics*. New York: Springer.
2. Lindgren, L.-E. (2007). *Computational welding mechanics: Thermomechanical and microstructural simulations*. Boca Raton, Florida: CRC Press.
3. Chiumenti, M. (1998). *Constitutive modeling and numerical analysis of thermo-mechanical phase change systems*. Dissertation, UPC, Barcelona.
4. Agelet de Saracibar, C., Cervera, M., & Chiumenti, M. (1999). On the formulation of coupled thermoplastic problems with phase-change. *International Journal of Plasticity, 15*(1), 1–34.
5. Cervera, M., Agelet de Saracibar, C., & Chiumenti, M. (1999). Thermo-mechanical analysis of industrial solidification processes. *International Journal for Numerical Methods in Engineering, 46*(9), 1575–1591.
6. Wessels, H., Weißenfels, C., & Wriggers, P. (2018). Metal particle fusion analysis for additive manufacturing using the stabilized optimal transportation meshfree method. *Computer Methods in Applied Mechanics and Engineering, 339*, 91–114.
7. Wessels, H. (2019). *Thermo-mechanical modeling for selective laser melting*. Ph.D. thesis, Hannover: Institut für Kontinuumsmechanik, Leibniz Universität Hannover.
8. Weißenfels, C., & Wriggers, P. (2018). Stabilization algorithm for the optimal transportation meshfree approximation scheme. *Computer Methods in Applied Mechanics and Engineering, 329*, 421–443.
9. Wriggers, P., Miehe, C., Kleiber, M., & Simo, J. C. (1992). On the coupled thermomechanical treatment of necking problems via finite element methods. *International Journal for Numerical Methods in Engineering, 33*(4), 869–883.
10. Lu, S. C. H., & Pister, K. S. (1975). Decomposition of deformation and representation of the free energy function for isotropic thermoelastic solids. *International Journal of Solids and Structures, 11*(7–8), 927–934.
11. Jiang, Q., & Wen, Z. (2011). *Thermodynamics of materials*. Springer Science & Business Media.
12. Miehe, C. (1988). *Zur numerischen Behandlung thermomechanischer Prozesse*. Dissertation, Leibniz Universität Hannover.
13. Simo, J. C., & Miehe, C. (1992). Associative coupled thermoplasticity at finite strains: Formulation, numerical analysis and implementation. *Computer Methods in Applied Mechanics and Engineering, 98*(1), 41–104.
14. Korelc, J., & Stupkiewicz, S. (2014). Closed-form matrix exponential and its application in finite-strain plasticity. *International Journal for Numerical Methods in Engineering, 98*(13), 960–987.

15. De Souza Neto, E. A., Perić, D., & Owen, D. R. J. (2008). *Computational methods for plasticity: Theory and applications*. Chichester, West Sussex, UK: Wiley.
16. Wessels, H., Bode, T., Weißenfels, C., Wriggers, P., & Zohdi, T. I. (2019). Investigation of heat source modeling for selective laser melting. *Computational Mechanics, 63*(5), 949–970.
17. Mercelis, P., & Kruth, J. P. (2006). Residual stresses in Selective Laser Sintering and Selective Laser Melting. *Rapid Prototyping Journal, 12*(5), 254–265.
18. Gu, D. D., Meiners, W., Wissenbach, K., & Poprawe, R. (2012). Laser Additive Manufacturing of metallic components: materials, processes and mechanisms. *International Materials Reviews, 57*(3), 133–164.

Computational Modelling of Flexoelectricity: State-of-the-art and Challenges

Xiaoying Zhuang, Binh Huy Nguyen, Srivilliputtur Subbiah Nanthakumar, Brahmanandam Javvaji, and Thai Quoc Tran

Talking to Prof. Wriggers is like reading a book rich of wisdom. The conversation with him is full of joy, insight and warmth. His thoughts and knowledge are like hands, they reached to your mind and lead you along. I am lucky that I spent the past years in his team and witness the learning and growth of myself. Sheng Ri Kuai Le (happy birthday in Chinese) to Prof. Wriggers.

Abstract Flexoelectricity is the polarization of dielectric materials under the gradient of strain. It is electromechanical coupling effect that manifests at micro/nanoscale. Flexoelectricity shares similarity to piezoelectricity, the linear polarization due to strain, but is also essentially different due to two facts. First is the size effects which is more prominent at nanoscale for flexoelectricity and the second the symmetries dislocation is different from piezoelectricity. Besides, under the dynamic loading, flexoelectric materials generate polarization waves having the magnitude proportional to the strain gradient, a phenomenon that is not observed in piezoelectric materials. It has a significant influence on the band-gap and dispersive behavior of meta-materials. Flexoelectricity has shown huge potentials in enabling technology such as self-powered nano devices and writing. Thus in the past decade, it has been intensively studied by various methodologies, theoretically and experimentally, from micro- to macroscopic continuum scale. In this report, we review the modeling of flexoelectricity at different length scales and current challenges remain to be solved. The characterization of flexoelectric coefficients from molecular dynamic simulation to continuum model remains the gap that needs to be bridged in a multiscale framework between different length-scales in flexolectric-based device modeling.

X. Zhuang (✉) · B. H. Nguyen · S. S. Nanthakumar · B. Javvaji · T. Q. Tran
Chair of Computational Science and Simulation Technology, Institute of Photonics, Faculty of Maths and Physics, Leibniz University Hannover, Appelstrasse 11, 30167 Hannover, Germany
e-mail: zhuang@iop.uni-hannover.de

F. Aldakheel et al. (eds.), *Current Trends and Open Problems in Computational Mechanics*, https://doi.org/10.1007/978-3-030-87312-7_57

1 Introduction

Advances in nanotechnology in the past two decades have allowed piezoelectric materials being produced at nanoscale. At the nanoscale, much higher strain gradients occur where the significance of flexoelectricity comes into play. Researchers observed that the reduced size of materials can result in a few orders of magnitude increase in voltage, which cannot be attributed to piezoelectricity alone. For example, in [1] a decrease in the thickness of a flexoelectric film from 3 m to 0.3 m has lead to an increase of two orders of magnitude in energy conversion efficiency. Such an effect has later been recognized due to flexoelectricity.

In small-scale structures such as micro- or nanoelectromechanical system (M-NEMS), the role of microstructure become dominant and lead to the size-effect, namely same specimen experiences different response at different dimensions. Computational homogenization technique can be employed such that macroscopic response is informed with microstructural properties [2]. The size-effect can be captured by enriching the continuum model with internal length scale as intensively studied by Cosserats theory [3], Mindlin's microelasticity theory [4], Toupin's couple stress theory [5], Eringen's micromorphic theory [6] or review collection [7]. However, for good reasons, we would want to particularly draw the attention to the deformation measure of strain gradient, which is a special case of micro-structure theory [8], at small-scale that lead to two consequences: (i) inversely proportional to the structural dimension and (ii) able to break centro-symmetry. Consequently, in small-scale dielectric solid material, the relationship between electrical polarization and strain gradient that characterize the so-called *flexoelectric* effect [9, 10], has became tangible and universal. Without being constrained to non-centrosymmetry, flexoelectricity offers a wider range of choice of electromechanical coupling materials that can open up unprecedented opportunities for flexible electronics [11], micro/nano-electromechanical systems (M/NEMS) [12], nano energy harvester [1] or in biological system [13]. Due to its promising potential, the characterization of flexoelectricity has been investigated through a wide spectrum of length-scale theoretically and experimentally. From the theoretical aspect, flexoelectricity theory has been developed from classical lattice mechanics [14, 15] and first ab inito principle [16, 17] to phenomenological and continuum theories. The experimental studies of flexoelectric effect, fueled by advanced nano-characterization technique, measured flexoelectric coefficients in many ceramic ferroelectric materials [18–21]. In-depth review of flexoelectricity progress can be found in variety of review articles [22–24].

In this report, we will give an overview on the the modeling of flexoelectricity with focus on the continuum approach. In contrast to the well-known piezoelectricity, whose theory has been well established, the continuum theory of flexoelectricity is recently developed [25–27]. Inherited from strain gradient theory, flexoelectricity results in fourth-order coupling partial differential equations that entails at least C^1-continuity for the approximation of displacement field. Several numerical remedies have been utilized to overcome this challenge including the maximum-local [28, 29], mixed finite element [30–32] and isogeometric analysis [33–36]. Another approach to capture the size effect is the micro-continuum field theory [37] based on the Cosserat theory [3] where the object of the study is the material particle including 6-DOFs instead of the material point with 3-DOFs, the theory allows the possibility of applying conventional C^0 finite element scheme taking into account the drill rotation and has been applied for flexoelectricity [38]. The constrained theory of micro-polar elasticity named couple stress theory [39–41] has been also employed to examine flexoelectric-continuum [42–45]. Different formats of elasticity theories considering the size effect come from different choices of the strain-gradient metric. However, in general, the final form of governing equations of these formulations are identical regardless of full strain-gradient tensor or rotational gradient tensor is employed. Hence, in this report for simplicity, the strain gradient effect is expressed in term of the second gradient of the deformation. In the following, we will present continuum modeling of flexoelectricity in various applications including soft dielectric, semi-conducting materials, optimization of nanogenerators in Sects. 2 and 3 followed by a brief overview of flexoelectric coefficients characterized by molecular dynamics in Sect. 5 and the discussion of bridging scales.

2 Continuum Model

Let us consider a dielectric solid occupying the volume $\mathcal{B}_0$ bounded by the boundary $\partial\mathcal{B}_0$ in the reference configuration. The dielectric body can be described by the displacement field $\mathbf{u} = \mathbf{x} - \mathbf{X}$ and electric potential (scalar) field ψ. vbx and $\mathbf{X}$ respectively denote the position vector in current and reference configuration such that they are related through the mapping $\mathbf{X} = \varphi(\mathbf{X})$. Subsequently, one can define kinematic variables such as the deformation gradient $\mathbf{F} = \nabla_{\mathbf{X}}\varphi$, the gradient of the deformation gradient $\mathbf{G} = \nabla_{\mathbf{X}}(\nabla_{\mathbf{X}}\varphi)$ and the material electric field $\mathbf{E} = -\nabla_{\mathbf{X}}\psi$. As an extension from strain gradient elasticity, the continuum theory of flexoelectric postulates the energy density function admits

$$W_0 = W_0(\mathbf{F}, \mathbf{G}, \mathbf{E}), \tag{1}$$

and the conjugated variables can be deduced, such as the first Piola-Kirchhoff stress $\mathbf{P} = \frac{\partial W_0}{\partial \mathbf{F}}$, the first Piola-Kirchhoff couple-stress and the material electric displacement $\mathbf{D} = -\frac{\partial W_0}{\partial \mathbf{E}}$. These fields obey the governing equations

$$\nabla_{\mathbf{X}} \cdot (\mathbf{P} - \nabla_{\mathbf{X}} \cdot \mathbf{Q}) + \mathbf{b} = 0 \text{ in } \mathcal{B}_0, \tag{2}$$

$$\nabla_{\mathbf{X}} \cdot \mathbf{D} = \rho^F, \text{ in } \mathcal{B}_0 \tag{3}$$

subjected to the boundary conditions

$$\boldsymbol{\varphi} = \bar{\boldsymbol{\varphi}} \text{ on } \partial\mathcal{B}_0^{\varphi}, \tag{4a}$$

$$\psi = \bar{\psi} \text{ on } \partial\mathcal{B}_0^{\psi}, \tag{4b}$$

$$\mathbf{t}^P = \bar{\mathbf{t}}^P \text{ on } \partial\mathcal{B}_0^{P}, \tag{4c}$$

$$\mathbf{t}^Q = \bar{\mathbf{t}}^Q \text{ on } \partial\mathcal{B}_0^{Q}, \tag{4d}$$

$$\nabla_{\mathbf{X}}^N \boldsymbol{\varphi} = \nabla_{\mathbf{X}}^N \bar{\boldsymbol{\varphi}} \text{ on } \partial\mathcal{B}_0^{F}, \tag{4e}$$

$$\rho^t = -\bar{\rho^t} \text{ on } \partial\mathcal{B}_0^{\rho^t}, \tag{4f}$$

in which $(\bar{\circ})$ denotes prescribed values on the boundary, i.e. $\bar{\boldsymbol{\varphi}}$ and $\bar{\mathbf{t}}^P$ are the prescribed mapping motion and traction on $\partial\mathcal{B}_0^{\varphi}$ and $\partial\mathcal{B}_0^{P}$, respectively, $\nabla_{\mathbf{X}}^N \bar{\boldsymbol{\varphi}}$ and $\bar{\mathbf{t}}^Q$ are the prescribed normal displacement derivative and double-traction on $\partial\mathcal{B}_0^{F}$ and $\partial\mathcal{B}_0^{Q}$, respectively, $\bar{\psi}$ and $\bar{\rho^t}$ are the prescribed electric potential and surface charge on $\partial\mathcal{B}_0^{\psi}$ and $\partial\mathcal{B}_0^{\rho^t}$, respectively, such that $\partial\mathcal{B}_0 \equiv \partial\partial\mathcal{B}_0^{\varphi} \cup \partial\mathcal{B}_0^{P} \equiv \partial\mathcal{B}_0^{F} \cup \partial\mathcal{B}_0^{Q} \equiv \partial\mathcal{B}_0^{\psi} \cup \partial\mathcal{B}_0^{\rho^t}$ and $\emptyset \equiv \partial\partial\mathcal{B}_0^{\varphi} \cap \partial\mathcal{B}_0^{P} \equiv \partial\mathcal{B}_0^{F} \cap \partial\mathcal{B}_0^{Q} \equiv \partial\mathcal{B}_0^{\psi} \cap \partial\mathcal{B}_0^{\rho^t}$. Here we remark that, for the sake of simplicity, we consider the surface double traction $\mathbf{t}^Q$, the normal derivative of displacement $\nabla_{\mathbf{X}}^N \boldsymbol{\varphi}$ and the surface charge ρ^t to be homogeneous on the respective boundaries.

Upon multiplying Eqs. (2) and (3) with test functions $\delta\boldsymbol{\varphi}$ and $\delta\psi$, respectively, then taking the integral over the domain and employing integration by part, the weak-form can be obtained

$$\int_{\mathcal{B}_0} \left[\mathbf{P} : \delta\mathbf{F} + \mathbf{Q} \vdots \delta\mathbf{G} - \mathbf{b} \cdot \delta\boldsymbol{\varphi} - \mathbf{D} \cdot \delta\mathbf{E} + \rho^F \delta\psi \right] \mathrm{d}V$$
$$- \int_{\partial\mathcal{B}_0} \bar{\mathbf{t}}^P \cdot \delta\boldsymbol{\varphi} \mathrm{d}A - \int_{\partial\mathcal{B}_0} \bar{\mathbf{t}}^Q \cdot \delta\nabla_{\mathbf{X}}^N \boldsymbol{\varphi} \mathrm{d}A + \int_{\partial\mathcal{B}_0} \rho^t \delta\psi \mathrm{d}A = 0. \tag{5}$$

In the case of finite deformation, Eq. (5) is further linearized and the resulting equations are approximated with suitable basis functions to obtain the system of linear equations in each Newton-Raphson iteration step. It should be remarked that the existence of strain gradient term entails the basis functions of at least C^1-continuity. A review on the numerical methods can be referred to [46].

3 Flexoelectric Effects

3.1 *Soft Dielectric Material*

With the rising interest in developing stretchable/flexible devices, nanoscale structures can undergo large deformation and high flexoelectricity can be generated with reduced size. Therefore, an important topic for numerical modeling is to support future development and design of the flexoelectric structure with is the prediction capability of large deformation induced nonlinear flexoelectric behavior. The flexoelectric effect has been studied in soft dielectric material [34]. Specifically, the dielectric material of circular shape with radius $R = 0.2$ mm embedded in air medium and subjected to electric field as schematically shown in figure. The material parameters of the dielectric droplet consist of: Young's modulus $E = 0.01$ MPa, Poisson's ratio $\nu = 0.49$ and relative permittivity $\epsilon_r = 10$ (the relative permittivity of the surrounding air is assumed to be 1). The problem is modeled by C^1-continuity quadratic NURBS basis functions. The advantage of IGA in controlling the continuity over different material surfaces is exploited in this work. Moreover, taking advantage of the problem symmetry, only a quarter of the domain is discretized and symmetric boundary conditions are applied on the bottom and left edges as shown in Fig. 1a. Note that in this work, the internal energy density is postulated instead of the electric Gibbs or enthalpy free energy. The equivalence of these two energy forms is compared in other work [46]. Now, due to the difference in dielectric permittivity

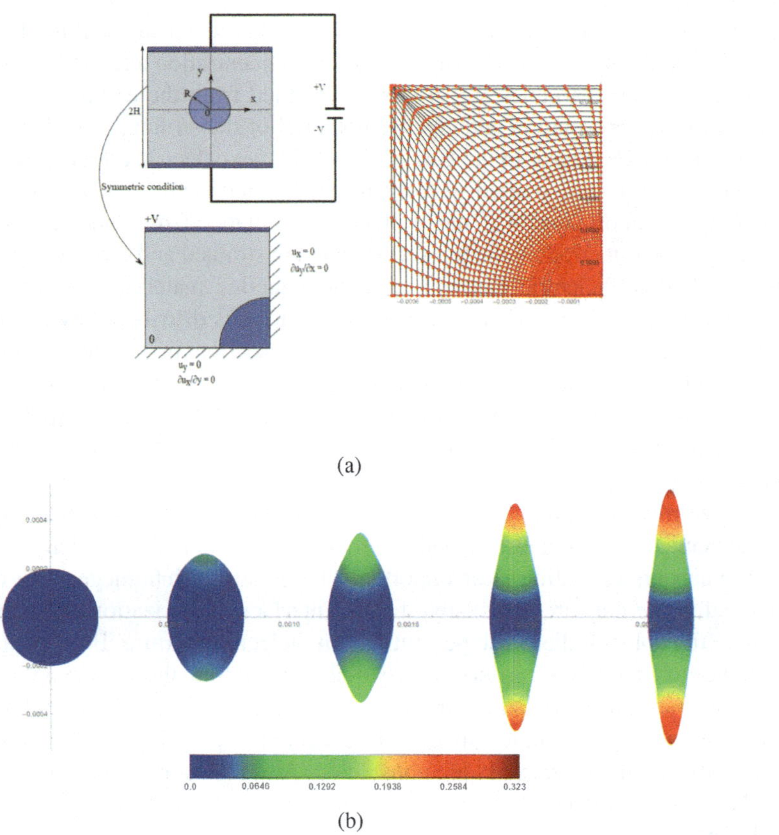

Fig. 1 Flexoelectric effect in soft dielectric material. **a** Schematic and NURBS discretization of a circular soft dielectric material under electric field. **b** Displacement u_y. Figures are adapted with permission from [34] Copyright © 2019, Elsevier

between the droplet and surrounding air, the surface of the droplet is induced by an electrical force as a consequence of the Maxwell stress such that the circular droplet deforms into a Taylor cone shape where higher-level of stress and electric potential concentrated at the two poles as shown in Fig. 1b A staggered scheme of this formulation can be found in [47] where an explicit-implicit isogeometric formulation was proposed which is possible to capture electromechanical instabilities of dielectric elastomers under extreme loading cases.

3.2 Semiconducting Material

Semiconducting materials are electronically similar to dielectric material, except having smaller energy band gap, and are ubiquitously used in nano-mechanical system. Unlike in dielectric material where the free charge carriers is negligible, semiconducting materials are even doped with purity to increase the feasibility of free charge carriers excitation. As a result, the body charge density will not be ignored, yet calculated based on the density of electron concentration and ionized donor concentration. These variables are in turn determined from the energy level which is related to the deformation and electric potential of the structure. In short, we have a non-linear source term in the Gauss's law. The flexoelectric effect in an semiconducting material was studied in [36] in which a nanowire with presumed square cross-section of dimension 50 nm $\times$ 50 nm and length 100 nm under lateral bending force in the y-direction on the top surface and constrained and electrically grounded bottom. Unlike dielectric material or semiconducting material with very low free charge carriers density, where the electric potential is distributed as in a capacitor with symmetric profile across the nanowire cross section, in semiconductor the conduction electrons tend to accumulate on the tensile side and deplete away from the compressive side. In addition, the flexoelectric effect also contributes to the potential difference of the two mentioned sides. Figure 2 shows the deformation and electric potential profile.

It is worthy to mention that the free charge density can be modulated via the electromechanical coupling effect (both piezoelectricity and flexoelectricity). Such free charge carriers migration is of importance especially in inhomegeneous dielectric material, since they are responsible for the interfacial polarization effect and might explain the colossal dielectric permittivity in dielectric mixture. By considering the existence of free charge carriers in lossy dielectric mixture through complex conductivity, we [35] have studied the frequency-dependence of the effective flexoelectric coefficient through numerical model. Note that in these models, only the bulk piezoelectricity and flexoelectricity are considered. Although surface piezoelectric effect has been investigated in numerical modeling work [48], its nature as well as relation with surface flexoelectricity are remained unexplored [22].

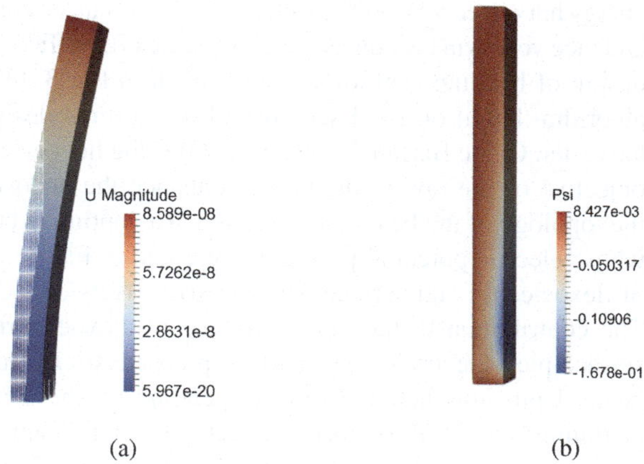

Fig. 2 Flexoelectric effect in semiconducting nanowire. **a** Displacement field. **b** Electric potential

3.3 Optimization of Energy Harvester

Flexoelectric materials are expected to replace piezoelectric materials in nano/micro electromechanical devices not only in terms of material choice but also because of the dominant role of flexoelectric effect at small scale. One typical example is the nano

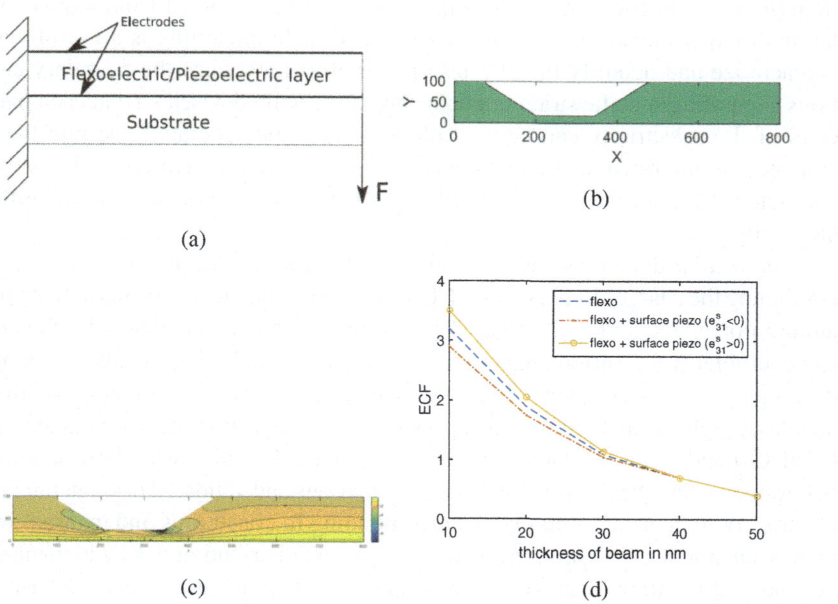

Fig. 3 Optimization of flexoelectric nano energy harvester. **a** Schematic of nano energy harvester. **b** Optimum geometry design. **c** Electric potential distribution. **d** Comparison of variation in optimal ECF with depth of nanobeam of optimal topology made of cubic BTO, for combination of flexoelectric and surface piezoelectric effects. Figures are adapted with permission from [32] Copyright © 2017, Elsevier

energy harvester, where the mechanical energy can be converted into electrical energy and vice versa via the converse flexoelectric effect. Taking one step beyond the trivial design of bending cantilever beam shown in Fig. 3, [32] has utilized optimization algorithm based on level sets to find the optimal design of a flexoelectric energy harvester. Cubic Barium Titanate (BTO) is the flexoelectric material considered, the objective of the optimization is to enhance the energy conversion, by modifying the topology of the beam cross-section. The optimized topology is shown in Fig. 3b whose electric potential profile is presented in Fig. 3c. Subsequently, the influence of flexoelectric and surface piezoelectric effects on ECF is also plotted in Fig. 3d. The contribution to the output voltage by flexoelectricity is higher compared to surface piezoelectricity. The surface piezoelectricity gains significance only as the beam depth falls below 30 nm. Depending on the sign of e^s_{31}, it either enhances or diminishes the flexoelectric effect. The significant finding of this work is that the presence of non-smooth surfaces in the beam leads to enhancement in energy conversion factor due to flexoelectricity.

4 Atomic-Scale Characterization of Flexoelectricity

To describe the relation between induced polarization and inhomogeneous deformation, the flexoelectric material parameters need to be extracted from atomic simulation or experimental measurement. Atomistic scale modelling is a useful way to characterize and quantify the physics of flexoelectricity since the boundary conditions generating specific strain gradient can be specially devised so that each parameters of flexoelectricity can be possibly extracted and seperated. Despite the discrepancy in order-of-magnitude between experimental and theoretical flexoelectric coefficients[22], atomic model still plays an important role in characterizing flexoelectricity.

A molecular dynamics based simulation scheme was developed in our team by combining the charge-dipole model (CDM) potential and the strong many-body interatomic potentials (INP) [49–51]. The developed scheme calculates the flexoelectric coefficients for various nano-material/structures including symmetry-breaking defect graphene sheet or bending two-dimensional materials of graphene allotropes, nitrides, graphene analogues of group-IV elements, transition metal dichalcogenides (TMDCs) and Janus TMDCs. For these materials, the unknown CDM parameters determined using the density functional calculations and validated through the piezoelectric coefficients. Essentially, in corresponding to constraints and external work, there is an atomic configuration and dipole moment distribution from which inhomogeneous deformation (averaged strain gradient) and polarization can be determined. In piezoelectric materials, the induced polarization consists of contributions from both the piezoelectric and flexoelectric effects. We proposed a parabolic bending scheme (shown in upper inset of Fig. 4a) that eliminates the piezoelectric contribution to the total polarization and enables directly measuring the flexoelectric coefficients. The linear variation of strain component ϵ_{yz} (see Fig. 4a) when applying the

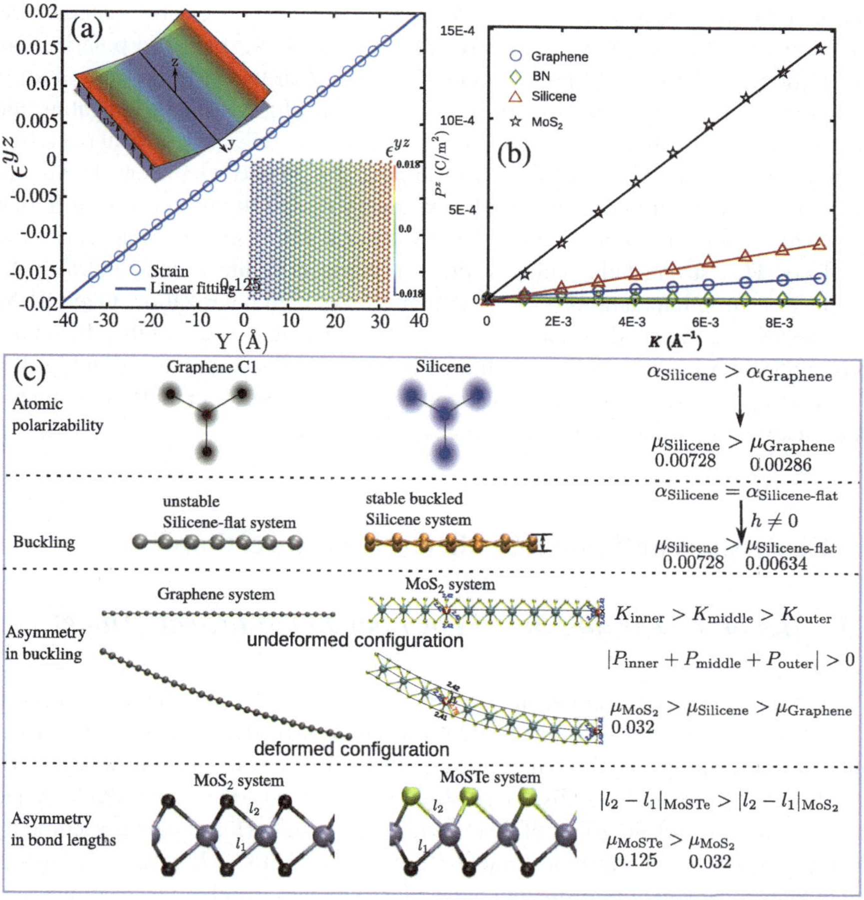

Fig. 4 **a** Binwise distribution of ϵ_{yz} along $y-$axis. Upper inset show the schematic for the bending deformation and lower inset shows the atomic configuration colored with strain component ϵ_{yz}. **b** Variation of polarization with strain gradient for graphene, boron nitride (BN), Silicene and MoS_2 TMDC monolayers. **c** The influence of various factors involved in rising the flexoelectric coefficients μ. The values respective to each material are also noted

displacement field $u_z = \frac{1}{2}Ky^2$ confirms that induced deformation is symmetric and the resulting polarization due to ϵ_{yz} canceled. Figure 4b shows the linear variation between the flexoelectric polarization and the strain gradient (K). The slope of this variation is noted as the flexoelectric coefficient (μ_{zyzy}) in units of nC/m. The flexoelectric coefficients are increasing in the order of boron nitride (BN), graphene, Silicene, MoS_2 and MoSTe. The weaker $\pi - \sigma$ electron interactions in BN tends to lower polarization and low flexoelectric coefficients compared to graphene. The high polarity of Silicon in Silicene over Carbon atoms in graphene induce more polarization in Silicene and hence the flexoelectric polarization is more. The small buckling between the successive atoms in Silicene improved the $\pi - \sigma$ electron interactions when compared to flat Silicene system, which rises the flexoelectric polarization. In addition to the increased polar nature, the bending of the MoS_2 system induces a local asymmetry by compressing the bond above Mo atom and elongating the bonds down to Mo atom. Such asymmetry creates differences in the local curvature (strain gradient) and boost the flexoelectric polarization and coefficient. In MoSTe Janus TMDCs, the initial bond length asymmetry due to the presence of different chalcogen atoms on top and bottom sides creates a strong electric field compared to MoS_2. That strong fields enhance the $\pi - \sigma$ electron interactions and also the bonding electron interactions ($\sigma - \sigma$) to induce very high polarization. Overall, the flexoelectric coefficient dependent on factors like the atomic polarizability, buckling, asymmetry in buckling and the asymmetry in bond lengths (summarized in Fig. 4c). The developed scheme and the atomistic insights highly useful in understanding the flexoelectric properties for the two-dimensional materials.

5 Discussions on Future Challenges and Directions

5.1 *Machine Learning Based Potential and Multiscale Model*

Though numerous contributions have been successfully made by the researchers in finding the geometrical shape dependent (different nanostructures, kirigami structures, designed polycrystalline monolayers) flexoelectric behavior for these materials. However, the continuous discovery of novel two-dimensional materials challenge the existing INPs for the study of flexoelectric properties. This can be overcome by the machine learning based moment tensor potentials (MTPs), where the training sets are prepared using the first-principle molecular dynamics simulations over hundreds of supercells. Later on, by using the coupled MTPs with CDM and performing the deformation simulations helps to predict the flexoelectric, piezoelectric properties as well as structural and thermal properties for the novel materials.

Although this approach is versatile and can be adapted in various material, its capability is typically limited to model up to fraction of micro-size structure, which might be insufficient for some flexoelectric-based application such as sensors, actuators or energy harvesters. Thus, this gap should be bridged by mean of combining or hybridizing atomic and continuum modeling for flexoelectricity.

It is worthy to note that phase field method has been employed to investigate meso-scale flexoelectric effect in ferroelectric materials, including the modification of domain structure [52, 53] or mechanical-driven polarization switching [54]. Notably, by linking intrinsic and extrinsic material properties from DFT calculation, the Helmholtz free energy coefficients of ferroelectric materials in phase-field model can be adjusted [55]. Furthermore, although a multiscale modeling with concurrent manner, in which MD simulation is coupled with FEM or coarse-grain MD, has been proposed for silicon micro-resonator [56], however until now there is no systematic multiscale framework for the modeling of flexoelectric coupling effect. Among several potential approaches, quasicontinuum (QC) method appears to be promising because of economical computational cost [57, 58]. However, in order to utilize QC in modeling flexoelectricity, extra care has to be paid to the long-range Coulomb interaction.

5.2 *Nonlinear Dynamic Flexoelectric Effect*

Some experimental and numerical efforts have been carried out to investigate the hysteresis behavior of the piezoelectric materials at nanoscale [59–61]. Although it was well-studied at the macro-scale, many different interpretations also were given in the efforts of interpreting the hysteresis loop at the micro-scale but there is a lack of theoretical approach to explain the mechanism of the phenomenon especially to capture its abnormal behavior near the surface of thin-film structures [62]. The dynamic flexoelectricity could be a possible solution for this issue. The influence of flexoelectricity on the dynamic hysteresis loop of micro-and nano-electro mechanical coupling systems has been got very limited attention in the literature until now. Considering dynamic flexoelectric effect requires the information of the micro-inertial effect [63], the dynamic of polarization and the flexoelectric dynamic effects [15, 64, 65] which can be determined from the analytical approach in some specific cases but still require the validation and verification from experimental data. To explain the aberrant hysteretic behavior, nonlinear behaviors such as large strain or large displacement should also be considered especially for extreme loading cases. The proposed approach can also be employed to study micro-actuators and micro-sensors which have a wide range of applications in our modern technology.

In this report, we have summarized the continuum and atomistic approach to model flexoelectricity. The necessity of bridging this gap is of importance for flexoelectric-based devices, yet a systematic multiscale framework for flexoelectricity is under development. Such multiscale flexoelectricity modeling might not only resolve the discrepancies in the theory but also provide a more reliable prediction for small-scale device designs.

Acknowledgements The first author would like to acknowledge the Sofja Kovalevskaja Programme from AvH (hosted by Prof. Wriggers) and the ERC Starting Grant COTOFLEXI (802205).

References

1. Deng, Q., Kammoun, M., Erturk, A., & Sharma, P. (2014). Nanoscale flexoelectric energy harvesting. *International Journal of Solids and Structures, 51*(18), 3218–3225.
2. Zohdi, T. I. & Wriggers, P. (2008). *An introduction to computational micromechanics*. Springer Science & Business Media.
3. Cosserat, E., & Cosserat, F. (1909). *Théorie des corps déformables*. A. Hermann et fils.
4. Mindlin, R. D. (1963). Microstructure in linear elasticity. Technical report, Columbia Univ New York Dept of Civil Engineering and Engineering Mechanics.
5. Toupin, R. A. (1962). Elastic materials with couple-stresses. *Archive for Rational Mechanics and Analysis, 11*(1), 385–414.
6. Cemal Eringen, A. (2012). *Microcontinuum field theories: I. foundations and solids*. Springer Science & Business Media.
7. Maugin, G. A., & Metrikine, A. V. (2010). *Mechanics of generalized continua.*
8. Mindlin, R. D., & Eshel, N. N. (1968). On first strain-gradient theories in linear elasticity. *International Journal of Solids and Structures, 4*(1), 109–124.
9. Tolpygo, K. B. (1963). Long wavelength oscillations of diamond-type crystals including long range forces. *Soviet Physics-Solid State, 4*(7), 1297–1305.
10. Kogan, S. M. (1964). Piezoelectric effect during inhomogeneous deformation and acoustic scattering of carriers in crystals. *Soviet Physics-Solid State, 5*(10), 2069–2070.
11. Nguyen, T. D., Mao, S., Yeh, Y.-W., Purohit, P. K., & McAlpine, M. C. (2013). Nanoscale flexoelectricity. *Advanced Materials, 25*(7), 946–974.
12. Bhaskar, U. K., Banerjee, N., Abdollahi, A., Wang, Z., Schlom, D. G., Rijnders, G., et al. (2016). A flexoelectric microelectromechanical system on silicon. *Nature nanotechnology, 11*(3), 263–266 (2016)
13. Ahmadpoor, F., & Sharma, P. (2015). Flexoelectricity in two-dimensional crystalline and biological membranes. *Nanoscale, 7*(40), 16555–16570.
14. Askar, A., Lee, P. C. Y., & Cakmak, A. S. (1970). Lattice-dynamics approach to the theory of elastic dielectrics with polarization gradient. *Physical Review B, 1*(8), 3525.
15. Tagantsev, A. K. (1985). Theory of flexoelectric effect in crystals. *Zhurnal Eksperimental'noi i Teoreticheskoi Fiziki, 88*(6), 2108–22.
16. Hong, J., & Vanderbilt, D. (2011). First-principles theory of frozen-ion flexoelectricity. *Physical Review B,84*(18), 180101.
17. Hong, J., & Vanderbilt, D. (2013). First-principles theory and calculation of flexoelectricity. *Physical Review B,88*(17), 174107.
18. Ma, W., & Eric Cross, L. (2001). Large flexoelectric polarization in ceramic lead magnesium niobate. *Applied Physics Letters, 79*(26), 4420–4422.
19. Ma, W., & Eric Cross, L. (2003). Strain-gradient-induced electric polarization in lead zirconate titanate ceramics. *Applied Physics Letters, 82*(19), 3293–3295.
20. Ma, W., & Eric Cross, L. (2005). Flexoelectric effect in ceramic lead zirconate titanate. *Applied Physics Letters, 86*(7), 072905.
21. Ma, W., & Eric Cross, L. (2006). Flexoelectricity of barium titanate. *Applied Physics Letters, 88*(23), 232902.
22. Wang, B., Yijia, G., Zhang, S., & Chen, L.-Q. (2019). Flexoelectricity in solids: Progress, challenges, and perspectives. *Progress in Materials Science, 106*, 100570.
23. Shujin Huang, L., Qi, W. H., Shu, L., Zhou, S., & Jiang, X. (2018). Flexoelectricity in dielectrics: Materials, structures and characterizations. *Journal of Advanced Dielectrics, 8*(02), 1830002.
24. Zubko, P., Catalan, G., & Tagantsev, A. K. (2013). Flexoelectric effect in solids. *Annual Review of Materials Research, 43*.
25. Maranganti, R., Sharma, N. D., & Sharma, P. (2006). Electromechanical coupling in non-piezoelectric materials due to nanoscale nonlocal size effects: Green's function solutions and embedded inclusions. *Physical Review B,74*(1), 014110.

26. ShuLing, H., & Shen, S. P. (2010). Variational principles and governing equations in nano-dielectrics with the flexoelectric effect. *Science China Physics, Mechanics and Astronomy, 53*(8), 1497–1504.
27. Shen, S., & Shuling, H. (2010). A theory of flexoelectricity with surface effect for elastic dielectrics. *Journal of the Mechanics and Physics of Solids, 58*(5), 665–677.
28. Abdollahi, A., Peco, C., Millan, D., Arroyo, M., & Arias, I. (2014). Computational evaluation of the flexoelectric effect in dielectric solids. *Journal of Applied Physics, 116*(9), 093502.
29. Abdollahi, A., Millán, D., Peco, C., Arroyo, M., & Arias, I. (2015). Revisiting pyramid compression to quantify flexoelectricity: A three-dimensional simulation study. *Physical Review B,91*(10), 104103.
30. Mao, S., Purohit, P. K., & Aravas, N. (2016). Mixed finite-element formulations in piezoelectricity and flexoelectricity. *Proceedings of the Royal Society A, 472*(2190), 20150879.
31. Deng, F., Deng, Q., Wenshan, Y., & Shen, S. (2017). Mixed finite elements for flexoelectric solids. *Journal of Applied Mechanics, 84*(8), 081004.
32. Nanthakumar, S. S., Zhuang, X., Park, H. S., & Rabczuk, T. (2017). Topology optimization of flexoelectric structures. *Journal of the Mechanics and Physics of Solids, 105*, 217–234.
33. Ghasemi, H., Park, H. S., & Rabczuk, T. (2017). A level-set based iga formulation for topology optimization of flexoelectric materials. *Computer Methods in Applied Mechanics and Engineering, 313*, 239–258.
34. Thai, T. Q., Rabczuk, T., & Zhuang, X. (2018.) A large deformation isogeometric approach for flexoelectricity and soft materials. *Computer Methods in Applied Mechanics and Engineering*.
35. Nguyen, B. H., Zhuang, X., & Rabczuk, T. (2018). Numerical model for the characterization of maxwell-wagner relaxation in piezoelectric and flexoelectric composite material. *Computers & Structures, 208*, 75–91.
36. Nguyen, B. H., Zhuang, X., & Rabczuk, T. (2018). Nurbs-based formulation for nonlinear electro-gradient elasticity in semiconductors. *Computer Methods in Applied Mechanics and Engineering*.
37. Cemal Eringen, A. (2002). *Nonlocal continuum field theories*. Springer Science & Business Media.
38. McBride, A. T., Davydov, D., & Steinmann, P. (2020). Modelling the flexoelectric effect in solids: A micromorphic approach. *Computer Methods in Applied Mechanics and Engineering, 371*, 113320.
39. Yang, F. A. C. M., Chong, A. C. M., Chun Lam, D. C., & Tong, P. (2002). Couple stress based strain gradient theory for elasticity. *International Journal of Solids and Structures, 39*(10), 2731–2743.
40. Hadjesfandiari, A. R., & Dargush, G. F. (2011). Couple stress theory for solids. *International Journal of Solids and Structures, 48*(18), 2496–2510.
41. Neff, P., Münch, I., Ghiba, I.-D., & Madeo, A. (2016). On some fundamental misunderstandings in the indeterminate couple stress model. a comment on recent papers of ar hadjesfandiari and gf dargush. *International Journal of Solids and Structures, 81*, 233–243.
42. Wang, G.-F., Shou-Wen, Yu., & Feng, X.-Q. (2004). A piezoelectric constitutive theory with rotation gradient effects. *European Journal of Mechanics-A/Solids, 23*(3), 455–466.
43. Ali R Hadjesfandiari. Size-dependent piezoelectricity. *International Journal of Solids and Structures*, 50(18):2781–2791, 2013.
44. Anqing, L., Shenjie, Z., Qi, L., & Xi, C. (2015). A flexoelectric theory with rotation gradient effects for elastic dielectrics. *Modelling and Simulation in Materials Science and Engineering, 24*(1), 015009.
45. Poya, R., Gil, A. J., Ortigosa, R., & Palma, R. (2019). On a family of numerical models for couple stress based flexoelectricity for continua and beams. *Journal of the Mechanics and Physics of Solids, 125*, 613–652.
46. Zhuang, X. (2020). Binh Huy Nguyen, Subbiah Srivilliputtur Nanthakumar, Thai Quoc Tran, Naif Alajlan, and Timon Rabczuk. *Computational modeling of flexoelectricity-a review. Energies, 13*(6), 1326.

47. Tran, Q. T., Zhuang, X., Harold Park, S., & Rabczuk, T. (2021). A staggered explicit-implicit isogeometric formulation for large deformation flexoelectricity. *Engineering Analysis with Boundary Element, 122*, 1–12.
48. Zhuang, X., Nanthakumar, S. S., & Rabczuk, T. (2020). A meshfree formulation for large deformation analysis of flexoelectric structures accounting for the surface effects. *Engineering Analysis with Boundary Elements, 120*, 153–165.
49. Javvaji, B., He, B., & Zhuang, X. (2018). The generation of piezoelectricity and flexoelectricity in graphene by breaking the materials symmetries. *Nanotechnology, 29*(22), 225702.
50. Zhuang, X., He, B., Javvaji, B., & Park, H. S. (2019). Intrinsic bending flexoelectric constants in two-dimensional materials. *Physical Review B, 99*(5), 054105.
51. Javvaji, B., He, B., Zhuang, X., & Park, H. S. (2019). High flexoelectric constants in janus transition-metal dichalcogenides. *Physical Review Materials, 3*(12), 125402.
52. Gu, Y., Li, M., Morozovska, A. N., Wang, Y., Eliseev, E. A., Gopalan, V., et al. (2014). Flexoelectricity and ferroelectric domain wall structures: Phase-field modeling and dft calculations. *Physical Review B, 89*(17), 174111.
53. Chen, H. T., Kah Soh, A., & Ni, Y. (2014). Phase field modeling of flexoelectric effects in ferroelectric epitaxial thin films. *Acta Mechanica, 225*(4–5), 1323–1333.
54. Yijia, G., Hong, Z., Britson, J., & Chen, L.-Q. (2015). Nanoscale mechanical switching of ferroelectric polarization via flexoelectricity. *Applied Physics Letters, 106*(2), 022904.
55. Völker, B., Marton, P., Elsässer, C., & Kamlah, M. (2011). Multiscale modeling for ferroelectric materials: A transition from the atomic level to phase-field modeling. *Continuum Mechanics and Thermodynamics, 23*(5), 435–451.
56. Rudd, R. E. (2001). Concurrent multiscale modeling of embedded nanomechanics. *MRS Online Proceedings Library Archive*, 677.
57. Knap, J., & Ortiz, M. (2001). An analysis of the quasicontinuum method. *Journal of the Mechanics and Physics of Solids, 49*(9), 1899–1923.
58. Miller, R. E., & Tadmor, E. B. (2002). The quasicontinuum method: Overview, applications and current directions. *Journal of Computer-Aided Materials Design, 9*(3), 203–239.
59. Ge, P., & Jouaneh, M. (1996). Tracking control of a piezoceramic actuator. *IEEE Transactions on control systems technology, 4*(3), 209–216.
60. Guo-Ying, G., Zhu, L.-M., Chun-Yi, S., Ding, H., & Fatikow, S. (2014). Modeling and control of piezo-actuated nanopositioning stages: A survey. *IEEE Transactions on Automation Science and Engineering, 13*(1), 313–332.
61. Damjanovic, D. (2006). *Chapter 4 - Hysteresis in Piezoelectric and Ferroelectric Materials*. Oxford: Academic Press.
62. Cao, H.-X., Lo, V. C., & Li, Z.-Y. (2006). Simulation of flexoelectricity effect on imprint behavior of ferroelectric thin films. *Solid state communications, 138*(8), 404–408.
63. Madeo, A., Neff, P., Aifantis, E. C., Barbagallo, G., & d'Agostino, M. V. (2017). On the role of micro-inertia in enriched continuum mechanics. *Proceedings of the Royal Society A: Mathematical, Physical and Engineering Sciences, 473*(2198), 20160722.
64. Harris, P. (1965). Mechanism for the shock polarization of dielectrics. *Journal of Applied Physics, 36*(3), 739–741.
65. Yudin, P. V., & Tagantsev, A. K. (2013). Fundamentals of flexoelectricity in solids. *Nanotechnology, 24*(43), 432001.

Peter Wriggers

We wish you a good health, further success in the years ahead and happy birthday!

F. Aldakheel et al. (eds.), *Current Trends and Open Problems in Computational Mechanics*, https://doi.org/10.1007/978-3-030-87312-7

The manufacturer's authorised representative in the EU is Springer Nature Customer Service Centre GmbH, Europaplatz 3, 69115 Heidelberg, Germany. If you have any concerns regarding our products, please contact ProductSafety@springernature.com

Printed and bound by CPI Group (UK) Ltd, Croydon, CR0 4YY
28/09/2026
02224067-0010